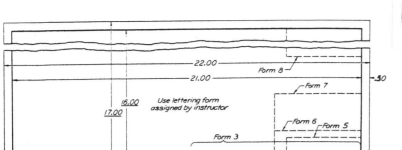

Fig. VI   Size C Sheet (17.00" × 22.00")

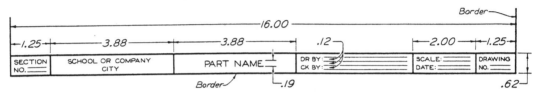

Fig. VII   Form 4. Title Block

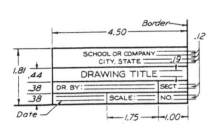

Fig. VIII   Form 5. Title Block

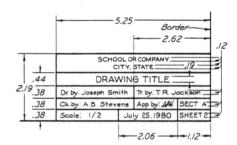

Fig. IX   Form 6. Title Block

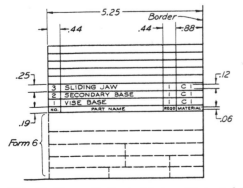

Fig. X   Form 7. Parts List or Material List

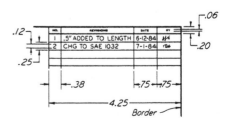

Fig. XI   Form 8. Revision Block

Library
SIAST Palliser Campus
P.O. Box 1420
Moose Jaw, SK
S6H 4R4

Phone: 694-3255

ENGINEERING
GRAPHICS

Seventh Edition

ENGINEERING GRAPHICS

FREDERICK E. GIESECKE

Late Professor Emeritus of Drawing, Texas A & M University

ALVA MITCHELL

Late Professor Emeritus of Engineering Drawing, Texas A & M University

HENRY CECIL SPENCER

Late Professor Emeritus of Technical Drawing
Formerly Director of Department Illinois Institute of Technology

IVAN LEROY HILL

Late Professor Emeritus of Engineering Graphics
Formerly Chairman of Department Illinois Institute of Technology

ROBERT OLIN LOVING

Professor Emeritus of Engineering Graphics
Formerly Chairman of Department Illinois Institute of Technology

JOHN THOMAS DYGDON

Professor Emeritus of Engineering Graphics
Formerly Chairman of Department,
and Director of the Division of Academic Services
and Office of Educational Services Illinois Institute of Technology

JAMES E. NOVAK

Assistant Professor of Engineering Graphics
Department of Civil and Architectural Engineering Illinois Institute of Technology

PRENTICE HALL
Upper Saddle River, NJ 07458

Library of Congress Cataloging-in-Publication Data

Engineering Graphics / Frederick E. Giesecke . . . [et al.]. – 7th ed.
    p.   cm.
    Includes bibliographical references and index.
    ISBN 0–13–030366–6
    1. Engineering Graphics.   I. Giesecke, Frederick E.
T467.E62 2000
642.3—dc21   00–20574
                             CIP

Acquisitions editor: *Eric Svendsen*
Editorial production/supervisor: *Rose Kernan*
Vice-president of Editorial Development, ECS: *Marcia Horton*
Managing editor: *David A. George*
Executive managing editor: *Vince O'Brien*
Vice-president of production and manufacturing: *David W. Riccardi*
Marketing Manager: *Danny Hoyt*
Copyeditor: *Pat Daly*
Manufacturing buyer: *Pat Brown*
Cover designer: *Jonathon Boylan*
Editorial assistant: *Kristen Blanko*
Art director: *Jonathon Boylan*
Insert designer: *Joe Sengotta*
Art manager: *Gus Vibal*
Project support: *Xiaohong Zhu*
Manager of formatting: *Jim Sullivan*
Composition: *Donna Marie Paukovits*

© 2000 by **PRENTICE-HALL, Inc.**
Simon & Schuster/A Viacom Company
Upper Saddle River, NJ 07458

Material printed from *Technical Drawing,* Eleventh Edition, Copyright 1999. Earlier editions of Technical Drawing Copyright 1933, 1936, 1940, 1949, © 1958, and Copyright © 1967, 1974, 1980, and 1986 by Macmillan Publishing Company. © Copyright 1991, 1997 by Prentice-Hall, Inc. Copyright renewed 1961 by Hulda G. Giesecke, R. Howard Mitchell, and Henry Cecil Spencer. Copyright renewed 1964 by Alma Giesecke Hodges, R. Howard Mitchell, and Henry C. Spencer. Copyright renewed 1968 by Henry Cecil Spencer, W. H. Mitchell, W. C. Mitchell, Alma Giesecke Hodges, Linda Giesecke Geren, and Minnie Giesecke Wight. Copyright renewed 1977 by Linda Giesecke Geren, Alma Giesecke Hodges, Minnie Giesecke Wight, R. Howard Mitchell, and Juanita M. Spencer.

The author and publisher of this book have used their best efforts in preparing this book. These efforts include the development, research, and testing of the theories and programs to determine their effectiveness. The author and publisher make no warranty of any kind, expressed or implied, with regard to these programs or the documentation contained in this book. The author and publisher shall not be liable in any event for incidental or consequential damages in connection with, or arising out of, the furnishing, performance, or use of these programs.

Printed in the United States of America
10 9 8 7 6 5 4 3 2

ISBN 0-13-030366-6

Prentice-Hall International (UK) Limited, *London*
Prentice-Hall of Australia Pty. Limited, *Sydney*
Prentice-Hall Canada, Inc., *Toronto*
Prentice-Hall Hispanoamericana, S.A., *Mexico*
Prentice-Hall of India Private Limited, *New Delhi*
Prentice-Hall of Japan, Inc., *Tokyo*
Pearson Education (Singapore) Pte. Ltd., *Singapore*
Editora Prentice-Hall do Brasil, Ltda., *Rio de Janeiro*
Prentice-Hall, Inc., *Upper Saddle River, New Jersey*

P R E F A C E

■ ABOUT THIS BOOK

Over the last sixty years, *Engineering Graphics* has taught well over a million students the practices and techniques of graphical communication. Both instructors and their students have come to depend on this text as the authority on the subject, and most students have used this book as a professional reference long after they have finished taking this course. Why has this book been so successful? First, the main goal of the text has always been to explain each principle so clearly that the student is certain to understand it, and to make the text interesting enough to encourage all students to read and study on their own. Secondly, this book has also had what are unquestionably the best, most detailed, and accurate set of drawings on the market. A student skilled with this book's drawings will have a full repertoire of graphical skills in his or her hands. Lastly, *Engineering Graphics* has continually sought to address the new technologies, and the skills that constantly change this field. By doing so, this text prepares students to enter the marketplace and face the challenges of a rapidly changing playing field.

■ THE SEVENTH EDITION

The Seventh Edition of *Engineering Graphics* continues to offer the best coverage of basic graphics principles available. Edition after edition, this text serves as the authoritative source on the subject. With this new version, we have acted upon the requests of over 30 reviewers to improve certain aspects of the book while preserving its core presentation. In particular, this new volume features:

- Greatly increased coverage of design process in chapter 14. This chapter now includes coverage of 3-D solid modeling, and parametric or constraint based modeling.
- Thoroughly revised chapter on manufacturing processes. We especially thank Professor Serope Kalpakjian of the Illinois Institute of Technology and the author of *Manufacturing Engineering and Technology* for his assistance with this chapter.
- Over 50 brand new problems — These problems feature parts that are not cast iron and that come from a variety of industries.
- Material on Instrumental Drawing and Lettering condensed to one chapter.
- New coverage of Geometric Dimensioning and Tolerancing.
- Extensive updating of text graphics to comply to most recent ANSI standards.
- New "Graphics Spotlights" feature that highlights a particular use of graphics in industry.
- A decrease in the overall size of the book — Instructors have told us that students are finding books too large. However, they want their students to have access to important reference material. To achieve both these goals, we have eliminated chapters on Graphs and Diagrams and Alignment Charts included in the last edition, and included them in Adobe pdf format on a free CD. This CD also contains over 30 animations of graphics concepts.
- Updated web site *http://www.prenhall.com/giesecke* now includes even more questions for review, as well as a new feature available for January of 2000

that lets professors create their own on-line syllabus. Using a simple interface, instructors build a graphic assignment list and use it to interact with their class over the web. This site also features links to other graphics related areas, a selection of animations available for class instruction or student review, an exploration of VRML technology, a case studies exploring how graphics communication is handled at an engineering company, and more. This site will be a constant resource to help instructors and students remain as current as possible.

- Eight Page Color Insert — To give students a feel for how color is often used in CAD software and other technological processes, but without overloading and obscuring the book's core content, we have included an eight page color insert.

■ SUPPLEMENTS

INSTRUCTOR'S MANUAL WITH SOLUTIONS/RESOURCE CD

This new manual prepared by Tom Kane of Pueblo Community College includes outlines, teaching tips, extra quiz questions, and other tools designed to aid in class preparation. In addition, this manual comes with a *new CD-ROM containing answer files for over 450 Giesecke drawing problems.* The problems are in both dwg and dwf form for easy electronic access and display. This CD also includes pdf files of all art in the text for quick integration in course web pages. Instructors have long asked for this supplement and we are happy to provide it with the new edition.

WORKBOOKS

Three workbooks with additional problems are available. These workbooks are fully class tested for effectiveness and relevance to the course. They range from having more traditional problems to more modern approaches.

- **Engineering Drawing, Problem Series 1** (ISBN — 0-13-658536-1): Contains traditional, mechanical workbook problems.
- **Engineering Drawing, Problem Series 2** (ISBN — 0-13-658881-6): Contains traditional problems with an emphasis on engineering concepts.
- **Engineering Drawing, Problem Series 3, New 2nd Edition** (ISBN — 0-13-025954-3): The new edition of a workbook by Paige Davis and Karen Juneau

of the Louisiana State University. This book contains modern problems, as well as an extensive CAD based project, and comes with its own disk of starter CAD files.

WORLDWIDE WEB SITE — http://www.prenhall.com/giesecke - AVAILABLE WITH SYLLABUS BUILDER FOR JANUARY 2000

In order to provide instructors and their students with the most exciting information available, Prentice Hall has created the Giesecke Web Site. This site now includes even more questions for review, as well as a new feature available for January of 2000 that lets professors create their own on-line syllabus. Using a simple interface, instructors build a graphic assignment list and use it to interact with their class over the web. This site features links to other graphics related areas, a selection of animations available for class instruction or student review, an exploration of VRML technology, a case study exploring how graphics communication is handled at an engineering company, and more. We hope it will be a constant resource to help instructors and students remain as current as possible.

PRENTICE HALL NEW YORK TIMES SUPPLEMENT

A bi-annual, free collection of articles excerpted from *The New York Times* covers areas of interest to freshman engineers and drafting students. Contact your Prentice Hall sales rep for a free supply of these supplements.

BUNDLES

To make the cost of purchasing several books for one course more manageable for students, Prentice Hall offers discounts when you purchase this book with several other Prentice Hall textbooks. Discounts range from up to 20% off the price of the two books purchased separately. At press time, you may bundle this text for discounts with several of our CAD books, including our new AutoCAD 2000 titles *Discovering AutoCAD 2000* by Dix/Riley, and *AutoCAD 2000 — One Step at a Time* (either the *Basics* or *Advanced* version) by Timothy Sean Sykes. You may also bundle this book with several books based on older releases of AutoCAD. Users of Pro/ENGINEER, I-DEAS, or SolidWorks may choose to bundle with books by Robert Rizza,

Sheryl Sorby, or Robert Lueptow. To request more specific and up-to-date pricing information, get ISBN's for ordering bundles, and learn more about Prentice Hall's offerings in graphics and CAD, either contact your Prentice Hall Sales Rep, or go to http://www.prenhall.com/cadgraphics/. For the name and number of your sales rep, please contact Prentice Hall Faculty Services at 1-800-526-0485.

■ ACKNOWLEDGMENTS

This edition underwent extensive reviewing. We would like to thank those who helped with their comments and suggestions:

Thomas Bledsaw, *ITT Educational Services, Inc.*
Jeff Cope, *Macon Technical Institute*
Catherine M. Ferman, *Scoolcraft College*
Wallace J. Franklin, *U.S. Merchant Marine Academy*
Bill Gordon, *Louisville Technical Institute*
Debora C. Edwards, *Appalachian State University*
Tim Hogue, *Oklahoma State University*
Jane Hook, *Salt Lake Community College*
Dean R. Kerste, *Monroe County Community College*
J.C. Malitzke, *Moraine Valley Community College*
William McCullough, *Triangle Tech*
Cliff Monroe, *Mission College*
Manuel Villalpando, *Texas State Technical College*
David L. Webb, *Salt Lake Community College*

This edition of *Engineering Graphics* is dedicated to the memory of our late coauthor, colleague and friend, Bob Loving, who passed away before its revision. His consummate knowledge of and intense interest in descriptive geometry are reflected in those chapters of this book.

John Thomas Dygdon
Tinley Park IL

James E. Novak
Illinois Institute of Technology
Chicago IL

C O N T E N T S

C H A P T E R 1

THE GRAPHIC LANGUAGE AND DESIGN

OBJECTIVES

After studying the material in this chapter, you should be able to:

1. Describe the role of the engineer on a design team.
2. List two types of drawings.
3. Explain why standards are important.
4. Draw examples of parallel and perspective projection.
5. Define *plane of projection* and *projectors*.
6. Identify uses of the graphic language.
7. Describe the differences between mechanical drawing and sketching.

OVERVIEW

A new machine, structure, or system must exist in the mind of the engineer or designer before it can become a reality. This original concept or idea is usually placed on paper or as an image on a computer screen and communicated to others by way of the *graphic language* in the form of freehand sketches. These first sketches are followed by other, more exact, sketches as the idea is developed more fully.

The engineer and drafter for the twenty-first century century must understand how to read and write in the graphic language. Everyone on the engineering and design team must be able to communicate quickly and accurately in order to compete in the world market. Like carpenters learning to use the tools of their trade, engineers, designers, and drafters must learn the tools of technical drawing. While CAD has replaced traditional drafting tools for many design teams, the basic concepts of the graphic language remain the same. Those students who can become proficient in graphic communication will succeed and add value to the employer who hires them.

■ INTRODUCTION

The old saying that "necessity is the mother of invention" is still true, and a new machine, structure, system, or device is the result of that need. If the new object is really needed or desired, people will buy it or use it as long as they can afford it.

Before a new object of any kind goes into production, certain questions must be answered: What is the potential market for this object? Can the object (device or system) be sold at a price that people are willing to pay? If the potential market is large enough and the estimated selling price seems reasonable, then the inventor, designer, or company officials may choose to proceed with development, production, and marketing plans for the new project.

A new machine, structure, or system, or even an improvement of an existing system, must exist in the mind of the inventor, engineer, or designer before it can become a reality. This original concept is usually placed on paper or as an image of a computer screen. It is then communicated to others by way of the *graphic language* in the form of freehand *idea sketches*, or *design sketches* (Figs. 1.1 and 5.1). As the idea is developed more fully, these original sketches are followed by other sketches, such as *computation sketches*.

Engineers and designers must be able to create idea sketches, calculate stresses, analyze motions, size parts, specify materials and production methods, make design layouts, and supervise the preparation of drawings and specifications that will control the numerous details of product manufacture, assembly, and maintenance. To perform or supervise these many tasks, engineers make liberal use of freehand sketches. They must be able to record and communicate ideas quickly to associate and support personnel. Both facility in freehand sketching (Chapter 3) and the ability to work with computer-controlled drawing techniques require a thorough knowledge of the graphic language. Engineers and designers who use a computer for drawing and design work must be proficient in drafting, designing, and conceptualizing.

Typical engineering and design departments are shown in Figs. 1.2 and 1.3. Such staffs include people who have considerable training and experience as well as recent graduates who are gaining experience. There is much to be learned on the job, and inexperienced people must to start at a low level and then advance to more responsible positions as they gain experience.

■ 1.1 THE GRAPHIC LANGUAGE

Although people around the world speak different languages, a universal graphic language has existed since the earliest of times. The earliest forms of writing were through picture forms, such as the Egyptian hieroglyphics (Fig. 1.4). Later these forms were simplified and became the abstract symbols used in our writing today.

A drawing is a *graphic representation* of a real thing, an idea, or a proposed design for later manufacture or construction. Drawings may take many forms, but the graphic method of representation is a basic natural form of communication of ideas that is universal and timeless.

■ 1.2 ARTISTIC AND TECHNICAL DRAWINGS

Graphic representation has been developed along two distinct lines, according to purpose: (1) artistic and (2) technical.

From the beginning of time, artists have used drawings to express aesthetic, philosophic, or other abstract ideas. People learned by conversing with their elders and by looking at sculptures, pictures, or drawings

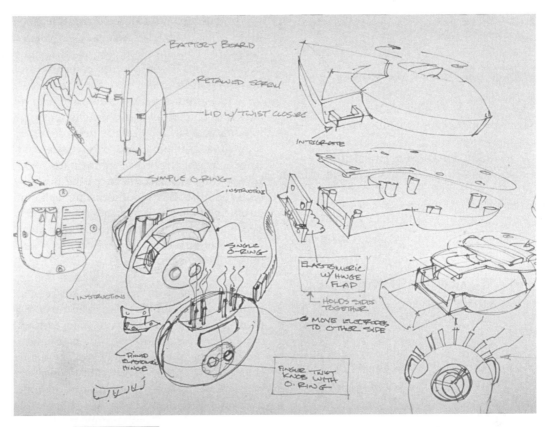

FIGURE 1.1 **An Initial Idea Sketch.** *Courtesy of Ratio Design Lab, Inc.*

FIGURE 1.2 **Engineers Work To Have an In-depth Understanding of a Product's Design.** *Courtesy of AC Engineering, Inc.*

FIGURE 1.3 **Part of a Typical Engineering Design Department.** *Courtesy of Hewlett-Packard Company.*

FIGURE 1.4   **Egyptian Hieroglyphics.**

in public places. Everybody could understand pictures, and they were a principal source of information.

The other line along which drawing has developed has been the technical. From the beginning of recorded history, people have used drawings to represent the design of objects to be built or constructed. No trace remains of these earliest drawings, but we know that drawings were used, for people could not have designed and built as they did without using fairly accurate drawings.

■ 1.3 EARLY TECHNICAL DRAWING

Perhaps the earliest known technical drawing in existence is the plan view for a design of a fortress drawn by the Chaldean engineer Gudea and engraved on a stone tablet (Fig. 1.5). It is remarkable how similar this plan is to those made by modern architects, although it was "drawn" thousands of years before paper was invented.

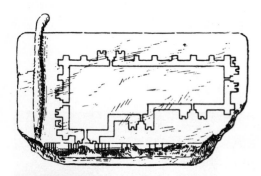

FIGURE 1.5   **Plan of a Fortress. This stone tablet is part of a statue now in the Louvre, in Paris, and is classified in the earliest period of Chaldean art, about 4000 B.C.** *From Transactions ASCE, May 1891.*

In museums we can see actual specimens of early drawing instruments. Compasses were made of bronze and were about the same size as those in current use. As shown in Fig. 1.6, the old compass resembled the dividers of today. Pens were cut from reeds.

The scriber-type compass gave way to the compass with a graphite lead shortly after graphic pencils were developed. At Mount Vernon we can see the drawing instruments used by the great civil engineer George Washington, bearing the date 1749. This set is very similar to the conventional drawing instruments used today. It consists of a divider and compass with pencil and pen attachments plus a ruling pen with parallel blades similar to the modern pens (Fig. 1.7).

The theory of projections of objects on imaginary planes of projection (to obtain *views*; see Chapter 6) apparently was not developed until the early part of the fifteenth century by the Italian architects Alberti, Brunelleschi, and others. It is well known that Leonardo da Vinci used drawings to record and transmit to others his ideas and designs for mechanical constructions, and many of these drawings are still in existence (Fig. 1.8). It is not clear whether Leonardo ever made mechanical drawings showing orthographic views as we now know them, but it is probable that he did. Leonardo's treatise on painting, published in 1651, is regarded as the first book ever printed on the theory of projection drawing; however, its subject was perspective and not orthographic projection.

■ 1.4 EARLY DESCRIPTIVE GEOMETRY

Descriptive geometry is the science of graphic representation and solution of spatial problems. The beginnings of descriptive geometry are associated with the problems encountered in designs for building construction and military fortifications of France in the eighteenth century. Gaspard Monge (1746–1818) is considered the "inventor" of descriptive geometry, although his efforts were preceded by publications on

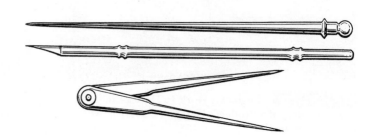

FIGURE 1.6   **Roman Stylus, Pen, and Compass.** *From Historical Note on Drawing Instruments, published by V & E Manufacturing Co.*

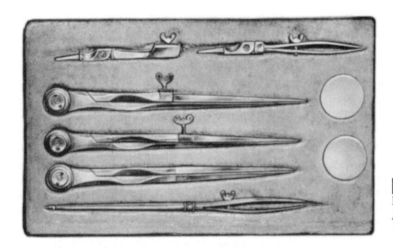

FIGURE 1.7   George Washington's Drawing Instruments. *From Historical Note on Drawing Instruments, published by V & E Manufacturing Co.*

FIGURE 1.8   An Arsenal, by Leonardo da Vinci. *Courtesy of the Bettmann Archive.*

stereotomy (the art or science of cutting solid bodies, especially stone, into desired shapes), architecture, and perspective, in which many of the principles were used. It was while he was a professor at the Polytechnic School in France near the close of the eighteenth century that Monge developed the principles of projection that are now the basis of our technical drawing. [A view of a part for a design is technically known as a **projection** (see §6.9; Chapter 6)]. These principles of descriptive geometry were soon recognized to be of such military importance that Monge was compelled to keep his principles secret until 1795, after which they became an important part of technical education in France and Germany and later in the United States. His book, *La Géométrie Descriptive*, is still regarded as the first text to expound the basic principles of projection drawing.

Monge's principles were brought to the United States from France in 1816 by Claude Crozet, an alumnus of the Polytechnic School and a professor at the United States Military Academy at West Point. He published the first text on the subject of descriptive geometry in the English language in 1821. In the years immediately following, these principles became a regular part of early engineering curricula at Rensselaer Polytechnic Institute, Harvard University, Yale University, and others. During the same period, the idea of manufacturing interchangeable parts in the early arms industries was being developed, and the principles of projection drawing were applied to these problems.

■ 1.5 MODERN TECHNICAL DRAWING

Perhaps the first text on technical drawing in this country was *Geometrical Drawing*, published in 1849 by William Minifie, a high school teacher in Baltimore. In 1850 the Alteneder family organized the first drawing instrument manufacturing company in the United States (Theo. Alteneder & Sons, Philadelphia). In 1876 the blueprint process was introduced at the Philadelphia Centennial Exposition. Up to this time the graphic language was more or less an art, characterized by fine-line drawings made to resemble copper-plate engraving, by the use of shade lines, and by the use of watercolor "washes." These techniques became unnecessary after the introduction of blueprinting, and drawings gradually were made less ornate to obtain the best results from this method of reproduction. This was the beginning of modern technical drawing. The graphic language now became a relatively exact method of representation, and the building of a work-

ing model as a regular preliminary to construction became unnecessary.

Up to about 1900, drawings everywhere were generally made in what is called *first-angle projection*, in which the top view was placed under the front view, the left-side view was placed at the right of the front view, and so on (§6.38). At this time in the United States, after a considerable period of argument pro and con, practice gradually settled on the present *third-angle projection*, in which the views are situated in what we regard as their more logical or natural positions. Today, third-angle projection is standard in the United States, but first-angle projection is still used throughout much of the world.

During the early part of the twentieth century, many books were published in which the graphic language was analyzed and explained in connection with its rapidly changing engineering design and industrial applications. Many of these writers were not satisfied with the term because they recognized that technical drawing was really a graphic language. Anthony's *An Introduction to the Graphic Language*, French's *Engineering Drawing*, and Giesecke et al., *Technical Drawing*, were all written with this point of view.

■ 1.6 DESIGN PROCESS

Design is the ability to combine ideas, scientific principles, resources, and often existing products into a solution for a problem. This ability to solve problems in design is the result of an organized and orderly approach to the problem known as the ***design process***. The design process is not the haphazard operation of an inventor working in a garage or basement, although it might well begin in that manner. Nearly all successful companies support a well-organized design effort, and the vitality of the company depends to a large extent on the planned output of its designers.

The design process leads to manufacturing, assembly, marketing, service, and the many activities necessary for a successful product, and it is composed of several phases. Although many industrial groups may identify them in their own particular way, one procedure for the design of a new or improved product follows these five stages:

1. Problem identification
2. Concepts and ideas
3. Compromise solutions
4. *Models* and/or *prototypes*
5. Production and/or working drawings

Ideally, the design moves through these stages, but as new information becomes available, it may be necessary to return to a previous stage and repeat a procedure. The design process will be discussed in detail in Chapter 14.

■ 1.7 DRAFTING STANDARDS

Modern technical drawing books tended to standardize the characters of the graphic language, to eliminate its provincialisms and dialects, and to give industry, engineering, and science a uniform, effective graphic language. Of prime importance in this movement in the United States has been the work of the American National Standards Institute (ANSI) with the American Society for Engineering Education, the Society of Automotive Engineers, and the American Society of Mechanical Engineers. As sponsors, they have prepared the *American National Standard Drafting Manual— Y14*, which is comprised of a number of separate sections that were published as approved standards as they were completed over a period of years (see Appendix 1).

These sections outline the most important idioms and usages in a form that is acceptable to the majority. They are considered the most authoritative guide to uniform drafting practices in this country today. The Y14 Standard gives the characters of the graphic language, and it remains for the textbooks to explain the grammar and the penmanship (see §1.9 for a definition of penmanship as it applies to technical drawing).

■ 1.8 DEFINITIONS

After this brief survey of the historical development of the graphic language, and before we begin a serious study of theory and applications, a few terms need to be defined.

DESCRIPTIVE GEOMETRY   This is the three-dimensional geometry forming the background for the practical applications of the graphic language and through which many of its problems may be solved graphically.

INSTRUMENTAL DRAWING OR MECHANICAL DRAWING   These terms properly apply only to a drawing made with drawing instruments. The use of "mechanical drawing" to denote all industrial drawings is unfortunate not only because such drawings are not always mechanically drawn, but also because that usage tends to belittle the broad scope of the graphic language by naming it superficially for its principal mode of execution.

COMPUTER GRAPHICS   This is the application of conventional computer techniques (with the aid of one of many graphic data processing systems available) to the analysis, modification, and finalizing of a graphical solution. The use of computers to produce technical drawings is called computer-aided design or computer-aided drafting (CAD) and also computer-aided design and drafting (CADD). (See Fig. 1.9.)

You can use CAD to create a useful database that accurately describes the three-dimensional geometry of the machine part, structure, or system you are designing. This database can be used to perform analysis, directly machine parts, or create illustrations for catalogs and service manuals.

ENGINEERING DRAWING AND ENGINEERING DRAFTING   These are broad terms widely used to denote the graphic language. However, since the graphic language is also used by a much larger group of people in diverse fields who are concerned with technical work or industrial production, these terms are not broad enough.

TECHNICAL DRAWING   This is a broad term that adequately suggests the scope of the graphic language. It is rightly applied to any drawing used to express technical ideas. This term has been used by various writers since Monge's time at least and is still widely used, mostly in Europe.

ENGINEERING GRAPHICS OR ENGINEERING DESIGN GRAPHICS   These terms are generally applied to drawings for technical use and have come to

FIGURE 1.9   **A CAD workstation.** *Courtesy of Digital Equipment Corporation.*

GRAPHICS SPOTLIGHT

From Art to Part

SINGLE DATABASE

You can use a single CAD database to design, document, analyze, create prototypes, and directly manufacture finished parts for your design. The term *art to part* is sometimes used to describe a CAD database being utilized for many or all of these purposes.

SKETCHING FREEDOM

Initial ideas for the design are frequently sketched freehand, as shown in Fig. A. While generating ideas for the design, it is important to be able to quickly generate creative ideas without the confines of using the computer. Sketching is still generally the best tool to help in this process.

INTELLIGENT MODELS

After generating the initial ideas, the best alternatives for the design are developed further. At this point, the engineer may create rough 3D drawing geometry like you see in Fig. B, perhaps using a parametric modeling software. *Parametric modeling* uses variables to constrain the shape of the geometry. Using parametric modeling the designer roughly sketches initial shapes and applies drawing dimensions and constraints to create models that have "intelligence." Later the de-

(B)

signer can change the dimensions and constraints as the design is refined so that new models do not have to be created for each design change. Realistic renderings of the model help you visualize the design.

OPTIMIZING THE DESIGN

You can export the refined model directly into a Finite Element Analysis (FEA) program to perform structural, thermal, and modal analysis as shown in Fig. C. The parametric model can easily be changed if the

(A)

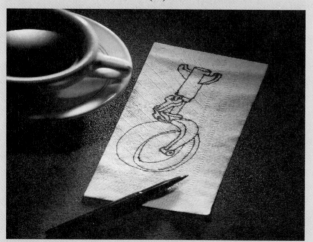

(C)

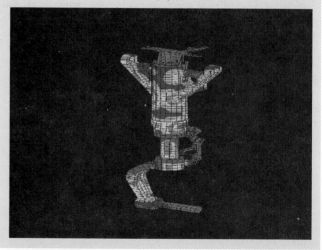

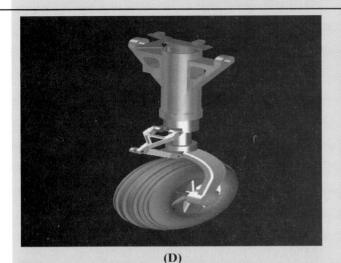

(D)

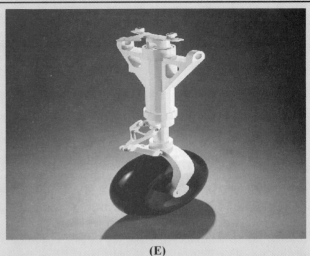

(E)

analysis shows that the initial design will not meet requirements. Simulation programs may even animate the performance and function of the system before a prototype is ever constructed. The tolerances and fits between mating parts can be checked within the parametric modeling and design software. Fig. D shows a shaded 3D model that closely resembles the final part.

RAPID PROTOTYPING

While refining the design ideas, engineers often work concurrently with manufacturing to determine the best ways to make and assemble the necessary parts. After several cycles of refining, analyzing, and synthesizing the best ideas, the final design is ready to go into production. Rapid prototyping systems allow parts to quickly be generated directly from the 3D models for mockup and testing. Fig. E shows the prototyped part.

When the design is approved the finished parts can be created using numerical controlled machines which get their tool paths directly from the 3D models.

GET NEW PRODUCTS TO MARKET QUICKLY

The necessary documentation for the design, manuals, brochures and other literature can be created directly from the same geometry used for design and manufacturing. Shortened design cycle time, improved communication, better opportunity to analyze and make design changes, are all advantages for companies using integrated CAD software for the design, documentation, and manufacture of their products.

mean that part of technical drawing that is concerned with the graphical representation of designs and specifications for physical objects and data relationships as used in engineering and science.

TECHNICAL SKETCHING   This is the freehand expression of the graphic language. Technical sketching is a valuable tool for engineers and others engaged in technical work because through it most technical ideas can be expressed quickly and effectively without the use of special equipment.

BLUEPRINT READING   This is the term applied to the "reading" of the language from drawings made by others. Actually, the blueprint process has now been replaced by other more efficient processes, but the term "blueprint reading" has been accepted through usage to mean the interpretation of all ideas expressed on technical drawings, whether or not the drawings are blueprints.

■ 1.9 WHAT ENGINEERING, SCIENCE, AND TECHNOLOGY STUDENTS SHOULD KNOW

From the dawn of history, the development of technical knowledge has been accompanied, and to a large extent made possible, by a corresponding graphic language. Today the intimate connection between engineering and science and the universal graphic language is more vital than ever before, and engineers, scientists, and technicians ignorant of or deficient in their field's principal mode of expression in technical field are professionally illiterate. Thus, training in the application of technical drawing is required in virtually every engineering school in the world.

The old days of fine-line drawings and of shading and "washes" are gone forever; artistic talent is no longer a prerequisite to learning the fundamentals of the graphic language. Instead, today's graphics student needs the same aptitudes, abilities, and computer skills that are needed in science and engineering courses.

The well-trained engineer, scientist, or technician must be able to make and read correct graphical representations of engineering structures, designs, and data relationships. This means that the individual must understand the fundamental principles, or the *grammar*, of the language and be able to execute the work with reasonable skill (which is *penmanship*).

Graphics students often try to excuse themselves for inferior results (usually caused by lack of application) by arguing that after graduation they do not expect to do any drafting at all. Such students presumptuously expect, immediately after graduation, to be accomplished engineers concerned with bigger things. They forget that first assignments may involve working with drawings and possibly revising drawings, either on a board or on a computer, under the direction of an experienced engineer. Entering the engineering profession via graphics provides an excellent opportunity to learn about the product, the company operations, and the supervision of others.

Even a young engineer who has not been successful in developing a skillful penmanship in the graphic language will have use for its grammar, since the ability to *read* a drawing will be of utmost importance. Furthermore, the engineering student is apt to overlook that, in practically all the subsequent courses taken in college, technical drawings will be encountered in most textbooks. The student is often called on by instructors to supplement calculations with mechanical drawings or sketches. Thus, a mastery of a course in technical drawing utilizing both traditional methods and computer systems (CAD) will aid materially, not only in professional practice after graduation but more immediately in other technical courses.

Besides the direct advantages of a serious study of the graphic language, many students learn the meaning of neatness, speed, and accuracy for the first time in a drawing course. These are basic and necessary habits for every successful engineer, scientist, and technician.

The ability to *think in three dimensions* is one of the most important requisites of successful scientists, designers, and engineers. Learning to visualize objects in space, to use the constructive imagination, is one of the principal values to be obtained from a study of the graphic language. Persons of extraordinary creative ability possess the ability to *visualize* to an outstanding degree. It is difficult to think of Edison, De Forest, or Einstein as being deficient in constructive imagination.

■ 1.10 PROJECTIONS

Behind every drawing of an object is a space relationship involving four imaginary things:

1. The observer's eye, or the *station point*
2. The object
3. The *plane of projection*
4. The *projectors*, also called *visual rays* or *lines of sight*.

For example, in Fig. 1.10a the drawing EFGH is the projection, on the plane of projection A, of the square

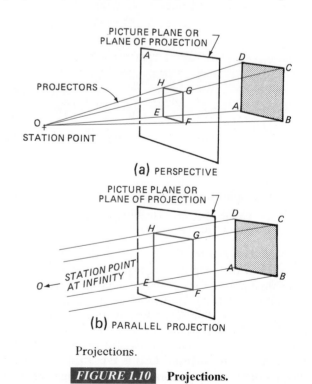

(a) PERSPECTIVE

(b) PARALLEL PROJECTION

Projections.

FIGURE 1.10   **Projections.**

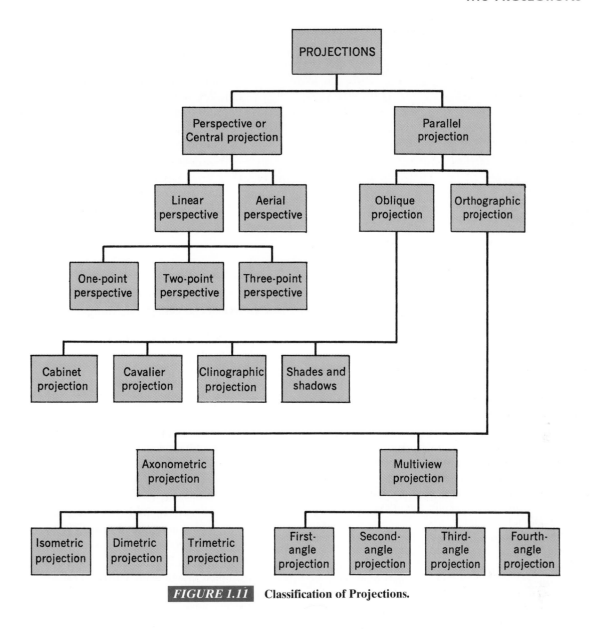

FIGURE 1.11   Classification of Projections.

ABCD as viewed by an observer whose eye is at the point O. The projection or drawing on the plane is produced by the points at which the projectors pierce the plane of projection (*piercing points*). In this case, where the observer is relatively close to the object and the projectors form a "cone" of projectors, the resulting projection is known as a *perspective*.

If the observer's eye is imagined as infinitely distant from the object and the plane of projection, the projectors will be parallel, as shown in Fig. 1.10b; hence, this type of projection is known as a *parallel projection*. If the projectors, in addition to being paral-lel to each other, are perpendicular (normal) to the plane of projection, the result is an *orthographic*, or right-angle, *projection*. If they are parallel to each other but oblique to the plane of projection, the result is an *oblique projection*.

These two main types of projection—perspective or central projection, and parallel, or central, projection—are further broken down into many subtypes, as shown in Fig. 1.11, and will be treated at length in the various chapters that follow.

A classification of the main types of projection according to their projectors is shown in Table 1.1.

Classes of Projection	Distance from Observer to Plane of Projection	Direction of Projectors
Perspective	Finite	Radiating from station point
Parallel	Infinite	Parallel to each other
Oblique	Infinite	Parallel to each other and oblique to plane of projection
Orthographic	Infinite	Perpendicular to plane of projection
Axonometric	Infinite	Perpendicular to plane of projection
Multiview	Infinite	Perpendicular to planes of projection

TABLE 1.1   **Classification by Projectors.**

■ KEY WORDS

GRAPHIC LANGUAGE

TECHNICAL DRAWING

DRAWING INSTRUMENTS

COMPUTER GRAPHICS

PERSPECTIVE

DESIGN PROCESS

DESIGN TEAM

DESCRIPTIVE GEOMETRY

STANDARDS

PROJECTION

PLANE OF PROJECTION

■ CHAPTER SUMMARY

• The members of the engineering design project team must be able to communicate among themselves and with the rest of the project team in order to contribute to the team's success.

• The graphic language is the universal language used by every engineering team designing and developing products throughout the world.

• There are two basic types of drawings: artistic and technical.

• Technical drawing is based on the universal principles of descriptive geometry, developed in the late eighteenth century in France.

• The design process is the ability to combine ideas, scientific principles, resources, and existing products into a solution for a problem. It consists of five specific stages.

• Every technical drawing is based on standards that prescribe what each symbol, line, and arc means.

• Computer running CAD software are the current tools used by drafters. However, the basic drawing principles are the same ones used for hundreds of years.

• Drawings are based on the projection of an image onto a plane of projection. There are two types of projection: parallel and perspective.

• Successful companies hire skilled people who can add value to their design team. The proper use of equipment and a thorough understanding of the graphic language are two essential skills employers require.

■ REVIEW QUESTIONS

1. What is the role of the engineer on the design team?
2. What is the difference between mechanical drawing and sketching?
3. Describe the main difference between parallel projection and perspective projection.
4. When is sketching an appropriate form of graphic communication?
5. Why are standards so important for members of the engineering design team?

6. What is the most important new tool used by drafters?

7. What is a plane of projection?

8. What are projectors and how are they drawn?

9. What is the design process?

10. What are the five phases of the design process?

CHAPTER 2

INTRODUCTION TO CAD

OBJECTIVES

After studying the material in this chapter, you should be able to:

1. List the basic components of a computer-aided drawing workstation.
2. Describe the relationship between computer-aided drawing (CAD) and computer-aided manufacturing (CAM).
3. List the major parts of a computer and describe their function.
4. Describe the purpose of a computer operating system.
5. List several input and output devices.
6. Describe ways in which a computer stores information.
7. Explain the differences between ROM and RAM.
8. Explain the differences between a bit and a byte.
9. Understand issues that affect the choice and the use of a CAD system.
10. Explain features common to most CAD software.

OVERVIEW

The use of electronic computers today in nearly every phase of engineering, science, business, and industry is well known. The computer has altered accounting and manufacturing procedures, as well as engineering concepts. The integration of computers into the manufacturing process from design to prototyping, manufacture, and marketing, is changing the methods used in the education and training of technicians, drafters, designers, and engineers.

Engineering, in particular, is a constantly changing field. As new theories and practices evolve, more powerful tools are developed and perfected to allow the engineer and designer to keep pace with the expanding body of technical knowledge. The computer has become an indispensable and effective tool for design and practical problem solving. New methods for analysis and design, the creation of technical drawings, and the solving of engineering problems, as well as the development of new concepts in automation and robotics, are the result of the influence of the computer on current engineering and industrial practice.

Computers are not new. Charles Babbage, an English mathematician, developed the idea of a mechanical digital computer in the 1830s, and many of the principles used in Babbage's design (Fig. 2.1) are the basis of today's computers. The computer has appeared in literature and science fiction to be a mysterious, uncompromising, often sinister, machine, but it is nothing more than a tool. It is capable of data storage, basic logical functions, and mathematical calculations. Computer applications have expanded human capabilities to such an extent that virtually every type of business and industry utilizes a computer directly or indirectly.

■ 2.1 COMPUTER SYSTEMS AND COMPONENTS

Engineers and drafters have used computers for many years to perform the mathematical calculations required in their work. Only recently, however, has the computer been accepted as a valuable tool in the preparation of technical drawings. Traditionally, drawings were made by using drafting instruments and applying ink, or graphite, to paper or film. Revisions and reproductions of these drawings were time consuming and often costly. Now the computer is used to produce, revise, store, and transmit original drawings. This

FIGURE 2.1   **A Working Model of Charles Babbage's "Difference Engine" Originally Designed in 1833.** *From the New York Public Library Picture Collection.*

method of producing drawings is called *computer-aided design* or *computer-aided drafting* (CAD) and *computer-aided design and drafting* (CADD). Since these and other comparable terms are used synonymously, and since industry and software creators are beginning to standardize, they will be referred to throughout this book simply as CAD.

Other terms, such as *computer-aided manufacturing* (CAM), *computer-integrated manufacturing* (CIM), and computer-assisted engineering (CAE), are often used in conjunction with the term "CAD." The term "CAD/CAM" refers to the integration of computers into the design and production process (see Fig. 2.2). The term "CAD/CAE/CAM/CIM" describes the use of computers in the total design and manufacturing process, from design to production, publishing of technical material, marketing, and cost accounting. The single concept that these terms refer to is the use of a computer and software to aid the designer or drafter in the preparation and completion of a task. (See Chapter 10 for a further discussion on CAD and manufacturing processes.)

Computer graphics is a very broad field. It covers the creation and manipulation of computer-generated images and may include areas in photography, busi-

FIGURE 2.2   **CAD/CAM Driven Machine Tool Cutting.** *Courtesy of David Sailors.*

ness, cartography, animation, publication, as well as drafting and design.

A complete computer system consists of *hardware* and *software*. The various pieces of physical equipment that comprise a computer system are known as hardware. The programs and instructions that permit the computer system to operate are classified as software. Computer programs are categorized as either *application programs* or *operating systems*. Operating systems, such as DOS, Windows (Fig. 2.3), and UNIX, are sets of instructions that control the operation of the computer and peripheral devices as well as the execution of specific programs. This type of program may also provide support for activities and programs (such as input/output [I/O] control, editing, storage assignment, data management, and diagnostics), assign drives for I/O devices, and provide

FIGURE 2.3   **Screenshot of the Windows 98 Operating System.** *Courtesy of Microsoft Corporation.*

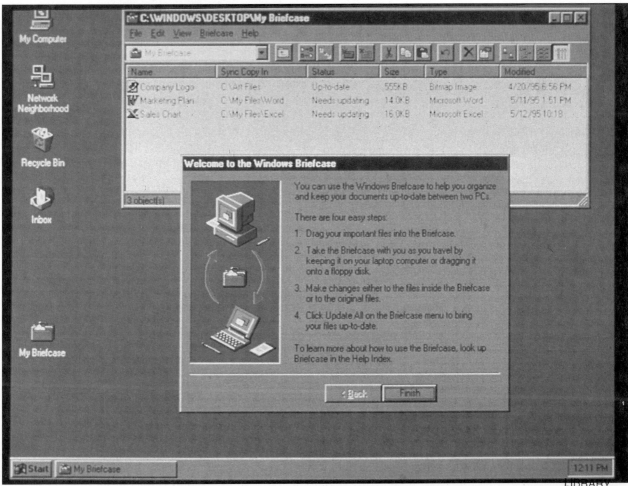

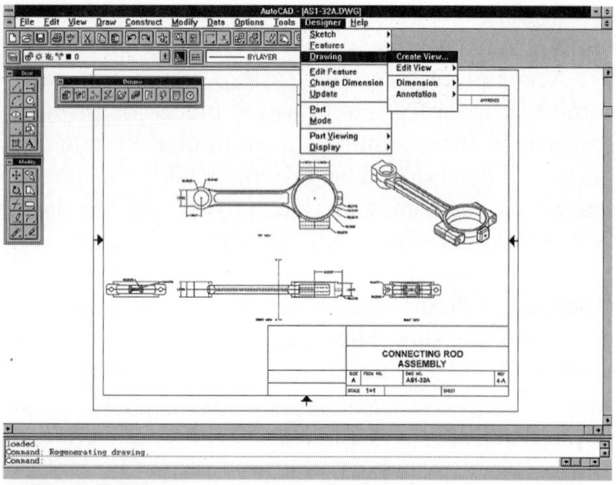

FIGURE 2.4   **AutoCAD is a Widely Used Drafting Application Program.** *This material has been reprinted with the permission from and under the copyright of Autodesk, Inc.*

FIGURE 2.5   **Advanced CAM Technology Used in High-Resolution Color Picture Tube Production.** *Courtesy of Zenith Electronics Corporation and Charlie Westerman.*

support for standard system commands and networking. Application programs are the link between specific system use and its related tasks—design, drafting, desktop publishing, etc.—and the general operating system program (Fig. 2.4).

■ 2.2 COMPUTER TYPES

Computers may be classified as one of two distinct types: *analog* or *digital*. An analog computer measures continuously without steps, whereas the digital computer counts by digits, going from one to two, three, and so on, in distinct steps. An electric wall clock with minute and hour hands and the radial speedometer on a car are examples of analog devices. An abacus and a digital watch are examples of digital devices. Digital computers are more widely used than analog computers because they are more flexible and can do a greater variety of jobs.

Analog computers are generally used for mathematical problem solving. This type of computer, which measures continuous physical properties, is often used to monitor and control electronic, hydraulic, or mechanical equipment. Digital computers have extensive applications in business and finance, engineering, numerical control, and computer graphics (Fig. 2.5).

Both types of computers have undergone great changes in appearance and in operation. Equipment that once filled the greater part of a large room has now been replaced by machines that occupy small

desktop areas. The single most important advancement in computer technology has been the development of the *integrated circuit* (IC). The IC chip has replaced thousands of components on the *printed circuit* (PC) board and made possible the development of microprocessors. The microprocessor is the processing unit of a computer. The difference in size between a PC board with individual components and an IC chip is shown in Fig. 2.6. The term "microminiaturization" is applied to advanced integrated circuit chip technology. The evolution of IC chip technology has led to the increased production of low-cost microcomputers. Microcomputers are largely responsible for the increase in use of computer-aided drafting systems in industry. Low-cost microcomputer CAD systems can now be cost justified by industrial users

FIGURE 2.6   Size Comparison of a PC Board and an IC Chip.

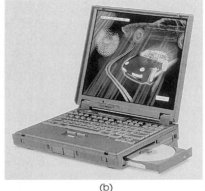
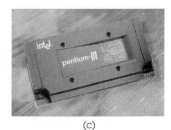

(a)

(b)

(c)

FIGURE 2.7 (a) Computer with Intel Pentium III. *Courtesy of Dell Corporation.* (b) Think pad. *Courtesy of IBM Corporation.* (c) Intel Pentium III chip. *Courtesy of Intel Corporation.*

(Figs. 2.7a-c). With the recent release of the Pentium III processor, CAD applications can run faster and more efficiently than ever before (See Figs. 2.7a and c). Thinkpad computers make mobile reference to CAD files not only possible, but commonplace (See Fig. 2.7b).

Since CAD systems utilize digital computers, we will restrict our discussion of computer types to digital computers.

■ 2.3 COMPUTER-AIDED DRAFTING

The first demonstration of the computer as a design and drafting tool was given at the Massachusetts Institute of Technology in 1963 by Dr. Ivan Sutherland. His system, called "Sketchpad," used a cathode ray tube and a light pen for graphic input to a computer. An earlier system, called SAGE, was developed in the 1950s for the Air Defense Command and used the light pen for data input. The first commercial computer-aided drafting system was introduced in 1964 by International Business Machines (IBM).

Many changes have taken place since the introduction of the first CAD system. The changes are due to the advent of the microprocessor, more sophisticated software (programs), and new industrial applications. In most cases, the drafter/engineer can create, revise, obtain prints (hard copy), and store drawings with relative ease, utilizing less space. CAD was originally used to aid in creating production drawings. The advent of three-dimensional CAD software made it apparent that a 3D computer model (Fig. 2.8) could assist not only in the manufacture of the part but also,

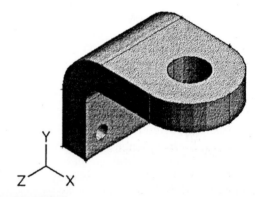

FIGURE 2.8 A CAD Solid Model of the Ball Bracket from a Trailor Hitch Assembly. *From* Machine Design: An Integrated Approach *by Robert Norton, © 1996. Reprinted by permission of Prentice-Hall, Inc., Upper Saddle River, NJ.*

along with its three-dimensional database, in testing the design with finite element analysis programs (Fig. 2.9), in developing technical manuals and other documentation that combine illustrations of the design with text from word processing programs, and in marketing (for which the 3D solid models can be used with a rendering and animation program). Increases in productivity and cost effectiveness are two advantages constantly stressed by CAD advocates. In addition, CAD stations can be linked either directly or through a *local area network* (LAN) to the manufacturing or production equipment, or they can be linked with *numerical control* (NC) equipment to program NC machines automatically in manufacturing operations or in robotics (Fig. 2.10).

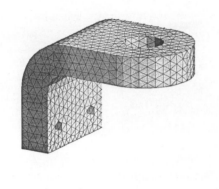

FIGURE 2.9   **An FEM Mesh Applied to the Solid Model of the Ball Bracket.** *From* **Machine Design: An Integrated Approach** *by Robert Norton, © 1996. Reprinted by permission of Prentice-Hall, Inc., Upper Saddle River, NJ.*

The primary users of CAD are in mechanical engineering and electronic design, civil engineering, and cartography. The design and layout of printed circuits are a principal application of CAD in the electronics industry, which, prior to 1976, was the largest CAD user. Mechanical engineering has since overtaken electronics and continues to expand its CAD applications and use. Continued expansion in mechanical design applications is expected because the design, analysis, and numerical control capabilities of CAD can be applied to a varied range of products and processes. Cartography, seismic data display, demographic analysis, urban planning, piping layouts, and especially architectural design also show growth in CAD use. A relatively new area in computer graphics is *image processing*,

FIGURE 2.10   **Computers Work with NC Machines in Modem Industry. This is an Example of a Computer Numerical Control (CNC) Four-Axis Turning Center.**

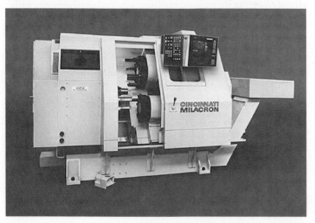

which includes animation, 35-mm slide preparation, photocolor enhancement, and font and character generation (used in television broadcasting and the graphic arts industry).

■ 2.4 CAD SYSTEM CONFIGURATIONS

All computer-aided drafting systems consist of similar hardware components (Fig. 2.11), such as input devices, a central processing unit, data storage devices, and output devices. For input devices, the system will have one or more of the following: a keyboard, mouse, trackball, digitizer/graphics tablet, and light pen. For output, The CAD system will include devices such as plotters, printers, and some type of monitor. The system must also have a data storage device, such as a tape drive, a hard (fixed) or soft (floppy) disk drive, or an optical disk drive. Finally, a computer or central processing unit (CPU) is needed to do all the numerical manipulations and to control all the other devices connected to the system.

Frequently, some devices are combined. For example, a terminal or workstation can contain the keyboard, monitor, disk drives, and a CPU all in the same cabinet. Such a combined device is often called a *workstation* (Fig. 2.12).

■ 2.5 CENTRAL PROCESSING UNIT

The CPU (Fig. 2.13), or computer, receives all data and manages, manipulates, and controls all functions of the CAD system. CAD systems use digital computers. All data must be converted into a binary form or

FIGURE 2.11   **Complete CAD Systems Need Input Devices, Output Devices, Storage Devices, and a Central Processing Unit.** *Courtesy of Hewlett-Packard Company.*

FIGURE 2.12 **A Powerful Computer Workstation.**
Courtesy of Sun Microsystems.

dicates the maximum word size that can be processed, as a unit, during an instruction cycle. Often, computers are categorized by their word length, such as 16-bit or 32-bit computers. The number of bits in the word length indicates the processing power of the computer (the larger the word length, the greater the processing power). A sequential group of adjacent bits in a computer is called a *byte*. The current industry standard is that 8 bits equal 1 byte. A byte represents a character that is operated on as a unit by the CPU. The length of a word on a majority of computer system is currently 4 bytes. This means that each word in any of these storage systems occupies a 32-bit storage location. The memory capacity of a computer is therefore expressed as a number of bytes rather than bits.

Inside the CPU is the brain of the computer—the microprocessor chip (Fig. 2.13). IBM-compatible computers have used the 386, 486, Pentium (P5), P6, II, and III chips. MacIntosh computers use the Motorola chips or PowerPC chip. Intel's Pentium III processor along with Apple's PowerPC G3 processor have kept pace with industries' every expanding need for speed and memory. Silicon Graphics and Sun computers use their own chips. The job of the microprocessor is to execute the instructions of the software programs.

The microprocessor chip's speed is rated in *megahertz* (MHz), and its power is rated in *millions of instructions per second* (MIPS). It is the largest chip in the CPU and is mounted on the main circuit board,

code for the computer to understand and accept. This code is called *binary coded decimal instructions* (BC-DI). This binary code uses a two-digit system, 1 and 0, to transmit all data through the circuits. The number 1 is the "on" signal; the 0 is the "off" signal. A *bit* is a binary digit. Bits are grouped, or organized, into larger instructions. Word length, which is expressed as bits, differs with various computers. The most common word lengths are 8, 16, and 32 bits. The word length in-

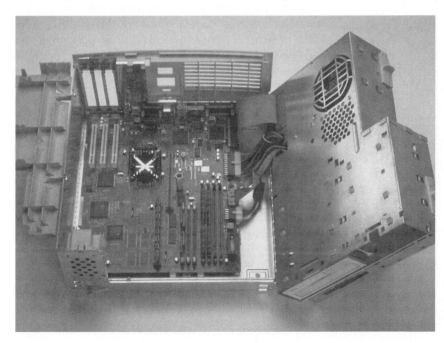

FIGURE 2.13 **Inside a CPU Box.**
Courtesy of Apple Computer.

called the *motherboard*. Attached to the motherboard are all the devices inside the CPU: the memory chips, power supply, connection ports (where the external cables are attached), internal modems, video cards, network interface cards, sound cards, and special hard drive controller cards.

Computer memory is stored on small circuit boards called *single inline memory modules* (SIMMs). Most of the SIMMs are for the main memory, called the *random access memory* (RAM). This memory is temporary. When you turn off the computer, the information in the memory is erased. The software programs and user files are stored in RAM when the program is actively running. The more RAM in a computer, the more programs you can run at one time. If you do not have enough memory, you may have a hard time running even one large program.

There is also permanent memory on the motherboard, which is called *read only memory* (ROM). When the computer is turned off, the ROM chips do not forget what is stored in them. The ROM chips contain basic operating system programs, like simple diagnostic programs that check the computer system to make sure all circuits and devices are operational when the computer is turned on. One type of ROM is called *flash* ROM, and it can be reprogrammed. Normal ROM cannot be reprogrammed.

FIGURE 2.14   **A Mother Board with 10 Bus Slots for Adding Capabilities Such as Video Display Cards.** *Courtesy of International Business Machines.*

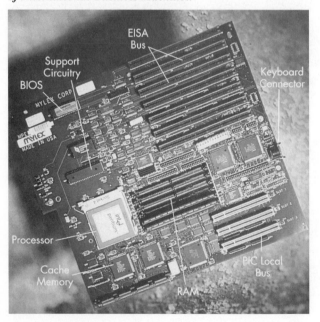

The motherboard contains slots for peripheral devices. (See Fig. 2.14.) Internal modems, video display cards, and network interface cards are typical cards that can be plugged into the motherboard. These cards talk to the microprocessor via an electrical pathway called the *bus*. The bus is like an expressway that allows electrical information to be shared between all the devices connected to the motherboard. Just like an expressway, computer busses have speed limits. They are rated in megahertz like the CPU. The faster the bus, the more quickly information can be transported into the computer.

The rear panel of the CPU contains a series of connector ports (plug receptacles). Different types of ports transfer data differently. Each corresponds to a matching cable that connects the CPU to an external peripheral device. Most printers, for example, use a 25-pin D connector to connect to the parallel port. The keyboard and mouse each have their own connection port (usually a small round port). Many of the cards inside the CPU have connection ports attached to the card that protrude through the back of the CPU. For example, the monitor connects to the port on the video display card. The modem telephone line connects to a port on the back of the internal modem. The LAN cable connects to the port on the back of the network interface card.

■ 2.6 DISPLAY DEVICES

Another major reason for the rapid growth in CAD systems is improvements in display devices. These display devices, commonly referred to as monitors (Fig. 2.15), utilize a wide variety of imaging principles. Each device has definite characteristics with regard to brightness, clarity, resolution, response time, and color. The purpose of any graphics display is to project an image on a screen. The image that is displayed may be alphanumeric (text symbols, letters, and/or numerals) or graphical (pictorial symbols and/or lines). Users of interactive CAD systems communicate directly or indirectly through graphics terminals. The information requested by the user may be displayed as animated figures, graphs, color-coded diagrams, or simply a series of lines.

Most interactive CAD systems use a raster scan monitor. *Raster scan* devices are similar to conventional television screens. These devices produce an image with a matrix of picture-element dots, called *pixels*, within a grid. Each pixel is either a light or dark image that appears on the screen as a dot. As in conventional

FIGURE 2.15   **21″ Monitor for CAD Hi-Res Images.**
Courtesy of NEC Technologies, Inc.

television, an electron beam is swept across the entire screen, line by line, top to bottom. This process is called raster scanning. A signal turns on or illuminates a pixel according to a pattern stored in memory. The screen is scanned around 60 times a second to update the image before the phosphor dims.

Most CAD computers use a 17-inch or 21-inch display. These sizes are measured diagonally across the front of the screen. The image on the monitor is generated by the video display card. These cards determine the resolution of the display and the number of colors. Standard *video graphics array* (VGA) resolution is 640×480 pixels. The more pixels per inch, the greater the resolution and the easier it is to read details on the monitor. Large-screen monitors can support up to 1600 pixels horizontally and 1200 pixels vertically.

In addition to resolution, video display cards can generate a range of colors. Normal color density is 256 colors. Photographic quality requires a color density of 16.7 million colors. When a monitor has a lot of pixels and many colors on the screen, it takes a lot of processing power and video memory to draw the image on the monitor. The microprocessor chip of the CPU cannot handle this load so the video display card often provides its own processor, called a video accelerator, and its own memory. Professional CAD computers usually have a very fast video accelerator that provides high resolution and a large number of colors. With a fast video accelerator, even the largest, high-resolution monitor can redraw in the blink of an eye. Slower video display cards can take up to a minute to redraw the screen. Most video display cards use a standard

15-pin VGA connector. However, larger monitors may require four separate cables.

Video monitors are rated by the speed with which they can refresh the screen. There are two ratings: horizontal refresh rate and vertical refresh rate. The faster the refresh rate, the easier the monitor is on the eyes. For example, if you are using a monitor with only a 60-Hz vertical refresh rate, you will begin to see a flicker on the screen after using it for several hours. Higher refresh rates reduce this annoying flicker. Quality large-screen monitors can cost as much as the CPU, but they are part of the user interface and can affect the long-term productivity of the operator.

■ 2.7 INPUT DEVICES

A CAD system may use one or a combination of input devices to create images on the display screen. Graphic input devices may be grouped into three categories: (1) *keyboard* and *touch sensitive*, (2) *time-dependent* devices, and (3) *coordinate-dependent* devices.

The keyboard is the universal input device by which data and commands are entered. A typical keyboard consists of alphanumeric character keys for keying in letters, numbers, and common symbols, such as #, &, and %; cursor control keys with words or arrows, printed on them, indicating directional movement of the screen cursor, and special function keys, which are used by some software programs for entering commands with a single keystroke. Many large mainframe-based CAD systems have used a special function keypad, or menu pad, that allows access to a command with a single keystroke (Fig. 2.16). Single-stroke command selection was considered so essential for cost effectiveness and ease of use that developers of mini- and microcomputer-based CAD systems included this feature of single and double keystroke command access into their program utilizing the CTRL, ALT, SHIFT, and function keys. Typically, a CAD system will use a keyboard for inputting commands and text, and another input device for cursor control.

A popular input device in use with both large and small CAD systems is a *mouse* (Fig. 2.17a-c). A mouse may be of the mechanical type or the optical type. A mechanical mouse uses a roller, or ball, on the underside of the device to detect movement. An optical mouse senses movement and position by bouncing a light off a special reflective surface. These optical mice track most reliably on a mousepad, but also can function on other surfaces, with the exception of glass. Most mice will have from one to three buttons on top

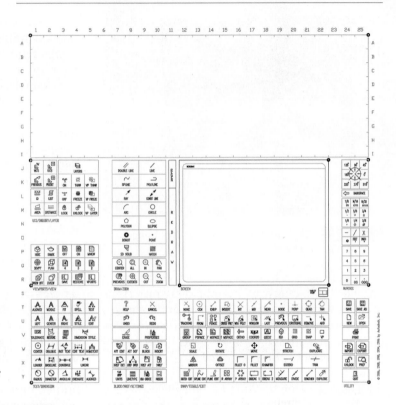

FIGURE 2.16   **An AutoCAD Menu Pad.** *This material has been reprinted with the permission from and under the copyright of Autodesk, Inc.*

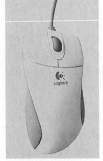  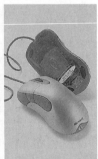

(a) Standard Mouse   (b) Cordless Mouse   (c) Microsoft IntelliEye Optical Mouse

FIGURE 2.17   **Mice (a–c).** *Courtesy of Logitech, Inc. and Microsoft, Inc.*

of them to select positions or commands. Microsoft's new IntelliEye optical mouse comes with two extra buttons. The advantages of a mouse include ease of use, small required working area, and is relatively low cost. A mouse cannot, however, be used to digitize existing drawings into a CAD format.

Digitizing tablets (Fig. 2.18) are another commonly used input device. They can be used to create an original CAD drawing or to convert an existing pen or pencil and paper drawing into a CAD drawing. Digitizing tables range in size from 80×110 to 360×480. Tablets larger than 360×480, called digitizing tables, are used primarily for converting existing drawings to a CAD system format. The *resolution* of digitizing tablets is important. This determines how small a movement the input device can detect (usually expressed in thousandths of an inch) and depends on the number of wires per inch in the tablet's grid system. The working area on a tablet can have areas that are used as menus to pick commands from the CAD system. Attached to the tablet will be either a puck or stylus. A *puck* is a small, hand-held device with a clear plastic extension (or window containing crosshairs) that transfers the location of the puck on the tablet-grid to the relative location on the screen. Single or multiple buttons on top of the puck are used to select points and/or commands. A *stylus* appears to be a ballpoint pen with an electronic cable attached to it. The tip of the stylus senses the position on the tablet grid and relays these coordinates to the computer. When the stylus is moved across the tablet, the screen cursor

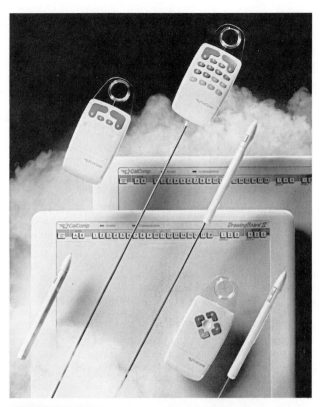

FIGURE 2.18 Tablets with Picking Devices—Puck and Stylus—Wired and Wireless. *Courtesy of CalComp Digitizer Division.*

FIGURE 2.19 Large Format Scanners for Capturing Documents Electronically. *Courtesy of CalComp Digitizer Division.*

moves correspondingly across the screen. The stylus also contains a pressure-sensitive tip that enables the user to select points or commands by pressing down on the stylus.

Existing paper drawings can also be converted to CAD drawings using a scanner (Fig. 2.19) to read the existing drawing. Scanners create raster or bitmapped files, so the scanned images need to be converted to vector (line drawing) formats before they are useful for CAD. There are a number of raster-to-vector conversion programs on the market that help automate this process.

One of the oldest input devices used on CAD systems is the *trackball* (Fig. 2.20a). Trackballs were used on many large mainframe-based CAD systems and were often incorporated into the keyboard. Now they are a popular input device on portable computer systems. A trackball consists of a ball nested in a holder or cup, much like the underside of a mechanical mouse, and from one to three buttons for entering coordinate data into the system. Within the holder are

sensors that pick up the movement of the ball. The ball is moved in any direction with the fingers or hand to control cursor movement on the CRT screen. Cursor speed and button functions can be set by the user. Figure 2.20b shows a thumb mouse, similar to a trackball, which is actually a 3-D controller. These controllers allow the user to manipulate objects on x, y, and z axes, as well as control pitch, roll and yaw movements. These 3-D controllers allow the user complete control of graphic objects in six degrees of freedom and are readily used in CAD applications.

A *joystick* (Fig. 2.21) is an input device more commonly used with video games today than with CAD systems. This device looks like a small rectangular box with a lever extending vertically form the top surface. This hand-controlled lever is used to manipulate the screen cursor and manually enter coordinate data. Buttons of the device can be used to enter coordinate location or specific commands. This device is inexpensive and requires a very small working area, but it sometimes presents a problem when precise cursor positioning is important.

The *light pen* is the oldest type of CAD input device currently in use (Fig. 2.22). It looks much like a ballpoint pen or the stylus on a digitized tablet. A light pen is a hand-held photosensitive device that works only with raster scan or vector refresh monitors and is used to identify displayed elements of a design or to specify a location on the screen where an action is to take place. The pen senses light created by the electron beam as it scans the surface of the CRT. When the

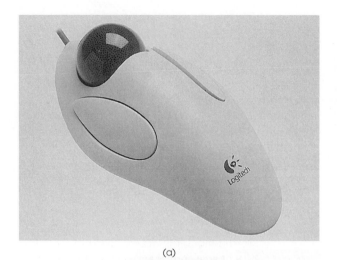

(a)

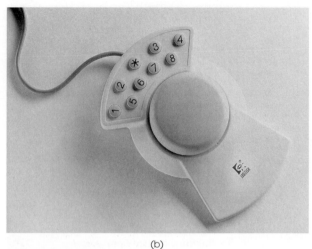

(b)

FIGURE 2.20 (a) A Trackball. *Photo Courtesy of Logitech, Inc.* (b) This Magellan 3-D Controller Allows the User to Manipulate Graphic Objects with *x, y, z*, Pitch, Roll and Yaw Movement. *Courtesy of Logitech, Inc.*

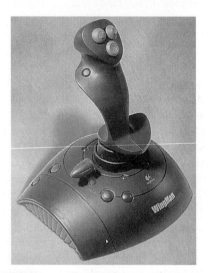

FIGURE 2.21 Joystick. *Courtesy of Logitech, Inc.*

FIGURE 2.22 Lightpen. *Courtesy of HEI, Inc.*

pen is held close to or touches the CRT screen, the computer can determine its location and position the cursor under the pen. Because this input device more closely emulates the traditional drafter's pencil or pen than other devices, it quickly gained popularity in the technical drawing field. It is popular for uses in which the user selects buttons or areas from the screen, because it is quick and easy to use.

Touch-panel displays, although not widely used with CAD systems, are another type of input device. These devices allow the user to touch an area on the screen with a finger to activate commands or functions. This area on the screen contains a pressure-sensitive mesh that picks up the location of the finger and transfers it to the computer, which activates the command associated with that specific location. These devices work well with function or command entry but are inaccurate for CAD coordinate entry.

Bar code readers, although not often used with CAD programs, can offer users a way of labeling, tracking, and storing data and diagrams fro future use. (See Fig. 2.23.)

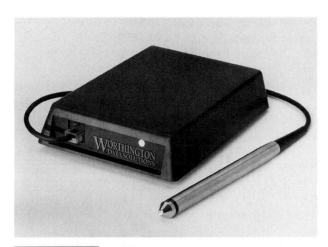

FIGURE 2.23 **Bar code reader** *Courtesy of Worthington Data Solutions.*

A relatively recent development in data input is *voice recognition* technology. This technology utilizes a combination of specialized integrated circuits and software to recognize spoken words. The system itself must first be "trained" by repeating commands into a microphone. The computer converts the operator's oral commands to digital form and then stores the characteristics of the operator's voice. When the operator gives an oral command, the system will check the sound against the words stored in its memory and then execute the command. A disadvantage of voice recognition systems is that the vocabulary supported by the system is limited. In addition, the memory required for storing complex sound or voice patterns can be extremely large. Access time, or time between the spoken word and command activation, may be several seconds. Finally, if the operator's voice changes in some manner or words have similar voice patterns, the system may not recognize the oral input.

■ 2.8 OUTPUT DEVICES

In most instances, the user of a CAD system will need a record of images that are stored on database files or displayed on the CRT. When an image is placed on paper, film, or other media, it is then referred to as *hard copy*. This hard copy can be produced by one of several types of output devices.

A commonly used device for the reproduction of computerized drawings is the *pen plotter*. Pen plotters may be classified as drum, flatbed, or microgrip. The *drum plotter* utilizes a long, narrow cylinder in combination with a movable pen carriage. The medium to be

drawn on (paper, vellum, or film) is mounted curved rather than flat and conforms to the shape of the cylinder drum. The drum rotates, moving the drawing surface, and provides one axis of movement, while the pen carriage moves the pen parallel to the axis of the cylinder and provides the other axis of movement. The combined movement of drum and pen allows circles, curves, and inclined lines to be drawn. The pen carriage on a drum plotter typically holds more than one pen, so varied line weights or multicolor plots can be drawn. These plotters accept up to E size paper in single sheets or in a roll, which is cut after the drawing is plotted.

Flatbed plotters (Fig. 2.24) differ from drum plotters in that the medium to be drawn on is mounted flat and held stationary by electrostatic or vacuum attraction while the pen carriage controls movement in box axes. The area of these plotters may be as small as A size (80×110) or larger than E size (340×440), and from one to eight pens may be used for varied line weights and multicolor plots.

Microgrip plotters (Fig. 2.25) have become one of the most widely used types of output devices. Their popularity is due to their adaptability to all types of computers, their size ranges, their low maintenance requirements, and their relatively inexpensive pricing. Microgrip plotters are similar to drum plotters in that the medium to be drawn on is moved in one axis while the pen moves along the other axis. These plotters get their name from the small rollers that grip the edges of the medium and move it back and forth under the pen carriage. These plotters range from A to E size, with single or multiple pen carriages, and may accept cut sheets or rolls.

FIGURE 2.24 **Flatbed Plotter.** *Courtesy of Houston Instruments, a division of CalComp Canada, Inc., Downsview, Canada.*

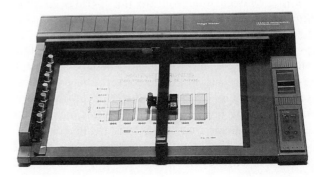

FIGURE 2.25   Microgrip Plotter. *Courtesy of Houston Instruments, a division of CalComp, Canada Inc., Downsview, Canada.*

All pen plotters are rated according to specified standards for accuracy, acceleration, repeatability, and speed. Accuracy is the amount of deviation in the geometry the pen plotter is supposed to draw (usually ranging from .0010 to .0050 in.). Acceleration is the rate at which the pen attains plotting speed and is expressed in *G*s (for gravitational force). Pen speed is important because slower speeds usually produce darker lines. The faster the pen attains a constant speed, the more consistent the line work will be. The ability of a plotter to retrace the same drawing over and over again is called repeatability. The deviation of the pen in redrawing the same line is the measure of repeatability and usually varies from .0010 to .0050. Plotter pen speed determines how fast the pen moves across the drawing medium. Most CAD software also allows the operator to adjust the pen speed to achieve maximum line quality and consistency. Slow pen speeds normally produce better quality plots than high pen speeds.

Other factors help determine the quality of a pen-plotted drawing. The variety of pens, inks, and drawing media available allows the operator to coordinate pen, ink, and paper to produce the most desirable hard copy.

Dot matrix or raster plotting is another method by which hard copy can be produced. These devices use a process called rasterization to convert images to a series of dots. The image is transferred optically (or sometimes by a laser) to the surface of the medium on a selenium drum that is electrostatically charged. Sometimes the image may be created by an array of nibs that electrically charge small dots on the medium.

A high-quality raster scan plotter can produce an image so fine and of such quality that it is not obvious how the image was produced unless examined under a magnifying glass.

Electrostatic plotters (Fig. 2.26) produce hard copy by placing an electrostatic charge on specially coated paper and having a toner, or ink, adhere to the charged area. Drawing geometry is converted through rasterization into a series of dots. These dots represent the charged area. Resolution of these plotters is determined by the number of dots per inch (dpi), usually ranging from 300 to 600 dpi. This type of plotter produces hard-copy drawings in single color or multicolors much faster than pen plotters, but the cost, power, and environmental requirements are also much greater.

An *ink jet printer/plotter* (Fig. 2.27) produces images by depositing droplets of ink on paper. These droplets correspond to the dots created by the rasterization process. This device places a charge on the ink rather than the paper, as in the electrostatic process. Ink jet plotters can produce good quality color-rendered images in addition to standard technical drawings.

Laser technology represents the newest evolution in plotter technology. A *laser printer/plotter* (Fig. 2.28 and 2.29) uses a beam of light to create images. This

FIGURE 2.26   Electrostatic Printer. *Courtesy of Houston Instruments, a division of CalComp, Canada Inc., Downsview, Canada.*

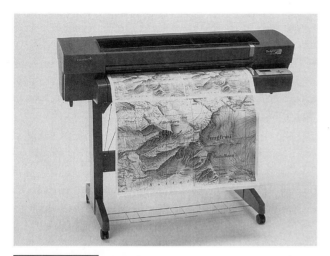

FIGURE 2.27 Inkjet Printer. *Photo courtesy of CalComp, Downsview, Ontario.*

FIGURE 2.29 Full Size Laserjet Printer. *Photo courtesy of CalComp.*

device utilizes electrostatic charging and raster scanning to produce a plotted image that is of very high quality.

Dot matrix printers (Fig. 2.30) produce images by one of two processes. They can produce images through impact on carbon or ink ribbon, or they may use a heat (thermal) process. Each method uses a number of pins in a specific configuration, such as 5×7, 7×9, or 9×9, set in the printer head. Machine commands control the sequence in which the pins strike the medium to produce an image. Dot matrix printers are normally less expensive than other types, but they are also associated with lower-quality printing.

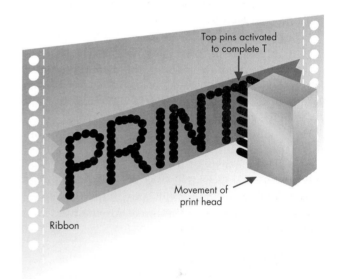

FIGURE 2.30 Dot Matrix Printer Character Formation—Each Character is Formed in a Matrix as the Print Head Moves Across the Paper. *Photo from Computers, 4/E by Long/Long. © 1996. Reprinted by permission of Prentice-Hall, Inc., Upper Saddle River, NJ.*

FIGURE 2.28 Laserjet Printer. *Courtesy of Hewlett-Packard Company.*

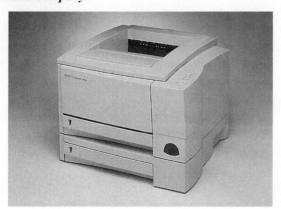

■ 2.9 DATA STORAGE DEVICES

Since all data kept in RAM will be lost when the computer is turned off, they must be saved, or stored, before the power is off. Data storage devices provide a place to save information permanently for later use. CAD programs, for example, are stored on a disk; when loaded (or activated), portions of the program go into RAM, which is temporary memory. While a drawing is being worked on, all data associated with that drawing are kept in the same temporary memory. Periodically the operator must save that drawing and all the associated data to a storage device before the program is exited or the power is shut off. Otherwise, all accumulated data from that work session will be lost. These storage devices can be considered electronic file cabinets.

Disk drives, optical drives, and magnetic tape are distinct categories of storage devices. Disk storage devices are the most commonly used method of data storage. *Disk drives* may be of the fixed (hard disk) variety, flexible (floppy) variety, or optical type. Disk drives file and read data in random order. This means that the device writes data to any portion of the disk that is empty, and it is able to locate data almost instantly because it has access to the whole disk at once. Disk drives are rated according to their type, access time, capacity, and transfer rate.

The *fixed disk drive*, or hard disk (Fig. 2.31) is the most common method of data storage. This type of drive uses an aluminum disk as the medium for storage. These drives may be internal, attached inside the computer case, or external, in a separate case of cabinet. A disk controller or controller card must be installed in the computer to allow the computer and drive to communicate or interface with each other. The storage capacity of these drives will range from 200 MB to several gigabytes. Access time is expressed in milliseconds (ms) and will range from 6 to 80 ms. The lower the number in milliseconds, the faster the access time.

FIGURE 2.32   **3.5 Inch Floppy Disk and Drive.** *Photo from* Computers, *4/E by Long/Long.* © *1996. Reprinted by permission of Prentice-Hall, Inc., Upper Saddle River, NJ.*

Floppy disk drives (Fig. 2.32) derive their name from the removable flexible plastic disks used in this device. The disks used in this drive are typically $3\frac{1}{2}''$ in diameter. The density of a disk refers to the amount of data the disk will hold. Typically, a $3\frac{1}{2}''$ disk, called microdiskette, will hold 1.44 MB of data in double-sided, high-density format. The floppy disk is inexpensive and convenient to use but holds less data and is slower than fixed disk drives.

Zip drives (Fig. 2.33) are high-capacity magnetic disk drives, similar to floppy drives, that can store up to 100 MB on a $3\frac{1}{2}''$ disk. The disks used in zip drives are a special high-capacity medium and are not the same as a standard floppy disk. Zip drives are popular because there are external parallel and SCSI transfer models available which can be used to transport large amounts of data from one machine to another. They

FIGURE 2.31   **Fixed or Hard Disk Drive Interior.** *Courtesy of Western Digital Corporation.*

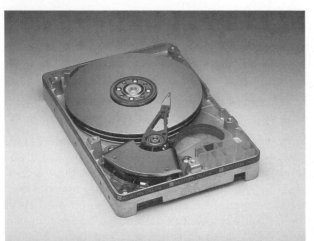

FIGURE 2.33   **Zip Disk and Drive.** *Courtesy of Iomega, Inc.*

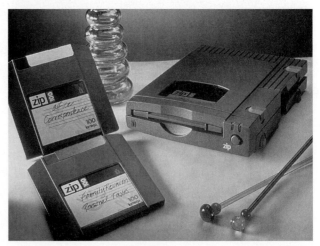

FIGURE 2.34   Jaz Drive and Disk. *Courtesy of Iomega Corporation.*

can also be used for effective short-term backup (disks have a shelf life of 10 years).

Jaz drives are similar to Zip drives but their disks can hold up to 2 GB of data (Fig. 2.34).

Similarly, *superdisks* can hold up to 120 MB of data while $\frac{1}{4}''$ *data cartridges* can hold up to 20 GB of information.

The ever-increasing need for larger storage capacity has spurred the development of new technologies. The newest technology in data storage uses lasers to read and write data and is generally termed optical storage. Optical storage media are capable of holding many gigabytes of data.

CD-ROM drives are a type of optical drive. Previously, CD-ROM disks were usually created from a master, so when you purchased a CD it already had the digital information written to it. Now you can store information on recordable CD-ROM systems, which produce a *write once read many* (WORM) disk. WORM devices allow data to be written to them, but the data become permanent on the disk and cannot be erased. This storage device is especially suited for archival purposes. CD-ROM drives use a laser to read and write data to a chemically coated aluminum disk (Fig. 2.35). The data is "burned" into the disk surface by a laser so the information becomes permanent (unlike magnetic storage) and the disk is removable. Recordable CD-ROM systems can store about 650 MB of information on a single disk.

FIGURE 2.35   CD-Rom and Disk. *Courtesy of NEC Technologies, Inc.*

FIGURE 2.36 **Optical Disk Cartridge and Optical Disk Drive.** *Courtesy of SyQuest Technology.*

Optical disk drives (Fig. 2.36) allow data to be erased and written over. These drives use a laser to change the state of optical magnetic media. These optical magnetic media can be changed again and again (write many read many). Because optical magnetic media are not sensitive to dust, like hard drives, optical disks can be removed from the drive and replaced with a new disk when additional storage capacity is needed.

Magnetic tape storage (Fig. 2.37) uses plastic tape coated with magnetic particles. A read/write head in the tape drive charges magnetic particles to store information on the tape. The data being sent are recorded as a series of charges along the tape. Once these particles are charged, they will remain charged until

FIGURE 2.37 **Tape Back-up System.** *Courtesy of Iomega, Inc.*

the head writes over them or they are demagnetized. Tape drives file and read data in sequential order. This means they must look through data in the order that the tape is wound or unwound. (This is similar to forwarding or rewinding a videotape to look for a specific scene, or an audiotape to play a particular song.) Gigabytes of data may be stored on some tape systems. These cassettes resemble audiocassettes in both size and appearance. Tape storage is essentially used for backing up data from a hard drive, or for archival purposes (since the tapes can be removed and stored for later use).

■ 2.10 CAD SOFTWARE

CAD software tells the computer how to interact with the flow of data entered by the user through an input device. For example, it lets the computer "use" formulas to solve complictaed questions requiring detailed analysis of large amounts of data, such as finding the center of gravity for a truncated cone. It also handles drawing processes, like creating many different models or views of the same object automatically (Fig. 2.38). Software helps organize data; it will find previously stored drawing symbols, and it will help create and archive new ones. Software can be used to count, measure, and direct devices to print or plot drawings, create a bill of materials, or exchange files with other programs. CAD software is extremely powerful and has been designed to serve all major branches of engineering. It will be an important tool in most engineering careers.

■ 2.11 COMMON CORE

CAD SOFTWARE All CAD software generates familiar geometric terminology for creating drawings. But even though the geometry is common and the procedures for construction are similar, every CAD software program will vary in operational procedures typically involving the basic hierarchy of command structure.

Three features are found in all CAD software. You can access these features interactively through basic commands and menu options.

1. Commands for geometry generators (basic geometric construction)
2. Functions to control viewing of drawing geometry
3. Modifiers for changing the drawing or editing variations in the drawing (rotate, mirror, delete, group, etc.).

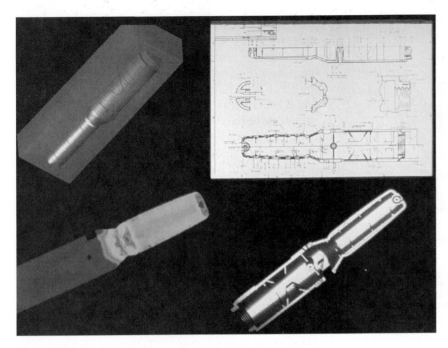

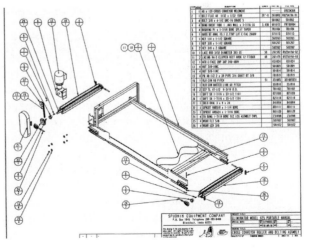

FIGURE 2.38   **CAD Software Easily Creates Different Outputs of the Same Object.** *Courtesy of SDRC, Milford, OH.*

The commands and menu options may often be selected in two basic ways: by typing or by picking using a tablet (digitizer) or mouse. The Cartesian coordinates may be accessed with the keyboard or the mouse. You may switch between these two input methods at any time to issue commands and select options. The sequence of selection is called the hierarchy of command structure and provides an ease of operation that is the basis for selecting one software program over another.

■ 2.12 CAD CAPABILITY CHECKLIST

CAD software can offer the following characteristics required for creating technical documents (Fig. 2.39):

1. Draw construction lines at any convenient spacing through any points, at any angles, and create tangent lines to one or more arcs.
2. Draw any type of line, such as visible, center, hidden, or section.
3. Draw circles and arcs of any size with given data.
4. Perform cross-hatching within specified boundaries.
5. Establish a scale or set a new scale within a drawing for various drawings within a document.
6. Calculate or list pertinent data of graphic construction, such as actual distances or angles.
7. Create a group of geometric figures for editing or copying.

8. Relocate drawing elements to any new position. Correct or change additions in stored documents.
9. Edit all or erase (delete) any part of a line, arc, or any geometric form on a drawing. Correct dimensions.
10. Make a mirrored image or create symmetrical forms.
11. Perform associative or datum unit dimensioning.

FIGURE 2.39   **CAD Assembly Drawing.** *Courtesy of Sputnik Equipment Corporation, Inc.*

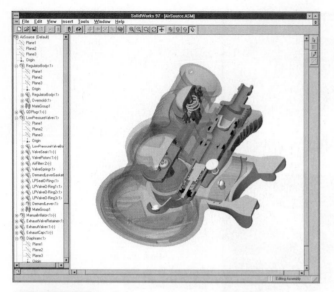

FIGURE 2.40   **Desktop PCs Running the Windows Operating System Can Now Provide Capabilities to Create Parametric Solid Models.**

12. Label drawings with notes and create title blocks and bills of material.
13. Save the entire drawing or any part for use on other documents and in other formats.
14. Create pictorials from three-view drawings.
15. Create orthographic views from a 3D model.
16. Retrieve and use stored drawings.

However, software programmers are constantly adding new capabilities and options to CAD software. For instance, many new programs offer very sophisticated 3D capabilities, often in a Windows environment (Fig. 2.40).

■ 2.13 DRAWING WITH CAD SOFTWARE

The geometry that is created, drawn, or generated with CAD programs is generally referred to as objects, entities, or elements. These geometric entities are individually constructed figures or groups of elements that consist of points, lines, arcs, circles, rectangles, polygons, splines, solids, surfaces, symbols or blocks, cross-hatching, dimensions, and notes. These basic building units are selected from the menu and constructed at specific graphic locations of the monitor screen by the CAD user. Many CAD systems can store standard drawing symbols (called blocks) and use overlays on the digitizer for retrieval.

■ 2.14 USING A CAD SYSTEM

After a CAD system has been installed, the beginning user must become thoroughly familiar with it and learn how to use it effectively. This will require learning some new skills as well as a different vocabulary. The CAD operator will need to learn to create accurate drawings and construct them to various scales. In addition, the operator will need to learn to interact effectively with the operating system to store and copy drawings and to routinely perform backups so that no data are lost. Most CAD manufacturers offer training programs and tutorials that will make the learning process much easier (Fig. 2.41). They will provide instruction and training manuals that give information and details about the operation of the system. These manuals can be used not only during the initial training period but also for reference purposes during later operation of the system.

Most experienced drafters have developed shorter or simpler methods for creating a drawing, such as using overlays or templates. CAD systems also have simplified methods for drawing. Some systems have symbol libraries that contain many of the frequently used symbols, such as electrical relays, switches, transformers, resistors, bolts, nuts, keys, piping, and architectural symbols. These symbols may be in a symbol library, or the symbol may be located on one of the templates in the library of templates. Most CAD systems allow users to customize their symbol libraries. The desired symbols must first be drawn by the user

FIGURE 2.41   **CAD Training Helps the User Learn the Software Quickly.** *Courtesy of Jeff Kaufman and FPG International.*

on the CAD system the same as they would appear on a drawing board. This process may initially take as much time as it would manually, but once the image has been entered in the computer database, it need never be drawn again. The symbol can easily be retrieved from the symbol library whenever required.

■ 2.15 SELECTING A CAD SYSTEM

As the number of manufacturers of computer system equipment has increased, there has been a corresponding decrease in the cost of these systems. Some CAD systems may contain only those features necessary to produce simple two-dimensional entities, while others have the capability to create true three-dimensional drawing objects. The manufacturers of some of these systems are eager to promote new or improved functions that may not be found on other systems. Almost daily the computer industry announces amazing advances.

Since the newer CAD systems are generally easier to operate, the words "user friendly" are frequently used by manufacturers to promote their systems. However, there has been a tendency to exaggerate what particular systems can do. A prospective user should therefore view all claims with some skepticism until proved, for they can be misleading and often lead to disappointment. Unfortunately, many firms have purchased a system only to discover that it did not perform as well as expected.

Before purchasing a CAD system, a careful, well-thought-out plan for selecting a system should be developed and followed. This plan can be divided into five phases:

1. Establish the need for a CAD system.
2. Survey and select system features.
3. Request CAD system demonstrations.
4. Review selected systems.
5. Select, purchase, and install a CAD system.

Let us now discuss each of these phases in greater detail.

ESTABLISH THE NEED FOR A CAD SYSTEM The first consideration in the selection process is to determine whether a CAD system is needed. All potential users of the CAD system should be consulted regarding how, when, and where a system would be used and whether it would be cost effective in their particular operations (see Fig. 2.42). Never purchase a system

FIGURE 2.42 **CAD Systems Can Be Expensive, So it is Important to Make Informed Decisions.** *Courtesy of Ron Chapple and FPG International.*

simply because it may be considered a first step into the future and your firm wants to project a progressive image. It is important to investigate and evaluate the time- and cost-saving claims of manufacturers by contacting firms that have CAD systems in operation. Contact as many firms as possible, especially those with a wide range of experience. Prepare a brief questionnaire to survey these firms, asking questions regarding costs, training periods, system operation, and so on. Also ask at what point after installation the system became cost effective. Evaluate the responses to these questions and compare them with your specific requirements. This information will assist you in deciding if a CAD system can be beneficial to your firm at present.

Undoubtedly, you will hear many spectacular claims about the productivity of CAD systems. The productivity, however, will depend on the type of engineering operation. For example, among the first companies to introduce and use CAD systems were the electronic industry manufacturers. They found that drafters were constantly redrawing many symbols (resistor, transistor, etc.) and standard hardware parts (nuts, screws, etc.). Although they used plastic templates, common in all drafting departments, a considerable amount of time was spent doing tedious, repetitive drawing tasks. Since much of the time was devoted to tracing or redrawing something that was previously drawn, it was determined that a CAD system would save as much as 50% of the drafting time compared to using traditional methods. It is important to remember that with CAD it is only necessary to draw something once, even though initially it may take as long, or perhaps longer, to create it on the system as on the drawing board. From that point on, however, you need only recall this information from data storage, which a CAD system can do with amazing speed. In this example the benefits of CAD are obvious.

It is difficult to place a direct monetary value on many benefits of using CAD. These benefits include things like shorter design cycle time, improved ability to visualize complex fits between parts, links to direct manufacturing, better ability to reuse existing drawing and designs, and improved analysis early in the design phase.

The process of determining the need for a CAD system can be time consuming and frustrating. Nevertheless, speed should be sacrificed to careful deliberation in this phase.

SURVEY AND SELECT SYSTEM FEATURES   It is generally agreed that software should be selected before hardware. However, some CAD systems are *turnkey* system—that is, a total system with software and hardware combined and inseparable. Therefore, you should examine all the features of any given system very carefully before being attracted by spectacular hardware. For example, some systems use dual monitors. The chance of being impressed by this feature may overshadow the question of whether one really needs the two displays.

Consider whether the system will be multipurpose or used strictly for CAD. Will other office operations, such as word processing or accounting, be done on this machine? The answer to this question may add or eliminate CAD programs based on their operating system software.

Investigate how well a system will exchange information or interface with other CAD or CAM systems. Many systems do not have this capability. One important consideration is the CAD system's ability to exchange information with other CAD systems and other engineering applications. One standard for such exchange is the initial graphics exchange specification (IGES). There are also a number of other common formats. Make sure that the system you select can export common file formats, particularly to other applications you are planning to use.

It is suggested that you survey the people who will use the system to determine what they think is desirable in a CAD system. From this survey, develop a checklist of hardware and software features that your future system should have.

REQUEST CAD SYSTEM DEMONSTRATIONS   After the checklist has been created and approved by all parties concerned, make arrangements to see the various CAD systems in operation. A list of vendors can be compiled from advertisements in trade journals, magazines, and so on, or from the various directories of computer graphics manufacturers that are published. Contact the vendors to arrange demonstrations. Explain to them exactly what you expect the system to do. If the vendors are completely aware of your requirements, they will be able to give a more realistic presentation. Most of all, be prepared to ask questions. With each succeeding demonstration, your questions will be more effective. Moreover, the answers that you receive will be more meaningful. It is a good idea to select your own project which you would like to have demonstrated. Some CAD systems run their own canned example well but may not be flexible enough to meet your needs. Having your own project also helps you to contrast the systems fairly. Some systems may have a fancy demonstration, while others will not look as good initially. You want to select a system that will perform well for your needs, not look good running a canned demonstration.

Some of the questions that you should ask are as follows:

1. What are the brand names of the equipment used?
2. Are service contracts available? If so, what are the details and the cost of the contract? In most instances, service contracts are very expensive. Are you required to buy the service contract? Doing so may not be cost effective compared to having repairs done as you need them. Can you afford to have some downtime if something goes wrong with your system? If not, you may want a service contract. It is also possible in some situations to obtain service contracts from third parties that specialize in this type of business. The service provided by these firms for preventive and downtime maintenance may be less costly than that offered by the original manufacturer.
3. What is the warranty period, or periods, if parts of the system are furnished by different vendors? Also, is the warranty paid for by the original vendor or is the cost shard by the user and vendor?

4. What types of CAD software are provided, and what operations can be performed?

5. What kind and length of training will be provided to the user staff? How long after installation will it take for the staff to be proficient enough for the system to be cost effective?

6. Does the vendor issue software updates? If so, how often is this done and what is the cost?

7. What is the reputation of the vendor for providing user support in the form of toll-free telephone assistance or other methods of communication? Is the vendor readily available when the user encounters software and/or hardware problems and requires assistance?

8. Who currently uses the particular system? Will the vendor provide a list of current users? Users that have several months' experience with a CAD system are best qualified to answer question 7.

A CAD system analysis worksheet can be helpful when demonstrations are presented. Items to be included on the worksheet may be arranged according to hardware and software specifications. The hardware features that should be listed will generally be included in the following five categories: (1) central processor, (2) input devices, (3) monitors, (4) data storage, and (5) output devices. You may want to create a worksheet similar to the one shown in Fig. 2.43 that you can use to track CAD system features during demonstrations. The worksheet may be expanded or modified to reflect particular requirements. You should also develop a similar worksheet listing CAD system software feature requirements.

With the completion of this phase, you will have achieved several important goals. Your knowledge and understanding of CAD systems will have increased, and you will be in a position to identify those system features that you require and can afford. In addition, you will have assembled a list of current users of CAD systems for future reference or exchange of information.

REVIEW SELECTED SYSTEMS   It is important at this stage that all of the collected information be organized and carefully reviewed. A list should be made of only those systems that merit further serious consideration. You may wish to request another demonstration of the particular systems that are on your revised list and to request additional information from current users regarding equipment performance, staff training, vendor

support, and so on. Remember, if you select a system made by a reputable and financially stable manufacturer, and purchase it from a dependable retailer, you will sacrifice nothing in terms of service and support. Throughout this chapter we have made every effort to feature or illustrate systems and equipment made by such firms.

FINAL SELECTION, PURCHASE, AND INSTALLATION OF A CAD SYSTEM   This last phase occurs when the final decision will be made regarding whether or not to purchase a CAD system. The decision invariably will depend on how you plan to use the system and how much you can afford to pay for it. You may want to consider a leasing arrangement, or lease-purchase arrangement. This can be effective for keeping the technology up to date. New technology is available constantly, and the lifetime of computer equipment is generally considered to be about three years. If your system cannot pay for itself in three years, perhaps it is not a good investment. You should plan to upgrade or purchase new equipment and software on a regular cycle. If the decision is made to acquire a system, determine costs, choose a delivery date, and arrange for installation and training.

■ 2.16 SUMMARY

The information presented in this chapter is intended to familiarize the student with the basic concepts, hardware, peripherals, and systems in CAD. It is not possible (nor was it intended) to present a comparison of CAD programs or all the commands used on CAD systems.

When possible, the instructor should arrange for students to visit nearby engineering and drafting departments that have CAD systems in operation. Those students who wish to obtain additional information on this subject should consult their school or local library.

The CAD system on the personal computer is replacing many drafting instruments, drafting tables, and drafting files. However, like no other drafting tool before, it raises engineering productivity without replacing the basic functions of the designer, engineer, and drafting technician. CAD developers, in their quest to harness computer technologies, have had a profound impact on the high-tech teams as they resolve problems in research, development, design, production, and operation (the five basic engineering functions).

The skills learned "on the board" are related and complementary to those needed by the CAD user.

FIGURE 2.43   Worksheet for Evaluating CAD System Hardware.

Item	Y/N	Size/Type	Comments	Cost
Central Processor				
memory (MB)				
word size (16/32 bit)				
cache				
speed (MHz/Mips)				
bus type (eisa, vesa, pci)				
expansion/upgrade				
Operating System				
32 bit				
multitasking				
software availability				
Data Input Devices				
mouse				
trackball				
digitizer				
light pen				
thumb wheel				
Display				
monochrome				
color				
screen size				
resolution				
Video Card				
memory				
software support				
dual display support				
Storage, Hard Drive				
type				
access time				
capacity				
expansion				
removable				
Storage, Floppy				
type				
access time				
capacity				
removable				
CD-ROM				
type				
speed				
capacity				
read/write				

FIGURE 2.43   (cont.)

Item	Y/N	Size/Type	Comments	Cost
Backup System				
type				
capacity				
speed				
automation				
Output Devices				
type				
provided with system?				
medium				
cost per sheet				
speed				
resolution/accuracy				
color				
Maintenance				

Learning the performance skills needed for creating drawings with CAD tools is time consuming and requires practice and manual dexterity. Both methods of drafting use simple and familiar geometric terminology for structuring the graphic production of technical documents, and both have the same goal—drawings that will meet industry standards.

The basic principles of drafting are common to traditional drafting and computer-aided drafting. The American National Standards Institute (ANSI) has well-established standards for shaping engineering drawings. Knowledge of drafting principles from the alphabet of lines to dimensioning and sectioning procedures, continues to be essential in shaping CAD documents. CAD can help you produce consistent lettering and regulate line work to improve the production of working drawings better than any other tool.

The CAD user is responsible for preparing engineering documents that are an integral part of the total manufacturing process (Fig. 2.44). The ability to interact with all forms of technical information increases the significant role of the drafting technician.

■ KEY WORDS

CAD

CAM

SOFTWARE

OPERATING SYSTEMS

HARDWARE

CPU

MICROPROCESSOR

MIPS

MEGAHERTZ (MHZ)

MOTHERBOARD

SIMM

RAM

WORM DISK

PORT

VIDEO DISPLAY MONITOR

MOUSE

DIGITIZING TABLET

PRINTER

PLOTTER

FLOPPY DISK

HARD DISK DRIVE

MODEM

MENU

WIRE FRAME

CD-ROM

ZIP DISK

JAZ DISK

SUPERDISKS

DATA CARTRIDGES

GRAPHICS SPOTLIGHT

Unlocking the Power of Solid Modeling

Successful use of solid modeling should increase the size of a company's market, increase its market share, and increase its profit margin. Early CAD systems basically only automated the drafting process, but solid modeling has the potential to affect the entire production process, from preliminary design through engineering and manufacturing. Ancillary functions, such as purchasing and marketing, can also be affected.

Solid modeling can create all the critical information for a product, and a company needs to take advantage of all the information contained in these models. Good implementation of solid modeling is marked by the wide-ranging use of the solid models in downstream applications. This maximizes return because it permits many operations to work from the original solid model rather than re-creating the design for each operation. Solid models should be the basis for virtual prototypes, engineering analyses, machine tool paths, purchase orders, marketing images, etc. Anywhere a design is re-created or its related data is retyped into another computer is a sign that solid modeling may be underutilized.

Although downstream use is an important consideration, there is nothing to be gained by insisting that all design be done with solid-modeling when various operations are just as well served by 2D processes. Another consideration in introducing a new technology is whether it affects existing bottlenecks in the production process. Unless the use of solid modeling (or any other technological innovation) helps to eliminate or decrease the bottlenecks, or constraints, in the overall process, it will not improve productivity or profits.

Introduction of solid modeling into company operations requires careful planning. First, there are many solid modeling systems on the market, and choice of the correct one is of paramount importance. Second, it will probably be necessary to undertake an expensive hardware upgrade because solid modeling requires larger workstations, better graphics, more memory, etc. than simpler programs. Third, everyone who will use the solid model should be given extensive training. Introducing solid modeling in a pilot project is also widely recommended. The gradual implementation of solid modeling on a project-by-project basis has been found to be more successful than a one-step installation throughout the company.

Small and medium-sized companies have an advantage over large companies in using solid modeling for maximum return. Large companies may have the resources and dedication to make it work, but the smaller companies have the most flexibility in terms of organizational structure.

Adapted from "Unlocking the Power of Solid Modeling" by Caren D. Potter, *Computer Graphics World*, Nov. 1995, Vol. 18, No. 11, p. S3(4). ©PennWell Publishing Company 1995.

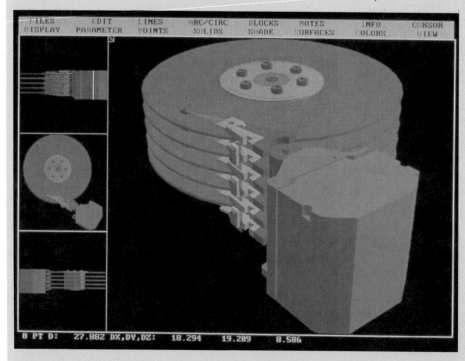

Multiview CAD Solid Model. *Courtesy of CADKEY.*

FIGURE 2.44 CAD Documents Supply a Multitude of Information which Engineers and Technicians Must Understand and Use. *Courtesy of SDRC, Milford, OH.*

■ CHAPTER SUMMARY

- Computers have revolutionized the drawing process. New technologies are constantly invented which make this process quicker, more versatile, and more powerful.

- CAD is the tool of choice for engineering design companies. The effective user of this tool requires an understanding of technical drawing fundamentals as well as training on the CAD software program.

- The microprocessor, RAM, and hard disk drive of the computer are essential components of a computer system. The keyboard and mouse are typical input devices. Printers and plotters output the drawing to paper for review and approval. The display monitor shows the drafter what is being drawn and offers command choices.

- CAD software can draw in three dimensions (width, height, and depth), unlike paper drawing which only consists of two dimensions in a single view.

- Different CAD packages have different operational procedures, and different strengths and weaknesses. Three features found in all CAD software are commands for geometry generators, functions to control the viewing of drawing geometry, and modifiers for changing the drawing or editing variations.

- Operating a CAD system typically has required extensive training. Newer CAD systems are becoming more user friendly, but one should not overestimate the claims CAD packages make. It is important to evaluate each package thoroughly and make an informed decision.

■ REVIEW QUESTIONS

1. What are the basic components of a computer-aided drawing (CAD) system?

2. Discuss the relationship between CAD and CAM in modern design and manufacturing facilities?

3. List the similarities and differences between a mouse and a digitizing pad with puck?

4. What are the main advantages of CAD over traditional drawing methods?

5. What is faster, a computer with a 100 Mhz micro-processor and a 500 MB hard disk, or a 500 Mhz com-puter with a 100 MB hard disk? Which will store more information?

6. What is the difference between RAM storage and hard disk storage? What computer parts are typically found on the motherboard?

7. What is the difference between plotting and printing?

8. What are the hardware and software specifications of your school's CAD system?

9. What are questions you should ask about any CAD system you consider buying?

■ CAD PROBLEMS

When necessary, refer to the appropriate sections of the chapter to check your answers.

Prob. 2.1 Define the following terms: computer system, hardware, software, analog, digital, computer graphics, CAD, CADD, and CAM.

Prob. 2.2 What are the principal components of a computer system? A CAD system? Draw a systems flowchart that illustrates the sequence of operations for each of the systems.

Prob. 2.3 Prepare a list of CAD system hardware components and give examples of each.

Prob. 2.4 Call a hardware company and compare the prices for three graphics monitors of different resolutions and sizes. Which purchase do you recommend? Give your reasoning.

Prob. 2.5 Prepare a list of possible data storage devices and determine total storage capabilities for each.

Prob. 2.6 Determine what would be your best storage de-vice, in terms of value, if your average CAD drawing file size is 1500k and you store 10–20 drawings per week.

Prob. 2.7 Arrange a visit to the computer center at your school, or to a local engineering design office, and prepare a written report on the use of computers in design and draft-ing at these facilities.

The following problems ask you to use CAD software to solve some typical geometric construction that is similar from one software program to another. All the problems have been prepared on a CAD system. Pre-pare the required CAD drawing problems, as shown with your CAD system, and produce a hard copy with a printer or plotter for approval.

Prob. 2.8 Prepare a list of modifying (or editing) com-mands available on the CAD system you will be using.

Prob. 2.9 The unknown distance KA in Fig. 2.45 has been determined and the angle measured using CAD. Re-create this problem with your CAD system, changing the 908 angle at H to 758; then determine the angles at K and A and the length of line KA.

Prob. 2.10 Prepare a revised version of the CAD drawing (Fig. 2.46) by increasing the radius 0.40 to 0.4375 and chang-ing the slot dimension 1.60 to 1.70.

Prob. 2.11 Prepare a detailed CAD drawing of the Safe-ty Key (Fig. 2.47) with the following changes: Correct the right-side view and add the missing dimension 0.40. Exam-ine the placement of dimensions and relocate where neces-sary. Change 1.12 to 1.25 and add the difference to dimension 4.70.

Prob. 2.12 Using a CAD system, determine the true length of lines AD and CD (Fig. 2.48) when the horizontal projection of point A is relocated to a new coordinate read-ing of (0, 3.125) and the horizontal projection of point D is relocated to a Cartesian coordinate of (1.75, 1.625). Revise the drawing using the "F" notation for the frontal projec-tions instead of the V notation, as shown. What is the new slope of line CD?

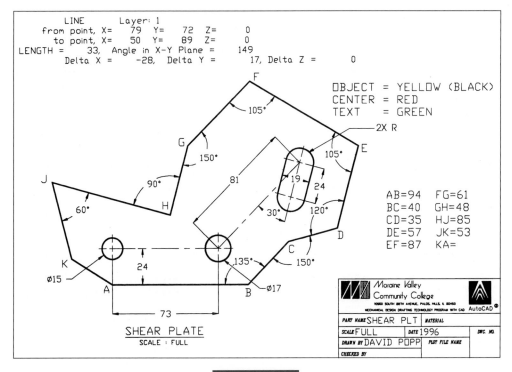

FIGURE 2.45

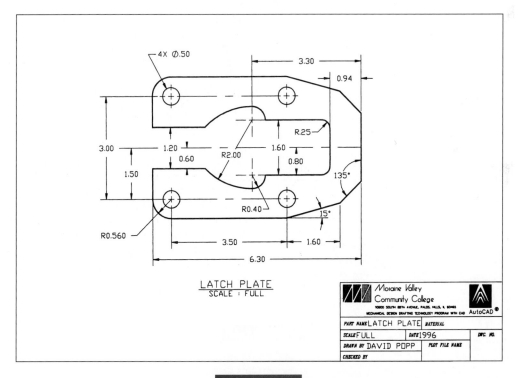

FIGURE 2.46

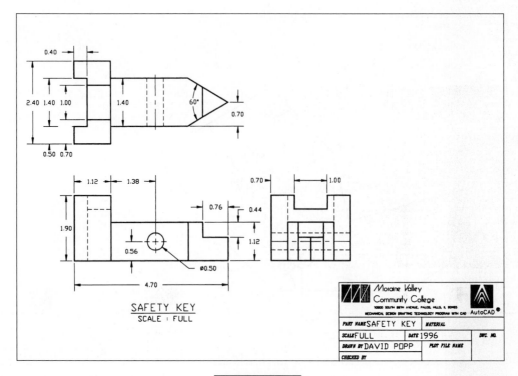

FIGURE 2.47

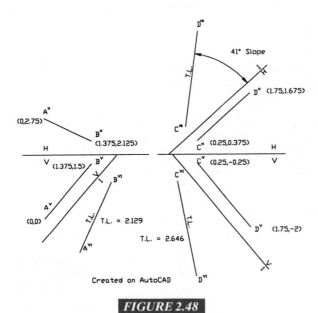

FIGURE 2.48

C H A P T E R   3

INSTRUMENT DRAWING, FREEHAND SKETCHING, AND LETTERING TECHNIQUES

OBJECTIVES

After studying the material in this chapter, you should be able to:

1. Identify the basic tools used by the drafter.
2. List the four objectives of drafting.
3. Describe the difference between the T-square, parallel rule, and drawing machine.
4. Identify various types of lines and how they are used.
5. Draw lines, arcs, and circles of specific size using drawing instruments.
6. Draw lines at specific angles.
7. Read and measure with the architect's scale, engineer's scale, and metric scale.
8. Draw irregular curves.
9. Identify several drawing media and standard sheet sizes.
10. Create freehand sketches using the correct sketching techniques.
11. Sketch parallel, perpendicular, and evenly spaced lines.
12. Sketch a circle and an arc of a given diameter.
13. Use techniques to keep your sketch proportionate.
14. Enlarge an object using grid paper.
15. Sketch various line types.
16. Add lettering to a sketch.

OVERVIEW

The traditional method of creating technical drawings is with drawing instruments. Since the eighteenth century, precision instruments have been the tools of drafters. Today, CAD software is another tool used by many drafters. The basic concepts of drawing and measuring lines and circles is the same for traditional and CAD drawing. The alphabet of lines and the meaning of line types is the same for traditional and CAD drawings. By understanding the basic principles of drawing, the properly trained drafter can create and modify any type of drawing. While some students may think that CAD software can replace the knowledge required to construct a drawing, this is not so. While CAD makes drawing easier, it does not replace the basic knowledge that enables a skilled drafter to manipulate either a pencil or CAD software.

Sketching technique is one of the most important skills for engineering visualization. Sketching is a quick way to communicate ideas with other members of the design team. A picture is often worth a thousand words (or 1K words, as it were). Sketching is a time-efficient way to plan out the drawing processes needed to create a complex object. Sketches act like a road map for the completion of a final paper or CAD drawing. When you sketch basic ideas ahead of time, you can often complete a drawing sooner and with fewer errors. Legible hand lettering is used on the sketch to specify important information.

INSTRUMENT DRAWING

■ 3.1 TYPICAL DRAWING EQUIPMENT

For many years the essential equipment for students in technical schools and for engineers and designers in professional practice remained unchanged. This equipment included a drawing board, T-square, triangles, an architects' or engineers' scale, and a professional-quality set of drawing instruments. Now, however, other equipment has come into general use, including the drafting machine, parallel-ruling straightedge, technical fountain pen, and, of course, the computer.

The basic items of drawing equipment are shown in Fig. 3.1. For best results, the drawing equipment you use should be of high grade. When you are ready to buy drawing instruments (item 3), you should talk to an experienced drafter or designer, or reliable dealer,

about your purchase because it is difficult for beginners to distinguish high-grade instruments from inferior instruments.

1. Drawing board (approximately 20″ × 24″), drafting table, or desk.
2. T-square (24″, transparent edge), drafting machine, or parallel-ruling edge (§§3.4, 3.51, and 3.52)
3. Set of instruments (§§3.28 and 3.29)
4. 45° triangle (8″ sides) (§3.11)
5. 30° × 60° triangle (10″ long side) (§3.11)
6. Ames Lettering Guide or lettering triangle
7. Architects' triangular scale (§3.24)
8. Engineers' triangular scale (§3.22)
9. Metric triangular scale (§3.20)
10. Irregular curve (§3.48)
11. Protractor (§3.13)
12. Mechanical pencils and/or thin-lead mechanical pencils and HB, F, 2H, and 4H to 6H leads, or drawing pencils (§3.7)
13. Drawing paper, tracing paper, tracing cloth, or films as required; backing sheet (drawing paper white, cream, or light green) to be used under drawings and tracings
14. Drafting tape (§3.6)
15. Templates
16. Calculator
17. Cleansing tissue or dust cloth
18. Eraser

■ 3.2 OBJECTIVES IN DRAWING

The following pages explain the correct methods for instrumental drawing. Students who practice and learn correct manipulation of their drawing instruments will eventually be able to draw correctly by habit and will be able to give their full attention to the problems at hand.

The following are the important objectives students should strive to attain:

1. *Accuracy.* No drawing is of maximum usefulness if it is not accurate. The engineer or designer cannot achieve success in professional employment if the habit of accuracy is not acquired.
2. *Speed.* Time is money in industry, and there is no demand for a slow drafter, technician, or engineer.

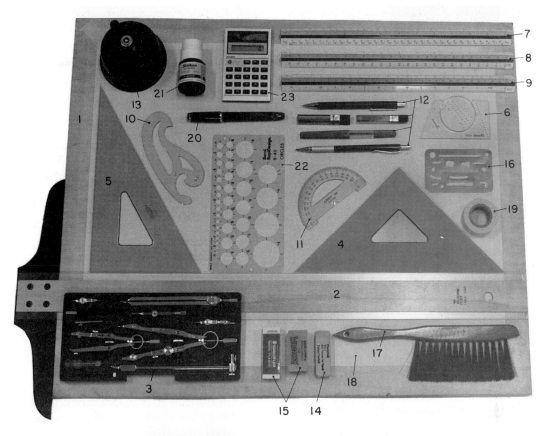

FIGURE 3.1   **Principal Items of Equipment.**

However, speed is not attained by hurrying; it is an unsought byproduct of *intelligent and continuous work*. It comes with study and practice

3. *Legibility*. Drafters, technicians, and engineers must remember that a drawing is a means of communication to others, and that it must be clear and legible to serve its purpose well. Care should be given to details, especially to lettering (discussed further at the end of this chapter).

4. *Neatness*. If a drawing is to be accurate and legible, it must also be clean. Untidy drawings are the result of sloppy and careless methods and will be unacceptable to an instructor or employer.

■ 3.3 DRAWING BOARDS

If the left edge of the drawing table top has a true straightedge and if the surface is hard and smooth (such as Masonite™), a drawing board is unnecessary,

provided that drafting tape is used to fasten the drawings. It is recommended that a backing sheet of heavy drawing paper be placed between the drawing and the table top.

In most cases a drawing board will be needed. Boards vary from 9″ × 12″ (for sketching and field work) up to 48″ × 72″ or larger. The recommended size for students is 20″ × 24″, which will accommodate the largest sheet likely to be used.

Drafters using drafting tape to hold paper in place, which in turn permits surfaces such as hardwood or other materials to be used for drawing boards.

For right-handed people, the left-hand edge of the board is the *working edge* because the T-square head slides against it (Fig. 3.2). (***Left-handers***: Place the head of the T-square on the right.) This edge must be straight, and you should test the edge with a T-square blade that has been tested and found straight (Fig. 3.3). If the edge of the board is not true, it should be replaced.

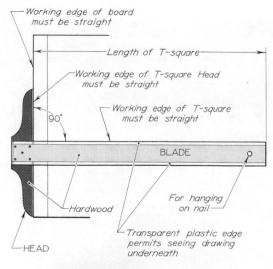

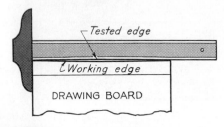

FIGURE 3.2   The T-square.

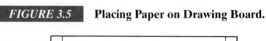

FIGURE 3.3   Testing the Working Edge of the Drawing Board.

3.4 T-SQUARES

The T-square is made of a long strip, called the *blade*, fastened rigidly at right angles to a shorter piece called the *head* (Fig. 3.2). The upper edge of the blade and the inner edge of the head are *working edges* and must be straight. The working edge of the head must not be convex, or the T-square will rock when the head is placed against the board. The blade should have transparent plastic edges and should be free of nicks along the working edge. Transparent edges are recommended, because they allow the drafter to see the drawing in the vicinity of the lines being drawn.

Do not use the T-square for any rough purpose. Never cut paper along its working edge, as the plastic is easily cut and even a slight nick will ruin the T-square.

3.5 TESTING AND CORRECTING T-SQUARES

To test the working edge of the head, see if the T-square rocks when the head is placed against a

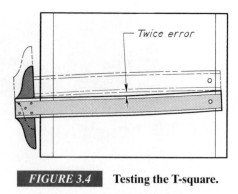

FIGURE 3.4   Testing the T-square.

straightedge, such as a drawing board working edge that has already been tested and found true. If the working edge of the head is not straight, the T-square should be replaced.

To test the working edge of the blade, draw a sharp line very carefully with a hard pencil along the entire length of the working edge; then turn the T-square over and draw the line again along the same edge (Fig. 3.4). If the edge is straight, the two lines will coincide; otherwise, the space between the lines will be twice the error of the blade.

It is difficult to correct a crooked T-square blade, and if the error is considerable, it may be necessary to discard the T-square and obtain another.

3.6 FASTENING PAPER TO THE BOARD

The drawing paper should be placed close enough to the working edge of the board to reduce to a minimum any error resulting from a slight "give," or bending, of the blade of the T-square. The paper should also be close enough to the upper edge of the board to permit space at the bottom of the sheet for using the T-square and supporting the arm while drawing (Fig. 3.5).

FIGURE 3.5   Placing Paper on Drawing Board.

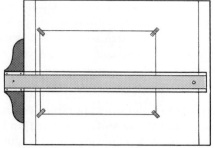

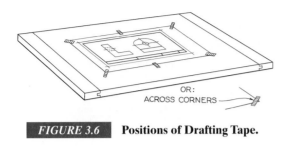

FIGURE 3.6 **Positions of Drafting Tape.**

Drafting tape is preferred for fastening the drawing to the board because it does not damage the board and it will not damage the paper if it is removed by *pulling it off slowly toward the edge of the paper.*

To fasten the paper in place, press the T-square head firmly against the working edge of the drawing board with the left hand, while the paper is adjusted with the right hand until the top edge coincides with the upper edge of the T-square. Then move the T-square to the position shown and fasten the upper left corner, then the lower right corner, and finally the remaining corners (Fig. 3.6). Large sheets may require additional fastening, whereas small sheets may require fastening only at the two upper corners.

Tracing paper should not be fastened directly to the board because small imperfections in the surface of the board will interfere with the line work. Always fasten a larger backing sheet of heavy drawing paper on the board first; then fasten the tracing paper over this sheet.

■ 3.7 DRAWING PENCILS

High-quality drawing pencils should be used in technical drawing never ordinary writing pencils (Fig. 3.7a).

Many makes of mechanical pencils are also available, together with refill leads of conventional size in all grades (Fig. 3.7b). Choose a holder that feels comfortable in your hand and that grips the lead firmly without slipping. Mechanical pencils have the advantages of maintaining a constant length of lead while permitting the use of a lead practically to the end, of being easily refilled with new leads, of affording a ready source for compass leads, of having no wood to be sharpened, and of easy sharpening of the lead by various mechanical pencil pointers.

Thin-lead mechanical pencils are available with 0.3-, 0.5-, 0.7-, or 0.9-mm-diameter drafting leads in several grades (Fig. 3.7c). These thin leads produce uniform width lines without sharpening, providing both a time savings and a cost benefit. Mechanical pencils are recommended as they are less expensive in the long run.

Drawing pencils are made of graphite with the addition of either a polymer binder or kaolin (clay) in varying amounts to make 18 grades from 9H, the hardest, to 7B, the softest. The uses of these different grades are described in Fig. 3.8. Note that small-diameter leads are used for the harder grades, whereas large-diameter leads are used to give more strength to the softer grades. Therefore, the degree of hardness in a wood pencil can be roughly judged by a comparison of diameters.

FIGURE 3.7 **Drawing Pencils.**

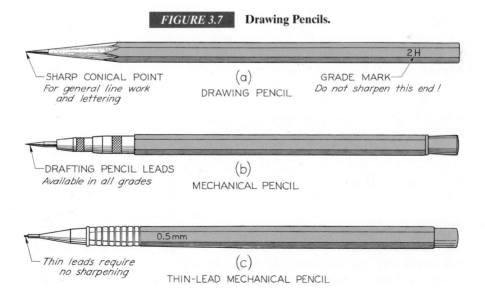

SHARP CONICAL POINT
*For general line work
and lettering*

(a)
DRAWING PENCIL

GRADE MARK
Do not sharpen this end!

2H

DRAFTING PENCIL LEADS
Available in all grades

(b)
MECHANICAL PENCIL

*Thin leads require
no sharpening*

(c)
THIN-LEAD MECHANICAL PENCIL

0.5mm

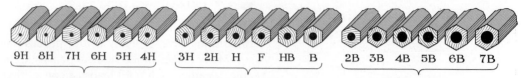

9H 8H 7H 6H 5H 4H	3H 2H H F HB B	2B 3B 4B 5B 6B 7B

Hard	Medium	Soft
The hard leads in this group (left) are used where extreme accuracy is required, as on graphical computations and charts and diagrams. The softer leads in this group (right) are sometimes used for line work on engineering drawings, but their use is restricted because the lines are apt to be too light.	These grades are for general purpose work in technical drawing. The softer grades (right) are used for technical sketching, for lettering, arrowheads, and other freehand work on mechanical drawings. The harder leads (left) are used for line work on machine drawings and architectural drawings. The H and 2H leads are widely used on pencil tracings for reproduction.	These leads are too soft to be useful in mechanical drafting. Their use for such work results in smudged, rough lines that are hard to erase, and the lead must be sharpened continually. These grades are used for art work of various kinds, and for full-size details in architectural drawing.

FIGURE 3.8   **Lead Grade Chart.**

Specifically formulated leads of carbon black particles in a polymer binder are also available in several grades for use on the polyester films now used quite extensively in industry (see §3.56).

In the selection of a grade of lead, the first consideration is the type of line work required. For light construction lines, guide lines for lettering, and accurate geometrical constructions or work in which accuracy is of prime importance, use a hard lead, such as 4H to 6H.

For mechanical drawings on drawing paper or tracing paper, the lines should be **black**, particularly for drawings to be reproduced. The lead chosen must be soft enough to produce jet black lines, but hard enough not to smudge too easily or permit the point to crumble under normal pressure. The same comparatively soft lead is preferred for lettering and arrowheads.

This lead will vary from F to 2H, depending on the paper and weather conditions. If the paper is hard, it will generally be necessary to use harder leads. For softer surfaces, softer leads can be used. On humid days, paper absorbs moisture from the atmosphere and becomes soft. This can be recognized because the paper expands and becomes wrinkled. It is necessary to select softer leads to offset the softening of the paper. If you have been using a 2H lead, for example, change to an F until the weather becomes drier.

■ 3.8 ALPHABET OF LINES

Each line on a technical drawing has a definite meaning and is drawn in a certain way. The line conventions endorsed by the American National Standards Institute, ANSI Y14.2M–1992, are presented in Fig. 3.9, together with illustrations of various applications.

Two widths of lines are recommended for use on drawings. The ratio of line widths should be approximately two-to-one. It is recommended the thin line width be 0.3 mm minimum, and the thick line width be 0.6 mm minimum. All required lines should be cleancut, dark, uniform throughout the drawing, and properly spaced for legible reproduction by all commonly used methods. Spacing between parallel lines may be exaggerated to a maximum of 3 mm .120 so there is no fill-in when the drawing is reproduced. The size and style of the drawing and the smallest size to which it is to be reduced govern the actual width of each line. The contrast between the two widths of lines should be distinct. Pencil leads should be hard enough to prevent smudging, but soft enough to produce the dense black lines necessary for quality reproduction.

When photoreduction and blowback are not necessary, as is the case for most drafting laboratory assignments, three weights of lines may improve the appearance and legibility of the drawing. The "thin lines" may be made in two widths regular thin lines for hidden lines, and stitch lines, and a somewhat thinner version for the other secondary lines (such as center lines, extension lines, dimension lines, leaders, section lines, phantom lines, and long-break lines).

For the "thick lines" visible, cutting plane, and short break use a relatively soft lead, such as F or H. All thin lines should be made with a sharp medium-grade lead, such as H or 2H. All lines (except construction lines) must be *sharp* and *dark*. Make construction

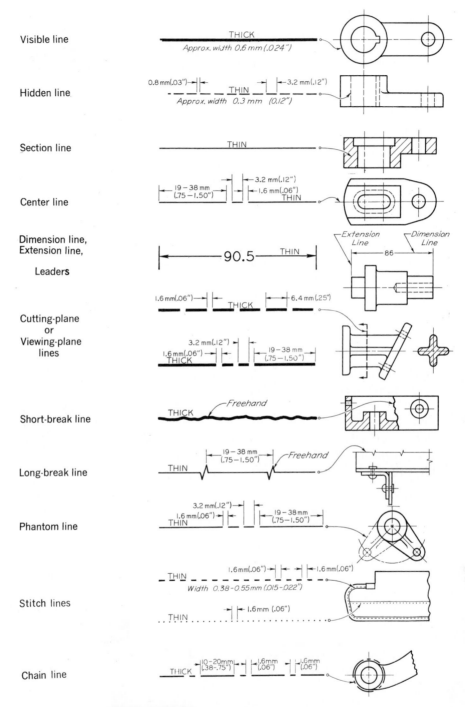

Visible line

Hidden line

Section line

Center line

Dimension line,
Extension line,

Leaders

Cutting-plane
or
Viewing-plane
lines

Short-break line

Long-break line

Phantom line

Stitch lines

Chain line

FIGURE 3.9 **Alphabet of Lines (Full Size).**

lines with a sharp 4H or 6H lead so thin that they can barely be seen at arm's length and need not be erased.

In Fig. 3.9, the ideal lengths of all dashes are indicated. You would do well to measure the first few hidden dashes and center-line dashes you make and thereafter to estimate the lengths carefully by eye. The line gage (Fig. 3.10) is a convenient reference for lines of various widths.

.005 in.	.007 in.	.010 in.	.012 in.	.014 in.	.020 in.	.024 in.	.028 in.	.031 in.	.039 in.	.047 in.	.055 in.	.079 in.
.13 mm	.18 mm	.25 mm	.30 mm	.35 mm	.50 mm	.60 mm	.70 mm	.80 mm	1.00 mm	1.20 mm	1.40 mm	2.00 mm

FIGURE 3.10 **Line Gage.** *Courtesy of Koh-I-Noor Rapidograph, Inc.*

■ 3.9 DRAWING HORIZONTAL LINES

To draw a horizontal line, press the head of the T-square firmly against the working edge of the board with your left hand; then slide your hand to the position shown in Fig. 3.11a so that the blade is pressed tightly against the paper. Lean the pencil in the direction of the line at an angle of approximately 60° with the paper, and draw the line from left to right (Fig. 3.11b). Keep the pencil in a vertical plane; otherwise, the line may not be straight (Fig. 3.11c). While drawing the line, let the little finger of the hand holding the pencil glide lightly on the blade of the T-square, and rotate the pencil slowly, except for the thin-lead pencils, between your thumb and forefinger to distribute the wear uniformly on the lead and maintain a symmetrical point. Thin-lead pencils should be held nearly vertical to the paper and not rotated. Also, pushing the thin-lead pencil from left to right, rather than pulling it, tends to minimize lead breakage.

When great accuracy is required, the pencil may be "toed in" to produce a perfectly straight line (Fig. 3.11d). *Left-handers:* In general, reverse the procedure just outlined. Place the T-square head against the right edge of the board, and with the pencil in the left hand, draw the line from right to left.

■ 3.10 DRAWING VERTICAL LINES

Use either the 45° triangle or the 30° × 60° triangle to draw vertical lines. Place the triangle on the T-square with the *vertical edge on the left,* as shown in Fig. 3.12a. With the left hand, press the head of the T-square against the board; then slide the hand to the position shown where it holds both the T-square and the triangle firmly in position. Draw the line upward, rotating the pencil slowly between the thumb and forefinger. (The only time it is advisable for right-handers to turn the triangle so that the vertical edge is on the right is when drawing a vertical line near the right end of the T-square. In this case, the line would be drawn downward.)

Lean the pencil in the direction of the line at an angle of approximately 608 with the paper and in a vertical plane (Fig. 3.12b). Meanwhile, the upper part of the body should be twisted to the right (Fig. 3.12c). (*Left-handers:* In general, reverse the foregoing procedure. Place the T-square head on the right and the ver-

FIGURE 3.11 **Drawing a Horizontal Line.**

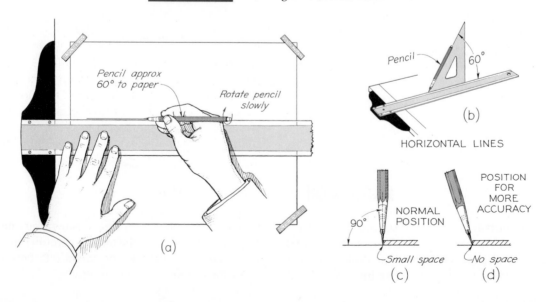

HORIZONTAL LINES

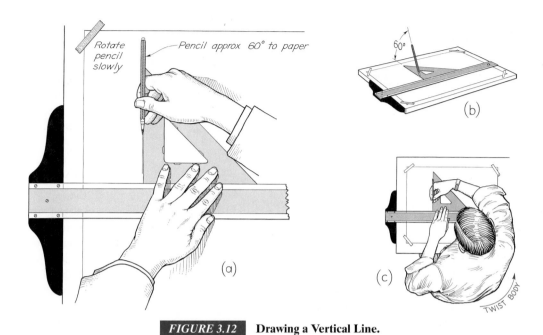

FIGURE 3.12   **Drawing a Vertical Line.**

tical edge of the triangle on the right; then, with the right hand, hold the T-square and triangle firmly together, and with the left hand draw the line upward.)

■ 3.11 TRIANGLES

Most inclined lines in mechanical drawing are drawn at standard angles with the *45° triangle* and the *30° × 60° triangle* (Fig. 3.13). The triangles are made of transparent plastic so that lines of the drawing can be seen through them. A good combination of triangles is the 30° × 60° triangle with a long side of 10″ and a 45° triangle with each side 8″ long.

FIGURE 3.13   **Triangles.**

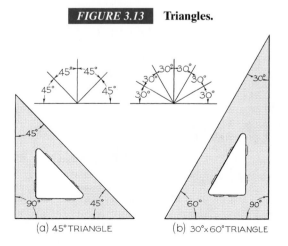

(a) 45°TRIANGLE   (b) 30°x 60°TRIANGLE

■ 3.12 DRAWING INCLINED LINES

The positions of the triangles for drawing lines at all of the possible angles are shown in Fig. 3.14. In the figure it is understood that the triangles in each case are resting on the blade of the T-square. Thus, it is possible to divide 360° into twenty-four 15° sectors with the triangles used singly or in combination. Note carefully the directions for drawing the lines, as indicated by the arrows, and that all lines in the left half are drawn *toward the center*, while those in the right half are drawn *away from the center*.

■ 3.13 PROTRACTORS

For measuring or setting off angles other than those obtainable with the triangles, the *protractor* is used. The best protractors, which produce the most accurate measurements, are made of nickel silver (Fig. 3.15a). For ordinary work, a plastic protractor is satisfactory and much cheaper (Fig. 3.15b). To set off angles with greater accuracy, use one of the methods presented in §4.19.

■ 3.14 DRAFTING ANGLES

There are a variety of devices that combine the protractor with triangles to produce great versatility of use. One such device is shown in Fig. 3.16.

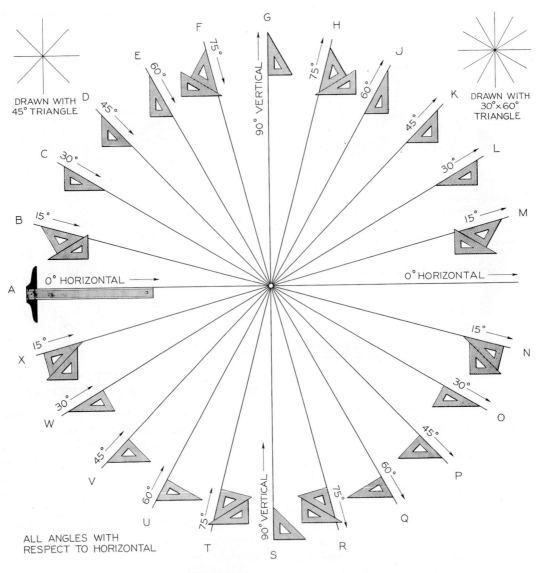

FIGURE 3.14 The Triangle Wheel.

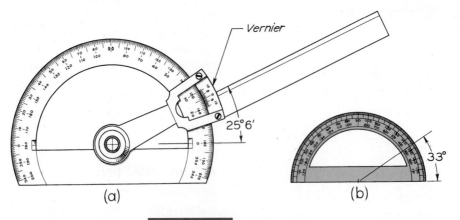

FIGURE 3.15 Protractors.

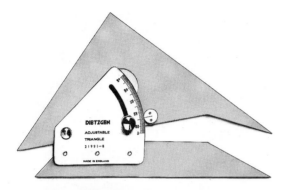

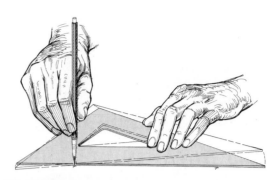

FIGURE 3.16   Adjustable Triangle.

FIGURE 3.17   To Draw a Pencil Line Through Two Points.

■ 3.15 DRAWING A LINE THROUGH TWO POINTS

To draw a line through two points, place the pencil vertically at one of the points (Fig. 3.17), and move the straightedge about the pencil point as a pivot until it lines up with the other point; then draw the line along the edge.

■ 3.16 DRAWING PARALLEL LINES

To draw a line parallel to a given line, move the triangle and T-square as a unit until the hypotenuse of the triangle lines up with the given line (Fig. 3.18a); then,

holding the T-square firmly in position, slide the triangle away from the line, and draw the required line along the hypotenuse (Figs. 3.18b and 3.18c).

Obviously, any straightedge, such as one of the triangles, may be substituted for the T-square in this operation, as shown in Fig. 3.18a.

■ 3.17 DRAWING PERPENDICULAR LINES

To draw a line perpendicular to a given line, move the T-square and triangle as a unit until one edge of the triangle lines up with the given line (Fig. 3.19a); then slide the triangle across the line and draws the required line (Figs. 3.19b and 3.19c).

FIGURE 3.18   To Draw a Line Parallel to a Given Line.

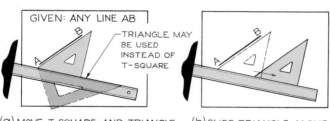

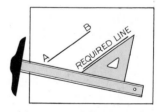

(a) MOVE T-SQUARE AND TRIANGLE TO LINE UP WITH AB

(b) SLIDE TRIANGLE ALONG T-SQUARE

(c) DRAW REQUIRED LINE PARALLEL TO AB

FIGURE 3.19   To Draw a Line Perpendicular to a Given Line.

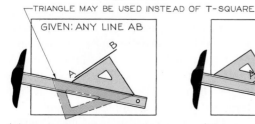

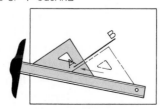

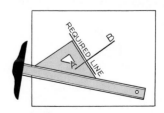

(a) MOVE T-SQUARE AND TRIANGLE TO LINE UP WITH AB

(b) SLIDE TRIANGLE ALONG T-SQUARE

(c) DRAW REQUIRED LINE PERPENDICULAR TO AB

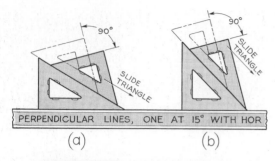

FIGURE 3.20 **Perpendicular Lines.**

To draw perpendicular lines when one of the lines makes 15° with horizontal, arrange the triangles as shown in Fig. 3.20.

■ 3.18 DRAWING LINES AT 30°, 60°, OR 45° WITH A GIVEN LINE

To draw a line making 30° with a given line, arrange the triangle as shown in Fig. 3.21. Angles of 60° and 45° may be drawn in a similar manner.

■ 3.19 SCALES

A drawing of an object may be the same size as the object (full size), or it may be larger or smaller than the object. The ratio of reduction or enlargement depends on the relative sizes of the object and of the sheet of paper on which the drawing is to be made. For example, a machine part may be half size; a building may be drawn size; a map may be drawn size; or a printed circuit board, may be drawn four times its size (Fig. 3.22).

Scales are instruments used in making technical drawings full size at a given enlargement or reduction. Figure 3.23 shows various types of scales, including (a) the *metric scale*, (b) the *engineers' scale*, (c) the *decimal scale*, (d) the *mechanical engineers' scale*, and (e) the *architects' scale*. A full-divided scale is one in which

FIGURE 3.21 **Line at 30° with Given Line.**

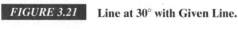
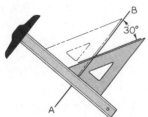

FIGURE 3.22 **Printed Circuit Board.** *United Nations/Guthrie.*

the basic units are subdivided throughout the length of the scale. The architects' scale is an open divided scale, one in which only the end unit is subdivided.

Scales are usually made of plastic or boxwood. The better wood scales have white plastic edges. Scales are either triangular or flat. The triangular scales have the advantage of combining many scales on one stick, but the user will waste much time looking for the required scale if a *scale guard* (Fig. 3.24) is not used. The scale guard marks the scale that is being used. Flat scales are almost universally used by professional drafters because of their convenience, but several flat scales are necessary to replace one triangular scale, and the total cost is greater.

■ 3.20 METRIC SYSTEM AND METRIC SCALES

The metric system is an international standard of measurement that, despite modifications over the past 200 years, has been the foundation of science and industry and is clearly defined. The modern form of the metric system is the International System of Units, commonly referred to as SI (from the French name, Le Système International d'Unités). The SI system was established in 1960 by international agreement and is now the international standard of measurement.

The metric scale is used when the meter is the standard for linear measurement. The meter was

(a) Metric Scale

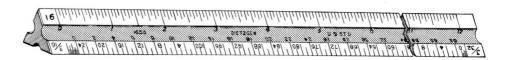

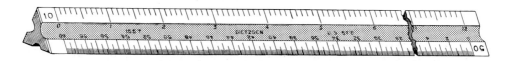

(b) Engineers' Scale

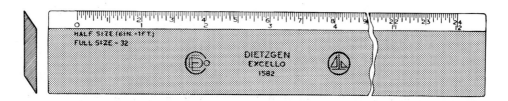

(c) Decimal Scale

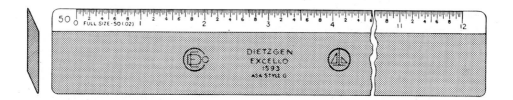

(d) Mechanical Engineers' Scale

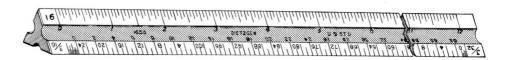

(e) Architects' Scale

FIGURE 3.23   **Types of Scales.**

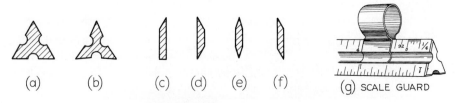

(a)   (b)   (c)   (d)   (e)   (f)   (g) SCALE GUARD

FIGURE 3.24   Scale Guard.

established by the French in 1791 with a length of one ten-millionth of the distance from the Earth's equator to the pole. The meter is equal to 39.37 inches or approximately 1.1 yards.

The metric system for linear measurement is a decimal system similar to our system of counting money. For example,

$$1 \text{ mm} = 1 \text{ millimeter} \left(\tfrac{1}{1000} \text{ of a meter}\right)$$

$$1 \text{ cm} = 1 \text{ centimeter} \left(\tfrac{1}{100} \text{ of a meter}\right)$$

$$= 10 \text{ mm}$$

$$1 \text{ dm} = 1 \text{ decimeter} \left(\tfrac{1}{10} \text{ of a meter}\right)$$

$$= 10 \text{ cm} = 100 \text{ mm}$$

$$1 \text{ m} = 1 \text{ meter}$$

$$= 100 \text{ cm} = 1000 \text{ mm}$$

$$1 \text{ km} = 1 \text{ kilometer} = 1000 \text{ m}$$

$$= 100,000 \text{ cm} = 1,000,000 \text{ mm}$$

The primary unit of measurement for engineering drawings and design in the mechanical industries is the millimeter (mm). Secondary units of measure are the meter (m) and the kilometer (km). The centimeter (cm) and the decimeter (dm) are rarely used.

In recent years, automotive and other industries have used a dual dimensioning system of millimeters and inches. Manufacturers of large agricultural machinery use all metric dimensions with the inch equivalents given in a table on the drawing.

Many of the dimensions in the illustrations and the problems in this text are given in metric units. Dimensions that are given in the customary units (inches and feet, either decimal or fractional) may be converted easily to metric values. In accordance with standard practice, the ratio 1 in. = 25.4 mm is used. Decimal equivalents tables can be found inside the front cover, and conversion tables are given in Appendix 31.

Metric scales are available in flat and triangular styles with a variety of scale graduations. The triangular scale illustrated in Fig. 3.35 has one full-size scale and five reduced-size scales, all full divided. By means

of these scales a drawing can be made full size, enlarged sized, or reduced sized. To specify the scale on a drawing see §3.26.

FULL SIZE   The 1:1 scale (Fig. 3.25a) is full size, and each division is actually 1 mm in width with the numbering of the calibrations at 10-mm intervals. The same scale is also convenient for ratios of 1:10, 1:100, 1:1000, and so on.

HALF SIZE   The 1:2 scale (Fig. 3.25a) is one-half size, and each division equals 2 mm with the calibration numbering at 20-unit intervals. This scale is also convenient for ratios of 1:20, 1:200, 1:2000, and so on.

The remaining four scales on this triangular metric scale include the typical scale ratios of 1:5, 1:25, 1:33⅓, and 1:75 (Figs. 3.25a and 3.25b). These ratios may also be enlarged or reduced as desired by multiplying or dividing by a factor of 10. Metric scales are

FIGURE 3.25   **Decimal Dimensions.**

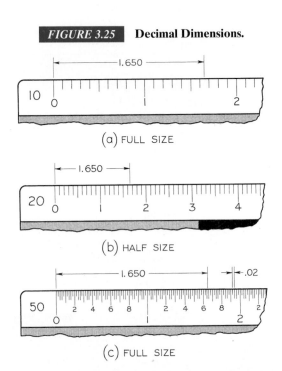

(a) FULL SIZE

(b) HALF SIZE

(c) FULL SIZE

also available with other scale ratios for specific drawing purposes.

The metric scale is used in map drawing and in drawing force diagrams or other graphical constructions that involve such scales as 1 mm = 1 kg and 1 mm = 500 kg.

■ 3.21 INCH-FOOT SCALES

Several scales that are based on the inch-foot system of measurement continue in domestic use today along with the metric system of measurement, which is accepted worldwide for science, technology, and international trade.

■ 3.22 ENGINEERS' SCALES

The *engineers' scale* is graduated in the decimal system. It is also frequently called the *civil engineers' scale* because it was originally used mainly in civil engineering. The name *chain scale* also persists because it was derived from the surveyors' chain composed of 100 links, used for land measurements.

The engineers' scale is graduated in units of 1 in. divided into 10, 20, 30, 40, 50, and 60 parts. Thus, the engineers' scale is convenient in machine drawing to set off dimensions expressed in decimals. For example, to set off 1.650″ full size, use the 10-scale and simply set off one main division plus $6\frac{1}{2}$ subdivisions (Fig. 3.26a).

FIGURE 3.26   **Metric Scales.**

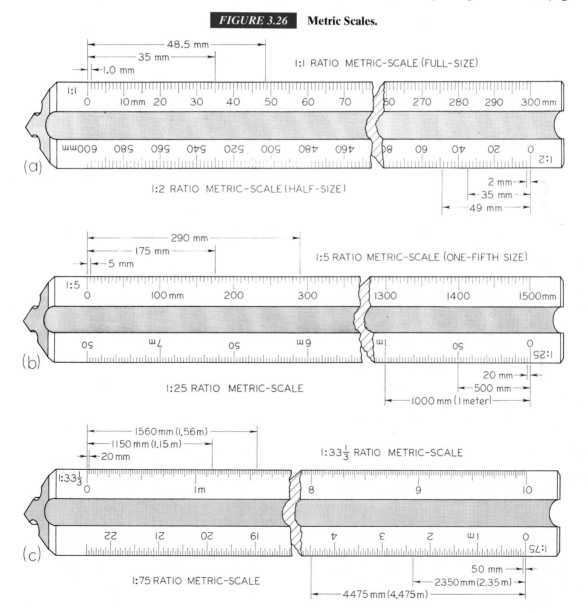

To set off the same dimension half size, use the 20-scale, since the 20-scale is exactly half the size of the 10-scale (Fig. 3.26b). Similarly, to set off a dimension quarter size, use the 40-scale.

The engineers' scale is also used in drawing maps to scales of $1'' = 50', 1'' = 500', 1'' = 5$ miles, and so on and in drawing stress diagrams or other graphical constructions to such scales as $1'' = 20$ lb and $1'' = 4000$ lb.

■ 3.23 DECIMAL SCALES

The increasing use of decimal dimensions has brought about the development of a scale specifically for that use. On the full-size scale, each inch is divided into fiftieths of an inch, or $.02''$ (Fig. 3.26c), and on the half- and quarter-size scales, the inches are compressed to half size or quarter size and then are divided into 10 parts, so that each subdivision stands for $.1''$.

The complete decimal system of dimensioning in which this scale is used is described in §12.10.

■ 3.24 ARCHITECTS' SCALES

The *architects' scale* is intended primarily for drawings of buildings, piping systems, and other large structures that must be drawn to a reduced scale to fit on a sheet of paper. The full-size scale is also useful in drawing relatively small objects, and for that reason the architects' scale has rather general usage.

The architects' scale has one full-size scale and ten overlapping reduced-sized scales. By means of these scales a drawing may be made to various sizes from full size to $\frac{1}{128}$ size. Note that, in all the reduced scales the major divisions represent feet, and their subdivisions represent inches and fractions thereof. Thus, the scale marked $\frac{3}{4}$ means $\frac{3}{4}$ inch = 1 foot, not $\frac{3}{4}$ inch = 1 inch; that is, one-sixteenth size, not three-fourths size. And the scale marked $\frac{1}{2}$ means $\frac{1}{2}$ inch + 1 foot, not $\frac{1}{2}$ inch = 1 inch, (that is, one twenty-fourth-size, not half size).

All the scales, from full size to $\frac{1}{128}$ size, are shown in Fig. 3.27. Some are upside down, just as they may occur in use. These scales are described as follows.

FULL SIZE Each division in the full-size scale is 0 (Fig. 3.27a). Each inch is divided first into halves, then quarters, eighths, and finally sixteenths, the dividing lines diminishing in length with each subdivision. To set off $\frac{1}{32}''$, estimate visually one half of $\frac{1}{16}''$; to set off $\frac{1}{64}''$, estimate one fourth of $\frac{1}{16}''$.

HALF SIZE Use the full-size scale, and divide every dimension mentally by two. (Do not use the $\frac{1}{2}''$ scale, which is intended for drawing to a scale of $\frac{1}{2}'' = 1'$, or one-twenty-fourth size.) To set off $1''$, measure $\frac{1}{4}''$; to set off $2''$, measure $1''$; to set off $\frac{6.5}{16}''$, measure $1\frac{1}{2}''$ (half of $30''$), then $\frac{1}{8}''$ (half of $\frac{1}{4}''$); to set off $2\text{-}\frac{13}{16}''$ (see Fig. 3.27), measure $1''$ then $\frac{13}{32}''$ ($\frac{6.5}{16}''$ or half of $\frac{13}{16}''$).

QUARTER SIZE Use the $3''$ scale in which $3'' = 1'$ (Fig. 3.27b). The subdivided portion to the left of zero, which represents 1 foot, is divided into inches, half inches, quarter inches, and eighth inches. The entire portion representing 1 foot actually measures 3 inches; therefore, $3'' = 1'$. To set off anything less than $12''$, start at zero and measure to the left.

To set off $10\frac{1}{8}''$, read off $9''$ from zero to the left and add $1\frac{1}{8}''$ and set off the total $10\frac{1}{8}''$, as shown. To set off more than $12''$ for example, $1'\text{-}\frac{9}{38}''$ (see your scale) find the $1'$ mark to the right of zero and the $9\frac{3}{8}''$ mark to the left of zero; the required distance is the distance between these marks and represents $1'\text{-}9\frac{3}{8}''$.

EIGHT SIZE Use the $1\frac{1}{2}''$ scale in which $1\frac{1}{2}'' = 1'$ (Fig. 3.27b). The subdivided portion of the right of zero represents $1'$ and is divided into inches, then half inches, and finally quarter inches. The entire portion, representing $1'$, actually is $1\frac{1}{2}''$; therefore, $1\frac{1}{2}'' = 1'$. To set off anything less than $12''$, start at zero and measure to the right.

DOUBLE SIZE Use the full-size scale, and multiply every dimension mentally by 2. To set off $1''$, measure $2''$; to set off $3\frac{1}{4}''$, measure $6\frac{1}{2}''$; and so on. The double-size scale is occasionally used to represent small objects. In such cases, a small actual-size outline view may be shown near the bottom of the sheet to help the shop worker visualized the actual size of the object.

OTHER SIZE The scales besides those just described are used chiefly by architects. Machine drawings are customarily made only double size, full size, half size, one-fourth size, or one-eighth size.

■ 3.25 MECHANICAL ENGINEERS' SCALES

The objects represented in machine drawing vary in size from small parts, an inch or smaller in size, to equipment or parts of large dimensions. By drawing these objects full size, half size, quarter size, or eighth size, the drawings will readily come within the limits of

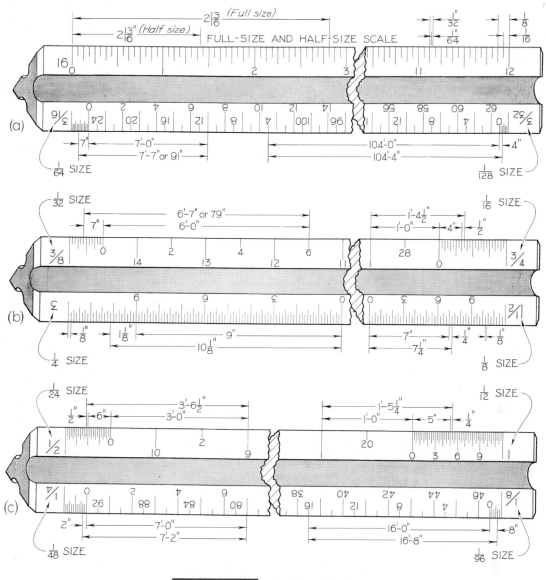

FIGURE 3.27 Architects' Scales.

the standard-size sheets. For this reason the mechanical engineers' scales are divided into units representing inches to full size, half size, quarter size, or eighth size (Fig. 3.27c). To make a drawing of an object to a scale of one-half size, for example, use the mechanical engineer's scale marked half size, which is graduated so that every $\frac{1}{2}''$ represents 1″. Thus, the half-size scale is simply a full-size scale compressed to one-half size.

These scales are also very useful in dividing dimensions. For example, to draw a $3\frac{11}{16}''$ diameter circle full size, we need half of $3\frac{11}{16}''$ to use as radius. Instead of using arithmetic to find half of $3\frac{11}{16}''$, it is easier to set off $3\frac{11}{16}''$ on the half-size scale.

Triangular combination scales are available that include the full- and half-size mechanical engineers' scales, several architects' scales, and an engineer's scale.

■ 3.26 SPECIFYING THE SCALE ON A DRAWING

For machine drawings, the scale indicates the ratio of the size of the drawn object to its actual size, irrespective of the unit of measurement used. The recommended practice is to letter FULL SIZE or 1:1; HALF SIZE or 1:2; and similarly for other reductions. Expansion or enlargement scales are given as 2:1 or 2:3; 3:1 or 3:3; 5:1 or 5:3; 10:1 or 10 3; and so on.

The various scale calibrations available on the metric scale and the engineers' scale provide almost unlimited scale ratios. The preferred metric scale ratios appear to be 1:1; 1:2; 1:5, 1:10, 1:20, 1:50, 1:100, and 1:200.

Map scales are indicated in terms of fractions, such as Scale $\frac{1}{62500}$, or graphically, such as

400   0   400   800 Ft

■ 3.27 ACCURATE MEASUREMENTS

Accurate drafting depends considerably on the correct use of the scale in setting off distances. Do not take measurements directly off the scale with the dividers or compass, as damage will result to the scale. Place the scale on the drawing with the edge parallel to the line on which the measurement is to be made and, with a sharp pencil having a conical point, make a short dash at right angles to the scale and opposite the correct graduation mark, as shown in Fig. 3.28a. If extreme accuracy is required, a tiny prick mark may be made at the required point with the needle point or stylus, (Fig. 3.28b),or with one leg of the dividers.

Avoid cumulative errors in the use of the scale. If a number of distances are to be set off end-to-end, all should be set off at one setting of the scale by adding each successive measurement to the preceding one, if possible. Avoid setting off the distances individually by moving the scale to a new position each time, since slight errors in the measurements may accumulate and give rise to a large error.

■ 3.28 DRAWING INSTRUMENTS

In technical drawing, accuracy, neatness, and speed are essential, §3.2. These objectives are not likely to be obtained with cheap or inferior drawing instruments. For the student or the professional drafter, it is advisable, and in the end more economical, to purchase the best instruments that can be afforded. Good instruments will satisfy the most rigid requirements, and the satisfaction, saving in time, and improved quality of work that good instruments can produce will more than justify the higher price.

Unfortunately, the qualities of high-grade instruments are not likely to be recognized by the beginner, who is not familiar with the performance characteristics required and who is apt to be attracted by elaborate sets containing a large number of shiny, low-quality instruments. Therefore, the student should obtain the advice of the drafting instructor, an experienced drafter, or a reliable dealer.

■ 3.29 GIANT BOW SETS

Giant bow sets contain various combinations of instruments, but all feature a large bow compass in place of the traditional large compass (Fig. 3.29). Most of the large bows are of the center-wheel type (Fig. 3.30a). Several manufacturers now offer different varieties of quick-acting bows. The large bow compass shown at (b) can be adjusted to the approximate setting by sim-

FIGURE 3.28   **Accurate Measurements.**

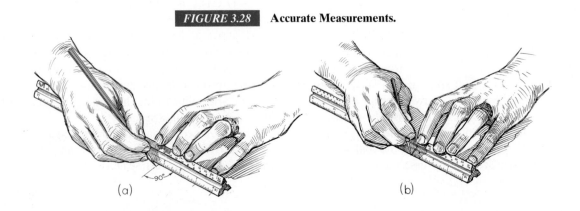

(a)   (b)

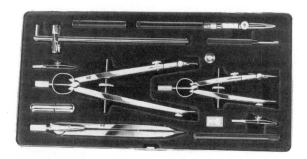

FIGURE 3.29 **Giant Bow Set.** *Courtesy of Frank Oppenheimer.*

ply opening or closing the legs in the same manner as for the other bow-style compass.

■ 3.30 COMPASSES

The compass, with pencil and inking attachments, is used for drawing circles of approximately 25 mm (10) radius or larger.

The giant bow compass (Figs. 3.29–3.31) has a socket joint in one leg that permits the insertion of either pencil or pen attachments. A *lengthening bar* or a *beam attachment* is often provided to increase the radius. Most of the large bows are of the center-wheel type (Fig. 3.30a). Several manufacturers now offer different varieties of quick-acting bows. The large bow compass shown in Fig. 3.30b can be adjusted to the approximate setting by simply opening or closing the legs in the same manner as for the other bow-style compass. For production drafting, in which it is necessary to make dense black lines to secure clear legible reproductions, the giant bow or an appropriate template is preferred. The large bow instrument is much sturdier than the traditional compass and is capable of taking the heavy pressure necessary to produce dense black lines without losing the setting.

■ 3.31 USING COMPASSES

The following instructions apply generally both to old style and giant bow compasses.

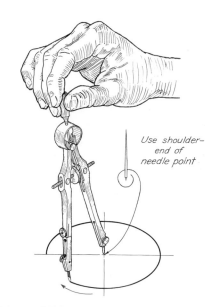

FIGURE 3.31 **Using the Giant Bow Compass.**

Most compass needle points have a plain end for use when the compass is converted into dividers and a shoulder end for use as a compass. Adjust the needle point with the shoulder end out and so that the small point extends *slightly* farther than the pencil lead or pen nib (Fig. 3.33d).

To draw a penciled circle, (1) set off the required radius on one of the center lines, (2) place the needle point at the exact intersection of the center lines, (3) adjust the compass to the required radius (25 mm or more), and (4) lean the compass forward and draw the circle clockwise while rotating the handle between the thumb and forefinger. To obtain sufficient weight of line, it may be necessary to repeat the movement several times.

Any error in radius will result in a doubled error in diameter; so it is best to draw a trial circle first on scrap paper or on the backing sheet and then check the diameter with the scale.

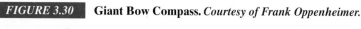

FIGURE 3.30 **Giant Bow Compass.** *Courtesy of Frank Oppenheimer.*

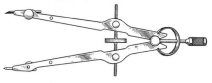

(a) CENTER-WHEEL

(b) QUICKACTING

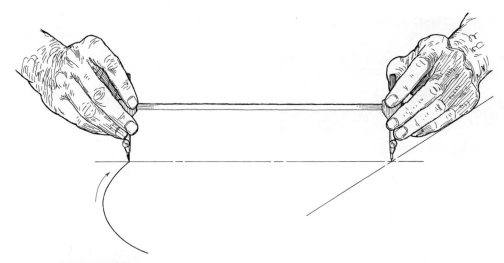

FIGURE 3.32 **Drawing a Circle of Large Radius with the Beam Compass.**

On drawings that have circular arcs and tangent straight lines, draw the acrs first, whether in pencil or in ink, as it is much easier to connect a straight line to an arc than the reverse.

For very large circles, a beam compass (discussed later in this section) is preferred, or use the lengthening bar to increase the compass radius. Use both hands, as shown in Fig. 3.32, but be careful not to jar the instrument and thus change the adjustment.

When using the compass to draw construction lines, use a 4H to 6H lead so that the lines will be very dim. For required lines, the arcs and circles must be black, and softer leads must be used. However, since heavy pressure cannot be exerted on the compass as it can on a pencil, it is usually necessary to use a compass lead that is one or two grades softer than the pencil used for the corresponding line work. For example, if an H lead is used for visible lines drawn with a pencil, then an F lead might be found suitable for the com-

pass work. The hard leads supplied with the compass are usually unsatisfactory for most line work except construction lines. In summary, use leads in the compass that will produce arcs and circles that *match* the straight pencil lines.

It is necessary to exert pressure on the compass to produce heavy "reproducible" circles, and this tends to enlarge the compass center hole in the paper, especially if there are a number of concentric circles. In such cases, use a horn center, or center tack, in the hole, and place the needle point of the compass in the center of the tack.

■ 3.32 SHARPENING THE COMPASS LEAD

Various forms of compass lead points are illustrated in Fig. 3.33. In Fig. 3.33a, a single elliptical face has been formed by rubbing on the sandpaper pad, as shown in Fig. 3.34. In Fig. 3.33b, the point is narrowed by small

FIGURE 3.33 **Compass Lead Points.**

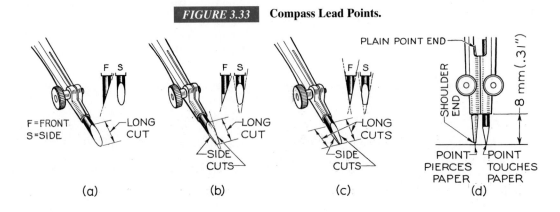

(a)   (b)   (c)   (d)

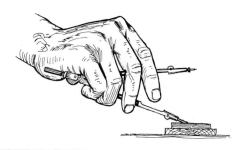

FIGURE 3.34 **Sharpening Compass Lead.**

side cuts. In Fig. 3.33c, two long cuts and two small side cuts have been made to produce a point similar to that on a screwdriver. In Fig. 3.33d, the cone point is prepared by chucking the lead in a mechanical pencil and shaping it in a pencil pointer. Avoid using leads that are too short to be exposed as shown.

In using the compass, *never use the plain end of the needle point*. Instead, use the shoulder end, as shown in Fig. 3.33d, adjusted so that the tiny needlepoint extends about halfway into the paper when the compass lead just touches the paper.

■ 3.33 BEAM COMPASSES

The beam compass, or trammel (Fig. 3.35), is used for drawing arcs or circles larger than can be drawn with the regular compass and for transferring distances too great for the regular dividers. Besides steel points, pencil and pen attachments are provided. The beams may

be made of nickel silver, steel, aluminum, or wood and are procurable in various lengths. A square nickel silver beam compass set is shown in Fig. 3.35a, and a set with the beam graduated in millimeters and inches is shown in Fig. 3.35b.

■ 3.34 DIVIDERS

Dividers, as the name implies, are used for *dividing* distances into a number of equal parts. They are also used for *transferring distances* or for *setting off* a series of equal distances. Dividers are similar to compasses in construction and are made in square, flat, and round forms.

The friction adjustment for the pivot joint should be loose enough to permit easy manipulation with one hand, as shown in Fig. 3.36. If the pivot joint is too tight, the legs of the divider tend to spring back instead of stopping at the desired point when the pressure of the fingers is released. To adjust tension, use a small screwdriver.

Many dividers are made with a spring and thumbscrew in one leg so that minute adjustments in the setting can be made by turning the small thumbscrew.

■ 3.35 USING DIVIDERS

Dividers are used for spaces of approximately 25 mm (10) or more. For spaces less than 25 mm, use the bow dividers (Fig. 3.40a). *Never use the large dividers for small spaces when the bow dividers can be used; the latter are more accurate.*

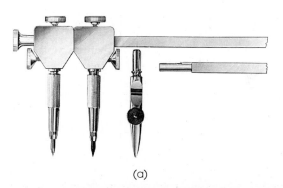

(a)

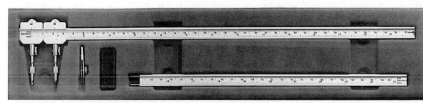

FIGURE 3.35 **Beam Compass Sets.** *(a) Courtesy of Frank Oppenheimer; (b) Courtesy of Tacro, Div. of A&T Importers, Inc.*

(b)

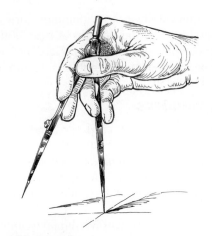

FIGURE 3.36   **Adjusting the Dividers.**

Dividing a given distance into a number of equal parts is a matter of trial and error (Fig. 3.36). Adjust the dividers with the fingers of the hand that holds them, to the approximate unit of division, estimated by eye. Rotate the dividers counterclockwise through 180°, and so on, until the desired number of units has been stepped off. If the last prick of the dividers falls short of the end of the line to be divided, increase the distance between the divider points proportionately. For example, to divide the line AB into three equal parts, the dividers are set by eye to approximately one-third the length AB. When it is found that the trial radius is too small, the distance between the divider points is increased by one-third the remaining distance. If the last prick of the dividers is beyond the end of the line, a similar decreasing adjustment is made.

Cumulative errors may result when dividers are used to set off a series of distances end to end. To set off a large number of equal divisions say, 15 first set off three equal large divisions and then divide each of these into five equal parts. Wherever possible in such cases, use the scale instead of the dividers (§3.27), or set off the total and then divide into the parts by means of the parallel-line method (§4.13).

■ 3.36 PROPORTIONAL DIVIDERS

For enlarging or reducing a drawing, proportional dividers are convenient (Fig. 3.38). They may also be

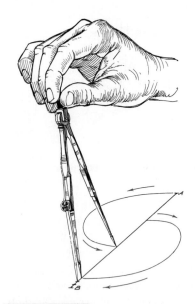

FIGURE 3.37   **Using the Dividers.**

used for dividing distances into a number of equal parts, or for obtaining a percentage reduction of a distance. For this purpose, points of division are marked on the instrument to secure the required subdivisions readily. Some instruments are calibrated to obtain special ratios, such as 1:square root of 2, the diameter of a circle to the side of an equal square, and feet to meters.

■ 3.37 BOW INSTRUMENTS

The bow instruments are classified as the *bow dividers*, *bow pen*, and *bow pencil*. A combination pen and pencil bow, usually with center-wheel adjustment, and separate instruments, with either side-wheel or center-wheel adjustment, are available (Figs. 3.39 and 3.40). The choice is a matter of personal preference.

■ 3.38 USING BOW INSTRUMENTS

Bow pencils and bow pens are used for drawing circles of approximately 25 mm (10) radius or smaller. Bow dividers are used for the same purpose as the large dividers, but they are used for smaller (approximately 25 mm or less) spaces and more accurate work.

Whether a center-wheel or side-wheel instrument is used, the adjustment should be made with the

FIGURE 3.38   **Proportional Dividers.**

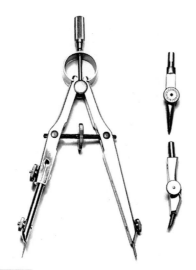

FIGURE 3.39 Combination Pen and Pencil Bow.
Courtesy of Frank Oppenheimer.

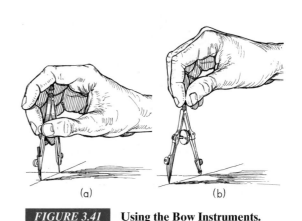

(a)   (b)

FIGURE 3.41 Using the Bow Instruments.

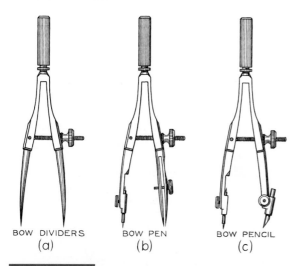

BOW DIVIDERS   BOW PEN   BOW PENCIL
(a)   (b)   (c)

FIGURE 3.40 Bow Instruments with Side Wheel.

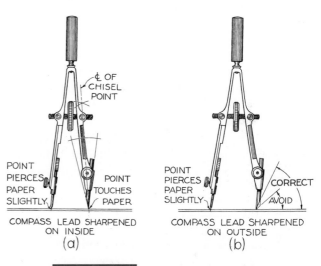

℄ OF CHISEL POINT

POINT PIERCES PAPER SLIGHTLY   POINT TOUCHES PAPER

COMPASS LEAD SHARPENED ON INSIDE
(a)

POINT PIERCES PAPER SLIGHTLY   CORRECT AVOID

COMPASS LEAD SHARPENED ON OUTSIDE
(b)

FIGURE 3.42 Compass-Lead Points.

thumb and finger of the hand that holds the instrument (Fig. 3.41a). The instrument is manipulated by twirling the head between the thumb and fingers (Fig. 3.41b).

The lead is sharpened in the same manner as for the large compass except that for small radii, the inclined cut may be turned *inside* if preferred (Fig. 3.42a). For general use, the lead should be turned on to the outside, as shown in Fig. 3.42b. In either case, always keep the compass lead sharpened. *Avoid stubby compass leads*, which cannot be properly sharpened. At least 6 mm (0) of lead should extend from the compass at all times.

In adjusting the needle point of the bow pencil or bow pen, be sure to have the needle extending slightly longer than the pen or the lead (Fig. 3.42b), the same as for the large compass.

In drawing small circles, greater care is necessary in sharpening and adjusting the lead and the needle point, and especially in accurately setting the desired radius. If a 6.35 mm (0) diameter circle is to be drawn, and if the radius is "off" only 0.8 mm (0), the total error on diameter is approximately 25%, which is far too much.

Appropriate templates may also be used for drawing small circles.

■ 3.39 DROP SPRING BOW PENCILS AND PENS

Drop spring bow pencils and pens (Fig. 3.43) are designed for drawing multiple identical small circles, such as drill holes or rivet heads. A central pin is made to move easily up and down through a tube to which the pen or pencil unit is attached. To use the instrument, hold the knurled head of the tube between your thumb and second finger, placing your first finger on top of the knurled head of the pin. Place the point of the pin at the desired center, lower the pen or pencil until it touches the paper, and twirl the instrument clockwise with your thumb and second finger. Then lift the tube independently of the pin, and finally lift the entire instrument.

■ 3.40 TO LAY OUT A SHEET

After the sheet has been attached to the board, as explained in §3.6, proceed as shown in Fig. 3.44 (see also Layout A-2, inside back cover).

1. Using the T-square, draw a horizontal *trim line* near the lower edge of the paper and then, using the triangle, draw a vertical trim line near the left edge of the paper. Both should be *light construction lines*.
2. Place the scale along the lower trim line with the full-size scale up. Draw short and light dashes *perpendicular* to the scale at the required distances (see Fig. 3.28a).

FIGURE 3.43   **Drop Spring Bow Instruments.**

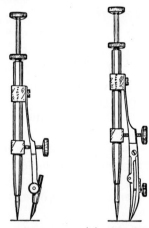

(a) DROP SPRING   (b) DROP SPRING
BOW PENCIL   BOW PEN

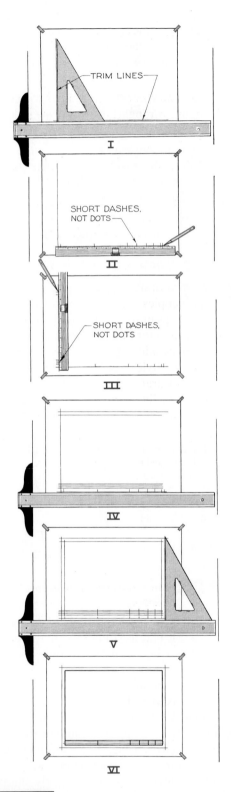

FIGURE 3.44   **To Lay Out a Sheet. Layout A–2; see inside front cover.**

3. Place the scale along the left trim line with the full-size scale to the left, and mark the required distances with short and light dashes perpendicular to the scale.

4. Draw horizontal construction lines with the aid of the T-square through the marks at the left of the sheet.

5. Draw vertical construction lines *from the bottom upward* along the edge of the triangle through the marks at the bottom of the sheet.

6. Retrace the border and the title strip to make them heavier. Notice that the layout is made independently of the edges of the paper.*

■ 3.41 TECHNIQUE OF PENCIL DRAWING

Most commercial drafting is executed in pencil. Most prints or photocopies are made from pencil tracings, and all ink tracings must be preceded by pencil drawings. It should therefore be evident that skill in drafting chiefly implies skill in pencil drawing.

Technique is a style or quality of drawing imparted by the individual drafter to the work. It is characterized by crisp black line work and lettering. Technique in lettering is discussed later in this chapter.

*In industrial drafting rooms the sheets are available, cut to standard sizes, with border and title strips already printed. Drafting supply houses can supply such papers, printed to order, to schools for little extra cost.

DARK ACCENTED LINES   The pencil lines of a finished pencil drawing or tracing should be very dark (Fig. 3.45). Dark crisp lines are necessary to give punch or snap to the drawing. (a) The ends of lines should be accented by a little extra pressure on the pencil. (b) Curves should be as dark as other lines. (c) and (d) Hidden-line dashes and center-line dashes should be carefully estimated as to length and spacing and should be of uniform width throughout their length.

Dimension lines, extension lines, section lines, and center lines also should be dark. The difference between these lines and visible lines is mostly in width there is very little difference, if any, in blackness.

A simple way to determine whether your lines on tracing paper or cloth are dense black is to hold the tracing up to the light. Lines that are not opaque black will not print clearly by most reproduction processes.

Construction lines should be made with a sharp, hard lead and *should be so light that need not be erased* when drawing is completed.

CONTRAST IN LINES   Contrast in pencil lines, like that in ink lines, should be mostly in *widths* of the lines, with little if any difference in the degree of darkness (Fig. 3.46). The visible lines should contrast strongly with the thin lines of the drawing. If necessary, draw over a visible line several times to get the desired thickness and darkness. A short retracing stroke backward (to the left), producing a jabbing action, results in a darker line.

FIGURE 3.45   **Technique of Lines (Enlarged).**

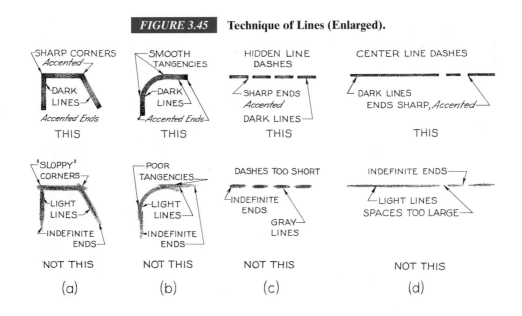

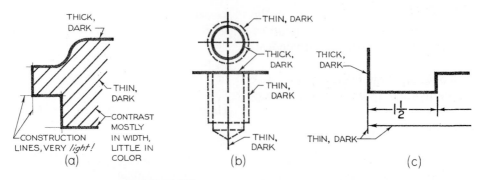

FIGURE 3.46   **Contrast of Lines (Enlarged).**

■ 3.42 PENCIL TRACING

While some pencil tracings are made of a drawing placed underneath the tracing paper (usually when a great deal of erasing and changing is necessary on the original drawing), most drawings today are made directly in pencil on tracing paper, pencil tracing cloth, films, or vellum. These are not tracings but pencil drawings, and the methods and technique are the same as previously described for pencil drawing.

In making a drawing directly on a tracing medium, a smooth sheet of heavy white drawing paper, a backing sheet, should be placed underneath. The whiteness of the backing sheet improves the visibility of the lines, and the hardness of the surface makes it possible to exert pressure on the pencil and produce dense black lines without excessive grooving of the paper.

All lines must be dark and cleanly drawn when drawings are to be reproduced.

■ 3.43 TECHNICAL FOUNTAIN PENS

Technical fountain pens (Fig. 3.47), with tube and needle point are available in several line widths. Many people prefer this type of pen because the line widths are fixed and it is suitable for freehand or mechanical lettering and line work. The pen requires an occasional filling and a minimum of skill to use. For uniform line work, the pen should be used perpendicular to the paper (Fig. 3.48). For best results, follow the manufacturer's recommendations for operation and cleaning.

FIGURE 3.47   **Technical Fountain Pen and Pen Set.** *Courtesy of Koh-I-Nor Rapidograph, Inc.*

■ 3.44 RULING PENS

The ruling pen (Fig. 3.49), is used to ink lines drawn with instruments, never to ink freehand lines or freehand lettering.

Ruling pens should be of the highest quality, with blades of high-grade tempered steel sharpened properly at the factory. The nibs should be sharp, but not sharp enough to cut the paper.

The *detail pen*, capable of holding a considerable quantity of ink, is extremely useful for drawing long heavy lines (Fig. 3.49b). This type of pen is preferred for small amounts of ink work because the pen is easily adjusted for the desired line width and is readily cleaned after use.

■ 3.45 USING RULING PENS

The ruling pen should lean at an angle of about 60° with the paper in the direction in which the line is being drawn and in a vertical plane containing the line (Fig. 3.50).

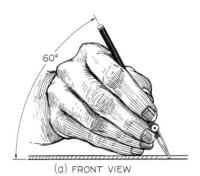
(a) FRONT VIEW

FIGURE 3.50   **Position of Hand in Using the Ruling Pen.**

The ruling pen is used in inking irregular curves, as well as straight lines. The pen should be held more nearly vertical when used with an irregular curve than when used with the T-square or a triangle. The ruling pen should lean only slightly in the direction in which the line is being drawn. The proper method for filling ruling pens is shown in Fig. 3.50.

■ 3.46 DRAWING INK

Drawing ink is composed chiefly of carbon in colloidal suspension and gum. The fine particles of carbon give the deep, black luster to the ink, and the gum makes it waterproof and quick to dry. The ink bottle should not be left uncovered, as evaporation will cause the ink to thicken.

Special drawing ink is available for use on acetate and polyester films. Such inks should not be used in technical fountain pens unless the pens are specifically made for acetate-based inks.

For removing dried waterproof drawing ink from pens or instruments, pen-cleaning fluids are available at dealers.

■ 3.47 TECHNIQUE OF INKING

The various widths of lines used for inked drawings or tracings are shown in Fig. 3.51. In inking a drawing or tracing (Fig. 3.52), proceed in the following order:

1. (a) Mark all tangent points in pencil directly on the drawing or tracing.
   (b) Indent all compass centers (with pricker or divider point).
   (c) Ink visible circles and arcs.
   (d) Ink hidden circles and arcs.
   (e) Ink irregular curves, if any.

FIGURE 3.48   **Using the Technical Fountain Pen.**

FIGURE 3.49   **Ruling Pens.**

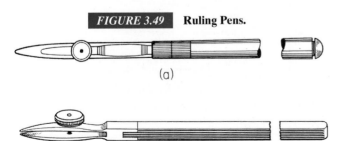
(a)

(b)

VISIBLE LINE

HIDDEN LINE

SECTION LINE

CENTER LINE

——————3.25——————
DIMENSION & EXTENSION LINES

CUTTING-PLANE LINES
OR
VIEWING-PLANE LINES

SHORT-BREAK LINE

LONG-BREAK LINE

PHANTOM LINE

STITCH LINE

FIGURE 3.51   Alphabet of Ink Lines (Full Size).

2. In (a) through (c), ink horizontal lines first, vertical lines second, and inclined lines last:

(a) Ink invisible straight lines.

(b) Ink hidden straight lines.

(c) Ink center lines, extension lines, dimension lines, leader lines, and section lines (if any).

3. (a) Ink arrowheads and dimension figures

(b) Ink notes titles, etc. (Pencil guide lines directly on the drawing or tracing.)

Some drafters prefer to ink center lines before indenting the compass centers because ink can go through the holes and cause blots on the back of the sheet.

When an ink blot is made, the excess ink should be taken up with a blotter, paper towel or tissue, and not allowed to soak into the paper. When the spot is thoroughly dry, the remaining ink can be erased easily.

For cleaning untidy drawings or for removing the original pencil lines from an inked drawing, a Pink Pearl or the Mars-Plastic eraser is suitable if used lightly.

When a gap in a thick ink line is made by erasing, the gap should be filled in with a series of fine lines

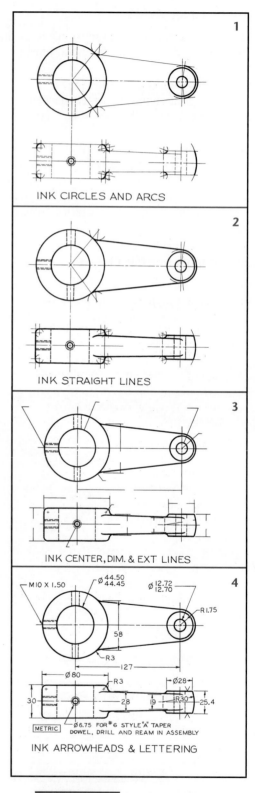

FIGURE 3.52   Order of Inking.

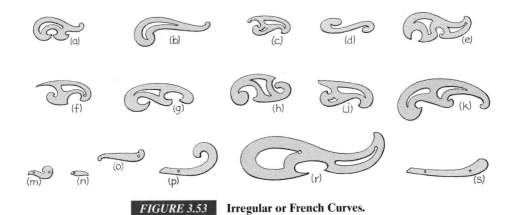

FIGURE 3.53   Irregular or French Curves.

that are allowed to run together. A single heavy line is difficult to match and is more likely to run and cause a blot.

■ 3.48 IRREGULAR CURVES

Drawing mechanical curves other than circles or circular arcs generally requires the use of an irregular or French curve. An irregular curve is a device for the *mechanical drawing of curved lines and should not be applied directly to the points* or used for purposes of producing an initial curve. Many different forms and sizes of curves are manufactured (Fig. 3.53).

The curves are composed largely of successive segments of the geometric curves, such as the ellipse, parabola, hyperbola, and involute. The best curves are made of transparent plastic. Among the many special types of curves available are hyperbolas, parabolas, ellipses, logarithmic spirals, ship curves, and railroad curves.

Adjustable curves are also available. Figure 3.54a consists of a core of lead, enclosed by a coil spring attached to a flexible strip. Figure 3.54b consists of a spline to which "ducks" (weights) are attached. The

spline can be bent to form any desired curve, limited only by the elasticity of the material. An ordinary piece of solder wire can be used very successfully by bending the wire to the desired curve.

■ 3.49 USING IRREGULAR CURVES

The proper use of the irregular curve requires skill, especially when the lines are to be drawn in ink (Fig. 3.55). After points have been plotted through which the curve is to pass, a light pencil line should be sketched freehand smoothly through the points.

To draw a mechanical line over the freehand line with an irregular curve, you match the various segments of the irregular curve with successive portions of the freehand curve and draw the line with pencil or ruling pen along the edge of the curve (Fig. 3.56). The irregular curve must match the sketched curve for some distance at each end beyond the segment to be drawn for any one setting of the curve so that successive sections of the curve will be tangent to each other, without any abrupt change in the curvature of the line (Fig. 3.56). In placing the irregular curve, the short-radius end of the curve should be turned toward

FIGURE 3.54   Adjustable Curves.

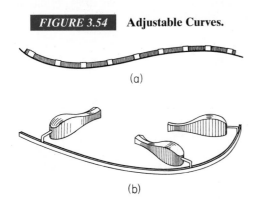

FIGURE 3.55   Using the Irregular Curves.

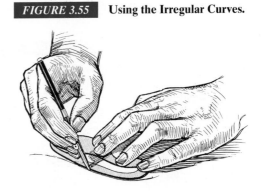

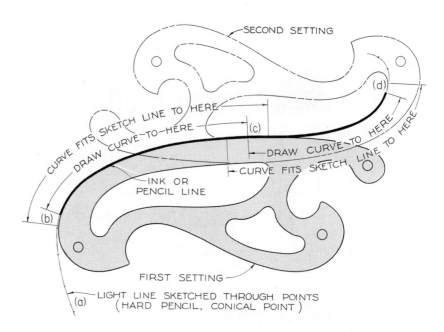

FIGURE 3.56   Settings of Irregular Curve.

the short-radius part of the curve to be drawn; that is, the portion of the irregular curve used should have the same curvilinear tendency as the portion of the curve to be drawn. This will prevent abrupt changes in direction.

The drafter should change the position of the drawing when necessary to avoid working on the lower side of the curve.

When plotting points to establish the path of a curve, it is desirable to plot more points, and closer together, where sharp turns in the curve occur.

Free curves may also be drawn with the compass, as shown in Fig. 4.42.

For symmetrical curves, such as an ellipse, use the same segment of the irregular curve in two or more opposite places (Fig. 3.57). For example, in Fig. 3.57a the irregular curve is matched to the curve and the line drawn from 1 to 2. Light pencil dashes are then drawn directly on the irregular curve at these points. (The curve will

take pencil marks well if it is lightly "frosted" by rubbing with a hard pencil eraser.) In Fig. 3.57b the irregular curve is turned over and matched so that the line may be drawn from 2 to 1. In similar manner, the same segment is used again in Figs. 3.57c and 3.57d. The ellipse is completed by filling in the gaps at the ends by using the irregular curve, or if desired, a compass.

■ 3.50 TEMPLATES

Templates are available for a great variety of specialized needs (Fig. 3.58). A template may be found for drawing almost any ordinary drafting symbols or repetitive features. The engineers' triangle (Fig. 3.58a) is useful for drawing hexagons or for bolt heads and nuts; the draftsquare (Fig. 3.58b) is convenient for drawing the curves on bolt heads and nuts, for drawing circles, thread forms, and so forth; and the chemistry stencil (Fig. 3.58c) is useful for drawing chemical apparatus in schematic form.

FIGURE 3.57   Symmetrical Figures.

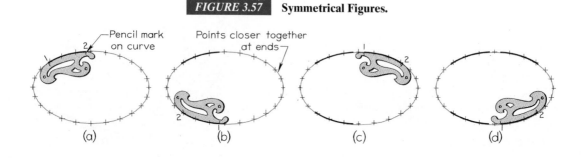

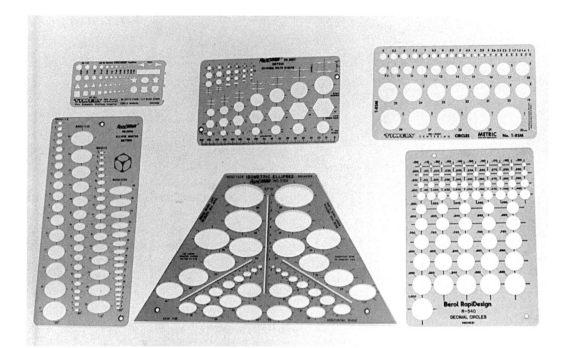

FIGURE 3.58   **Templates.**

Ellipse templates (§4.53), are perhaps more widely used than any other type. Circle templates are useful for drawing small circles quickly and for drawing fillets and rounds: such templates are used extensively in tool and die drawings.

■ 3.51 DRAFTING MACHINES

The drafting machine is an ingenious device that replaces the T-square, triangles, scales, and protractor (Figs. 3.59 and 3.60). The links, or bands, are arranged so that the controlling head is always in any desired fixed position regardless of where it is placed on the board; thus, the horizontal straightedge will remain horizontal if so set. The controlling head is graduated in degrees (including a vernier on certain machines), which allows the straightedges, or scales, to be set and locked at any angle. There are automatic stops at the more frequently used angles, such as 15°, 30°, 45°, 60°, 75°, and 90°.

The chief advantage of the drafting machine is that it speeds up drafting. Since its parts are made of metal, their accurate relationships are not subject to change, whereas T-squares, triangles, and working edges of drawing boards must be checked and corrected frequently. Drafting machines for left-handers are available from the manufacturers.

FIGURE 3.59   **Drafting Machine.** *Courtesy of VEMCO Corporation.*

■ 3.52 PARALLEL-RULING STRAIGHTEDGE

For large drawings, the long T-square becomes unwieldy, and considerable inaccuracy may result from the "give" or swing of the blade. In such a case the parallel-ruling straightedge is recommended (Fig. 3.61). The ends of the straightedge are controlled by a system of cords and pulleys that permit the straightedge

FIGURE 3.60 **Adjustable Drafting Table with Track Drafting Machine.** *Courtesy of Keuffel & Esser Co.*

FIGURE 3.61 **Parallel-Ruling Straightedge.**

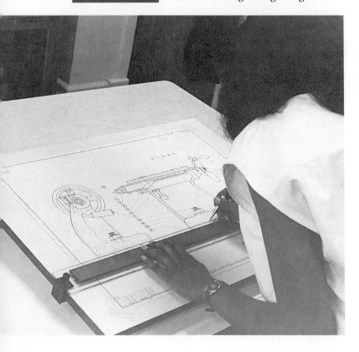

to be moved up or down on the board while maintaining a horizontal position.

■ 3.53 DRAWING PAPERS

Drawing paper, or *detail paper*, is used whenever a drawing is to be made in pencil but not for reproduction. From working drawings and for general use, the preferred paper is light cream or buff in color, and it is available in rolls of widths 24″ and 36″ and in cut sheets of standard sizes, such as 8.5″ × 11″, 11″ × 17″, 17″ × 22″, and so on. Most industrial drafting rooms use standard sheets with printed borders and title strips (§3.57). Since the cost for printing adds so little to the price per sheet, many schools have also adopted printed sheets.

The best drawing papers have up to 100% pure rag stock; they have strong fibers that afford superior erasing qualities, folding strength, and toughness; and they will not discolor or grow brittle with age. The paper should have a fine grain or tooth that will pick up the graphite and produce clean, dense black lines. However, if the paper is too rough, it will wear down the pencil excessively and will produce ragged, grainy lines. The paper should have a hard surface so that it will not groove too easily when pressure is applied to the pencil.

For ink work, as for catalog and book illustrations, white papers are used. The better papers, such as Bristol Board and Strathmore, come in several thicknesses, such as 2-ply, 3-ply, and 4-ply.

■ 3.54 TRACING PAPERS

Tracing paper is a thin transparent paper on which drawings are made for the purpose of reproducing by blueprinting or by other similar processes. Tracings are usually made in pencil but may also be made in ink. Most tracing papers will take pencil or ink, but some are especially suited to one or to the other.

Tracing papers called *vellums* have been treaded with oils, waxes, or similar substances to render them more transparent; other tracing papers are not so treated, but may be quite transparent due to the high quality of the raw materials and the methods of manufacture. Some treated papers deteriorate rapidly with age, becoming brittle within a few months, but some excellent vellums are available. Untreated papers made entirely of good rag stock will last indefinitely and will remain tough.

■ 3.55 TRACING CLOTH

Tracing cloth is a thin transparent muslin fabric (cotton, not linen as commonly supposed) sizes with a starch compound or plastic to provide a good working surface for pencil or ink. It is much more expensive than tracing paper. Tracing cloth is available in rolls of standard widths, such as 30″, 36″, and 42″, and also in sheets of standard sizes, with or without printed borders and title forms.

For pencil tracings, special pencil tracing cloths are available. Many concerns make their drawings in pencil directly on this cloth, dispensing entirely with the preliminary pencil drawing on detail paper, thus saving a great deal of time. These cloths generally have a surface that will produce dense black lines when hard pencils are used. Hence, these drawings do not easily smudge and will stand up well to handling.

■ 3.56 POLYESTER FILMS AND COATED SHEETS

Polyester film is a superior drafting material available in rolls and sheets of standard size. It is made by bonding a matte surface to one or both sides of a clear polyester sheet. The transparency and printing qualities are very good, the matte drawing surface is excellent for pencil or ink, erasures leave no ghost marks, and the film has high dimensional stability. Its resistance to cracking, bending, and tearing makes it virtually indestructible, if given reasonable care. The film has rapidly replaced cloth and is competing with vellum in some applications. Some companies have found it more economical to make their drawings directly in ink on the film.

Large coated sheets of aluminum (which provides a good dimensional stability) are often used in the aircraft and auto industry for full-scale layouts that are scribed into the coating with a steel point rather than a pencil. The layouts are reproduced from the sheets photographically.

■ 3.57 STANDARD SHEETS

Two systems of sheet sizes, together with length, width, and letter designations, are listed by ANSI, as shown in the accompanying table.

The use of the basic sheet size, 8.5″ × 11.0″ or 210 mm × 297 mm, and multiples thereof permits filing of small tracings and of folded prints in standard files with or without correspondence. These sizes can be cut without waste from the standard rolls of paper, cloth, or film.

For layout designations, title blocks, revision blocks, and a list of materials blocks (see inside the front cover of this book).

Nearest International Size[a] (millimeter)	Standard U.S. Size[a] (inch)
A4 210 × 297	A 8.5 × 11.0
A3 297 × 420	B 11.0 × 17.0
A2 420 × 594	C 17.0 × 22.0
A1 594 × 841	D 22.0 × 34.0
A0 841 × 1189	E 34.0 × 44.0

[a] ANSI Y14.1m-1992.

■ 3.58 THE COMPUTER AS A DRAFTING TOOL

Many of you will be using a CAD system as your drafting tool. Drawings created using a computer are basically the same as drawings created by hand. Accuracy, speed, and the ability to understand spatial and visual information, are equally important in instrumental drawing and in using a CAD system. Drawings created using a CAD system should follow the proper drafting standards so that they can be easily interpreted. Most CAD drawings are plotted on standard sheet sizes and to similar scales as hand prepared instrumental drawings. You still need to master the concepts and standards for orthographic and pictorial projections in order to use a CAD system effectively to create models and drawings.

An advantage of using CAD is that the system contains commands for easily drawing perfectly straight uniform lines and other geometric elements. Also the various styles of lines can be quickly represented by the CAD system (Fig. 3.62). Though it will take you some time to learn the command structure of your CAD system, you would take as long to learn instrumental drawing techniques for preparing neat accurate drawings. Keeping your drawing files organized and following conventions for naming the drawings so that you can find them on the CAD system is also an important consideration. Even when using a CAD system, skill in freehand sketching is still necessary to quickly get your ideas down on paper.

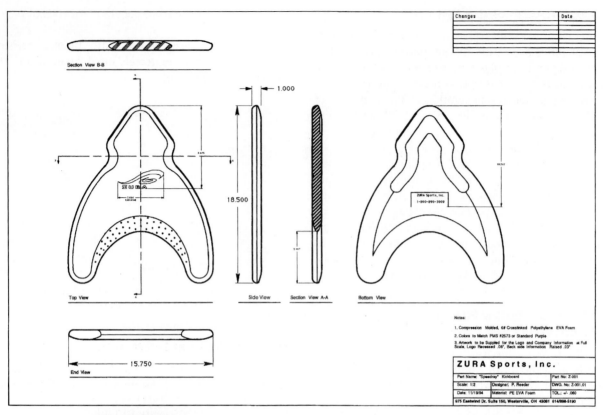

FIGURE 3.62 **A Drawing Created Using CAD.** *Courtesy of Zura Sports, Inc.*

FREEHAND SKETCHING*

■ 3.59 TECHNICAL SKETCHING

Freehand sketches are a helpful way to organize your thoughts and record ideas. They provide a quick, low-cost way to explore various solutions to a problem so that the best choice can be made. Investing too much time in doing a scaled layout before exploring your options through sketches can be a costly mistake. Sketches are also used to clarify information about changes in design or provide information on repairing existing equipment.

The degree of precision needed in a given sketch depends on its use. Quick sketches to supplement verbal descriptions may be rough and incomplete. Sketches that are supposed to convey important and precise information should be drawn as carefully and accurately as possible.

The term freehand sketch does not mean a sloppy drawing. As shown in Figure 3.63, a freehand sketch shows attention to proportion, clarity, and correct line widths.

■ 3.60 SKETCHING MATERIALS

One advantage of freehand sketching is that it requires only pencil, paper, and eraser. Small notebooks or sketch pads are useful in the field (when working at a site) or when an accurate record is needed. Graph paper can be helpful in making neat sketches like the one in Figure 3.64. Paper with 4, 5, 8, or 10 squares per inch is convenient for maintaining correct proportions.

Find a style of pencil that suits your use. Figure 3.7 (see page 47) shows three styles which are all good for preparing sketches. Automatic mechanical pencils (shown as C in the illustration) come in .3-mm, .5-mm, .7-mm, and .9-mm leads that advance automatically and are easy to use. The .5-mm lead is a good general size, or you can use a .7-mm lead for thick lines and .3-mm for thin lines. The lead holder shown as part B requires a special sharpener, so it is not usually suit-

---

* Freehand sketching is discussed again, at length, in detail in Chapter 5 within the discussion on "Technical Sketching Shape Description."

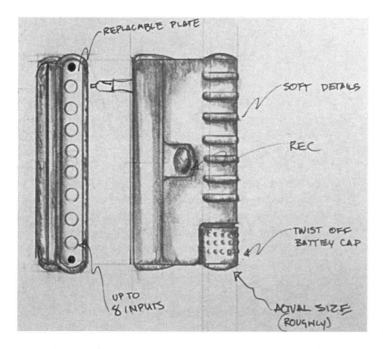

FIGURE 3.63 Great Ideas Often Start as Freehand Sketches Made on Scratch Paper. *Courtesy of ANATech, Inc.*

■ 3.61 TYPES OF SKETCHES

Technical sketches of 3-D objects are usually one of four standard types of projection, shown in Figure 3.65:

- Multiview projection
- Axonometric (isometric) projection
- Oblique projection
- Perspective sketches

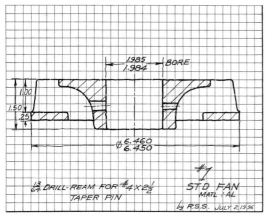

FIGURE 3.64 Sketch on Graph Paper.

FIGURE 3.65 Types of Projection.

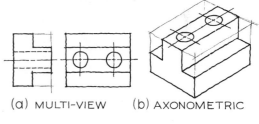

(a) MULTI-VIEW   (b) AXONOMETRIC

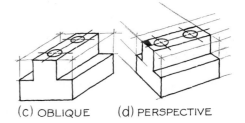

(c) OBLIQUE   (d) PERSPECTIVE

able for work in the field. Plain wooden pencils work great. They are inexpensive and make it easy to produce thick or thin lines by the amount you sharpen them.

A sketch pad of plain paper with a master grid sheet showing through underneath works well as a substitute for grid paper. You can create your own master grid sheets for different sketching purposes using CAD. Specially ruled isometric paper is available for isometric sketching.

Figure 3.8 (see page 48) shows the grades of lead and their uses. Use soft pencils, such as HB or F, for freehand sketching. Soft vinyl erasers are recommended.

Multiview projection shows one or more necessary views. You will learn multiview projection in Chapter 6. Axonometric, oblique, and perspective sketches are methods of showing the object pictorially in a single view. They will be discussed in Chapters 16, 17, 18.

■ 3.62 SCALE

Sketches are not usually made to a specific scale. Sketch objects in their correct proportions as accurately as possible by eye. Grid paper helps you sketch the correct proportions by providing a ready-made scale (by counting squares). The size of the sketch is up to you, depending on the complexity of the object and the size of the paper available. Sketch small objects oversize to show the details clearly.

■ 3.63 TECHNIQUE OF LINES

The main difference between an instrument drawing and a freehand sketch is in the style or technique of the lines. A good freehand line is not expected to be precisely straight or exactly uniform, as is a CAD or instrument-drawn line. Freehand lines show freedom and variety. Freehand construction lines are very light, rough lines. All other lines should be dark and clean.

■ 3.64 STYLES OF LINES

Each line on a technical drawing has a definite meaning. Drawings use two different line widths thick and thin and different line styles indicate the meaning of the line. Aperson reading a drawing depends on line styles to communicate whether a line is visible or hidden, if it represents a center axis, or if its purpose is to convey dimension information. Without making these distinctions, drawings would become a confusing jumble of lines. To make your drawings clear and easy to read, make the contrast between the two widths of lines distinct. Thick lines such as visible lines and cutting plane lines should be twice as thick as thin lines. Thin lines are used for construction lines, hidden lines, dimension lines, extension lines, center lines, and phantom lines. Figure 3.9 (see page 49) shows the different styles of lines that you will be using. All lines except for construction lines should be sharp and dark. Construction lines should be very light so that they are not visible (or are barely visible) in the completed drawing. Figures 3.66 and 3.67 show examples of technique for sketching using different line patterns.

■ 3.65 SKETCHING CIRCLES, ARCS, AND ELLIPSES

Small circles and arcs can be sketched in one or two strokes without any preliminary blocking in. Sketching arcs is similar to sketching circles. In general, it is easier to sketch arcs by holding your pencil on the inside of the curve. In sketching arcs, look closely at the actual geometric constructions and carefully approximate all points of tangency so that the arc touches a line or other entity at the right point. Circle templates also make it easy to sketch accurate circles of various sizes.

If a circle is tipped away from your view, it appears as an ellipse. Figure 3.68 shows a coin viewed so

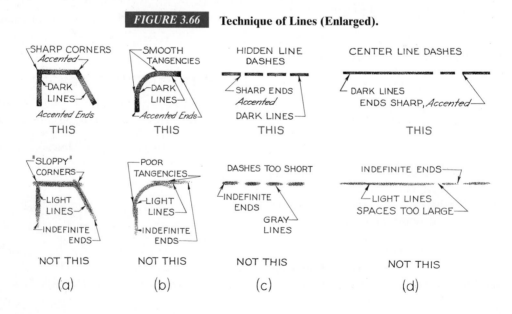

FIGURE 3.66 **Technique of Lines (Enlarged).**

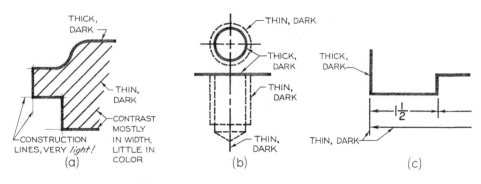

FIGURE 3.67 **Contrast of Lines (Enlarged).**

FIGURE 3.68 **Circle Viewed as an Ellipse.**

that it appears as an ellipse. You can learn to sketch small ellipses with a free arm movement similar to the way you sketch circles, or you can use ellipse templates to help you easily sketch ellipses. These templates are usually grouped according to the amount a circular shape the would be rotated to form the ellipse. They provide a number of sizes of ellipses on each template, but usually only one or a couple typical rotations.

■ 3.66 MAINTAINING PROPORTIONS

The most important rule in freehand sketching is to keep the sketch in proportion. No matter how brilliant the technique or how well-drawn the small details, if the proportions are bad, the sketch will be of little use. To keep your sketch in proportion, first determine the relative proportions of the height to the width and lightly block them in. Then lightly block in the medium-size areas and the small details.

Compare each new estimated distance with already-established distances. One way to estimate dis-

tances is to mark an arbitrary unit on the edge of a card or strip of paper. Then see how many units wide and how many units high the object is.

To sketch an object with many curves to a different scale, use the squares method. On the original picture, rule accurate grid lines to form squares of any convenient size. It is best to use a scale and some convenient spacing, such as $\frac{1}{2}$ inch or 10 mm. On the new sheet, rule a similar grid, marking the spacing of the lines proportional to the original, but reduced or enlarged as needed. Draw the object's contours in and across the new grid lines to match the original as closely as you can by eye.

LETTERING

Lettered text is often necessary to completely describe an object or to provide detailed specifications. Lettering should be legible, be easy to create, and use styles acceptable for traditional drawing and CAD drawing.

■ 3.67 FREEHAND LETTERING

Most engineering lettering is single-stroke Gothic font. A font is the name for a particular shape of letters. Figure 3.69 shows some common fonts. Most hand-drawn notes are lettered $\frac{1}{8}''$ high and are drawn within light horizontal guidelines. CAD notes are typed from the keyboard and sized according to the plotted size of the drawing.

■ 3.68 LETTERING STANDARDS

The modern styles of letters were derived from the design of Roman capital letters, whose origins date all the way back to Egyptian hieroglyphics. The term

ABCDEFGH
abcdefghij
ABCDEFGHI
abcdefghijklm

Roman *All letters having elementary strokes "accented" or consisting of heavy and light lines, are classified as Roman.*

Italic- All slanting letters are classified as Italics- These may be further designated as Roman-Italics, Gothic Italics or Text Italic.

FIGURE 3.69   **Serif and Sans-Serif Lettering.**

Roman refers to any letter that has wide downward strokes, thin connecting strokes, and ends terminating in spurs called serifs. In the late 19th century, the development of technical drawing created a need for a simplified, legible alphabet that could be drawn quickly with an ordinary pen. Single-stroke Gothic sans-serif (meaning without serifs or spurs) letters are used today because they are very legible.

■ 3.69 COMPUTER LETTERING

Lettering is a standard feature available in computer graphics programs. Using CAD software, you can add titles, notes, and dimensioning information to a drawing. Several fonts and a variety of sizes may be selected. When modifications are required, it is easy to make appropriate lettering changes on the drawing by editing existing text.

CAD drawings typically use a Gothic style of lettering, but often use a Roman style of lettering for titles. When adding lettering to a CAD drawing, a good rule of thumb is not to use more than two fonts within the same drawing. You may want to use one font for the titles and a different one for notes and other text. However, you may have a couple different sizes of lettering in the drawing and perhaps some slanted lettering all using the same font. It is sometimes tempting to use many different fonts in a drawing because of the wide variety available on CAD systems, but drawings that use too many different fonts have been jokingly referred to as having a ransom note style of lettering.

■ 3.70 LETTERING TECHNIQUE

Lettering is more similar to freehand drawing than it is to writing, so the six fundamental drawing strokes and their directions are basic to lettering. Horizontal strokes are drawn left to right. Vertical, inclined, and curved strokes are drawn downward. If you are left-handed, you can use a system of strokes similar to the sketching strokes that work for you.

Lettering ability has little relationship to writing ability. You can learn to letter neatly even if you have terrible handwriting. There are three necessary aspects of learning to letter:

- Proportions and forms of the letters (to make good letters, you need to have a clear mental image of their correct shape)
- Composition the spacing of letters and words
- Practice

■ 3.71 VERTICAL LETTERS AND NUMERALS

The proportions of vertical capital letters and numerals are shown in Figure 3.70 in a grid six units high. Numbered arrows indicate the order and direction of strokes. The widths of the letters can be easily remembered: The letter l and the numeral 1 are only a pencil width. The W is eight grid units wide ($1\frac{1}{3}$ times its height) and is the widest letter in the alphabet. All the other letters or numerals are either five or six grid units wide, and it is easy to remember the six-unit letters because when assembled they spell TOM Q. VAXY. This means that most letters are as wide as they are tall, which is probably wider than your usual writing. All numerals except the 1 are five units wide.

Lowercase letters are rarely used in engineering sketches except for lettering large volumes of notes. Vertical lowercase letters are used on map drawings, but very seldom on machine drawings. Lowercase letters are shown in Figure 3.71. The lower part of the letter is usually two-thirds the height of the capital letter.

■ 3.72 INCLINED LETTERS AND NUMERALS

Inclined capital letters and numerals, shown in Figure 3.72, are similar to vertical characters, except for the slope. The slope of the letters is about 68° from the horizontal. While you may practice drawing slanted hand-lettering at approximately this angle, it is important in CAD drawings to always set the amount of incline for the letters at the same value within a drawing so that the lettering is consistent. Inclined lowercase letters, shown in Figure 3.73, are similar to vertical lowercase letters.

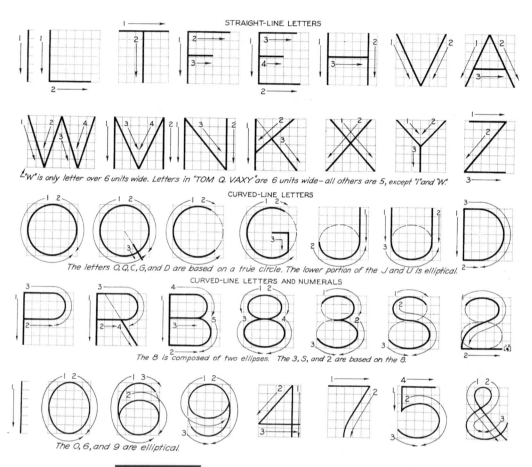

STRAIGHT-LINE LETTERS

"W" is only letter over 6 units wide. Letters in "TOM Q. VAXY" are 6 units wide – all others are 5, except 'I' and 'W'.

CURVED-LINE LETTERS

The letters O, Q, C, G, and D are based on a true circle. The lower portion of the J and U is elliptical.

CURVED-LINE LETTERS AND NUMERALS

The 8 is composed of two ellipses. The 3, S, and 2 are based on the 8.

The 0, 6, and 9 are elliptical.

FIGURE 3.70 **Vertical Capital Letters and Numerals.**

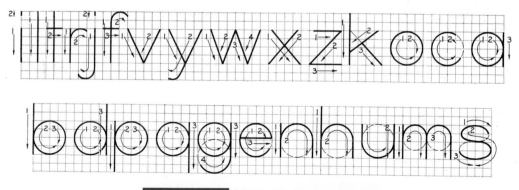

FIGURE 3.71 **Vertical Lowercase Letters.**

FIGURE 3.72   Inclined Capital Letters and Numerals.

FIGURE 3.73   Inclined Lowercase Letters.

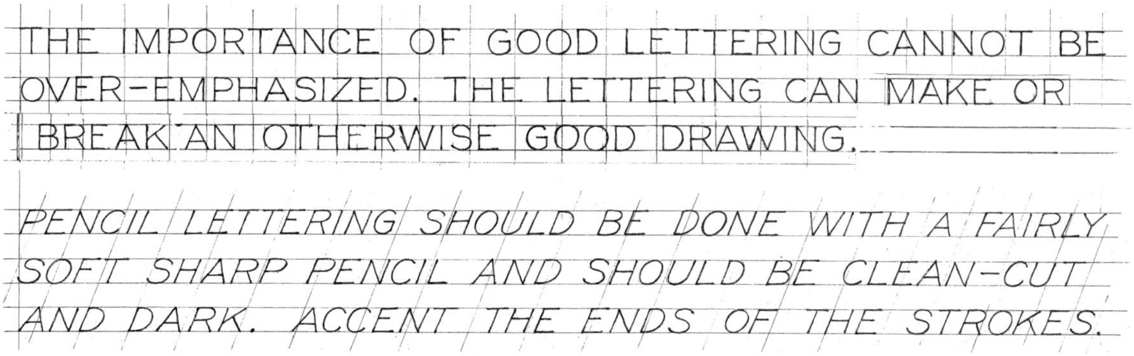

THE IMPORTANCE OF GOOD LETTERING CANNOT BE
OVER-EMPHASIZED. THE LETTERING CAN MAKE OR
BREAK AN OTHERWISE GOOD DRAWING.
PENCIL LETTERING SHOULD BE DONE WITH A FAIRLY
SOFT SHARP PENCIL AND SHOULD BE CLEAN-CUT
AND DARK. ACCENT THE ENDS OF THE STROKES.

FIGURE 3.74 **Pencil Lettering (Full Size).**

3.73 GUIDELINES

Use extremely light horizontal guidelines to keep letter height uniform, as is shown in Figure 3.74. Capital letters are commonly made $\frac{1}{8}''$ (3.2 mm) high, with the space between lines of lettering being from three-fifths to full height of the letters. Lettering size may vary depending on the size of the sheet. Do not use vertical guidelines to space the letters this should be done by eye while lettering. Use a vertical guideline at the beginning of a row of text to help you line up the left edges of the following rows, or use randomly spaced vertical guidelines to help you maintain the correct slant.

A simple method of spacing horizontal guidelines is to use a scale and set off a series of $\frac{1}{8}''$ spaces, making both the letters and the spaces between lines of letters $\frac{1}{8}''$ high. Another quick method of creating guidelines is to use a guideline template like the Berol Rapidesign 925 shown in Figure 3.75.

When large and small capitals are used in combination, the small capitals should be three-fifths to two-thirds as high as the large capitals.

3.74 GUIDELINES FOR WHOLE NUMBERS AND FRACTIONS

Beginners should use guidelines for whole numbers and fractions. Draw five equally spaced guidelines for whole numbers and fractions, as shown in Figure 3.76. Fractions are twice the height of the corresponding whole numbers. Make the numerator and the denominator each about three-fourths as high as the whole number to allow enough space between them and the

REVERSE WITH INKING PEN
3/36
5/32
1/8
3/32
1/4
3/16
5/32
1/8
Berol.RapiDesign.
R-925
LETTERING AID

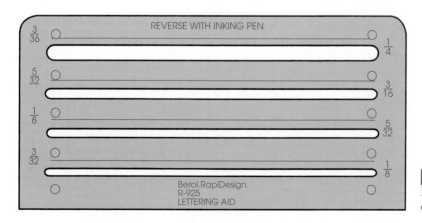

FIGURE 3.75 **The Berol Rapidesign 925 template is used to quickly create guidelines for lettering.**

Always leave ample clear spaces
Equal Spaces
(a)
(b)
(c)
SCALE

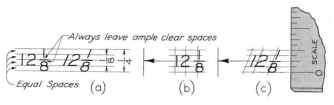

FIGURE 3.76 **Guide Lines for Dimension Figures.**

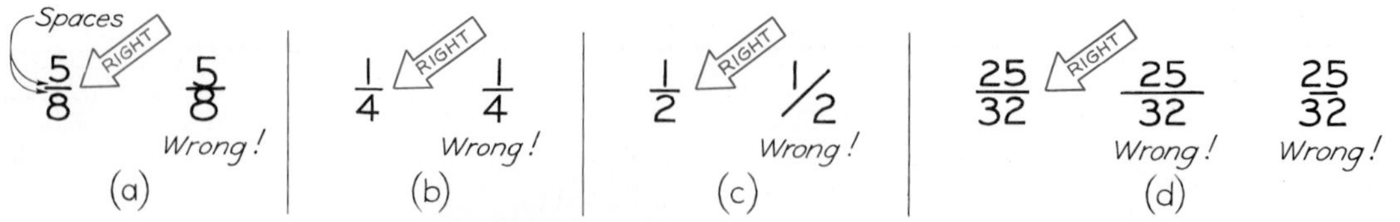

$$\frac{5}{8} \quad \frac{5}{8} \quad \Bigg| \quad \frac{1}{4} \quad \frac{1}{4} \quad \Bigg| \quad \frac{1}{2} \quad \frac{1}{2} \quad \Bigg| \quad \frac{25}{32} \quad \frac{25}{32} \quad \frac{25}{32}$$

Spaces / RIGHT

Wrong! (a)   Wrong! (b)   Wrong! (c)   Wrong! Wrong! (d)

FIGURE 3.77   **Common Errors.**

fraction bar. For dimensioning, the most commonly used height for whole numbers is $\frac{1}{8}''$ (3.2 mm), and for fractions $\frac{1}{4}''$ (6.4 mm), as shown in the figure.

Some of the most common errors in lettering fractions are shown in Figure 3.77. To make fractions appear correctly:

- Never let numerals touch the fraction bar.
- Center the denominator under the numerator.
- Never use an inclined fraction bar, except when lettering in a narrow space, as in a parts list.
- Make the fraction bar slightly longer than the widest part of the fraction.

■ 3.75 SPACING OF LETTERS AND WORDS

Uniform spacing of letters is a matter of equalizing spaces by eye. The background areas between letters, not the distances between them, should be approximately equal. Equal distances from letter to letter causes the letters to appear unequally spaced. Equal background areas between letters results in an even and pleasing spacing.

Some combinations, such as LT and VA, may even have to be slightly overlapped to secure good spacing. In some cases the width of a letter may be decreased. For example, the lower stroke of the L may be shortened when followed by A. These pairs of letters that need to be spaced extra closely to appear correctly are called kerned pairs in typesetting.

Space words well apart, but space letters closely within words. Make each word a compact unit well separated from adjacent words. For either uppercase or lowercase lettering, make the spaces between words approximately equal to a capital O. Be sure to have space between rows of letters, usually equal to the letter height. Rows spaced too closely are hard to read. Rows that are too far apart do not appear related.

■ 3.76 TITLES

In most cases, the title and related information are lettered in title boxes or title strips, which may be printed directly on the drawing paper or polyester film, as shown in Figure 3.78. The main drawing title is usually centered in a rectangular space, which is easy to do in CAD. When lettering by hand, arrange the title symmetrically about an imaginary centerline, as shown in Figure 3.79. In any kind of title, the most important words are given most prominence by making the lettering larger, heavier, or both. Other data, such as scale and date, can be smaller.

■ 3.77 WEB SITES FOR FURTHER INFORMATION

Check the sites below for engineering graphics supplies and equipment:

- http://www.reprint-draphix.com/
- http://www.eclipse.net/~essco/draft/draft.htm
- http://www.seventen.com/art_eng/index.html

These sites feature typography information:

- http://www.graphic-design.com/type/
- http://www.webcom.com/cadware/letease2.html

To find other sites like these, use keywords like reprographic supplies or engineering type fonts.

FIGURE 3.78   **Centering Title in Title Box.**

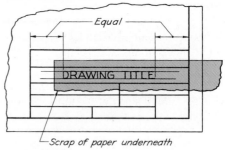

Equal

DRAWING TITLE

Scrap of paper underneath

FIGURE 3.79   **Balanced Machine-Drawing Title.**

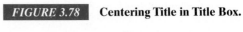

TOOL GRINDING MACHINE
TOOL REST SLIDE
SCALE : FULL SIZE
AMERICAN MACHINE COMPANY
NEW YORK CITY

DRAWN BY ____ CHECKED BY ____

■ KEY WORDS

CAD	CONSTRUCTION LINES
SCALE	OBLIQUE
ALPHABET OF LINES	GRID PAPER
PARALLEL	HIDDEN LINES
VERTICAL	CENTERLINES
RADIUS	SHADING
DRAWING MEDIA	SKETCH
PROTRACTOR	PROPORTIONS
TRIANGLE	LETTERING
T-SQUARE	GOTHIC
IRREGULAR CURVE	ROMAN
LINE TYPE	SERIF
HORIZONTAL	INCLINED
DIAMETER	STABILITY
TEMPLATE	SPACING
COMPASS	GUIDELINES
DIVIDERS	TITLE BLOCK
PERPENDICULAR	KERNED PAIRS
FREEHAND SKETCH	

■ CHAPTER SUMMARY

- An understanding of the basic principles of drawing is required to draw with either a pencil or with CAD software.

- The line weight (thickness) and type (dashed or solid) has specific meaning in all technical drawings. This is called the alphabet of lines and is essential knowledge for every drafter.

- Both CAD and traditional drawing have specific methods for drawing lines, arcs, and circles. Proper understanding of the elements of this basic geometry is essential for both mechanical and CAD drawing.

- Every drawing tool, including every CAD software program, requires careful study of the tools and procedures for using the tools. Proper use of each tool facilitates the creation of neat, accurate drawings. Improper use of a tool creates sloppy, inaccurate drawings.

- The proper sizing of a drawing requires complete understanding of the use of scales. Paper drawings are scaled before they are drawn. CAD drawings are scaled when they are printed.

- Complex circles and arcs are difficult to draw using either CAD software or a mechanical compass. The prescribed techniques for either tool require understanding of the proper technique and practice with the appropriate tool.

- There are many drawing and printing media used in the creation of traditional drawings and the printing of CAD drawings. Each media type has specific advantages. Drawing and printing media are chosen based on the cost, durability, image quality, and reproduction capability.

- Sketching is a quick way of visualizing and solving a drawing problem. It is an effective way of communicating with all members of the design team.

- There are special techniques for sketching lines, circles, and arcs. These techniques should be practiced so they become second nature.

- Moving your thumb up or down the length of a pencil at arms length is an easy method for estimating proportional size.

- Using a grid makes sketching in proportion an easy task. Grid paper comes in a variety of types, including square grid and isometric grid.

- You can sketch circles by constructing a square and locating the four tangent points where the circle touches the square.

- A sketched line does not need to look like a mechanical line. The main distinction between instrumental drawing and freehand sketching is the character or technique of the line work.

- Freehand sketches are made to proportion, but not necessarily to a particular scale.
- Notes and dimensions are added to sketches using uppercase letters drawn by hand.

- The standard shapes of letters used in engineering drawing have been developed to be legible and quick to produce.

■ REVIEW QUESTIONS

1. What tools are used to draw straight lines?

2. What tools are used to draw arcs and circles?

3. Draw the alphabet of lines and label each line.

4. Describe the proper technique for erasing a line using an erasing shield.

5. Why is the pencil pulled and never pushed when drawing lines?

6. Which architect's scale represents a size ratio of 1 : 24? Which metric scale represents a half size? Which engineering scale would be used for full size?

7. Which scale type is the only one to use fractions of an inch?

8. Is the bevel of a compass lead sharpened on the inside or outside surface?

9. What are the minimum number of points that you should connect when using an irregular curve?

10. What are the main advantages of polyester film as a drawing media?

11. What are the four standard types of projections?

12. What are the advantages of using grid paper for sketching?

13. What is the correct technique for sketching a cricle or an arc?

14. Sketch the alphabet of lines. Which lines are thick? Which are thin? Which are very light and should not reproduce when copied?

15. What is the advantage of sketching an object first before drawing it using CAD?

16. What is the difference between proportion and scale?

17. What font provides the shape of standard engineering lettering?

18. Describe the characteristics of good freehand lettering.

19. Why must guidelines always be used for lettering?

20. How are sketches used in the design process?

■ DRAWING PROBLEMS

The constructions in Figs. 3.80 to 3.90 are to be drawn in pencil on Layout A–2 (see the inside front cover of this book). The steps in drawing this layout are shown in Fig. 3.44. Draw all construction lines *lightly*, using a hard lead (4H to 6H), and all required lines dense black with a softer lead (F to H). If construction lines are drawn properly—that is, *lightly*—they need not be erased in the final drawing.

If the layout is to be made on the A4 size sheet, width dimensions for title-strip forms will need to be adjusted to fit the available space.

The pencil drawings of Figs. 3.85 to 3.90 should be done on tracing paper or vellum; then prints should be made to show the effectiveness of the student's technique. If ink tracings are required, the originals may be drawn on film or on detail paper and then traced on vellum or tracing cloth. For any assigned problem, the instructor may require that all dimensions and notes be lettered to afford further lettering practice.

Since many of the problems in this chapter are of a general nature, they can also be solved on most computer graphic systems. If a system if available, the instructor may choose to assign specific problems to be completed by this method.

Prob. 3.1 Using Layout A–2 or A4–2 (adjusted), divide working space into six equal rectangles and draw visible lines, as shown in Fig. 3.80. Draw construction lines AB through centers C at right angles to required lines; then along each construction line, set off 0.50″ spaces and draw required visible lines. Omit dimensions and instructional notes.

Problems in convenient form for solution may be found in *Technical Drawing Problems*, Series 1, by Giesecke, Mitchell, Spencer, Hill, Dygdon, and Novak; *Technical Drawing Problems*, Series 2, by Spencer, Hill, Dygdon, and Novak; and *Technical Drawing Problems*, Series 3, by Spencer, Hill, Dygdon, and Novak; all designed to accompany this text and published by Prentice Hall.

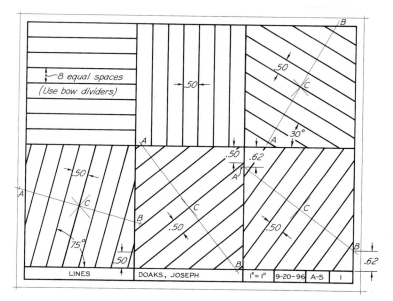

FIGURE 3.80 Using Layout A-2 or A4-2 (adjusted), divide working space into six equal rectangles and draw visible lines as shown. Draw construction lines AB through centers C at right angles to required lines; then along each construction line, set off 0.50″ spaces and draw required visible lines. Omit dimensions and instructional notes.

Prob. 3.2 Using Layout A–2 or A4–2 (adjusted), divide working space into six equal rectangles and draw lines as shown in Fig. 3.81. In the first two spaces, draw conventional lines to match those in Fig. 3.9. In remaining spaces, locate centers C by diagonals, and then work constructions out from them. Omit the metric dimensions and instructional notes.

Prob. 3.3 Using Layout A–2 or A4–2 (adjusted), draw views in pencil, as shown in Fig. 3.82. Omit all dimensions.

Prob. 3.4 Using Layout A–2 or A4–2 (adjusted), draw figures in pencil, as shown in Fig. 3.83. Use bow pencil for all arcs and circles within it radius range. Omit all dimensions.

Prob. 3.5 Using Layout A–2 or A4–2 (adjusted), draw views in pencil, as shown in Fig. 3.84. Use bow pencil for all arcs and circles within its radius range. Omit all dimensions.

Prob. 3.6 Using Layout A–2 or A4–2 (adjusted), draw in pencil the friction plate in Fig. 3.85. Omit dimensions and notes.

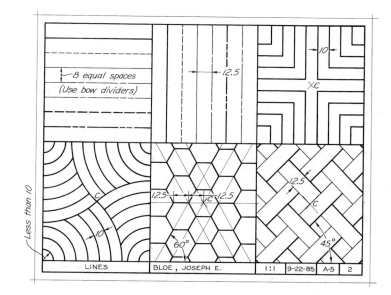

FIGURE 3.81 Using Layout A–2 or A4–3 (adjusted), divide working space into six equal rectangles, and draw lines as shown. In first two spaces, draw conventional lines to match those in Fig. 3.9. In remaining spaces, locate centers C by diagonals, and then work constructions out from them. Omit the metric dimensions and instructional notes.

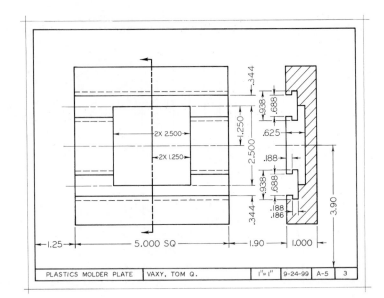

FIGURE 3.82   Using Layout A–2 or A4–2 (adjusted), draw views in pencil as shown. Omit all dimensions.

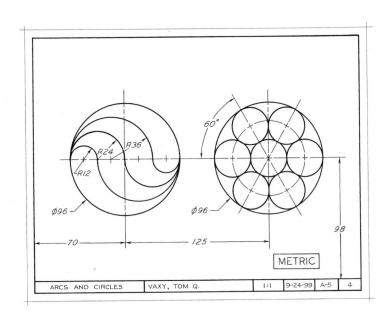

FIGURE 3.83   Using Layout A–2 or A4–3 (adjusted), draw figures in pencil as shown. Use bow pencil for all arcs and circles within its radius range. Omit all dimensions.

Prob. 3.7   Using Layout A–2 or A4–2 (adjusted), draw views in pencil of the seal cover in Fig. 3.86. Omit dimensions and notes.

Prob. 3.8   Using Layout A–2 or A4–2 (adjusted), draw in pencil the Geneva cam in Fig. 3.87. Omit dimensions and notes.

Prob. 3.9   Using Layout A–2 or A4–2 (adjusted), draw accurately in pencil the shear plate in Fig. 3.88. Give length of KA. Omit other dimensions and notes.

Prob. 3.10   Using Layout A–2 or A4–2 (adjusted), draw in pencil the ratchet wheel in Fig. 3.89. Omit dimensions and notes.

Prob. 3.11   Using Layout A–2 or A4–2 (adjusted), draw in pencil the latch plate in Fig. 3.90. Omit dimensions and notes.

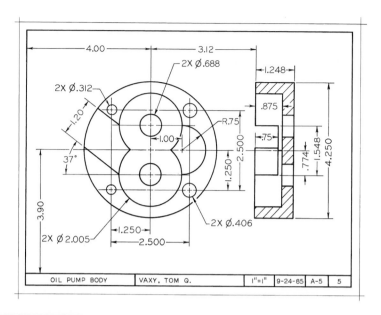

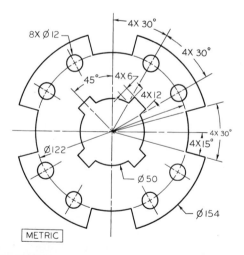

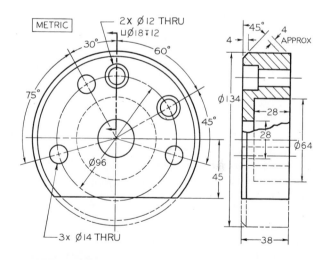

FIGURE 3.84   Using Layout A–2 or A4–2 (adjusted), draw views in pencil as shown. Use bow pencil for all arcs and circles within its radius range. Omit all dimensions.

FIGURE 3.85   Friction Plate. Using Layout A–2 or A4–2 (adjusted), draw in pencil. Omit dimensions and notes.

FIGURE 3.86   Seal Cover. Using Layout A–2 or A4–2 (adjusted), draw in pencil. Omit dimensions and notes.

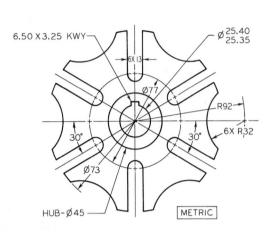

FIGURE 3.87 Geneva Cam. Using Layout A–2 or A4–2 (adjusted), draw in pencil. Omit dimensions and notes.

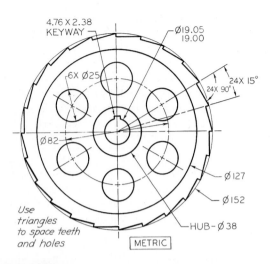

FIGURE 3.88 Ratchet Wheel. Using Layout A–2 or A4–2 (adjusted), draw in pencil. Omit dimensions and notes.

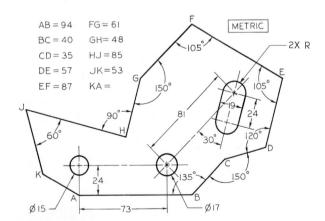

FIGURE 3.89 Shear Plate. Using Layout A–2 or A4–2 (adjusted), draw accurately in pencil. Give length of KA. Omit other dimensions and notes.

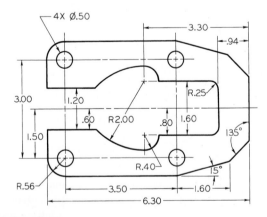

FIGURE 3.90 Latch Plate. Using Layout A–2 or A4–2 (adjusted), draw in pencil. Omit dimensions and notes.

■ LETTERING PROBLEMS

Layouts for lettering problems are given in Figs. 3.91 through 3.94. Draw complete horizontal and vertical or inclined guide lines *very lightly*. Draw the vertical or inclined guide lines through the full height of the lettered area of the sheet. For practice in ink lettering, the last two lines and the title strip on each sheet may be lettered in ink, if assigned by the instructor. Omit all dimensions.

Prob. 3.12 As shown in Fig. 3.91, lay out sheet, add vertical or inclined guide lines, and fill in vertical or inclined capital letters as assigned. For decimal-inch and millimeter equivalents of given dimensions, see table inside of back cover.

Prob. 3.13 As shown in Fig. 3.92, lay out sheet, add vertical or inclined guide lines, and fill in vertical or inclined capital letters as assigned. For decimal-inch and millimeter equivalents of given dimensions, see table inside of back cover.

Prob. 3.14 As shown in Fig. 3.93, lay out sheet, add vertical or inclined guide lines, and fill in vertical or inclined capital letters as assigned. For decimal-inch and millimeter equivalents of given dimensions, see table inside of back cover.

Prob. 3.15 As shown in Fig. 3.94, lay out sheet, add vertical or inclined guide lines, and fill in vertical or inclined capital letters as assigned. For decimal-inch and millimeter equivalents of given dimensions, see table inside of back cover.

Lettering sheets in convenient form for lettering practice may be found in *Engineering Drawing Problems*, Series 1, by Giesecke, Mitchell, Spencer, Hill, Dygdon, and Novak; *Engineering Drawing Problems*, Series 2, by Spencer, Hill, Dygdon, and Novak; and *Engineering Drawing Problems*, Series 3 by Davis and Juneau; all designed to accompany this text and published by Prentice Hall.

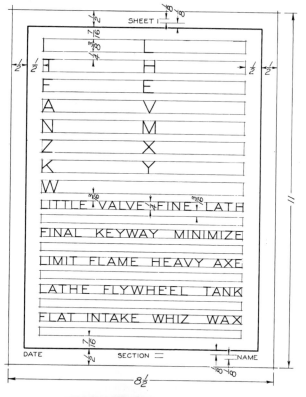

FIGURE 3.91 Prob. 3.12.

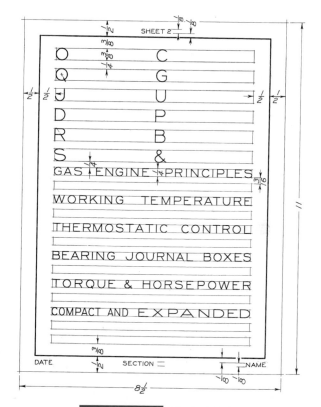

FIGURE 3.92 Prob. 3.13.

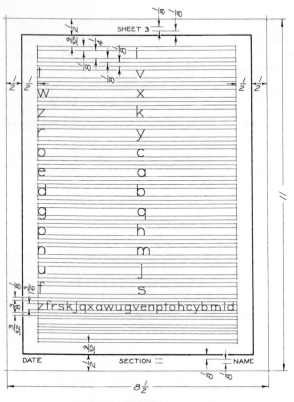

FIGURE 3.93   Prob. 3.14.

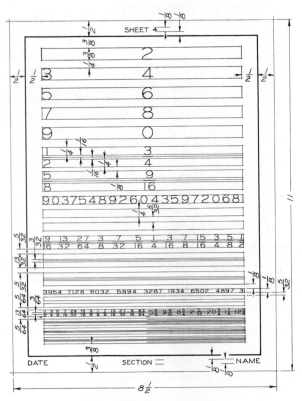

FIGURE 3.94   Prob. 3.15.

C H A P T E R   4

GEOMETRIC CONSTRUCTIONS

OBJECTIVES

After studying the material in this chapter, you should be able to:

1. Identify and draw four different types of lines.
2. Identify and draw any size angle.
3. Identify and draw four different types of triangles and five types of parallelograms.
4. Identify and draw a regular triangle, square, hexagon, and octagon.
5. Divide a space into equal parts; divide an angle in half; divide a line into equal parts.
6. Draw lines and arcs tangent to each other.
7. Identify and draw several conic sections.
8. Define the meaning of a tangent point and construct its location on any tangent construction.

OVERVIEW

All traditional drawing and CAD drawing techniques are based on the construction of basic geometric elements. A point, line, arc, and circle are the basic elements used to create the most complex drawings. The geometry of drawing is the same for traditional drawing and CAD drawing. The drafting student must understand basic geometric construction techniques in order to draw on paper or with CAD, or to apply these geometric techniques to solving problems.

Pure geometry problems may be solved with only the compass and a straightedge, and in some cases these methods may be used to advantage in technical drawing. However, drafters and designers have available the T-square, triangles, dividers, and other equipment, that can often yield accurate results more quickly by what we may term "preferred methods." Therefore, many of the solutions in this chapter are practical adaptations of the principles of pure geometry.

This chapter is designed to present definitions of terms and geometric constructions of importance in technical drawing, suggest simplified methods of construction, point out practical applications, and afford opportunity for practice in accurate instrumental drawing.

■ 4.1 POINTS AND LINES

A *point* represents a location in space or on a drawing and has no width, height, or depth (Fig. 4.1). A point is represented by the intersection of two lines (Fig. 4.1a), by a short crossbar on a line (Fig. 4.1b), or by a small cross (Fig. 4.1c). Never represent a point by a simple dot on the paper.

A **line** is defined by Euclid as "that which has length without breadth." A *straight line* is the shortest distance between two points and is commonly referred to simply as a "line." If the line is indefinite in extent, the length is a matter of convenience, and the endpoints are not fixed (Fig. 4.1d). If the endpoints of the line are significant, they must be marked by means of small mechanically drawn crossbars (Fig. 4.1e). Other common terms are illustrated in Figs. 4.1f to 4.1h. Either straight lines or curved lines are parallel if the shortest distance between them remains constant. The common symbol for parallel lines is ||, and for perpendicular lines it is ⊥ (singular) or ⊥s (plural). Two perpendicular lines may be marked with a "box" to indicate perpendicularity, as shown in Fig. 4.1k. Such symbols may be used on sketches, but not on production drawings.

■ 4.2 ANGLES

An angle is formed by two intersecting lines. A common symbol for angle is | (singular) or | s (plural) (Fig. 4.2).

There are 360 degrees (360°) in a full circle, as shown in Fig. 4.2a. A degree is divided into 60 minutes (60′), and a minute is divided into 60 seconds (60″). Thus, 37° 26′ 10″ is read 37 degrees, 26 minutes, and 10 seconds. When minutes alone are indicated, the number of minutes should be preceded by 0°, as in 0° 20′.

The different kinds of angles are illustrated in Fig. 4.2b to 4.2e. Two angles are *complementary* (Fig. 4.2f) if they total 90°, and are *supplementary* (Fig. 4.2g) if, they total 180°. Most angles used in technical drawing can be drawn easily with the T-square or straightedge and triangles. To draw odd angles, use a protractor. For considerable accuracy, use a *vernier protractor*, or the tangent, sine, or chord methods (§4.19).

■ 4.3 TRIANGLES

A triangle is a plane figure bounded by three straight sides, and the sum of the interior angles is always 180° (Fig. 4.3). A right triangle (Fig. 4.3d) has one 90° angle, and the square of the hypotenuse is equal to the sum of the squares of the two sides (Fig. 4.3c). As shown in Fig. 4.3f, any triangle inscribed in a semicircle is a right triangle if the hypotenuse coincides with the diameter.

■ 4.4 QUADRILATERALS

A *quadrilateral* is a plane figure bounded by four straight sides (Fig. 4.4). If the opposite sides are parallel, the quadrilateral is also a *parallelogram*.

■ 4.5 POLYGONS

A *polygon* is any plane figure bounded by straight lines (Fig. 4.5). If the polygon has equal angles and equal sides, it can be inscribed in or circumscribed around a circle and is called a *regular polygon*.

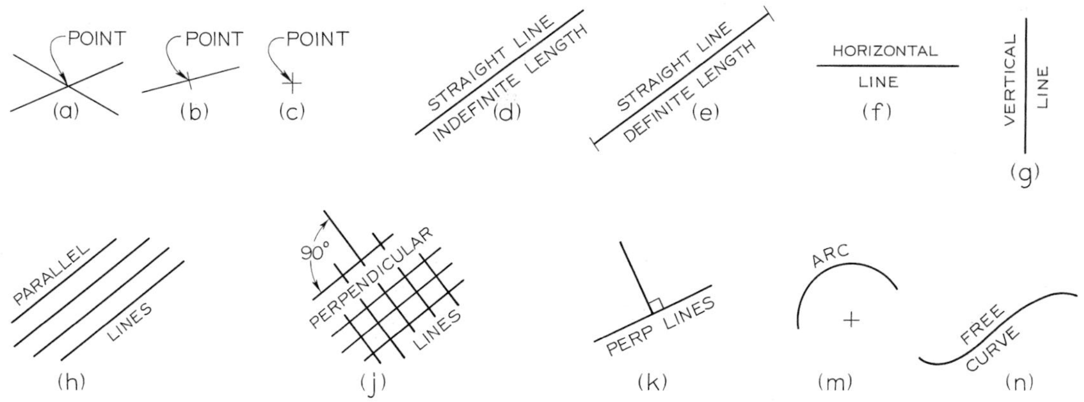

FIGURE 4.1 Points and Lines.

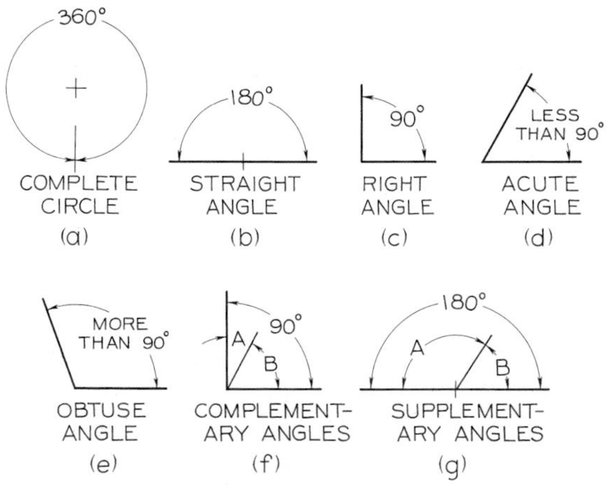

FIGURE 4.2 Angles.

FIGURE 4.3 Triangles.

■ 4.6 CIRCLES AND ARCS

A *circle* is a closed curve, all points of which are the same distance from a point called the center (Fig. 4.6a). *Circumference* refers to the circle or to the distance around the circle. This distance equals the diameter multiplied by π (called pi) or 3.1416. Other definitions are illustrated in Figs. 4.6b through 4.6e.

■ 4.7 SOLIDS

Solids bounded by plane surfaces are called *polyhedra* (Fig. 4.7). The surfaces are called *faces*, and if the faces are equal regular polygons, the solids are called *regular polyhedra*.

A *prism* has two bases, which are parallel equal polygons, and three or more lateral faces, which are parallelograms. A triangular prism has a triangular base; a rectangular prism has rectangular bases; and so on. If the bases are parallelograms, the prism is a parallelpiped. A right prism has faces and lateral edges perpendicular to the bases; an oblique prism has faces and laterals edge oblique to the bases. If one end is cut off to form an end and not parallel to the bases, the prism is said to be truncated.

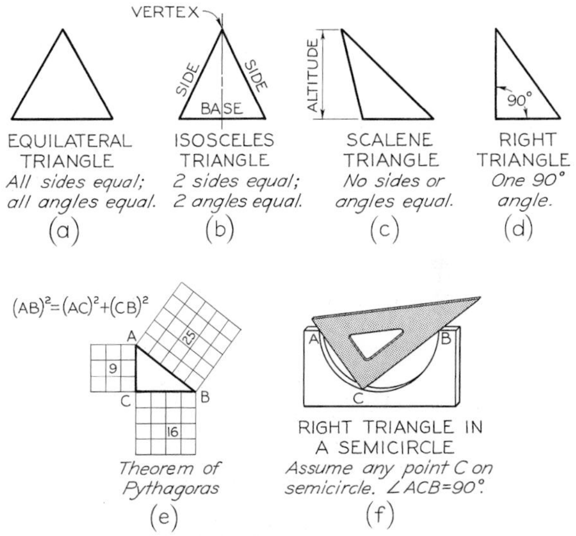

A *pyramid* has a polygon for a base and triangular lateral faces intersecting at a common point called the *vertex*. The center line from the center of the base to the vertex is the *axis*. If the axis is perpendicular to the base, the pyramid is a right pyramid; otherwise it is an oblique pyramid. A triangular pyramid has a triangular base; a square pyramid has a square base; and so

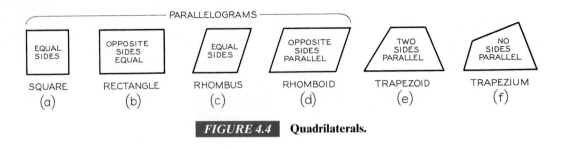

FIGURE 4.4   Quadrilaterals.

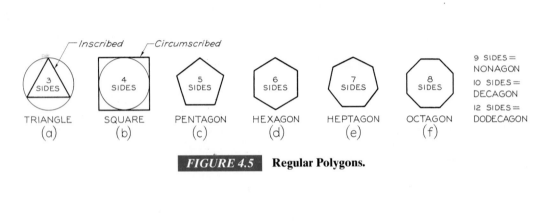

FIGURE 4.5   Regular Polygons.

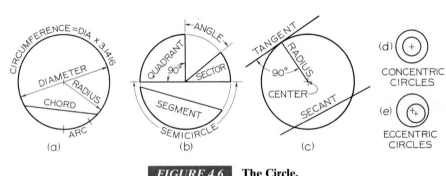

FIGURE 4.6   The Circle.

on. If a portion near the vertex has been cut off, the pyramid is *truncated*, or it is referred to as a *frustum*.

A **cylinder** is generated by a straight line, called the *generatrix*, moving in contact with a curved line and always remaining parallel to its previous position or to the axis. Each position of the generatrix is called an element of the cylinder.

A **cone** is generated by a straight line moving in contact with a curved line and passing through a fixed point, the *vertex* of the cone. Each position of the generatrix is an element of the cone.

A **sphere** is generated by a circle revolving about one of its diameters. This diameter becomes the axis of the sphere, and the ends of the axis are *poles* of the sphere.

A **torus**, which is shaped like a doughnut, is generated by a circle (or other curve) revolving about an axis that is eccentric to the curve.

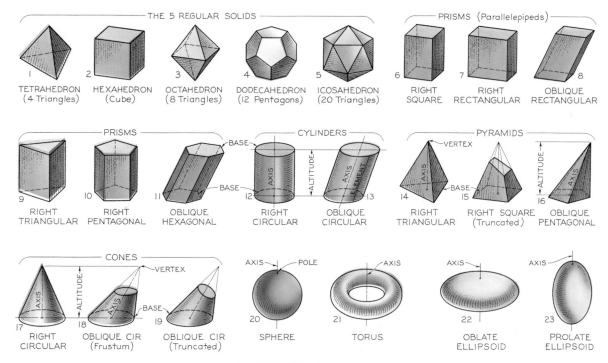

FIGURE 4.7 Solids.

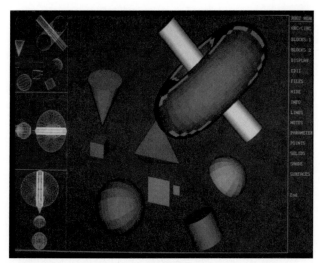

Examples of Solids Created with CAD. *Courtesy of American Small Business Computers.*

An oblate or prolate **ellipsoid** is generated by revolving an ellipse about its minor or major axis, respectively. For more information on working with solids, see Chapter 14 for an in-depth discussion on solid modeling.

■ 4.8 BISECTING A LINE OR CIRCULAR ARC

Figure 4.8a shows the given line or arc AB to be bisected.

I. From A and B draw equal arcs with radius greater than half AB.

II. and III. Join intersections D and E with a straight line to locate center C.

■ 4.9 BISECTING A LINE WITH TRIANGLE AND T-SQUARE

As shown in Fig. 4.9, from endpoints A and B, draw construction lines at 30°, 45°, or 60° with the given line; then through their intersection, C, draw a line perpendicular to the given line to locate the center D, as shown.

■ 4.10 BISECTING AN ANGLE

Figure 4.10a shows the given angle BAC to be bisected.

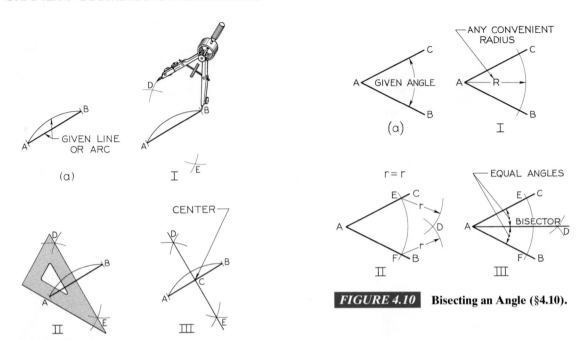

FIGURE 4.10   Bisecting an Angle (§4.10).

FIGURE 4.8   Bisecting a Line or a Circular Arc (§4.8).

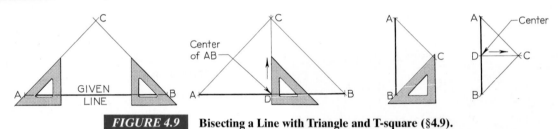

FIGURE 4.9   Bisecting a Line with Triangle and T-square (§4.9).

I. Strike large arc R.

II. Strike equal arcs r with radius slightly larger than half BC, to intersect at D.

III. Draw line AD, which bisects the angle.

■ 4.11 TRANSFERRING AN ANGLE

Figure 4.11a shows the given angle BAC to be transferred to the new position at A′B′.

I. Use any convenient radius R, and strike arcs from centers A and A′.

II. Strike equal arcs r, and draw side A′C′.

■ 4.12 DRAWING A LINE THROUGH A POINT AND PARALLEL TO A LINE

With given point P as center, and any convenient radius R, strike arc CD to intersect the given line AB at E (Fig. 4.12a). With E as center and the same radius, strike arc R′ to intersect the given line at G. With PG as radius and E as center, strike arc r to locate point H. The line PH is the required line.

FIGURE 4.11   Transferring an Angle (§4.11).

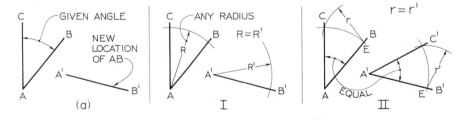

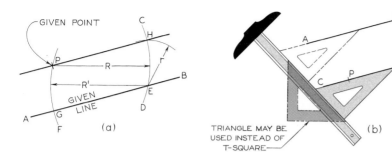

FIGURE 4.12   Drawing a Line
Through a Point Parallel to a Line (§4.12).

PREFERRED METHOD   As shown in Fig. 4.12b, move the triangle and T-square as a unit until the triangle lines up with given line AB; then slide the triangle until its edge passes through the given point P. Draw CD, the required parallel line.

■ 4.13 DRAWING A LINE PARALLEL TO A LINE AND AT A GIVEN DISTANCE

Let AB be the line and CD the given distance. As shown in Fig. 4.13a, draw two arcs with points E and F near A and B, respectively, as centers, and CD as radius. The line GH, tangent to the arcs, is the required line.

PREFERRED METHOD   With any point E of the line as center and CD as radius, strike an arc JK (Fig. 4.13b). Move the triangle and T-square as a unit until the triangle lines up with the given line AB; then slide the triangle until its edge is tangent to the arc JK, and draw the required line GH.

As shown in Fig. 4.13c, with centers selected at random on the curved line AB and with CD as radius, draw a series of arcs; then draw the required line tangent to these arcs.

■ 4.14 DIVIDING A LINE INTO EQUAL PARTS

There are two methods for dividing a line into equal parts. The first method is shown in Fig. 4.14:

I.   Draw a light construction line at any convenient angle from one end of line.

II.   With dividers or scale, set off from intersection of lines as many equal divisions as needed (in this case, three).

III.   Connect last division point to the other end of line, using triangle and T-square, as shown.

IV.   Slide triangle along T-square and draw parallel lines through other division points, as shown.

The second method for dividing a line into equal parts is shown in Fig. 4.15:

I.   Draw vertical construction line at one end of given line.

II.   Set zero of scale at other end of line.

III.   Swing scale up until third unit falls on vertical line, and make tiny dots at each point, or prick points with dividers.

FIGURE 4.13   Drawing a Line Parallel to a Line at a Given Distance (§4.13).

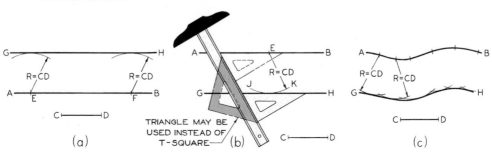

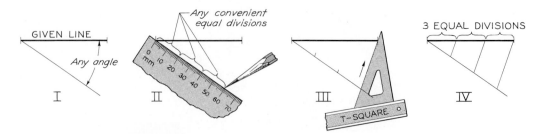

FIGURE 4.14   Dividing a Line into Equal Parts (§4.14).

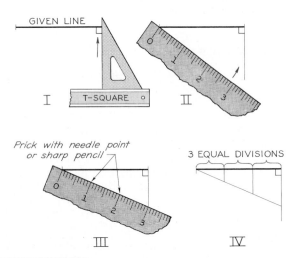

FIGURE 4.15   Dividing a Line into Equal Parts (§4.14).

IV. Draw vertical construction lines through each point.

Some practical applications of this method are shown in Fig. 4.16.

■ 4.15 DIVIDING A LINE INTO PROPORTIONAL PARTS

Let it be required to divide the line AB into three parts proportional, to 2, 3, and 4.

PREFERRED METHOD   Draw a vertical line from point B (Fig. 4.17a). Select a scale of convenient size for a total of nine units and set the zero of the scale at A. Swing the scale up until the ninth unit falls on the vertical line. Along the scale, set off points for 2, 3, and 4 units, as shown. Draw vertical lines through these points.

Draw a line CD parallel to AB and at any convenient distance (Fig. 4.17b). On this line, set off 2, 3, and 4 units, as shown. Draw lines through the ends of the two lines to intersect at the point O. Draw lines through O and the points 2 and 5 to divide AB into the required proportional parts.

Given AB, divide into proportional parts, in this case proportional to the square of x, where $x = 1$, 2, 3, ... (Fig. 4.17c). Set zero of scale at end of line and set off divisions 4, 9, 16, ... Join the last division to the other end of the line, and draw parallel lines as shown. This method may be used for any power of x.

FIGURE 4.16   Practical Applications of Dividing a Line into Equal Parts (§4.14).

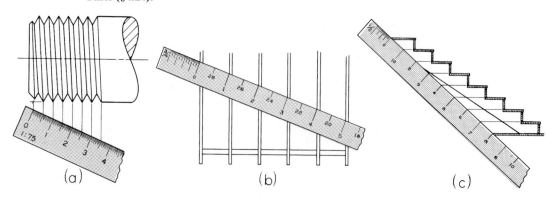

(a)   (b)   (c)

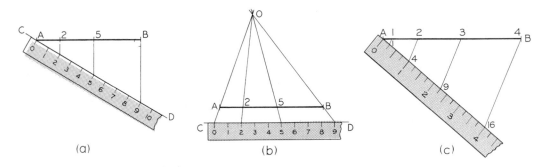

FIGURE 4.17   Dividing a Line into Proportional Parts (§4.15).

■ 4.16 DRAWING A LINE THROUGH A POINT AND PERPENDICULAR TO A LINE

The line AB and a point P (Fig. 4.18) are given.

WHEN THE POINT IS NOT ON THE LINE   From P, draw any convenient inclined line, as PD (Fig. 4.18a). Find center C of line PD, and draw arc with radius CP. The line EP is the required perpendicular.

With P as center, strike an arc to intersect AB at C and D (Fig. 4.18b). With C and D as centers and radius slightly greater than half CD, strike arcs to intersect at E. The line PE is the required perpendicular.

WHEN THE POINT IS ON THE LINE   With P as center and any radius, strike arcs to intersect AB at D and G (Fig. 4.18c). With D and G as centers and radius slightly greater than half DG, strike equal arcs to intersect at F. The line PF is the required perpendicular.

As shown in Fig. 4.18d, select any convenient unit of length (for example, 6 mm or $\frac{1}{4}''$). With P as center and 3 units as radius, strike an arc to intersect the given line at C. With P as center and 4 units as radius, strike arc DE. With C as center and 5 units as radius,

strike an arc to intersect DE at F. The line PF is the required perpendicular.

This method makes use of the 3–4–5 right triangle and is frequently used in laying off rectangular foundations of large machines, buildings, or other structures. For this purpose a steel tape may be used and distances of 30′, 40′, and 50′ measured as the three sides of the right triangle.

PREFERRED METHOD   Move the triangle and T-square as a unit until the triangle lines up with AB (Fig. 4.18e); then slide the triangle until its edge passes through the point P (whether P is on or off the line), and draw the required perpendicular.

■ 4.17 DRAWING A TRIANGLE WITH SIDES GIVEN

Given the sides A , B, and C, as shown in Fig. 4.19a:

I.   Draw one side, as C, in desired position, and strike arc with radius equal to side A.

II.   Strike arc with radius equal to side B.

III.   Draw sides A and B from intersection of arcs, as shown.

FIGURE 4.18   Drawing a Line Through a Point and Perpendicular to a Line (§4.16).

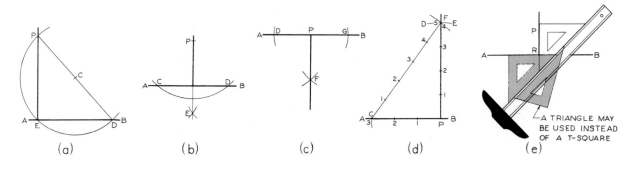

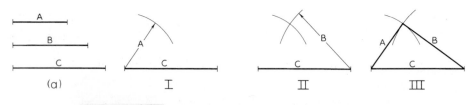

FIGURE 4.19   Drawing a Tringle with Sides Given (§4.17).

■ 4.18 DRAWING A RIGHT TRIANGLE WITH HYPOTENUSE AND ONE SIDE GIVEN

Given sides S and R (Fig. 4.20), with AB as a diameter equal to S, draw a semicircle. With A as center and R as radius, draw an arc intersecting the semicircle at C. Draw AC and CB to complete the right triangle.

■ 4.19 LAYING OUT AN ANGLE

Many angles can be laid out directly with the triangle, or they may be laid out with the protractor. Other methods, for which considerable accuracy is required, are discussed next (Fig. 4.21).

TANGENT METHOD   The tangent of angle θ is y/x and $y = x \tan \theta$. To construct the angle, assume a convenient value for x, preferably 10 units of convenient length (Fig. 4.21a). (The larger the unit, the more accurate will be the construction.) Find the tangent of angle θ in a table of natural tangents, multiply by 10, and set off $y = 10 \tan \theta$.

> **EXAMPLE**   To set off $31\frac{1}{2}°$, find the natural tangent of $31\frac{1}{2}°$, which is 0.6128. Then
>
> $$y = 10 \text{ units} \times 0.6128 = 6.128 \text{ units.}$$

SINE METHOD   Draw line x to any convenient length, preferably 10 units (Fig. 4.21b). Find the sine of angle θ in a table of natural sines, multiply by 10, and strike arc $R = 10 \sin \theta$. Draw the other side of the angle tangent to the arc, as shown.

> **EXAMPLE**   To set off $25\frac{1}{2}°$, find the natural sine of $25\frac{1}{2}°$, which is 0.4304. Then
>
> $$R = 10 \text{ units} \times 0.4305 = 4.305 \text{ units.}$$

CHORD METHOD   Draw line x to any convenient length, and draw arc with any convenient radius R—say 10 units (Fig. 4.21c). Find the chordal length C in a table of chords (see a machinists' handbook), and multiply the value by 10 since the table is made for a radius of 1 unit.

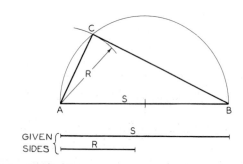

FIGURE 4.20   Drawing a Right Triangle (§4.18).

FIGURE 4.21   Laying Out Angles (§4.19).

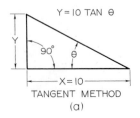

TANGENT METHOD
(a)

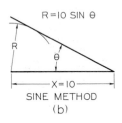

SINE METHOD
(b)

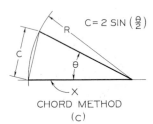

CHORD METHOD
(c)

EXAMPLE To set off 43° 20′, the chordal length C for 1 unit radius, as given in a table of chords equals 0.7384; and if $R = 10$ units, then $C = 7.384$ units.

If a table is not available, the chord C may be calculated by the formula $C = 2 \sin \theta/2$.

EXAMPLE Half of 43° 20′ = 21° 40′. The sine of 21° 40′ = 0.3692. $C = 2 \times 0.3692 = 0.7384$ for a 1 unit radius. For a 10 unit radius, $C = 7.384$ units.

■ 4.20 DRAWING AN EQUILATERAL TRIANGLE

Side AB is given. With A and B as centers and AB as radius, strike arcs to intersect at C (Fig. 4.22a). Draw lines AC and BC to complete the triangle.

PREFERRED METHOD Draw lines through points A and B, making angles of 60° with the given line and intersecting C (Fig. 4.22b).

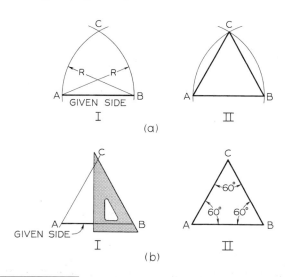

FIGURE 4.22 Drawing an Equilateral Triangle (§4.20).

■ 4.21 DRAWING A SQUARE

One side AB (Fig. 4.23a) is given. Through point A, draw a perpendicular (see Fig. 4.18c). With A as center and AB as radius, draw the arc to intersect the perpendicular at C. With B and C as centers and AB as radius, strike arcs to intersect at D. Draw lines CD and BD.

PREFERRED METHOD One side AB (Fig. 4.23b) is given. Using the T-square or parallel straightedge and 45° triangle, draw lines AC and BD perpendicular to AB and the lines AD and BC at 45° with AB. Draw line CD.

PREFERRED METHOD Given the circumscribed circle (distance "across corners"), draw two diameters at right angles to each other (Fig. 4.23c). The intersections of these diameters with the circle are vertexes of an inscribed square.

PREFERRED METHOD Given the inscribed circle (Fig. 4.23d) (distance "across flats," as in drawing bolt heads), use the T-square (or parallel straightedge) and 45° triangle and draw the four sides tangent to the circle.

■ 4.22 DRAWING A REGULAR PENTAGON

Given the circumscribed circle, do the following:

PREFERRED METHOD Divide the circumference of the circle into five equal parts with the dividers, and join the points with straight lines (Fig. 4.24a).

GEOMETRICAL METHOD
As shown in Fig. 4.24b:

I. Bisect radius OD at C.
II. With C as center and CA as radius, strike arc AE. With A as center and AE as radius, strike arc EB.

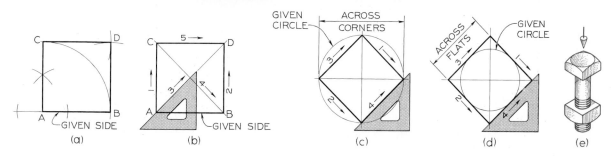

FIGURE 4.23 Drawing a Square (§4.21).

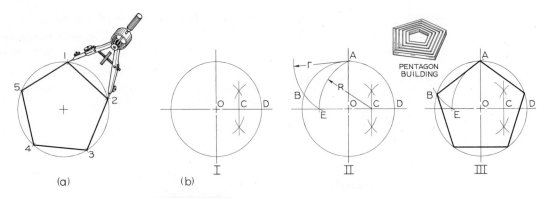

FIGURE 4.24   Drawing a Pentagon (§4.22).

III. Draw line AB; then set off distances AB around the circumference of the circle, and draw the sides through these points.

■ 4.23 DRAWING A HEXAGON

The circumscribed circle (Fig. 4.25) as given. Each side of a hexagon is equal to the radius of the circumscribed circle (Fig. 4.25a). Therefore, using the compass or dividers and the radius of the circle, set off the six sides of the hexagon around the circle, and connect the points with straight lines. As a check on the accura-

cy of the construction, make sure that opposite sides of the hexagon are parallel.

PREFERRED METHOD   This construction (Fig. 4.25b) is a variation of the one shown in Fig. 4.25a. Draw vertical and horizontal center lines. With A and B as centers and radius equal to that of the circle, draw arcs to intersect the circle at C, D, E, and F, and complete the hexagon as shown.

Given the circumscribed circle (distance "across corners") (Figs. 4.26a and 4.26b), draw vertical and horizontal center lines, and then diagonals AB and CD

FIGURE 4.25   Drawing a Hexagon (§4.23).

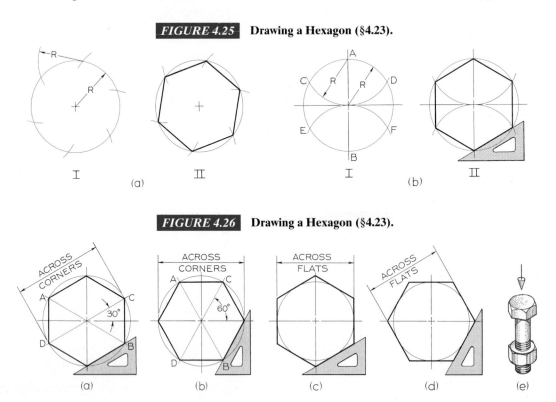

FIGURE 4.26   Drawing a Hexagon (§4.23).

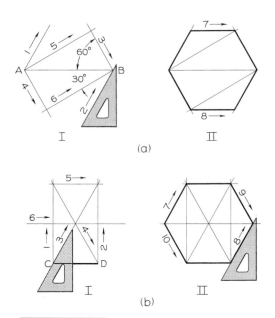

FIGURE 4.27   Drawing a Hexagon (§4.23).

at 30° or 60° with horizontal; finally, with the 30° × 60° triangle and the T-square, draw the six sides as shown.

Given the inscribed circle (distance "across flats") (Figs. 4.26c and 4.26d), draw vertical and horizontal center lines; then with the 30° × 60° triangle and the T-square or straightedge, draw the six sides tangent to the circle. This method is used in drawing bolt heads and nuts. For maximum accuracy, diagonals may be added, as in Figs. 4.26a and 4.26b.

As shown in Fig. 4.27, use the 30° × 60° triangle and the T-square or straightedge; draw lines in the order shown in Fig. 4.27a, where the distance AB ("across corners") is given or, as shown in Fig. 4.27b, where a side CD is given.

■ 4.24 DRAWING AN OCTAGON

Given an inscribed circle, or distance "across flats" (Fig. 4.28a), use a T-square or straightedge and a 45° triangle to draw the eight sides tangent to the circle, as shown.

Given a circumscribed square, or distance "across flats" (Fig. 4.28b), draw diagonals of square; then with the corners of the given square as centers and with half the diagonal as radius, draw arcs cutting the sides as shown in I. Using a T-square and 45° triangle, draw the eight sides, as shown in II.

■ 4.25 TRANSFERRING PLANE FIGURES BY GEOMETRIC METHODS

TRANSFERRING A TRIANGLE TO A NEW LOCATION Set off any side of the given triangle (Fig. 4.29a), such as AB, in the new location (Fig. 4.29b). With the ends of the line as centers and the lengths of the other sides of the given triangle as radii, strike two arcs to intersect at C. Join C to A and B to complete the triangle.

TRANSFERRING A POLYGON BY THE TRIANGLE METHOD Divide the polygon into triangles as shown, and transfer each triangle as explained previously (Fig. 4.29c).

TRANSFERRING A POLYGON BY THE RECTANGLE METHOD Circumscribe a rectangle abut the given polygon (Fig. 4.29d). Draw a congruent rectangle in the new location and locate the vertexes of the polygon by transferring location measurements a, b, c, and so on along the sides of the rectangle to the new rectangle. Join the points thus found to complete the figure.

TRANSFERRING IRREGULAR FIGURES Figures composed of rectangular and circular forms are readily transferred by enclosing the elementary features in rectangles and determining centers of arcs and circles (Fig. 4.29e). These may then be transferred to the new location.

TRANSFERRING FIGURES BY OFFSET MEASUREMENTS *Offset location measurements* are frequently useful in transferring figures composed of free curves (Fig. 4.29f). When the figure has been enclosed by a rectangle, the sides of the rectangle are used as reference lines for the location of points along the curve.

FIGURE 4.28   Drawing an Octagon (§4.24). *Photo by Laima Druskis.*

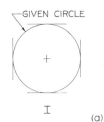
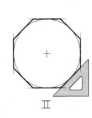
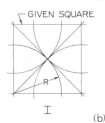
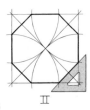

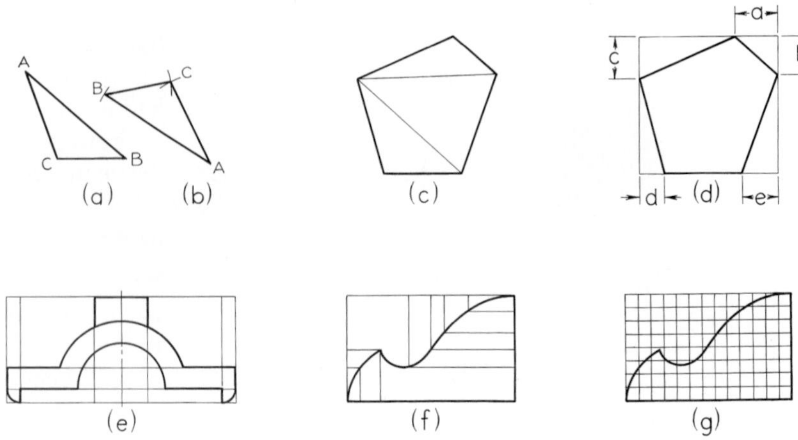

FIGURE 4.29   **Transferring a Plane Figure (§4.25).**

TRANSFERRING FIGURES BY A SYSTEM OF SQUARES   Figures involving free curves are easily copied, enlarged, or reduced by the use of a system of squares (Fig. 4.29g). For example, to enlarge a figure to double size, draw the containing rectangle and all small squares double their original size. Then draw the lines through the corresponding points in the new set of squares.

■ 4.26 TRANSFERRING DRAWINGS BY TRACING-PAPER METHODS

To transfer a drawing to an opaque sheet, the following procedures may be used.

PRICKED-POINT METHOD   Lay tracing paper over the drawing to be transferred. With a sharp pencil, make a small dot directly over each important point of the drawing. Encircle each dot so it is not lost. Remove the tracing paper, place it over the paper to receive the transferred drawing, and maneuver the tracing paper into the desired position. With a needle point (such as a point of the dividers), prick through each dot. Remove the tracing paper and connect the pricked points to reproduce the lines of the original drawing.

To reproduce arcs or circles, it is only necessary to transfer the center and one point on the circumference. To duplicate a free curve, transfer as many pricked points on the curve as desired.

TRACING METHOD   Lay tracing paper over the drawing to be transferred, and make a pencil tracing of it. Turn the tracing paper over and mark over the lines with short strokes of a soft pencil to provide a coating of graphite over every line. Turn tracing face up and fasten in position where drawing is to be transferred. Trace over all lines of the tracing, using a hard pencil. The graphite on the back acts as a carbon paper and will produce dim but definite lines. Heavy in the dim lines to complete the transfer.

If one half of a symmetrical object has been drawn (Fig. 4.30), such as the ink bottle in I, the other half may be drawn with the aid of tracing paper, as follows:

I.   Trace the half already drawn.

II.   Turn tracing paper over and maneuver to the position for the right half. Then trace over the lines freehand or mark over the lines with short strokes, as shown.

III.   Remove the tracing paper, revealing the dim imprinted lines for the right half. Heavy in these lines to complete the drawing.

■ 4.27 ENLARGING OR REDUCING A DRAWING

The construction shown in Fig. 4.31a is an adaptation of the parallel-line method (Figs. 4.12 and 4.13); it may

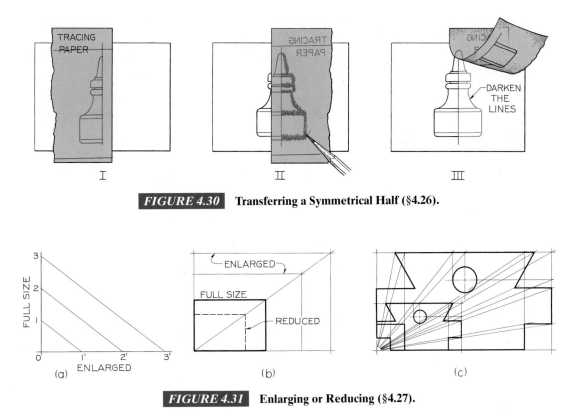

FIGURE 4.30   **Transferring a Symmetrical Half (§4.26).**

FIGURE 4.31   **Enlarging or Reducing (§4.27).**

be used to enlarge or reduce any group of dimensions to the same ratio. Thus, if full-size dimensions are laid off along the vertical line, the enlarged dimensions will appear along the horizontal line, as shown.

To enlarge or reduce a rectangle (say, a photograph), a simple method is to use the diagonal, as shown in Fig. 4.31b. A simple method of enlarging or reducing a drawing is to make use of radial lines (Fig. 4.31c). The original drawing is placed underneath a

sheet of tracing paper, and the enlarged or reduced drawing is made directly on the tracing paper.

■ 4.28 DRAWING A CIRCLE THROUGH THREE POINTS

I.   Let A, B, and C be the three given points not in a straight line (Fig. 4.32a). Draw lines AB and BC,

FIGURE 4.32   **Finding Center of Circle (§§4.28 and 4.29).**

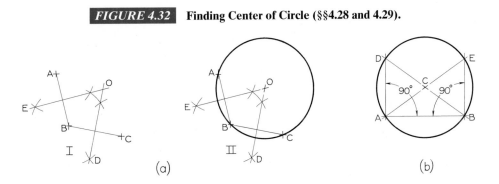

which will be chords of the circle. Draw perpendicular bisectors EO and DO intersecting at O (see Fig. 4.8).

II. With center at O, draw required circle through the points.

■ 4.29 FINDING THE CENTER OF A CIRCLE

Draw any chord AB, preferably horizontal as shown (Fig. 4.32b). Draw perpendiculars from A and B, cutting the circle at D and E. Draw diagonals DB and EA whose intersection C will be the center of the circle. This method uses the principle that any right triangle inscribed in a circle cuts off a semicircle (see Fig. 4.3f).

Another method, slightly longer, is to reverse the procedure of Fig. 4.32a. Draw any two nonparallel chords and draw perpendicular bisectors. The intersection of the bisectors will be the center of the circle.

■ 4.30 DRAWING A CIRCLE TANGENT TO A LINE AT A GIVEN POINT

Given a line AB and a point P on the line (Fig. 4.33a):

I. At P, erect a perpendicular to the line.

II. Set off the radius of the required circle on the perpendicular.

III. Draw a circle with radius CP.

■ 4.31 DRAWING A TANGENT TO A CIRCLE THROUGH A POINT

PREFERRED METHOD   Given point P on the circle (Fig. 4.34a), move the T-square and triangle as a unit until one side of the triangle passes through the point P and the center of the circle; then slide the triangle until the other side passes through point P, and draw the required tangent.

Given point P outside the circle (Fig. 4.34b), move the T-square and triangle as a unit until one side of the triangle passes through point P and, by inspection, is tangent to the circle; then slide the triangle until the other side passes through the center of the circle, and lightly mark the point of tangency T. Finally, move the triangle back to its starting position, and draw the required tangent.

In both constructions either triangle may be used. Also, a second triangle may be used in place of the T-square.

FIGURE 4.33   Drawing a Circle Tangent to a Line (§4.30).

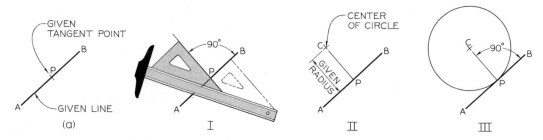

FIGURE 4.34   Drawing a Tangent to a Circle Through a Point (§4.31).

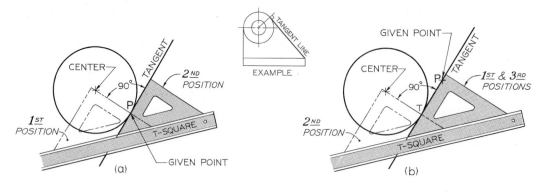

■ 4.32 DRAWING TANGENTS TO TWO CIRCLES

Move the triangle and T-square as a unit until one side of the triangle is tangent, by inspection, to the two circles (Figs. 4.35a and 4.35b); then slide the triangle until the other side passes through the center of one circle, and lightly mark the point of tangency. Then slide the triangle until the slide passes through the center of the other circle, and mark the point of tangency. Finally, slide the triangle back to the tangent position, and draw the tangent lines between the two points of tangency. Draw the second tangent line in a similar manner.

■ 4.33 DRAWING AN ARC TANGENT TO A LINE OR ARC AND THROUGH A POINT

Given line AB, point P, and radius R (Fig. 4.36a), draw line DE parallel to the given line and distance R from it. From P draw arc with radius R, cutting line DE at C, the center of the required tangent arc.

Given line AB, with tangent point Q on the line and point P (Fig. 4.36b), draw PQ, which will be a chord of the required arc. Draw perpendicular bisector DE, and at Q erect a perpendicular to the line to intersect DE at C, the center of the required tangent arc.

Given arc with center Q, point P, and radius R (Fig. 4.36c), from P, strike arc with radius R. From Q,

FIGURE 4.35 Drawing Tangents to Two Circles (§4.32).

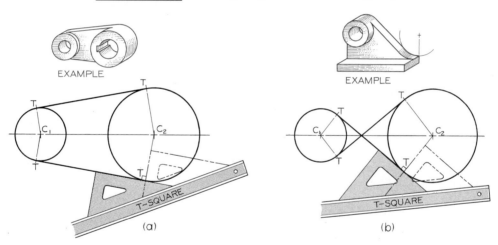

(a) (b)

FIGURE 4.36 Tangents (§4.33).

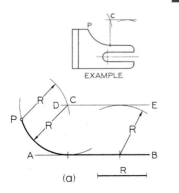

(a)

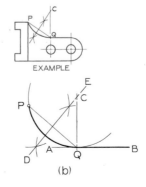

(b)

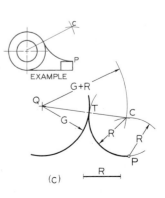

(c)

strike arc with radius equal to that of the given arc plus R. The intersection C of the arcs is the center of the required tangent arc.

■ 4.34 DRAWING AN ARC TANGENT TO TWO LINES AT RIGHT ANGLES

I. Two lines are given at right angles to each other (Fig. 4.37a).

II. With given radius R, strike arc intersecting given lines at tangent points T.

III. With given radius R again, and with points T as centers, strike arcs intersecting at C.

IV. With C as center and given radius R, draw the required tangent arc.

For small radii, such as $\frac{1}{8}R$ for fillets and rounds, it is not practicable to draw complete tangency constructions. Instead, draw a 45° bisector of the angle and locate the center of the arc by trial along this line (Fig. 4.37b).

Note that the center C can be located by intersecting lines parallel to the given lines, as shown in Fig. 4.13b. The circle template can also be used to draw the arcs R for the parallel line method of Fig. 4.13b. While the circle template is convenient to use for small radii up to about $\frac{5}{8}''$ or 16 mm, the diameter of the template circle is precisely equal twice the required radius.

■ 4.35 DRAWING AN ARC TANGENT TO TWO LINES AT ACUTE OR OBTUSE ANGLES

I. Two lines intersecting not making 90° with each other (Fig. 4.38a and 4.38b) are given.

II. Draw lines parallel to given lines, at distance R from them, to intersect at C, the required center.

III. From C, drop perpendiculars to the given lines, respectively, to locate tangent points T.

IV. With C as center and with given radius R, draw the required tangent arc between the points of tangency.

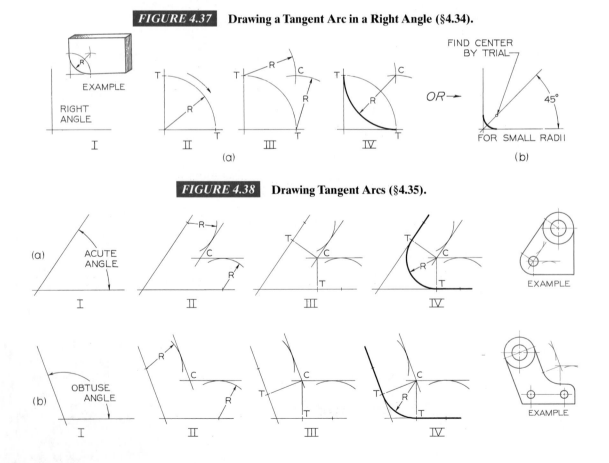

FIGURE 4.37   **Drawing a Tangent Arc in a Right Angle (§4.34).**

FIGURE 4.38   **Drawing Tangent Arcs (§4.35).**

■ 4.36 DRAWING AN ARC TANGENT TO AN ARC AND A STRAIGHT LINE

I. An arc with radius G and a straight line AB (Figs. 4.39a and 4.39b) are given.

II. Draw a straight line and an arc parallel, respectively, to the given straight line and arc at the required radius distance R from them, to intersect at C, the required center.

III. From C, drop a perpendicular to the given straight line to obtain one point of tangency T. Join the

centers C and O with a straight line to locate the other point of tangency T.

IV. With center C and given radius R, draw the required tangent arc between the points of tangency.

■ 4.37 DRAWING AN ARC TANGENT TO TWO ARCS

I. Arcs with centers A and B and required radius R (Figs. 4.40a and 4.40b) are given.

II. With A and B as centers, draws arcs parallel to the

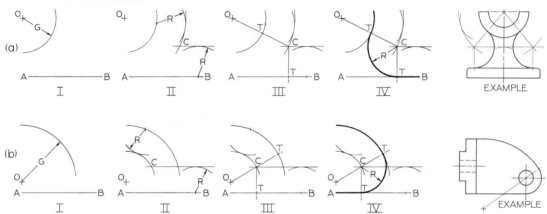

FIGURE 4.39   Drawing an Arc Tangent to an Arc and a Straight Line (§4.36).

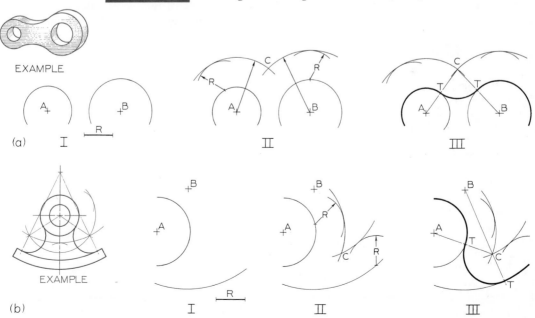

FIGURE 4.40   Drawing an Arc Tangent to Two Arcs (§4.37).

given arcs and at a distance R from them; their intersection C is the center of the required tangent arc.

III. Draw lines of centers AC and AC to locate points of tangency T, and draw the required tangent arc between the points of tangency, as shown.

■ 4.38 DRAWING AN ARC TANGENT TO TWO ARCS AND ENCLOSING ONE OR BOTH

REQUIRED ARC ENCLOSES BOTH GIVEN ARCS
With A and B as centers, strike arcs HK − r (given radius minus radius of small circle) and HK − R (given radius minus radius of large circle) intersecting at G, the center of the required tangent arc. Lines of centers GA and GB (extended) determine points of tangency T (Fig. 4.41a).

REQUIRED ARC ENCLOSES ONE GIVEN ARC
With C and D as centers, strike arcs HK + r (given radius plus radius of small circle) and HK − R (given radius minus radius of large circle) intersecting at G, the center of the required tangent arc. Lines of centers GC and GD (extended) determine points of tangency T (Fig. 4.41b).

■ 4.39 DRAWING A SERIES OF TANGENT ARCS CONFORMING TO A CURVE

First sketch lightly a smooth curve as desired (Fig. 4.42). By trial, find a radius R and a center C, produc-

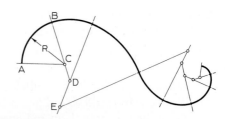

FIGURE 4.42 Drawing a Series of Tangent Arcs Conforming to a Curve (§4.39).

ing an arc AB that closely follows that portion of the curve. The successive centers D, E, and so on will be on lines joining the centers with the points of tangency, as shown.

■ 4.40 DRAWING AN OGEE CURVE

CONNECTING TWO PARALLEL LINES   Let NA and BM be the two parallel lines. Draw AB, and assume inflection point T (at midpoint if two equal arcs are desired) (Fig. 4.43a). At A and B, erect perpendiculars AF and BC. Draw perpendicular bisectors of AT and BT. The intersections F and C of these bisectors and the perpendiculars, respectively, are the centers of the required tangent arcs.

Let AB and CD be the two parallel lines, with point B as one end of the curve and R the given radii (Fig. 4.43b). At B, erect perpendicular to AB, make BG = R, and draw the arc as shown. Draw line SP parallel to CD at distance R from CD. With center G, draw the arc of radius 2R, intersecting line SP at O. Draw perpendicular OJ to locate tangent point J, and

FIGURE 4.41   Drawing an Arc Tangent to Two Arcs and Enclosing One or Both (§4.38).

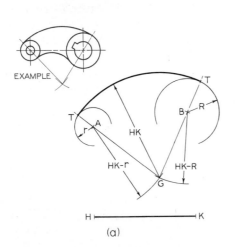

(a)

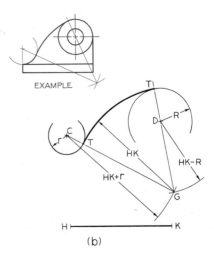

(b)

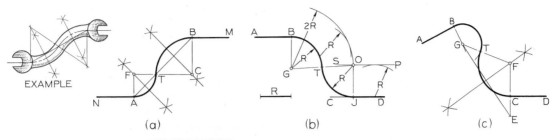

FIGURE 4.43   **Drawing an Ogee Curve (§4.40).**

join centers G and O to locate point of tangency T. Using centers G and O and radius R, draws the two tangent arcs as shown.

CONNECTING TWO NONPARALLEL LINES Let AB and CD be the two nonparallel lines (Fig. 4.43c). Erect perpendicular to AB at B. Select point G on the perpendicular so that BG equals any desired radius, and draw the arc as shown. Erect perpendicular to CD at C and make CE = BG. Join G to E and bisect it. The intersection F of the bisector and the perpendicular CE, extended, is the center of the second arc. Join centers of the two arcs to locate tangent point T, the inflection point of the curve.

■ 4.41 DRAWING A CURVE TANGENT TO THREE INTERSECTING LINES

Let AB, BC, and CD be the given lines (Fig. 4.44a and 4.44b). Select point of tangency P at any point on line

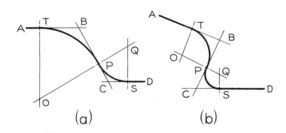

(a)   (b)

FIGURE 4.44   **Drawing Two Curves Tangent to Three Intersecting Lines (§4.41).**

BC. Make BT equal to BP, and CS equal to CP, and erect perpendiculars at the points P, T, and S. Their intersections O and Q are the centers of the required tangent arcs.

■ 4.42 RECTIFYING A CIRCULAR ARC

To *rectify* an arc is to lay out its true length along a straight line. The constructions are approximate, but well within the range of accuracy of drawing instruments.

TO RECTIFY A QUADRANT OF A CIRCLE, AB Draw AC tangent to the circle BC at 60° to AC, as shown (Fig. 4.45a). The line AC is almost equal to the arc AB; the difference in length is about 1 in 240.

TO RECTIFY ARC, AB Draw tangent at B (Fig. 4.45b). Draw chord AB and extend it to C, making BC equal to half AB. With C as center and radius CA, strike the arc AD. The tangent BD is slightly shorter than the given arc AB. For an angle of 45° the difference in length is about 1 in 2866.

Use the bow dividers and, beginning at A, set off equal distances until the division point nearest B is reached (Fig. 4.45c). At this point, reverse the direction and set off an equal number of distances along the tangent to determine point C. The tangent BC is slightly shorter than the given arc AB. If the angle subtended

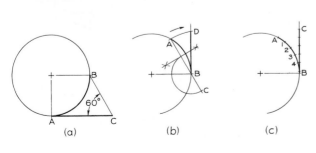

(a)   (b)   (c)

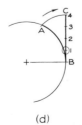

(d)

FIGURE 4.45   **Rectifying Circular Arcs (§§4.42 and 4.43).**

by each division is 10°, the error is approximately 1 in 830. *

■ 4.43 SETING OFF A GIVEN LENGTH ALONG A GIVEN ARC

To transfer distances from the tangent line to the arc, reverse the preceding method (Fig. 4.45c).

To set off the length BC along the arc BA, draw BC tangent to the arc at B (Fig. 4.45d). Divide BC into four equal parts. With center at 1, the first division point, and radius 1–C, draw the arc CA. The arc BA is practically equal to BC for angles less than 30°. For 45° the difference is approximately 1 in 3232, and for 60° it is about 1 in 835.

■ 4.44 THE CONIC SECTIONS

The conic sections are curves produced by planes intersecting a right circular cone (Fig. 4.46). Four types of curves are produced: the **circle**, **ellipse**, **parabola**, and **hyperbola**, according to the position of the planes, as shown. (Also see §19.7.) These curves were studied in detail by the ancient Greeks and are of great interest in mathematics, as well as in technical drawing. For equations, see any text on analytic geometry.

* If the angle θ subtending an arc of radius R is known, the length of the arc is

$$2\pi R \, \frac{\theta}{360°} = 0.01745R\theta.$$

■ 4.45 ELLIPSE CONSTRUCTION

The long axis of an ellipse is the major axis and the short axis is the minor axis (Fig. 4.47a). The foci E and F are found by striking arcs with radius equal to half the major axis and with center at the end of the minor axis. Another method is to draw a semicircle with the major axis as diameter, and then to draw GH parallel to the major axis and GE and HF parallel to the minor axis as shown.

An **ellipse** is generated by a point moving so that the sum of its distances from two points (the foci) is constant and equal to the major axis. As shown in Fig. 4.47b, an ellipse may be constructed by placing a looped string around the foci E and F and around C, one end of the minor axis, and moving the pencil point P along its maximum orbit while the string is kept taut.

■ 4.46 DRAWING A FOCI ELLIPSE

Let AB be the major axis and CD the minor axis (Fig. 4.48). This method is the geometrical counterpart of the pin-and-string method. Keep the construction very light, as follows:

I.   To find foci E and F, strike arcs R with radius equal to half the major axis and with centers at the ends of the minor axis.

II.   Between E and O on the major axis, mark at random a number of points (spacing those on the left more closely), equal to the number of points de-

---

FIGURE 4.46   Conic Sections (§4.44).

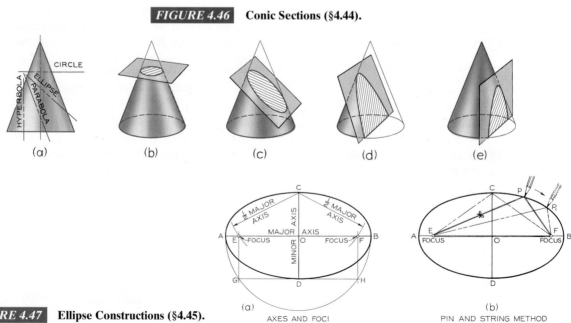

(a)   (b)   (c)   (d)   (e)

FIGURE 4.47   Ellipse Constructions (§4.45).

(a) AXES AND FOCI

(b) PIN AND STRING METHOD

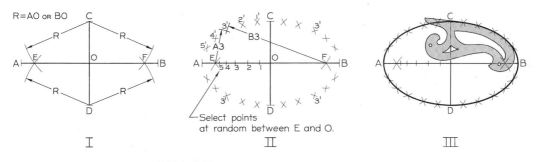

FIGURE 4.48 Drawing a Foci Ellipse (§4.46).

sired in each quadrant of the ellipse. In this figure, five points were deemed sufficient. For large ellipses, more points should be used—enough to ensure a smooth, accurate curve. Begin construction with any one of these points, such as 3. With E and F as centers and radii A–3 and B–3, respectively (from the ends of the major axis to point 3), strike arcs to intersect at four points 3', as shown. Using the remaining points 1, 2, 4, and 5, for each find four additional points on the ellipse in the same manner.

III. Sketch the ellipse lightly through the points; then heavy in the final ellipse with the aid of an irregular curve.

■ 4.47 DRAWING A TRAMMEL ELLIPSE

A "long trammel" or a "short trammel" may be prepared from a small strip of stiff paper or thin cardboard (Fig. 4.49). In both cases, set off on the edge of the trammel distances equal to the semimajor (AC or BC) and semiminor (DC or EC) axes. With the short trammel, these distances overlap; with the long trammel they are end to end. To use either method, place the trammel so that two of the points are on the respective axes, as shown; the third point will then be on the curve and can be marked with a small dot. Find additional points by moving the trammel to other positions, always keeping the two points exactly on the respective axes. Extend the axes to use the long trammel. Find enough points to ensure a smooth and symmetrical ellipse. Sketch the ellipse lightly through the

points; then heavy in the ellipse with the aid of an irregular curve.

■ 4.48 DRAWING A CONCENTRIC-CIRCLE ELLIPSE

If a circle is viewed so that the line of sight is perpendicular to the plane of the circle, as shown for the silver dollar in Fig. 4.50a, the circle will appear as a circle, in true size and shape. If the circle is viewed at an angle, as shown in Fig. 4.50b, it will appear as an ellipse. If the circle is viewed edgewise, it appears as a straight line, as shown in Fig. 4.50c. The case shown in Fig. 4.50b is the basis for the construction of an ellipse by the concentric-circle method, which follows. (Keep the construction very light.)

I. Draw circles on the major and minor axes, using them as diameters; draw any diagonal XX through center O. From the points X, where the diagonal intersects the large circle, draw lines XE parallel to the minor axis, and from the points H, where the diagonal intersects the small circle, draw lines HE parallel to the major axis. The intersections E are points on the ellipse. Two additional points, S and R, can be found by extending lines XE and HE, giving a total of four points from the one diagonal XX.

II. Draw as many additional diagonals as needed to provide a sufficient number of points for a smooth and symmetrical ellipse, each diagonal accounting for four points on the ellipse. Notice that where the curve is sharpest (near the ends of the ellipse),

GRAPHICS SPOTLIGHT

Virtual Reality on the World Wide Web

You can access 3D virtual reality sites on the World Wide Web (WWW) using special add-on software. One of the popular formats for showing virtual reality models on the web is VRML, which stands for Virtual Reality Modeling Language.

THE BEGINNINGS OF THE WEB

When the Internet was first started, its main use was to allow text and data to be stored on various hosts in a consistent format so that the information could be shared effectively. Originally it was difficult to use as you had to keep track of how to get to where the data was stored. A pioneer, Tim Berners-Lee, working at CERN, developed the World Wide Web creating the Universal Resource Locators (URLs) which are used to tell someone where to go and how to access your data on the Web, although you still have to remember the URL.

VRML STANDARD

By 1994 several groups were already working on building 3D visualization tools for the WWW. There was the possibility that there would be no clear standard for how 3D information would be stored and accessed. Tim Berners-Lee and Dave Raggett organized a session to discuss virtual reality interfaces to the World Wide Web and from this session VRML was conceived. The need for virtual reality tools to have a common language to describe 3D scenes and WWW links similar to HTML (hypertext markup language) was a common goal. In 1995 VRML 1.0 was developed. It allows you to move around static scenes, examine objects, and jump to links within the scene. The VRML 2.0 standard allows dynamic object behaviors, multi-user interaction, as well as animation, sound, and video. Using a VRML browser you can interact with VRML web sites in a way that is fun and visual. This is especially important for 3D information, like engineering designs. Software packages like *ANSYS/AutoFEA—Validation* include ways to export your models for VRML browsers so that you can use them on the WWW to interact with other members of your engineering team.

LIVE3D BROWSER FOR NETSCAPE

There are a number of different virtual reality browsers that you can use in order to view and interact with 3D worlds on the WWW. Live3D is a browser provided by Netscape. You may want to look for others such as Black Sun's Cybergate, Vream's WIRL, Superscape's Viscape, among many others. Live3D may not be quite as easy to control movement as some of the others, however it is a good starting point to view virtual reality worlds.

DYNOJET RESEARCH'S SITE USES VRML

One site you may find interesting to visit is Dynojet Research Incorporated's web site. It contains VRML models showing Dynojet Dynamometers which are used to diagnose motorcycle and car performance (see Fig. A). The dynamometer uses computerized technology to simulate road testing in the shop. In the VRML car dynamometer model, you can use Live3D to even walk into the pit and see the equipment under the dyno. In the VRML model shown below, you can walk all around the dyno and even view the calendars on the wall. Bruce Olson, CAD Engineer at Dynojet says, "Notice that the back wheel of the motorcycle is actually turning." It is animated, and as you are walking around the motorcycle dynamometer, you can see that the wheel turns just as it would on the real thing.

Another site that uses VRML that you may find interesting is Worlds, Inc.'s Alpha World, where you can interact with other uses, if you enroll as a citizen. In Alpha World you can have an *avatar*, a name from Hindu mythology meaning an incarnation of a god, as your 3D presence on the WWW. You can recreate yourself in almost any shape you wish. Your avatar can chat with other avatars, and you can even create your own buildings, grass, and other structures. (Hurry, real estate is hard to find!) Alpha World is only one of many such 3D interactive worlds.

INTRANET

Many companies are using VRML on their Intranet, a version of Internet that can only be accessed from inside the company's firewall. This can be a useful way for engineering groups to share and access up to date information, which may be sensitive or not ready for publication on the WWW.

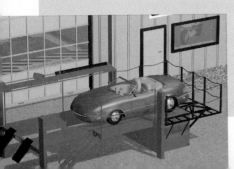

(A) Copyright Dynojet Research.
Rendering by Bruce A. Olson.

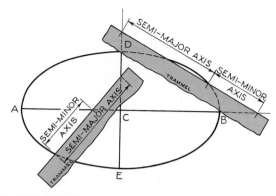

FIGURE 4.49 Drawing a Trammel Ellipse (§4.47).

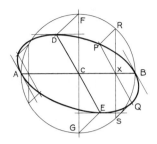

FIGURE 4.51 Oblique-Circle Ellipse (§4.49).

the points are constructed closer together to determine the curve better.

III. Sketch the ellipse lightly through the points, and then heavy in the final ellipse with the aid of an irregular curve.[†]

■ 4.49 DRAWING AN ELLIPSE ON CONJUGATE DIAMETERS: THE OBLIQUE-CIRCLE METHOD

Let AB and DE be the given conjugate diameters (Fig. 4.51). *Two diameters are conjugate when each is parallel to the tangents at the extremities of the other*. With center at C and radius CA, draw a circle; draw the diameter GF perpendicular to AB, and draw lines joining points D and F and points G and E.

Assume that the required ellipse is an oblique projection of the circle just drawn; the points D and E of the ellipse are the oblique projections of the points F and G of the circle, respectively; similarly, the points P

and Q are the oblique projections of the points R and S, respectively. The points P and Q are determined by assuming point X at any point on AB and drawing the lines RS and PQ, and RP and SQ, parallel, respectively, to GF and DE and FD and GE.

Determine at least five points in each quadrant (more for larger ellipses) by assuming additional points on the major axis and proceeding as explained for point X. Sketch the ellipse lightly through the points; then heavy in the final ellipse with the aid of an irregular curve.

■ 4.50 DRAWING A PARALLELOGRAM ELLIPSE

Given the major and minor axes, or the conjugate diameters AB and CD, draw a rectangle or parallelogram with sides parallel to the axes, respectively (Fig. 4.52a and 4.52b). Divide AO and AJ into the same number of equal parts, and draw *light* lines through these points from the ends of the minor axis, as shown. The intersection of like-numbered lines will be points on the ellipse. Locate points in the remaining three quadrants in a similar manner. Sketch the ellipse lightly through the points; then heavy in the final ellipse with the aid of an irregular curve.

[†] In Fig. 4.50, part I, the ordinate EZ of the ellipse is to the corresponding ordinate XZ of the circle as b is to a, where b represents the semiminor axis and a the semimajor axis. Thus, the area of the ellipse is equal to the area of the circumscribed circle multiplied by b/a; hence, it is equal to πab.

FIGURE 4.50 Drawing a Concentric-Circle Ellipse (§4.48).

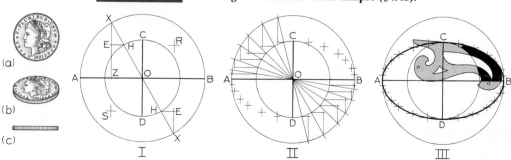

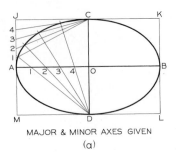

MAJOR & MINOR AXES GIVEN
(a)

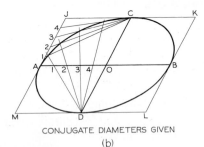

CONJUGATE DIAMETERS GIVEN
(b)

FIGURE 4.52 Parallelogram Ellipse (§4.50).

■ 4.51 FINDING THE AXES OF AN ELLIPSE WITH CONJUGATE DIAMETERS GIVEN

Conjugate diameters AB and CD and the ellipse are given (Fig. 4.53a). With intersection O of the conjugate diameters (center of ellipse) as center, and any convenient radius, draw a circle to intersect the ellipse in four points. Join these points with straight lines, as shown; the resulting quadrilateral will be a rectangle whose sides are parallel, respectively, to the required major and minor axes. Draw the axes EF and GH parallel to the sides of the rectangle.

An ellipse only is given (Fig. 4.53b). To find the center of the ellipse, draw a circumscribing rectangle or parallelogram about the ellipse; then draw diagonals to intersect at center O as shown. The axes are then found as shown in Fig. 4.53a.

Conjugate diameters AB and CD only are given (Fig. 4.53c). With O as center and CD as diameter, draw a circle. Through center O and perpendicular to CD, draw line EF. From points E and F, where this perpendicular intersects the circle, draw lines FA and EA to form angle FAE. Draw the bisector AG of this angle. The major axis JK will be parallel to this bisector, and the minor axis LM will be perpendicular to it. The length AH will be one half the major axis, and HF one half the minor axis. The resulting major and minor axes are JK and LM, respectively.

■ 4.52 DRAWING A TANGENT TO AN ELLIPSE

CONCENTRIC-CIRCLE CONSTRUCTION To draw a tangent at any point on an ellipse, such as E, draw the ordinate at E to intersect the circle at V (Fig. 4.54a). Draw a tangent to the circumscribed circle at V (§4.31), and extend it to intersect the major axis extended at G. The line GE is the required tangent.

To draw a tangent from a point outside the ellipse, such as P, draw the ordinate PY and extend it. Draw DP, intersecting the major axis at X. Draw FX and extend it to intersect the ordinate through P at Q. Then, from similar triangles QY:PY = OF:OD. Draw a tangent to the circle from Q, (§4.31), find the point of tangency R, and draw the ordinate at R to intersect the ellipse at Z. The line ZP is the required tangent. As a check on the drawing, the tangents RQ and ZP should intersect at a point on the major axis extended. Two tangents to the ellipse can be drawn from point P.

FOCI CONSTRUCTION To draw a tangent at any point on the ellipse, such as point 3, draw the focal radii E–3 and F–3, extend one, and bisect the exterior angle, as shown in Fig. 4.54b. The bisector is the required tangent.

To draw a tangent from any point outside the ellipse, such as point P, with center at P and radius PF,

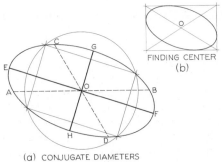

FINDING CENTER
(b)

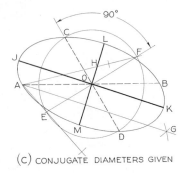

FIGURE 4.53 Finding the Axes of an Ellipse (§4.51).

(a) CONJUGATE DIAMETERS AND ELLIPSE ARE GIVEN

(c) CONJUGATE DIAMETERS GIVEN

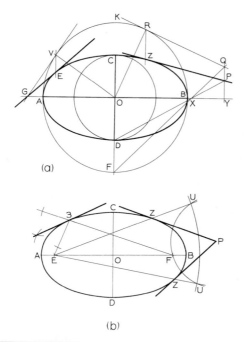

(a)

(b)

FIGURE 4.54 **Tangents to an Ellipse (§4.52).**

strike an arc as shown. With center at E and radius AB, strike an arc to intersect the first arc at points U. Draw the lines EU to intersect the ellipse at the points Z. The lines PZ are the required tangents.

■ 4.53 ELLIPSE TEMPLATES

To save time in drawing ellipses and to ensure uniform results, ellipse templates are often used (Fig. 4.55a). These are plastic sheets with elliptical openings in a wide variety of sizes, and they usually come in sets of six or more sheets.

Ellipse guides are usually designated by the ellipse angle, the angle at which a circle is viewed to appear as an ellipse. In Fig. 4.55b, the angle between the line of sight and the edge view of the plane of the circle is found to be about 49°; hence the 50° ellipse template is indicated. Ellipse templates are generally available in ellipse angles at 5° intervals, such as 15°, 20°, and 25°. The 50° template provides a variety of sizes of 50° ellipses, and it is only necessary to select the one that fits. If the ellipse angle is not easily determined, you can always look for the ellipse that is approximately as long and as "fat" as the ellipse to be drawn.

A simple construction for finding the ellipse angle when the views are not available is shown in Fig. 4.55c. Using center O, strike arc BF; then draw CE parallel to the major axis. Draw diagonal OE, and measure angle EOB with a protractor. Use the ellipse template nearest to this angle; in this case a 35° template is selected.

Since it is not feasible to have ellipse openings for every exact size that may be required, it is often necessary to use the template somewhat in the manner of an irregular curve. For example, if the opening is too long and too "fat" for the required ellipse, one end may be drawn and then the template may be shifted slightly to draw the other end. Similarly, one long side may be drawn and then the template may be shifted slightly to draw the opposite side. In such cases, leave gaps between the four segments, to be filled in freehand or with the aid of an irregular curve. When the differences between the ellipse openings and the required ellipse are small, it is only necessary to lean the pencil slightly outward or inward from the guiding edge to offset the differences.

For inking the ellipses, a technical fountain pen is recommended (Fig. 4.55d).

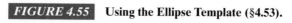

FIGURE 4.55 **Using the Ellipse Template (§4.53).**

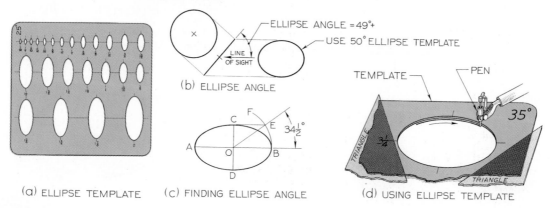

(a) ELLIPSE TEMPLATE   (b) ELLIPSE ANGLE   (c) FINDING ELLIPSE ANGLE   (d) USING ELLIPSE TEMPLATE

■ 4.54 DRAWING AN APPROXIMATE ELLIPSE

For many purposes, particularly where a small ellipse is required, the approximate circular-arc method is perfectly satisfactory (Fig. 4.56). Such an ellipse is sure to be symmetrical and may be quickly drawn.

Given axes AB and CD,

I. Draw line AC. With O as center and OA as radius, strike the arc AE. With C as center and CE as radius, strike the arc EF.

II. Draw perpendicular bisector GH of the line AF; the points K and J, where it intersects the axes, are centers of the required arcs.

III. Find centers M and L by setting off OL = OK and OM = OJ. Using centers K, L, M, and J, draw circular arcs as shown. The points of tangency T are at the junctures of the arcs on the lines joining the centers.

■ 4.55 DRAWING A PARABOLA

The curve of intersection between a right circular cone and a plane parallel to one of its elements, is a parabola (see Fig. 4.46d). The parabola is used to reflect surfaces for light and sound, for vertical curves in highways, for forms of arches, and approximately for forms of the curves of cables for suspension bridges. It is also used to show the bending moment at any point on a uniformly loaded beam or girder.

A *parabola* is generated by a point moving so that its distances from a fixed point, the focus, and from a fixed line, the directrix, remain equal.

Focus F and directrix AB are given. A parabola may be generated by a pencil guided by a string (Fig. 4.57a). Fasten the string at F and C; its length is GC. The point C is selected at random; its distance from G depends on the desired extent of the curve. Keep the

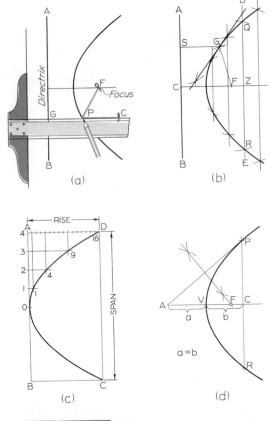

FIGURE 4.57   Drawing a Parabola (§4.55).

string taut and the pencil against the T-square, as shown.

Given focus F and directrix AB, draw a line DE parallel to the directrix and at any distance CZ from it (Fig. 4.57b). With center at F and radius CZ, strike arcs to intersect the line DE in the points Q and R, which are points on the parabola. Determine as many additional points as are necessary to draw the parabola ac-

FIGURE 4.56   Drawing an Approximate Ellipse (§4.54).

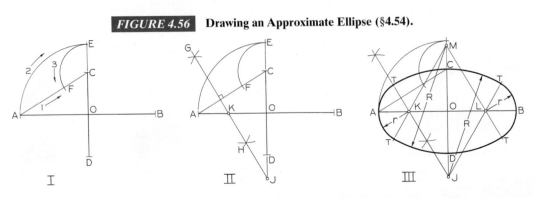

curately, by drawing additional lines parallel to line AB and proceeding in the same manner.

A tangent to the parabola at any point G bisects the angle formed by the focal line FG and the line SG perpendicular to the directrix.

Given the rise and span of the parabola (Fig. 4.57c), divide AO into any number of equal parts, and divide AD into a number of equal parts amounting to the square of that number. From line AB, each point on the parabola is offset by a number of units equal to the square of the number of units from point O. For example, point 3 projects 9 units (the square of 3). This method is generally used for drawing parabolic arches.

To find the focus, F, given points P, R, and V of a parabola (Fig. 4.57d), draw a tangent at P, making a = b. Draw perpendicular bisector of AP, which intersects the axis at F, the focus of the parabola.

Draw a parabola given rectangle or parallelogram ABCD (Figs. 4.58a and 4.58b). Divide BC into any even number of equal parts, divide the sides AB and DC each into half as many parts, and draw lines as shown. The intersections of like-numbered lines are points on the parabola.

■ 4.56 JOINING TWO POINTS BY A PARABOLIC CURVE

Let X and Y be the given points (Fig. 4.59). Assume any point O, and draw tangents XO and YO. Divide

XO and YO into the same number of equal parts, number the division points as shown, and connect corresponding points. These lines are tangents of the required parabola and form its envelope. Sketch a light smooth curve, and then heavy in the curve with the aid of an irregular curve.

These parabolic curves are more pleasing in appearance than circular arcs and are useful in machine design. If the tangents OX and OY are equal, the axis of the parabola will bisect the angle between them.

■ 4.57 DRAWING A HYPERBOLA

The curve of intersection between a right circular cone and a plane making an angle with the axis smaller than that made by the elements is a hyperbola (see Fig. 4.46e). A *hyperbola* is generated by a point moving so that the difference of its distances from two fixed points, the foci is constant and equal to the transverse axis of the hyperbola.

Let F and F′ be the foci and AB the transverse axis (Fig. 4.60a). The curve may be generated by a pencil guided by a string, as shown. Fasten a string at F′ and C; its length is FC minus AB. The point C is chosen at random; its distance from F depends on the desired extent of the curve.

Fasten the straightedge at F. If it is revolved about F, with the pencil point moving against it and with the string taut, the hyperbola may be drawn as shown.

To construct the curve geometrically, select any point X on the transverse axis produced (Fig. 4.60b). With centers at F and F′ and BX as radius, strike the arcs DE. With the same centers, F and F′, and AX as radius, strike arcs to intersect the arcs first drawn in the points Q, R, S, and T, which are points of the required hyperbola. Find as many additional points as are necessary to draw the curves accurately by selecting other points similar to point X along the transverse axis and proceeding as described for point X.

To draw the tangent to a hyperbola at a given point P, bisect the angle between the focal radii FP and F′P. The bisector is the required tangent.

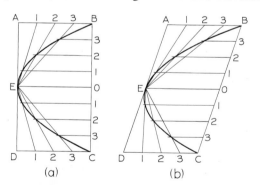

(a)   (b)

FIGURE 4.58   **Drawing a Parabola (§4.55).**

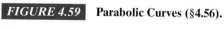

FIGURE 4.59   **Parabolic Curves (§4.56).**

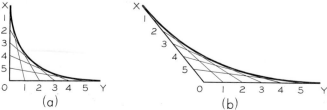
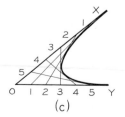

(a)   (b)   (c)

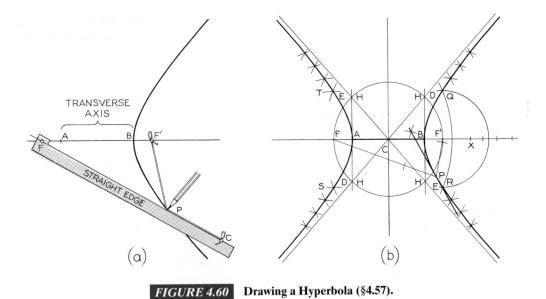

FIGURE 4.60 Drawing a Hyperbola (§4.57).

To draw the asymptotes HCH of the hyperbola, draw a circle with the diameter FF′ and erect perpendiculars to the transverse axis at the points A and B to intersect the circle in the points H. The lines HCH are the required asymptotes.

■ 4.58 DRAWING AN EQUILATERAL HYPERBOLA

Let the asymptotes OB and OA, at right angles to each other, and the point P on the curve be given (Fig. 4.61).

In an equilateral hyperbola, the asymptotes, which are at right angles to each other, may be used as the axes to which the curve is referred. If a chord of the hyperbola is extended to intersect the axes, the intercepts between the curve and the axes are equal (Fig. 4.61a). For example, a chord through given point P intersects the axes at points 1 and 2, intercepts P–1 and 2–3 are equal, and point 3 is a point on the hyperbola. Likewise, another chord through P provides equal intercepts P–1′ and 3′–2′, and point 3′ is a point on the curve. Not all chords need be drawn through given point P, but as new points are established on the curve, chords may be drawn through them to obtain more

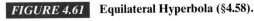

FIGURE 4.61 Equilateral Hyperbola (§4.58).

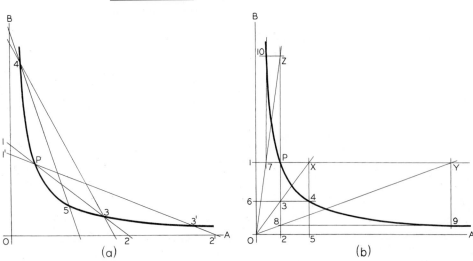

points. After enough points are found to ensure an accurate curve, the hyperbola is drawn with the aid of an irregular curve.

In an equilateral hyperbola, the coordinates are related so their products remain constant. Through given point P, draw lines 1–P–Y and 2–P–Z parallel, respectively, to the axes (Fig. 4.61b). From the origin of coordinates O, draw any diagonal intersecting these two lines at points 3 and X. At these points draw lines parallel to the axes, intersecting at point 4, a point on the curve. Likewise, another diagonal from O intersects the two lines through P at points 8 and Y, and lines through these points parallel to the axes intersect at point 9, another point on the curve. A third diagonal similarly produces point 10 on the curve, and so on. Find as many points as necessary for a smooth curve, and draw the parabola with the aid of an irregular curve. It is evident from the similar triangles O–X–5 and O–3–2 that lines P–1 × P–2 = 4–5 × 4–6.

The equilateral hyperbola can be used to represent varying pressure of a gas as the volume varies,

since the pressure varies inversely with the volume; that is, pressure × volume is constant.

■ 4.59 DRAWING A SPIRAL OF ARCHIMEDES

To find points on the curve, draw lines through the pole C, making equal angles with each other, such as 30° angles (Fig. 4.62). Beginning with any one line, set off any distance, such as 2 mm or $\frac{1}{16}$″; set off twice that distance on the next line, three times on the third, and so on. Through the points thus determined, draw a smooth curve, using irregular curve.

■ 4.60 DRAWING A HELIX

A *helix* is generated by a point moving around and along the surface of a cylinder or cone with a uniform angular velocity about the axis, and with a uniform linear velocity about the axis, and with a uniform velocity in the direction of the axis (Fig. 4.63). A cylindrical helix is generally known simply as a helix. The distance measure parallel to the axis traversed by the point in one revolution is called the lead.

If the cylindrical surface on which a helix is generated is rolled out onto a plane, the helix becomes a straight line (Fig. 4.63a). The portion below the helix becomes a right triangle, the altitude of which is equal to the lead of the helix; the length of the base is equal to the circumference of the cylinder. Such a helix, therefore, can be defined as the shortest line that can be drawn on the surface of a cylinder connecting two points not on the same element.

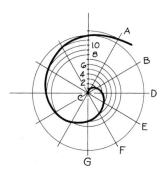

FIGURE 4.62   **Spiral of Archimedes (§4.59).**

FIGURE 4.63   **Helix (§4.60).**

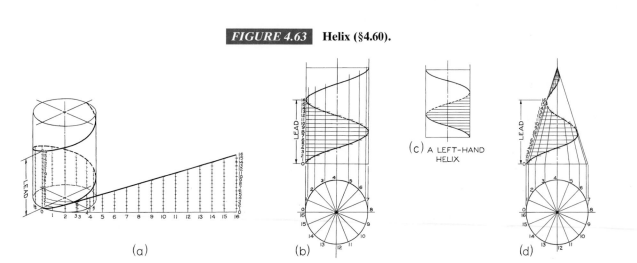

(a)   (b)   (c) A LEFT-HAND HELIX   (d)

To draw the helix, draw two views of the cylinder on which the helix is generated (Fig. 4.63b). Divide the circle of the base into any number of equal parts. On the rectangular view of the cylinder, set off the lead and divide it into the same number of equal parts as the base. Number the divisions as shown (in this case 16). When the generating point has moved one sixteenth of the distance around the cylinder, it will have risen one sixteenth of the lead; when it has moved halfway around the cylinder, it will have risen half the lead; and so on. Points on the helix are found by projecting up from point 1 in the circular view to line 1 in the rectangular view, from point 2 in the circular view to line 2 in the rectangular view, and so on.

Figure 4.63b is a right-hand helix. In a left-hand helix (Fig. 4.63c), the visible portions of the curve are inclined in the opposite direction—that is, downward to the right. The helix shown in Fig. 4.63b can be converted into a left-hand helix by interchanging the visible and hidden lines.

The helix finds many applications in industry, as in screw threads, worm gears, conveyors, spiral stairways, and so on. The stripes of a barber pole are helical in form.

The construction for a right-hand conical helix is shown in Fig. 4.63d.

■ 4.61   DRAWING AN INVOLUTE

An *involute* is the path of a point on a string as the string unwinds from a line, polygon, or circle.

TO DRAW AN INVOLUTE OF A LINE   Let AB be the given line. With AB as radius and B as center, draw the semicircle AC (Fig. 4.64a). With AC as radius and A as center, draw the semicircle CD. With BD as radius and B as center, draw the semicircle DE. Continue similarly, alternating centers between A and B, until a figure of the required size is completed.

TO DRAW AN INVOLUTE OF A TRIANGLE   Let ABC be the given triangle. With CA as radius and C as center, strike the arc AD (Fig. 4.64b). With BD as radius and B as center, strike the arc DE. With AE as radius and A as center, strike the arc EF. Continue similarly until a figure of the required size is completed.

TO DRAW AN INVOLUTE OF A SQUARE   Let ABCD be the given square. With DA as radius and D as center, draw the 90° arc AE (Fig. 4.64c). Proceed as for the involute of a triangle until a figure of the required size is completed.

TO DRAW AN INVOLUTE OF A CIRCLE   A circle may be regarded as a polygon with an infinite number of sides (Fig. 4.64d). The involute is constructed by dividing the circumference into a number of equal parts, drawing a tangent at each division point, setting off along each tangent the length of the corresponding circular arc (see Fig. 4.45c), and drawing the required curve through the points set off on the several tangents.

An involute can be generated by a point on a straight line that is rolled on a fixed circle (Fig. 4.64e). Points on the required curve may be determined by setting off equal distances 0–1, 1–2, 2–3, and so on, along the circumference, drawing a tangent at each division point, and proceeding as explained for Fig. 4.64d.

The involute of a circle is used in the construction of involute gear teeth. In this system, the involute forms the face and a part of the flank of the teeth of gear wheels; the outlines of the teeth of racks are straight lines.

■ 4.62   DRAWING A CYCLOID

A *cycloid* generated by a point P in the circumference of a circle that rolls along a straight line (Fig. 4.65).

FIGURE 4.64   Involutes (§4.61).

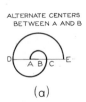

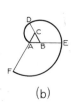

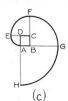

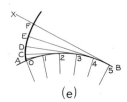

(a)   (b)   (c)   (d)   (e)

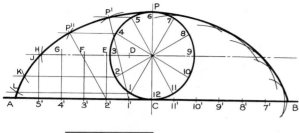

FIGURE 4.65 Cycloid (§4.62).

Given the generating circle and the straight line AB tangent to it, make the distances CA and CB each equal to the semicircumference of the circle (see Fig. 4.45e). Divide these distances and the semicircumference into the same number of equal parts (six, for instance) and number them consecutively, as shown. Suppose the circle rolls to the left; when point 1 of the circle reaches point 1' of the line, the center of the circle will be at D, point 7 will be the highest point of the circle, and the generating point 6 will be at the same distance from the line AB as point 5 is when the circle is in its central position. Hence, to find the point P', draw a line through point 5 parallel to AB and intersect it with an arc drawn from the center D with a radius equal to that of the circle. To find point P", draw a line through point 4 parallel to AB, and intersect it with an arc drawn from the center E, with a radius equal to that of the circle. Points J, K, and L are found in a similar manner.

Another method that may be employed is shown in the right half of Fig. 4.64. With center at 11' and the chord 11–6 as radius, strike an arc. With 10' as center and the chord 10–6 as radius, strike an arc. Continue similarly with centers 9', 8', and 7'. Draw the required cycloid tangent to these arcs.

Either method may be used; however, the second is the shorter one and is preferred. It is evident, from the tangent arcs drawn in the manner just described, that the line joining the generating point and the point of contact for the generating circle is a normal of the cycloid. The lines 1'–P" and 2'–P', for instance, are normals; this property makes the cycloid suitable for the outlines of gear teeth.

■ **4.63 DRAWING AN EPICYCLOID OR A HYPOCYCLOID**

If the generating point P is on the circumference of a circle that rolls along the convex side of a larger circle, the curve generated is an epicycloid (Fig. 4.66a). If the circle rolls along the concave side of a larger circle, the curve generated is a hypocycloid (Fig. 4.66b). These curves are drawn in a manner similar to the cycloid (Fig. 4.65). Like the cycloid, these curves are used to form the outlines of certain gear teeth and are, therefore, of practical importance in machine design.

■ **4.64 COMPUTER GRAPHICS**

Through the use of various application programs and routines available in computer graphics, it is possible to establish accurately the various geometric constructions shown in this chapter. CAD programs are particularly well suited for repetitive operations, such as dividing a line into a number of equal parts, and for generating lines representing mathematical curves, such as the hyperbola and parabola. Examples of CAD-produced geometric shapes and surfaces are shown in Fig. 4.67.

FIGURE 4.66 Epicycloid and Hypocycloid (§4.63).

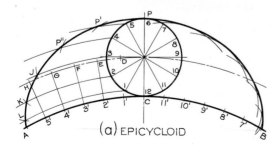

(a) EPICYCLOID

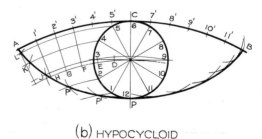

(b) HYPOCYCLOID

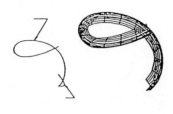
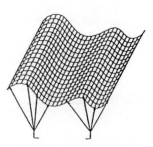

SURFACE OF REVOLUTION   SWEPT SURFACE   MESH SURFACE

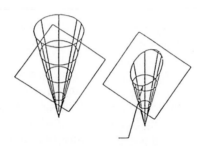
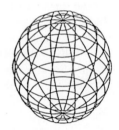

RULED SURFACE   PLANE/SURFACE INTERSECTION   SPHERE

FIGURE 4.67   **Geometric Shapes and Surfaces Produced with TRI-CAD System.** *Courtesy of Lodgrafix, Inc.* **(§4.64).**

■ **KEY WORDS**

POINT	LINE
ARC	CIRCLE
TANGENT	TANGENT POINT
PARALLEL	PERPENDICULAR
POLYGON	REGULAR POLYGON
BISECT	CONIC SECTION
CONSTRUCTION LINES	CENTER
RADIUS	DIAMETER
CONCENTRIC	RIGHT ANGLE

■ CHAPTER SUMMARY

- Understanding basic geometric construction techniques is fundamental to the success of both traditional drawing and CAD drawing.

- All drawings are made up of points, lines, arcs, and circles drawn at various sizes and constructed in specific location to each other. Geometric construction cannot be approximated.

- Every point, line, arc and circle must be precisely located. CAD cannot draw correctly if you cannot tell the program where you want an element drawn.

- The advantage that CAD provides in geometric construction is drawing precision. Only the drafter knows where a point, line, arc, or circle needs to be drawn, and more importantly, *how* to draw it. A poor paper drafter will most likely be a poor CAD drafter.

- One of the more difficult types of geometric construction are tangencies. There are many different types of tangencies and each requires a special drawing technique.

- Knowing when to draw a line parallel or perpendicular to another line is an important part of geometric construction.

- Unless the drafter knows the basics of geometric construction, it will be difficult to create a drawing using either instruments or CAD.

■ REVIEW QUESTIONS

1. What are the four basic conic sections and how are they cut from a cone?

2. How can you divide a line into equal parts? A space into equal parts?

3. How many ways can an arc be tangent to one line? To two lines? To a line and an arc? To two arcs? Draw an example of each.

4. Draw an equilateral triangle, a right triangle, and an isosceles triangle.

5. Given a line, draw another line (1) parallel to the first, (2) perpendicular to the first. Then draw a horizontal line through the intersection of the lines. Finally, draw a vertical line through the intersection.

6. Draw a regular square, hexagon, and octagon so that the distance from flat to flat is 2 inches.

7. List the different methods for drawing an approximation of an ellipse. Pick one and draw an ellipse with a major diameter of 6 inches and a minor diameter of 3 inches.

8. Demonstrate the technique for drawing a line parallel to a given line using only a pencil, triangle, and straight edge.

9. Demonstrate the technique for drawing a line perpendicular to a given line using only a pencil, triangle, and straight edge.

■ GEOMETRIC CONSTRUCTION PROBLEMS

Geometric constructions should be done very accurately, using a hard pencil (2H to 4H) with a long, sharp, conical point. Draw given and required lines dark and medium in thickness, and draw construction lines *very light*. Do not erase construction lines. Indicate points and lines as described in §4.1.

In your assignments from the following problems, use Layout A–2 (see inside of front cover) divided into four parts, as shown in Fig. 4.68, or Layout A4–2 (adjusted). Additional sheets with other problems selected from Figs. 4.69–4.80 and drawn on the same sheet layout may be assigned.

Many problems are dimensioned in the metric system. Your instructor may ask you to convert these remaining problems to metric measure. (See the inside front cover for decimal and millimeter equivalents.)

Set up each problem so as to make the best use of the space available, to present the problem to best advantage, and to produce a pleasing appearance. Letter the principal points of all constructions in a manner similar to the various illustrations in this chapter.

Since many of the problems in this chapter are of a general nature, they can also be solved on most computer graphics systems. If a system is available, your instructor may choose to assign specific problems to be completed by this method.

The first four problems are shown in Fig. 4.68.

Prob. 4.1   Draw an inclined line AB 65 mm long and bisect it (see Fig. 4.8).

Prob. 4.2   Draw any angle with vertex at C. Bisect it (see Fig. 4.10), and transfer one half to a new position at D (see Fig. 4.11).

Prob. 4.3   Draw in inclined line EF and assume distance GH = 42 mm. Draw a line parallel to EF and at the distance GH from it (see Fig. 4.13a).

Prob. 4.4   Draw the line JK 95 mm long and divide it into five equal parts using dividers. Draw a line LM 58 mm long and divide it into three equal parts by the parallel-line method (see Fig. 4.15).

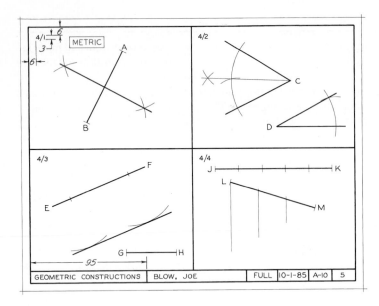

FIGURE 4.68   (Probs. 4.1–4.4) Geometric Constructions. Layout A–2 or A4–2 (adjusted).

Prob. 4.5   Draw a line OP 92 mm long and divide it into three proportional parts to in the ratio of 3:5:9 (see Fig. 4.17a).

Prob. 4.6   Draw a line 87 mm long and divide it into parts proportional to the square of x, where $x = 1, 2, 3,$ and 4 (see Fig. 4.17c).

Prob. 4.7   Draw a triangle with sides of 76 mm, 85 mm, and 65 mm (see Fig. 4.19). Bisect the three interior angles (see Fig. 4.10). The bisectors should meet at a point. Draw the inscribed circle with the point as center.

Prob. 4.8   Draw a right triangle having a hypotenuse 65 mm and one leg 40 mm (see Figs. 4.3 and 4.20), and draw a circle through the three vertexes (see Fig. 4.32).

Prob. 4.9   Draw an inclined line QR 84 mm long. Select a point P on the line 32 mm form Q, and erect a perpendicular (see Fig. 4.18c). Assume a point S 45.5 mm from the line, and erect a perpendicular from S to the line (see Fig. 4.18b).

Prob. 4.10   Draw two lines making an angle of $35\frac{1}{2}°$ with each other using the tangent method (see Fig. 4.21a). Check with a protractor.

Prob. 4.11   Draw two lines making an angle of 33° 16′ with each other using the sine method (see Fig. 4.21b). Check with a protractor.

Prob. 4.12   Draw an equilateral triangle having 63.5-mm sides (see Fig. 4.22a). Bisect the interior angles (see Fig. 4.10). Draw the inscribed circle using the intersection of the bisectors as center.

Prob. 4.13   Draw inclined line TU 55 mm long, and then draw a square on TU as a given side (see Fig. 4.23a).

Prob. 4.14   Draw a 54-mm diameter-circle (lightly); inscribe a square in the circle and circumscribe a square on the circle (see Figs. 4.23c and 4.23d).

Prob. 4.15   Draw a 65-mm-diameter circle (lightly). Find the vertexes of a regular inscribed pentagon (see Fig. 4.24a), and join the vertexes to form a five-pointed star.

Prob. 4.16   Draw a 65-mm-diameter circle (lightly). Inscribe a hexagon (see Fig. 4.25b), and circumscribe a hexagon (see Fig. 4.26d).

Prob. 4.17   Draw a square (lightly) with 63.5-mm sides (see Fig. 4.23b), and inscribe an octagon (see Fig. 4.28b).

Prob. 4.18   Draw a triangle similar to that in Fig. 4.29a, with sides 50 mm, 38 mm, and 73 mm long. Transfer the triangle to a new location and turned 180° similar to that in Fig. 4.29b. Check by the pricked-point method (§4.29).

Prob. 4.19   In the center of your space, draw a rectangle 88 mm wide and 61 mm high. Show construction for reducing this rectangle first to 70 mm wide and then to 58 mm wide (see Fig. 4.31b).

Prob. 4.20   Draw three points arranged approximately as those in Fig. 4.32a), and draw a circle through the three points.

Prob. 4.21   Draw a 58-mm-diameter circle. Assume a point S on the left side of the circle and draw a tangent at that point (see Fig. 4.34a). Assume a point T to the right of the circle 50 mm from its center. Draw two tangents to the circle through the point (see Fig. 4.34b).

Prob. 4.22   Draw a horizontal center line through your space. Then draw two circles with 50-mm-diameter and 38-mm-diameter, respectively, with centers 54 mm apart. Locate the circles so that the construction will be centered in the space. Draw "open-belt" tangents to the circles (see Fig. 4.35a).

Prob. 4.23 Do the same as for Prob. 4.21 except draw "crossed-belt" tangents to the circle (see Fig. 4.35b).

Prob. 4.24 Draw a vertical line VW 33 mm from the left side of your space. Assume point P 44 mm farther to the right and 25 mm down from top of space. Draw a 56-mm-diameter circle through P, tangent to VW (see Fig. 4.36a).

Prob. 4.25 Draw a vertical line XY 35 mm from the left side of your space. Assume point P 44 mm farther to the right and 25 mm down from the top of the space. Assume point Q on line XY and 50 mm from P. Draw a circle through P and tangent to XY at Q (see Fig. 4.36b).

Prob. 4.26 Draw a 64-mm-diameter circle with center C 16 mm directly to left of center of space. Assume point P at the lower right and 60 mm from C. Draw an arc with 25-mm-radius through P and tangent to the circle (see Fig. 4.36c).

Prob. 4.27 Draw a vertical line and a horizontal line, each 65 mm long (see Fig. 4.37, part I). Draw an arc with a 38 mm radius, tangent to the lines.

Prob. 4.28 Draw a horizontal line 20 mm up from the bottom of your space. Select a point on the line 50 mm from the left side of space, and through it draw a line upward to the right at 60° to horizontal. Draw arcs with a 35 mm radius within an obtuse angle and acute angle, respectively, tangent to the two lines (see Fig. 4.38).

Prob. 4.29 Draw two intersecting lines making an angle of 60° with each other similar to Fig. 4.38a. Assume a point P on one line at a distance of 45 mm from the intersection. Draw an arc tangent to both lines with one point of tangency at P (see Fig. 4.33).

Prob. 4.30 Draw vertical line AB 32 mm from the left side of your space. Draw an arc of 42 mm radius with its center 75 mm to the right of the line and in the lower right portion of your space. Draw an arc of 25 mm radius tangent to AB and to the first arc (see Fig. 4.39).

Prob. 4.31 With centers 20 mm up from the bottom of your space and 86 mm apart, draw arcs of radii 44 mm and 24 mm, respectively. Draw an arc of 32 mm radius tangent to the two arcs (see Fig. 4.40).

Prob. 4.32 Draw two circles as in Prob. 4.22. Draw an arc of 70 mm radius tangent to the upper sides of, and enclosing, the circles (see Fig. 4.41a). Draw an arc of 50 mm radius tangent to the circles but enclosing only the smaller circle (see Fig. 4.41b).

Prob. 4.33 Draw two parallel inclined lines 45 mm apart. Choose a point on each line and connect them with an ogee curve tangent to the two parallel lines (see Fig. 4.43a).

Prob. 4.34 Draw an arc of 54 mm radius that subtends an angle of 90°. Find the length of the arc by two methods (see Figs. 4.45a and 4.45c). Calculate the length of the arc and compare with the lengths determined graphically. (See first-note in §4.42).

Prob. 4.35 Draw a major axis 102 mm long (horizontally) and a minor axis 64 mm long, with their intersection at the enter of the space. Draw an ellipse by the foci method with at least five points in each quadrant (see Fig. 4.48).

Prob. 4.36 Draw axes as in Prob. 4.35, and draw an ellipse by the trammel method (see Fig. 4.49).

Prob. 4.37 Draw axes as in Prob. 4.35, and draws an ellipse by the concentric-circle method (see Fig. 4.50).

Prob. 4.38 Draw axes as in Prob. 4.35, and draw an ellipse by the parallelogram method (see Fig. 4.52a).

Prob. 4.39 Draw conjugate diameters intersecting at the center of your space. Draw 88-mm-diameter horizontally, and 70-mm-diameter at 60° with horizontal. Draw an oblique-circle ellipse (see Fig. 4.51). Find at least five points in each quadrant.

Prob. 4.40 Draw conjugate diameters as in Prob. 4.39, and draw the ellipse by the parallelogram method (see Fig. 4.52b).

Prob. 4.41 Draw axes as in Prob. 4.35, and draw an approximate ellipse (see Fig. 4.56).

Prob. 4.42 Draw a parabola with a vertical axis and the focus 12 mm from the directrix (see Fig. 4.57b). Find at least nine points on the curve.

Prob. 4.43 Draw a hyperbola with a horizontal transverse axis 25 mm long and the foci 38 mm apart (see Fig. 4.60b). Draw the asymptotes.

Prob. 4.44 Draw a horizontal line near the bottom of the space, and a vertical line near the left side of the space. Assume point P 16 mm to the right of the vertical line and 38 mm above the horizontal line. Draw an equilateral hyperbola through P and with reference to the two lines as asymptotes. Use either method shown in Fig. 4.61.

Prob. 4.45 Using the center of the space as the pole, draw a spiral of Archimedes with the generating point moving in a counterclockwise direction and away from the pole at the rate of 25 mm in each convolution (see Fig. 4.62).

Prob. 4.46 Through center of your space, draw a horizontal center line, and on it construct a right-hand helix 50 mm diameter, 64 mm long, and with a lead of 25 mm (see Fig. 4.63). Draw only a half-circular end view.

Prob. 4.47 Draw the involute of an equilateral triangle with 15-mm sides (see Fig. 4.64b).

Prob. 4.48 Draw the involute of a 20-mm-diameter circle (see Fig. 4.64d).

Prob. 4.49 Draw a cycloid generated by a 30-mm-diameter circle rolling along a horizontal straight line (see Fig. 4.65).

Prob. 4.50 Draw and epicycloid generated by a 38-mm-diameter circle rolling along a circular arc with a radius of 64 mm (see Fig. 4.66a).

Prob. 4.51 Draw a hypocycloid generated by a 38-mm-diameter circle rolling along a circular arc with a radius of 64 mm (see Fig. 4.66b).

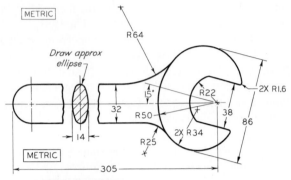

FIGURE 4.69 **(Prob. 4.52) Spanner.**

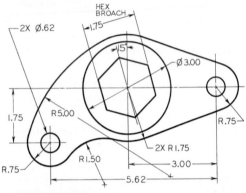

FIGURE 4.70 **(Prob. 4.53) Rocker Arm.**

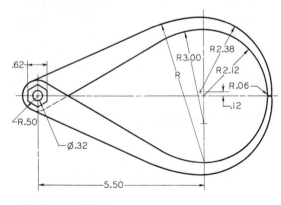

FIGURE 4.71 **(Prob.4.54) Outside Caliper.**

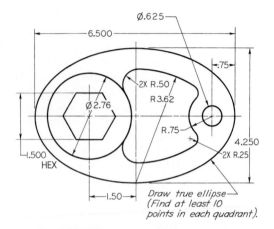

FIGURE 4.72 **(Prob. 4.55) Special Cam.**

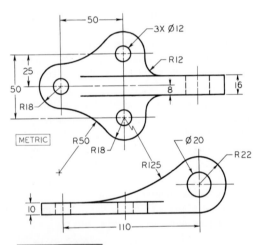

FIGURE 4.73 **(Prob. 4.56) Boiler Stay.**

Prob. 4.52 Using Layout A–2 or A4–2 (adjusted), draw the spanner in Fig. 4.69. Omit dimensions and notes unless assigned.

Prob. 4.53 Using Layout A–2 or A4–2 (adjusted), draw the rocker arm in Fig. 4.70. Omit dimensions and notes unless assigned.

Prob. 4.54 Using Layout A–2 or A4–2 (adjusted), draw the outside caliper in Fig. 4.71. Omit dimensions and notes unless assigned.

Prob. 4.55 Using Layout A–2 or A4–2 (adjusted), draw the special cam in Fig. 4.72. Omit dimensions and notes unless assigned.

Prob. 4.56 Using Layout A–2 or A4–2 (adjusted), draw the boiler stay in Fig. 4.73. Omit dimensions and notes unless assigned.

Prob. 4.57 Using Layout A–2 or A4–2 (adjusted), draw the shaft hanger casting in Fig. 4.74. Omit dimensions and notes unless assigned.

Prob. 4.58 Using Layout A–2 or A4–2 (adjusted), draw the shift lever in Fig. 4.75. Omit dimensions and notes unless assigned.

Prob. 4.59 Using Layout A–2 or A4–2 (adjusted), draw the gear arm in Fig. 4.76. Omit dimensions and notes unless assigned.

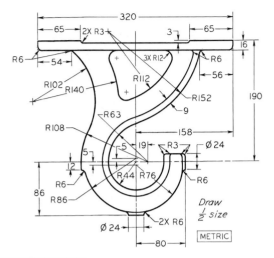

FIGURE 4.74 (Prob. 4.57) Shaft Hanger Casting.

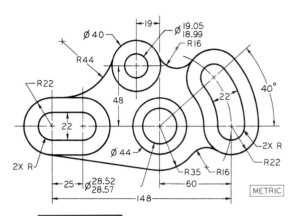

FIGURE 4.76 (Prob. 4.59) Gear Arm.

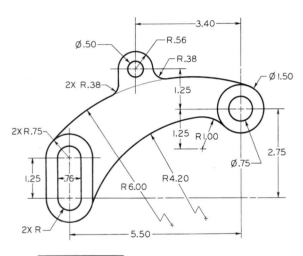

FIGURE 4.75 (Prob. 4.58) Shift Lever.

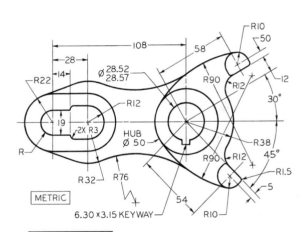

FIGURE 4.77 (Prob. 4.60) Form Roll Lever.

Prob. 4.60 Using Layout A–2 or A4–2 (adjusted), draw the form roll lever in Fig. 4.77. Omit dimensions and notes unless assigned.

Prob. 4.61 Using Layout A–2 or A4–2 (adjusted), draw the press base in Fig. 4.78. Omit dimensions and notes unless assigned.

Prob. 4.62 Using Layout A–2 or A4–2 (adjusted), draw the special S-wrench in Fig. 4.79. Omit dimensions and notes unless assigned.

Prob. 4.63 Using Layout A–2 or A4–2 (adjusted), draw the photo floodlight reflector in Fig. 4.80. Omit dimensions and notes unless assigned.

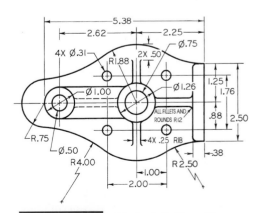

FIGURE 4.78 (Prob. 4.61) Press Base.

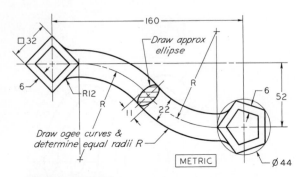

FIGURE 4.79   (Prob. 4.62) Special S-Wrench.

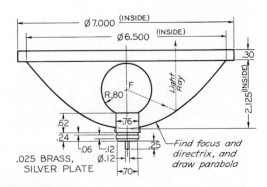

FIGURE 4.80   (Prob. 4.63) Photo Floodlight Reflector.

C H A P T E R   5

TECHNICAL SKETCHING AND SHAPE DESCRIPTION

OBJECTIVES

After studying the material in this chapter, you should be able to:

1. Read any type of sketch and understand the object depicted in the sketch.

2. Create sketches of both two- and three-dimensional objects using the correct sketching methods.

3. Demonstrate how to sketch parallel, perpendicular, and evenly spaced lines.

4. Demonstrate how to sketch a circle and arc of a given diameter.

5. Demonstrate how to estimate and compare dimensions of an object using a pencil at arms length.

6. Sketch an object to scale using grid drawing media.

7. Prepare a sketch of an object to be drawn using CAD showing all tangents, centers, and sizes.

8. Sketch an object on isometric grid paper.

9. Sketch the primary views of an orthographic drawing in proper orientation and alignment.

10. Show examples of the correct use of the alphabet of sketch lines.

11. Describe the difference between a sketch and a drawing.

OVERVIEW

Proper sketching technique is one of the most important skills that a traditional or CAD drafter must master. Sketching is a formal process of visualizing three-dimensional objects in preparation for drawing on paper or with CAD. Sketching is a quick way to communicate ideas with other members of the design team. (The old Chinese saying that "one picture is worth a thousand words" is not without foundation.) Sketching is an excellent way of planning out the drawing process necessary to effectively create a complex object. A properly drawn sketch will act like a road map for the completion of the final paper or CAD drawing. Drafters who sketch solutions before placing lines on paper or the CAD screen can often complete a drawing sooner and with fewer errors than those who cannot sketch effectively. Sketching is an excellent way of learning how to represent three-dimensional objects on a two-dimensional surface.

■ 5.1 TECHNICAL SKETCHING

Freehand sketches are of great value to designers in organizing their thoughts and recording their ideas. Sketching is an effective and economical means of formulating various solutions to a given problem so that a choice can be made between them. Time can be lost if the designer starts to do a scaled layout before adequate preliminary study with the aid of sketches. Information about changes in design or covering replacement of broken parts or lost drawings is usually conveyed through sketches.

The degree of perfection required in a given sketch depends on its intended use. Quick sketches done to supplement oral descriptions may be rough and incomplete. On the other hand, sketches that are supposed to convey important and precise information to engineers, technicians, or skilled workers should be executed as carefully and exactly as possible.

The term "freehand sketch" does not mean a crude or sloppy freehand drawing in which no particular effort has been made. On the contrary, as shown in Fig. 5.1, a freehand sketch should be made with care and with attention to proportion, clarity, and correct line widths.

■ 5.2 SKETCHING MATERIALS

One advantage of freehand sketching is that it requires only pencil, paper, and eraser. When sketches are made in the field and an accurate record is required, small notebooks or sketching pads are frequently used. Graph paper can be helpful, especially to someone who cannot sketch reasonably well without guide lines. Paper with 4, 5, 8, or 10 squares per inch is recommended. Such paper is convenient for maintaining correct proportions (Fig. 5.2).

Sketching pads of plain tracing paper may be accompanied by a master cross-section sheet, which shows clearly through a transparent sheet placed on top of it. (A master cross-section sheet can also be drawn with instruments.) A specially ruled isometric paper is available for isometric sketching (see Fig. 5.26).

Soft pencils, such as HB or F, should be used for freehand sketching. Use a mechanical pencil, and sharpen it to a conical point. Use this sharp point for center lines, dimension lines, and extension lines. For visible lines, hidden lines, and cutting-plane lines, round off the point slightly to produce the desired thickness of line (Fig. 5.3). For carefully made sketches, two soft erasers are recommended—Pink Pearl and a Mars-Plastic.

■ 5.3 TYPES OF SKETCHES

Since technical sketches are made of three-dimensional objects, the form of the sketch conforms approximately to one of the four standard types of projection. In multiview projection (Fig. 5.4a), the object is described by its necessary views (§§5.14–5.16). As shown in Figs. 5.4b through 5.4d, the object may also be shown pictorially in a single view, by *axonometric* (*isometric*), *oblique*, or *perspective sketches*, respectively (see §§5.9–5.13).

■ 5.4 SCALE

Sketches usually are *not made to any scale*. Objects should be sketched in their correct proportions as accurately as possible, by eye. However, cross-section paper provides a ready scale (by counting squares) that may be used to assist in sketching to correct proportions. The size of the sketch is purely optional, depending on the complexity of the object and the size of paper available. Small objects are often sketched oversize to show the necessary details clearly.

■ 5.5 TECHNIQUE OF LINES

The chief difference between an instrument drawing and a freehand sketch lies in the character or *tech-*

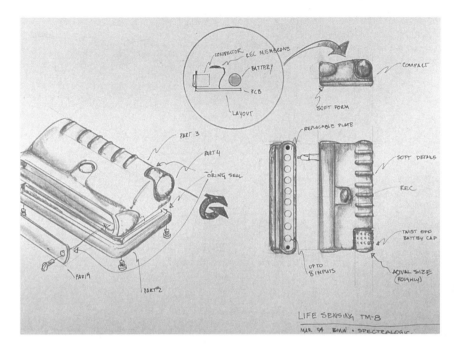

FIGURE 5.1 **Great Ideas Often Start as Freehand Sketches Made on Scratch Paper.** *Courtesy of ANATech, Inc.*

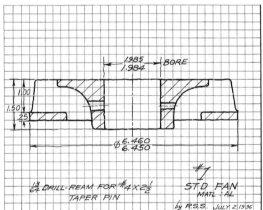

FIGURE 5.2 **Sketch on Graph Paper.**

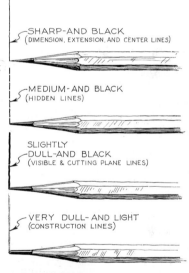

FIGURE 5.3 **Pencil Points.**

nique of the lines. A good freehand line is not expected to be as rigidly straight or exactly uniform as an instrument line. While the effectiveness of an instrument line lies in exacting uniformity, the quality of a freehand line lies in its *freedom* and *variety* (Figs. 5.5 and 5.7).

Conventional lines drawn instrumentally and freehand are shown in Fig. 5.6. The freehand construction line is a very light, rough line in which some strokes may overlap. All other lines should be dark and clean-cut. Accent the ends of all dashes, and maintain a sharp contrast between the line thicknesses. In particular, make visible lines **heavy** so the outline will stand out clearly, and make hidden lines, center lines, dimension lines, and extension lines *thin*.

■ 5.6 SKETCHING STRAIGHT LINES

Since most lines on an average sketch are straight lines, you should learn to make them correctly. Hold your pencil naturally, about $1\frac{1}{2}''$ back from the point, and approximately at right angles to the line to be drawn. Draw horizontal lines from left to right with a

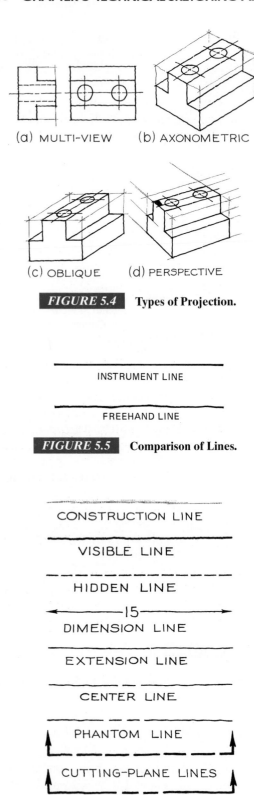

(a) MULTI-VIEW   (b) AXONOMETRIC

(c) OBLIQUE   (d) PERSPECTIVE

FIGURE 5.4   Types of Projection.

INSTRUMENT LINE

FREEHAND LINE

FIGURE 5.5   Comparison of Lines.

CONSTRUCTION LINE

VISIBLE LINE

HIDDEN LINE

←―――15―――→
DIMENSION LINE

EXTENSION LINE

CENTER LINE

PHANTOM LINE

CUTTING-PLANE LINES

FIGURE 5.6   Sketch Lines.

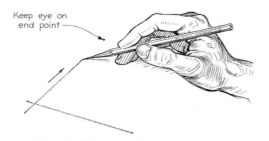

Keep eye on
end point

(a)   POOR—SHOWS TIGHT GRIP
ON PENCIL—DOES NOT CONTINUE ALONG
STRAIGHT PATH—IS AN ATTEMPT TO
IMITATE MECHANICAL LINES.

(b)   BETTER—SHOWS FREE HANDLING OF
PENCIL—CONTINUES ALONG STRAIGHT PATH-
SLIGHT WIGGLES DO NOT DETRACT.

(c)   BEST—HAS EFFECTIVENESS OF (b), PLUS
SNAP ADDED BY OCCASIONAL GAPS—
EASIER TO DRAW STRAIGHT.

FIGURE 5.7   Drawing Horizontal Lines.

Keep eye on
end point

FIGURE 5.8   Drawing Vertical Lines.

free and easy wrist-and-arm movement (Fig. 5.7). Draw vertical lines downward with finger and wrist movements (Fig. 5.8).

Inclined lines may be made to conform in direction to horizontal or vertical lines by shifting position with respect to the paper or by turning the paper slightly; hence, they may be drawn with the same general movements (Fig. 5.9).

In sketching long lines, mark the ends of the line with light dots. Then move the pencil back and forth between the dots in long sweeps, always keeping your eye on the dot toward which the pencil is moving. The point of the pencil should touch the paper lightly, with each successive stroke correcting the defects of the preceding strokes. When the path of the line has been established sufficiently, apply a little more pressure, replacing the trial series with a distinct line. Then dim the line with a soft eraser and draw the final line clean-

FIGURE 5.9   **Drawing Inclined Lines.**

cut and dark, now keeping your eye on the point of the pencil.

An easy method for blocking in horizontal or vertical lines is to hold your hand and pencil rigidly and glide your fingertips along the edge of the pad or board, as shown in Fig. 5.10a. Another method is to mark the distance on the edge of a card or a strip of paper and transfer this distance at intervals (Fig. 5.10b); then draw the final line through these points. A third method is to hold your pencil as shown in the lower part of Fig. 5.10b and to make distance marks on the paper at intervals by tilting the lead down to the paper. All of these methods of transferring distances are substitutes for dividers and have many uses in sketching.

To find the midpoint of a line AB (Fig. 5.10c), hold the pencil in your left hand with your thumb gauging the estimated half-distance. Try this distance on the left and then on the right until you locate the center by trial and mark it C. Another method is to mark the total distance AB on the edge of a strip of paper and then to fold the paper to bring points A and B together, thus locating center C at the crease. To find quarter points, the folded strip can be folded once more.

■ 5.7 SKETCHING CIRCLES AND ARCS

Small circles and arcs can be easily sketched in one or two strokes without any preliminary blocking in.

One method of sketching a larger circle (Fig. 5.11) is to first sketch lightly the enclosing square, mark the midpoints of the sides, draw light arcs tangent to the sides of the square, and then heavy in the final circle. Another method (Fig. 5.12) is to sketch the two center lines, add light 45° radial lines, sketch light arcs across the lines at the estimated radius distance from the center, and finally sketch the required circle heavily. Dim all construction lines with a soft eraser before heavying in the final circle.

An excellent method, particularly for large circles (Fig. 5.13a), is to mark the estimated radius on the edge of a card or scrap of paper, to set off from the center as many points as desired, and to sketch the final heavy circle through these points.

Clever drafters will prefer the method shown in Fig. 5.13b and 5.13c, in which the hand is used as a compass. You place the tip of your little finger or the knuckle joint of your little finger at the center, "feed" the pencil out to the desired radius, hold this position rigidly, and carefully revolve the paper with the other hand, as shown. If you are using a sketching pad, place the pad on your knee and revolve the entire pad.

In Fig. 5.13d, two pencils are held rigidly like a compass and the paper is slowly revolved.

Methods of sketching arcs (Fig. 5.14) are adaptations of those used for sketching circles. In general, it is easier to sketch arcs with your hand and pencil on the concave side of the curve. In sketching tangent arcs,

FIGURE 5.10   **Blocking in Horizontal and Vertical Lines.**

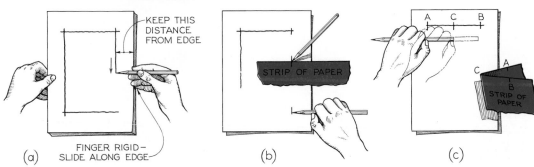

(a)   FINGER RIGID—SLIDE ALONG EDGE   KEEP THIS DISTANCE FROM EDGE   (b)   STRIP OF PAPER   (c)

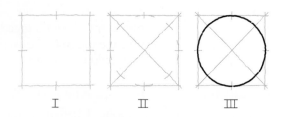

FIGURE 5.11 **Sketching a Circle.**

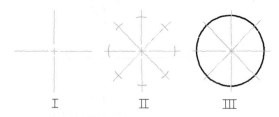

FIGURE 5.12 **Sketching a Circle.**

always keep in mind the actual geometric constructions, carefully approximating all points of tangency.

■ 5.8 SKETCHING ELLIPSES

If a circle is viewed obliquely, it appears as an ellipse (see Fig. 5.50b). With a little practice, you can learn to sketch small ellipses with a free arm movement (Fig. 5.15a). Hold the pencil naturally, rest the weight on the upper part of the forearm, and move the pencil rapidly above the paper in the elliptical path desired; then lower the pencil to describe several light overlapping ellipses, as shown in Fig. 5.15a, part I. Lighten all lines with a soft eraser and heavy in the final ellipse (Fig. 5.15a, part II).

Another method (Fig. 5.15b) is to sketch lightly the enclosing rectangle (Fig. 5.15b, part I), mark the midpoints of the sides, and sketch light tangent arcs, as shown. Then complete the ellipse lightly (Fig. 5.15b,

FIGURE 5.13 **Sketching Circles.**

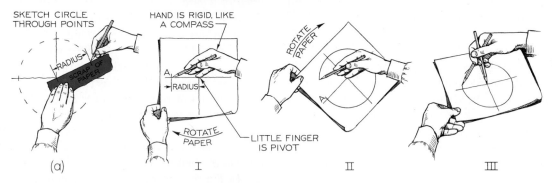

FIGURE 5.14 **Sketching Arcs.**

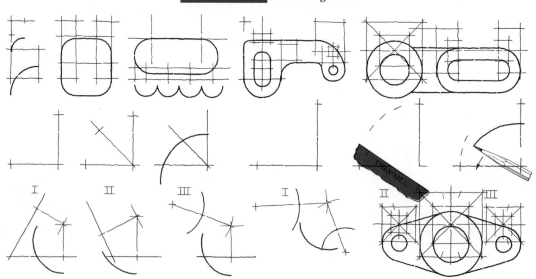

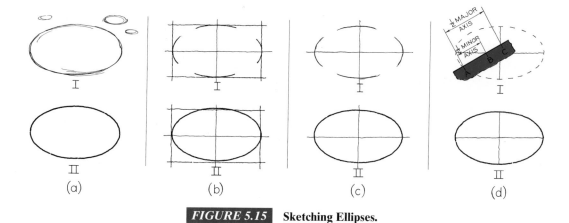

FIGURE 5.15 Sketching Ellipses.

part II), lighten all lines with a soft eraser, and heavy in the final ellipse. The same general procedure shown in Fig. 5.15b may be used in sketching the ellipse on the given axes (Fig. 5.15c).

The trammel method (Fig. 5.15d), is excellent for sketching large ellipses. Prepare a trammel on the edge of a card or strip of paper, move it to different positions, and mark points on the ellipse at A. (The trammel method is explained in §5.50.) Sketch the final ellipse through the points, as shown. For sketching isometric ellipses, see §5.12.

■ 5.9 MAINTAINING PROPORTIONS

The most important rule in freehand sketching is *keep the sketch in proportion*. No matter how brilliant the technique or how well the small details are drawn, if the proportions—especially the large overall proportions—are bad, the sketch will be bad. First, the relative proportions of the height to the width must be

carefully established; then as you proceed to the medium-sized areas and the small details, constantly compare each new estimated distance with already established distances.

If you are working from a given picture, such as the utility cabinet in Fig. 5.16a, you must first establish the relative width compared to the height. One way is to use the pencil as a measuring stick, as shown. In this case, the height is about $1\frac{3}{4}$ times the width. Then

1. Sketch the enclosing rectangle in the correct proportion (Fig. 5.16I). In this case, the sketch is to be slightly larger than the given picture.

2. Divide the available drawer space into three parts with the pencil by trial (Fig. 5.16II). Sketch light diagonals to locate centers of drawers, and block in drawer handles. Sketch all remaining details.

3. Dim all construction with a soft eraser, and heavy in all final lines (Fig. 5.16III).

FIGURE 5.16 Sketching a Utility Cabinet.

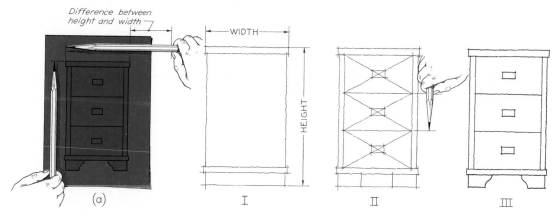

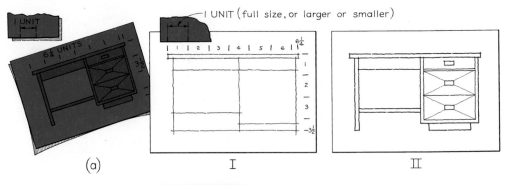

FIGURE 5.17 Sketching a Desk.

Another method of estimating distances is illustrated in Fig. 5.17. On the edge of a card or strip of paper, mark an arbitrary unit. Then see how many units wide and how many units high the desk is. If you are working from the actual object, you can use a scale, a piece of paper, or the pencil itself as a unit to determine the proportions.

To sketch an object composed of many curves to the same scale or to a larger or smaller scale, the method of "squares" is recommended (Fig. 5.18). On the given picture, rule accurate grid lines to form squares of any convenient size. It is best to use a scale and some convenient spacing, such as either .50″ or 10 mm. On the new sheet rule a similar grid, marking the spacing of the lines proportional to the original, but reduced or enlarged as desired. Make the final sketch by drawing the lines in and across the grid lines as in the original, as near as you can estimate by eye.

In sketching from an actual object, you can easily compare various distances on the object by using the pencil to compare measurements, as shown in Fig. 5.19. While doing this, do not change your position, and always hold your pencil at arm's length. The length sighted can then be compared in a similar manner with any other dimension of the object. If the object is small, such as a machine part, you can compare distances in the manner of Fig. 5.16, by actually placing the pencil against the object itself.

In establishing proportions, the blocking-in method is recommended, especially for irregular shapes. The steps for blocking in and completing the sketch of a shaft hanger are shown in Fig. 5.20. As always, first give attention to the main proportions, next to the general sizes and direction of flow of curved shapes, and finally to the snappy lines of the completed sketch.

In making sketches from actual machine parts, it is necessary to use the measuring tools used in the

FIGURE 5.18 Squares Method.

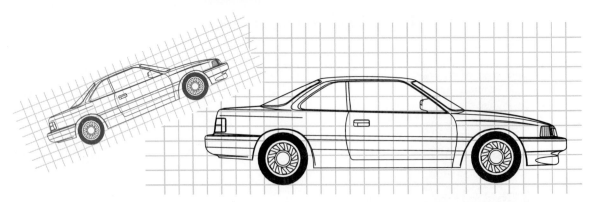

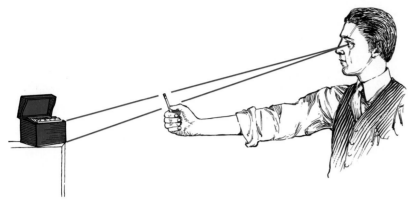

FIGURE 5.19   **Estimating Dimensions.**

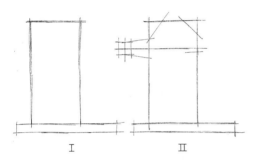

I   II

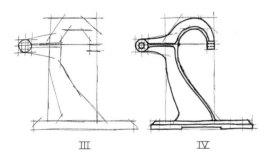

III   IV

FIGURE 5.20   **Blocking in an Irregular Object (Shaft Hanger).**

shop, especially those needed to determine dimensions that must be relatively accurate. For a discussion of these methods, see Chapter 10.

■ 5.10 PICTORIAL SKETCHING

We will now examine several simple methods of preparing pictorial sketches that will be of great assistance in learning the principles of multiview projection. A detailed and more technical treatment of pictorial drawing is given in Chapters 16 through 18.

■ 5.11 ISOMETRIC SKETCHING

Isometric sketching is one of several simple methods of preparing pictorial sketches that will be of great assistance in learning the principles of multiview projection. A detailed and more technical treatment of pictorial drawing is given in Chapters 16 through 18.

To make an *isometric sketch* from an actual object, hold the object in your hand and tilt it toward you, as shown in Fig. 5.21a. In this position, the front

FIGURE 5.21   **Isometric Sketching.**

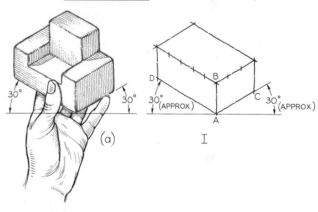

(a)   I

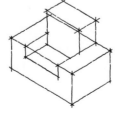

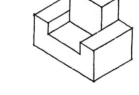

II   III

corner will appear vertical, and the two receding bottom edges and those parallel to them, respectively, will appear at about 30° with horizontal, as shown. The steps in sketching are as follows:[*]

1.   Sketch the enclosing box lightly, making AB vertical and AC and AD approximately 30° with horizontal (Fig. 5.21I). These three lines are the *isometric axes*. Make AB, AC, and AD approximately proportional in length to the actual corresponding edges on the object. Sketch the remaining lines parallel, respectively, to these three lines.
2.   Block in the recess and the projecting block (Fig. 5.21II).
3.   Lighten all construction lines with a soft eraser, and heavy in all final lines (Fig. 5.21III).

■ 5.12 ISOMETRIC ELLIPSES

When objects with cylindrical or conical shapes are placed in isometric or other oblique positions, the circles are seen at an angle and appear as ellipses (Fig. 5.22a).

[*] The angle of the receding lines may be less than 30° (say, 20° or 15°). Although the result will not be an isometric sketch, the sketch may be more pleasing and effective in some cases.

The most important consideration in sketching isometric ellipses is as follows: *The major axis of the ellipse is always at right angles to the center line of the cylinder, and the minor axis is at right angles to the major axis and coincides with the center line.*

Two views of a block with a large cylindrical hole are shown in Fig. 5.23a. The steps in sketching the object are as follows:

1.   Sketch the block and the enclosing parallelogram for the ellipse, making the sides of the parallelogram parallel to the edges of the block and equal in length to the diameter of the hole (Fig. 5.23I). Draw diagonals to locate the center of the hole, and then draw center lines AB and CD. Points A, B, C, and D will be midpoints of the sides of the parallelogram, and the ellipse will be tangent to the sides at those points. The major axis will be on the diagonal EF, which is at right angles to the center line of the hole, and the minor axis will fall along the short diagonal. Sketch long, flat elliptical sides CA and BD, as shown.
2.   Sketch short, small-radius arcs CB and AD to complete the ellipse (Fig. 5.23II). Avoid making the ends of the ellipse "squared off" or pointed like a football.
3.   Sketch lightly the parallelogram for the ellipse

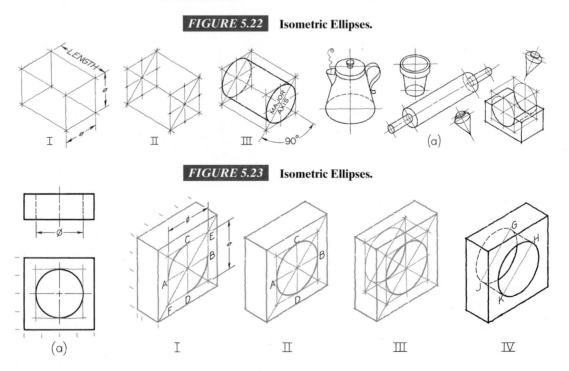

FIGURE 5.22   Isometric Ellipses.

FIGURE 5.23   Isometric Ellipses.

that lies in the back plane of the object, and sketch the ellipse in the same manner as the front ellipse (Fig. 5.23III).

4. Draw lines GH and JK tangent to the two ellipses (Fig. 5.23IV). Dim all construction with a soft eraser, and heavy in all final lines.

Another method for determining the back ellipse is shown in Fig. 5.24.

1. Select points at random on the front ellipse and sketch "depth lines" equal in length to the depth of the block (Fig. 5.24I).
2. Sketch the ellipse through the ends of the lines (Fig. 5.24II).

Two views of a bearing with a semicylindrical opening are shown in Fig. 5.25a. The steps in sketching are as follows:

1. Block in the object, including the rectangular space for the semicylinder (Fig. 5.25I).
2. Block in the box enclosing the complete cylinder (Fig. 5.25II). Sketch the entire cylinder lightly.
3. Dim all construction lines, and heavy in all final lines, showing only the lower half of the cylinder (Fig. 5.25III).

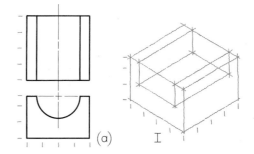

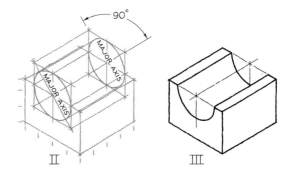

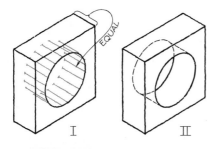

FIGURE 5.24   Isometric Ellipses.

FIGURE 5.25   Sketching Semiellipses.

■ 5.13 SKETCHING ON ISOMETRIC PAPER

Two views of a guide block are shown in Fig. 5.26a. The steps in sketching illustrate not only the use of isometric paper, but also the sketching of individual planes or faces to build up a pictorial visualization from the given views.

1. Sketch the isometric of the enclosing box, counting off the isometric grid spaces to equal the cor-

FIGURE 5.26   Sketching on Isometric Paper.

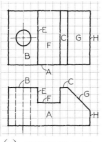

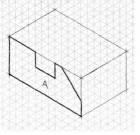

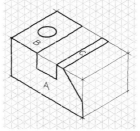

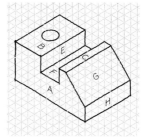

(a) GUIDE BLOCK   I   II   III

responding squares on the given views (Fig. 5.26I). Sketch surface A, as shown.

2. Sketch additional surfaces B and C and the small ellipse (Fig. 5.26II).

3. Sketch additional surfaces E, F, G, and H to complete the sketch (Fig. 5.26III).

■ 5.14 OBLIQUE SKETCHING

Another simple method for sketching pictorially is oblique sketching. Hold the object in your hand (Fig. 5.27a).

1. Block in the front face of the bearing, as if you were sketching a front view (Fig. 5.27I).

2. Sketch receding lines parallel to each other and at any convenient angle (say, 30° or 45°) with the horizontal (Fig. 5.27II). Cut off the receding lines at the correct depth. These lines may be full length, but three quarters or one half size produces a more natural appearance. If the lines are full length, the sketch is a *cavalier sketch*. If half size, the sketch is a *cabinet sketch* (see §18.4).

3. Lighten all construction lines with a soft eraser and heavy in the final lines (Fig. 5.27III).

Oblique sketching is not a very desirable method for sketching an object that has circular shapes in or parallel to more than one plane of the object because ellipses result when circular shapes are viewed obliquely. Therefore, place the object with most or all of the circular shapes toward you so that they will appear as true circles and arcs in oblique sketching, as shown in Fig. 5.27a.

■ 5.15 OBLIQUE SKETCHING ON GRAPH PAPER

Ordinary graph paper is convenient for oblique sketching. Two views of a bearing bracket are shown in Fig. 5.28a. The dimensions are determined simply by counting the squares.

1. Sketch lightly the enclosing box construction (Fig. 5.28I). Sketch the receding lines at 45° diagonally through the squares. To establish the depth at a reduced scale, sketch the receding lines diagonally

FIGURE 5.27 **Sketching in Oblique.**

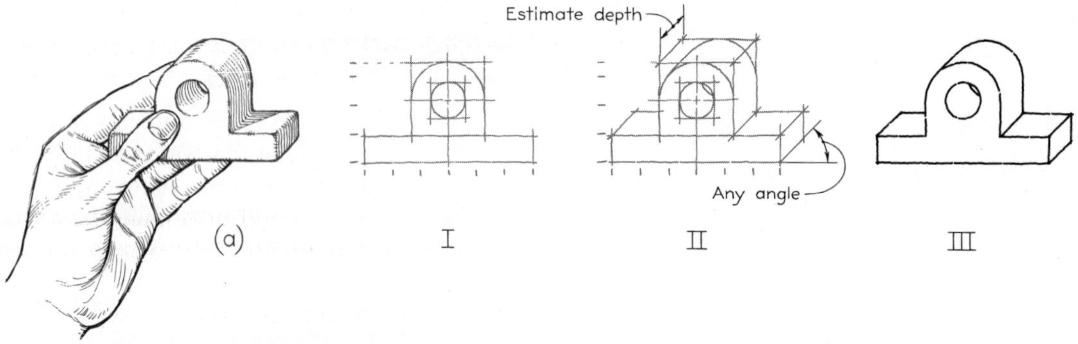

FIGURE 5.28 **Oblique Sketching on Cross-section Paper.**

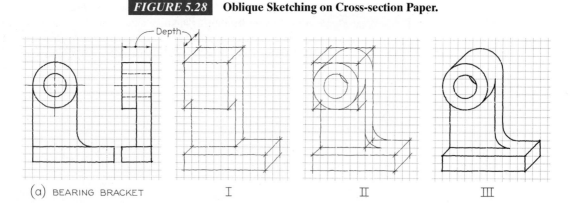

(a) BEARING BRACKET   I   II   III

through half as many squares as the given number shown in Fig. 5.28a.

2. Sketch all arcs and circles (Fig. 5.28II).

3. Heavy in all final lines (Fig. 5.28III).

■ 5.16 PERSPECTIVE SKETCHING

The bearing sketched in oblique in Fig. 5.27 can easily be sketched in *one-point perspective*—that is, with one vanishing point (Fig. 5.29).

1. Sketch the true front face of the object just as in oblique sketching (Fig. 5.29I). Select the vanishing point (VP) for the receding lines. In most cases, it is desirable to place VP above and to the right of the picture, as shown, although it can be placed anywhere in the vicinity of the picture. However, if the vanishing point is placed too close to the center, the lines will converge too sharply, and the picture will be distorted.

2. Sketch the receding lines toward VP (Fig. 5.29II).

3. Estimate the depth to look good, and sketch in the back portion of the object (Fig. 5.29III). Note that the back circle and arc will be slightly smaller than the front circle and arc.

4. Lighten all construction lines with a soft eraser, and heavy in all final lines (Fig. 5.27IV). Note the similarity between the perspective sketch and the oblique sketch in Fig. 5.27.

Two-point perspective (two vanishing points) is the most true to life of all pictorial methods, but it requires some natural sketching ability or considerable practice for best results. The simple method shown in Fig. 5.30 can be used successfully by the nonartistic student.

1. Sketch the front corner of the desk in true height, and locate two vanishing points (VPL and VPR) on a horizon line (eye level) (Fig. 5.30I). The distance CA may vary—the greater it is, the higher the eye

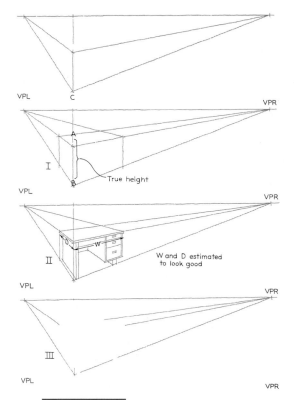

FIGURE 5.30   **Two-Point Perspective.**

level will be and the more we will be looking down on top of the object. A good rule of thumb is to make C–VPL one third to one fourth of C–VPR.

2. Estimate depth and width, and sketch the enclosing box (Fig. 5.30II).

3. Block in all details (Fig. 5.30III). Note that all parallel lines converge toward the same vanishing point.

4. Lighten the construction lines with a soft eraser as necessary, and heavy in all final lines (Fig. 5.30IV). Make the outlines thicker and the inside lines thinner, especially where they are close together.

FIGURE 5.29   **Sketching in One-Point Perspective.**

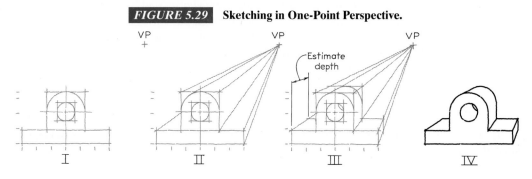

■ 5.17 VIEWS OF OBJECTS

A pictorial drawing or a photograph shows an object as it *appears* to the observer, but not as it *is*. Such a picture cannot describe the object fully, no matter which direction it is viewed from, because it does not show the exact shapes and sizes of the several parts.

In industry, a complete and clear description of the shape and size of an object to be made is necessary to make certain that the object is manufactured exactly as intended by the designer. To provide this information, a number of *views*, systematically arranged, are used. This system of views is called *multiview projection*. Each view provides certain definite information if the view is from a direction perpendicular to a principal face or side of the object. For example, an observer looking perpendicularly toward one face of an object obtains a true view of the shape and size of that side (Fig. 5.31a). The view as seen by the observer is shown in Fig. 5.31b. (The observer is theoretically at an infinite distance from the object.)

The three principal dimensions of an object are *width*, *height*, and *depth* (Fig. 5.31a). In technical drawing, these fixed terms are used for dimensions taken in these directions, regardless of the shape of the object. The terms "length" and "thickness" are not used because they cannot be applied in all cases. Note that the front view in Fig. 5.31b shows only the height and width of the object and not the depth. In fact, *any one view of a three-dimensional object shows only two dimensions; the third dimension is found in an adjacent view*.

■ 5.18 REVOLVING THE OBJECT

In addition to the three views just described, other views can be obtained by revolving the object, as shown in Fig. 5.32. First, hold the object in the front-view position (Fig. 5.32a). To get the *top view* (Fig. 5.32b), revolve the object to bring the *top of the object up and toward you*. To get the *right-side view* (Fig. 5.32c), revolve the object to bring the *right side toward you*. To obtain views of any of the other sides, merely turn the object to bring those sides toward you.

The top, front, and right-side views, arranged closer together, are shown in Fig. 5.32d. These are called the *three regular views* because they are the views most frequently used.

At this stage we can consider spacing between views as purely a matter of appearance. The views should be spaced well apart, yet close enough to appear related to each other. The space between the front and top views may or may not be equal to the space between the front and side views. If dimensions (Chapter 11) are to be added to the sketch, adequate space for them will have to be left between views.

An important advantage of a view over a photograph of an object is that hidden features can be clearly shown by means of hidden lines. In Fig. 5.32d, surface 7–8–9–10 in the front view appears as a visible line 5–6 in the top view and as a hidden line 15–16 in the side view. Also, hole A, which appears as a circle in the front view, shows as hidden lines 1–4 and 2–3 in the top view, and 11–12 and 13–14 in the side view. For a complete discussion of hidden lines, see §5.24. Also note the use of center lines for the hole (see §5.25).

■ 5.19 THE SIX VIEWS

Any object can be viewed from six mutually perpendicular directions, as shown in Fig. 5.33a. These six views may be drawn if necessary, as shown in Fig. 5.33b. Except as explained in §6.8, the six views are always arranged as shown, which is the American National Standard arrangement. The *top, front,* and *bottom views* align vertically, while the *rear, left-side, front,* and *right-side views* align horizontally. To draw a view out of place is a serious error and is generally regarded as one of the worst possible mistakes in drawing.

Note that height is shown in the rear, left-side, front, and right-side views; width is shown in the rear, top, front, and bottom views; and depth is shown in the four views that surround the front view—namely, the left-side, top, right-side, and bottom views. Each view shows two of the principal dimensions. Note also that in the four views that surround the front view, the front of the object faces toward the front view.

Adjacent views are reciprocal. If the front view in Fig. 5.33 is imagined to be the object itself, the right-side view is obtained by looking toward the right side of the front view, as shown by the arrow RS. Likewise, if the right-side view is imagined to be the object, the front view is obtained by looking toward the left side of the right-side view, as shown by the arrow F. The same relation exists between any two adjacent views.

Obviously, the six views may be obtained either by shifting the object with respect to the observer (Fig. 5.32) or by shifting the observer with respect to the object (Fig. 5.33). Another illustration of the second

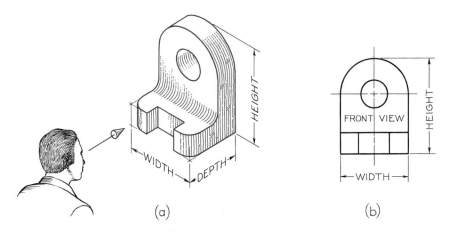

(a)   (b)

FIGURE 5.31   **Front View of an Object.**

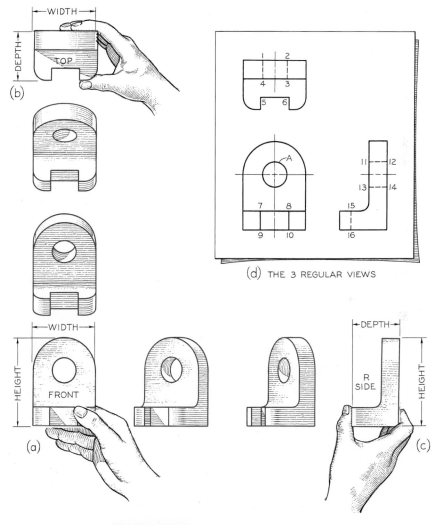

(d) THE 3 REGULAR VIEWS

FIGURE 5.32   **The Three Regular Views.**

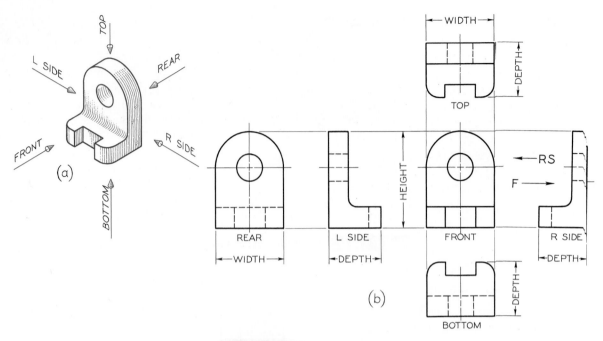

FIGURE 5.33 The Six Views.

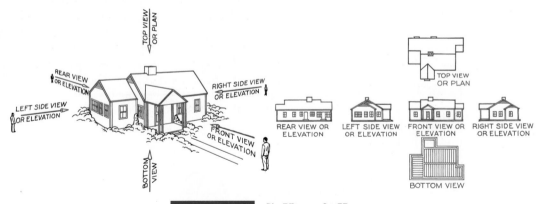

FIGURE 5.34 Six Views of a House.

method is given in Fig. 5.34, which show six views of a house. The observer can walk around the house and view its front, sides, and rear and can imagine the top view as seen from an airplane and the bottom or "worm's-eye view" as seen from underneath. Notice the use of the terms "plan," for the top view, and "elevation," for all views showing the height of the build-

ing. These terms are regularly used in architectural drawing and occasionally with reference to drawings in other fields.[*]

---

[*] Architects frequently draw the views of a building on separate sheets because of the large sizes of the drawings.

■ 5.20 ORIENTATION OF FRONT VIEW

Six views of a compact automobile are shown in Fig. 5.35. The view chosen for the front view in this case is the side, not the front of the automobile. In general, the front view should show the object in its operating position, particularly of familiar objects (such as the house shown and the automobile). A machine part is often drawn in the position it occupies in the assembly. However, in most cases this is not important, and the drafter may assume the object to be in any convenient position. For example, an automobile connecting rod is usually drawn horizontally on the sheet (see Fig. 14.28). Also, it is customary to draw screws, bolts, shafts, tubes, and other elongated parts in a horizontal position.

■ 5.21 NECESSARY VIEWS

A drawing for use in production should contain *only* those views needed for a clear and complete shape description of the object. These minimum required views are referred to as the **necessary views**. In selecting views, the drafter should choose those that best show essential contours or shapes and have the least number of hidden lines.

As shown in Fig. 5.36a, three distinctive features of this object need to be shown on the drawing: (1) rounded top and hole, seen from the *front*; (2) rectangular notch and rounded corners, seen from the *top*; and (3) right angle with filleted corner, seen from the *side*.

Another way to choose required views is to eliminate unnecessary views. Figure 5.36b shows a thumbnail sketch of the six views. Both the front and rear views show the true shapes of the hole and the rounded top, but the front view is preferred because it has no hidden lines. Therefore, the rear view (which is seldom needed) is crossed out. Both the top and bottom views show the rectangular notch and rounded corners, but the top view is preferred because it has fewer hidden lines. Both the right-side and left-side views show the right angle with the filleted corner. In fact, in this case the side views are identical, except reversed. In such instances, it is customary to choose the right-side view.

The necessary views, then, are the three remaining views: the top, front, and right-side views. These are the three regular views referred to in connection with Fig. 5.32. More complicated objects may require more than three views or special views, such as partial views.

■ 5.22 TWO-VIEW DRAWINGS

Often only two views are needed to clearly describe the shape of an object. In Fig. 5.37a, the right-side view shows no significant contours of the object and is crossed out. In Fig. 5.37b, the top and front views are identical, so the top view is eliminated. In Fig. 5.37c, all necessary information is given in the front and top views so the side view is unnecessary.

The question often arises: What are the absolute minimum views required? For example, in Fig. 5.38,

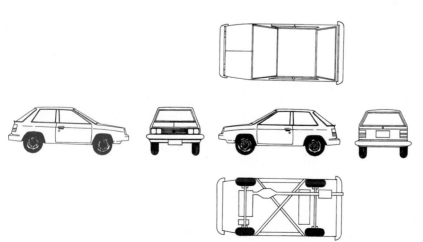

FIGURE 5.35 Six Views of a Compact Automobile.

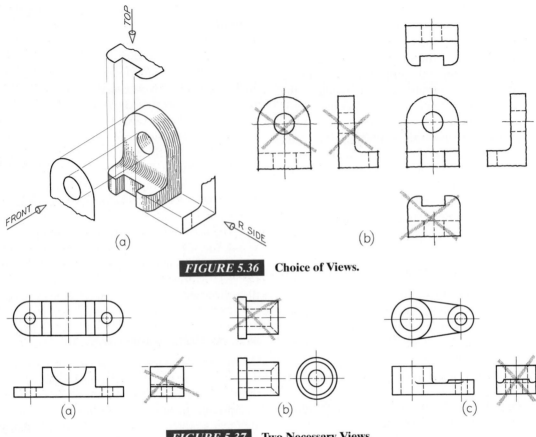

FIGURE 5.36   **Choice of Views.**

FIGURE 5.37   **Two Necessary Views.**

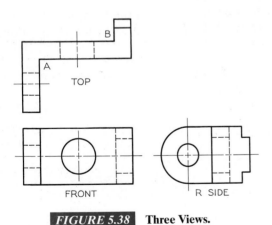

FIGURE 5.38   **Three Views.**

(top view) are square and not filleted. In this example, all three views are necessary.

If an object requires only two views and the left-side and right-side views are equally descriptive, the right-side view is customarily chosen (Fig. 5.39). If contour A were omitted, then the presence of slot B would make it necessary to choose the left-side view in preference to the right-side view.

If an object requires only two views and the top and bottom views are equally descriptive, the top view is customarily chosen (Fig. 5.40).

If only two views are necessary and the top view and right-side view are equally descriptive, the combination chosen is the one that spaces best on the paper (Fig. 5.41).

■ **5.23 ONE-VIEW DRAWINGS**

Frequently, a single view supplemented by a note or lettered symbols is sufficient to describe clearly the shape of a relatively simple object. In Fig. 5.42a, one view of the shim plus a note indicating the thickness

the top view might be omitted, leaving only the front and right-side views. However, it is more difficult to "read" the two views or visualize the object, because the characteristic "Z" shape of the top view is omitted. In addition, one must assume that corners A and B

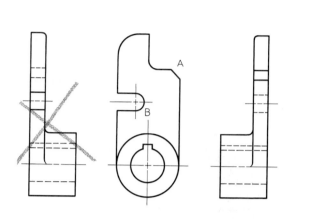

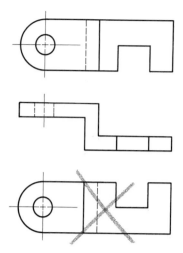

FIGURE 5.39   Choice of Right-Side View.

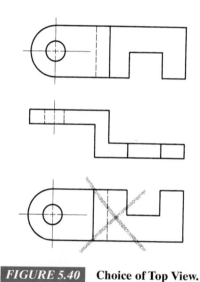

FIGURE 5.40   Choice of Top View.

as 0.25 mm is sufficient. In Fig. 5.42b, the left end is 65 mm square, the next portion is 49.22 mm diameter, the next is 31.75 mm diameter, and the portion with the thread is 20 mm diameter, as indicated in the note.

Nearly all shafts, bolts, screws, and similar parts should be represented by single views in this manner.

■ 5.24 HIDDEN LINES

Correct and incorrect practices in drawing hidden lines are illustrated in Fig. 5.43. In general, a hidden line should join a visible line except when it causes the visible line to extend too far, as shown in Fig. 5.43a. In other words, *leave a gap whenever a hidden line is a continuation of a visible line.* Hidden lines should intersect to form L and T corners (Fig. 5.43b). A hidden line preferably should "jump" a visible line when possible (Fig. 5.43c). Parallel hidden lines should be drawn so that the dashes are staggered, as in bricklaying (Fig. 5.43d). When two or three hidden lines meet at a point, the dashes should join, as shown for the bottom of the drilled hole in Fig. 5.43e and for the top of a countersunk hole in Fig. 5.43f. The example in Fig. 5.43g is similar to that in Fig. 5.43a; hidden lines should not join visible lines when this makes the visible line extend too far. Correct and incorrect methods of drawing hidden arcs are shown in Fig. 5.43h.

FIGURE 5.41   Choice of Views to Fit Paper.

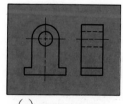

(a) PREFERRED

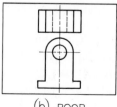

(b) POOR

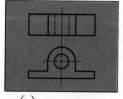

(c) PREFERRED

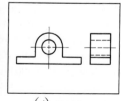

(d) POOR

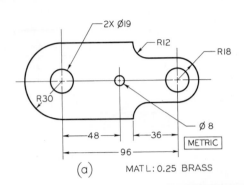

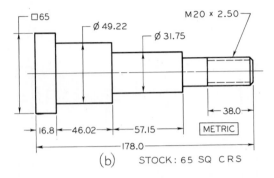

FIGURE 5.42 One-View Drawings.

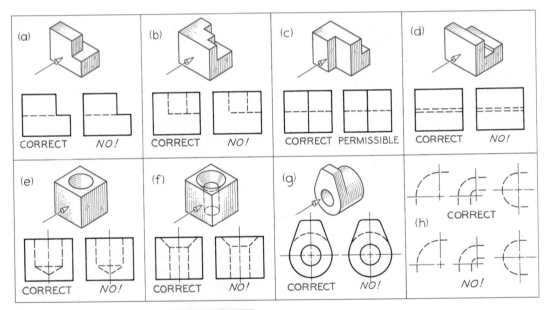

FIGURE 5.43 Hidden-Line Practices.

Poorly drawn hidden lines can easily spoil a drawing. Dashes should be about 5 mm long and spaced about 1 mm apart, by eye. Accent the beginning and end of each dash by pressing down on the pencil, whether drawn freehand or mechanically.

As far as possible, views should be chosen that show features with visible lines. Hidden lines should be used where necessary to make the drawing clear. Hidden lines not needed for clarity should be omitted so as not to clutter the drawing and also to save time. The beginner, however, should be cautious about leaving out hidden lines until experience shows when they can be safely omitted.

■ 5.25 CENTER LINES

Center lines (symbol: ₵) are used to indicate axes of symmetrical objects or features, bolt circles, and paths of motion. Typical applications are shown in Fig. 5.44. As shown in Fig. 5.44a, a single center line is drawn in the longitudinal view and crossed center lines in the circular view. The small dashes should cross at the in-

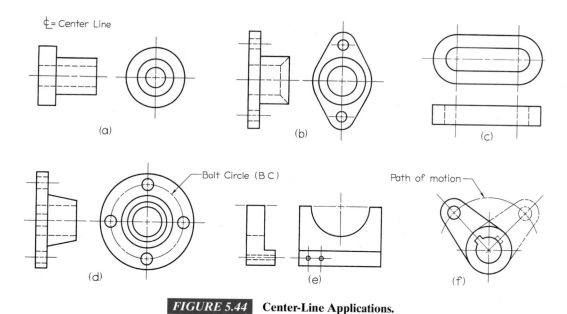

₵ = Center Line

(a)

(b)

(c)

Bolt Circle (B C)

(d)

(e)

Path of motion

(f)

FIGURE 5.44 Center-Line Applications.

tersections of center lines. Center lines should extend uniformly about 8 mm outside the feature for which they are drawn.

The long dashes of center lines may vary from 20 to 40 mm or more in length, depending on the size of the drawing. The short dashes should be about 5 mm long, with spaces about 2 mm. Center lines should always start and end with long dashes. Short center lines, especially for small holes, may be made solid (Fig. 5.44e). Always leave a gap when a center line forms a continuation of a visible or hidden line. Center lines should be thin enough to contrast well with the visible and hidden lines, but dark enough to reproduce well.

Center lines are useful mainly in dimensioning and should be omitted from unimportant rounded or filleted corners and other shapes that are self-locating.

■ 5.26 SKETCHING TWO VIEWS

A sketch of the support block in Fig. 5.45a requires only two views. The steps in sketching are as follows:

1. Block in lightly the enclosing rectangles for the two views (Fig. 5.45I). Sketch horizontal lines 1 and 2 to establish the height of the object, while making spaces A approximately equal. Sketch vertical lines 3, 4, 5, and 6 to establish the width and depth in correct proportion to the already

FIGURE 5.45 Sketching Two Views of a Support Block.

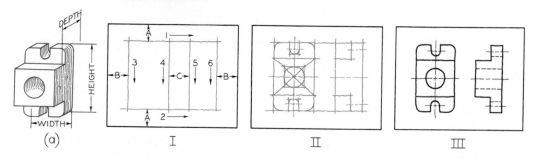

(a)

I

II

III

G R A P H I C S   S P O T L I G H T

Sketching and Parametric Modeling

THE DESIGN PROCESS

Using CAD parametric modeling in many ways mirrors the design process (see Chapter 14). To get the rough ideas down, the designer starts by making hand sketches. Then as the ideas are refined, more accurate drawings are created either with instruments or using CAD. Necessary analysis is performed and in response the design may change. The drawings are revised as needed to meet the new requirements. Eventually the drawings are approved so that the parts may be manufactured.

ROUGH SKETCHES

Using parametric modeling software, initially the designer roughly sketches the basic on the screen. These sketches do not have to have perfectly straight lines or accurate corners. The software interprets the sketch much as you would interpret a rough sketch given to you by a colleague. If the lines are nearly horizontal or vertical, the software assumes that you meant them thus. If the line appears to be perpendicular it is assumed that it is.

CONSTRAINING THE SKETCH

Using a parametric CAD system, you can start by sketching on the computer screen as though you were sketching freehand. Then the two-dimensional sketch is refined by adding geometric constraints, which tell how to interpret the sketch and by adding parametric dimensions, which control the size of sketch geometry. Once the sketch is refined, it can be

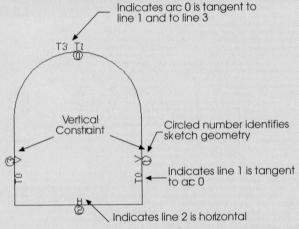

(B) Constrained Sketch

created as a 3D feature to which other features can be added. As the design changes the dimensions and constraints that control the sketch geometry can be changed, and the parametric model will update to reflect the new design.

When you are creating sketches by hand or for parametric modeling, think about the implications of the geometry you are drawing. Does the sketch imply that lines are perpendicular? Are the arcs you have drawn intended to be tangent or intersecting? When you are creating a parametric model, the software makes assumptions about how you intend to constrain the geometry based on your sketch. You can remove, change, or add new constraints as you wish.

AutoCAD Mechanical Desktop software contains AutoCAD Designer parametric design software. Using it you can create a rough sketch like the one you see in Fig. A.

You select the Profile command to have the software constrain the sketch automatically. The results of profiling the sketch are shown in Fig. B. The symbols show the constraints that were assumed.

The dialog box shown on the next page in Fig. C lists types of geometric constraints you can use to control the sketch geometry in AutoCAD Designer. The dialog box labeled Fig. D on the left shows the constraints that you can use to control the way parts fit together in an assembly.

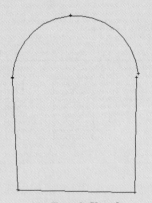

(A) Rough Sketch

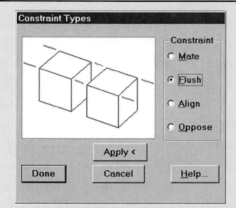

(C) Assembly Constraints

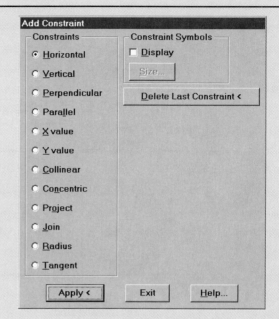

(D) Sketch Constraints

When you have completed the parametric model, you have an "intelligent" part. When design changes are necessary, you can change a dimension or constraint causing the model to automatically update. Orthographic drawings with correctly shown hidden lines and dimensions can be generated auto-matically. Or the part can be exported for rapid proto-typing or manufacture.

established height, while making spaces B approximately equal and space C equal to or slightly less than space B.

2. Block in smaller details, using diagonals to locate the center (Fig. 5.45II). Sketch lightly the circle and arcs.

3. Lighten all construction lines with a soft eraser, and heavy in all final lines (Fig. 5.45III).

■ 5.27 SKETCHING THREE VIEWS

A sketch of a lever bracket that requires three views is shown in Fig. 5.46a. The steps in sketching the three views are as follows:

1. Block in the enclosing rectangles for the three views (Fig. 5.46I). Sketch horizontal lines 1, 2, 3, and 4 to establish the height of the front view and the depth of the top view, making spaces A approximately equal and space C equal to or slightly less than one space A. Sketch vertical lines 5, 6, 7, and 8 to establish the width of the top and front views and the depth of the side view. Make sure that this is in correct proportion to the height, while making spaces B approximately equal and space D equal to or slightly less than one space B. Note that spaces C and D are not necessarily equal, but are independent of each other. Similarly, spaces A and B are not necessarily equal. To transfer the depth dimension from the top view to the side view, use the edge of a card or strip of paper, as shown, or transfer the distance by using the pencil as a measuring stick (see Figs. 5.10b and 5.10c). Note that *the depth in the top and side views must always be equal.*

2. Block in all details lightly (Fig. 5.46II).

3. Sketch all arcs and circles lightly (Fig. 5.46III).

4. Lighten all construction lines with a soft eraser (Fig. 5.46IV).

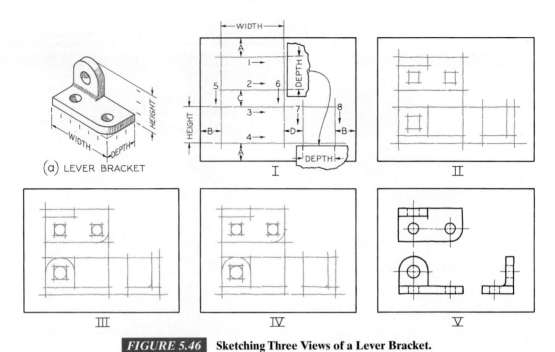

FIGURE 5.46   Sketching Three Views of a Lever Bracket.

5. Heavy in all final lines so that the views will stand out clearly (Fig. 5.46V).

■ 5.28 ALIGNMENT OF VIEWS

Errors in arranging the views are so commonly made by students that it is necessary to repeat this: The views must be drawn in accordance with the American National Standard arrangement shown in Fig. 5.33. Figure 5.47a shows an offset guide that requires three views. These three views, correctly arranged, are shown in Fig. 5.47b. The top view must be directly above the front view, and the right-side view directly to the right of the front view—not out of alignment, as in Fig. 5.47c. Also, never draw the views in reversed positions, with the bottom over the front view or the right-side to the left of the front view (Fig. 5.47d), even though the views do line up with the front view.

■ 5.29 MEANING OF LINES

A visible line or a hidden line has three possible meanings (Fig. 5.48): (1) intersection of two surfaces, (2) edge view of a surface, and (3) contour view of a curved surface. Since *no shading is used on a working drawing*, it is necessary to examine all the views to determine the meaning of the lines. For example, the line AB at the top of the front view might be regarded as the edge view of a flat surface if we look at only the

front and top views and do not observe the curved surface on top of the object, as shown in the right-side view. Similarly, the vertical line CD in the front view might be regarded as the edge view of a plane surface if we look at only the front and side views. However, the top view shows that the line represents the intersection of an inclined surface.

■ 5.30 PRECEDENCE OF LINES

Visible lines, hidden lines, and center lines often coincide on a drawing, and the drafter must know which line to show. A visible line always takes precedence over (covers up) a center line or a hidden line, as shown at A and B in Fig. 5.49. A hidden line takes precedence over a center line, as shown at C. Note that at A and C the ends of the center line are shown, but are separated from the view by short gaps.

■ 5.31 COMPUTER GRAPHICS

Preliminary sketches are usually done on paper with a pencil or pen. More detailed CAD drawings usually follow the initial sketches. Finished CAD drawings should apply the same rules for arranging views, clearly depicting the subject of the drawing, using the proper line patterns and line weights, and following all of the necessary standards as manually created drawings (Fig. 5.50).

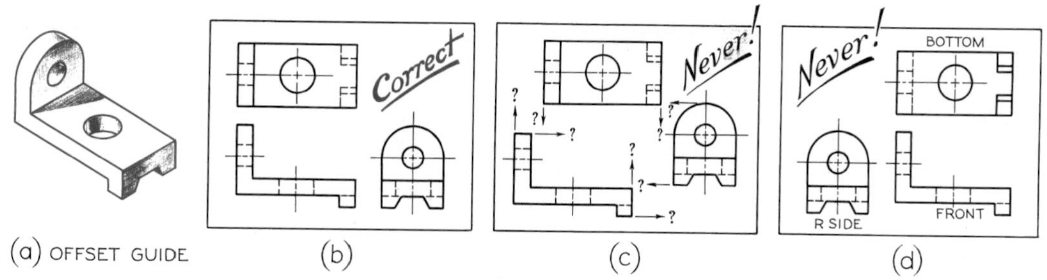

(a) OFFSET GUIDE

(b)

Correct

(c)

Never!

(d)

Never!

BOTTOM

R SIDE

FRONT

FIGURE 5.47   Position of Views.

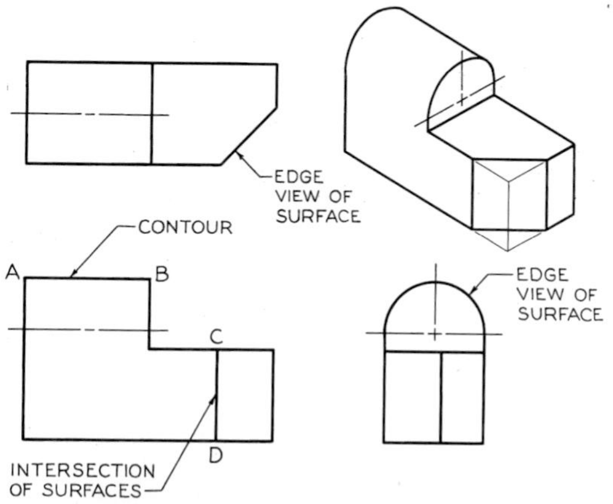

EDGE
VIEW OF
SURFACE

CONTOUR

A   B

C

D

INTERSECTION
OF SURFACES

EDGE
VIEW OF
SURFACE

FIGURE 5.48   Meaning of Lines.

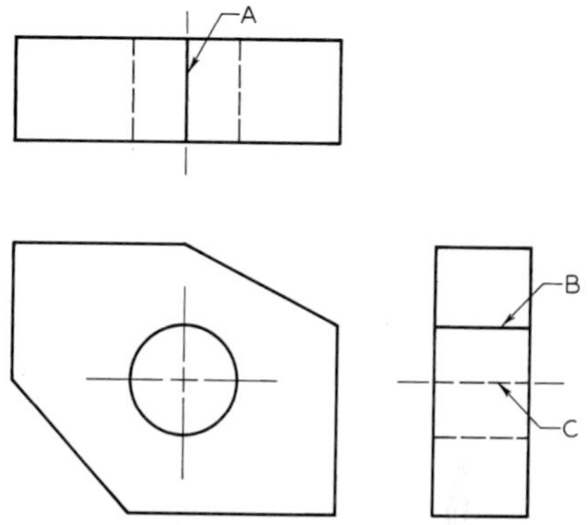

A

B

C

FIGURE 5.49   Precedence of Lines.

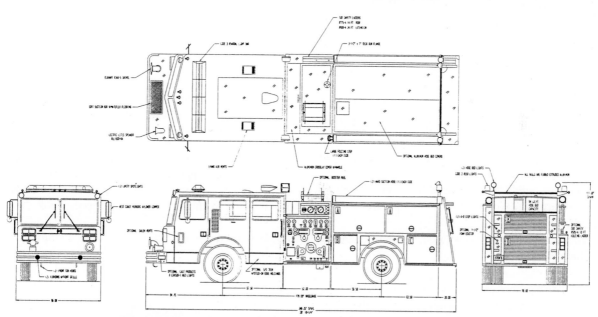

FIGURE 5.50   Multiview CAD Assembly Drawing of a MAXIM Fire
Truck. *Courtesy of CADKEY.*

■ KEY WORDS

CONSTRUCTION LINES	PICTORIAL
ORTHOGRAPHIC PROJECTION	OBLIQUE
ISOMETRIC	GRID PAPER
PRIMARY VIEWS	HIDDEN LINES
VISIBLE LINES	CENTER LINES
PRECEDENCE OF LINES	SHADING
ALIGNMENT OF VIEWS	SKETCH
SHAPE DESCRIPTION	PROPORTIONS

■ CHAPTER SUMMARY

- Sketching is one of the most important skills that a traditional or CAD drafter must learn to become effective.
- Sketching is a quick way of visualizing and solving a drawing problem. It is an effective way of communicating with all members of the design team.
- There are special techniques for sketching lines, circles, and arcs. These techniques should be practiced so they become second nature.
- Moving your thumb up or down the length of a pencil at arms length is an easy method for estimating proportional size.
- Using a sketching grid makes measuring an easy task. Grid paper comes in a variety of types, including square grid and isometric grid.

- You can sketch circles by constructing a square and locating the four tangent points where the circle touches the square. This same technique can be applied to isometric circles by drawing isometric "squares" (parallelograms).
- You can sketch objects using the same view orientation and alignment that traditional drawing conventions demand.
- A sketched line must not look like a mechanical line. The main distinction between instrumental drawing and freehand sketching is the character or technique of the linework.
- Freehand sketches are made to proportion, but not necessarily to a particular scale.

■ REVIEW QUESTIONS

1. What are the four standard types of projections?
2. What are the advantages of using grid paper for sketching?
3. What is the correct technique for sketching a circle or arc?
4. Sketch the alphabet of lines. Which lines are thick? Which are thin? Which are very light and will not reproduce when copied?
5. When is isometric grid paper used? When is square grid paper used?
6. What type of three-dimensional drawing can easily be drawn on square grid paper?
7. What is the advantage of sketching an object first before drawing it using CAD?
8. Why are tangent points so important when sketching, drawing on paper with instruments, or using CAD?
9. What is the difference between proportion and scale?
10. Which view should be sketched directly below the TOP view?
11. Why should center lines, visible lines, and hidden lines line up between views when sketching?

■ SKETCHING PROBLEMS

Figures 5.52 and 5.53 present a variety of objects from which the student is to sketch the necessary views. Using 8.5″ × 11.0″ graph paper, sketch a border and title strip and divide the sheet into two parts (Fig. 5.51). Sketch two assigned problems per sheet. On the problems in Fig. 5.52, "ticks" are given that indicate .50″ or .25″ spaces. Thus, measurements may be easily spaced off on graph paper having .12″ or .25″ grid spacings.

If desired, the "ticks" on the problems in Fig. 5.52 may be used to indicate 10-mm and 5-mm spaces. Thus, metric measurements may be easily utilized on appropriate metric-grid graph paper.

On the problems in Fig. 5.53, no indications of size are given. The student is to sketch the necessary views of assigned problems to fit the spaces comfortably, as shown in Fig. 5.51. It is suggested that the student prepare a small paper scale, making the divisions equal to those on the paper scale in Prob. 5.1. This scale can be used to determine the approximate sizes. Let each division equal either .50″ or 10 mm on your sketch.

Missing-line and missing-view problems are given in Figs. 5.54 and 5.55, respectively. These are to be sketched, two problems per sheet, in the arrangement shown in Fig. 5.51. If the instructor so assigns, the missing lines or views may be sketched with a colored pencil. The problems given in Figs. 5.54 and 5.55 may be sketched in isometric on isometric paper or in oblique on graph paper.

Since many of the problems in this chapter are of a general nature, they can also be solved on most computer graphics systems. If a system is available, the instructor may choose to assign specific problems to be completed by this method.

For all of the following problems, use 8.5″ × 11.0″ graph paper or plain paper. Sketch a border and title strip, and divide the sheet into two parts, as shown in Fig. 5.51. Sketch two assigned problems per sheet.

Prob. 5.1–5.24 (Fig. 5.52) Sketch necessary views using Layout A–1 or A4–1 adjusted (freehand). The "ticks" on the figures indicate .50″ or .25″ spaces. Thus, measurements may be easily spaced off on graph paper having .12″ or .25″ grid spacings. The "ticks" may also be used to indicate 10-mm or 5-mm spaces, so that metric measurements may be easily utilized on appropriate metric-grid graph paper. In these problems, all holes are through holes.

Prob. 5.25–5.48 (Fig. 5.53) Sketch necessary views using Layout A–1 or A4–1 adjusted (freehand). No indication of size is given in these problems. Sketch the necessary views to fit the spaces comfortably. Prepare a small paper scale, making the divisions equal to those on the paper scale in Prob. 5.25, and apply to the problems to obtain approximate sizes. Let each division equal either .50″ or 10 mm on your sketch. For Prob. 5.41–5.48, study §§6.34–6.36.

Prob. 5.49–5.72 (Fig. 5.54) Sketch given views using Layout A–1 or A4–1 adjusted (freehand). Add missing lines. The squares may be either .25″ or 5 mm. The problems may be sketched in isometric on isometric paper or in oblique on graph paper.

Prob. 5.73–5.105 (Fig. 5.55) Using Layout A–1 or A4–1 adjusted (freehand), sketch the two given views and add the missing views, as indicated. The squares may be either .25″ or 5 mm. The given views are either front and right-side views or front and top views. Hidden holes with center lines are drilled holes. Sketch in isometric on isometric paper or in oblique on cross-section paper.

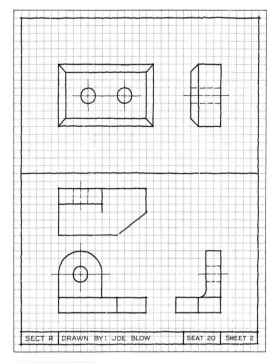

FIGURE 5.51 **Multiview Sketch (Layout A–1).**

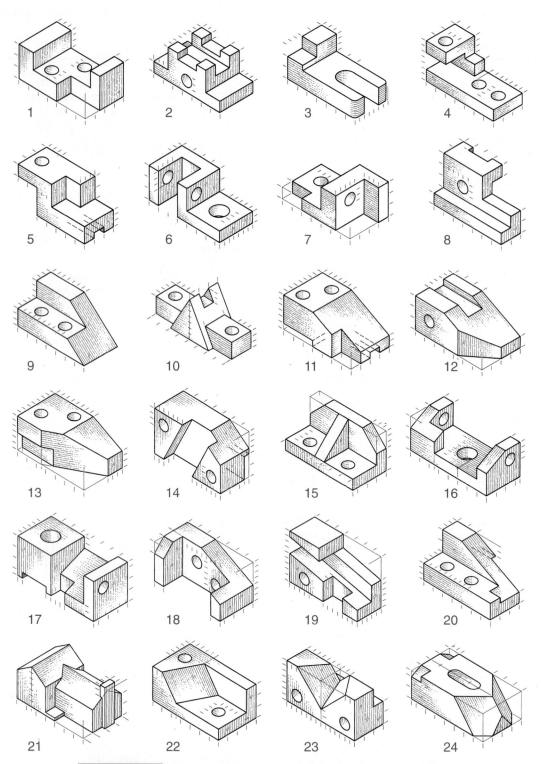

FIGURE 5.52 Multiview Sketching Problems. Sketch necessary views, using Layout A–1 (see Fig. 5.51) or A4–1 (see front cover endpapers) adjusted (freehand), on graph paper or plain paper, two problems per sheet as in Fig. 5.51. The units shown may be either .50″ and .25″ or 10 mm and 5 mm. See instructions on page 159. All holes are through holes.

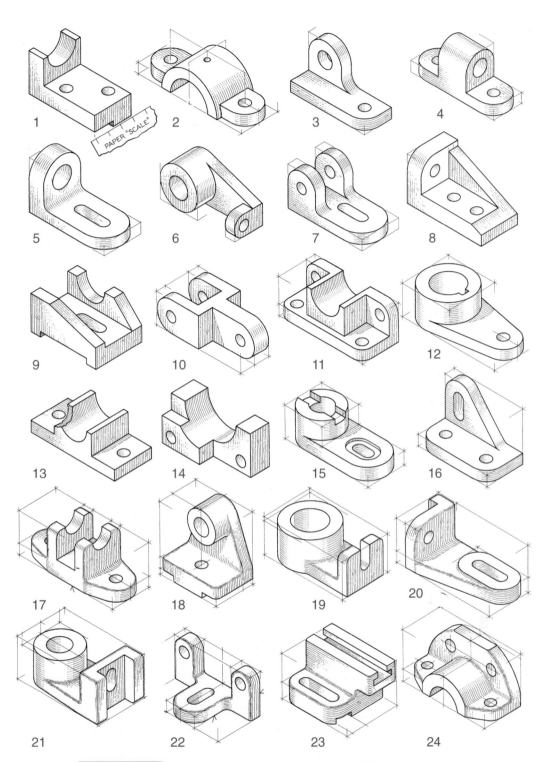

FIGURE 5.53 Multiview Sketching Problems. Sketch necessary views, using Layout A–1 or A4–1 adjusted (freehand), on graph paper or plain paper, two problems per sheet as in Fig. 5.51. Prepare paper scale with divisions equal to those in Prob. 1, and apply to problems to obtain approximate sizes. Let each division equal either .50″ or 10 mm on your sketch. See instructions on page 159. For Probs. 17–24, study §§6.34–6.36.

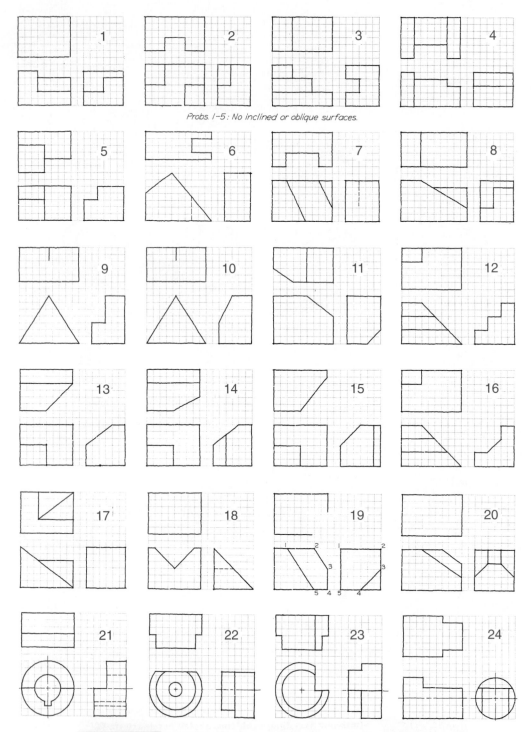

Probs. 1–5 : No inclined or oblique surfaces.

FIGURE 5.54 Missing-Line Sketching Problems. (1) Sketch given views, using Layout A–1 or A4–1 adjusted (freehand), on graph paper or plain paper, two problems per sheet as in Fig. 5.51. Add missing lines. The squares may be either .25″ or 5 mm. See instructions on page 159. (2) Sketch in isometric on isometric paper or in oblique on cross-section paper.

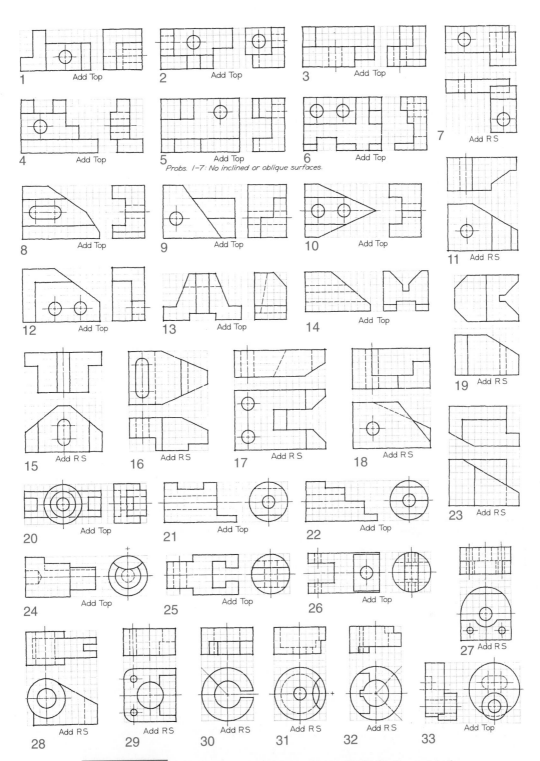

1   Add Top
2   Add Top
3   Add Top

4   Add Top
5   Add Top
6   Add Top
7   Add R S

Probs. 1–7: No inclined or oblique surfaces.

8   Add Top
9   Add Top
10   Add Top
11   Add R S

12   Add Top
13   Add Top
14   Add Top

15   Add R S
16   Add R S
17   Add R S
18   Add R S
19   Add R S

20   Add Top
21   Add Top
22   Add Top
23   Add R S

24   Add Top
25   Add Top
26   Add Top
27   Add R S

28   Add R S
29   Add R S
30   Add R S
31   Add R S
32   Add R S
33   Add Top

FIGURE 5.55   **Third-View Sketching Problems. (1) Using Layout A–1 or A4–1 adjusted (freehand), on graph paper or plain paper, two problems per sheet as in Fig. 5.51, sketch the two given views and add the missing views, as indicated. The squares may be either .25″ or 5 mm. See instructions on page 159. The given views are either front and right-side views or front and top views. Hidden holes with center lines are drilled holes. (2) Sketch in isometric on isometric paper or in oblique on cross-section paper.**

C H A P T E R   6

MULTIVIEW PROJECTION

OBJECTIVES

After studying the material in this chapter, you should be able to:

1. Draw the six standard views of an object.
2. Draw any three views using proper conventions, placement, and alignment.
3. Transfer height, width, or depth dimensions between views.
4. Layout a three-view drawing so it is centered on the drawing medium.
5. Apply conventional practices to the revolution of ribs, spokes, and webs.
6. Identify and draw visible and hidden lines in all six standard views.
7. Identify and project surfaces appearing in all views.
8. Identify surface shapes, and determine edge views of surfaces.
9. Draw and project normal, inclined, and oblique surfaces in all views.
10. Draw positive and negative cylinders in all views.
11. Plot conic sections and irregular curves in all views.
12. Understand drawing conventions for hole treatments and machine processes.
13. Describe the concept of first and third angle projections.
14. Understand and use fold lines or miter lines to create new views.

OVERVIEW

A view of an object is known technically as a **projection**. A projection is a view conceived to be drawn or projected onto a plane known as the **plane of projection**. A system of views of an object formed by projectors from the object perpendicular to the desired planes of projection is known as **orthographic**, or **multiview**, **projection** (see ANSI/ASME Y14.3M–1994). This system of required views provides for the shape description of the object.

Drawing conventions describe the projection process so that all technical drawings can be created and interpreted in the same way. Conventions assure that no ambiguity exists between the drawer's intent and the reader's interpretation. Understanding surface orientations can help students visualize the six standard views of projection. Surfaces are either normal, inclined, or oblique. The endpoints of lines define surface boundaries. Conic sections and irregular curves can be approximated by mapping definable lines from one view to another. Various manufacturing processes for holes, fillets, and rounds are described in specific ways on technical drawings.

■ 6.1 PROJECTION METHOD

The method of viewing an object to obtain a multiview projection is illustrated for a front view in Fig. 6.1a. Between the observer and the object, a transparent plane or pane of glass representing a plane of projection is located parallel to the front surfaces of the object. The outline on the plane of projection shows how the object appears to the observer. Theoretically, the observer is at an infinite distance from the object so that the *lines of sight* are parallel.

In more precise terms, this view is obtained by drawing perpendicular lines, or *projectors*, from all points on the edges or contours of the object to the plane of projection (Fig. 6.1b). The infinite number of collective piercing points of these projectors form lines on the pane of glass (Fig. 6.1c). As shown, a projector from point 1 on the object pierces the plane of projection at point 7, which is a view or projection of the point. The same procedure applies to point 2, whose projection is point 9. Since 1 and 2 are endpoints of a straight line on the object, the projections 7 and 9 are joined to give the projection of the line 7–9. Similarly, if the projections of the four corners 1, 2, 3, and 4 are

found, the projections 7, 9, 10, and 8 may be joined by straight lines to form the projection of the rectangular surface.

The same procedure can be applied to curved lines—for example, the top curved contour of the object. A point, 5, on the curve is projected to the plane at 6. The projection of an infinite number of such points (a few are shown in Fig. 6.1b) on the plane of projection results in the projection of the curve. If this procedure of projecting points is applied to all edges and contours of the object, a complete view or projection of the object results. This view is necessary in the shape description because it shows the true curvature of the top and the true shape of the hole.

A similar procedure may be used to obtain the top view (Fig. 6.2a). This view is necessary in the shape description because it shows the true angle of the inclined surface. In this view, the hole is invisible and its extreme contours are represented by hidden lines, as shown.

The right-side view (Fig. 6.2b) is necessary because it shows the right-angled characteristic shape of the object and the true shape of the curved intersection. Note how the cylindrical contour at the top of the object appears when viewed from the side. The extreme, or contour, element 1–2 on the object is projected to give the line 3–4 on the view. The hidden hole is also represented by projecting the extreme elements.

The plane of projection on which the front view is projected is called the *frontal plane;* that on which the top view is projected, the *horizontal plane*; and that on which the side view is projected, the *profile plane*.

■ 6.2 THE GLASS BOX

If planes of projection are placed parallel to the principal faces of the object, they form a "glass box," as shown in Fig. 6.3a. Notice that the observer is always on the outside looking in, so the object is seen through the planes of projection. Since the glass box has six sides, six views of the object can be obtained.

Note that the object has three principal dimensions: *width, height,* and *depth.* These are fixed terms used for dimensions in these directions, regardless of the shape of the object (see §5.17).

To show the views of a solid, or three-dimensional, object on a flat sheet of paper, it is necessary to unfold the planes so that they will all lie in the same plane (Fig. 6.3b). All planes except the rear plane are hinged

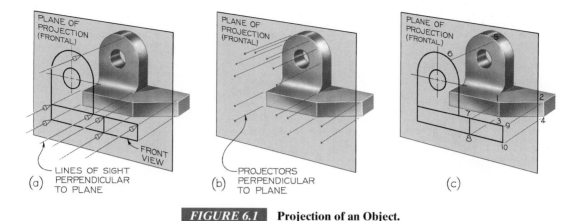

(a) LINES OF SIGHT PERPENDICULAR TO PLANE

(b) PROJECTORS PERPENDICULAR TO PLANE

(c)

FIGURE 6.1 Projection of an Object.

FIGURE 6.2 Top and Right-Side Views.

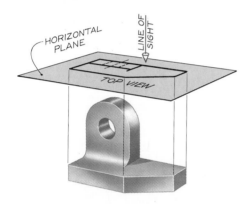

TOP VIEW
(a)

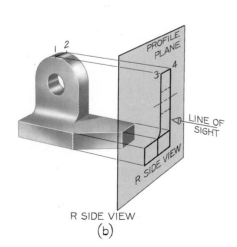

R SIDE VIEW
(b)

to the frontal plane; the rear plane is hinged to the left-side plane, except as explained in §6.8. Each plane revolves outwardly from the original box position until it lies in the frontal plane, which remains stationary. The hinge lines of the glass box are known as *folding lines*.

The positions of these six planes after they have been revolved are shown in Fig. 6.4. Carefully identify each of these planes and corresponding views with its original position in the glass box, and repeat this mental procedure, if necessary, until the revolutions are thoroughly understood.

Observe that in Fig. 6.3b lines extend around the glass box from one view to another on the planes of projection. These are the *projections of the projectors* from points on the object to the views. For example, the projector 1–2 is projected on the horizontal plane at 7–8 and on the profile plane at 16–17. When the top plane is folded up, lines 9–10 and 7–8 will become vertical and line up with 10–6 and 8–2, respectively. Thus, 9–10 and 10–6 form a single straight line 9–6, and 7–8 and 8–2 form a single straight line 7–2, as shown in Fig. 6.4. This explains why the top view is the same width as the front view and why it is placed directly above the front view. The same relation exists between the front and bottom views. Therefore, *the front, top, and bottom views all line up vertically and are the same width.*

In Fig. 6.3b, when the profile plane is folded out, lines 4–13 and 13–15 become a single straight line 4–15, and lines 2–16 and 16–17 become a single straight line 2–17, as shown in Fig. 6.4. The same relation exists between the front, left-side, and rear views. Therefore, *the rear, left-side, front, and right-side views all line up horizontally and are the same height.*

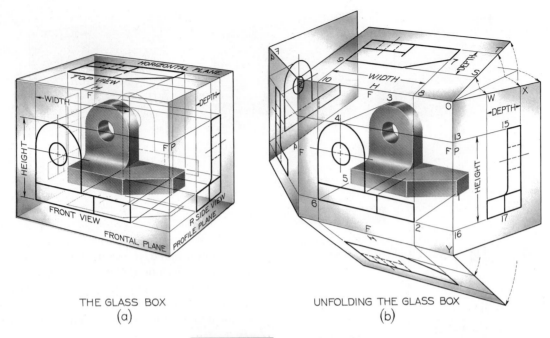

THE GLASS BOX
(a)

UNFOLDING THE GLASS BOX
(b)

FIGURE 6.3   The Glass Box.

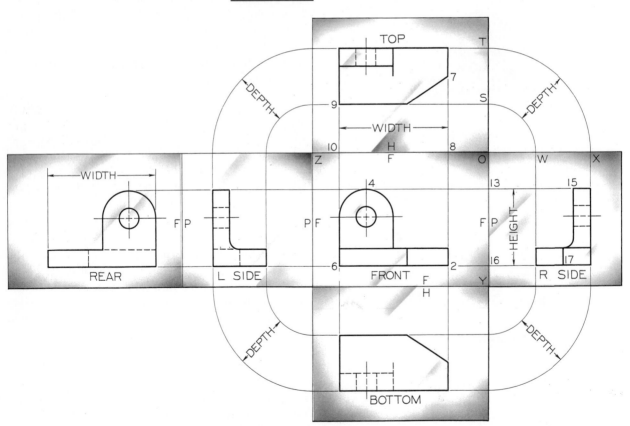

FIGURE 6.4   The Glass Box Unfolded.

Note that in Fig. 6.3b lines OS and OW and lines ST and WX are respectively equal. These lines of equal length are shown in the unfolded position in Fig. 6.4. Thus, it is seen that the top view must be the same distance from the folding line OZ as the right-side view is from the folding line OY. Similarly, the bottom view and left-side view are the same distance from their respective folding lines as are the right-side view and the top view. Therefore, *the top, right-side, bottom, and left-side views are all equidistant from the respective folding lines and are the same depth.* Note that in these four views that surround the front view, the front surfaces of the object are faced inward or toward the front view. Observe also that the left-side and right-side views and the top and bottom views are the reverse of each other in outline shape. Similarly, the rear and front views are the reverse of each other.

■ 6.3 FOLDING LINES

The front, top, and right views of the object shown in the previous figures are shown in Fig. 6.5a with folding lines between the views. These folding lines correspond to the hinge lines of the glass box, as we have seen. The H/F folding line, between the top and front views, is the intersection of the horizontal and frontal planes. The F/P folding line, between the front and side views, is the intersection of frontal and profile planes (see Figs. 6.3 and 6.4).

The distances X and Y, from the front view to the respective folding lines, are not necessarily equal, since they depend on the relative distances of the object from the horizontal and profile planes. However, as explained in §6.2, distances D_1, from the top and side views to the respective folding lines, must always be

equal. Therefore, the views may be any desired distance apart, and the folding lines may be drawn anywhere between them, as long as distances D_1 are kept equal and the folding lines are at right angles to the projection lines between the views.

It will be seen that distances D_2 and D_3, respectively, are also equal and that the folding lines H/F and F/P are in reality reference lines for making equal *depth* measurements in the top and side views. Thus, any point in the top view is the same distance from H/F as the corresponding point in the side view is from F/P.

While it is necessary to understand the folding lines, particularly because they are useful in solving graphical problems in descriptive geometry, they are as a rule omitted in industrial practice as shown in Fig. 6.5b. Again, the distances between the top and front views and between the side and front views are not necessarily equal. Instead of using the folding lines as reference lines for setting off depth measurements in the top and side views, we may use the front surface A of the object as a reference line. In this way, D_1, D_2, and all other depth measurements are made to correspond in the two views in the same manner as if folding lines were used.

■ 6.4 TWO-VIEW INSTRUMENTAL DRAWING

The complete structure of some objects can be shown with only two views. For example, the necessary views of the operating arm in Fig. 6.6a include only the front and top views, as shown by the arrows.

To draw [full size with instruments in Layout A–2 (inside back cover)] the necessary views of the operating arm in Fig. 6.6, do the following:

FIGURE 6.5 Folding Lines.

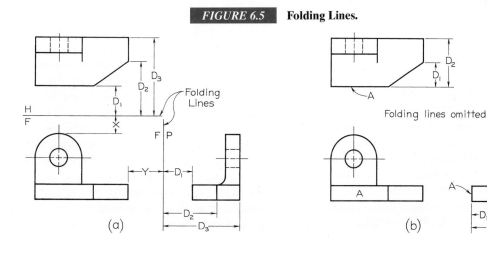

(a)

(b)

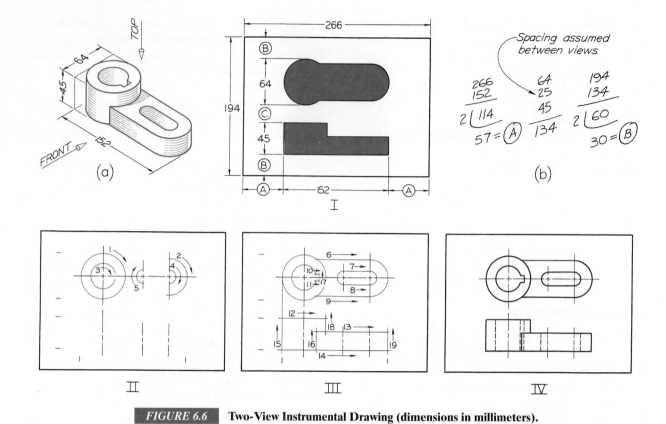

FIGURE 6.6 Two-View Instrumental Drawing (dimensions in millimeters).

1. Determine the spacing of the views (Fig. 6.6I). The width of the front and top views is approximately 152 mm (6″; 25.4 mm = 1″), and the width of the working space is approximately 266 mm (10 $\frac{1}{2}$″). As shown in Fig. 6.6b, subtract 152 mm from 266 mm and divide the result by 2 to get the value of space A. To set off the spaces, place the scale horizontally along the bottom of the sheet and make short vertical marks.

   The depth of the top view is approximately 64 mm (2 $\frac{1}{2}$″) and the height of the front view is 45 mm (1 $\frac{3}{4}$″), while the height of the working space is 194 mm (7 $\frac{5}{8}$″). Assume a space C—say, 25 mm (1″)—between views that will look well and that will provide sufficient space for dimensions, if any.

   Add 64 mm, 25 mm, and 45 mm, subtract the total from 194 mm, and divide the result by 2 to get the value of space B. To set off the spaces, place the scale vertically along the left side of the sheet with the full-size scale on the left, and make short marks perpendicular to the scale.

2. Locate center lines from spacing marks (Fig. 6.6II). Construct arcs and circles lightly.

3. Draw horizontal and then vertical construction lines in the order shown (Fig. 6.6III). Allow construction lines to cross at corners.

4. Add hidden lines and heavy in all final lines, clean-cut and dark (Fig. 6.6IV).

■ 6.5 TRANSFERRING DEPTH DIMENSIONS

Since all depth dimensions in the top and side views must correspond point for point, accurate methods of transferring these distances must be used.

Professional drafters transfer dimensions between the top and side views either with dividers or a scale (Figs. 6.7a and 6.7b). The scale method is especially convenient when a drafting machine is used because both vertical and horizontal scales are readily available. Beginners may find it convenient to use a 45° miter line to project dimensions between top and side views (Fig. 6.7c). Note that the right-side view may be moved to the right or left, or the top view may be moved upward or downward, by shifting the 45° line accordingly. It is not necessary to draw continuous

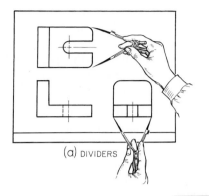
(a) DIVIDERS

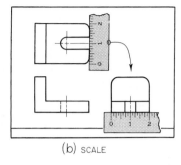
(b) SCALE

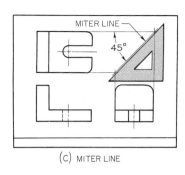
(c) MITER LINE

FIGURE 6.7   **Transferring Depth Dimensions.**

lines between the top and side views via the miter line. Instead, make short dashes across the miter line and project from these.

The 45° miter-line method is also convenient for transferring a large number of points, as when plotting a curve (see Fig. 6.35).

■ 6.6 PROJECTING A THIRD VIEW

Figure 6.8a is a pictorial drawing of an object that has three necessary views. Each corner of the object is numbered. Figure 6.8I shows the top and front views, with each corner properly numbered. If a point is *visible* in a given view, the number is placed *outside* the corner, but if the point is hidden, the numeral is placed *inside* the corner. For example, in Fig. 6.8I, point 1 is visible in both views and the number is therefore placed outside the corners in both views. However, point 2 is visible in the top view and the number is placed outside, while in the front view it is hidden and is placed inside.

This system, in which points are identified by the same numbers in all views, is useful in projecting known points in two views to unknown positions in a third view. Note that this numbering system assigns the same number to a given point in all views and should not be confused with the system used in Fig. 6.23 and elsewhere, in which a point has different numbers in each view.

Before starting to project the right-side view in Fig. 6.8, try to visualize the view as seen in the direction of the arrow in Fig. 6.8a. Then construct the right-side view point by point, using a hard pencil and very light lines.

As shown in Fig. 6.8I, locate point 1 in the side view by projecting from point 1 in the top view and point 1 in the front view. In Fig. 6.8II, project points 2,

3, and 4 in a similar manner to complete the vertical end surface of the object. In Fig. 6.8III, project points 5 and 6 to complete the side view of the inclined surface 5–6–2–1. This completes the right-side view, since invisible points 9, 10, 8, and 7 are directly behind visible corners 5, 6, 4, and 3, respectively.

As shown in Fig. 6.8IV, the drawing is completed by heavying in the lines in the right-side view.

■ 6.7 THREE-VIEW INSTRUMENTAL DRAWING

The complete structure of some objects, such as that in Fig. 6.9a, requires three views.

To draw (full size with instruments on Layout A–2), the necessary views of the V-block in Fig. 6.9a, do the following:

1. Determine the spacing of the views (Fig. 6.9I). The width of the front view is 108 mm and the depth of the side view is 58 mm, while the width of the working space is 266 mm. Assume a space C between views (say, 32 mm) that will look well and will allow sufficient space for dimensions, if any.

   As shown in Fig. 6.9b, add 108 mm, 32 mm, and 58 mm, subtract the total from 266 mm, and divide the result by 2 to get the value of space A. To set off these horizontal spacing measurements, place the scale along the bottom of the sheet and make short vertical marks.

   The depth of the top view is 58 mm and the height of the front view is 45 mm, while the height of the working space is 194 mm. Assume a space D between views (say, 25 mm). As explained in §6.3, space D need not be the same as space C. Add 58 mm, 25 mm, and 45 mm, subtract the total from 194 mm, and divide the result by 2 to get the value

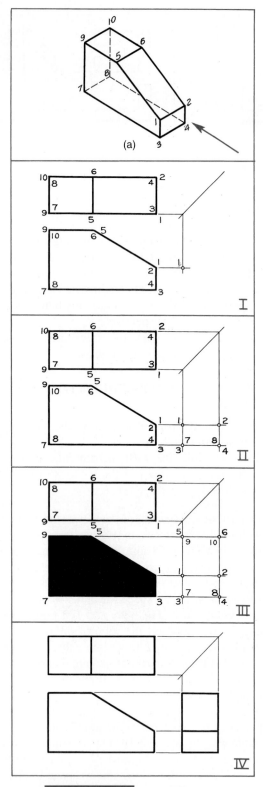

FIGURE 6.8   **Use of Numbers.**

of space B. To set off these vertical spacing measurements, place the scale along the left side of the sheet with the scale used on the left, and make short marks perpendicular to the scale. Allow space for written dimensions, if required.

2. Locate the center lines from the spacing marks (Fig. 6.9II). Construct lightly the arcs and circles.

3. Draw horizontal, then vertical, then inclined construction lines, in the order shown (Fig. 6.9III). Allow construction lines to cross at the corners. Do not complete one view at a time; construct the views simultaneously. A convenient method for transferring the hole diameter from the top view to the side view is to use the compass with the same setting used for drawing the hole.

4. Add hidden lines and heavy in all final lines, clean-cut and dark (Fig. 6.9IV).

■ 6.8 ALTERNATE POSITIONS OF VIEWS

If three views of a wide, flat object are drawn using the conventional arrangement of views, a large wasted space is left on the paper (Fig. 6.10a). In such cases, the profile plane may be considered hinged to the horizontal plane instead of the frontal plane (Fig. 6.10b). This places the side view beside the top view, which results in better spacing and sometimes makes the use of a reduced scale unnecessary.

It is also permissible in extreme circumstances to place the side view across horizontally from the bottom view. In this case the profile plane is considered hinged to the bottom plane of projection. Similarly, the rear view may be placed directly above the top view or under the bottom view, if necessary. As a result, the rear plane is considered hinged to the horizontal or bottom plane, as the case may be, and then rotated into coincidence with the frontal plane.

■ 6.9 PARTIAL VIEWS

A view may not need to be complete but may show only what is necessary for the clear description of the object (Fig. 6.11). Such a view is a partial view. A break line may be used to limit the partial view (Fig. 6.11a); the contour of the part shown may limit the view (Fig. 6.11b); if symmetrical, a half-view may be drawn on one side of the center line (Fig. 6.11c); or a partial view, "broken out," may be drawn (Fig. 6.11d). The half view shown in Figs. 6.11c and 6.11d should be the

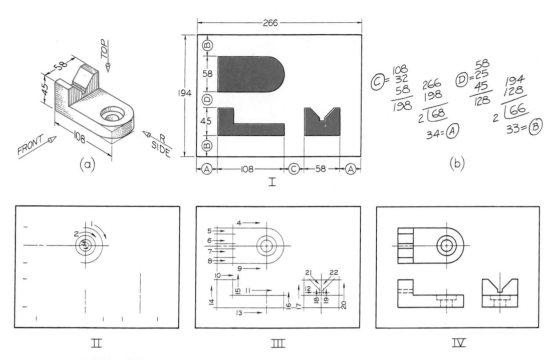

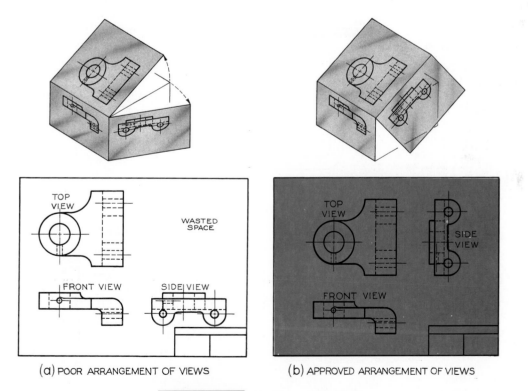

FIGURE 6.9   Three-View Instrumental Drawing (dimensions in millimeters).

FIGURE 6.10   Position of Side View.

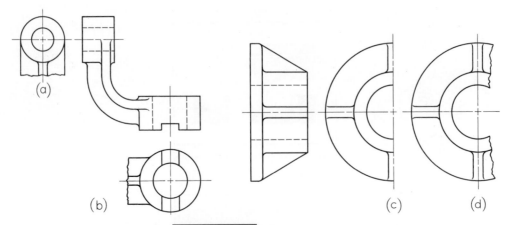

FIGURE 6.11 **Partial Views.**

near side, as shown. For half views in connection with sections, see Fig. 7.33.

Do not place a break line where it will coincide with a visible or hidden line.

Occasionally the distinctive features of an object are on opposite sides. In either complete side view there will be a considerable overlapping of shapes. In such cases two side views are often the best solution (Fig. 6.12). The views are partial views, and in both,

certain visible and invisible lines have been omitted for clarity.

■ 6.10 REVOLUTION CONVENTIONS

Regular multiview projections are sometimes awkward, confusing, or actually misleading. For example, Fig. 6.13a shows an object that has three triangular ribs, three holes equally spaced in the base, and a key-

FIGURE 6.12 **Incomplete Side Views.**

FIGURE 6.13 **Revolution Conventions.**

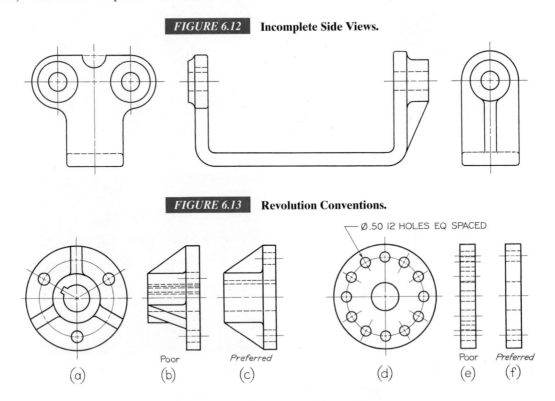

way. The right-side view (Fig. 6.13b) is a regular projection and is not recommended. The lower ribs appear in a foreshortened position, the holes do not appear in their true relation to the rim of the base, and the keyway is projected as a confusion of hidden lines.

The conventional method shown in Fig. 6.13c is preferred, not only because it is simpler to read, but also because it requires less drafting time. Each of the features mentioned has been revolved in the front view to lie along the vertical center line, from where it is projected to the correct side view (Fig. 6.13c).

Figures 6.13d and 6.13e show regular views of a flange with many small holes. The hidden holes in Fig. 6.13e are confusing and take unnecessary time to draw. The preferred representation in Fig. 6.13f shows the holes revolved, and the drawing is clear.

Another example is given in Fig. 6.14. As shown in Fig. 6.14a, a regular projection produces a confusing foreshortening of the inclined arm. To preserve the appearance of symmetry about the common center, the lower arm is revolved to line up vertically in the front view so that it projects true length in the side view (Fig. 6.14b).

Revolutions of the type discussed here are frequently used in connection with sectioning. Such sectional views are called *aligned sections* (see §7.13).

■ 6.11 REMOVED VIEWS

A removed view is a complete or partial view removed to another place on the sheet so that it no longer is in direct projection with any other view (Fig. 6.15). Such a view may be used to show some feature of the object more clearly, possibly to a larger scale, or to save drawing a complete regular view. A viewing-plane line is used to indicate the part being viewed; the arrows at the corners show the direction of sight (see §7.5). The removed views should be labeled VIEW A–A or VIEW B–B and so on; the letters refer to those placed at the corners of the viewing-plane line.

■ 6.12 VISUALIZATION

As stated in §1.9, the ability to *visualize* or *think in three dimensions* is one of the most important abilities of successful engineers, designers, and scientists. In practice, this means the ability to study the views of an object and to form a mental picture of it—to *visualize* its three-dimensional shape. To the designer it means the ability to *synthesize*, or form, a mental picture before the object even exists and the ability to express this image in terms of views. The engineer is the master planner in the construction of new equipment, structures, or processes. The ability to visualize and to

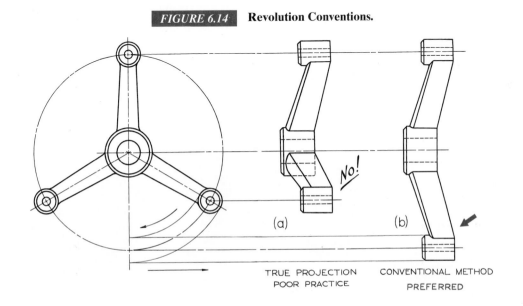

FIGURE 6.14   **Revolution Conventions.**

(a) (b)

TRUE PROJECTION
POOR PRACTICE

CONVENTIONAL METHOD
PREFERRED

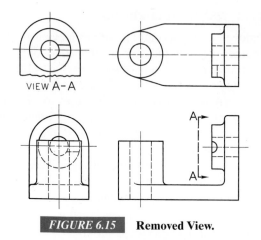

FIGURE 6.15 Removed View.

1. The front view (Fig. 6.16I) shows that the object is L-shaped, the height and width of the object, and the thickness of the members. The meaning of the hidden and center lines is not yet clear; nor do we yet know the depth of the object.

2. The top view (Fig. 6.16II) tells us that the horizontal member is rounded at the right end and has a round hole. Some kind of slot is indicated at the left end. The depth and width of the object are shown.

3. The right-side view (Fig. 6.16III) tells us that the left end of the object has rounded corners at the top and has an open-end slot in a vertical position. The height and depth of the object are shown.

Thus, each view provides certain definite information regarding the shape of the object. All views must be considered to visualize the object completely.

use the graphic language as a means of communication or recording of mental images is indispensable.

Even experienced engineers and designers cannot look at a multiview drawing and instantly visualize the object represented (except for the simplest shapes) any more than we can grasp the ideas on a book page merely at a glance. It is necessary to *study* the drawing, to read the lines in a logical way, and to piece together the little things until a clear idea of the whole emerges. How this is done is described in §§6.13–6.32.

■ 6.13 VISUALIZING THE VIEWS

Figure 6.16 illustrates a method of reading drawings that is essentially the reverse of the mental process used in obtaining the views by projection. The given views of an angle bracket are shown in Fig. 6.16a.

■ 6.14 MODELS

One of the best aids to visualization is an actual model of the object. Such a model need not be made accurately to scale and may be made of any convenient material, such as modeling clay, soap, wood, styrofoam, or any material that can easily be shaped, carved, or cut.

A typical example of the use of soap or clay models is shown in Fig. 6.17. Three views of an object are given in Fig. 6.17a, and the student is to supply a missing line. The model is carved as shown in Figs. 6.17I–6.17III. The "missing" line discovered in the process is added to the drawing (Fig. 6.17b).

FIGURE 6.16 Visualizing from Given Views.

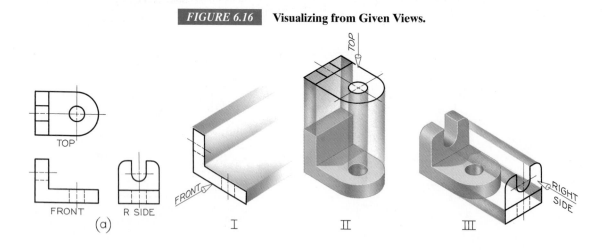

FIGURE 6.17   Use of Model to Aid Visualization.

Some typical examples of soap models are shown in Fig. 6.18.

■ 6.15 SURFACES, EDGES, AND CORNERS

To analyze and synthesize multiview projections, the component elements that make up most solids must be considered. A *surface* (plane) may be bounded by straight lines, curves, or a combination of them. A surface may be *frontal, horizontal,* or *profile,* accord-ing to the plane of projection to which it is parallel (see §6.1).

If a plane surface is perpendicular to a plane of projection, it appears as a line or *edge view* (EV) (Fig. 6.19a). If it is parallel, it appears as a *true size* (TS) sur-face (Fig. 6.19b). If it is situated at an angle, it appears as a *foreshortened* (FS) surface (Fig. 6.19c). Thus, *a plane surface always projects as a line or a surface.*

The intersection of two plane surfaces produces an *edge*, or a straight line. Such a line is common to both surfaces and forms a boundary line for each. If an edge

FIGURE 6.18   Soap Models.

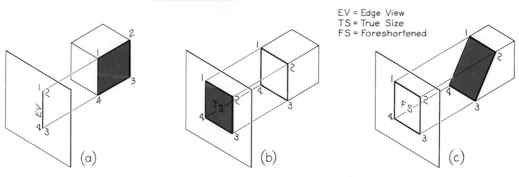

FIGURE 6.19   Projections of Surfaces.

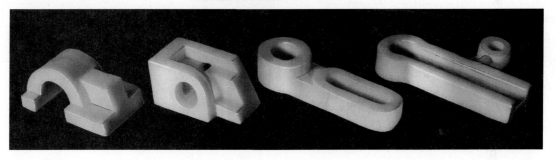

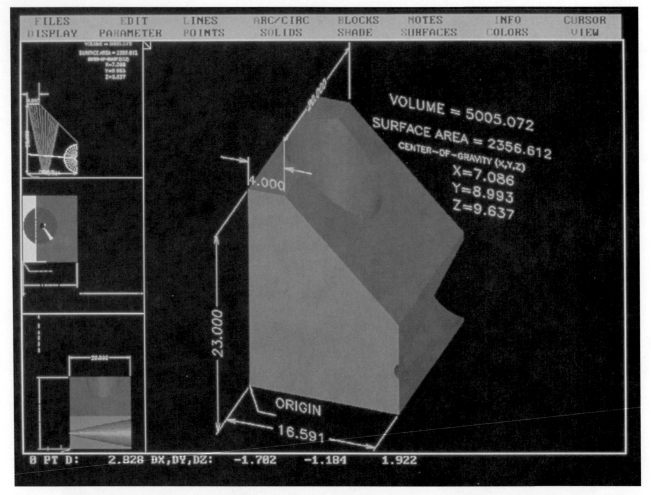

A CAD Solid Model and Several Surface Projections. *Courtesy of American Small Business Computers.*

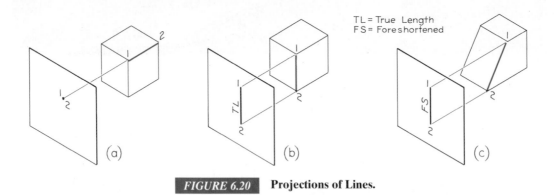

FIGURE 6.20   Projections of Lines.

is perpendicular to a plane of projection, it appears as a point (Fig. 6.20a); otherwise it appears as a line, as in Figs.6.20b and 6.20c. If it is parallel to the plane of projection, it shows true length (Fig. 6.20b); if not parallel,

it appears foreshortened (Fig. 6.20c). Thus, a *straight line always projects as a straight line or as a point.* A line may be *frontal, horizontal,* or *profile,* according to the plane of projection to which it is parallel.

A *corner*, or point, is the common intersection of three or more surfaces or edges. A point appears as a point in every view.

■ 6.16 ADJACENT AREAS

Consider the top view in Fig. 6.21a. Lines divide the view into three areas. Each area must represent a surface *at a different level*. Surface A may be high and surfaces B and C lower (Fig. 6.21b). Or B may be lower than C (Fig. 6.21c). Or B may be highest, with C and A each lower (Fig. 6.21d). Or one or more surfaces may be inclined (Fig. 6.21e). Or one or more surfaces may be cylindrical (Fig. 6.21f), and so on. *No two adjacent areas can lie in the same plane.*

The same reasoning can apply, of course, to the adjacent areas in any given view. Since an area (surface) in a view can be interpreted in several different ways, other views must be observed to determine which interpretation is correct.

■ 6.17 SIMILAR SHAPES OF SURFACES

If a surface is viewed from several different positions, it will in each case been seen to have a certain number of sides and to have a certain characteristic shape. An L-shaped surface (Fig. 6.22a) will appear as an L-shaped figure in every view in which it does not appear as a line. A T-shaped surface (Fig. 6.22b), a U-shaped surface (Fig. 6.22c), or a hexagonal surface

(Fig. 6.22d) will in each case have the same number of sides and the same characteristic shape in every view in which it appears as a surface.

This repetition of shapes is one of our best methods for analyzing views.

■ 6.18 READING A DRAWING

Suppose we want to visualize the object shown by three views in Fig. 6.23. Since no lines are curved, we know that the object is made up of plane surfaces.

Surface 2-3-10-9-6-5 in the top view is an L-shaped surface of six sides. It appears in the side view at 16-17-21-20-18-19 and is L-shaped and six-sided. No such shape appears in the front view, but we note that points 2 and 5 line up with 11 in the front view, points 6 and 9 line up with 13, and points 3 and 10 line up with 15. Evidently, line 11-15 in the front view is the edge view of the L-shaped surface.

Surface 11-13-12 in the front view is triangular in shape, but no corresponding triangles appear in either the top or the side view. We note that point 12 lines up with 8 and 4 and that point 13 lines up with 6 and 9. However, surface 11-13-12 of the front view cannot be the same as surface 4-6-9-8 in the top view because the former has three sides and the latter has four. Obviously, the triangular surface appears as line 4-6 in the top view and as line 16-19 in the side view.

Surface 12-13-15-14 in the front view is trapezoidal in shape. But there are no trapezoids in the top

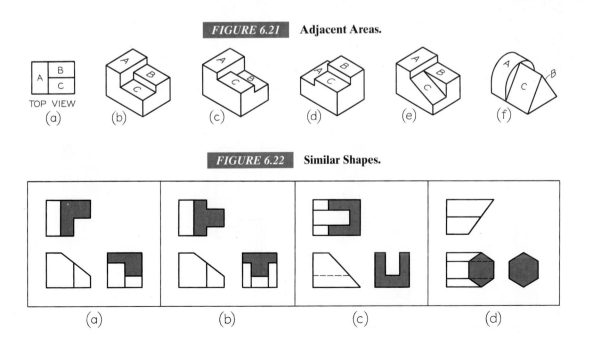

FIGURE 6.21   **Adjacent Areas.**

TOP VIEW
(a) (b) (c) (d) (e) (f)

FIGURE 6.22   **Similar Shapes.**

(a) (b) (c) (d)

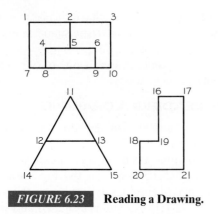

FIGURE 6.23   **Reading a Drawing.**

and side views, so the surface evidently appears in the top view as line 7–10 and in the side view as line 18–20.

The remaining surfaces can be identified in the same manner, where it will be seen that the object is bounded by seven plane surfaces, two of which are rectangular, two triangular, two L-shaped, and one trapezoidal.

Note that the numbering system used in Fig. 6.23 is different from that in Fig. 6.8 in that different numbers are used for all points and there is no significance in a point being inside or outside a corner.

■ 6.19 NORMAL SURFACES

A *normal surface* is a plane surface that is parallel to a plane of projection. It appears in true size and shape on the plane to which it is parallel, and as a vertical or a horizontal line on adjacent planes of projection.

In Fig. 6.24 four stages in machining a block of steel to produce the final tool block in Fig 6.24IV are shown. All surfaces are normal surfaces. In Fig. 6.24I, normal surface A is parallel to the horizontal plane

FIGURE 6.24   **Machining a Tool Block—Normal Surfaces and Edges.**

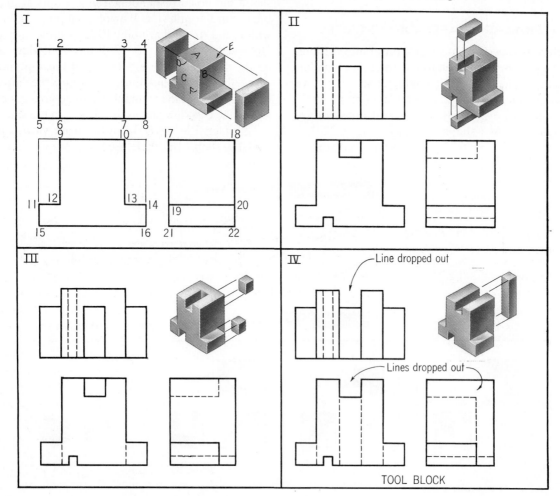

and appears true size in the top view at 2–3–7–6, as line 9–10 in the front view, and as line 17–18 in the side view. Normal surface B is parallel to the profile plane and appears true size in the side view at 17–18–20–19, as line 3–7 in the top view, and as line 10–13 in the front view. Normal surface C, an inverted T-shaped surface, is parallel to the frontal plane and appears true size in the front view at 9–10–13–14–16–15–11–12, as line 5–8 in the top view, and as line 17–21 in the side view.

All other surfaces of the object may be visualized in a similar manner. In the four stages of Fig. 6.24, observe carefully the changes in the views produced by the machining operations, including the introduction of new surfaces, new visible edges, and hidden edges and the dropping out of certain lines as the result of a new cut.

The top view in Fig. 6.24I is cut by lines 2–6 and 3–7, which means that there are three surfaces, 1–2–6–5, 2–3–7–6, and 3–4–8–7. In the front view, surface 9–10 is seen to be the highest, and surfaces 11–12 and 13–14 are at the same lower level. In the side view, both of these latter surfaces appear as one line 19–20. Surface 11–12 might appear as a hidden line in the side view, but surface 13–14 appears as a visible line 19–20, which covers up the hidden line and takes precedence over it (see §5.30).

■ 6.20 NORMAL EDGES

A **normal edge** is a line that is perpendicular to a plane of projection. It appears as a point on the plane of projection to which it is perpendicular and as a line in true length on adjacent planes of projection. In Fig. 6.24I, edge D is perpendicular to the profile plane of projection and appears as point 17 in the side view. It is parallel to the frontal and horizontal planes of projection and is shown true length at 9–10 in the front view and 6–7 in the top view. Edges E and F are perpendicular, respectively, to the frontal and horizontal planes of projection, and their views may be similarly analyzed.

■ 6.21 INCLINED SURFACES

An **inclined surface** is a plane surface that is perpendicular to one plane of projection but inclined to adjacent planes. An inclined surface projects as a straight line on the plane to which it is perpendicular; it appears foreshortened (FS) on planes to which it is inclined, with the degree of foreshortening being proportional to the angle of inclination.

Figure 6.25 shows four stages in machining a locating finger, producing several inclined surfaces. In Fig. 6.25I, inclined surface A is perpendicular to the horizontal plane of projection and appears as line 5–3 in the top view. It is shown as a foreshortened surface in the front view at 7–8–11–10 and in the side view at 12–13–16–15. Note that the surface is more foreshortened in the side view than in the front view because the plane makes a greater angle with the profile plane of projection than with the frontal plane of projection.

In Fig. 6.25III, edge 23–24 in the front view is the edge view of an inclined surface that appears in the top view as 21–2–3–22 and in the side view as 25–14–27–26. Note that 25–14 is equal in length to 21–22 and that the surface has the same number of sides (four) in both views in which it appears as a surface.

In Fig. 6.25IV, edge 29–23 in the front view is the edge view of an inclined surface that appears in the top view as visible surface 1–21–22–5–18 and in the side view as invisible surface 25–14–32–31–30. While the surface does not appear true size in any view, it does have the same characteristic shape and the same number of sides (five) in the views in which it appears as a surface.

To obtain the true size of an inclined surface, it is necessary to construct an auxiliary view (Chapter 8) or to revolve the surface until it is parallel to a plane of projection (Chapter 9).

■ 6.22 INCLINED EDGES

An **inclined edge** is a line that is parallel to a plane of projection but inclined to adjacent planes. It appears true length on the plane to which it is parallel and foreshortened on adjacent planes, with the degree of foreshortening being proportional to the angle of inclination. The true-length view of an inclined line is always inclined, while the foreshortened views are either vertical or horizontal lines.

In Fig. 6.25I, inclined edge B is parallel to the horizontal plane of projection and appears true length in the top view at 5–3. It is foreshortened in the front view at 7–8 and in the side view at 12–13. Note that plane A produces two normal edges and two inclined edges.

In Figs. 6.25III and 6.25IV, some of the sloping lines are not inclined lines. In Fig. 6.25III, the edge that appears in the top view at 21–22, in the front view at

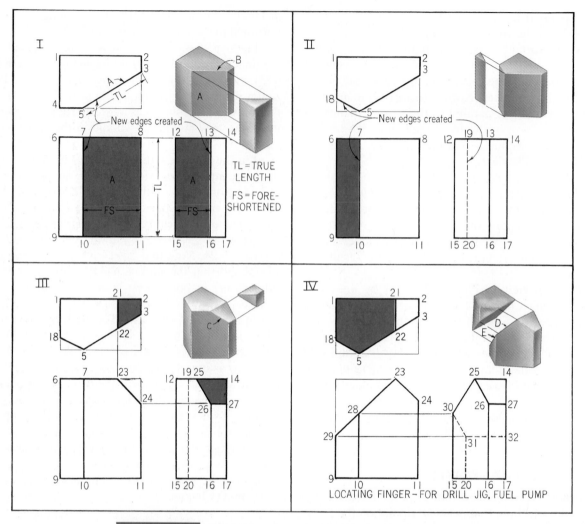

FIGURE 6.25   Machining a Locating Finger—Inclined Surfaces.

23–24, and in the side view at 14–27 is an inclined line. However, the edge that appears in the top view at 22–23, in the front view at 23–24, and in the side view at 25–26 is not an inclined line by the definition given here. Actually, it is an oblique line (see §6.24).

■ 6.23 OBLIQUE SURFACES

An *oblique surface* is a plane that is oblique to all planes of projection. Since it is not perpendicular to any plane, it cannot appear as a line in any view. Since it is not parallel to any plane, it cannot appear true size in any view. Thus, an oblique surface always appears as a foreshortened surface in all three views.

In Fig. 6.26II, oblique surface C appears in the top view at 25–3–6–26 and in the front view at 29–8–31–30. What are its numbers in the side view? Note that any surface appearing as a line in any view cannot be an oblique surface. How many inclined surfaces are there? How many normal surfaces?

To obtain the true size of an oblique surface, it is necessary to construct a secondary, auxiliary view (see §§8.21 and 8.22) or to revolve the surface until it is parallel to a plane of projection (see §9.11).

■ 6.24 OBLIQUE EDGES

An *oblique edge* is a line that is oblique to all planes of projection. Since it is not perpendicular to any plane, it cannot appear as a point in any view. Since it is not parallel to any plane, it cannot appear true length in

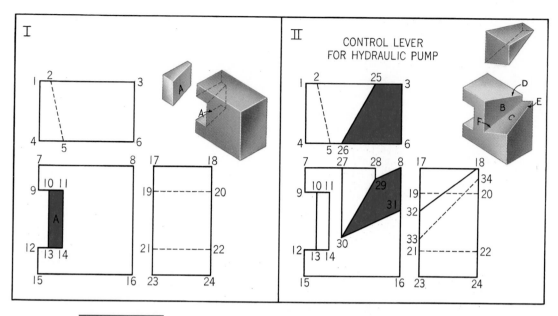

Machining a Control Lever—Inclined and Oblique Surfaces.

any view. An oblique edge appears foreshortened and in an inclined position in every view.

In Fig. 6.26II, oblique edge F appears in the top view at 26–25, in the front view at 30–29, and in the side view at 33–34.

■ 6.25 PARALLEL EDGES

If a series of parallel planes is intersected by another plane, the resulting lines of intersection are parallel (Fig. 6.27a). In Fig. 6.27b, the top plane of the object intersects the front and rear planes, producing the parallel edges 1–2 and 3–4. If two lines are parallel in space, their projections in any view are parallel. The example in Fig. 6.27b is a special case in which the two lines appear as points in one view and coincide as a single line in another and should not be regarded as an exception to the rule. Note that even in the pictorial drawings the lines are shown parallel. Parallel inclined lines are shown in Fig. 6.27c, and parallel oblique lines in Fig. 6.27d.

Figure 6.28 shows three views of an object after a plane has been passed through the points A, B, and C. As shown in Fig. 6.28b, only points that lie in the same plane are joined. In the front view, join points A and C, which are in the same plane, extending the line to P on the vertical front edge of the block extended. In the side view, join P to B, and in the top view, join B to A.

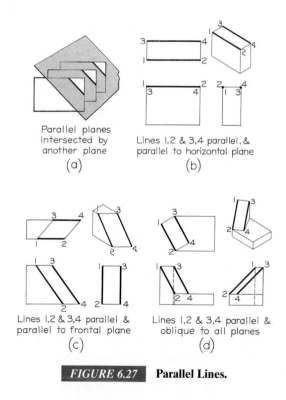

Parallel planes intersected by another plane (a)

Lines 1,2 & 3,4 parallel, & parallel to horizontal plane (b)

Lines 1,2 & 3,4 parallel & parallel to frontal plane (c)

Lines 1,2 & 3,4 parallel & oblique to all planes (d)

FIGURE 6.27 **Parallel Lines.**

Complete the drawing by applying this rule: *Parallel lines in space will be projected as parallel lines in any view.* The remaining lines are thus drawn parallel to lines AP, PB, and BA.

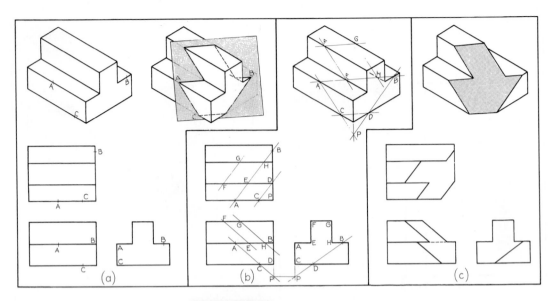

FIGURE 6.28 **Oblique Surface.**

■ 6.26 ANGLES

If an angle is in a normal plane—that is, parallel to a plane of projection—the angle will be shown true size on the plane of projection to which it is parallel (Fig. 6.29a). If the angle is in an inclined plane, it may be projected either larger or smaller than the true angle, depending on its position (Figs. 6.29b and 6.29c). In Fig. 6.29b, the 45° angle is shown *oversize* in the front view, while in Fig. 6.29c the 60° angle is shown *undersize* in both views.

A 90° angle will be projected true size, even if it is in an inclined plane, provided that one leg of the angle is a normal line (Fig. 6.29d). In this figure, the 60° angle is projected *oversize* and the 30° angle *undersize*. Study these relations, using your own 30° × 60° triangle as a model.

■ 6.27 CURVED SURFACES

Rounded surfaces are common in engineering practice because they are easily formed on the lathe, the drill press, and other machines using the principle of rotation either of the "work" or of the cutting tool. The most common rounded surfaces are the cylinder, cone, and sphere, a few of whose applications are shown in Fig. 6.30. For other geometric solids, see Fig. 4.7.

■ 6.28 CYLINDRICAL SURFACES

Three views of a right-circular cylinder, the most common type, are shown in Fig. 6.31a. The single cylindri-

cal surface is intersected by two plane (normal) surfaces, forming two curved lines of intersection or circular edges (the bases of the cylinder). These circular edges are the only actual edges on the cylinder. Figure 6.31b shows a cylindrical hole in a right square prism.

The cylinder is represented on a drawing by its circular edges and the contour elements. An *element* is a straight line on the cylindrical surface, parallel to the axis, as shown in the pictorial view of the cylinder in Fig. 6.31a. In both parts of Fig. 6.31, the circular edges of the cylinder appear in the top views as circles A, in the front views as horizontal lines 5–7 and 8–10, and in the side views as horizontal lines 11–13 and 14–16.

The contour elements 5–8 and 7–10 in the front views appear as points 3 and 1 in the top views. The contour elements 11–14 and 13–16 in the side views appear as points 2 and 4 in the top views.

In Fig. 6.32, four possible stages in machining a cap are shown, producing several cylindrical surfaces. In Fig. 6.32I, the removal of the two upper corners forms cylindrical surface A, which appears in the top view as surface 1–2–4–3, in the front view as arc 5, and in the side view as surface 8–9–Y–X.

In Fig. 6.32II, a large reamed hole shows in the front view as circle 16, in the top view as cylindrical surface 12–13–15–14, and in the side view as cylindrical surface 17–18–20–19.

In Fig. 6.32III, two drilled and counterbored holes are added, producing four more cylindrical surfaces

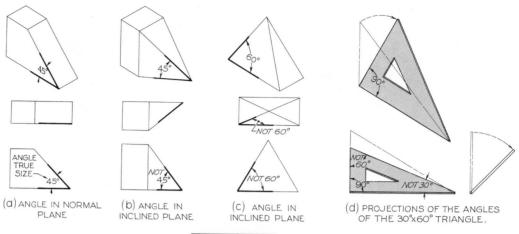

(a) ANGLE IN NORMAL PLANE

(b) ANGLE IN INCLINED PLANE

(c) ANGLE IN INCLINED PLANE

(d) PROJECTIONS OF THE ANGLES OF THE 30°x60° TRIANGLE.

FIGURE 6.29   **Angles.**

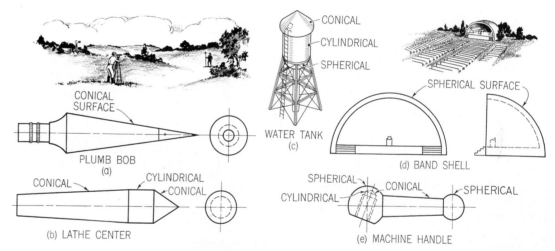

CONICAL SURFACE

PLUMB BOB
(a)

(b) LATHE CENTER

WATER TANK
(c)

(d) BAND SHELL

(e) MACHINE HANDLE

FIGURE 6.30   **Curved Surfaces.**

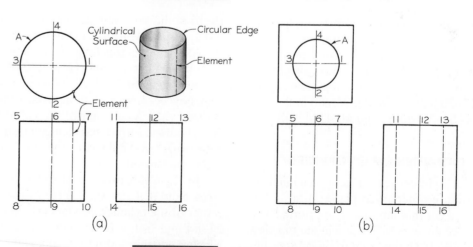

(a)

(b)

FIGURE 6.31   **Cylindrical Surfaces.**

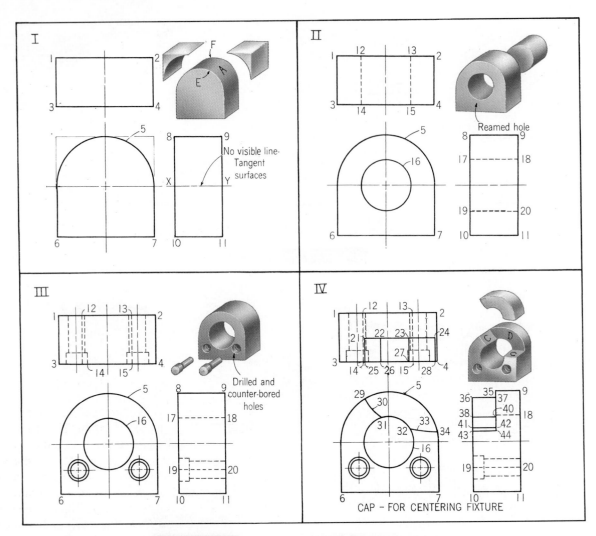

FIGURE 6.32   Machining a Cap—Cylindrical Surfaces.

and two normal surfaces. The two normal surfaces are those at the bottoms of the counterbores.

In Fig. 6.32IV, a cylindrical cut is added, producing two cylindrical surfaces that appear edgewise in the front view as arcs 30 and 33, in the top view as surfaces 21–22–26–25 and 23–24–28–27, and in the side view as surfaces 36–37–40–38 and 41–42–44–43.

■ 6.29 DEFORMITIES OF CYLINDERS

In shop practice, cylinders are usually machined or formed to introduce other, often plane, surfaces. Figure 6.33a shows a cut that introduces two normal surfaces. One surface appears as line 3–4 in the top view, as surface 6–7–10–9 in the front view, and as line 13–16 in the side view. The other appears as line

15–16 in the side view, as line 9–10 in the front view, and as surface 3–4, arc 2 in the top view.

All elements touching arc 2, between 3 and 4 in the top view, become shorter as a result of the cut. For example, element A, which shows as a point in the top view, now becomes CD in the front view and 15–17 in the side view. As a result of the cut, the front half of the cylindrical surface has changed from 5–8–12–11 to 5–6–9–10–7–8–12–11 (front view). The back half remains unchanged.

In Fig. 6.33b, two cuts introduce four normal surfaces. Note that surface 7–8 (top view) is through the center of the cylinder, producing in the side view line 21–24 and in the front view surface 11–14–16–15, which is equal in width to the diameter of the cylinder. Surface 15–16 (front view) is read in the top view as

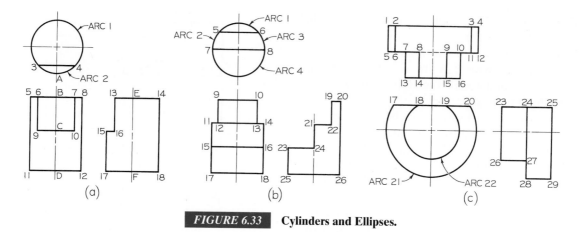

FIGURE 6.33   Cylinders and Ellipses.

7-8-ARC 4. Surface 11–14 (front view) is read in the top view as 5-6-ARC 3-8-7-ARC 2.

In Fig. 6.33c, two cylinders on the same axis are shown, intersected by a normal surface parallel to the axis. Surface 17–20 (front view) is 23–25 in the side view and 2-3-11-9-15-14-8-6 in the top view. A common error is to draw a visible line in the top view between 8 and 9. However, this would produce two surfaces 2-3-11-6 and 8-9-15-14 not in the same plane. In the front view, the larger surface appears as line 17–20 and the smaller as line 18–19. These lines coincide; hence, they are all one surface, and there can be no visible line joining 8 and 9 in the top view.

The vertical surface that appears in the front view at 17-18-ARC 22-19-20-ARC 21 appears as a line in the top view at 5–12, which explains the hidden line 8–9 in the top view.

■ 6.30 CYLINDERS AND ELLIPSES

If a cylinder is cut by an inclined plane, as in Fig. 6.34a, the inclined surface is bounded by an ellipse. The ellipse appears as circle 1 in the top view, as straight line 2–3 in the front view, and as ellipse ADBC in the side view. Note that circle 1 in the top view would remain a circle regardless of the angle of the cut. If the cut is 45° with horizontal, the ellipse will appear as a circle in the side view (see phantom lines) since the major and minor axes in that view would be equal. Finding the true size and shape of the ellipse requires an auxiliary view

FIGURE 6.34   Deformities of Cylinders.

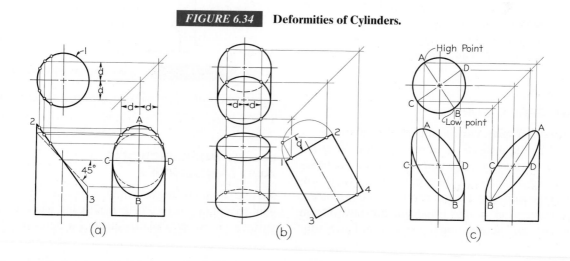

with the line of sight perpendicular to surface 2–3 in the front view (see §8.12).

Since the major and minor axes AB and CD are known, the ellipse can be drawn by any of the methods in Figs. 4.48–4.50 and 4.52a (true ellipses) or with the aid of an ellipse template.

If a cylinder is tilted forward its bases or circular edges 1–2 and 3–4 (side view) become ellipses in the front and top views (Fig. 6.34b). Points on the ellipses can be plotted from the semicircular end view of the cylinder, as shown, with distances d being equal. Since the major and minor axes for each ellipse are known, the ellipses can be drawn with the aid of an ellipse template, or by any of the true ellipse methods, or by the approximate method.

If a cylinder is cut by an oblique plane, the elliptical surface appears as an ellipse in two views (Fig. 6.34c). In the top view, points A and B are selected, diametrically opposite, as the high and low points in the ellipse, and CD is drawn perpendicular to AB. These are the projections of the major and minor axes, respectively, of the actual ellipse in space. In the front and side views, points A and B are assumed at the desired altitudes. Since CD appears true length in the top view, it will appear horizontal in the front and side views, as shown. These axes in the front and side views are the conjugate axes of the ellipses. The ellipses may be drawn on these axes by the methods shown in Figs. 4.51 and 4.52b or by trial with the aid of an ellipse template.

The intersection of a plane and a quarter-round molding is shown in Fig. 6.35a, and intersection with a cove molding is shown in Fig. 6.35b. In both figures, assume points 1, 2, 3, ... at random in the side views in which the cylindrical surfaces appear as curved lines, and project the points to the front and top views, as shown. A sufficient number of points should be used to ensure smooth curves. Draw the final curves through the points with an irregular curve.

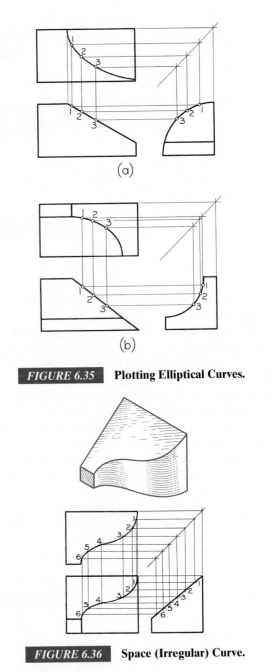

FIGURE 6.35 **Plotting Elliptical Curves.**

FIGURE 6.36 **Space (Irregular) Curve.**

■ 6.31 SPACE CURVES

The views of a space curve (an irregular cuve) are established by the projections of points along the curve (Fig. 6.36). In this figure, any points 1, 2, 3, ... are selected along the curve in the top view and then projected to the side view (or the reverse), and points are located in the front view by projecting downward from the top view and across from the side view. The resulting curve in the front view is drawn with an irregular curve.

■ 6.32 INTERSECTIONS AND TANGENCIES

No line should be drawn where a curved surface is tangent to a plane surface (Fig. 6.37a), but when a curved surface *intersects* a plane surface, a definite edge is

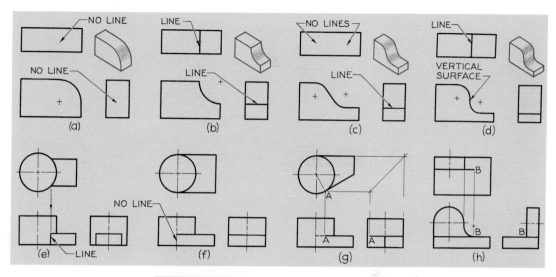

FIGURE 6.37   Intersections and Tangencies.

formed (Fig. 6.37b). If curved surfaces are arranged as in Fig. 6.37c, no lines appear in the top view. If the surfaces are arranged as in Fig. 6.37d, a vertical surface in the front view produces a line in the top view. Other typical intersections and tangencies of surfaces are shown in Figs. 6.37e–6.37h. To locate the point of tangency A in Fig. 6.37g, refer to Fig. 4.34b.

The intersection of a small cylinder with a large cylinder is shown in Fig. 6.38a. The intersection is so small that it is not plotted; a straight line is used instead. In Fig. 6.38b, the intersection is larger, but still not large enough to justify plotting the curve. The curve is approximated by drawing an arc whose radius r is the same as radius R of the large cylinder.

The intersection in Fig. 6.38c is significant enough to justify constructing the true curve. Points are selected at random in the circle in the side or top view, and these are then projected to the other two views to locate points on the curve in the front view, as shown. Depending on the size of the intersection, a sufficient number of points should be used to ensure a smooth

and accurate curve. Draw the final curve with the aid of an irregular curve.

In Fig. 6.38d, the cylinders are the same diameter. The figure of intersection consists of two semiellipses that appear as straight lines in the front view.

If the intersecting cylinders are holes, the intersections will be similar to those for the external cylinders in Fig. 6.38 (see also Fig. 7.34d).

In Fig. 6.39a, a narrow prism intersects a cylinder, but the intersection is insignificant and is ignored. In Fig. 6.39b, the prism is larger, and the intersection is noticeable enough to warrant construction, as shown. In Fig. 6.39c and 6.39d, a keyseat and a small drilled hole, respectively, are shown; in both cases the intersection is not important enough to construct.

■ 6.33 HOW TO REPRESENT HOLES

The correct methods of representing most common types of machined holes are shown in Fig. 6.40. Instructions to the machinist are given in the form of

FIGURE 6.38   Intersections of Cylinders.

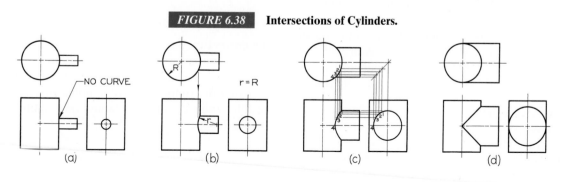

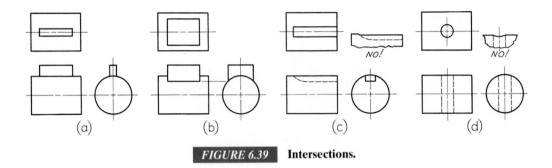

FIGURE 6.39 Intersections.

notes, and the drafter represents the holes in conformity with these specifications. In general, the notes tell the machine operator what to do and in which order it is to be done. Hole sizes are always specified by diameter—never by radius. For each operation specified, the diameter is given first, followed by the method—such as drill, ream, and so on—as shown in Figs. 6.40a and 6.40b.

The size of the hole may be specified as a diameter without the specific method (such as drill, ream, and so on) since the selection of the method will depend on available production facilities (Figs. 6.40h–6.40j).

A drilled hole is a *through hole* if it goes through a member. If the hole has a specified depth, as shown in Fig. 6.40a, the hole is called a *blind hole*. The depth includes the cylindrical portion of the hole only. The point of the drill leaves a conical bottom in the hole, drawn approximately with the 30° × 60° triangle, as shown.[*]

A through-drilled or reamed hole is drawn as shown in Fig. 6.40b. The note tells how the hole is to be produced—in this case by reaming. Note that tolerances are ignored in actually laying out the diameter of a hole.

In Fig. 6.40c, a hole is drilled and then the upper part is enlarged cylindrically to a specified diameter and depth. In Fig. 6.40d, a hole is drilled and then the upper part is enlarged conically to a specified angle and diameter. The angle is commonly 82° but is drawn 90° for simplicity. In Fig. 6.40e, a hole is drilled and then the upper part is enlarged cylindrically to a specified diameter. The depth usually is not specified, but is

---

[*] For drill sizes, see Appendix 16 ("Twist Drill Sizes"). For abbreviations, see Appendix 4.

left to the shop to determine. For average cases, the depth is drawn 1.5 mm ($\frac{1}{16}''$).

■ 6.34 FILLETS AND ROUNDS

A rounded interior corner is called a *fillet*, and a rounded exterior corner is called a *round* (Fig. 6.41a). Sharp corners should be avoided in designing parts to be cast or forged not only because they are difficult to produce but also because, in the case of interior corners, they are a source of weakness and failure (see §10.5 for manufacturing processes involved).

Two intersecting rough surfaces produce a rounded corner (Fig. 6.41b). If one of these surfaces is machined (Fig. 6.41c), or if both surfaces are machined (Fig. 6.41d), the corner becomes sharp. Therefore, in drawings, a rounded corner means that both intersecting surfaces are rough, while a sharp corner means that one or both surfaces are machined. On working drawings, fillets and rounds are never shaded. The presence of the curved surfaces is indicated only where they appear as arcs, except as shown in Fig. 6.45.

Fillets and rounds should be drawn with the filleted corners of a triangle, a special fillets and rounds template, or a circle template.

■ 6.35 RUNOUTS

The correct method of representing fillets in connection with plane surfaces tangent to cylinders is shown in Fig. 6.42. These small curves are called *runouts*. Note that the runouts F have a radius equal to that of the fillet and a curvature of about one eighth of a circle (Fig. 6.42d).

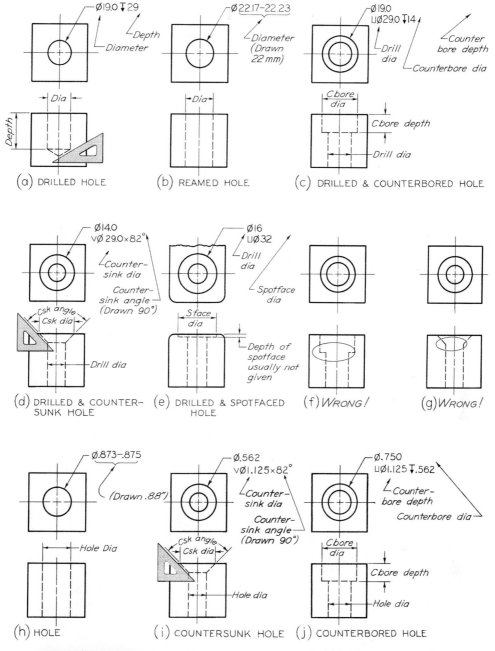

FIGURE 6.40   How to Represent Holes. Dimensions for (a)–(e) in metric.

Typical filleted intersections are shown in Fig. 6.43. The runouts from Figs. 6.43a–6.43d differ because of the different shapes of the horizontal intersecting members. In Figs. 6.43e and 6.43f, the runouts differ because the top surface of the web in Fig. 6.43e is flat, with only slight rounds along the edge, while the top surface of the web in Fig. 6.43f is considerably rounded. When two different sizes of fillets intersect, as in

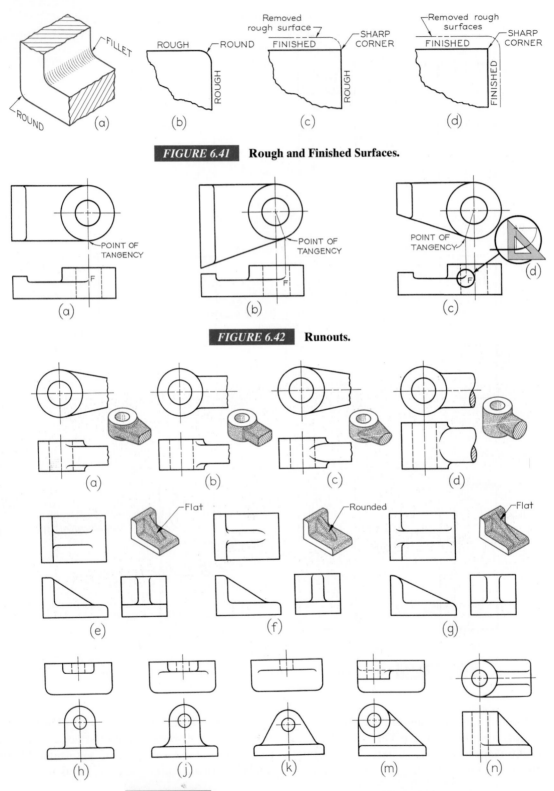

FIGURE 6.41 **Rough and Finished Surfaces.**

FIGURE 6.42 **Runouts.**

FIGURE 6.43 **Conventional Fillets, Rounds, and Runouts.**

Figs. 6.43g and 6.43j, the direction of the runout is dictated by the larger fillet, as shown.

■ 6.36 CONVENTIONAL EDGES

Rounded and filleted intersections eliminate sharp edges and sometimes make it difficult to present a clear shape description. In fact, in some cases, true projection may actually be misleading, as in Fig. 6.44a, where the side view of the railroad rail is quite blank. A clearer representation results if lines are added for rounded and filleted edges, as shown in Figs. 6.44b and 6.44c. The added lines are projected from the actual intersections of the surfaces as if the fillets and rounds were not present.

Figure 6.45 shows top views for each given front view. The upper top views are nearly devoid of lines that contribute to the shape descriptions, while the lower top views, in which lines are used to represent the rounded and filleted edges, are quite clear. Note, in the lower top views in Figs. 6.45a and 6.45c, the use of small Ys where rounded or filleted edges meet a rough surface. If such an edge intersects a finished surface, no Y is shown.

■ 6.37 RIGHT-HAND AND LEFT-HAND PARTS

In industry, many individual parts are located symmetrically so they can function in pairs. These opposite parts are often exactly alike (for example, the wheel covers used on the left and right sides of an automobile). In fact, whenever possible, for economy's sake the designer will design identical parts for use on both the right and left. But opposite parts often cannot be exactly alike, such as a pair of gloves or a pair of shoes. Similarly, the right-front fender of an automobile cannot be the same shape as the left-front fender. Therefore, a left-hand part is not simply a right-hand part turned around; the two parts will be mirror images and not interchangeable.

A left-hand part is referred to as an LH part, and a right-hand part as an RH part. In Fig. 6.46a, the image in the mirror is the "other hand" of the part shown. If

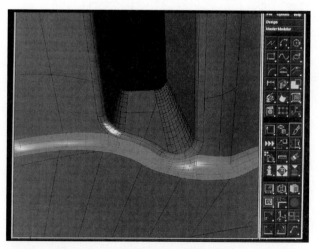

A 3D CAD Representation of Fillets. *Courtesy of SDRC, Milford, OH.*

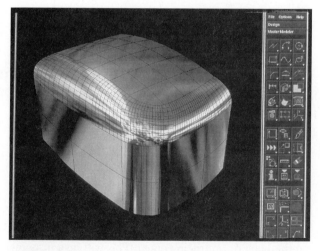

A 3D CAD Representation of Rounds. *Courtesy of SDRC, Milford, OH.*

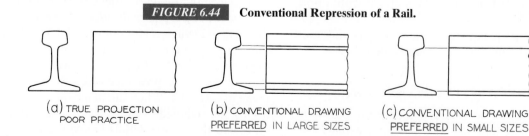

| FIGURE 6.44 | Conventional Repression of a Rail. |

(a) TRUE PROJECTION POOR PRACTICE

(b) CONVENTIONAL DRAWING PREFERRED IN LARGE SIZES

(c) CONVENTIONAL DRAWING PREFERRED IN SMALL SIZES

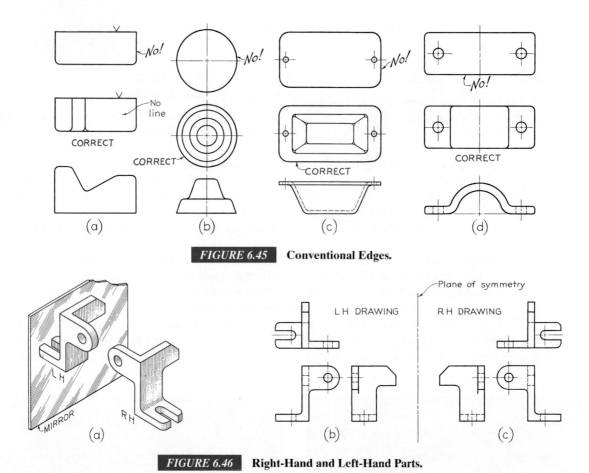

FIGURE 6.45 **Conventional Edges.**

FIGURE 6.46 **Right-Hand and Left-Hand Parts.**

the part in front of the mirror is an RH part, the image shows the LH part. No matter how the object is turned, the image will show the LH part. Figures 6.46b and 6.46c show LH and RH drawings of the same object; these drawings are also symmetrical with respect to a reference-plane line between them.

If you hold a drawing faced against a windowpane or on a light table so that the lines can be seen through the paper, you can trace the reverse image of the part on the back or on tracing paper. This will be a drawing of the opposite part.

It is customary to draw only one of two opposite parts and to label the one that is drawn with a note, such as LH PART SHOWN, RH OPPOSITE. If the opposite-hand shape is not clear, a separate drawing must be made for it and must be properly identified.

■ 6.38 FIRST-ANGLE PROJECTION

If the vertical and horizontal planes of projection are considered indefinite in extent and intersecting at 90°

with each other, the four dihedral angles produced are the *first, second, third,* and *fourth* angles (Fig. 6.47a). The profile plane intersects these two planes and may extend into all angles. If the object is placed below the horizontal plane and behind the vertical plane, as in the glass box (see Fig. 6.3), the object is said to be in the third angle. In this case, as we have seen, the observer is always "outside, looking in," so that for all views the lines of sight proceed from the eye *through the planes of projection and to the object.*

If the object is placed above the horizontal plane and in front of the vertical plane, the object is in the first angle. In this case, the observer always looks *through the object and to the planes of projection.* Thus, the right-side view is still obtained by looking toward the right side of the object, the front by looking toward the front, and the top by looking down toward the top; but the views are projected from the object onto a plane in each case. When the planes are unfolded (Fig. 6.47b), the right-side view falls at the left of the front view, and the top view falls below the front view, as

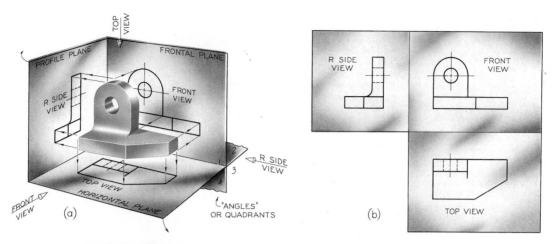

FIGURE 6.47 First-Angle Projection. An object that is above the horizontal plane and in front of the vertical plane is in the first angle. An observer looks through the object to the planes of projection.

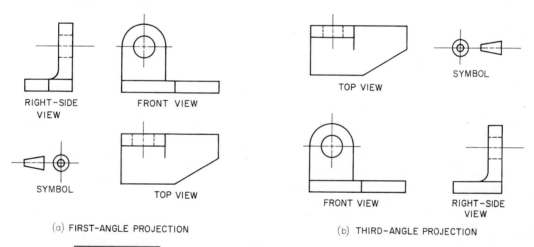

(a) FIRST-ANGLE PROJECTION

(b) THIRD-ANGLE PROJECTION

FIGURE 6.48 First-Angle Projection Compared to Third-Angle Projection.

shown. A comparison between first-angle orthographic projection and third-angle orthographic projection is shown in Fig. 6.48. The front, top, and right-side views shown in Fig. 6.47b for first-angle projection are repeated in Fig. 6.48a. The front, top, and right-side views for third-angle projection of Fig. 6.4 are repeated in Fig. 6.48b. Ultimately, the only difference between third-angle and first-angle projection is the arrangement of the views. Still, confusion and possibly manufacturing errors may result when the user reading a first-angle drawing thinks it is a third-angle drawing, or vice versa. To avoid misunderstanding, international projection symbols, shown in Fig. 6.48, have been developed to distinguish between first-an-

gle and third-angle projections on drawings. On drawings where the possibility of confusion is anticipated, these symbols may appear in or near the title box.

In the United States and Canada (and, to some extent, in England), third-angle projection is standard, while in most of the rest of the world, first-angle projection is used. First-angle projection was originally used all over the world, including the United States, but it was abandoned around 1890.

■ 6.39 COMPUTER GRAPHICS

You can use CAD to create 2D multiview projections in a way similar to creating a multiview drawing on

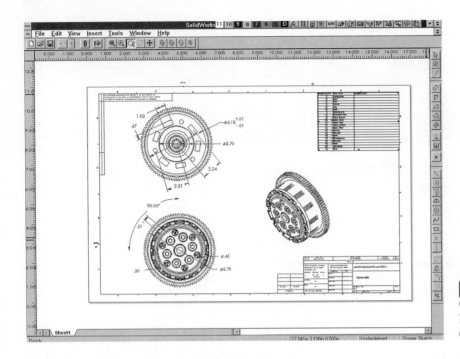

FIGURE 6.49 **Computer-Generated Multiview and Pictorial Drawing.** *Courtesy of SolidWorks Corporation.*

paper. You can also create 3D models of a part using CAD. From the 3D model, you can create 2D orthographic views automatically. This can save time in creating many drawings because you only have to create the 3D model once to generate as many different views as you want. It is important to develop an understanding of projection in order to use 3D CAD effectively. When you are looking at a 3D wireframe drawing on your computer screen, it can be difficult to tell from which direction you are viewing the model. This is because some objects may look identical on the computer screen even when viewed from different directions. If you do not develop an understanding of orthographic views, it can be very easy to lose yourself in the 3D space of your drawing (Fig. 6.49).

■ KEY WORDS

PROJECTION	THIRD-ANGLE PROJECTION
PROJECTORS	PLANE OF PROJECTION
MITER LINE	ORTHOGRAPHIC
FOLD LINES	THE GLASS BOX
DEPTH	ALIGNMENT OF VIEWS
HEIGHT	INCLINED
WIDTH	OBLIQUE
REVOLUTION	EDGES
NORMAL	HOLE CONVENTIONS
SURFACES	FILLETS AND ROUNDS
CYLINDERS	

■ CHAPTER SUMMARY

- Orthographic drawings are the result of projecting the image of a three-dimensional object onto one of six standard planes of projection. The planes of projection intersect each other at fold lines. The six standard planes of projection are often thought of as a "glass box."

- Each view in an orthographic projection is aligned with an adjacent view. The principal views most often used are top, front, and right side.

- To help project or verify surfaces you can label them with letters, and the corners of surfaces can be labeled with numbers. There are normal, inclined, and oblique surfaces. Normal surfaces appear true size in one principal view and as an edge in the other two principal views. Inclined surfaces appear as an edge view in one of the three

principal views. Oblique surfaces do not appear in edge view in any of the principal views.

- Conic sections and irregular curves must be plotted by identifying points on the object. The points can be projected to approximate the boundaries of the curved surface.

- Drawing conventions define usual practices for the representation of features such as holes, bosses, ribs, webs, spokes, fillets, and rounds.

- Creating CAD drawings involves applying the same concepts as paper drawing. The main difference is that the paper surface is replaced by the monitor screen, and CAD software can draw a line faster and more accurately than most drafters.

■ REVIEW QUESTIONS

1. Sketch a representation of third angle projection.
2. List the six principal views of projection.
3. Sketch the top, front, and right side views of an object having normal, inclined, and oblique surfaces.
4. In a drawing that shows the top, front, and right side view, which two views show depth? Which view shows depth vertically? Which view shows depth horizontally?
5. What is the definition of a normal surface? An inclined surface? An oblique surface?
6. Describe the procedure necessary to center a three-view drawing on a sheet of paper.
7. Describe the procedure necessary to center a CAD drawing on a sheet of paper.
8. How is the computer monitor used to view a CAD program similar to a sheet of paper?
9. What dimensions are the same between the top and front view: width, height, or depth? Between the front and right side view? Between the top and right side view?
10. List two ways of transferring depth between the top and right side views.
11. If surface A contained corners 1, 2, 3, 4, and surface B contained corners 3, 4, 5, 6, what is the name of the line where surfaces A and B intersect?
12. If the top view of an object shows a drilled through hole, how many hidden lines would be necessary in the front view to describe the hole?

■ MULTIVIEW PROJECTION PROBLEMS

The following problems are intended primarily to afford practice in instrumental drawing, but any of them may be sketched freehand on graph paper or plain paper. Sheet layouts such as those in Figs. 6.50 and 6.51 or inside the back cover are suggested, but your instructor may prefer a different sheet size or arrangement. Use metric or decimal-inch as assigned.

If your instructor requires dimensions, you should study §§11.1–11.25. In the following problems, often it is not possible for the dimensions to be shown in the preferred places in the standard manner. In doing the problems, you are expected to move dimensions to the preferred locations and to conform to the dimensioning practices recommended in Chapter 11.

For additional problems, see Fig. 8.29. Draw top views instead of auxiliary views.

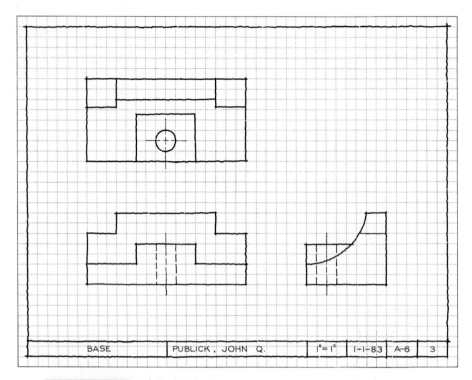

| BASE | PUBLICK , JOHN Q. | | 1"=1" | 1-1-83 | A-6 | 3 |

FIGURE 6.50 Suggested Layout for Freehand Sketch (Layout A–2 or A4–2 adjusted).

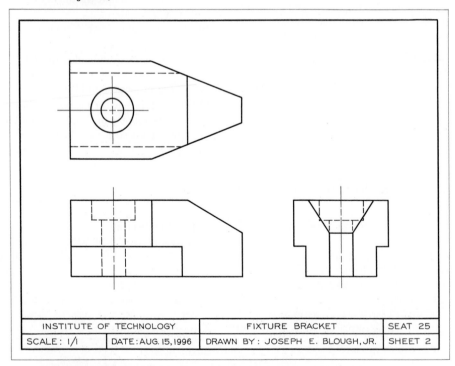

| INSTITUTE OF TECHNOLOGY | | FIXTURE BRACKET | SEAT 25 |
| SCALE : 1/1 | DATE: AUG. 15, 1996 | DRAWN BY : JOSEPH E. BLOUGH, JR. | SHEET 2 |

FIGURE 6.51 Suggested Layout for Mechanical Drawing (Layout A–3 or A4–3 adjusted).

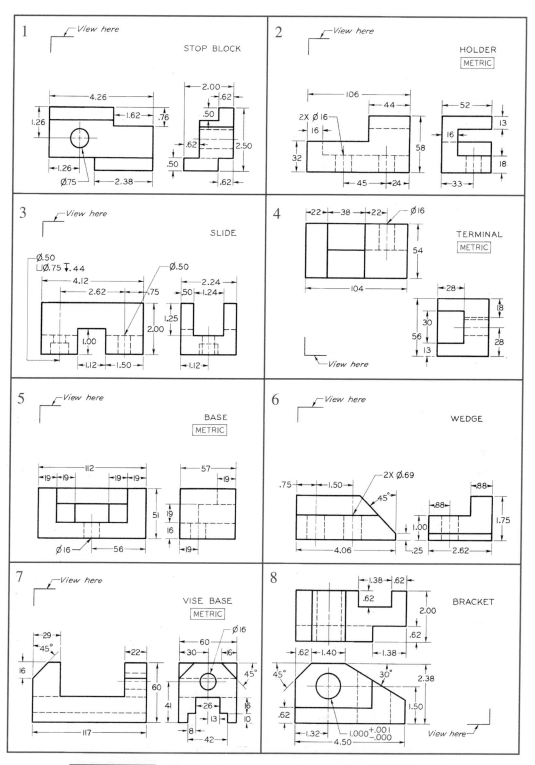

FIGURE 6.52 **Missing-View Problems. Using Layout A–2 or 3 or Layout A4–2 or 3 (adjusted), sketch or draw with instruments the given views, and add the missing view, as shown in Figs. 6.50 and 6.51. If dimensions are required, study §§11.1–11.25. Use metric or decimal-inch dimensions as assigned by the instructor. Move dimensions to better locations where possible. In Probs. 1–5, all surfaces are normal surfaces.**

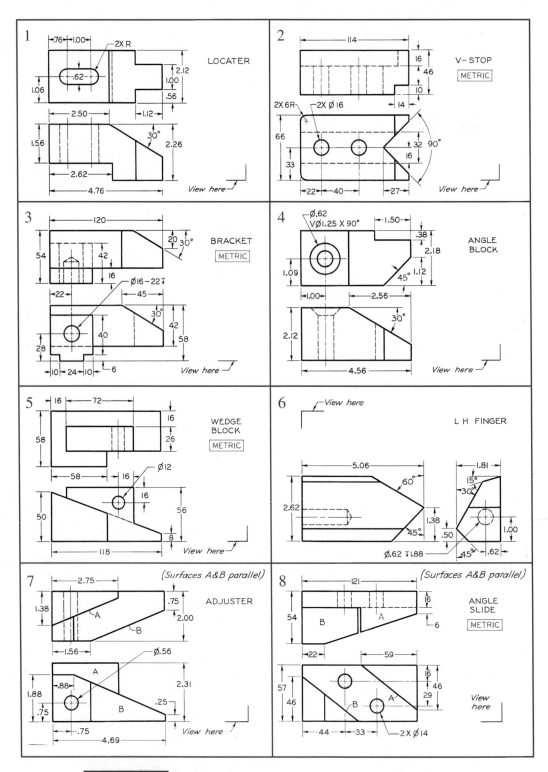

FIGURE 6.53 Missing-View Problems. Using Layout A–2 or 3 or Layout A4–2 or 3 (adjusted), sketch or draw with instruments the given views, and add the missing view, as shown in Figs. 6.50 and 6.51. If dimensions are required, study §§11.1–11.25. Use metric or decimal-inch dimensions as assigned by the instructor. Move dimensions to better locations where possible.

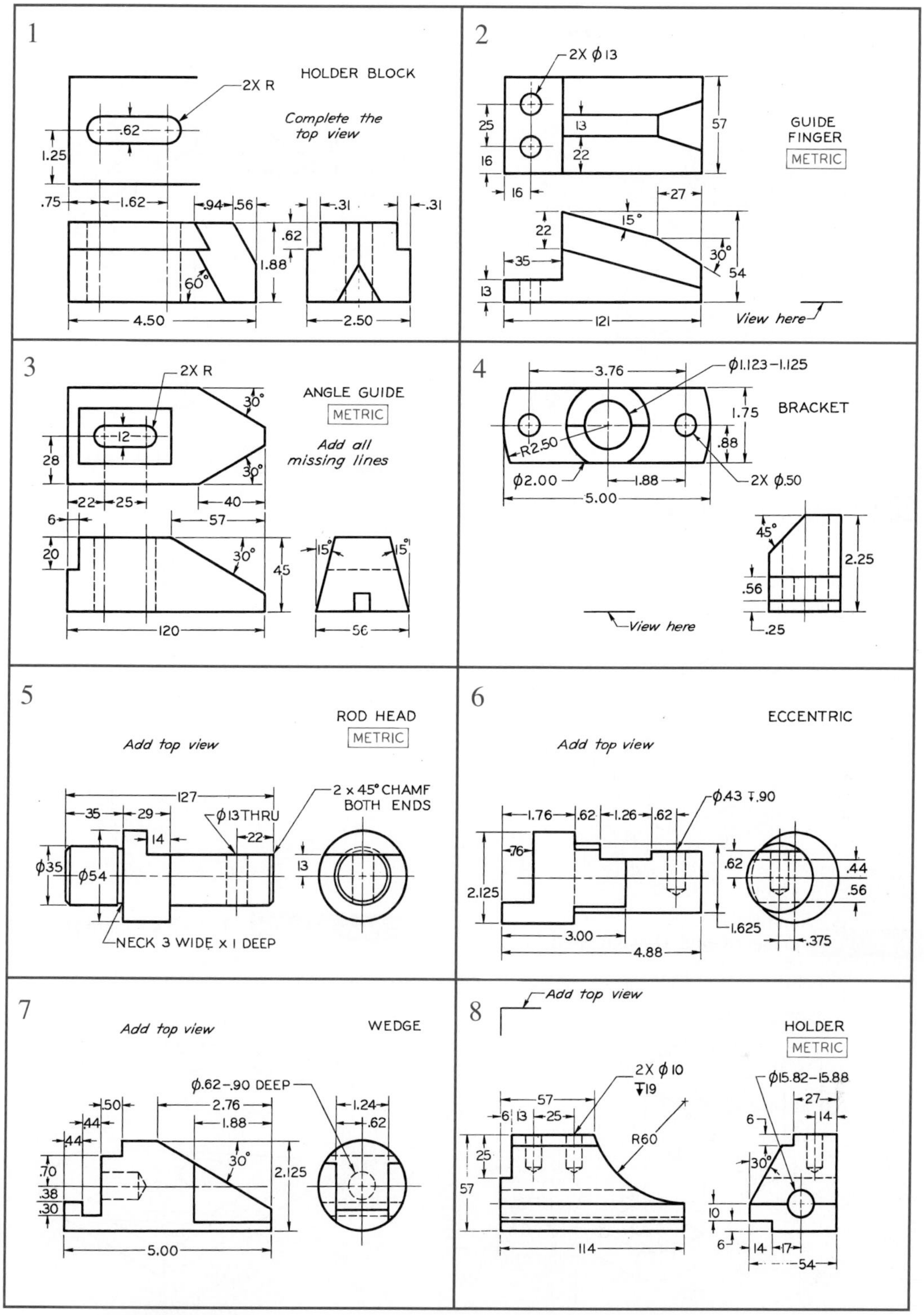
1
HOLDER BLOCK
Complete the top view
2X R
.62
1.25
.75
1.62
.94
.56
.31
.31
.62
1.88
60°
4.50
2.50
2
2X ϕ 13
25
13
16
22
57
GUIDE FINGER
METRIC
16
27
15°
22
30°
35
54
13
121
View here
3
2X R
ANGLE GUIDE
METRIC
Add all missing lines
30°
12
28
30°
22
25
40
6
57
20
30°
45
15°
15°
120
56
4
3.76
ϕ1.123–1.125
BRACKET
1.75
.88
R2.50
ϕ2.00
1.88
2X ϕ.50
5.00
45°
2.25
.56
.25
View here
5
ROD HEAD
METRIC
Add top view
127
35
29
ϕ13THRU
2 x 45° CHAMF BOTH ENDS
14
22
13
ϕ35
ϕ54
NECK 3 WIDE x 1 DEEP
6
ECCENTRIC
Add top view
ϕ.43 ↧.90
1.76
.62
1.26
.62
.76
.62
.44
.56
2.125
1.625
3.00
.375
4.88
7
Add top view
WEDGE
ϕ.62–.90 DEEP
.50
2.76
1.24
.44
1.88
.62
.44
30°
.70
2.125
.38
.30
5.00
8
Add top view
HOLDER
METRIC
2X ϕ 10
↧19
ϕ15.82–15.88
57
27
6
13
25
14
6
R60
25
30°
57
10
6
14
17
114
54

FIGURE 6.54 **Missing-View Problems. Using Layout A–2 or 3 or Layout A4–2 or 3 (adjusted), sketch or draw with instruments the given views, and add the missing view, as shown in Figs. 6.50 and 6.51. If dimensions are required, study §§11.1–11.25. Use metric or decimal-inch dimensions as assigned by the instructor. Move dimensions to better locations where possible.**

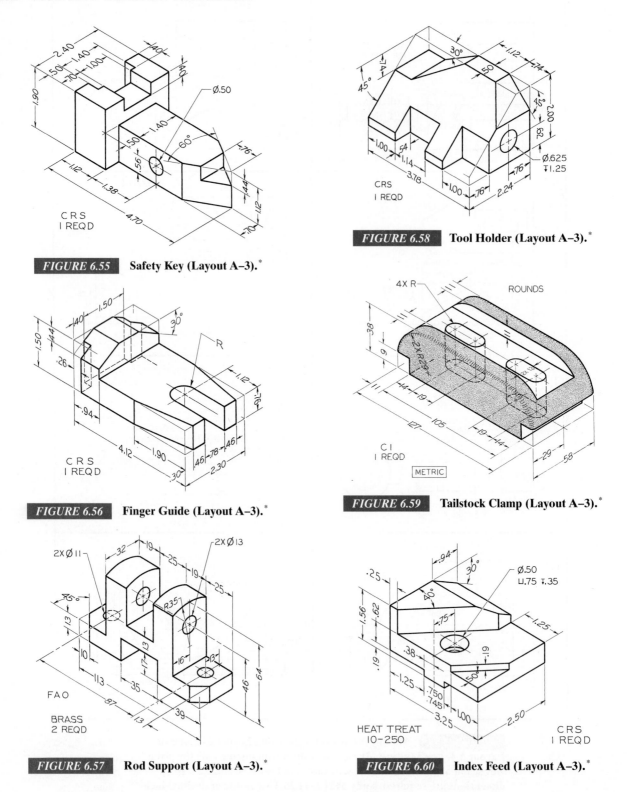

FIGURE 6.55 Safety Key (Layout A–3).*

FIGURE 6.56 Finger Guide (Layout A–3).*

FIGURE 6.57 Rod Support (Layout A–3).*

FIGURE 6.58 Tool Holder (Layout A–3).*

FIGURE 6.59 Tailstock Clamp (Layout A–3).*

FIGURE 6.60 Index Feed (Layout A–3).*

*Draw or sketch necessary views. Layout A4–3 (adjusted) may be used. If dimensions are required, study §§11.1–11.25. Use metric or decimal-inch dimensions as assigned by the instructor.

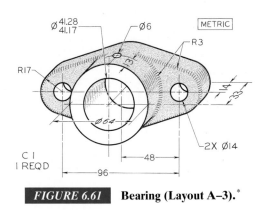

FIGURE 6.61   Bearing (Layout A–3).[*]

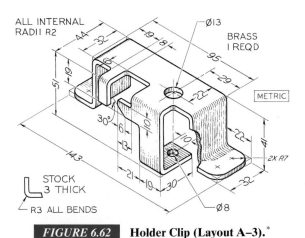

FIGURE 6.62   Holder Clip (Layout A–3).[*]

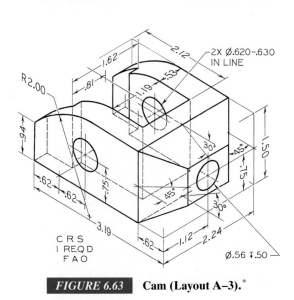

FIGURE 6.63   Cam (Layout A–3).[*]

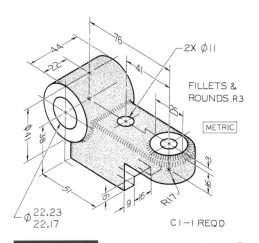

FIGURE 6.64   Index Arm (Layout A–3).[*]

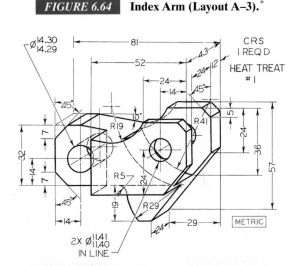

FIGURE 6.65   Roller Lever (Layout A–3).[*]

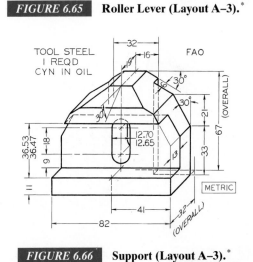

FIGURE 6.66   Support (Layout A–3).[*]

[*] Draw or sketch necessary views. Layout A4–3 (adjusted) may be used. If dimensions are required, study §§11.1–11.25. Use metric or decimal-inch dimensions as assigned by the instructor.

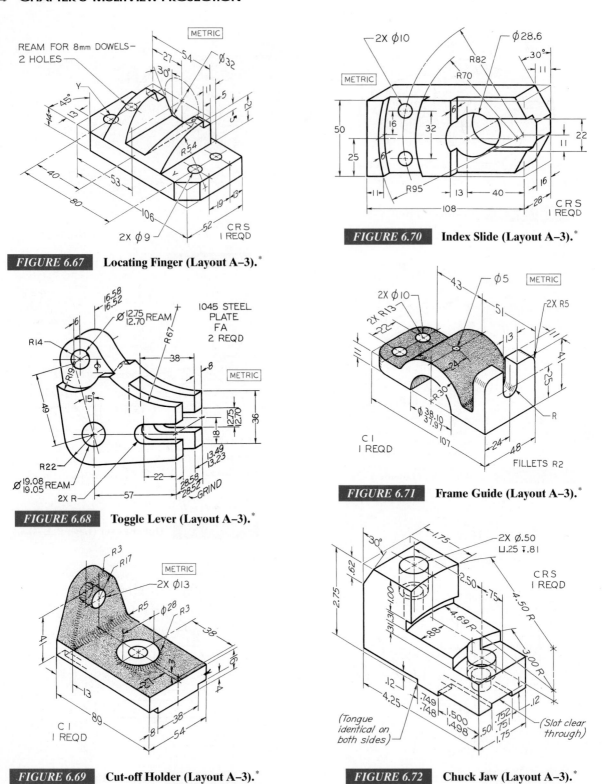

FIGURE 6.67　Locating Finger (Layout A–3).[*]

FIGURE 6.70　Index Slide (Layout A–3).[*]

FIGURE 6.68　Toggle Lever (Layout A–3).[*]

FIGURE 6.71　Frame Guide (Layout A–3).[*]

FIGURE 6.69　Cut-off Holder (Layout A–3).[*]

FIGURE 6.72　Chuck Jaw (Layout A–3).[*]

[*] Draw or sketch necessary views. Layout A4–3 (adjusted) may be used. If dimensions are required, study §§11.1–11.25. Use metric or decimal-inch dimensions as assigned by the instructor.

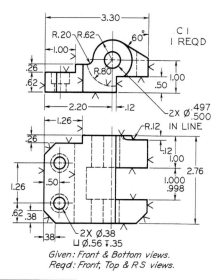

Given: Front & Bottom views.
Reqd: Front, Top & RS views.

FIGURE 6.73 **Hinge Bracket (Layout A–3).**[*]

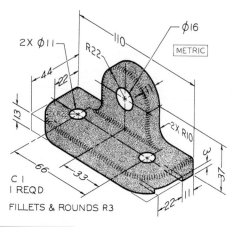

FIGURE 6.76 **Cross-feed Stop (Layout A–3).**[*]

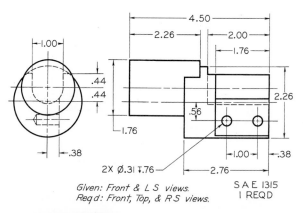

Given: Front & LS views.
Reqd: Front, Top, & RS views.

S A E 1315
1 REQD

FIGURE 6.74 **Tool Holder (Layout A–3).**[*]

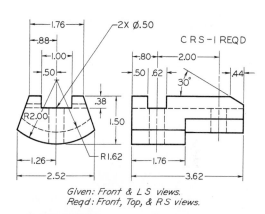

Given: Front & L S views.
Reqd: Front, Top, & RS views.

FIGURE 6.77 **Cross Cam (Layout A–3).**[*]

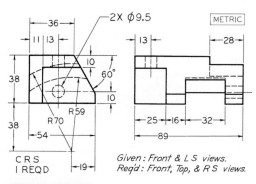

Given: Front & L S views.
Reqd: Front, Top, & RS views.

FIGURE 6.75 **Shifter Block (Layout A–3).**[*]

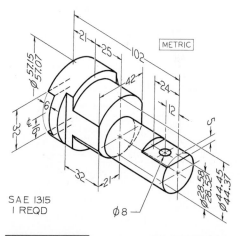

SAE 1315
1 REQD

FIGURE 6.78 **Roller Stud (Layout A–3).**[*]

[*] Draw or sketch necessary views. Layout A4–3 (adjusted) may be used. If dimensions are required, study §§11.1–11.25. Use metric or decimal-inch dimensions as assigned by the instructor.

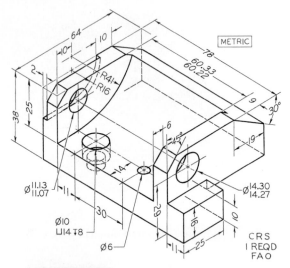

FIGURE 6.79   **Hinge Block (Layout A–3).**[*]

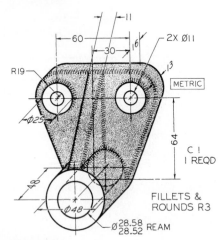

FIGURE 6.80   **Feed Rod Bearing (Layout A–3).**[*]

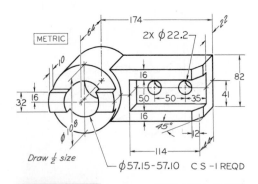

FIGURE 6.81   **Lever Hub (Layout A–3).**[*]

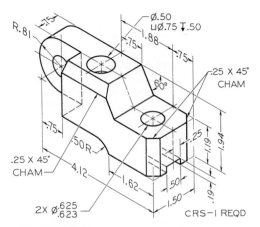

FIGURE 6.82   **Vibrator Arm (Layout A–3).**[*]

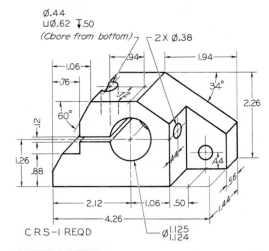

FIGURE 6.83   **Clutch Lever (Layout A–3).**[*]

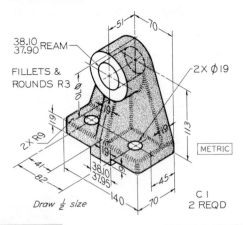

FIGURE 6.84   **Counter Bearing Bracket (Layout A–3).**[*]

[*] Draw or sketch necessary views. Layout A4–3 (adjusted) may be used. If dimensions are required, study §§11.1–11.25. Use metric or decimal-inch dimensions as assigned by the instructor.

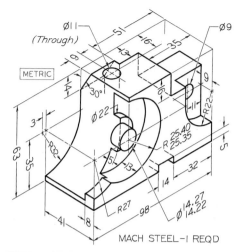

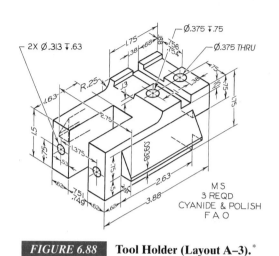

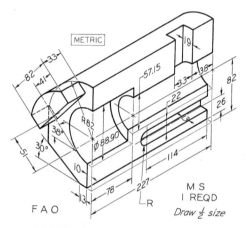

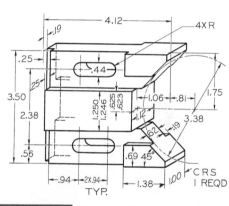

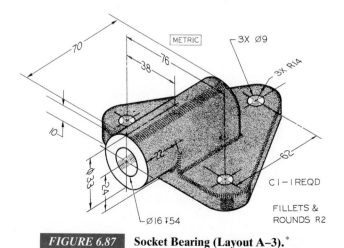

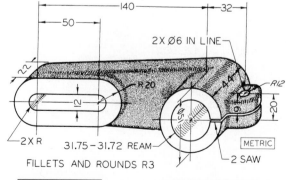

FIGURE 6.85   Tool Holder (Layout A–3).*

FIGURE 6.88   Tool Holder (Layout A–3).*

FIGURE 6.86   Control Block (Layout A–3).*

FIGURE 6.89   Locating V-Block (Layout A–3).*

FIGURE 6.87   Socket Bearing (Layout A–3).*

FIGURE 6.90   Anchor Bracket (Layout A–3).*

* Draw or sketch necessary views. Layout A4–3 (adjusted) may be used. If dimensions are required, study §§11.1–11.25. Use metric or decimal-inch dimensions as assigned by the instructor.

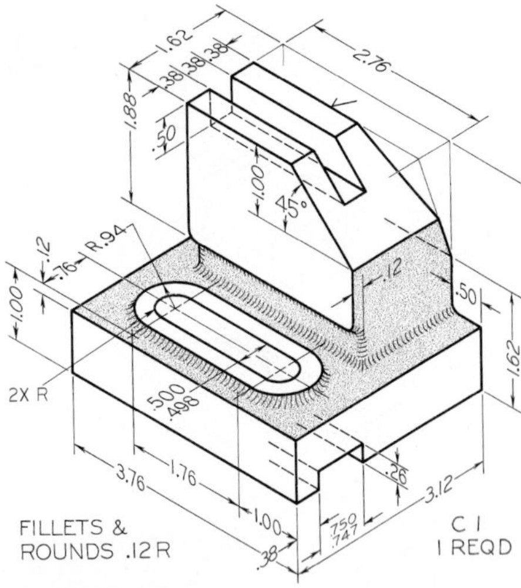

FIGURE 6.91 **Door Bearing (Layout B–3).**[*]

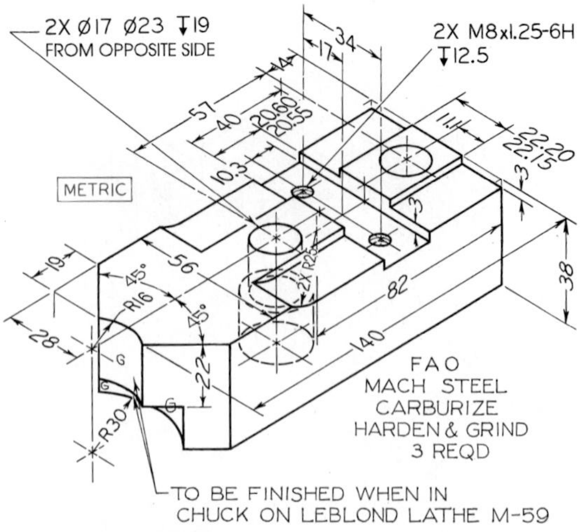

FIGURE 6.94 **Chuck Jaw (Layout B–3).**[*]

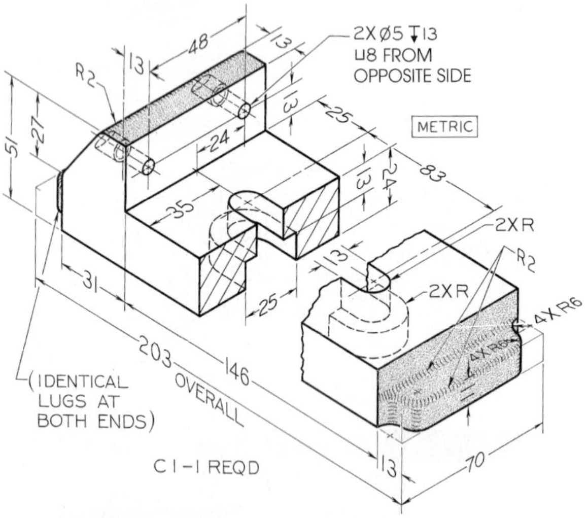

FIGURE 6.92 **Vise Base (Layout B–3).**[*]

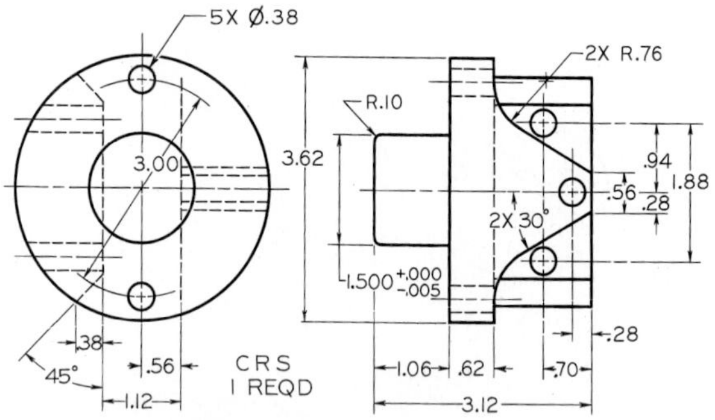

Given: Front and L Side views.
Reqd: Front, R Side, and Top views.

FIGURE 6.95 **Holder (Layout B–3).**[*]

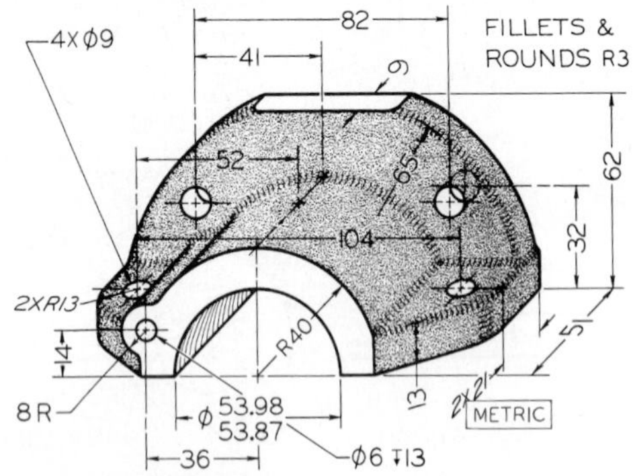

FIGURE 6.93 **Dust Cap (Layout B–3).**[*]

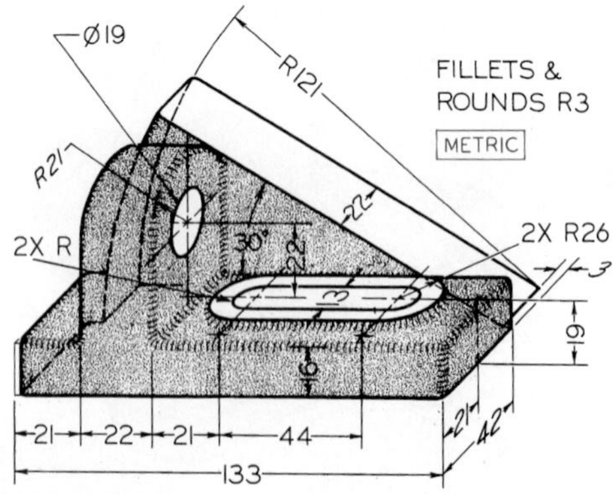

FIGURE 6.96 **Centering Wedge (Layout B–3).**[*]

[*] Draw or sketch necessary views. Layout A4–3 (adjusted) may be used. If dimensions are required, study §§11.1–11.25. Use metric or decimal-inch dimensions as assigned by the instructor.

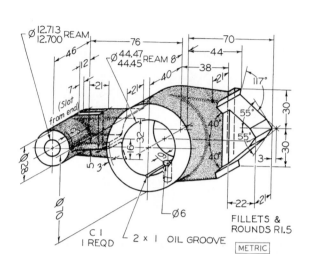

FIGURE 6.97 Motor Switch Lever. Draw or sketch necessary views (Layout B–3 or A3–3).*

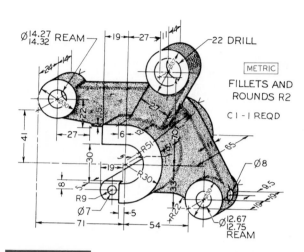

FIGURE 6.98 Socket Form Roller—LH. Draw or sketch necessary views (Layout B–4 or A3–4 adjusted).*

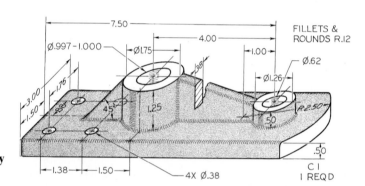

FIGURE 6.99 Stop Base. Draw or sketch necessary views (Layout B–3 or A3–3).*

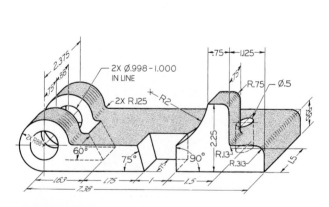

FIGURE 6.100 Hinge Base. Draw or sketch necessary views (Layout B–3 or A3–3).*

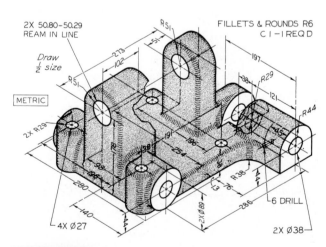

FIGURE 6.101 Automatic Stop Base. Draw or sketch necessary views (Layout C–3 or A2–3).*

---

* If dimensions are required, study §§11.1–11.25. Use metric or decimal-inch dimensions as assigned by the instructor.

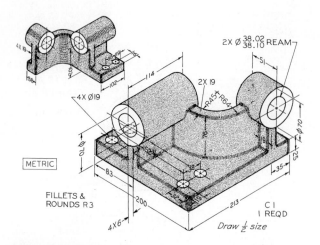

FIGURE 6.102 Lead Screw Bracket. Draw or sketch necessary views (Layout C–3 or A2–3).*

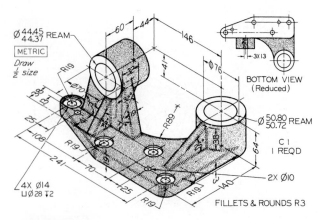

FIGURE 6.103 Lever Bracket. Draw or sketch necessary views (Layout C–3 or A2–3).*

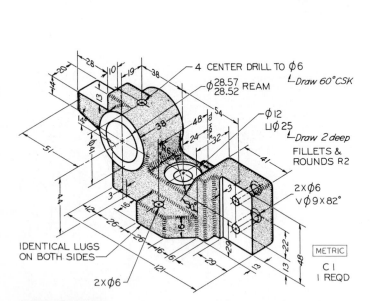

FIGURE 6.104 Gripper Rode Center. Draw or sketch necessary views (Layout B–3 or A3–3).*

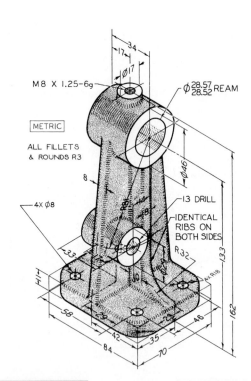

FIGURE 6.105 Bearing Bracket. Draw or sketch necessary views (Layout B–3 or A3–3).*

* If dimensions are required, study §§11.1–11.25. Use metric or decimal-inch dimensions as assigned by the instructor.

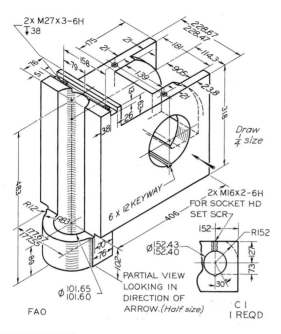

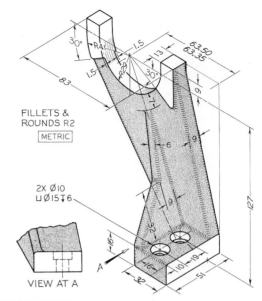

FIGURE 6.106 Link Arm Connector. Draw or sketch necessary views (Layout B–3 or A3–3).*

FIGURE 6.108 LH Shifter Fork. Draw or sketch necessary views (Layout B–3 or A3–3).*

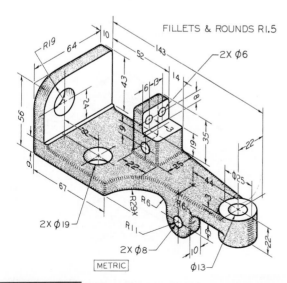

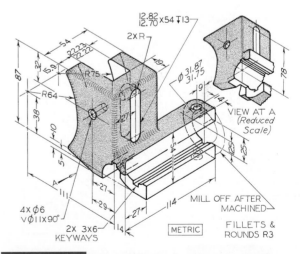

FIGURE 6.107 Mounting Bracket. Draw or sketch necessary views (Layout B–3 or A3–3).*

FIGURE 6.109 Gear Shift Bracket. Draw or sketch necessary views (Layout C–4).*

* If dimensions are required, study §§11.1–11.25. Use metric or decimal-inch dimensions as assigned by the instructor.

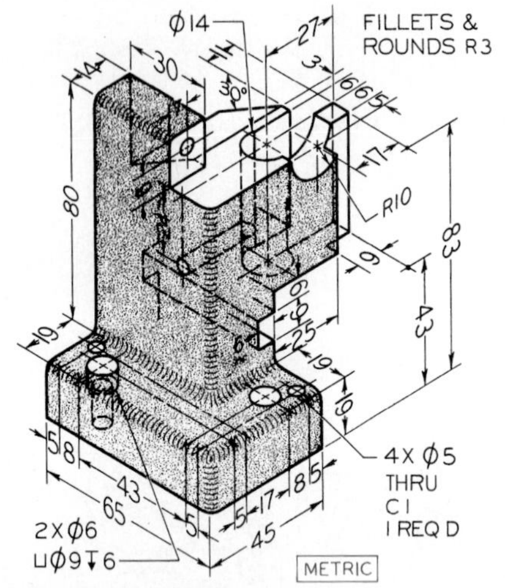

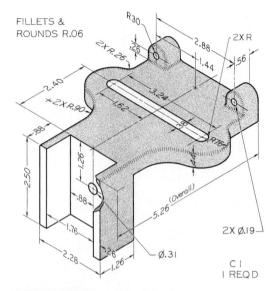

FIGURE 6.110 **Fixture Base (Layout C–4).**[*]

FIGURE 6.112 **Tension Bracket (Layout C–4).**[*]

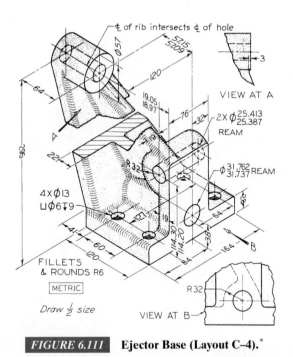

FIGURE 6.111 **Ejector Base (Layout C–4).**[*]

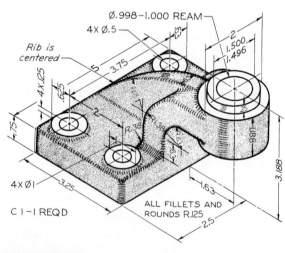

FIGURE 6.113 **Offset Bearing (Layout C–4 or A2–4).**[*]

[*] Draw or sketch necessary views. Layout A2–4 may be used. If dimensions are required, study §§11.1–11.25. Use metric or decimal-inch dimensions as assigned by the instructor.

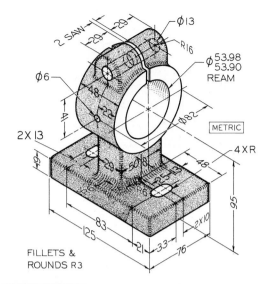

FILLETS &
ROUNDS R3

FIGURE 6.114 Feed Guide (Layout C–4 or A2–4). *

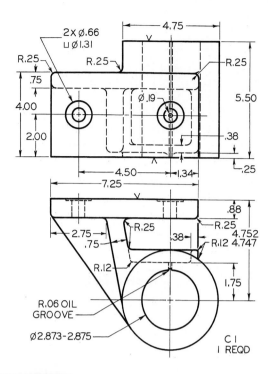

FIGURE 6.115 Feed Shaft Bracket. Given: Front and top views. Required: Front, top, and right-side views, half size (Layout B–3 or A3–3). *

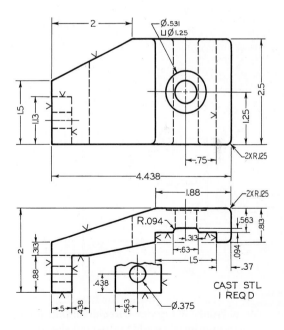

FIGURE 6.116 Trip Lever. Given: Front, top, and partial side views. Required: Front, bottom, and left-side views, drawn completely (Layout B–3 or A3–3). *

* Draw or sketch necessary views. Layout A2–4 may be used. If dimensions are required, study §§11.1–11.25. Use metric or decimal-inch dimensions as assigned by the instructor.

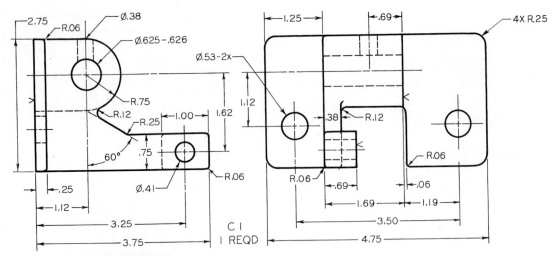

FIGURE 6.117 Knurl Bracket Bearing. Given: Front and left-side views. Required: Take front as top view on new drawing, and add front and right-side views (Layout B–3 or A3–3).*

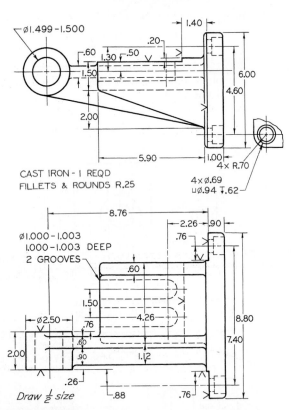

FIGURE 6.118 Horizontal Bracket for Broaching Machine. Given: Front and top views. Required: Take top as front view in new drawing; then add top and left-side views (Layout C–4 or A2–4).*

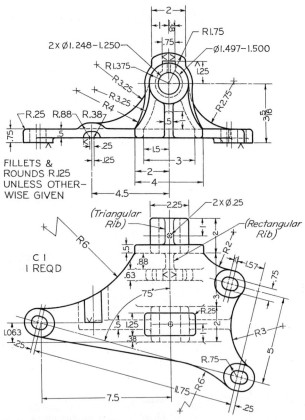

FIGURE 6.119 Boom Swing Bearing for a Power Crane. Given: Front and bottom views. Required: Front, top, and left-side views (Layout C–4 or A2–4).*

* Draw or sketch necessary views. If dimensions are required, study §§11.1–11.25. Use metric or decimal-inch dimensions as assigned by the instructor.

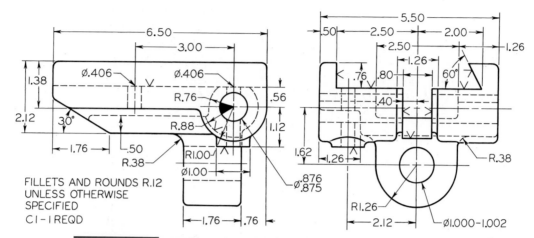

FILLETS AND ROUNDS R.12
UNLESS OTHERWISE
SPECIFIED
CI - I REQD

FIGURE 6.120   Sliding Nut for Mortiser. Given: Top and right-side views. Required: Front, top, and left-side views, full size (Layout C–4 or A2–4). *

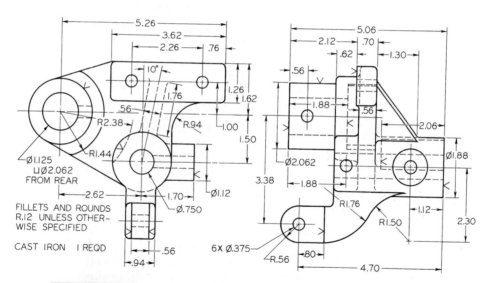

FILLETS AND ROUNDS
R.12 UNLESS OTHER-
WISE SPECIFIED

CAST IRON   I REQD

FIGURE 6.121   Power Feed Bracket for Universal Grinder. Given: Front and right-side views. Required: Front, top, and left-side views, full size (Layout C–4 or A2–4). *

* Draw or sketch necessary views. If dimensions are required, study §§11.1–11.25. Use metric or decimal-inch dimensions as assigned by the instructor.

C H A P T E R 7

SECTIONAL VIEWS

OBJECTIVES

After studying the material in this chapter, you should be able to:

1. Understand the meaning of sections and cutting-plane lines.
2. Identify seven types of sections.
3. Draw a sectional view, given a two-view drawing.
4. Demonstrate the proper techniques for sectioning ribs, webs, and spokes.
5. Demonstrate the proper technique for aligned sections.
6. Demonstrate correct hidden-line practices.
7. Draw correct conventional break symbols for elongated objects.
8. Recognize and draw the correct section-lining symbols for 10 different materials.

OVERVIEW

The basic method of representing objects by views, or projections, has been explained in previous chapters. By means of a limited number of carefully selected views, the external features of the most complicated designs can be fully described. However, we often need to show interiors that cannot be illustrated clearly by hidden lines. We show such interiors by slicing through the object much as we cut through an apple or melon. A cutaway view of the part is then drawn; such views are called **sectional views**, **cross sections**, or simply *sections*.

Drafters use sectional views to improve the clarity of complex objects when internal surfaces result in too many hidden lines. There are many types of sectional views, and a sectional view may replace one of the primary views. The cutting-plane line shows where the object is imagined to be cut. Section lining shows the solid parts of the object that have been in contact with the cutting plane. Hidden lines are not usually found in sectional views. Special conventions are used to make a sectional view easy to understand. Creating a sectional view is one of the more complicated operations for a CAD program. CAD users need to understand thoroughly the concepts of sectional views to create a sectional view with CAD software. See ANSI/ASME Y14.2M–1992 and Y14.3M–1994 for complete standards for multiview and sectional-view drawings.

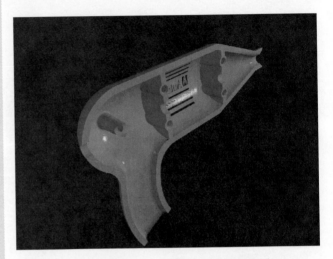

A Section Created with CAD. *This material has been reprinted with the permission from and under the copyright of Autodesk, Inc.*

tage in clarity of the sectional view. The left-side view would naturally be omitted. In the front view, the cutting plane appears as a line, called a **cutting-plane line** (see §7.5). The arrows at the ends of the cutting-plane line indicate the direction of sight for the sectional view.

To obtain the sectional view, the right half is only *imagined* to be removed and is not actually shown removed anywhere except in the sectional view itself. In the sectional view, the section-lined areas are those portions that have been in actual contact with the cutting plane. Those areas are *cross-hatched* with thin parallel section lines spaced carefully by eye. In addition, the visible parts behind the cutting plane are shown but are not cross-hatched.

As a rule, the location of the cutting plane is obvious from the section itself, and, therefore, the cutting-plane line is omitted. It is shown in Fig. 7.2 for illustration only. Cutting-plane lines should, of course, be used wherever necessary for clarity, as in Figs. 7.21, 7.22, 7.24, and 7.25.

■ 7.1 SECTIONING

To produce a sectional view, a **cutting plane** (§7.5) is assumed to be passed through the part (Fig. 7.1a). The cutting plane is then removed, and the two halves are drawn apart, exposing the interior construction (Fig. 7.1b). In this case, the direction of sight is toward the left half, as shown, and for purposes of the section, the right half is mentally discarded. The sectional view will be in the position of a right-side view.

■ 7.2 FULL SECTIONS

The sectional view obtained by passing the cutting plane fully through the object is called a **full section** (Fig. 7.2c). A comparison of this sectional view with the left-side view (Fig. 7.2a) emphasizes the advan-

■ 7.3 LINES IN SECTIONING

A correct front view and sectional view are shown in Figs. 7.3a and 7.3b. In general, *all visible edges and contours behind the cutting plane should be shown*; otherwise a section will appear to be made up of disconnected and unrelated parts, as shown in Fig. 7.3c.

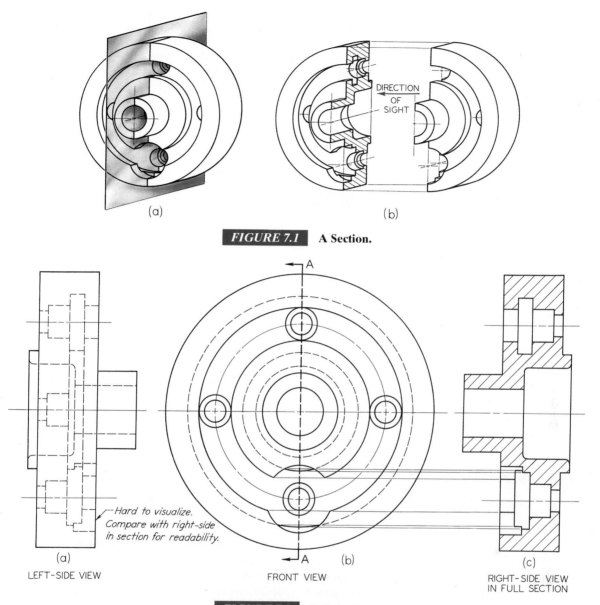

(a)   (b)

FIGURE 7.1   A Section.

LEFT-SIDE VIEW (a)

*Hard to visualize.
Compare with right-side
in section for readability.*

FRONT VIEW (b)

A

A

RIGHT-SIDE VIEW
IN FULL SECTION (c)

FIGURE 7.2   Full Section.

Occasionally, however, visible lines behind the cutting plane are not necessary for clarity and should be omitted.

Sections are used primarily to replace hidden-line representation; and, as a rule, *hidden lines should be omitted in sectional views*. As shown in Fig. 7.3d, the hidden lines do not clarify the drawing; they tend to confuse, and they take unnecessary time to draw. Sometimes hidden lines are necessary for clarity and should be used in such cases, especially if their use will make it possible to omit a view (Fig. 7.4).

A section-lined area is always completely bounded by a visible outline—never by a hidden line, as in Fig. 7.3e, since in every case the cut surfaces and their boundary lines will be visible. Also, a visible line can never cut across a section-lined area.

In a sectional view of an object, alone or in assembly, the section lines in all sectioned areas must be parallel, not as shown in Fig. 7.3f. The use of section lining in opposite directions is an indication of different parts, as when two or more parts are adjacent in an assembly drawing.

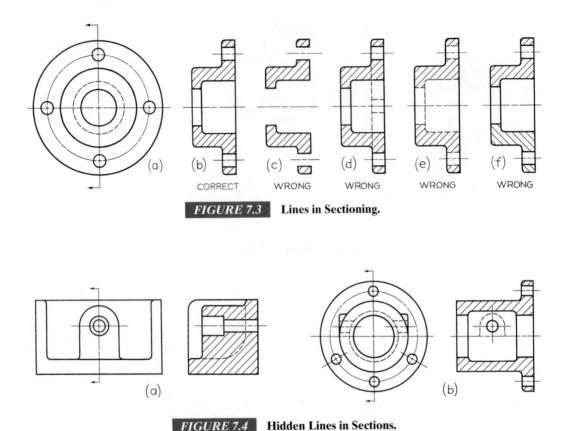

FIGURE 7.3 Lines in Sectioning.

FIGURE 7.4 Hidden Lines in Sections.

■ 7.4 SECTION LINING

Section-lining symbols (Fig. 7.5) have been used to indicate specific materials. These symbols represent general material types only, such as cast iron, brass, and steel. Now, however, because there are so many different types of materials, and each has so many subtypes, a general name or symbol is not enough. For example, there are hundreds of different kinds of steel. Since detailed specifications of material must be lettered in the form of a note or in the title strip, the general-purpose (cast-iron) section lining may be used for all materials on detail drawings (single parts).

Section-lining symbols may be used in assembly drawings in cases where it is desirable to distinguish different materials; otherwise, the general-purpose symbol is used for all parts.

CAD programs usually include a library that allows the user to select from a variety of section-lining patterns, making it easy to indicate various types of material rather than using the generic cast-iron symbol for all sectioned parts.

The correct method of drawing section lines is shown in Fig. 7.6a. Draw the section lines with a sharp, medium-grade pencil (H or 2H) with a conical point. Always draw the lines at 45° with horizontal as shown, unless there is some advantage in using a different angle. Space the section lines as evenly as possibly by eye from approximately 1.5 mm ($\frac{1}{16}$″) to 3 mm ($\frac{1}{8}$″) or more apart, depending on the size of the drawing or of the sectioned area. *For most drawings, space the lines about 2.5 mm ($\frac{3}{32}$″) or slightly more apart.* As a rule, space the lines as generously as possible and yet close enough to distinguish clearly the sectioned areas.

After the first few lines have been drawn, look back repeatedly at the original spacing to avoid gradually increasing or decreasing the intervals (Fig. 7.6b). Beginners almost invariably draw section lines too close together (Fig. 7.6c). This is very tedious because with small spacing the least inaccuracy in spacing is obvious.

Section lines should be uniformly thin, never varying in thickness, as in Fig. 7.6d. There should be a marked contrast in thickness of the visible outlines

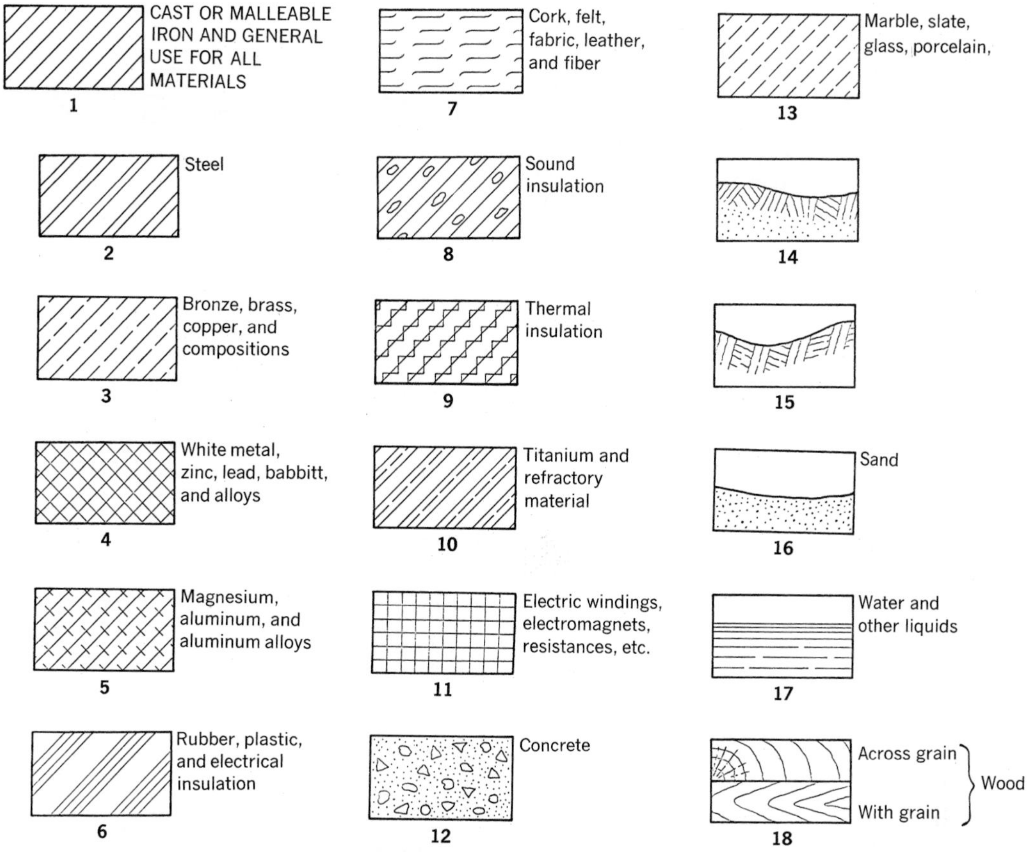

FIGURE 7.5 **Symbols for Section Lining.**

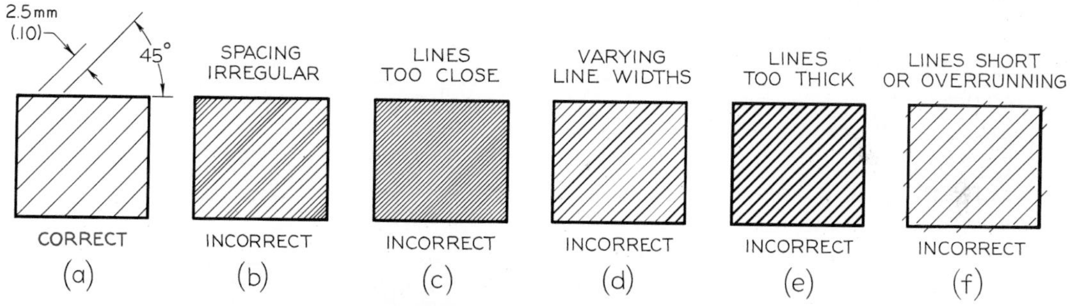

FIGURE 7.6 **Section-Lining Technique.**

and the section lines. Section lines should not be too thick, as in Fig. 7.6e. Also avoid running section lines beyond the visible outlines or stopping the lines too short, as in Fig. 7.6f.

If section lines drawn at 45° with horizontal would be parallel or perpendicular (or nearly so) to a prominent visible outline, the angle should be changed to 30°, 60°, or some other angle (Fig. 7.7).

Dimensions should be kept off sectioned areas, but when this is unavoidable the section lines should be omitted where the dimension figure is placed (see Fig. 11.13).

Section lines may be drawn adjacent to the boundaries of the sectioned areas (outline sectioning), provided that clarity is not sacrificed.

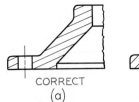
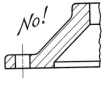
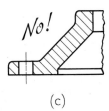

CORRECT
(a)

(c)

FIGURE 7.7   **Direction of Section Lines.**

■ 7.5 THE CUTTING PLANE

The cutting plane is indicated in a view adjacent to the sectional view (Fig. 7.8). In this view, the cutting plane appears edgewise as a line called the cutting-plane line. Alternate styles of cutting-plane lines are also shown in Fig. 7.9. The form shown in Fig. 7.9a is composed of equal dashes each about 6 mm ($\frac{1}{4}$″) or more long plus the arrowheads. This form without the dashes between the ends is especially desirable on complicated drawings. The form shown in Fig. 7.9b, composed of alternate long dashes and pairs of short dashes plus

the arrowheads, has been in general use for a long time. Both lines are drawn the same thickness as visible lines. Arrowheads indicate the direction in which the cutaway object is viewed.

Capital letters are used at the ends of the cutting-plane line when necessary to identify the cutting-plane line with the indicated section. This most often occurs in the case of multiple sections (see Fig. 7.25) or removed sections (see Fig. 7.21).

Sectional views occupy normal projected positions in the standard arrangement of views. In Fig. 7.8a, the

FIGURE 7.8   **Cutting Planes and Sections.**

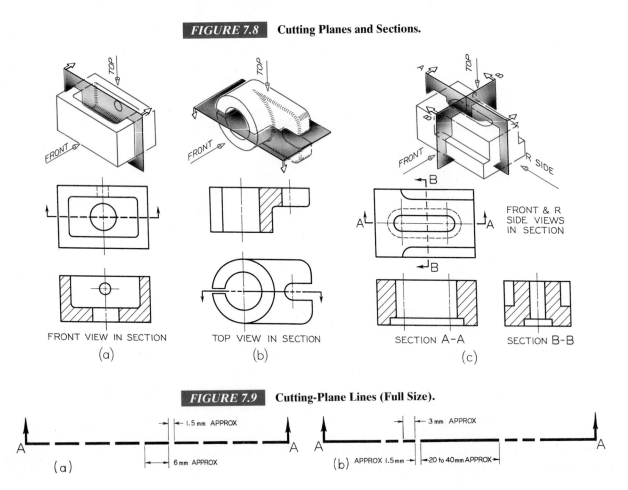

FRONT VIEW IN SECTION
(a)

TOP VIEW IN SECTION
(b)

SECTION A-A   SECTION B-B
(c)

FIGURE 7.9   **Cutting-Plane Lines (Full Size).**

A ————— 1.5 mm APPROX ————— A   A ————— 3 mm APPROX ————— A

(a)   6 mm APPROX   (b)   APPROX 1.5 mm   20 to 40 mm APPROX

cutting plane is a frontal plane (see §6.15) and appears as a line in the top view. The front half of the object (lower half in the top view) is imagined removed. The arrows at the ends of the cutting-plane line point in the direction of sight for a front view—that is, away from the front view or section. Note that the arrows do not point in the direction of withdrawal of the removed portion. The resulting full section may be referred to as the front view in section since it occupies the front view position.

In Fig. 7.8b, the cutting plane is a horizontal plane (see §6.15) and appears as a line in the front view. The upper half of the object is imagined removed. The arrows point toward the lower half in the same direction of sight as for a top view, and the resulting full section is a top view in section.

In Fig. 7.8c, two cutting planes are shown, one a frontal plane and the other a profile plane (see §6.15), both of which appear edgewise in the top view. *Each section is completely independent of the other and drawn as if the other were not present.* For section A–A, the front half of the object is imagined removed. The back half is then viewed in the direction of the arrows for a front view, and the resulting section is a front view in section. For section B–B, the right half of the object is imagined removed. The left half then is viewed in the direction of the arrows for a right-side view, and the resulting section is a right-side view in section. The cutting-plane lines are preferably drawn through an exterior view

(in this case the top view, as shown) instead of a sectional view.

The cutting-plane lines in Fig. 7.8 are shown for purposes of illustration only. They are generally omitted in cases such as these, in which the location of the cutting plane is obvious. When a cutting-plane line coincides with a center line, the cutting-plane line takes precedence.

Correct and incorrect relations between cutting-plane lines and corresponding sectional views are shown in Fig. 7.10.

■ 7.6 VISUALIZING A SECTION

Figure 7.11a shows two views of an object to be sectioned; it has a drilled and counterbored hole. The cutting plane is assumed to pass through the horizontal center line in the top view, and the front half of the object (lower half of the top view) is imagined removed. A pictorial drawing of the remaining back half is shown in Fig. 7.11b. The two cut surfaces produced by the cutting plane are 1-2-5-6-10-9 and 3-4-12-11-7-7. However, the corresponding section shown in Fig. 7.11c is incomplete because certain visible lines are missing.

If the section is viewed in the direction of sight (Fig. 7.11b), arcs A, B, C, and D will be visible. As shown in Fig. 7.11d, these arcs appear as straight lines 2–3, 6–7, 5–8, and 10–11. These lines may also be accounted for in other ways. The top and bottom surfaces of the object appear in the section as lines

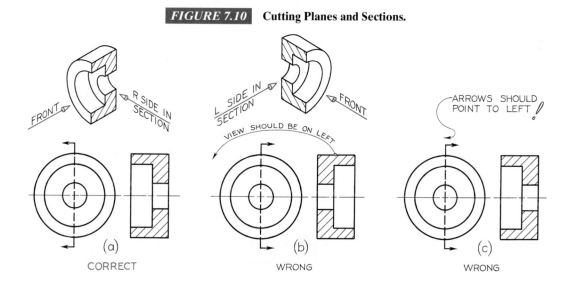

FIGURE 7.10 Cutting Planes and Sections.

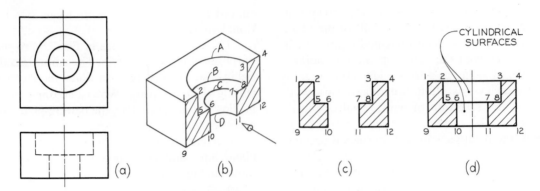

FIGURE 7.11 Visualizing a Section

1–4 and 9–12. The bottom surface of the counter-bore appears in the section as line 5–8. Also, the semicylindrical surfaces for the back half of the counterbore and of the drilled hole will appear as rectangles in the section at 2–3–8–5 and 6–7–11–10.

The front and top views of a collar are shown in Fig. 7.12a, and a right-side view in full section is required. The cutting plane is understood to pass along the center lines AD and EL. If the cutting plane were drawn, the arrows would point to the left in conformity with the direction of sight (see arrow) for the right-side view. The right side of the object is imagined removed, and the left half is viewed in the direction of the arrow, as shown pictorially in Fig. 7.12d. The cut surfaces appear edgewise in the top and front views along AD and EL; since the direction of sight for the

FIGURE 7.12 Drawing a Full Section.

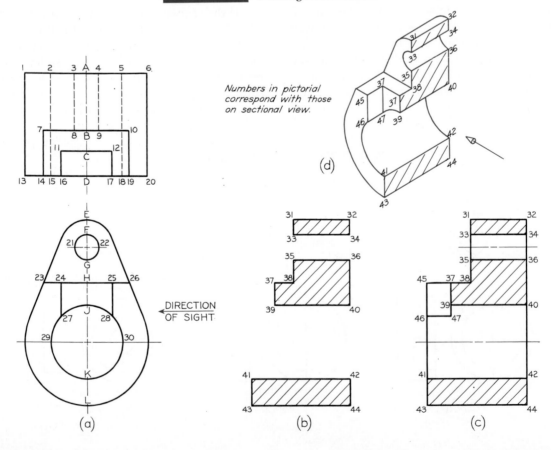

section is at right angles to them, they appear in true size and shape in the sectional view. Each sectioned area is completely enclosed by a boundary of visible lines. In addition to the cut surfaces, the sectional view shows all visible parts behind the cutting plane. No hidden lines are shown.

Whenever a surface of the object (plane or cylindrical) appears as a line and is intersected by a cutting plane that also appears as a line, a new edge (line of intersection) is created that will appear as a *point* in that view. Thus, in the front view, the cutting plane creates new edges appearing as points at E, F, G, H, J, K, and L. In the sectional view (Fig. 7.12b), these are horizontal lines 31–32, 33–34, 35–36, 37–38, 39–40, 41–42, and 43–44.

Whenever a surface of an object appears as a surface (that is, not as a line) and is cut by a cutting plane that appears as a line, a new edge is created that appears as a line in the view, coinciding with the cutting-plane line, and as a line in the section.

In the top view, D is the **point view** of the vertical line KL in the front view and 41–43 in the section (Fig. 7.12b). Point C is the point view of the vertical line HJ in the front view and 37–39 in the section. Point B is the point view of two vertical lines EF and GH in the front view, and 31–33 and 35–38 in the section. Point A is the point view of three vertical lines EF, GJ, and KL in the front view, and 32–34, 36–40, and 42–44 in the section. This completes the boundaries of three sectioned areas 31–32–34–33, 35–36–40–39–37–38, and 41–42–44–43. It is only necessary now to add the visible lines beyond the cutting plane.

The semicylindrical left half F–21–G of the small hole (front view) is visible as a rectangle in the sec-

tions at 33–34–36–35 (Fig. 7.12c). The two semicircular arcs appear as straight lines in the section at 33–35 and 34–36.

Surface 24–27, a line in the front view, appears as line 11–16 in the top view and as surface 45–37–47–46, true size, in the section (Fig. 7.12c).

Cylindrical surface J–29–K, an arc in the front view, appears in the top view as 2–A–C–11–16–15 and in the section as 46–47–39–40–42–41. Thus, arc 27–29–K (front view) appears in the section (Fig. 7.12c) as straight line 46–41; and arc J–29–K appears as straight line 40–42.

All cut surfaces here are part of the same object, and the section lines must all run in the same direction, as shown.

■ 7.7 HALF SECTIONS

If a cutting plane passes halfway through an object, the result is a **half section** (Fig. 7.13). A half section has the advantage of exposing the interior of one half of the object and retaining the exterior of the other half. Its usefulness is, therefore, largely limited to symmetrical objects. It is not widely used in detail drawings (single parts) because of this limitation of symmetry and also because of difficulties in dimensioning internal shapes that are shown in part only in the sectioned half (Fig. 7.13b).

In general, hidden lines should be omitted from both halves of a half section. However, they may be used in the unsectioned half if necessary for dimensioning.

The greatest usefulness of the half section is in assembly drawing, in which it is often necessary to show both internal and external construction on the same view, but without dimensioning.

FIGURE 7.13 Half Section.

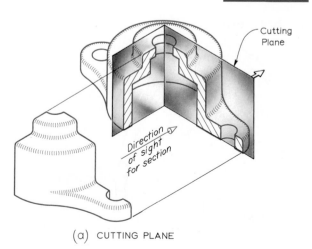

(a) CUTTING PLANE

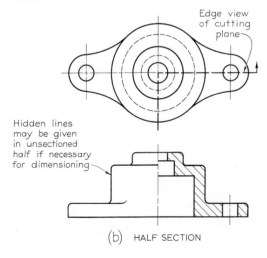

(b) HALF SECTION

As shown in Fig. 7.13b, a center line is used to separate the halves of the half section. The American National Standards Institute recommends a center line for the division line between the sectioned half and the unsectioned half of a half-sectional view, although in some cases the same overlap of the exterior portion, as in a broken-out section, is preferred (see Fig. 7.33b). Either form is acceptable.

■ 7.8 BROKEN-OUT SECTIONS

It often happens that only a partial section of a view is needed to expose interior shapes. Such a section, limited by a break line, is called a **broken-out section**. In Fig. 7.14, a full or half section is not necessary, and a small broken-out section is sufficient to explain the construction. In Fig. 7.15, a half section would have caused the removal of half the keyway. The keyway is preserved by breaking out around it. In this case, the section is limited partly by a break line and partly by a center line.

■ 7.9 REVOLVED SECTIONS

The shape of the cross section of a bar, arm, spoke, or other elongated object may be shown in the longitudinal view by means of a revolved section (Fig. 7.16). **Revolved sections** are made by assuming a plane perpendicular to the center line or axis of the bar or other object, as shown in Fig. 7.17a, and then revolving the plane through 90° about a center line at right angles to the axis (Fig. 7.17b and 7.17c).

The visible lines adjacent to a revolved section may be broken out if desired, as shown in Figs. 7.16k and 7.17.

The superimposition of the revolved section requires the removal of all original lines covered by it

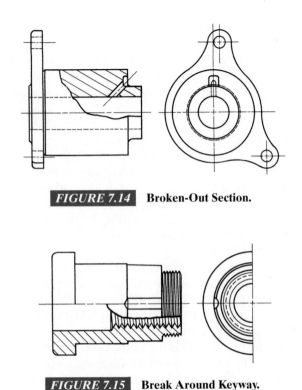

FIGURE 7.14 Broken-Out Section.

FIGURE 7.15 Break Around Keyway.

(Fig. 7.19). The true shape of a revolved section should be retained after the revolution of the cutting plane, regardless of the direction of the lines in the view (Fig. 7.20).

■ 7.10 REMOVED SECTIONS

A **removed section** is a section that is not in direct projection from the view containing the cutting plane—that is, it is not positioned in agreement with the standard arrangement of views. This displacement from

FIGURE 7.16 Revolved Sections.

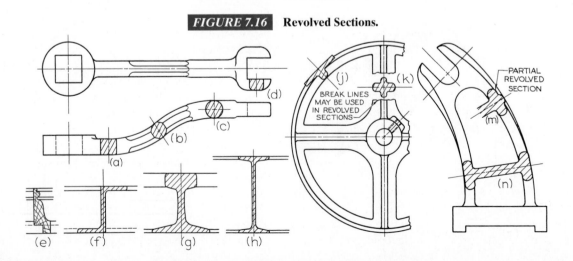

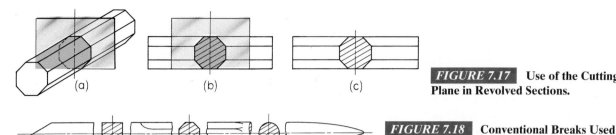

FIGURE 7.17 Use of the Cutting Plane in Revolved Sections.

FIGURE 7.18 Conventional Breaks Used with Revolved Sections.

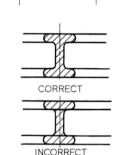

CORRECT

INCORRECT

FIGURE 7.19 A Common Error in Drawing Revolved Sections.

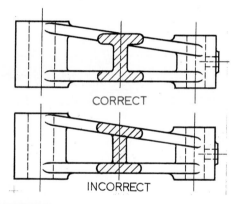

CORRECT

INCORRECT

FIGURE 7.20 A Common Error in Drawing Revolved Sections.

the normal projection position should be made without turning the section from its normal orientation.

Removed sections should be labeled, such as SEC-TION A–A and SECTION B–B, corresponding to the letters at the ends of the cutting-plane line (Fig. 7.21). They should be arranged in alphabetical order from left to right on the sheet. Section letters should be used in alphabetical order, but letters I, O, and Q should not be used because they are easily confused with the numeral 1 or the zero.

A removed section is often a partial section. Such a removed section is frequently drawn to an enlarged scale (Fig. 7.22). This is often desirable to show small detail clearly and to provide sufficient space for di-

mensioning. In such a case the enlarged scale should be indicated below the section title.

A removed section should be placed so that it no longer lines up in projection with any other view. It should be separated clearly from the standard arrangement of views (see Fig. 12.9). Whenever possible, removed sections should be on the same sheet as the regular views. If a section must be placed on a different sheet, cross-references should be given on the related sheets. A note should be given below the section title, such as

SECTION B–B ON SHEET 4, ZONE A3

A similar note should be placed on the sheet on which the cutting-plane line is shown, with a leader pointing to the cutting-plane line and referring to the sheet on which the section will be found. Sometimes it is convenient to place removed sections on center lines extended from the section cuts (Fig. 7.23).

■ 7.11 OFFSET SECTIONS

In sectioning through irregular objects, it is often desirable to show features that do not lie in a straight line by "offsetting" or bending the cutting plane. Such a section is called an **offset section**. In Fig. 7.24a the cutting plane is offset in several places to include the hole at the left end, one of the parallel slots, the rectangular recess, and one of the holes at the right end. The front portion of the object is then imagined to be removed (Fig. 7.24b). The path of the cutting plane is shown by the cutting-plane line in the top view (Fig. 7.24c), and the resulting offset section is shown in the front view. The offsets or bends in the cutting plane are all 90° and are *never shown in the sectional view*.

Figure 7.24 also illustrates how hidden lines in a section eliminate the need for an additional view. In this case, an extra view would be needed to show the small boss on the back if hidden lines were not shown.

Figure 7.25 shows an example of multiple offset sections. Notice that the visible background shapes without hidden lines appear in each sectional view.

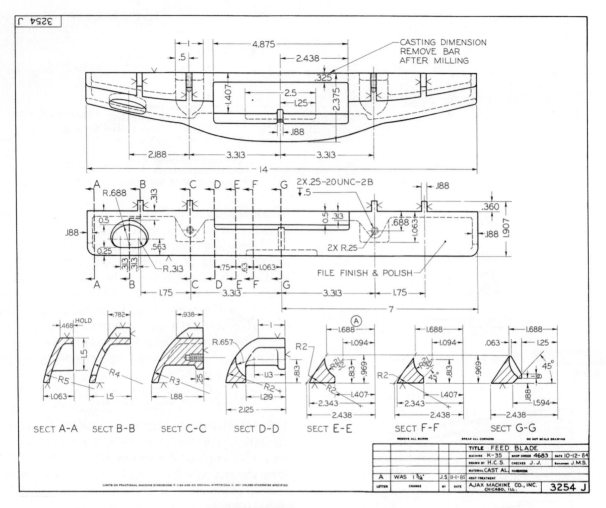

FIGURE 7.21 Removed Sections.

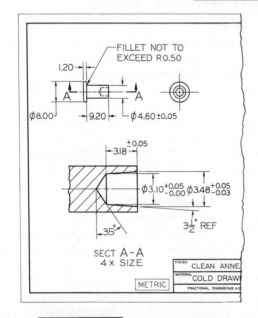

FIGURE 7.22 Removed Section.

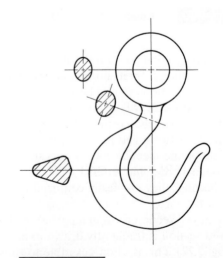

FIGURE 7.23 Removed Sections.

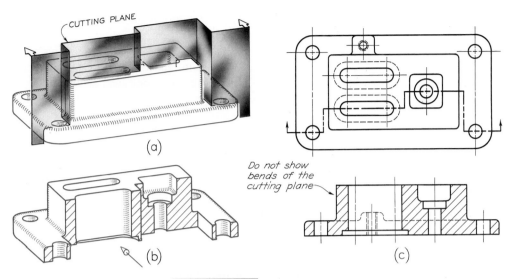

FIGURE 7.24 Offset Section.

■ 7.12 RIBS IN SECTIONS

To avoid a false impression of thickness and solidity, ribs, webs, gear teeth, and other similar flat features are not sectioned even though the cutting plane passes along the center plane of the feature. For example, in Fig. 7.26, the cutting plane A–A passes flatwise through the vertical web, or rib, and the web is not section lined (Fig. 7.26a). *Such thin features should not be section lined, even though the cutting plane passes through them.* The incorrect section is shown in Fig. 7.26b. Note the false impression of thickness or solidity resulting from section lining the rib.

If the cutting plane passes *crosswise* through a rib or any thin member, as does the plane B–B, the member should be section lined in the usual manner, as shown in the top view (Fig. 7.26c).

If a rib is not section lined when the cutting plane passes through it flatwise, it is sometimes difficult to tell whether the rib is actually present, as, for example, ribs A in Figs. 7.27a and 7.27b. It is difficult to distinguish spaces B as open spaces and spaces A as ribs. In such cases, double-spaced *section lining* of the ribs should be used (Fig. 7.27c). This consists simply in continuing alternate section lines through the ribbed areas, as shown.

■ 7.13 ALIGNED SECTIONS

To include in a section certain angled elements, the cutting plane may be bent to pass through those features. The plane and features are then imagined to be revolved into the original plane. For example, in Fig. 7.28, the cutting plane was bent to pass through the

FIGURE 7.25 Three Offset Sections.

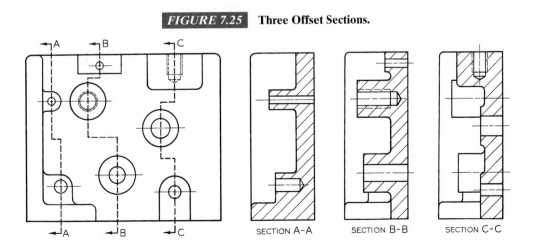

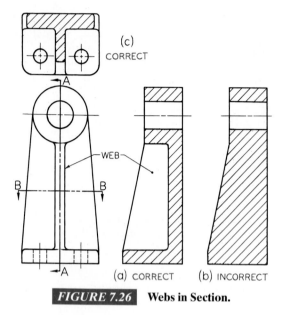

(c) CORRECT

WEB

(a) CORRECT   (b) INCORRECT

FIGURE 7.26   **Webs in Section.**

angled arm and then revolved to a vertical position (aligned), from where it was projected across to the sectional view.

In Fig. 7.29 the cutting plane is bent to include one of the drilled and counterbored holes in the sectional view. The correct section view in Fig. 7.29b gives a clearer and more complete description than does the section in Fig. 7.29c, which was taken along the vertical center line of the front view—that is, without any bend in the cutting plane. In such cases, the angle of revolution should always be less than 90°.

The student is cautioned *not to revolve* features when clearness is not improved. In some cases revolving the features results in a loss of clarity. Examples in

which revolution should not be used are Fig. 7.40 and Probs. 17 and 18.

In Fig. 7.30a, the projecting lugs are not sectioned for the same reason that ribs are not sectioned. In Fig. 7.30b, the projecting lugs are located so that the cutting plane passes through them crosswise; therefore, they are sectioned.

Another example involving rib sectioning and aligned sectioning is shown in Fig. 7.31. In the circular view, the cutting plane is offset in circular-arc bends to include the upper hole and upper rib, the keyway and center hole, the lower rib, and one of the lower holes. These features are imagined to be revolved until they line up vertically and are then projected from that position to obtain the section shown in Fig. 7.31b. Note that the ribs are not sectioned. If a regular full section of the object were drawn without using the conventions discussed here, the resulting section (Fig. 7.31c) would be incomplete and confusing and would take more time to draw.

In sectioning a pulley or any spoked wheel (Fig. 7.32a), it is standard practice to revolve the spokes if necessary (if there is an odd number) and not to section line the spokes (Fig. 7.32b). If the spoke is sectioned, the section gives a false impression of continuous metal (Fig. 7.32c). If the lower spoke is not revolved, it will be foreshortened in the sectional view, in which it presents an "amputated" and wholly misleading appearance.

Figure 7.32 also illustrates correct practice in omitting visible lines in a sectional view. Notice that spoke B is omitted in Fig. 7.32b. If it is included, as shown in Fig. 7.32c, the spoke is foreshortened, difficult and time-consuming to draw, and confusing to the reader of the drawing.

FIGURE 7.27   **Alternate Section Lining.**

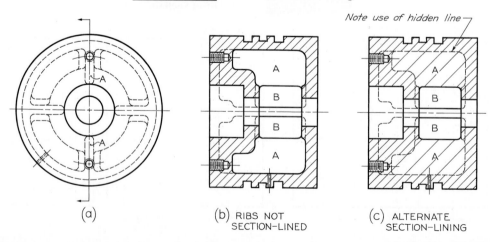

(a)

(b) RIBS NOT SECTION–LINED

(c) ALTERNATE SECTION–LINING

Note use of hidden line

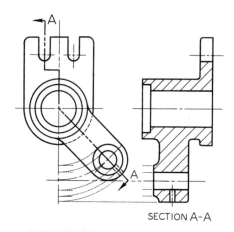

SECTION A-A

FIGURE 7.28 Aligned Section.

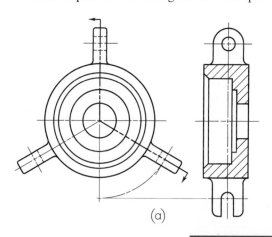

FIGURE 7.29 Aligned Section.

7.14 PARTIAL VIEWS

If space is limited on the paper or if it is necessary to save drafting time, **partial views** may be used in connection with sectioning (Fig. 7.33). Half views are shown in Figs. 7.33a and 7.33b in connection with a full section and a half section, respectively. Note that in each case the back half of the object in the circular view is shown, in conformity with the idea of removing the front portion of the object to expose the back portion for viewing in section (see also §6.9).

Another method of drawing a partial view is to break out much of the circular view, retaining only those features that are needed for minimum representation (Fig. 7.33c).

7.15 INTERSECTIONS IN SECTIONING

Where an intersection is small or unimportant in a section, it is standard practice to disregard the true pro-

jection of the figure of intersection, as shown in Figs. 7.34a and 7.34c. Larger figures of intersection may be projected, as shown in Fig. 7.34b, or approximated by circular arcs, as shown for the smaller hole in Fig. 7.34d. Note that the larger hole K is the same diameter as the vertical hole. In such cases the curves of intersection (ellipses) appear as straight lines, as shown (see also Figs. 6.38 and 6.39).

7.16 CONVENTIONAL BREAKS

To shorten a view of an elongated object, conventional breaks are recommended, as shown in Fig. 7.35. For example, the two views of a garden rake in Fig. 7.36a are drawn to a small scale to fit them on the paper. In Fig. 7.36b, the handle is "broken," a long central portion is removed, and the rake is then drawn to a larger scale, producing a much clearer delineation.

Parts to be broken must have the same section throughout, or if tapered they must have a uniform

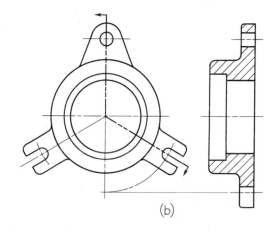

(a)   (b)

FIGURE 7.30 Aligned Sections.

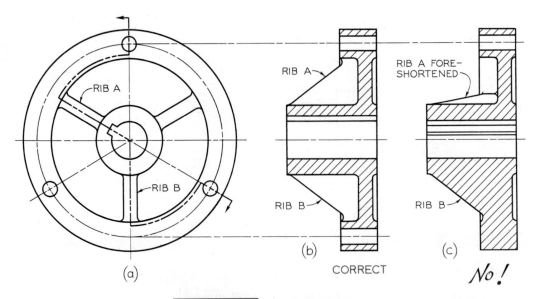

FIGURE 7.31 Symmetry of Ribs.

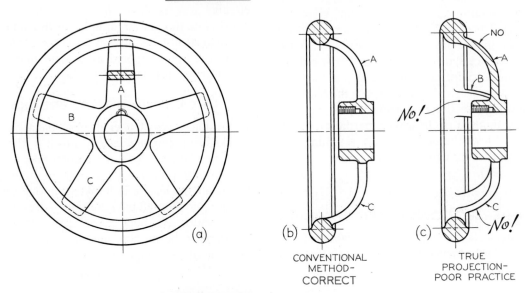

FIGURE 7.32 Spokes in Section.

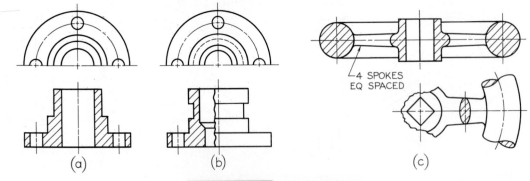

FIGURE 7.33 Partial Views.

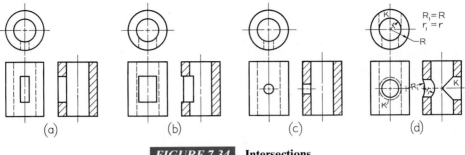

FIGURE 7.34   Intersections.

taper. Note in Fig. 7.36b that the full-length dimension is given, just as if the entire rake were shown.

The breaks used on cylindrical shafts or tubes are often referred to as "S-breaks" and in industrial drafting are usually drawn entirely freehand or partly freehand and partly with an irregular curve or compass. The results of these methods are often very crude, especially when attempted by beginners. Simple construction methods for students or industrial drafters are shown in Figs. 7.37 and 7.38 and will always produce a professional result. Excellent S-breaks are also obtained with an S-break template.

Breaks for rectangular metal and wood sections are always drawn freehand, as shown in Fig. 7.35. (See also Fig. 7.18, which illustrates the use of breaks in connection with revolved sections.)

■ 7.17 COMPUTER GRAPHICS

USING CAD

You create 2D and 3D sectional views using CAD. Most CAD systems have a Hatch command to generate the section lining and hatch patterns to fill an area automatically. A wide variety of hatch patterns are

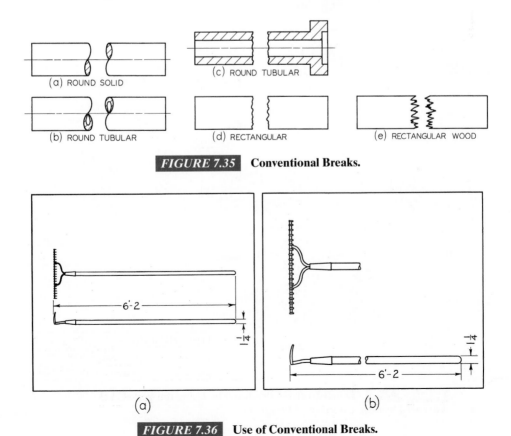

FIGURE 7.35   Conventional Breaks.

(a)

(b)

FIGURE 7.36   Use of Conventional Breaks.

generally available to show materials such as steel, bronze, sand, concrete, and many more.

Creating a full-section view from a 3D model is generally very easy. You only need to define the cutting plane. Often the hatching for the cut surfaces is generated automatically. Sectioned views other than full sections can be more difficult to create. To create good sectional drawings using CAD, you should have a clear understanding of the standards for showing section views (Fig. 7.39).

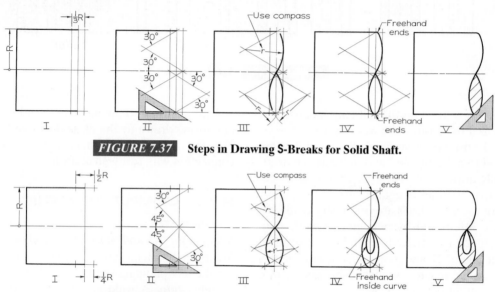

FIGURE 7.37 Steps in Drawing S-Breaks for Solid Shaft.

FIGURE 7.38 Steps in Drawing S-Breaks for Tubing.

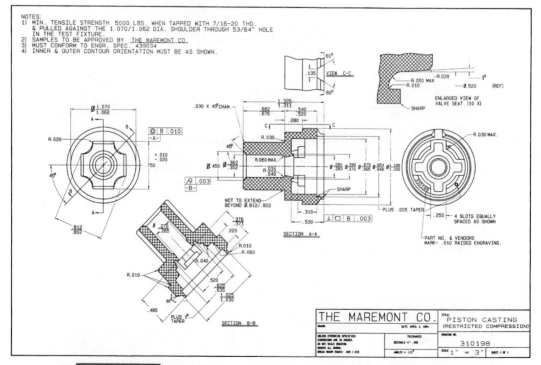

FIGURE 7.39 Detail Drawing Produced by Using the VersaCAD Advanced System. *Courtesy of VersaCAD.*

■ KEY WORDS

SECTIONAL VIEW	SECTION LINES
CUTTING-PLANE LINE	FULL AND HALF SECTION
OFFSET SECTION	BROKEN-OUT SECTION
REVOLVED SECTION	REMOVED SECTION
AUXILIARY SECTION	CONVENTIONS
RIBS, SPOKES, AND WEBS	ALIGNED SECTIONS
CONVENTIONAL BREAKS	

■ CHAPTER SUMMARY

- Sectional views show internal details without the need for hidden lines.
- Objects are imagined to be cut apart along the cutting-plane line.
- Along the cutting plane, the part of the object that is solid (cut by the cutting plane) is shown with section lines (thin lines drawn at 45°).
- In a sectional view, many hidden lines are replaced by object lines because the internal surfaces are exposed when the object is imagined to be cut open.

- The section-lining symbol denotes the material of the object.
- Ribs, webs, and spokes are not shown with section lining.
- Symmetrical features like spokes and webs are revolved so the sectional view appears symmetrical.
- Conventional breaks are used to represent various objects in shortened form.

■ REVIEW QUESTIONS

1. What does the cutting-plane line represent?
2. Sketch the section-line symbols for 10 different materials.
3. List seven different types of sections and sketch an example of each.
4. Which sectional views are used to *replace* an existing primary view? Which sectional views are used in addition to the primary views?
5. How much of an object is imagined to be cut away in a half section?
6. What type of line is used to show the boundary of a broken-out section?
7. Why are hidden lines generally omitted in a sectional view?
8. Why are some symmetrical features, like spokes and webs, revolved in the sectional view?
9. Why is a rib outlined with object lines and not filled with section lining?

■ SECTIONING PROBLEMS

Any of the following problems (Figs. 7.40–7.71) may be drawn freehand or with your instruments, as assigned by the instructor. However, the problems in Fig. 7.40 are especially suitable for sketching on 8.5" × 11.0" graph paper with appropriate grid squares. Two problems can be drawn on one sheet, using Layout A–1 similarly to Fig. 5.51, with borders drawn freehand. If desired, the problems may be sketched on plain drawing paper. Before making any sketches, the student should study carefully §§5.1–5.9.

The problems in Figs. 7.41–7.60 are intended to be drawn with instruments, but may be drawn freehand if desired. If metric or decimal dimensions are required, you should first study §§11.1–11.25.

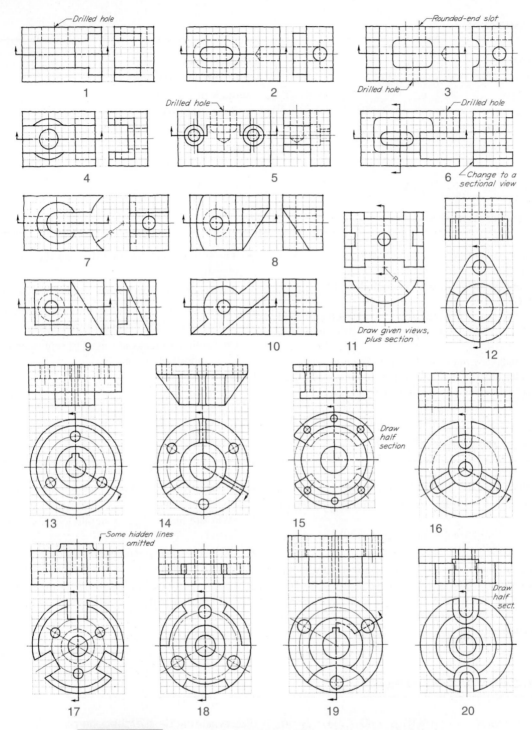

FIGURE 7.40 Freehand Sectioning Problems. Using Layout A–1 or A4–1 adjusted (freehand) on graph paper or plain paper, two problems per sheet, sketch views with sections as indicated. Each grid square = 6 mm ($\frac{1}{4}''$). In Probs. 1–10, top and right-side views are given. Sketch front sectional views and then move right-side views to line up horizontally with front sectional views. Omit cutting planes except in Probs. 5 and 6.

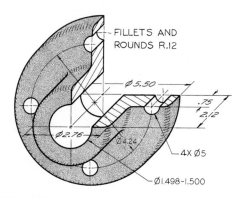

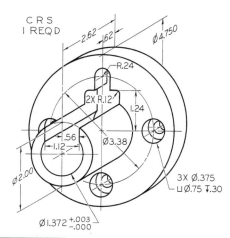

FIGURE 7.41 Bearing. Draw necessary views, with full section (Layout A–3).*

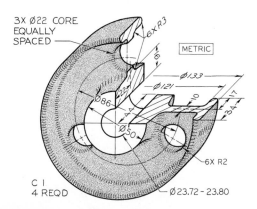

FIGURE 7.44 Centering Bushing. Draw necessary views, with full section (Layout A–3).*

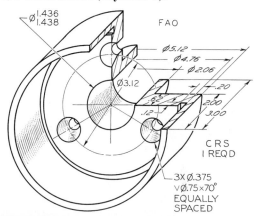

FIGURE 7.42 Truck Wheel. Draw necessary views, with half section (Layout A–3).*

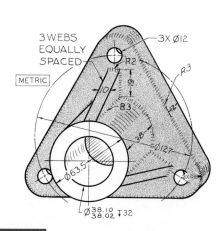

FIGURE 7.45 Special Bearing. Draw necessary views, with full section (Layout A–3).*

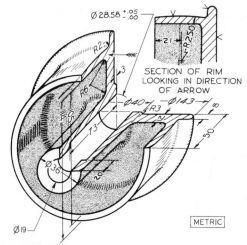

FIGURE 7.43 Column Support. Draw necessary views, with full section (Layout A–3).*

FIGURE 7.46 Idler Pulley. Draw necessary views, with full section (Layout A–3).*

*Layout A4–3 (adjusted) may be used. If dimensions are required, study §§11.1–11.25. Use metric or decimal-inch dimensions as assigned by the instructor.

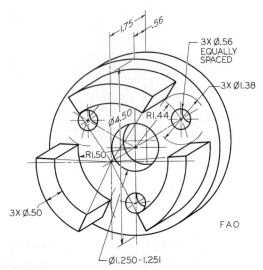

FIGURE 7.47 Cup Washer. Draw necessary views, with full section (Layout A–3 or A4–3 adjusted).*

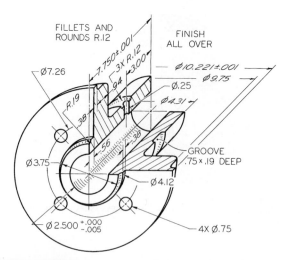

FIGURE 7.50 Bearing. Draw necessary views, with half section. Scale: half size (Layout B–4 or A3–4 adjusted).*

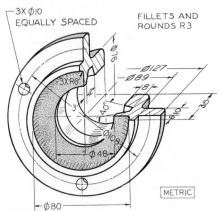

FIGURE 7.48 Fixed Bearing Cup. Draw necessary views, with full section (Layout A–3 or A4–3 adjusted).*

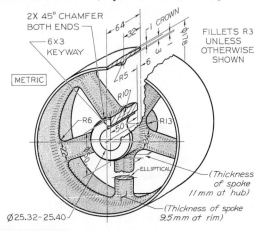

FIGURE 7.51 Pulley. Draw necessary views, with full section, and revolved section of spoke (Layout B–4 or A3–4 adjusted).*

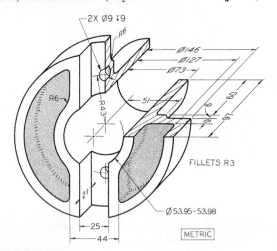

FIGURE 7.49 Stock Guide. Draw necessary views, with half section (Layout B–4 or A3–4 adjusted).*

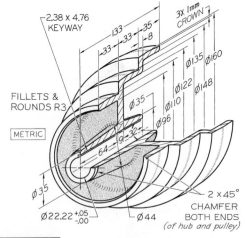

FIGURE 7.52 Step-Cone Pulley. Draw necessary views, with full section (Layout B–4 or A3–4 adjusted).*

* If dimensions are required, study §§11.1–11.25. Use metric or decimal-inch dimensions as assigned by the instructor.

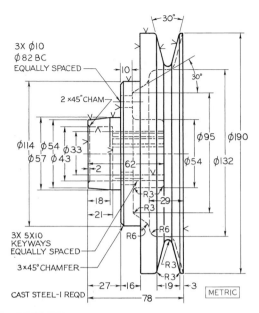

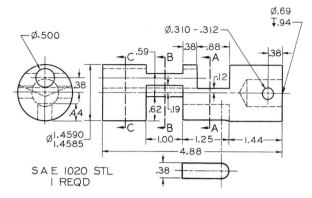

FIGURE 7.54 Operating Valve. Given: Front, left-side, and partial bottom views. Required: Front, right-side, and full bottom views, plus indicated removed sections (Layout B–4).*

FIGURE 7.53 Sheave. Draw two views, including half section (Layout B–4).*

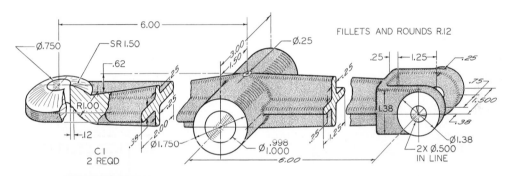

FIGURE 7.55 Rocker Arm. Draw necessary views, with revolved sections (Layout B–4).*

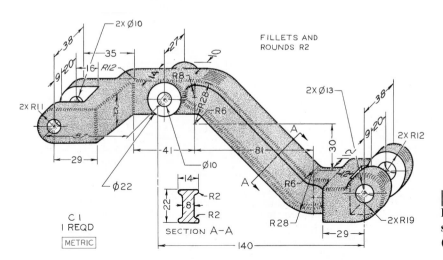

FIGURE 7.56 Dash Pot Lifter. Draw necessary views, using revolved section instead of removed section (Layout B–4).*

*Layout A3–4 (adjusted) may be used. If dimensions are required, study §§11.1–11.25. Use metric or decimal-inch dimensions as assigned by the instructor.

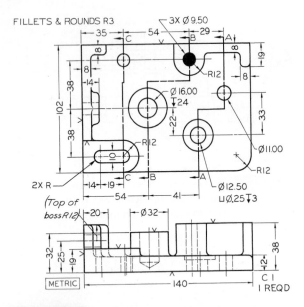

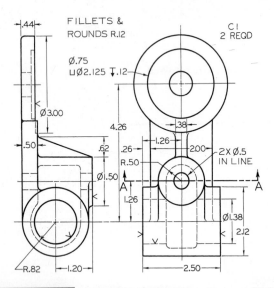

FIGURE 7.57 Adjuster Base. Given: Front and top views. Required: Front and top views and sections A–A, B–B, and C–C. Show all visible lines (Layout B–4).[*]

FIGURE 7.58 Mobile Housing. Given: Front and left-side views. Required: Front view, right-side view in full section, and removed section A–A (Layout B–4).[*]

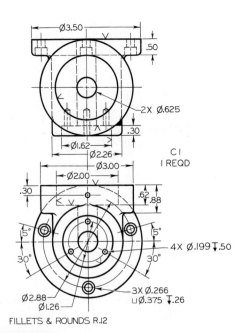

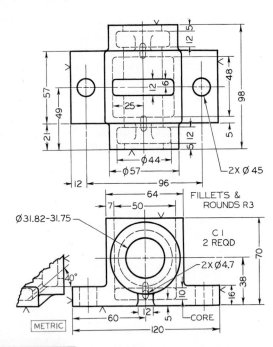

FIGURE 7.59 Hydraulic Fitting. Given: Front and top views. Required: Front and top views and right-side view in full section (Layout B–4).[*]

FIGURE 7.60 Auxiliary Shaft Bearing. Given: Front and top views. Required: Front and top views and right-side view in full section (Layout B–4).[*]

[*] If dimensions are required, study §§11.1–11.25. Use metric or decimal-inch dimensions as assigned by the instructor.

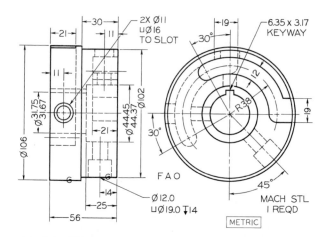

FIGURE 7.61 Traverse Spider. Given: Front and left-side views. Required: Front and right-side views and top view in full section (Layout B–4 or A3–4 adjusted).[*]

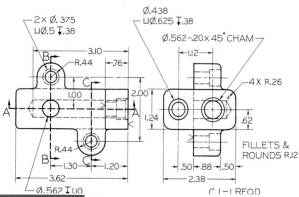

FIGURE 7.63 Bracket. Given: Front and right-side views. Required: Take front as new top; then add right-side view, front view in full section A-A, and sections B-B and C-C (Layout B–4 or A3–4 adjusted).[*]

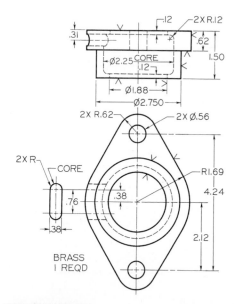

FIGURE 7.62 Gland. Given: Front, top, and partial left-side views. Required: Front view and right-side view in full section (Layout A–3 or A4–3 adjusted).[*]

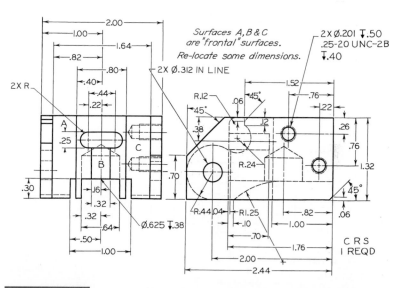

FIGURE 7.64 Cocking Block. Given: Front and right-side views. Required: take front as new top view; then add new front view, and right-side view in full section. Draw double size on Layout C–4 or A2–4 adjusted.[*]

[*] Layout A3–4 (adjusted) may be used. If dimensions are required, study §§11.1–11.25. Use metric or decimal-inch dimensions as assigned by the instructor.

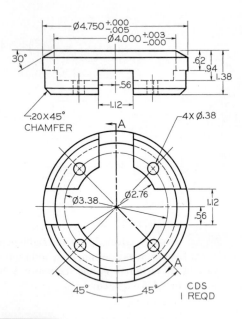

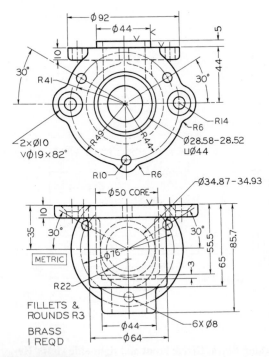

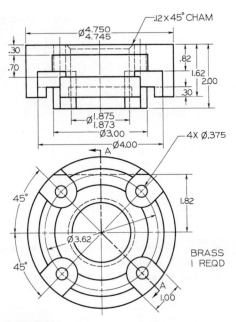

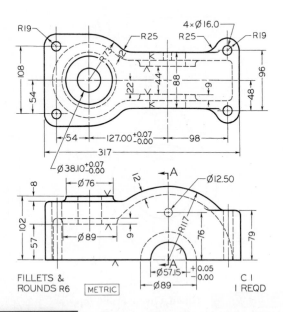

FIGURE 7.65   Packing Ring. Given: Front and top views. Required: Front view and section **A–A** (Layout A–3 or A4–3 adjusted).[*]

FIGURE 7.67   Oil Retainer. Given: Front and top views. Required: Front view and section **A–A** (Layout B–4 or A3–4 adjusted).[*]

FIGURE 7.66   Strainer Body. Given: Front and bottom views. Required: Front and top views and right-side view in full section (Layout C–4 or A2–4).[*]

FIGURE 7.68   Gear Box. Given: Front and top views. Required: Front in full section, bottom view, and right-side section A–A. Draw half size on Layout B–4 or A3–4 (adjusted).[*]

[*] If dimensions are required, study §§11.1–11.25. Use metric or decimal-inch dimensions as assigned by the instructor.

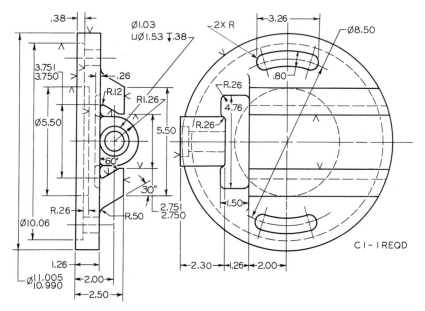

FIGURE 7.69 Slotted Disk for Threading Machine. Given: Front and left-side views. Required: Front and right-side views and top full-section view. Draw half size on Layout B–4 or A3–4 (adjusted). [*]

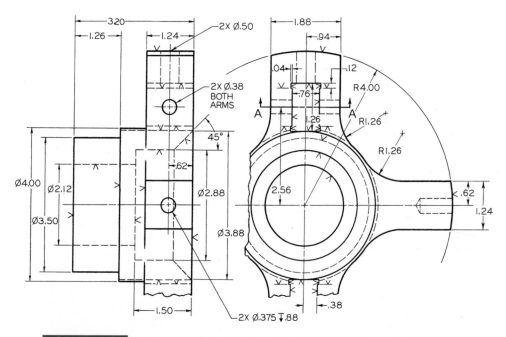

FIGURE 7.70 Web for Lathe Clutch. Given: Partial front and left-side views. Required: Full front view, right-side view in full section, and removed section A-A (Layout C–4 or A2–4). [*]

[*] If dimensions are required, study §§11.1–11.25. Use metric or decimal-inch dimensions as assigned by the instructor.

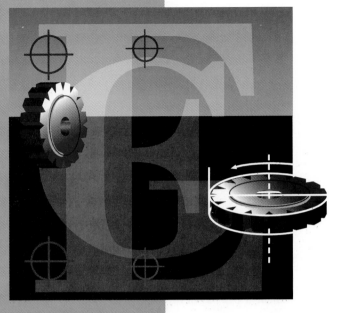

C H A P T E R 8

AUXILIARY VIEWS

OBJECTIVES

After studying the material in this chapter, you should be able to:

1. Create an auxiliary view from any orthographic projection using drawing instruments or CAD.

2. Draw folding lines or reference-plane lines between any two adjacent views.

3. Construct depth, height, or width auxiliary views.

4. Plot curves in auxiliary views.

5. Construct partial auxiliary views.

6. Create auxiliary sectional views.

7. Find the true length of an oblique line by constructing an auxiliary view.

8. Find the true size of an oblique plane by constructing auxiliary views.

OVERVIEW

Inclined planes and oblique lines do not appear true length or true size in any of the principal planes of projection. To determine the true length of an oblique line or the true size of an inclined plane, an auxiliary view must be created. The techniques for creating an auxiliary view are the same using traditional drawing or CAD. A line of sight and reference plane are defined. With traditional drawing, the view is manually created along line-of-sight projectors. With CAD drawing, the computer generates the view automatically if a 3D model of the object was originally created.

Many objects are shaped such that their principal faces cannot be assumed to be parallel to the regular planes of projection. For example, in Fig. 8.1a, the base of the design for the bearing is shown in its true size and shape, but the rounded upper portion is situated at an angle with the planes of projection and does not appear in its true size and shape in any of the three regular views. To show the true circular shapes, it is necessary to assume a direction of sight perpendicular to the planes of those curves, as shown in Fig. 8.1b. The resulting view is known as an *auxiliary view*. This view, together with the top view, completely describes the object. The front and right-side views are not necessary.

■ 8.1 DEFINITIONS

Any view obtained by a projection on a plane other than the horizontal, frontal, and profile projection planes is an **auxiliary view**. A *primary auxiliary view* is projected onto a plane that is perpendicular to one of the principal planes of projection and is inclined to the other two. A *secondary auxiliary view* is projected from a primary auxiliary view onto a plane that is inclined to all three principal projection planes (see §8.19).

■ 8.2 THE AUXILIARY PLANE

In Fig. 8.2a, the object shown has an inclined surface that does not appear in its true size and shape in any regular view. The auxiliary plane is assumed parallel to the inclined surface P—that is, perpendicular to the line of sight, which is at right angles to that surface. The auxiliary plane is then perpendicular to the frontal plane of projection and hinged to it.

When the horizontal and auxiliary planes are unfolded to appear in the plane of the front view (Fig.

8.2b), the folding lines represent the hinge lines joining the planes. The drawing is simplified by retaining the folding lines (H/F and F/1) and omitting the planes (Fig. 8.2c). As will be shown later, the folding lines may themselves be omitted in the actual drawing. The inclined surface P is shown in its true size and shape in the auxiliary view; the long dimension of the surface is projected directly from the front view and the depth from the top view.

The positions of the folding lines depend on the relative positions of the planes of the glass box. If the horizontal plane is moved upward, the distance Y is increased. If the frontal plane is brought forward, the distances X are increased but remain *equal*. If the auxiliary plane is moved to the right, the distance Z is increased. Note that both the top and auxiliary views show the *depth* of the object.

■ 8.3 DRAWING AN AUXILIARY VIEW: FOLDING-LINE METHOD

As shown in Fig. 8.2c, the folding lines are the hinge lines of the glass box. Distances X must be equal since they both represent the distance of the front surface of the object from the frontal plane of projection.

Although distances X must remain equal, distances Y and Z, from the front view to the respective folding lines, may or may not be equal.

The steps in drawing an auxiliary view with the aid of the folding lines, shown in Fig. 8.3, are described as follows.

1. The front and top views are given (Fig. 8.3I). It is necessary to draw an auxiliary view showing the true size and shape of inclined surface P. Draw the folding line H/F between the views at right angles to the projection lines. Distances X and Y may or may not be equal, as desired.

NOTE In the following steps, manipulate the triangle (either triangle), as shown in Fig. 8.4, to draw lines parallel or perpendicular to the inclined face.

2. Assume a direction of sight perpendicular to surface P (Fig. 8.3II). Draw light projection lines from the front view parallel to the arrow, or perpendicular to surface P.

3. Draw folding line F/1 for the auxiliary view at right angles to the projection lines and at any convenient distance from the front view (Fig. 8.3III).

4. Draw the auxiliary view using the numbering system explained in §6.6 (Fig. 8.3IV). Locate all

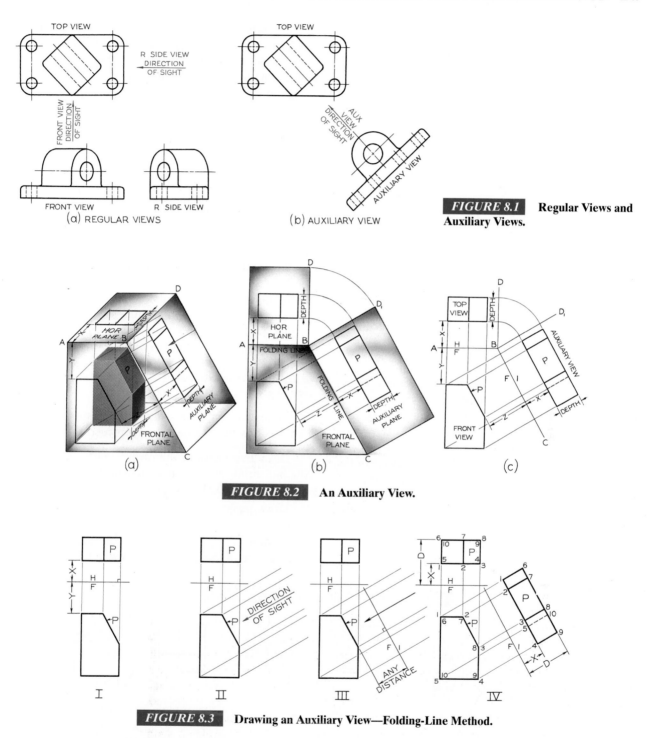

FIGURE 8.1 **Regular Views and Auxiliary Views.**

FIGURE 8.2 **An Auxiliary View.**

FIGURE 8.3 **Drawing an Auxiliary View—Folding-Line Method.**

points the same distances from folding line F/1 as they are from folding line H/F in the top view. For example, points 1 to 5 are distance X from the folding lines in both the top and auxiliary views,

and points 6 to 10 are distance D from the corresponding folding lines. Since the object is viewed in the direction of the arrow, it will be seen that edge 5–10 will be hidden in the auxiliary view.

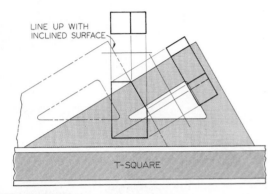

LINE UP WITH
INCLINED SURFACE

T-SQUARE

FIGURE 8.4 **Drawing Parallel or Perpendicular Lines.**

■ 8.4 REFERENCE PLANES

In the auxiliary views shown in Figs. 8.2c and 8.3, the folding lines represent the edge views of the frontal plane of projection. In effect, the frontal plane is used as a **reference plane**, or *datum plane*, for transferring distances (depth measurements) from the top view to the auxiliary view.

Instead of using one of the planes of projection as a reference plane, it is often more convenient to assume a reference plane inside the glass box parallel to the plane of projection and touching or cutting through the object. For example, in Fig. 8.5a, a reference plane is assumed to coincide with the front surface of the object. This plane appears edgewise in the top and auxiliary views, and the two reference lines are then used in the same manner as folding lines. Dimensions D in the top and auxiliary views are equal. The advantage of the reference-plane method is that fewer measurements are required since some points of the object lie in the reference plane.

The reference plane may coincide with the front surface of the object (Fig. 8.5a); it may cut through the object if the object is symmetrical (Fig. 8.5b); it may coincide with the back surface of the object (Fig. 8.5c); or it may cut through any intermediate point of the object.

The reference plane should be assumed in the position most convenient for transferring distances with respect to it. Remember the following:

1. Reference lines, like folding lines, are always at right angles to the projection lines between the views.
2. A reference plane appears as a line in two alternate views, never in adjacent views.
3. Measurements are always made at right angles to the reference lines or parallel to the projection lines.
4. In the auxiliary view, all points are at the same distances from the reference line as the corresponding points are from the reference line in the alternate view, or the *second previous view*.

■ 8.5 DRAWING AN AUXILIARY VIEW: REFERENCE-PLANE METHOD

The object in Fig. 8.6a is numbered as explained in §6.6. To draw the auxiliary view, proceed as follows:

1. Draw two views of the object, and assume an arrow indicating the direction of sight for the auxiliary view of surface A (Fig. 8.6I).
2. Draw projection lines parallel to the arrow (Fig. 8.6II).
3. Assume a reference plane coinciding with the back surface of the object, as shown in Fig. 8.6a. Draw

FIGURE 8.5 **Position of the Reference Plane.**

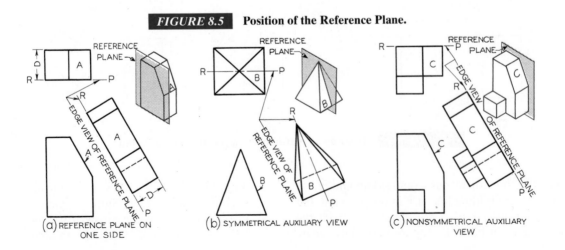

(a) REFERENCE PLANE ON ONE SIDE

(b) SYMMETRICAL AUXILIARY VIEW

(c) NONSYMMETRICAL AUXILIARY VIEW

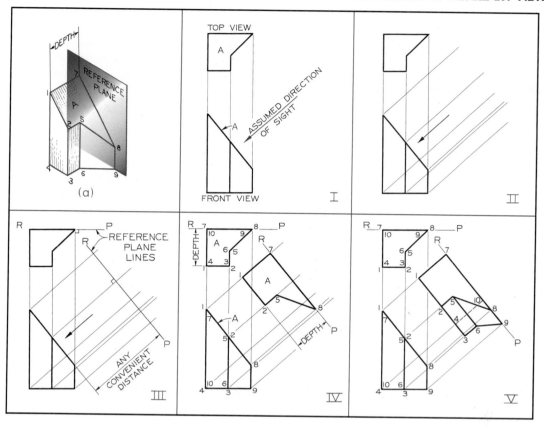

FIGURE 8.6 Drawing an Auxiliary View—Reference-Plane Method.

reference lines in the top and auxiliary views at right angles to the projection lines (Fig. 8.6III). *These are the edge views of the reference plane.*

4. Draw auxiliary view of surface A (Fig. 8.6IV). It will be true size and shape because the direction of sight was taken perpendicular to that surface. Transfer depth measurements from the top view to the auxiliary view with dividers or scale. Each point in the auxiliary view will be on its projection line from the front view and the same distance from the reference line as it is in the top view to the corresponding reference line.

5. Complete the auxiliary view by adding other visible edges and surfaces of the object (Fig. 8.6V). Each numbered point in the auxiliary view lies on its projection line from the front view and is the same distance from the reference line as it is in the top view. Note

that two surfaces of the object appear as lines in the auxiliary view.

■ 8.6 CLASSIFICATION OF AUXILIARY VIEWS

Auxiliary views are classified and named according to the principal dimensions of the object shown in the auxiliary view. For example, the auxiliary view in Fig. 8.6 is a *depth auxiliary view* because it shows the principal dimension of the object, *depth*. Any auxiliary view projected from the front view, also known as a front adjacent view, will show the depth of the object and is a depth auxiliary view.

Similarly, any auxiliary view projected from the top view, also known as a top adjacent view, is a height auxiliary view; and any auxiliary view projected from the side view (either side), also known as a side adjacent view, is a width auxiliary view. For examples of height auxiliary views, see Figs. 8.1b and 8.13b. Depth auxiliary views are illustrated in Figs. 8.27 and 8.33.

■ 8.7 DEPTH AUXILIARY VIEWS

An infinite number of auxiliary planes can be assumed perpendicular to, and hinged to, the frontal plane (F) of projection. Five such planes are shown in Fig. 8.7a; the horizontal plane is included to show that it is similar to the others. All these views show the principal dimension, depth; therefore, all the auxiliary views are depth auxiliary views.

The unfolded auxiliary planes are shown in Fig. 8.7b, which also shows how the depth dimension may be projected from the top view to all auxiliary views. The arrows indicate the directions of sight for the several views, and the projection lines are respectively parallel to these arrows. The arrows may be assumed but need not be actually drawn since the projection lines determine the direction of sight. The folding lines are perpendicular to the arrows and the corresponding projection lines. Since the auxiliary planes can be assumed at any distance from the object, it follows that the folding lines may be any distance from the front view.

The complete drawing, with the outlines of the planes of projection omitted, is shown in Fig. 8.7c. This shows the drawing as it would appear on paper, in which use is made of reference planes as described in §8.4; all depth dimensions are measured perpendicular to the reference line in each view.

Note that the front view shows the height and the width of the object, but not the depth. The depth is shown in all views that are projected from the front view; thus, this rule: *The principal dimension shown in an auxiliary view is the one not shown in the adjacent view from which the auxiliary view was projected.*

■ 8.8 HEIGHT AUXILIARY VIEWS

An infinite number of auxiliary planes can be assumed perpendicular to, and hinged to, the horizontal plane (H) of projection, several of which are shown in Fig. 8.8a. The front view and all the auxiliary views show the principal dimension, height. Therefore, all the auxiliary views are height auxiliary views.

The unfolded projection planes are shown in Fig. 8.8b, and the complete drawing, with the outlines of the planes projection omitted, is shown in Fig. 8.8c. All reference lines are perpendicular to the corresponding projection lines, and all height dimensions are measured parallel to the projection lines, or perpendicular to the reference lines, in each view. Note that in the view projected from, which is the top view, the only dimension *not shown* is height.

■ 8.9 WIDTH AUXILIARY VIEWS

An infinite number of auxiliary planes can be assumed perpendicular to, and hinged to, the profile plane (P) of projection, several of which are shown in Fig. 8.9a. The front view and all the auxiliary views show the principal dimension, width. Therefore, all the auxiliary views are width auxiliary views.

The unfolded planes are shown in Fig. 8.9b, and the complete drawing, with the outlines of the planes of projection omitted, is shown in Fig. 8.9c. All refer-

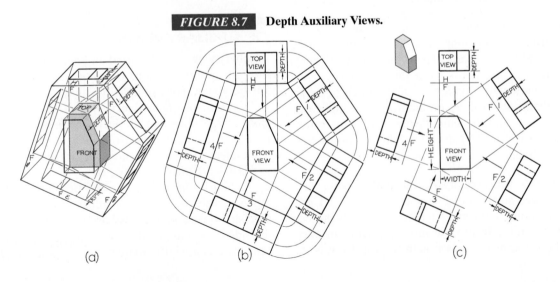

FIGURE 8.7 **Depth Auxiliary Views.**

(a)   (b)   (c)

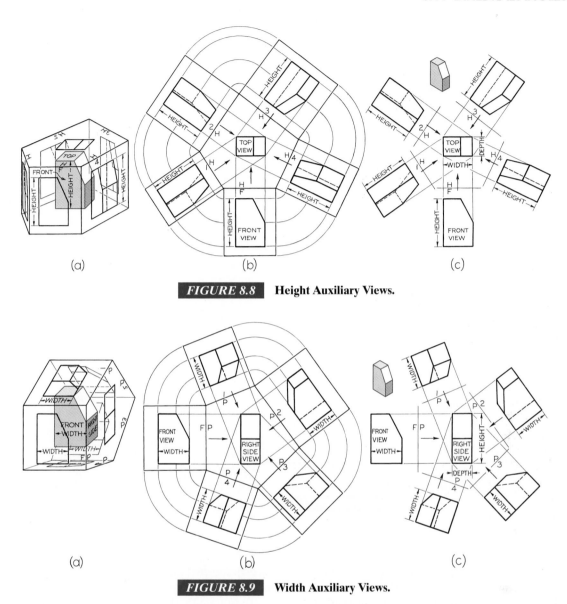

FIGURE 8.8   Height Auxiliary Views.

FIGURE 8.9   Width Auxiliary Views.

ence lines are perpendicular to the corresponding projection lines, and all width dimensions are measured parallel to the projection lines, or perpendicular to the reference lines, in each view. Note that in the right-side view from, which the auxiliary views are projected, the only dimension *not shown* is width.

■ 8.10 REVOLVING A DRAWING

In Fig. 8.10a is a drawing showing top, front, and auxiliary views. Figure 8.10b shows the drawing revolved, as indicated by the arrows, until the auxiliary view and the front view line up horizontally. Although the views remain exactly the same, the names of the views are changed if drawn in this position. The auxiliary view now becomes a right-side view, and the top view becomes an auxiliary view. Some students find it easier to visualize and draw an auxiliary view when revolved to the position of a regular view in this manner. In any case, it should be understood that an auxiliary view basically is like any other view.

■ 8.11 DIHEDRAL ANGLES

The angle between two planes is a **dihedral angle**. One of the principal uses of auxiliary views is to show dihedral

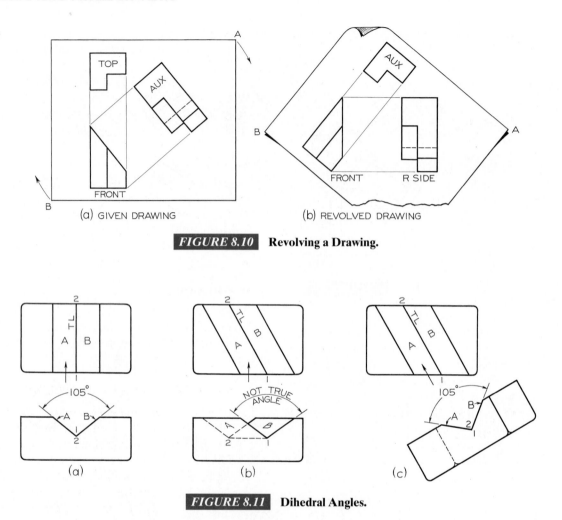

FIGURE 8.10 Revolving a Drawing.

FIGURE 8.11 Dihedral Angles.

angles in true size, mainly for dimensioning purposes. In Fig. 8.11a, a block is shown with a V-groove situated so that the true dihedral angle between inclined surfaces A and B is shown in the front view.

Assume a line in a plane. For example, draw a straight line on a sheet of paper; then hold the paper to view the line as a point. You will observe that when the line appears as a point, the plane containing the line appears as a line. Thus, this rule applies: *To get the edge view of a plane, find the point view of any line in that plane.*

In Fig. 8.11a, line 1–2 is the line of intersection of planes A and B. Now, line 1–2 lies in both planes at the same time; therefore, a point view of this line will show both planes as lines, and the angle between them is the dihedral angle between the planes. *To get the true an-*

gle between two planes, find the point view of the line intersection of the planes.

In Fig. 8.11b, the line of intersection 1–2 does not appear as a point in the front view; as a result, planes A and B do not appear as lines, and the true dihedral angle is not shown. Assuming that the actual angle is the same as in Fig. 8.11a, does the angle show larger or smaller than in Fig. 8.11a? The angle does not appear true size in Fig. 8.11b because the direction of sight (see arrow) is not parallel to the line of intersection 1–2.

In Fig. 8.11c, the direction-of-sight arrow is taken parallel to the line 1–2, producing an auxiliary view in which line 1–2 appears as a point, planes A and B appear as lines, and the true dihedral angle is shown. *To draw a view showing a true dihedral angle, assume the direction of sight parallel to the line of intersection between the planes of the angle.*

■ 8.12 PLOTTED CURVES

As described in §6.30, when a cylinder is cut by an inclined plane, the inclined surface is elliptical in shape. In Fig. 6.34a, such a surface is produced, but the ellipse does not show true size and shape because the plane of the ellipse is not seen at right angles in any view.

In Fig. 8.12a, the line of sight is taken perpendicular to the edge view of the inclined surface, and the resulting ellipse is shown in true size and shape in the auxiliary view. The major axis is found by direct projection from the front view, and the minor axis is equal to the diameter of the cylinder. The left end of the cylinder (a circle) will appear as an ellipse in the auxiliary view, the major axis of which is equal to the diameter of the cylinder.

Since this is a symmetrical object, the reference plane is assumed to be located through the center, as shown. To plot points on the ellipses, select points on the circle of the side view, and project them across to the inclined surface or to the left-end surface, and then upward to the auxiliary view. In this manner, two points can be projected each time, as shown for points 1–2, 3–4, and 5–6. Distances *a* are equal and are transferred from the side view to the auxiliary view. A sufficient number of points must be projected to establish the curves accurately. Use the irregular curve.

Since the major and minor axes are known, any of the true ellipse methods in Figs. 4.48–4.50 and 4.52a may be used. If an approximate ellipse is adequate, the

method in Fig. 4.56 can be used. The quickest and easiest method using instruments is to use an ellipse template (§4.53).

In Fig. 8.12b, the auxiliary view shows the true size and shape of the inclined cut through a piece of molding. The method of plotting points is similar to that explained for the ellipse in Fig. 8.12a.

■ 8.13 REVERSE CONSTRUCTION

To complete the regular views, it is often necessary to construct an auxiliary view first. For example, in Fig. 8.13a, the upper portion of the right-side view cannot be constructed until the auxiliary view is drawn and points are established on the curves and then projected back to the front view, as shown.

In Fig. 8.13b, the 60° angle and the location of line 1–2 in the front view are given. To locate line 3–4 in the front view and the lines 2–4, 3–4, and 4–5 in the side view, it is necessary to first construct the 60° angle in the auxiliary view and project back to the front and side views, as shown.

■ 8.14 PARTIAL AUXILIARY VIEWS

The use of an auxiliary view often makes it possible to omit one or more regular views and thus to simplify the shape description, as shown in Fig. 8.1b. In Fig. 8.14 three complete auxiliary-view drawings are shown. To make such drawings takes a great deal of time, particularly when ellipses are involved, as is so often the case. The completeness of detail may add nothing to clearness or may even detract from it because of the clutter of lines. However, in these cases, some portion of

FIGURE 8.12   **Plotted Curves.**

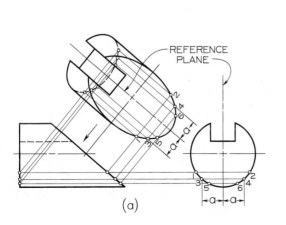

(a)

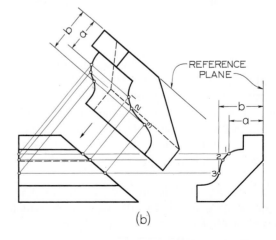

(b)

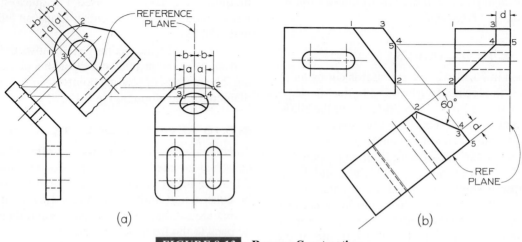

FIGURE 8.13 Reverse Construction.

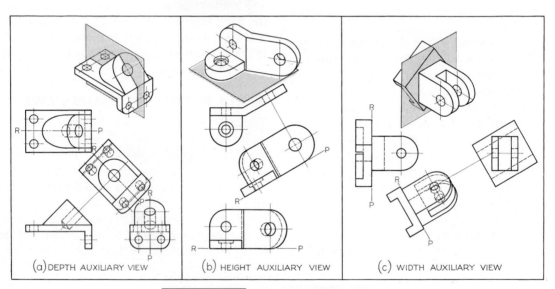

(a) DEPTH AUXILIARY VIEW   (b) HEIGHT AUXILIARY VIEW   (c) WIDTH AUXILIARY VIEW

FIGURE 8.14 Primary Auxiliary Views.

every view is needed—no view can be completely eliminated.

As described in §6.9, partial views are often sufficient, and the resulting drawings are considerably simplified and easier to read. Similarly, as shown in Fig. 8.15, partial regular views and partial auxiliary views are used with the same result. Usually a break line is used to indicate the imaginary break in the views. *Do not draw a break line coinciding with a visible line or a hidden line*.

To clarify the relationship of views, the auxiliary views should be connected to the views from which they are projected, either with a center line or with one or two projection lines. This is particularly important for partial views, which are often small and appear to be "lost" and not related to any view.

■ 8.15 HALF AUXILIARY VIEWS

If an auxiliary view is symmetrical, and if it is necessary to save space on the drawing or to save time in drafting, only half of the auxiliary view may be drawn (Fig. 8.16). In this case, half of a regular view is also shown, since the bottom flange is also symmetrical (see §§6.9 and 7.14). Note that in each case the *near half* is shown.

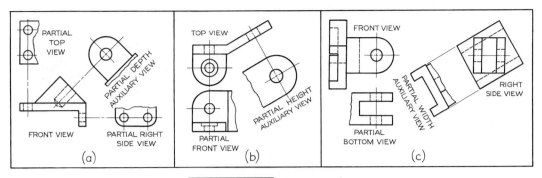

FIGURE 8.15   **Partial Views.**

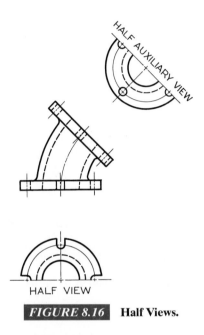

FIGURE 8.16   **Half Views.**

■ 8.16 HIDDEN LINES IN AUXILIARY VIEWS

In practice, hidden lines should be omitted in auxiliary views, unless they are needed for clearness. The beginner, however, should show all hidden lines, especially if the auxiliary view of the entire object is shown. Later, in advanced applications, it will become clearer as to when hidden lines can be omitted.

■ 8.17 AUXILIARY SECTIONS

An auxiliary section is simply an auxiliary view in section. In Fig. 8.17a, note the cutting-plane line and the terminating arrows that indicate the direction of sight for the auxiliary section. Observe that the section lines

are drawn at approximately 45° with visible outlines. In an auxiliary section drawing, the entire portion of the object behind the cutting plane may be shown (Fig. 8.17a), or the cut surface alone may be shown (Fig. 8.17b and 8.17c).

An auxiliary section through a cone is shown in Fig. 8.18. This is one of the conic sections (§4.44), in this case a parabola. The parabola may be drawn by other methods (Figs. 4.57 and 4.58), but the method shown here is by projection. In Fig. 8.18, elements of the cone are drawn in the front and top views. These intersect the cutting plane at points 1, 2, 3, and so on. These points are established in the top view by projecting upward to the top views of the corresponding elements. In the auxiliary section, all points on the parabola are the same distance from the reference plane RP as they are in the top view.

A typical example of an auxiliary section in machine drawing is shown in Fig. 8.19. Here, there is not sufficient space for a revolved section, although a removed section could have been used instead of an auxiliary section.

■ 8.18 TRUE LENGTH OF LINE: AUXILIARY-VIEW METHOD

A line will show in true length when projected to a projection plane parallel to the line.

Figure 8.20 shows how to find the true length of the hip rafter 1–2 by means of a depth auxiliary view.

1.   Assume an arrow perpendicular to 1–2 (front view) indicating the direction of sight, and place the H/F folding line as shown.

2.   Draw the F/1 folding line perpendicular to the arrow and at any convenient distance from 1–2 (front view), and project the points 1 and 3 toward it.

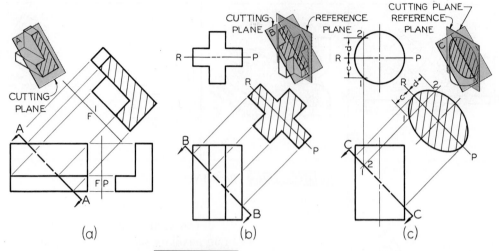

FIGURE 8.17 Auxiliary Sections.

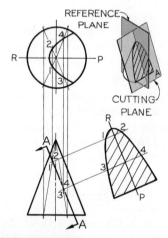

FIGURE 8.18 Auxiliary Section.

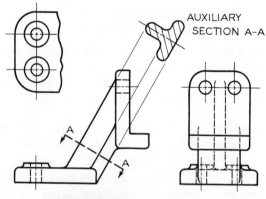

FIGURE 8.19 Auxiliary Section.

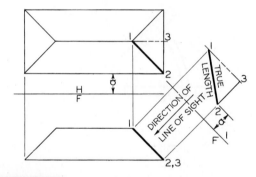

FIGURE 8.20 True Length of a Line by Means of an Auxiliary View.

3. Set off the points 1 and 2 in the auxiliary view at the same distance from the folding line as they are in the top view. The triangle 1–2–3 in the auxiliary view shows the true size and shape of the roof section 1–2–3, and the distance 1–2 in the auxiliary view is the true length of the hip rafter 1–2.

To find the true length of a line by revolution, see §9.10.

■ 8.19 SUCCESSIVE AUXILIARY VIEWS

Up to this point we have dealt with **primary auxiliary views**—that is, single auxiliary views projected from one of the regular views. In Fig. 8.21, auxiliary view 1 is a primary auxiliary view projected from the top view.

From primary auxiliary view 1, a **secondary auxiliary view** 2 can be drawn; then from it a **third auxiliary**

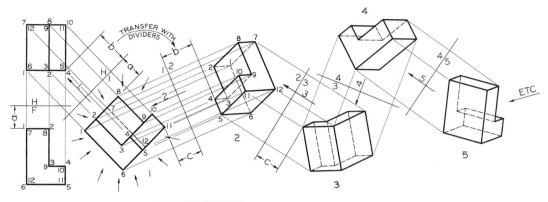

FIGURE 8.21 Successive Auxiliary Views.

view 3, and so on. An infinite number of such successive auxiliary views may be drawn. However, secondary auxiliary view 2 is not the only one that can be projected from primary auxiliary view 1. As shown by the arrows around view 1, an infinite number of secondary auxiliary views, with different lines of sight, may be projected. *Any auxiliary view projected from a primary auxiliary view is a secondary auxiliary view.* Furthermore, any succeeding auxiliary view may be used to project an infinite series of views from it.

In this example, folding lines are more convenient than reference-plane lines. In auxiliary view 1, all numbered points of the object are the same distance from folding line H/1 as they are in the front view from folding line H/F. These distances, such as distance a, are transferred from the front view to the auxiliary view.

To draw the secondary auxiliary view 2, disregard the front view and center attention on the sequence of three views: the top view, view 1, and view 2. Draw arrow 2 toward view 1 in the direction desired for view 2, and draw light projection lines parallel to the arrow. Draw folding line 1/2 perpendicular to the projection lines and at any convenient distance from view 1. Locate all numbered points in view 2 from folding line 1/2 at the same distances they are in the top view from folding line H/1, using the dividers to transfer distances. For example, transfer distance b to locate points 4 and 5. Connect points with straight lines, and determine visibility. The corner nearest the observer (11) for view 2 will be visible, and the one farthest away (1) will be hidden, as shown.

To draw views 3, 4, and so on, repeat this procedure, remembering that each time we will be concerned only with a sequence of three views. In drawing

any auxiliary view, the paper may be revolved to make the last two views line up as regular views.

■ 8.20 USES OF AUXILIARY VIEWS

Generally, auxiliary views are used to show the true shape or true angle of features that appear distorted in the regular views. Basically, auxiliary views have the following four uses:

1. True length of line, TL (§8.18).
2. Point view of line (§8.11).
3. Edge view of plane, EV (§8.21).
4. True size of plane, TS (§8.21).

■ 8.21 TRUE SIZE OF AN OBLIQUE SURFACE: FOLDING-LINE METHOD

A typical use for a secondary auxiliary view is to show the true size and shape of an oblique surface, such as surface 1–2–3–4 in Fig. 8.22. In this case folding lines are used, but the same results can be obtained with reference lines. Proceed as follows:

1. Draw the primary auxiliary view showing surface 1–2–3–4 as a line. As explained in §8.11, the edge view (EV) of a plane is found by getting the point of view of a line in that plane. To get the point view of a line, the line of sight must be assumed parallel to the line. Therefore, draw arrow P parallel to lines 1–2 and 3–4, which are true length (TL) in the front view, and draw projection lines parallel to the arrow. Draw folding line H/F between the top and front views and F/1 between the front and auxiliary views, perpendicular to the respective projection lines. All points in the auxiliary view will be the same distance from the folding line F/1 as they are in the top view from folding line

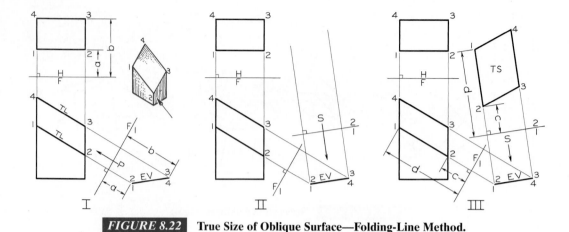

FIGURE 8.22 True Size of Oblique Surface—Folding-Line Method.

H/F. Lines 1–2 and 3–4 will appear as points in the auxiliary view, and plane 1–2–3–4 will therefore appear *edgewise*—that is, as a line.

2. Draw arrow S perpendicular to the edge view of plane 1–2–3–4 in the primary auxiliary view, and draw projection lines parallel to the arrow. Draw folding line 1/2 perpendicular to these projection lines and at a convenient distance from the primary auxiliary view.

3. Draw the secondary auxiliary view. Locate each point (transfer with dividers) the same distance from the folding line 1/2 as it is in the front view to the folding line F/1—as, for example, dimensions c and d. The true size (TS) of the surface 1–2–3–4 will be shown in

the secondary auxiliary view since the direction of sight, arrow S, was taken perpendicular to it.

■ 8.22 TRUE SIZE OF AN OBLIQUE SURFACE: REFERENCE-PLANE METHOD

Figure 8.23a shows the steps in drawing an auxiliary view in which triangular surface 1–2–3 appears in true size and shape. For the true size of the surface to appear in the secondary auxiliary view, arrow S must be assumed perpendicular to the edge view of that surface; it is therefore necessary to have the edge view of surface 1–2–3 in the primary auxiliary view first. To do this, the direction of sight, arrow P, must be parallel to

FIGURE 8.23 True Size of an Oblique Surface—Reference-Plane Method.

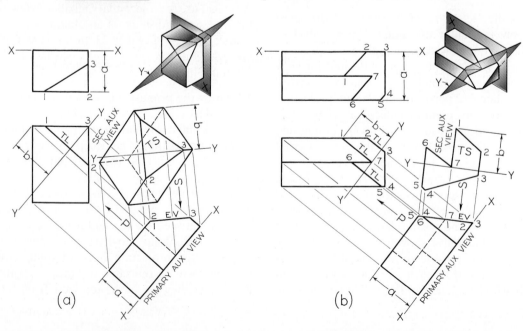

(a)   (b)

a line in surface 1–2–3 that appears true length (TL) in the front view. Therefore, arrow P is drawn parallel to line 1–2 of the front view, line 1–2 will appear as a point in the primary auxiliary view, and surface 1–2–3 must therefore appear edgewise in that view.

In this case it is convenient to use reference lines and to assume the reference plane X (for drawing the primary auxiliary view) coincides with the back surface of the object, as shown. For the primary auxiliary view, all depth measurements, as a in the figure, are transferred with dividers from the top view with respect to the reference line X–X.

For the secondary auxiliary view, reference plane Y is assumed to be cutting through the object for convenience in transferring measurements. All measurements perpendicular to Y–Y in the secondary auxiliary view are the same as between the reference plane and the corresponding points in the front view. Note that corresponding measurements must be *inside* (toward the central view in the sequence of three views) or *outside* (away from the central view). For example, dimension b is on the side of Y–Y *away* from the primary auxiliary view in both places.

Figure 8.23b shows how to determine the true size and shape of surface 1–2–3–4–5–6–7 without drawing the complete secondary auxiliary view. The method is similar to that just described.

■ 8.23 SECONDARY AUXILIARY VIEW, OBLIQUE DIRECTION OF SIGHT GIVEN

In Fig. 8.24 two views of a block are given, with two views of an arrow indicating the lines of sight for the desired secondary auxiliary views of the object. To draw these views, proceed as follows:

1. *Draw a primary auxiliary view of both the object and the assumed arrow,* which will show the true length *of the arrow.* To do this, assume a horizontal reference plane X–X in the front and auxiliary views, as shown. Then assume a direction of sight perpendicular to the given arrow. In the front view, the butt end of the arrow is a distance a higher than the arrow point, and this distance is transferred to the primary auxiliary view, as shown. All *height* measurements in the auxiliary view correspond to those in the front view.

2. *Draw a secondary auxiliary view* that shows the arrow as a point. This can be done because the arrow appears true length in the primary auxiliary view and projection lines for the secondary auxiliary view are drawn parallel to it. Draw reference line Y–Y for the secondary auxiliary view perpendicular to these pro-

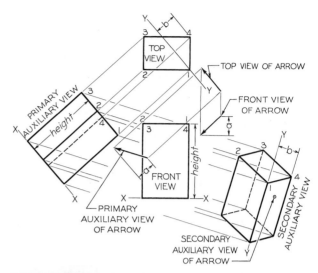

FIGURE 8.24 Secondary Auxiliary View with Oblique Direction of Sight Given.

jection lines. In the top view, draw Y–Y perpendicular to the projection lines to the primary auxiliary view. All measurements, such as b, with respect to Y–Y correspond in the secondary auxiliary view and the top view.

Note that the secondary auxiliary views of Figs. 8.23a and 8.24 have considerable pictorial value. These are *trimetric projections* (see §16.30). However, the direction of sight could be assumed, in the manner of Fig. 8.24, to produce either isometric or dimetric projections. If the direction of sight is assumed parallel to the diagonal of a cube, the resulting view is an *isometric projection* (§16.3).

A typical application of a secondary auxiliary view is shown in Fig. 8.25. All views are partial views, except the front view. The partial secondary auxiliary view illustrates a case in which break lines are not needed. Note the use of an auxiliary section to show the true shape of the arm.

■ 8.24 ELLIPSES

As shown in §6.30, a circle viewed obliquely appears as an ellipse. This often occurs in successive auxiliary views because of the variety of directions of sight. In Fig. 8.26a, the hole appears as a true circle in the top view. The circles appear as straight lines in the primary auxiliary view and as ellipses in the secondary auxiliary view. In the latter, the major axis AB of the ellipse is parallel to the projection lines and equal in length to the true diameter of the circle in the top view. The minor axis CD is perpendicular to the major axis, and

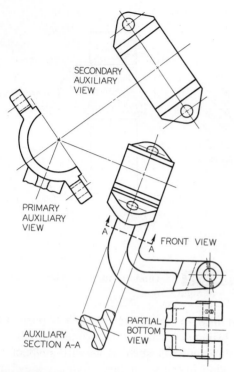

FIGURE 8.25 **Secondary Auxiliary View—Partial Views.**

its foreshortened length is projected from the primary auxiliary view.

The ellipse can be completed by projecting points, such as 1 and 2, symmetrically located about the reference plane RP coinciding with CD, with distances *a* equal in the top and secondary auxiliary views as shown, and finally, after a sufficient number of points have been plotted, by using an irregular curve.

Since the major and minor axes are easily found, any of the true-ellipse methods of Figs. 4.48–4.50 and 4.52a may be used, or an approximate ellipse (Fig. 4.56) may be found sufficiently accurate for a particular drawing. Or the ellipses may be easily and rapidly drawn with the aid of an ellipse template. The "angle" of ellipse to use is one that most closely matches the angle between the direction-of-sight arrow and the plane (EV) containing the circle, as seen in this case in the primary auxiliary view. Here the angle is $36\frac{1}{2}°$, so a 35° ellipse is selected.

In Fig. 8.26b, successive auxiliary views are shown in which the true circular shapes appear in the secondary auxiliary view, and the elliptical projections in the front and top views. It is necessary to construct the circular shapes in the secondary auxiliary view, then to project plotted points back to the primary auxiliary view, the front view, and finally to the top view, as shown in the figure for points 1, 2, 3, and 4. The final curves are then drawn with the aid of an irregular curve.

If the major and minor axes are found, any of the true-ellipse methods may be used; or better still, an ellipse template may be employed. The major and minor axes are easily established in the front view, but in the top view, they are more difficult to find. The major axis AB is at right angles to the center line GL of the hole, and equal in length to the true diameter of the hole. The minor axis ED is at right angles to the major axis. Its length is found by plotting several points in the vicinity of one end of the minor axis or by using descriptive geometry to find the angle between the line of sight and the inclined surface, and by this angle selecting the ellipse guide required.

FIGURE 8.26 **Ellipses.**

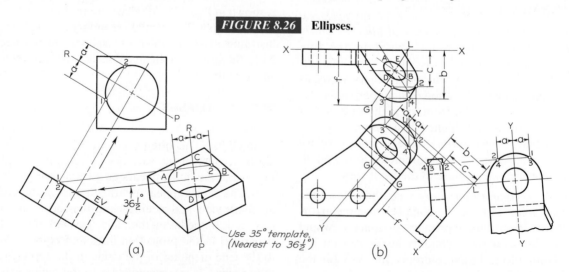

(a)

(b)

■ 8.25 COMPUTER GRAPHICS

USING CAD

Using 3D CAD, any view can be generated in one or two steps, eliminating the need to project auxiliary views manually. It is still very important to have a clear understanding which line of sight will produce a true size view or a view which shows a true dihedral angle. When measuring or dimensioning a view from a CAD screen, if the surface or angle is not true size the automatic dimension from the CAD system will be that of the apparent, or projected, distance. Incorrectly dimensioned dihedral angles can be a common error in CAD drawings created by inexperienced operators.

■ KEY WORDS

AUXILIARY VIEW	FOLDING LINES
REFERENCE PLANE	PARTIAL AUXILIARY VIEW
PRIMARY AUXILIARY	SECONDARY AUXILIARY
GLASS BOX	TRUE LENGTH
DIHEDRAL ANGLE	INCLINED SURFACE
TRUE SIZE	OBLIQUE SURFACE
OBLIQUE LINE	

■ CHAPTER SUMMARY

- An auxiliary view can be used to create a projection that shows the true length of an oblique line or true size of an inclined plane.

- Primary and secondary auxiliary views can also create the edge view of an oblique plane, the point view of an oblique line, and the true size of an oblique plane.

- An auxiliary view can be automatically created using CAD if the original object was drawn as a three-dimensional model.

- Folding lines or reference plane lines represent the edge views of projection planes.

- Points are projected between views parallel to the line of sight and perpendicular to the reference plane lines or folding lines.

- A principal use of auxiliary views is to show dihedral angles in true size.

- Curves are plotted as points in auxiliary views and are connected with an irregular curve drawing instrument or spline CAD function.

- An auxiliary view can also be drawn as a sectional view. The same sectional view conventions apply when the section is also an auxiliary view.

- A secondary auxiliary view can be constructed from a previously drawn (primary) auxiliary view.

■ REVIEW QUESTIONS

1. What is meant by *true length*? By *true size*?
2. Why is a true-length line always parallel to an adjacent reference plane line?
3. If an auxiliary view is drawn from the front view, its depth dimensions would be the same as in what other views?
4. Describe one method for transferring depth between views.
5. What is the difference between a complete auxiliary view and a partial auxiliary view?
6. How are curves drawn in an auxiliary view?
7. How many auxiliary views are necessary to draw the true size of an inclined plane? Of an oblique plane?
8. What is the angle between the reference-plane lines and the direction-of-sight lines?

■ AUXILIARY-VIEW PROBLEMS

The problems in Figs. 8.27–8.59 are to be drawn with instruments or freehand. If partial auxiliary views are not assigned, the auxiliary views are to be complete views of the entire object, including all necessary hidden lines.

It is often difficult to space properly the views of an auxiliary-view drawing. In some cases it may be necessary to make a trial blocking out on a preliminary sheet before starting the actual drawing. Al-lowances for dimensions must be made if metric or decimal dimensions are to be included. In such a case, the student should study §§11.1–11.25.

Since many of the problems in this chapter are of a general nature, they can also be solved on most computer graphics systems. If a system is available, your instructor may choose to assign specific problems to be completed by this method.

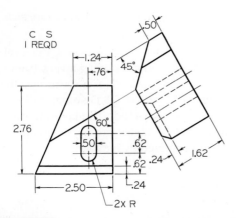

FIGURE 8.27 RH Finger. Given: Front and auxiliary
views. Required: Complete front, auxiliary, left-side, and top
views (Layout A–3 or A4–3 adjusted).

FIGURE 8.27 RH Finger. Given: Front and auxiliary
views. Required: Complete front, auxiliary, left-side, and top
views (Layout A–3 or A4–3 adjusted).

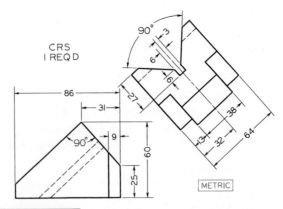

FIGURE 8.28 V-Block. Given: Front and auxiliary
views. Required: Complete front, top, and auxiliary views
(Layout A–3 or A4–3 adjusted).

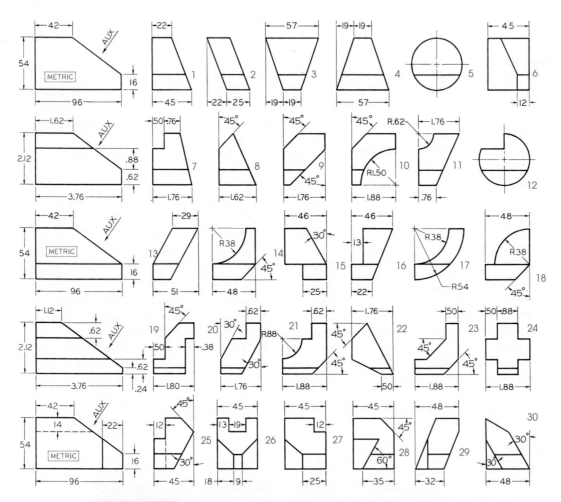

FIGURE 8.29 Auxiliary View Problems. Make freehand sketch or
instrument drawing of selected problem as assigned. Draw given front and
right-side views, and add incomplete auxiliary view, including all hidden lines
(Layout A–3 or A4–3 adjusted). If assigned, design your own right-side view
consistent with given front view, and then add complete auxiliary view.

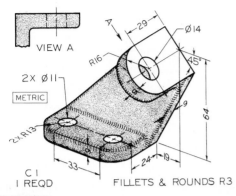

FIGURE 8.30   Anchor Bracket. Draw necessary views or partial views (Layout A–3 or A4–3 adjusted).*

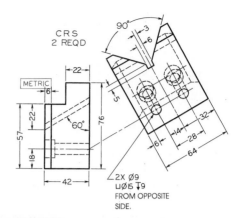

FIGURE 8.33   Guide Block. Given: Right-side and auxiliary views. Required: Right-side, auxiliary, plus front and top views—all complete (Layout B–3 or A3–3).*

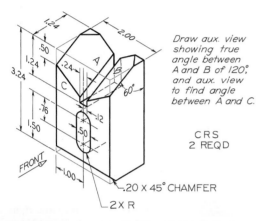

FIGURE 8.31   Centering Block. Draw complete front, top, and right-side views, plus indicated auxiliary views (Layout B–3 or A3–3).*

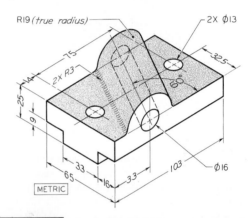

FIGURE 8.34   Angle Bearing. Draw necessary views, including a complete auxiliary view (Layout A–3 or A4–3 adjusted).*

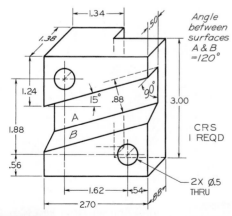

FIGURE 8.32   Clamp Slide. Draw necessary views completely (Layout B–3 or A3–3).*

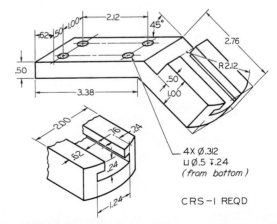

FIGURE 8.35   Guide Bracket. Draw necessary views or partial views (Layout B–3 or A3–3).*

* If dimensions are required, study §§11.1–11.25. Use metric or decimal-inch dimensions as assigned.

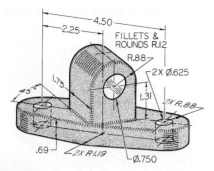

FIGURE 8.36 Rod Guide. Draw necessary views, including complete auxiliary view showing true shape of upper rounded portion (Layout B–4 or A3–4 adjusted).*

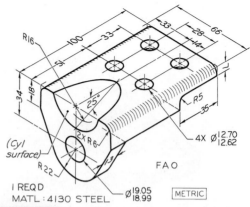

I REQD
MATL : 4130 STEEL

FIGURE 8.39 Angle Guide. Draw necessary views, including a partial auxiliary view of cylindrical recess (Layout B–4 or A3–4 adjusted).*

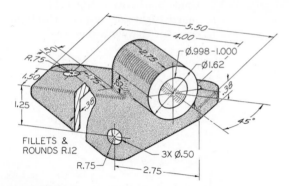

FIGURE 8.37 Brace Anchor. Draw necessary views, including partial auxiliary view showing true shape of cylindrical portion (Layout B–4 or A3–4 adjusted).*

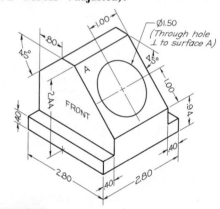

FIGURE 8.40 Holder Block. Draw front and right-side views (2.80″ apart) and complete auxiliary view of entire object showing true shape of surface A and all hidden lines (Layout A–3 or A4–3 adjusted).*

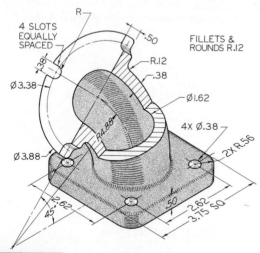

FIGURE 8.38 45° Elbow. Draw necessary views, including a broken section and two half views of flanges (Layout B–4 or A3–4 adjusted).*

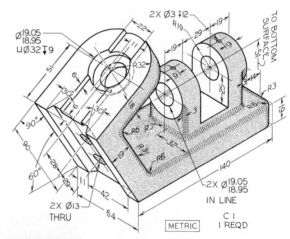

FIGURE 8.41 Control Bracket. Draw necessary views, including partial auxiliary views and regular views (Layout C–4 or A2–4).*

*If dimensions are required, study §§11.1–11.25. Use metric or decimal-inch dimensions as assigned.

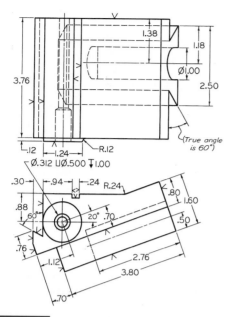

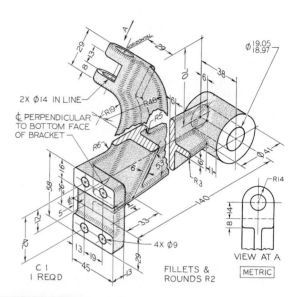

FIGURE 8.42 Tool Holder Slide. Draw given views, and add complete auxiliary view showing true curvature of slot on bottom (Layout B–4 or A3–4 adjusted).*

FIGURE 8.44 Guide Bearing. Draw necessary views and partial views, including two partial auxiliary views (Layout C–4 or A2–4).*

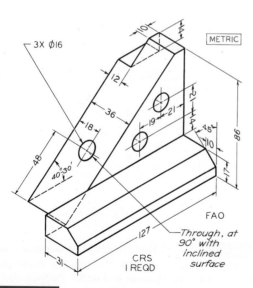

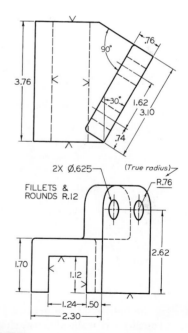

FIGURE 8.43 Adjuster Block. Draw necessary views, including complete auxiliary view showing true shape of inclined surface (Layout B–4 or A3–4 adjusted).*

FIGURE 8.45 Drill Press Bracket. Draw given views and add complete auxiliary view showing true shape of inclined face (Layout B–4 or A3–4 adjusted).*

* If dimensions are required, study §§11.1–11.25. Use metric or decimal-inch dimensions as assigned.

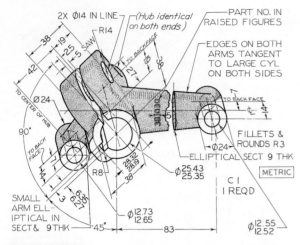

FIGURE 8.46 Brake Control Lever. Draw necessary views and partial views (Layout B–4 or A3–4 adjusted).*

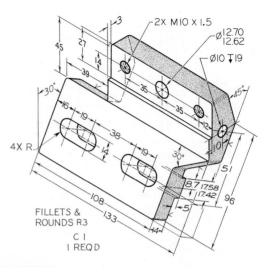

FIGURE 8.48 Cam Bracket. Draw necessary views or partial views as needed. (Layout B–4 or A3–4 adjusted).*

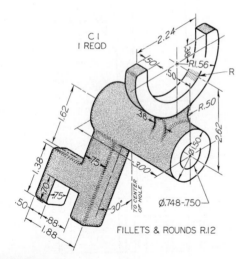

FIGURE 8.47 Shifter Fork. Draw necessary views, including partial auxiliary view showing true shape of inclined arm (Layout B–4 or A3–4 adjusted).*

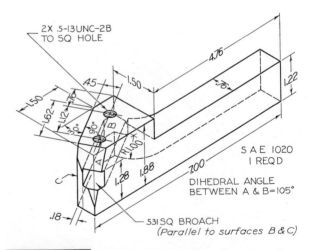

FIGURE 8.49 RH Tool Holder. Draw necessary views, including partial auxiliary views showing 105° angle and square hole true size. (Layout B–4 or A3–4 adjusted).*

---

† If dimensions are required, study §§11.1–12.25. Use metric or decimal-inch dimensions as assigned.

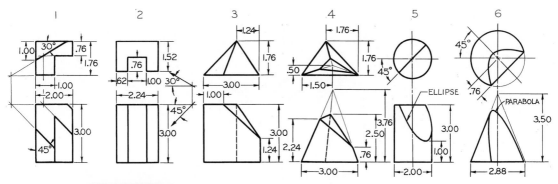

FIGURE 8.50 Draw secondary auxiliary views, complete, which (except Prob. 2) will show the true sizes of the inclined surfaces. In Prob. 2 draw secondary auxiliary view as seen in direction of arrow (Layout B–3 or A3–3).[*]

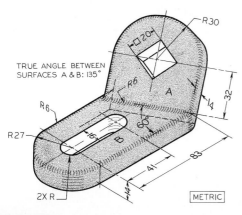

FIGURE 8.51 Control Bracket. Draw necessary views including primary and secondary auxiliary views so that the latter shows true shape of oblique surface A (Layout B–4 or A3–4 adjusted).[*]

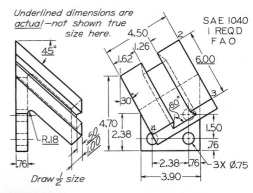

FIGURE 8.53 Dovetail Slide. Draw complete given views and auxiliary views, including view showing true size of surface 1–2–3–4 (Layout B–4 or A3–4 adjusted).[*]

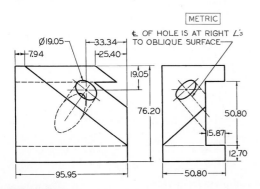

FIGURE 8.52 Holder Block. Draw given views and primary and secondary auxiliary views so that the latter shows true shape of oblique surface (Layout B–4 or A3–4 adjusted).[*]

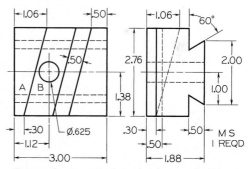

Draw primary aux. view showing angle between planes A and B; then secondary auxiliary view showing true size of surface A.

FIGURE 8.54 Dovetail Guide. Draw given views plus complete auxiliary views as indicated (Layout B–4 or A3–4 adjusted).[*]

[*] If dimensions are required, study §§11.1–11.25. Use metric or decimal-inch dimensions as assigned.

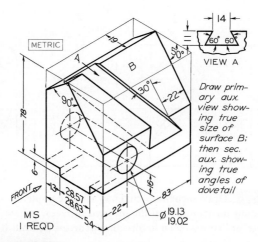

FIGURE 8.55   Adjustable Stop. Draw complete front and auxiliary views plus partial right-side view. Show all hidden lines (Layout C–4 or A2–4).[*]

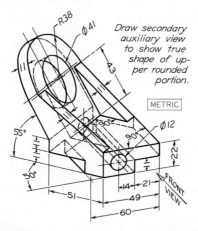

FIGURE 8.56   Tool Holder. Draw complete front view, and primary and secondary auxiliary views as indicated (Layout B–4 or A3–4 adjusted).[*]

[*] If dimensions are required, study §§11.1–11.25. Use metric or decimal-inch dimensions as assigned.

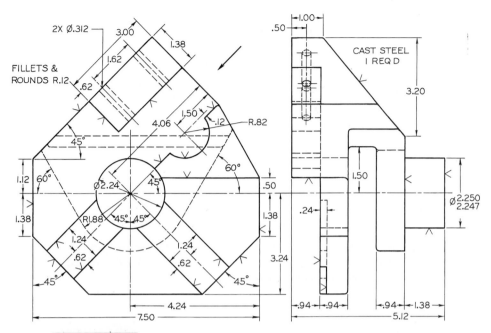

FIGURE 8.57 Box Tool Holder for Turret Lathe. Given: Front and right-side views. Required: Front and left-side views, and complete auxiliary view as indicated by arrow (Layout C–4 or A2–4).*

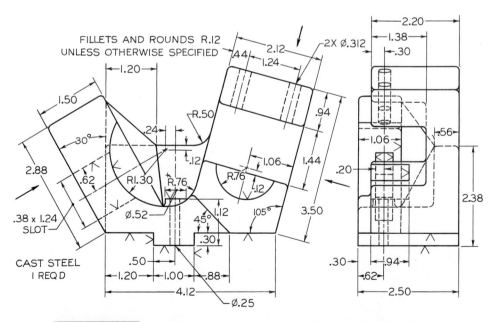

FIGURE 8.58 Pointing Tool Holder for Automatic Screw Machine. Given: Front and right-side views. Required: Front view and three partial auxiliary views (Layout C–4 or A2–4).*

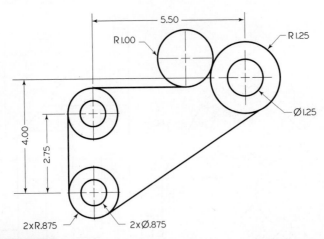

FIGURE 8.59 Print Roller. Given: Right-side view. Design your own front and auxiliary view (Use Layout A–3 or A4–3 adjusted). If assigned, use CAD to create a partial auxiliary view.

C H A P T E R 9

REVOLUTIONS

OBJECTIVES

After studying the material in this chapter, you should be able to:

1. Revolve an object about any principal axis of revolution.
2. Revolve an object to show the true length and true size of lines and planes.
3. Construct multiple revolutions to determine the true size of an oblique plane.
4. Revolve a point about a normal axis.
5. Revolve a circle to create an ellipse.
6. Revolve a three-dimensional object in space using CAD.

OVERVIEW

Revolution, like auxiliary-view projection, is a method of determining the true length and true size of inclined and oblique lines and planes. To obtain an auxiliary view, the observer changes position with respect to the object, as shown by the arrow in Fig. 9.1a. The auxiliary view shows the true size and shape of surface A. The same view of the object can be obtained by *moving the object* with respect to the observer, as shown in Fig. 9.1b. Here the object is revolved until surface A appears in its true size and shape in the right-side view. Revolution determines true length and true size without creating another view. Instead, revolution positions an object in space to create true size and shape.

■ 9.1 AXIS OF REVOLUTION

In Fig. 9.1b the axis of revolution is assumed perpendicular to the frontal plane of projection. The view in which the axis of revolution appears as a point (in this case the front view) *revolves but does not change shape*; in the views in which the axis is shown as a line in true length, the *dimensions of the object parallel to the axis do not change*.

To make a revolution drawing, the view on the plane of projection that is perpendicular to the axis of revolution is drawn first, since it is the only view that remains unchanged in size and shape. This view is drawn revolved either *clockwise* or *counterclockwise* about a point that is the end view, or point view, of the axis of revolution. This point may be assumed at any convenient position on or outside the view. The other views are then projected from this view.

The axis of revolution is usually considered perpendicular to one of the three principal planes of projection. Thus, an object may be revolved about an axis perpendicular to the horizontal, frontal, or profile planes of projection, and the views drawn in the new positions. Such a process is called a **primary revolution**. If this drawing is then used as a basis for another revolution, the operation is called **successive revolutions**. Obviously, this process may be continued indefinitely, which reminds us of the succession of auxiliary views shown in Fig. 8.21.

■ 9.2 REVOLUTION ABOUT AXIS PERPENDICULAR TO FRONTAL PLANE

A primary revolution is illustrated in Fig. 9.2. An imaginary axis XY is assumed (Fig. 9.2I), about which the object is to revolve to the desired position. In this case the axis is selected perpendicular to the frontal plane of projection, and during the revolution all points of the object describe circular arcs parallel to that plane. The axis may pierce the object at any point or may be exterior to it. In Fig. 9.2II, the front view is drawn *revolved*, but not changed in shape, through the angle desired (30° in this case), and the top and side views are obtained by projecting from the front view.

The *depth* of the top view and the side view is found by projecting from the top view of the first unrevolved position (Fig. 9.2I) *because the depth, since it is parallel to the axis, remains unchanged*. If the front view of the revolved position is drawn directly without first drawing the normal unrevolved position, the depth of the object, as shown in the revolved top and side views, may be drawn to known dimensions.

Note the similarity between the top and side views in Fig. 9.2II and some of the auxiliary views of Fig. 8.7c.

■ 9.3 REVOLUTION ABOUT AXIS PERPENDICULAR TO HORIZONTAL PLANE

A revolution about an axis perpendicular to the horizontal plane of projection is shown in Fig. 9.3. An imaginary axis XY is assumed perpendicular to the top plane of projection (Fig. 9.3I), and the top view is drawn revolved (but not changed in shape) to the desired position (30° in this case) (Fig. 9.3II); the other views are obtained by projecting from this view. During the revolution, all points of the object describe circular arcs parallel to the horizontal plane. The *heights* of all points in the front and side views in the revolved position remain unchanged, since they are measured parallel to the axis, and may be drawn by projecting from the initial front and side views in Fig. 9.3I.

Note the similarity between the front and side views in Fig. 9.3II and some of the auxiliary views of Fig. 8.8c.

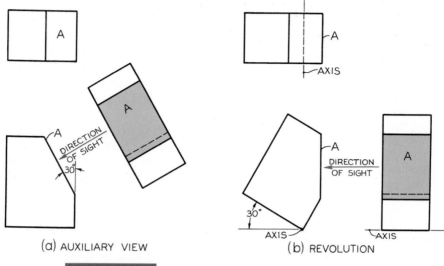

(a) AUXILIARY VIEW   (b) REVOLUTION

FIGURE 9.1   **Auxiliary View and Revolution Compared.**

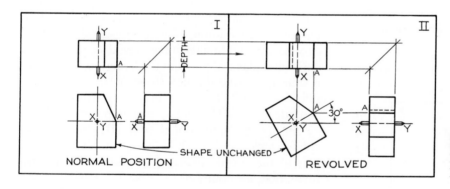

FIGURE 9.2   **Primary Revolution About an Axis Perpendicular to Frontal Plane.**

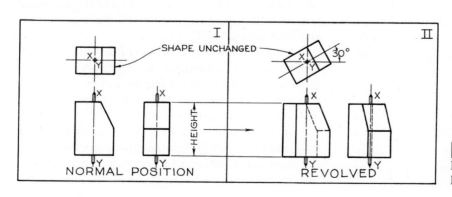

FIGURE 9.3   **Primary Revolution About an Axis Perpendicular to Horizontal Plane.**

■ 9.4 REVOLUTION ABOUT AXIS PERPENDICULAR TO PROFILE PLANE

A revolution about an axis XY perpendicular to the profile plane of projection is illustrated in Fig. 9.4. During the revolution, all points of the object describe circular arcs parallel to the profile plane of projection. The widths of the top and front views in the revolved position remain unchanged, since they are measured parallel to the axis and may be obtained by projection from the top and front views in Fig. 9.4I, or may be set off by direct measurement.

Note the similarity between the top and front views in Fig. 9.4II and some of the auxiliary views of Fig. 8.9c.

■ 9.5 SUCCESSIVE REVOLUTIONS

It is possible to draw an object in an infinite number of revolved positions by making successive revolutions. Such a procedure (Fig. 9.5), limited to three or four stages, offers excellent practice in multiview projection. While it is possible to make several revolutions of a simple object without the aid of a system of numbers, it is absolutely necessary in successive revolutions to assign a number or a letter to every corner of the

FIGURE 9.4   **Primary Revolution About an Axis Perpendicular to Profile Plane.**

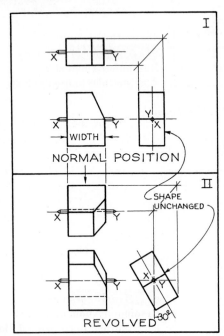

object (see §6.6). The numbering or lettering must be consistent in the various views of the several stages of revolution. Figure 9.5 shows four sets of multiview drawings numbered I, II, III, and IV, respectively. They represent the same object in different positions with reference to the planes of projection.

In Fig. 9.5I, the object is represented in its normal position with its faces parallel to the planes of projection. In (II), the object is shown revolved clockwise through an angle of 30° about an axis perpendicular to the frontal plane. The drawing in (III) is placed under (I) so that the side view, whose width remains unchanged, can be projected from space I to space II as shown.

During the revolution, all points of the object describe circular arcs parallel to the frontal plane of projection and remain at the same distance from that plane. The side view, therefore, may be projected from the side view in (I) and the front view in (II). The top view may be projected in the usual manner from the front and side views in (II).

In (III), the object from (II) has been revolved counterclockwise through an angle of 30° about an axis perpendicular to the horizontal plane of projection. During the revolution, all points describe *horizontal* circular arcs and remain at the same distance from the horizontal plane of projection. The top view from (II) is shown revolved through 30° in (III). The front and side views are obtained by projecting from the front and side views of (II) and from the top view of (III).

In Fig. 9.5IV, the object from (III) has been revolved clockwise through 15° about an axis perpendicular to the profile plane of projection. During the revolution, all points of the object describe circular arcs parallel to the profile plane of projection and remain at the same distance from that plane. The side view is copied (see §4.25) from the side view of (III), but revolved through 15°. The front and top views are projected from the side view of (IV), and from the top and front views of (III).

Another convenient method of copying a view in a new revolved position is to use tracing paper, as described in §4.26. Either a tracing can be made and transferred by rubbing, or the prick points may be made and transferred.

In Figs. 9.5III and IV, each view is an axonometric projection (see §16.2). An isometric projection can be obtained by revolution, as shown in Fig. 16.3, and a dimetric projection (see §16.28) can be constructed in a

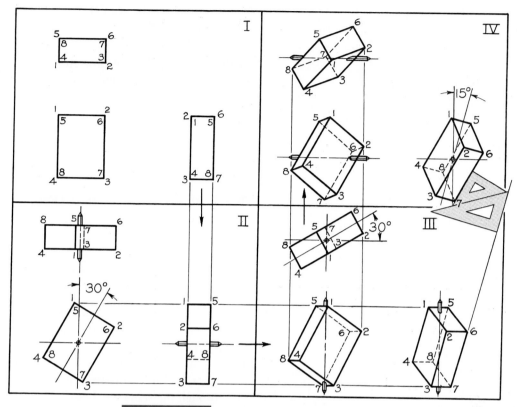

FIGURE 9.5 Successive Revolutions of a Prism.

similar manner. If neither an isometric nor a dimetric projection is specifically sought, successive revolution will produce a trimetric projection (see §16.30), as shown in Fig. 9.5.

■ 9.6 REVOLUTION OF A POINT: NORMAL AXIS

Examples of the revolution of a point about a straight-line axis are often found in design problems that involve pulleys, gears, cranks, linkages, and so on. For example, in Fig. 9.6a, as the disk is revolved, point 3 moves in a circular path lying in a plane perpendicular to the axis 1–2. This relationship is represented in the two views in Fig. 9.6b. Note in this instance that the axis is normal or perpendicular to the frontal plane of projection, resulting in a front view that shows a point view of the axis and a true-size view of the circular path of revolution for point 3. The top view shows the path of revolution in edge view and perpendicular to the true-length view of the axis. Similar two-view relationships would occur if the axis were perpendicular

or normal to either the horizontal or profile planes of projection.

The clockwise revolution through 150° for point 3 is illustrated in Fig. 9.6c.

■ 9.7 REVOLUTION OF A POINT: INCLINED AXIS

In Fig. 9.7a, the axis of revolution for point 3 is positioned parallel to the frontal plane and inclined to the horizontal and profile projection planes. Since the axis 1–2 is true length in the front view, the edge view of the path of revolution can be located as in Fig. 9.7b. To establish the circular path of revolution for point 3, an auxiliary view showing the axis in point view is required (Fig. 9.7c). The required revolution of point 3 (in this case, 210°) is now performed in this circular view. The revolved position of the point is projected to the given front and top views, as shown.

Note the similarity of the relationships of the front view and auxiliary view and the constructions shown in Fig. 9.6c.

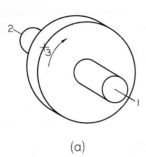

(a)

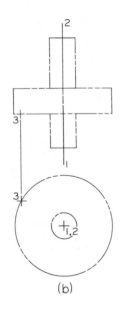

(b)

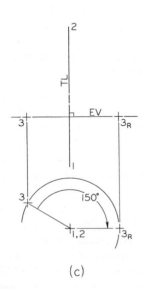

(c)

FIGURE 9.6 **Revolution of a Point About a Normal Axis.**

■ 9.8 REVOLUTION OF A POINT: OBLIQUE AXIS

In Fig. 9.8a, the axis of revolution for point 3 is oblique to all principal planes of projection and, therefore, is shown neither in true length nor as a point view in the top, front, or profile views. To establish the necessary true length and point view of the axis 1–2 in adjacent views, two successive auxiliary views are needed (Fig. 9.8b). The required revolved position of point 3 can now be located and then projected back to complete the given front and top views.

■ 9.9 REVOLUTION OF A LINE

The procedure for the revolution of a line about an axis is very similar to that required for the revolution of a point (see § 9.6). All points on a line must revolve through the same angle or the revolved line becomes altered.

In Fig. 9.9a, the line 1–2 is to be revolved through 150° about the inclined axis 3–4. Since the axis 3–4 is given in true length in the top view, an auxiliary view is required to provide a point view of the axis (Fig. 9.9b). The necessary revolution can then be made about point view 3–4. To ensure that all points on the line rotate through the same number of degrees, a construction circle is drawn tangent to line 1–2 and a perpendicular through the tangency point becomes the reference for measuring the angle of rotation. The circular arc paths for points 1 and 2 locate the points 1_R and 2_R, as the revolved position of the line is drawn perpendicular to the radial line extending under the 150° arc of revolution and tangent to the smaller circle. The alternate-position line is used to distinguish the revolved-position line from the original given line.

■ 9.10 TRUE LENGTH OF A LINE: REVOLUTION METHOD

If a line is parallel to one of the planes of projection, its projection on that plane is equal in length to the line (see Fig. 6.20). In Fig. 9.10a, the element AB of the cone is oblique to the planes of projection; therefore, its projections are foreshortened. If AB is revolved about the axis of the cone until it coincides with either of the contour elements (for example, AB_R), it will be shown in its true length in the front view because it will then be parallel to the frontal plane of projection.

Likewise, in Fig. 9.10b, the edge of the pyramid CD is shown in its true length CD_R when it has been revolved about the axis of the pyramid until it is paral-

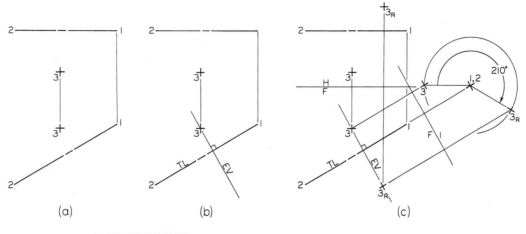

(a)   (b)   (c)

FIGURE 9.7   Revolution of a Point About an Inclined Axis.

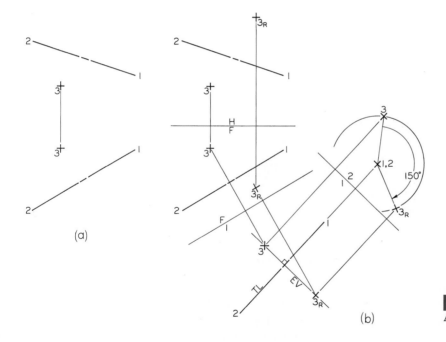

FIGURE 9.8   Revolution of a Point About an Oblique Axis.

lel to the frontal plane of projection. In Fig. 9.10c, the line EF is shown true length at EF_R when it has been revolved about a vertical axis until it is parallel to the frontal plane of projection.

The true length of a line may also be found by constructing a right triangle or a true-length diagram (Fig. 9.10d) whose base is equal to the top view of the line and whose altitude is the difference in elevation of the ends. The hypotenuse of the triangle is equal to the true length of the line.

In these cases the lines are revolved until parallel to a plane of projection. The true length of a line may also be found by leaving the line stationary but shifting the position of the observer—that is, using auxiliary views (see §8.18).

■ 9.11 TRUE SIZE OF A PLANE SURFACE: REVOLUTION METHOD

If a surface is parallel to one of the planes of projection, its projection on that plane is true size (see

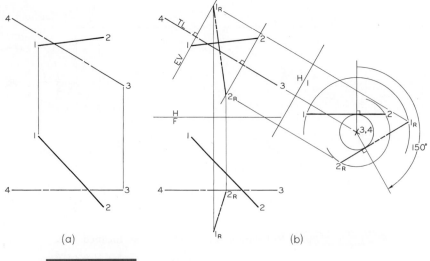

FIGURE 9.9   Revolution of a Line About an Inclined Axis.

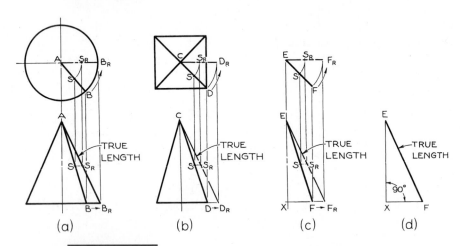

FIGURE 9.10   True Length of a Line—Revolution Method.

Fig. 6.19). In Fig. 9.11a, the inclined surface 1–2–3–4 is foreshortened in the top and side views and appears as a line in the front view. Line 2–3 is taken as the axis of revolution, and the surface is revolved clockwise in the front view to the position 4_R–3 and projected to the side view at 4_R–1_R–2–3, which is the true size of the surface. In this case the surface was revolved until parallel to the profile plane of projection.

In Fig. 9.11b, triangular surface 1–2–3 is revolved until parallel to the horizontal plane of projection so that the surface appears true size in the top view, as shown.

In Fig. 9.11c, the true size of the oblique surface 1–2–3–4–5 cannot be found by a simple primary revolution. The true size can be found by two successive revolutions or by a combination of an auxiliary view and a primary revolution (as in Fig. 9.11c). First, draw an auxiliary view that will show the edge view (EV) of the plane (see Fig. 9.22). Second, revolve the edge view of the surface until it is parallel to the folding line F/1, as shown. All points in the front view, except those in the axis of revolution line 4–5, will describe circular arcs parallel to the folding line F/1. These arcs will appear in the front view as lines parallel to the folding line, such as 2–2_R and 3–3_R. The true size of the surface is found by connecting the points with straight lines.

■ 9.12 REVOLUTION OF CIRCLES

As shown in Figs. 4.50a to 4.50c, a circle, when viewed obliquely, appears as an ellipse. In that case the coin is

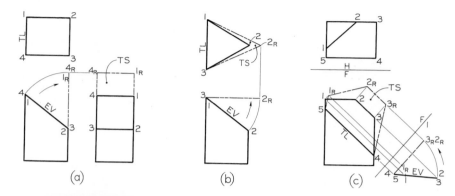

FIGURE 9.11 True Size of a Plane Surface—Revolution Method.

revolved by the fingers. The geometric construction of this revolution is shown in Fig. 9.12a. In the front view the circle appears as ACBD, and in the side view as line CD, which is really the edge view of the plane containing the circle. In the side view, CD (the side view of the circle) is revolved through any desired acute angle to $C_R D_R$.

To find points on the ellipse, draw a series of horizontal lines across the circle in the front view. Each line will cut the circle at two points, as 1 and 2. Project these points across to the vertical line representing the unrevolved circle; then revolve each point and project horizontally to the front view to establish points on the ellipse. Plot as many points as necessary to secure a smooth curve. Figure 9.12b shows an application of this construction to the representation of a revolved object with a large hole.

■ 9.13 COUNTERREVOLUTION

The reverse procedure to revolution is *counterrevolution*. For example, if the three views in Fig. 9.2–II are given, the object can be drawn in the unrevolved position in Fig. 9.2–I by counterrevolution. The front

view is simply counterrevolved back to its normal upright position, and the top and side views are drawn as shown. Similarly, in Fig. 9.5, the object may be counterrevolved from its position in Fig. 9.5–IV to its unrevolved position in Fig. 9.5–I by simply reversing the process.

In practice, it sometimes becomes necessary to draw the views of an object located on or parallel to a given oblique surface. In such an oblique position, it is very difficult to draw the views of the object because of the foreshortening of lines. The work is greatly simplified by counterrevolving the oblique surface to a simple position, completing the drawing, and then revolving to the original given position.

An example is shown in Fig. 9.13. Assume that the oblique surface 8–4–3–7, shown in three views in Fig. 9.13–I, is given and that it is required to draw the three views of a prism 13 mm high with the given oblique surface as its base. Revolve the surface about any horizontal axis XX, perpendicular to the side view, until the edges 8–4 and 3–7 are horizontal, as shown in Fig. 9.13–II. Then revolve the surface about any vertical axis YY, which appears as a point in the top view, until the edges 8–7 and 4–3 are parallel to the frontal plane,

FIGURE 9.12 Revolution of a Circle.

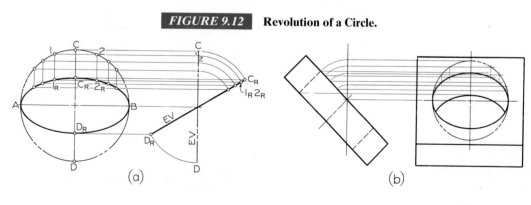

With a slow surfer's drawl, John Gillis describes his creation as "unreal," an adjective not generally used by most inventors. But then, most inventors don't surf. Mr. Gillis, who makes surfboards in a Sydney suburb in Australia, hopes that his invention, the Computer Enhanced Technology board-shaping machine, changes the sport. Traditionally, surfboard shapers spend hours in poorly ventilated bays, hand-carving surfable shapes out of slabs of polyurethane foam or wood. To be a successful shaper, you need an intimate knowledge of how a surfboard interacts with waves. You also must stoop and squat continually and tolerate a lifetime of breathing foam dust.

Using Mr. Gillis's machine, along with software designed by Ian Pearce, a British surfer, shapers forgo these last two requirements. They create boards on a PC using Mr. Pearce's Deadly Accurate Template, or D.A.T., a computer-assisted design and modeling system. The information is then fed into the Computer Enhanced Technology machine, which will carve a board with one-hundredth-of-a-millimeter accuracy. The machine can also act as a board scanner, creating exact duplicates of existing boards. This means that wherever there is a PC, there's a shaping bay.

Shaping machines have been around for years and similar computer-aided technology is being developed independently in California, France, and Brazil. Mr. Gillis says the advantage of such machines is that they allow the user to design from scratch in the computer, not just replicate an existing model.

In the early 1990s, when Mr. Gillis was shaping sailboards on the Hawaiian island of Maui, he found that once he built a great board, it was almost impossible to reproduce. "Slight changes in the board made major differences," he explained. "We couldn't get it by hand."

After he returned to Australia in 1992, he stumbled across Mr. Pearce's software, which was being used to design board templates that were printed out on a blotter. Hoping Mr. Pierce would rewrite the software for his machine, Mr. Gillis rang the developer and invited him to Australia. "He came out here for four weeks and the surf was good," Mr. Gillis said with a shrug. "It inspired him to do the job."

There are those detractors, including the legendary surfer Greg Noll, who said shaping machines "took the heart and soul" out of shaping. Mr. Gillis says soul has nothing to do with shaping. "The chemicals in a board, have they got soul?" he asked. "The soul we're talking about here is when you go surfing."

Adapted from "*The Art of Making Surfboards Has Become More of a Science*," by Denis Faye, *The New York Times*, May 27, 1999.

as shown in Fig. 9.13–III. In this position the given surface is perpendicular to the frontal plane, and the front and top views of the required prism can be drawn, as shown by phantom lines in the figure, because the edges 4–1 and 3–2, for example, are parallel to the frontal plane and, therefore, are shown in their true lengths, 13 mm. When the two views in (III) have been drawn, counterrevolve the object from (III) to (II) and then from (II) to (I) to find the required views of the given object in (I).

■ 9.14 COMPUTER GRAPHICS

Computer graphics programs provide the user with simple and fast ways of revolving objects about any desired axis. Successive revolutions can be easily accomplished using CAD. Computer-generated revolutions also enable a drafter to readily depict circular features, which appear elliptical when viewed obliquely (Fig. 9.14).

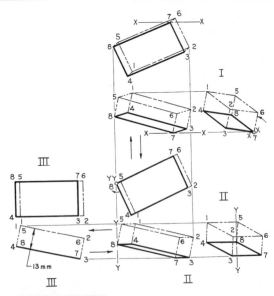

FIGURE 9.13   **Counterrevolution of a Prism.**

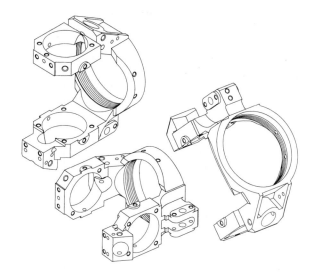

FIGURE 9.14 Pictorial Drawings With Different Viewing Angles Created by Using Computervision Designer System. The system provides complete flexibility to manipulate the original illustration to display any desired viewing orientation. *Courtesy of Computervision Corporation, a subsidiary of Prime Computer, Inc.*

■ KEY WORDS

AXIS OF REVOLUTION

NORMAL AXIS

INCLINED AXIS

OBLIQUE AXIS

SUCCESSIVE REVOLUTIONS

COUNTERREVOLUTION

CLOCKWISE

■ CHAPTER SUMMARY

- Revolution moves an object in space, to reveal what would normally be an auxiliary view of the object in a primary view (top, front, right side).

- The main purpose of revolution is to reveal the true length and true size of inclined and oblique lines and planes in a primary view.

- A normal axis is perpendicular to one of the principal planes of projection (front, top, right side).

- Three-dimensional objects can be revolved at will with CAD. Traditional drawing methods require projection to locate revolved points and lines. Both drawing tools use the same principles of revolution.

- Circles can be revolved to create ellipses or circles that will appear as ellipses when foreshortened on an inclined or oblique plane.

- When solving for true shape, the axis of revolution is drawn perpendicular to the plane of projection that shows the inclined or oblique surface in edge view.

■ REVIEW QUESTIONS

1. What is the purpose of revolution?

2. What is the axis of revolution? What determines where the axis is drawn?

3. How can an ellipse be created by revolution?

4. What is a normal axis? An inclined axis?

5. What are successive revolutions?

■ REVOLUTION PROBLEMS

In Figs. 9.15–9.19 are problems covering primary revolutions, successive revolutions, and counterrevolutions.

Since many of the problems in this chapter are of a general nature, they can also be solved on most computer graphics systems. If a system is available, the instructor may choose to assign specific problems to be completed by this method.

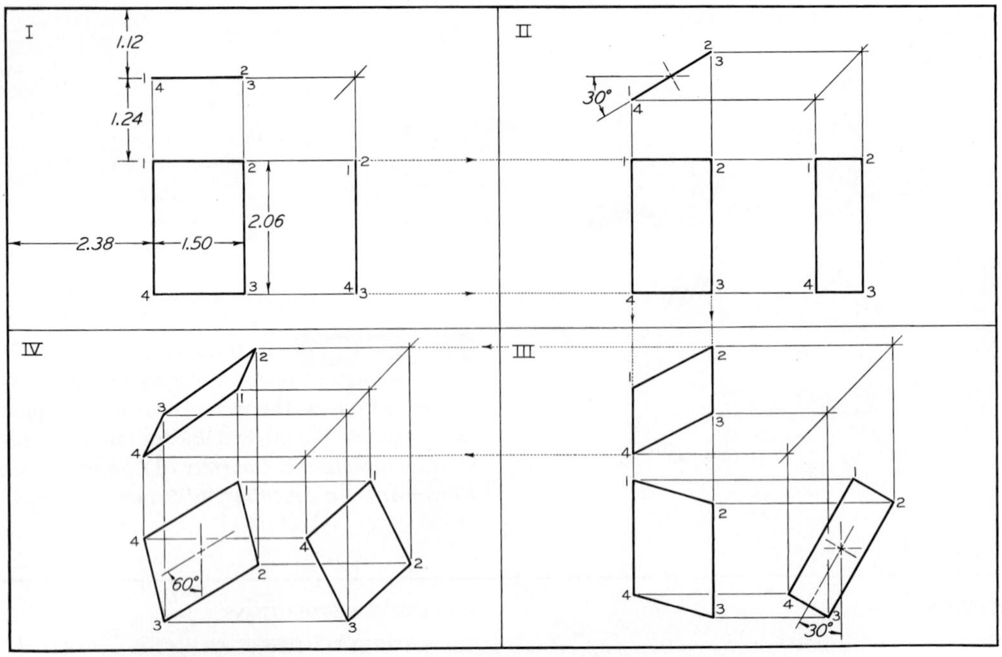

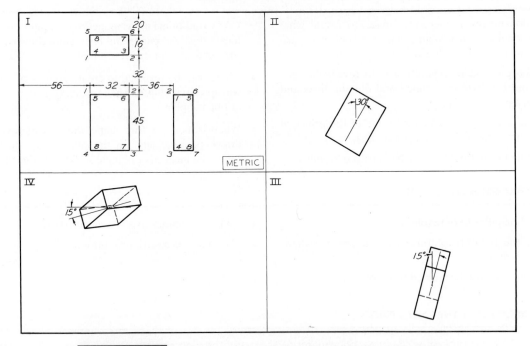

FIGURE 9.15   Using size B or A3 sheet, divide working area into four equal parts, as shown. Draw given views of rectangle, and then the primary revolution in space II, followed by successive revolutions in spaces III and IV. Number points as shown. Omit dimensions. Use Form 3 title box.

FIGURE 9.16   Using size B or A3 sheet, divide working area into four equal parts, as shown. Draw given views of prism as shown in space I; then draw three views of the revolved prism in each succeeding space, as indicated. Number all corners. Omit dimensions. Use Form 3 title box.

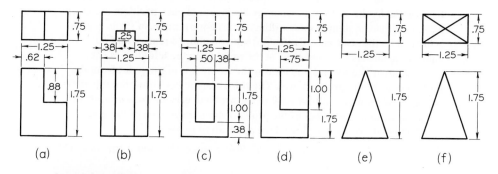

(a) (b) (c) (d) (e) (f)

FIGURE 9.17 Using Layout B–4 or A3–4 (adjusted) sheet, divide into four equal parts as in Fig. 9.15. In the upper two spaces, draw a simple revolution as in Fig. 9.2 and in the lower two spaces, draw a simple revolution as in Fig. 9.3, but for each problem use a block assigned from Fig. 9.17.

Alternative Assignment: Using Layout B–4 or A3–4 (adjusted), divide into four equal parts as in Fig. 9.15. In the two left-hand spaces, draw a simple revolution as in Fig. 9.4, but use an object assigned from Fig. 9.17. In the two right-hand spaces, draw another simple revolution as in Fig. 9.4, but use a different object taken from Fig. 9.17 and revolve through 45° instead of 30°.

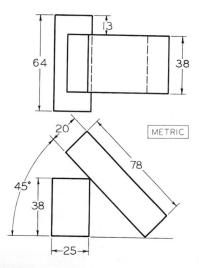

FIGURE 9.18 Using Layout A–2 or A–3 or Layout A4–2 or A4–3 (adjusted), draw three views of the blocks but revolved 30° clockwise about an axis perpendicular to the top plane of projection. Do not change the relative positions of the blocks.

FIGURE 9.19 *(opposite)* **Use Layout A–1 or A4–1 (adjusted), and divide the working area into four equal areas for four problems per sheet to be assigned by the instructor. Data for the layout of each problem are given by a coordinate system in metric dimensions. For example, in Prob. 1, point 1 is located by the scale coordinates (28 mm, 38 mm, 76 mm). The first coordinate locates the front view of the point from the left edge of the problem area. The second one locates the front view of the point from the bottom edge of the problem area. The third one locates either the top view of the point from the bottom edge of the problem area or the side view of the point from the left edge of the problem area. Inspection of the given problem layout will determine which application to use.**

1. Revolve clockwise point 1(28, 38, 76) through 210° about the axis 2(51, 58, 94)–3(51, 8, 94).
2. Revolve point 3(41, 38, 53) about the axis 1(28, 64, 74)–2(28, 8, 74) until point 3 is at the farthest distance behind the axis.
3. Revolve point 3(20, 8, 84) about the axis 1(10, 18, 122)–2(56, 18, 76) through 210° and to the rear of line 1–2.
4. Revolve point 3(5, 53, 53) about the axis 1(10, 13, 71)–2(23, 66, 71) to its extreme position to the left in the front view.
5. Revolve point 3(15, 8, 99) about the axis 1(8, 10, 61)–2(33, 25, 104) through 180°.
6. By revolution find the true length of line 1(8, 48, 64)–2(79, 8, 119). Scale: 1:100.
7. Revolve line 3(30, 38, 81)–4(76, 51, 114) about axis 1(51, 33, 69)–2(51, 33, 122) until line 3–4 is shown true length and below the axis 1–2. Scale: 1:20.
8. Revolve line 3(53, 8, 97)–4(94, 28, 91) about the axis 1(48, 23, 81)–2(91, 23, 122) until line 3–4 is in true length and above the axis.
9. Revolve line 3(28, 15, 99)–4(13, 30, 84) about the axis 1(20, 20, 97)–2(43, 33, 58) until line 3–4 is level an above the axis.

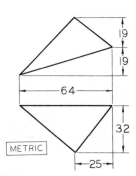

FIGURE 9.20 **Using Layout B–3 or A3–3, draw three views of a right prism 38 mm high that has as its lower base the triangle shown above. See §9.13.**

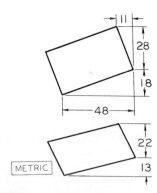

FIGURE 9.21 **Using Layout B–3 or A3–3, draw three views of a right pyramid 51 mm high, having as its lower base the parallelogram shown above. See §9.13.**

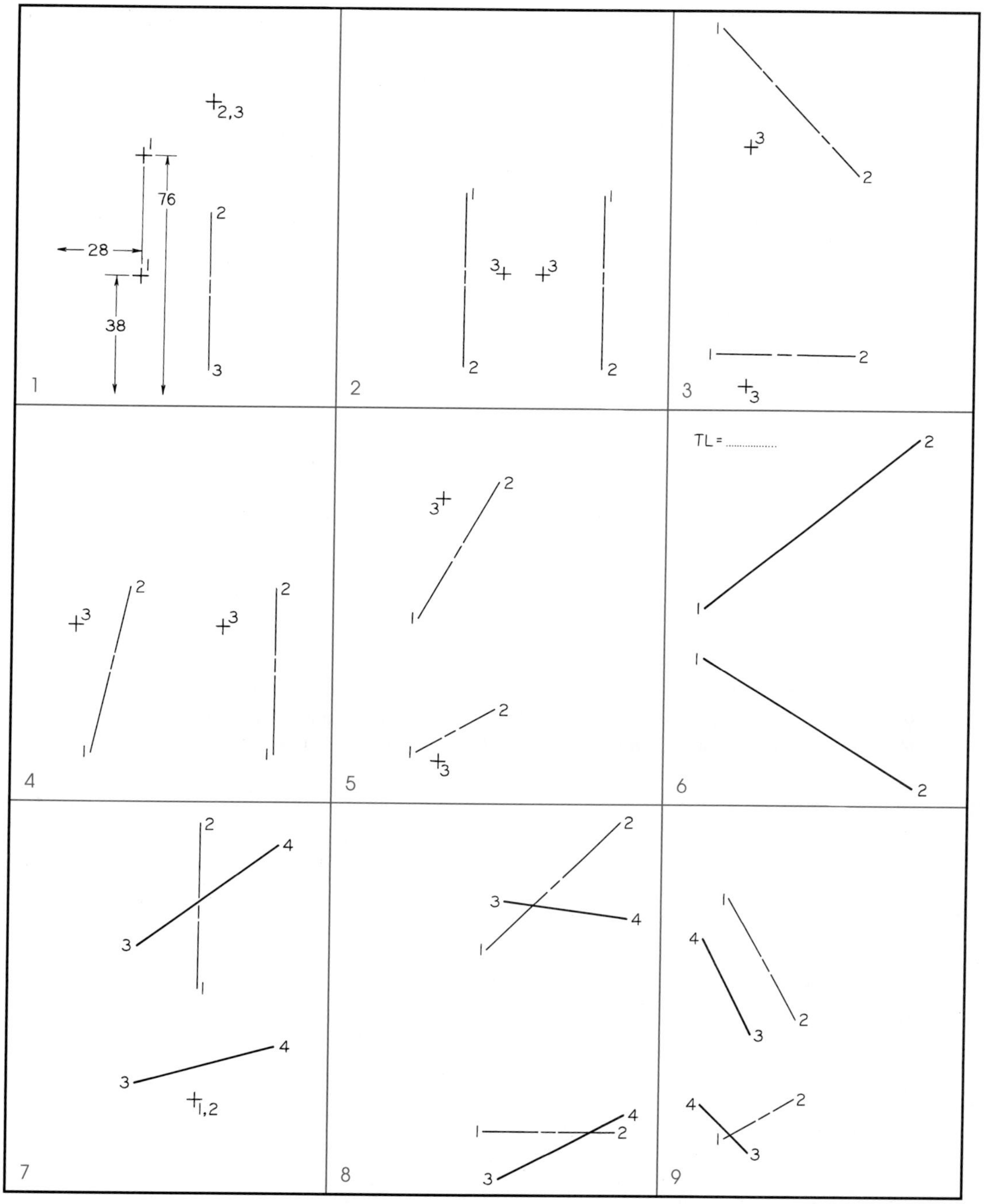
1
2,3
1
76
2
28
1
38
3
2
1
3
3
1
2
2
3
1
3
2
1
2
3
4
2
3
1
3
2
1
1
5
2
3
1
2
1
3
6
TL = ...
2
1
1
2
7
2
4
3
1
3
4
1,2
8
2
3
4
1
1
4
2
3
9
1
4
2
3
4
2
1
3

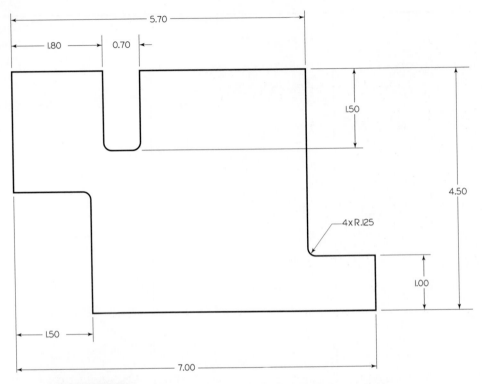

FIGURE 9.22 Using Layout A–2 or A–3 or a computer graphic system (CAD), design a three-dimensional view of this mounting plate revolved through 45° clockwise about an axis perpendicular to the top plane.

C H A P T E R 1 0

MANUFACTURING DESIGN AND PROCESSES

OBJECTIVES

After studying the material in this chapter, you should be able to:

1. Define *manufacturing* as we know it today.

2. Discuss the importance of the initial design stages in product development and manufacture.

3. Define *concurrent engineering* and show how this systematic approach integrates the design and manufacturing processes.

4. Explain the importance of computer-aided design, engineering, and manufacture in product development.

5. Explain the importance of prototypes and rapid prototyping in modern manufacturing processes.

6. Define *design for manufacture* and *design for manufacture and assembly*.

7. Explain the importance of proper material selection based on their properties, cost and availability, appearance, service life, and recycle potentials.

8. List the manufacturing processes typically used today.

9. Explain the importance of dimensional accuracy and surface finish.

10. List the typical measuring devices used in manufacturing.

11. Discuss the benefits of computer-integrated manufacturing.

OVERVIEW

Take a few moments and inspect various objects around you: your pen, watch, calculator, telephone, chair, and light fixtures. You will soon realize that all these objects had a different shape at one time. You could not find them in nature as they appear in your room. They have been transformed from various raw materials and assembled into the shapes that you now see.

Some objects are made of a single part, such as nails, bolts, wire or plastic coat hangers, metal brackets, and forks. However, most objects such as aircraft jet engines (invented in 1939), ball-point pens (1938), toasters (1926), washing machines (1910), air conditioners (1928), refrigerators (1931), photocopiers (1949), all types of machines and thousands of other products are made of an assembly of several parts made from a variety of materials. All are made by various processes that we call "manufacturing."

Manufacturing, in its broadest sense, is the process of converting raw materials into products. It encompasses (1) product design, (2) selection of raw materials, and (3) selection of processes by which manufacturing of goods take place, using various production methods and techniques.

Manufacturing is the backbone of any industrialized nation. Its importance is emphasized by the fact that, as an economic activity, it comprises approximately 20% to 30% of the value of all goods and services produced. The level of manufacturing activity is directly related to the economic health of a country. Generally, the higher the level of manufacturing activity in a country, the higher is the standard of living of its people.

Manufacturing also involves activities in which the manufactured product is itself used to make other products. Examples are large presses to shape sheet metal for car bodies, machinery to make bolts and nuts, and sewing machines for making clothing. An equally important aspect of manufacturing activities is servicing and maintaining this machinery during its useful life.

Engineering drawings, whether created with drawing instruments or CAD, are detailed instructions for manufacturing the described objects. The drawings must give information regarding shape, size, materials, finish, and, sometimes, the manufacturing process required. This chapter provides engineers with information about terms and processes used in manufacturing; information that will assist them with their drawings.

■ 10.1 "MANUFACTURING" DEFINED

The word "manufacturing" is derived from the Latin *manu factus*, meaning "made by hand." In the modern sense, manufacturing involves making products from raw materials by various processes, machinery, and operations, following a well-organized plan for each activity required. The word product means something that is produced, and the words product and production first appeared sometime during the 15th century. The word production is often used interchangeably with the word manufacturing. Whereas manufacturing engineering is the term used widely in the United States to describe this area of industrial activity, the equivalent term in other countries is production engineering.

Because a manufactured item has undergone a number of processes in which a piece of raw material has become a useful product, it has a value—defined as monetary worth or marketable price. For example, as the raw material for ceramics, clay has a certain value as mined. When the clay is used to make a ceramic cutting tool or electrical insulator, value is added to the clay. Similarly, a wire coat hanger or a nail has a value over and above the cost of a piece of wire from which it is made. Thus manufacturing has the important function of adding value.

Manufacturing may produce discrete products, meaning individual parts, or continuous products. Nails, gears, balls for bearings, beverage cans, and engine blocks are examples of discrete parts, even though they are mass produced at high production rates. On the other hand, a spool of wire, metal or plastic sheet, tubes, hose, and pipe are continuous products, which may be cut into individual lengths and thus become discrete parts.

Manufacturing is generally a complex activity involving a wide variety of resources and activities such as:

- Product design
- Purchasing
- Marketing
- Machinery and tooling
- Manufacturing
- Sales
- Process planning
- Production control
- Shipping

- Materials
- Support services
- Customer service

Manufacturing activities must be responsive to several demands and trends:

1. A product must fully meet design requirements and product specifications and standards.

2. A product must be manufactured by the most environmentally friendly and economical methods.

3. Quality must be built into the product at each stage, from design to assembly, rather than relying on quality testing after the product is made. Furthermore, quality should be appropriate to the product's use.

4. In a highly competitive environment, production methods must be sufficiently flexible so as to respond to changing market demands, types of products, production rates, production quantities, and on-time delivery to the customer.

5. New developments in materials, production methods, and computer integration of both technological and managerial activities in a manufacturing organization must constantly be evaluated with a view to their appropriate, timely, and economic implementation.

6. Manufacturing activities must be viewed as a large system, each part of which is interrelated to others. Such systems can be now modeled in order to study the effect of factors such as changes in market demands, product design, and materials. Various other factors and production methods affect product quality and cost.

7. A manufacturing organization must constantly strive for higher levels of quality and productivity (defined as the optimum use of all its resources: materials, machines, energy, capital, labor, and technology). Output per employee per hour in all phases must be maximized. Zero-based part rejection and waste are also an integral aspect of productivity.

■ 10.2 THE DESIGN PROCESS AND CONCURRENT ENGINEERING

The *design process* for a product first requires a clear understanding of the functions and the performance expected of that product. The product may be new, or it may be a revised version of an existing product. We all have observed, for example, how the design and style of radios, toasters, watches, automobiles, and washing machines have changed. The market for a product and its anticipated uses must be defined clearly, with the assistance of sales personnel, market analysts, and others in the organization. Product design is a critical activity because it has been estimated that 70% to 80% of the cost of product development and manufacture is determined at the initial design stages. The design process is discussed at length in Chapter 14.

Traditionally, design and manufacturing activities have taken place sequentially rather than concurrently or simultaneously (Fig. 10.1a). Designers would spend considerable effort and time in analyzing components and preparing detailed part drawings; these drawings would then be forwarded to other departments in the organization, such as materials departments where, for example, particular alloys and vendor sources would be identified. The specifications would then be sent to a manufacturing department where the detailed drawings would be reviewed and processes selected for efficient production. While this approach seems logical and straightforward, in practice it has been found to be extremely wasteful or resources.

In theory, a product can flow from one department in an organization to another and directly to the marketplace, but in practice there are usually difficulties encountered. For example, a manufacturing engineer may wish to taper the flange on a part to improve its castability, or a different alloy may be desirable, thus necessitating a repeat of the design analysis stage to ensure that the product will still function satisfactorily. These iterations, also shown in Fig. 10.1a, are certainly wasteful of resources but, more importantly, of time.

There is a great desire, originally driven by the consumer electronics industry, to bring products to market as quickly as possible. The rationale is that products introduced early enjoy a greater percentage of the market and hence profits, and have a longer life before obsolescence (clearly a concern with consumer electronics). For these reasons, concurrent engineering, also called simultaneous engineering, has come to the fore.

A more modern product development approach is shown in Fig. 10.1b. While there is a general product flow from market analysis to design to manufacturing, there are recognized iterations which occur in the process. The main difference to the more modern

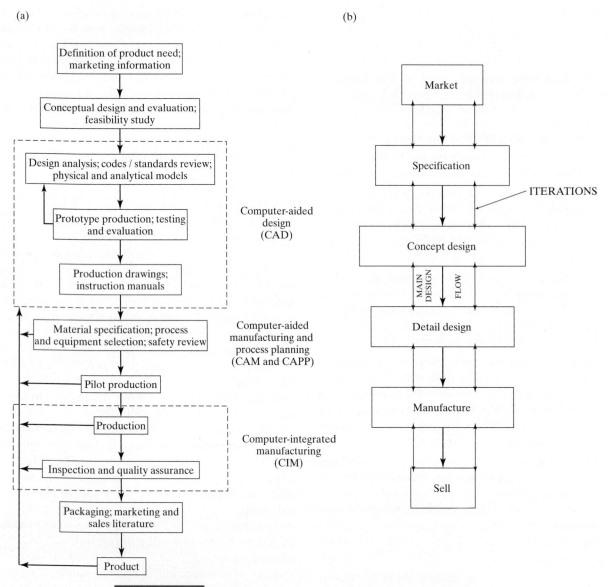

(a)   (b)

FIGURE 10.1   Design and manufacturing activities traditionally have taken place sequentially rather than concurrently. With CAM, the design process can work simultaneously with the manufacturing process.

approach is that all disciplines are involved in the early design stages, so that the iterations which naturally occur result in less wasted effort and lost time. A key to the approach is the now well-recognized importance of communication between and within disciplines. That is, while there must be communication between engineering and marketing and service functions, so too must there be avenues of interactions between engineering sub-disciplines, for example, design for manufacture, design recyclability and design for safety.

The design process begins with the development of an original product concept. An innovative approach to design is highly desirable—and even essential —at this stage for the product to be successful in the marketplace (see Chapter 14). Innovative approaches can also lead to major savings in material and production costs. The design engineer or product designer must be knowledgeable of the interrelationships among materials, design, and manufacturing, as well as the overall economics of the operation.

Concurrent engineering is a systematic approach integrating the design and manufacture of products with the view of optimizing all elements involved in the life cycle of the product. *Life cycle* means that all aspects of a product (such as design, development, production, distribution, use, and its ultimate disposal and recycling) are considered simultaneously. The basic goals of concurrent engineering are to minimize product design and engineering changes and the time and costs involved in taking the product from design concept to production and introduction of the product into the marketplace.

The philosophy of *life cycle engineering* requires that the entire life of a product be considered in the design stage (i.e., the design, production, distribution, use and disposal/recycling must be considered simultaneously). Thus a well-designed product is functional (design stage), well manufactured (production), packaged so that it safely arrives to the end user or customer (distribution), functions effectively for its intended life and has components which can be easily replaced for maintenance or repair (use), and can be disassembled so that components can be recycled (disposal).

Although the concept of concurrent engineering appears to be logical and efficient, its implementation can take considerable time and effort when those using it either do not work as a team or fail to appreciate its real benefits. It is apparent that for concurrent engineering to succeed it must:

a. Have the full support of the upper management,

b. Have multifunctional and interactive teamwork, including support groups, and

c. Utilize all available technologies.

There are numerous examples of the benefits of concurrent engineering. An automotive company, for example, has reduced the number of components in an engine by 30%, and as a result has decreased its weight by 25% and cut manufacturing time by 50%. The concurrent engineering concept can be implemented not only in large organizations but in smaller companies as well. This is particularly noteworthy in view of the fact that 98% of U.S. manufacturing establishments have fewer than 500 employees.

For both large and small companies, product design often involves preparing analytical and physical models of the product, as an aid to studying factors such as forces, stresses, deflections, and optimal part shape. The necessity for such models depends on product complexity. Today, constructing and studying analytical models is simplified through the use of computer-aided design (CAD), engineering (CAE), and manufacturing (CAM) techniques.

■ 10.3 COMPUTER-AIDED DESIGN AND PRODUCT DEVELOPMENT

Computer-aided design (CAD) allows the designer to conceptualize objects more easily without having to make costly illustrations, models, or prototypes. These systems are now capable of rapidly and completely analyzing designs, from a simple bracket to complex structures. For example, the two-engine Boeing 777 passenger airplane was designed completely by computer (paperless design) with AutoCAD 2000 workstations linked to eight computers (Fig. 10.2). The airplane was constructed directly from the CAD/CAM software developed and no prototypes or mockups were built, unlike previous models.

■ 10.4 COMPUTER-AIDED ENGINEERING ALLOWS FOR FUTURE MODIFICATION

Using *computer-aided engineering*, the performance of structures subjected to static or fluctuating loads and various temperatures can now be simulated, analyzed, and tested efficiently, accurately, and more quickly

FIGURE 10.2   **Boeing 777 in Flight.** *Courtesy of Boeing Co.*

than ever. The information developed can be stored, retrieved, displayed, printed, and transferred anywhere in the organization. Designs can be optimized and modifications can be made directly and easily at any time.

■ 10.5 COMPUTER-AIDED ENGINEERING LINKS ALL PHASES OF MANUFACTURING

Computer-aided manufacturing (CAM) involves all phases of manufacturing by utilizing and processing further the large amount of information on materials and processes collected and stored in the organization's database. Computers now assist manufacturing engineers and others in organizing tasks such as programming numerical control of machines; programming robots for material handling and assembly; designing tools, dies, and fixtures; and maintaining quality control.

On the basis of the models developed using the above mentioned techniques, the product designer selects and specifies the final shape and dimensions of the product, its dimensional accuracy and surface finish, and the materials to be used. The selection of materials is often made with the advice and cooperation of materials engineers, unless the design engineer is also experienced and qualified in this area. An important design consideration is how a particular component is to be assembled into the final product. Lift the hood of your car and observe how hundreds of components are put together in a limited space.

■ 10.6 THE ROLE OF PROTOTYPES AND RAPID PROTOTYPING IN PRODUCT DEVELOPMENT

The next step in the production process is to make and test a *prototype*, that is, an original working model of the product. An important development is *rapid prototyping*, which relies on CAD/CAM and various manufacturing techniques (using metallic or non-metallic materials) to quickly produce prototypes in the form of a solid physical model of a part and at low cost. For example, prototyping new automotive components by traditional methods of shaping, forming, machining, etc. costs hundreds of millions of dollars a year; some components may take a year to produce. Rapid prototyping can cut these costs as well as development times significantly. These techniques are being advanced further so that they can be used for low-volume economical production of actual parts.

Tests of prototypes must be designed to simulate as closely as possible the conditions under which the product is to be used. These include environmental conditions such as temperature and humidity, as well as the effects of vibration and repeated use and misuse of the product. Computer-aided engineering techniques are now capable of comprehensively and rapidly performing such simulations. During this stage, modifications in the original design, materials selected, or production methods may be necessary. After this phase has been completed, appropriate process plans, manufacturing methods (Table 10.1), equipment, and tooling are selected with the cooperation of manufacturing engineers, process planners, and others involved in production.

■ 10.7 DESIGN FOR MANUFACTURE, ASSEMBLY, DISASSEMBLY, AND SERVICE

As we have seen, design and manufacturing must be intimately interrelated; they should never be viewed as separate disciplines or activities. Each part or component of a product must be designed so that it not only meets design requirements and specifications, but also can be manufactured economically and efficiently. This approach improves productivity and allows a manufacturer to remain competitive.

This broad view has become recognized as the area of *design for manufacture* (DFM). It is a comprehensive approach to production of goods and integrates the design process with materials, manufacturing methods, process planning, assembly, testing, and quality assurance. Effectively implementing design for manufacture requires that designers have a fundamental understanding of the characteristics, capabilities, and limitations of materials, manufacturing processes, and related operations, machinery, and equipment. This knowledge includes characteristics such as variability in machine performance, and dimensional accuracy and surface finish of the workpiece, processing time, and the effect of processing method on part quality.

Designers and product engineers must be able to assess the impact of design modifications on manufacturing process selection, assembly, inspection, tools and dies, and product cost. Establishing quantitative relationships is essential in order to optimize the design for ease of manufacturing and assembly at minimum product cost (also called producibility). Computer-aided design, engineering, manufacturing,

Shape of Feature	Production Method
Flat surfaces	Rolling, planing, broaching, milling, shaping, grinding
Parts with cavities	End milling, electrical-discharge machining, electrochemical machining, ultrasonic machining, casting
Parts with sharp features	Permanent mold casting, machining, grinding, fabricating, powder metallurgy
Thin hollow shapes	Slush casting, electroforming, fabricating
Tubular shapes	Extrusion, drawing, roll forming, spinning, centrifugal casting
Tubular parts	Rubber forming, expanding with hydraulic pressure, explosive forming, spinning
Curvature on thin sheets	Stretch forming, peen forming, fabricating, assembly
Opening in thin sheets	Blanking, chemical blanking, photochemical blanking
Cross-sections	Drawing, extruding, shaving, turning, centerless grinding
Square edges	Fine blanking, machining, shaving, belt grinding
Small holes	Laser, electrical discharge machining, electrochemical machining
Surface textures	Knurling, wire brushing, grinding, belt grinding, shot blasting, etching, deposition
Detailed surface feature	Coining, investment casting, permanent-mold casting, machining
Threaded parts	Thread cutting, thread rolling, thread grinding, chasing
Very large parts	Casting, forging, fabricating, assembly
Very small parts	Investment casting, machining, etching, powder metallurgy, nanofabrication, micromachining

TABLE 10.1   **Shapes and Some Common Methods of Production.**

and process planning techniques, using powerful computer programs, have become indispensable to those conducting such analysis. New developments include expert systems, which have optimization capabilities, thus expediting the traditional iterative process in design optimization.

After individual parts have been manufactured, they have to be assembled into a product. Assembly is an important phase of the overall manufacturing operation and requires consideration of the ease, speed, and cost of putting parts together. Also, many products must be designed so that disassembly is possible, enabling the products to be taken apart for maintenance, servicing, or recycling of their components. Because assembly operations can contribute significantly to product cost, *design for assembly* (DFA) as well as design for disassembly are now recognized as important aspects of manufacturing. Typically, a product that is easy to assemble is also easy to disassemble. The latest trend now includes design for service, ensuring that individual parts or sub-assemblies in a product are easy to reach and service.

Methodologies and computer software (CAD) have been developed for DFA utilizing 3-D conceptual designs and solid models. In this way, subassembly and assembly times and costs are minimized while maintaining product integrity and performance; the system also improves the product's ease of disassembly. The trend now is to combine design for manufacture and design for assembly into the more comprehensive *design for manufacture and assembly* (DFMA) which recognizes the inherent interrelationships between design and manufacturing.

There are several methods of assembly, such as using fasteners, adhesives, or by welding, soldering, and brazing, each with its own characteristics and requiring different operations. The use of a bolt and nut, for example, requires preparation of holes that must match in location and size. Hole generation requires operations such as drilling or punching, which take additional time, require separate operations, and produce scrap. On the other hand, products assembled with bolts and nuts can be taken apart and reassembled with relative ease.

Parts can also be assembled with adhesives. This method, which is being used extensively in aircraft and automobile production, does not require holes. However, surfaces to be assembled must match properly and be clean because joint strength is adversely affected by the presence of contaminants such as dirt, dust, oil, and moisture. Unlike mechanical fastening, adhesively joined components, as well as those that are welded, are not usually designed to be taken apart and reassembled, hence are not suitable for the important purposes of recycling individual parts in the product.

Parts may be assembled by hand or by automatic equipment and robots. The choice depends on factors such as the complexity of the product, the number of parts to be assembled, the protection required to prevent damage or scratching of finished surfaces of the parts, and the relative costs of labor and machinery required for automated assembly.

■ 10.8 MATERIAL SELECTION

An ever-increasing variety of materials is now available, each having its own characteristics, applications, advantages, and limitations. The following are the general types of materials used in manufacturing today either individually or in combination.

- Ferrous metals: carbon, alloy, stainless, and tool and die steels.
- Nonferrous metals: aluminum, magnesium, copper, nickel, titanium, superalloys, refractory metals, beryllium, zirconium, low-melting alloys, and precious metals.
- Plastics: thermoplastics, thermosets, and elastomers.
- Ceramics, glass ceramics, glasses, graphite, diamond, and diamond-like materials.
- Composite materials: reinforced plastics, metal-matrix and ceramic-matrix composites. These are also known as engineered materials.
- *Nanomaterials*, shape-memory alloys, amorphous alloys, superconductors, and various other materials with unique properties.

As new materials are developed, the selection of appropriate materials becomes even more challenging. Aerospace structures, as well as products such as sporting goods, have been at the forefront of new material usage. The trend has been to use more titanium and composites for the airframe of commercial aircraft, with a gradual decline of the use of aluminum and steel. There are constantly shifting trends in the usage of materials in all products, driven principally by economic considerations as well as other considerations.

■ 10.9 PROPERTIES OF MATERIALS

When selecting materials for products, we first consider their mechanical properties: strength, toughness, ductility, hardness, elasticity, fatigue, and creep. The strength-to-weight and stiffness-to-weight ratios of material are also important, particularly for aerospace and automotive applications. Aluminum, titanium, and reinforced plastics, for example, have higher ratios than steels and cast irons. The mechanical properties specified for a product and its components should, of course, be for the conditions under which the product is expected to function. We then consider the physical properties of density, specific heat, thermal expansion and conductivity, melting point, and electrical and magnetic properties.

Chemical properties also play a significant role in hostile as well as normal environments. Oxidation, corrosion, general degradation of properties, toxicity, and flammability of materials are among the important factors to be considered. In some commercial airline disasters, for example, many deaths have been caused by toxic fumes from burning nonmetallic materials in the aircraft cabin.

Manufacturing properties of materials determine whether they can be cast, formed, machined, welded, and heat treated with relative ease (Table 10.2). The method(s) used to process materials to the desired shapes can adversely affect the product's final properties, service life, and its cost.

■ 10.10 COST AND AVAILABILITY OF MATERIALS

Cost and availability of raw and processed materials and manufactured components are major concerns in manufacturing. Competitively, the economic aspects of material selection are as important as the technological considerations of properties and characteristics of materials.

If raw or processed materials or manufactured components are not available in the desired shapes, dimensions and quantities, substitutes and/or additional processing will be required, which can contribute significantly to product cost. For example, if we need a round bar of a certain diameter and it is not available

Alloy	Castability	Weldability	Machinability
Aluminum	E	F	G-E
Copper	F-G	F	F-G
Gray cast iron	E	D	G
White cast iron	G	VP	VP
Nickel	F	F	F
Steels	F	E	F
Zinc	E	D	E

E, excellent; G, good; F, fair; D, difficult; VP, very poor

TABLE 10.2   **General Manufacturing Characteristics of Various Alloys.**

in standard form, then we have to purchase a larger rod and reduce its diameter by some means, such as machining, drawing through a die, or grinding. It should be noted, however, that a product design can be modified to take advantage of standard dimensions of raw materials, thus avoiding additional manufacturing costs.

Reliability of supply, as well as demand, affects cost. Most countries import numerous raw materials that are essential for production. The United States, for example, imports the majority of raw materials such as natural rubber, diamond, cobalt, titanium, chromium, aluminum, and nickel from other countries. The broad political implications of such reliance on other countries are self-evident.

Different costs are involved in processing materials by different methods. Some methods require expensive machinery, others require extensive labor, and still others require personnel with special skills, a high level of education, or specialized training.

■ 10.11 APPEARANCE, SERVICE LIFE, AND RECYLING

The appearance of materials after they have been manufactured into products influences their appeal to the consumer. Color, feel, and surface texture are characteristics that we all consider when making a decision about purchasing a product.

Time- and service-dependent phenomena such as wear, fatigue, creep, and dimensional stability are important. These phenomena can significantly affect a product's performance and, if not controlled, can lead to total failure of the product. Similarly, compatibility of materials used in a product is important. Friction and wear, corrosion, and other phenomena can shorten

a product's life or cause it to fail prematurely. An example is galvanic corrosion between mating parts made of dissimilar metals.

Recycling or proper disposal of materials at the end of their useful service lives has become increasingly important in an age when we are more conscious of preserving resources and maintaining a clean and healthy environment. Note, for example, the use of biodegradable packaging materials or recyclable glass bottles and aluminum beverage cans. The proper treatment and disposal of toxic wastes and materials are also a crucial consideration.

■ 10.12 MANUFACTURING PROCESSES

Before preparing a drawing for the production of a part, the drafter/designer should consider what manufacturing processes are to be used. These processes will determine the representation of the detailed feature s of the part, the choice of dimensions, and the machining of processing accuracy. Many processes are used to produce parts and shapes (Table 10.1) and there is usually more than one method of manufacturing a part from a given material. The broad categories of processing methods for materials are:

a. *Casting*: Expendable molds (i.e., sand casting) and permanent molds (Fig. 10.3).

b. *Forming and shaping*: Rolling, forging, extrusion, drawing, sheet forming, powder metallurgy, and molding (Fig. 10.4a-d).

c. *Machining*: Turning, boring, drilling, milling, planing, shaping, broaching, grinding, ultrasonic machining; chemical, electrical, and electrochemical machining; and high-energy beam machining (Fig. 10.5a-g).

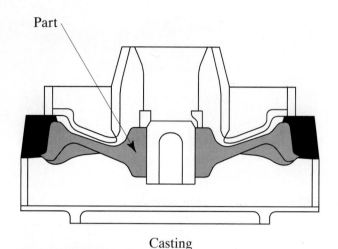

Part

Casting

FIGURE 10.3 **This casting mold is an example of a permanent mold.**

d. *Joining*: Welding, brazing, soldering, diffusion bonding, adhesive bonding, and mechanical joining (Fig. 10.6a-b). Welding procedures and the use of standardized symbols for their representation are described in Chapters 25.

e. *Finishing*: Honing, lapping, polishing, burnishing, deburring, surface treating, coating, and plating.

Selection of a particular manufacturing process, or a series of processes, depends not only on the shape to be produced but also on many other factors pertaining to material properties (Table 10.2). Brittle and hard materials, for example, cannot be shaped easily, whereas they can be cast or machined by several methods. The manufacturing process usually alters the properties of materials. Metals that are formed at room temperature, for example, become stronger, harder, and less ductile than they were before processing.

FIGURE 10.4 **Examples of forming and shaping methods.**

(a)

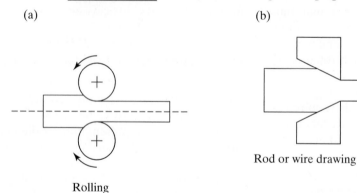

Rolling

(b)

Rod or wire drawing

(c)

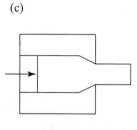

Extrusion

(d)

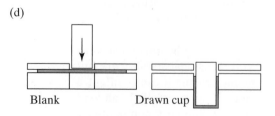

Blank   Drawn cup

Deep Drawing

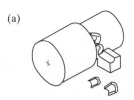

(a)

Turning

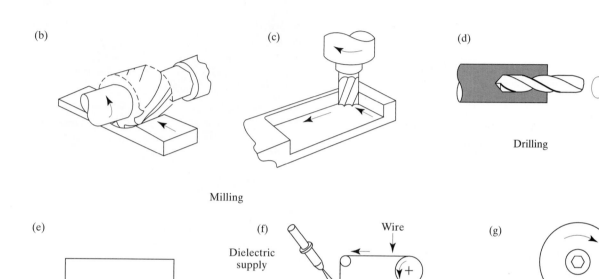

(b)

(c)

(d)

Drilling

Milling

(e)

(f)

Dielectric supply

Wire

(g)

Broaching

Grinding

Wire electrical-discharge machining

FIGURE 10.5 **Examples of different types of machining.**

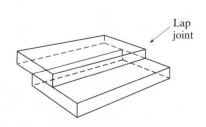

Lap joint

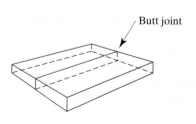

Butt joint

Joining

FIGURE 10.6 **Example of joining method.**

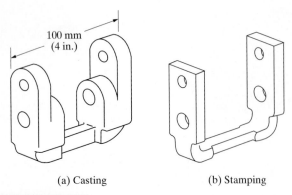

(a) Casting   (b) Stamping

FIGURE 10.7   **Two steel mounting brackets (a) made by casting, (b) made by stamping.**

Two steel mounting brackets are shown in Fig. 10.7, one made by casting, and the other by stamping of sheet metal. Note that there are some differences in the designs, although the parts are basically alike. Each of these two manufacturing processes has its own advantages and limitations, as well as production rates and manufacturing cost.

Manufacturing engineers are constantly being challenged to find new solutions to manufacturing problems and cost reduction. For a long time, for example, sheet metal parts were cut and fabricated by traditional tools, punches, and dies. Although they are still widely used, some of these operations are now being replaced by laser cutting techniques (Fig. 10.8). With advances in computer technology, we can automatically control the path of the laser, thus increasing the capability for producing a wide variety of shapes accurately, repeatedly, and economically.

■ 10.13 DIMENSIONAL ACCURACY AND SURFACE FINISH

Size, thickness, and shape complexity of the part have a major bearing on the manufacturing process selected to produce it. Flat parts with thin cross-sections, for example, cannot be cast properly. Complex parts cannot be formed easily and economically, whereas they may be cast or fabricated from individual pieces.

Tolerances and surface finish obtained in hot-working operations cannot be as good as those obtained in cold-working (room temperature) operations because dimensional changes, warpage, and surface oxidation occur during processing at elevated temperatures. Some casting processes produce a better surface finish than others because of the different types of mold materials used and their surface finish.

FIGURE 10.8   **Cutting sheet metal with a laser beam.** *Courtesy of Rofin-Sinar, Inc., and Manufacturing Engineering Magazine, Society of Manufacturing Engineers.*

The size and shape of manufactured products vary widely. For example, the main landing gear for a twin-engine, 400-passenger Boeing 777 jetliner is 4.3 m (14 ft) high, with three axles and six wheels, made by forging and machining processes (Fig. 10.2). At the other extreme is the generation of a 0.05-mm (0.002-in.) diameter hole at one end of a 0.35-mm (0.014-in.) diameter needle (Fig. 10.9), using a process called electrical-discharge machining. The hole is burr-free and has a location accuracy of ±0.003 mm (0.0001 in.).

Another small-scale manufacturing example is given in Fig. 10.10, which shows microscopic gears as small as 100 m (0.004 in.) in diameter. These gears have possible applications such as powering *microrobots* to repair human cells, *microknives* in surgery, and camera shutters for precise photography. The gears are made by a special electroplating and x-ray etching technique of metal plates coated with a polymer film. The center hole in these gears is so small that a human hair cannot pass through it. Such small-scale operations are called *nanotechnology* and *nanofabrication* ("nano" meaning one billionth).

Ultraprecision manufacturing techniques and machinery are now being developed and are coming into more common use. For machining mirrorlike surfaces,

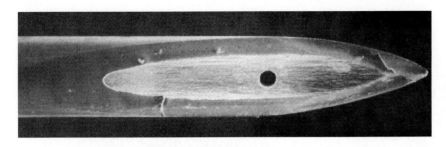

FIGURE 10.9 A 0.05-mm hole produced in a needle, using the electrical-discharge machining process. *Courtesy of Derata Corporation.*

Human Hair

FIGURE 10.10 Microscopic gear with a diameter on the order of 100μm, made by a special etching process. *Courtesy of Wisconsin Center for Applied Microelectronics, University of Wisconsin–Madison.*

for example, the cutting tool is a very sharp diamond tip and the equipment has very high stiffness and must be operated in a room where the temperature is controlled within 1 °C. Highly sophisticated techniques such as molecular-beam epitaxy and scanning-tunneling engineering are being implemented to obtain accuracies on the order of the atomic lattice (0.1 nm; 10^{-8} in.).

■ 10.14 MEASURING DEVICES USED IN MANUFACTURING

Although the machinist uses various measuring devices depending on the kind of dimensions (fractional, decimal, or metric) shown on the drawing, it is evident that to dimension correctly, the engineering designer must have at least a working knowledge of the common measuring tools. The machinists' *steel rule*, or *scale*, is a commonly used measuring tool in the shop (Fig. 10.11a). The smallest division on one scale of this rule is 0, and such a scale is used for common fractional dimensions. Also, many machinists' rules have a decimal scale with

the smallest division of .010, which is used for dimensions given on the drawing by the decimal system. For checking the nominal size of outside diameters, the *outside spring caliper* and steel scale are used, as shown in Figs. 10.11b and 10.11c. Likewise, the *inside spring caliper* is used for checking nominal dimensions, as shown in Figs. 10.11d and 10.11e. Another use for the outside caliper (Fig. 10.11f) is to check the nominal distance between holes (center to center). The *combination square* may be used for checking height (Fig. 10.11g) and for a variety of other measurements. Measuring devices are also available that have metric scales.

For dimensions that require more precise measurements, the *vernier caliper* (Figs. 10.11h and 10.11j) or the *micrometer caliper* (Fig. 10.11k) may be used. It is common practice to check measurements to 0.025 mm (.0010) with these instruments, and in some instances they are used to measure directly to 0.0025 mm (.00010).

Many of the measuring devices discussed here have been supplemented with newer, more sophisticated tools. Computerized measuring devices have broadened the range of accuracy previously attainable. Figure 10.12 illustrates an ultraprecision electronic digital readout micrometer and caliper that contain integral microprocessors. In addition to the hand-held printer/recorder providing a hard-copy output of measurements, the printer also calculates and lists statistical mean, minimum, and maximum values, as well as standard deviation.

Most measuring devices in manufacturing are adjustable so they can each be employed to measure any size within their range of designed usage. There is also a need for measuring devices designed to be used for only one particular dimension. These are called *fixed gages* because their setting is fixed and cannot be changed.

A common type of fixed gage consists of two carefully finished rounds. One might think of each of these rounds as being 25.4 mm (1.000) in diameter and 38 mm (1.5000) long. Let one of these diameters be slightly larger than the other. One can see that, for a

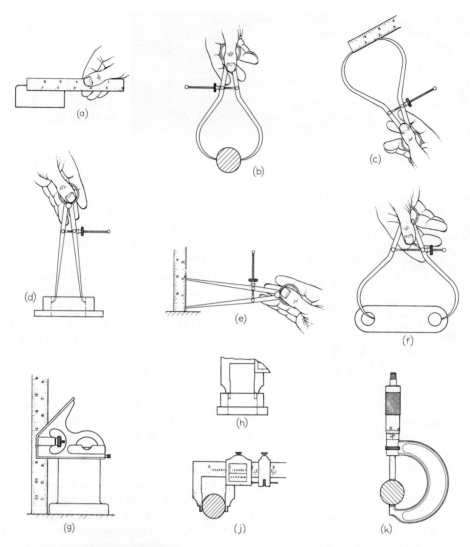

FIGURE 10.11 Measuring Devices Used by the Machinist.

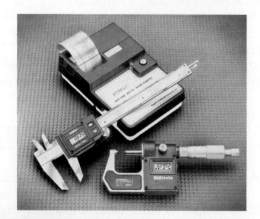

FIGURE 10.12 Computerized Measurement System. *Courtesy of Fred V. Fowler Co., Inc.*

certain range of hole sizes, the smaller round will enter the hole but the larger will not. If the larger round diameter is made slightly greater than the largest acceptable hole diameter and if the diameter of the smaller round is made slightly less than the smallest acceptable hole diameter, then the large round *will never go* into any acceptable hole but the small round *will go* into any acceptable hole. A fixed gage consisting of two such rounds is called a "go–no go" gage. There are, of course, many kinds of "go–no go" gages.

The subject of gages and gaging is a specialized field and involves so many technical considerations that many large companies employ highly trained workers to attend to nothing but this one feature of their operations.

■ 10.15 OPERATIONAL AND MANUFACTURING COSTS

The design and cost of tooling, the lead time required to begin production, and the effect of workpiece material on tool and die life are major considerations. Depending on its size, shape, and expected life, the cost of tooling can be substantial. For example, a set of steel dies for stamping sheet-metal fenders for automobiles may cost about $2 million.

For parts made from expensive materials, the lower the scrap rate, the more economical the production process will be; thus, every attempt should be made for *zero-base waste*. Because it generates chips, machining may not be more economical than forming operations, all other factors being the same.

Availability of machines and equipment and operating experience within the manufacturing facility are also important cost factors. If they are not available, some parts may have to be manufactured by outside firms. Automakers, for example, purchase many parts from outside vendors, or have them made by outside firms according to their specifications.

The number of parts required (quantity) and the required production rate (pieces per hour) are important in determining the processes to be used and the economics of production. Beverage cans or transistors, for example, are consumed in numbers and at rates much higher than telescopes and propellers for ships.

The operation of machinery has significant environmental and safety implications. Depending on the type of operation, some processes adversely affect the environment, such as the use of oil-base lubricants in hot metalworking processes. Unless properly controlled, such processes can cause air, water, and noise pollution. The safe use of machinery is another important consideration, requiring precautions to eliminate hazards in the workplace.

■ 10.16 CONSEQUENCES OF IMPROPER SELECTION OF MATERIALS AND PROCESSES

Numerous examples of product failure can be traced to improper selection of material or manufacturing processes or improper control of process variables. A component or a product is generally considered to have failed when:

- It stops functioning (broken shaft, gear, bolt, cable, or turbine blade).

- It does not function properly or perform within required specification limits (worn bearings, gears, tools, and dies).

- It becomes unreliable or unsafe for further use (frayed cable in a winch, crack in a shaft, poor connection in a printed-circuit board, or delamination of a reinforced plastic component).

■ 10.17 NET-SHAPE MANUFACTURING

Since not all manufacturing operations produce finished parts, additional operations may be necessary. For example, a forged part may not have the desired dimensions or surface finish; thus additional operations such as machining or grinding may be necessary. Likewise, it may be difficult, impossible, or economically undesirable to produce a part with holes using just one manufacturing process, thus necessitating additional processes such as drilling. Also, the holes produced by a particular manufacturing process may not have the proper roundness, dimensional accuracy, or surface finish, thus creating a need for additional operations such as honing.

Finishing operations can contribute significantly to the cost of a product. Consequently, the trend has been for *net-shape* or *near-net-shape manufacturing*, in which the part is made as close to the final desired dimensions, tolerances, surface finish, and specifications as possible. Typical examples of such manufacturing methods are near-net-shape forging and casting of parts, stamped sheet-metal parts, injection molding of plastics, and components made by powder-metallurgy techniques.

■ 10.18 COMPUTER-INTEGRATED MANUFACTURING

The major goals of automation in manufacturing facilities are to integrate various operations to improve productivity, increase product quality and uniformity, minimize cycle times, and reduce labor costs. Beginning in the 1940s, automation has accelerated because of rapid advances in control systems for machines and in computer technology.

Few developments in the history of manufacturing have had a more significant impact than computers. Computers are now used in a very broad range of applications, including control and optimization of manufacturing processes, material handling, assembly, automated inspection and testing of products, as well as inventory control and numerous management activities. Beginning with computer graphics and computer-aided design and manufacturing, the use of computers has been extended to *computer-integrated manufacturing* (CIM). Computer-integrated manufacturing is particularly effective because of its capability for:

- Responsiveness to rapid changes in market demand and product modification.
- Better use of materials, machinery and personnel, and reduced inventory.
- Better control of production and management of the total manufacturing operation.
- High-quality products at low cost.

The major applications of computers in manufacturing are:

a. *Computer numerical control (CNC)* is a method of controlling the movements of machine components by direct insertion of coded instructions in the form of numerical data. Numerical control was first implemented in the early 1950s and was a major advance in automation of machines.

b. *Adaptive control (AC).* The parameters in a manufacturing process are adjusted automatically to optimize production rate and product quality, and to minimize cost. Parameters such as forces, temperatures, surface finish, and dimensions of the part are monitored constantly. If they move outside the acceptable range, the system adjusts the process variables until the parameters again fall within the acceptable range.

c. *Industrial robots.* Introduced in the early 1960s, industrial robots (Figs. 10.13 and 10.14) have been replacing humans in operations that are repetitive, boring, and dangerous, thus reducing the possibility of human error, decreasing variability in product quality, and improving productivity. Robots with sensory perception capabilities are being developed (*intelligent robots*), with movements that simulate those of humans.

d. *Automated handling of materials.* Computers have allowed highly efficient handling of materials and products in various stages of completion (*work in progress*), such as from storage to machines, from machine to machine, and at the points of inspection, inventory, and shipment.

e. *Automated and robotic assembly systems* are replacing costly assembly by operators. Products are designed or redesigned so that they can be assembled more easily by machine (Fig. 10.15).

f. *Computer-aided process planning (CAPP)* is capable of improving productivity in a plant by opti-

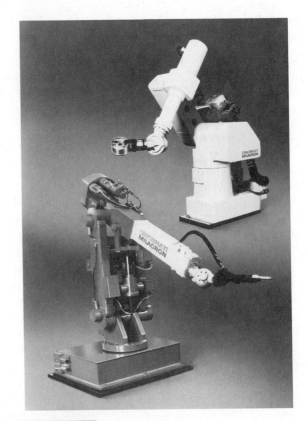

FIGURE 10.13 **Industrial Robots.** *Courtesy of Cincinnati Milacron.*

mizing process plans, reducing planning costs, and improving the consistency of product quality and reliability. Functions such as cost estimating and work standards (time required to perform a certain operation) can also be incorporated into the system.

g. *Group technology (GT).* The concept of group technology is that parts can be grouped and produced by classifying them into families, according to similarities in design and similarities in manufacturing processes to produce the part. In this way, part designs and process plans can be standardized and families of parts can be produced efficiently and economically.

h. *Just-in-time production (JIT).* The principal of JIT is that supplies are delivered just in time to be used, parts are produced just in time to be made into subassemblies and assemblies, and products are finished just in time to be delivered to the customer. In this way, inventory-carrying costs are low, part defects are detected right away, productivity is increased, and high-quality products are made at low cost.

FIGURE 10.14   Robotic Welding on Ford Automobile Assembly Line. *Courtesy of Ford Motor Co.*

FIGURE 10.15   Automated Manufacturing System. *Courtesy of Cargill Detroit.*

Digital Polish for Factory Floors

Without question, this has been the decade in which software moved irrevocably into the industrial designer's world. In corporate offices across the country, engineers have been booting up programs that let them tinker, in three dimensions, with every permutation and combination of a product's design. Be the item as lowly as a dinner plate or as complex as a Boeing 777 jet, the goal has always been the same: save time and money in getting products to market.

It was just a matter of time before engineers would aim their computers at designing and refining the assembly lines on which those products are made. Apparently, that time has come:

- These days, before it approves a design for a new car or van, the Ford Motor Company checks the plan against computer models of its factory floor. Often, through subtle changes like relocating a few seams or shaving a few millimeters from the length of a fender, Ford can lop weeks off the time it takes to prepare an old plant to make a new car.

- The Dow Chemical Company now uses computers to simulate its methods for making plastics, running what-if scenarios to fine-tune the temperatures, pressures and rates at which it feeds in raw materials. Dow can now switch production among 15 different grades of plastics in minutes, with almost no wasted material. Before computer modeling, the process took two hours and yielded lots of useless byproducts.

- The International Business Machines Corporation, the world's largest computer maker and an early convert to factory-simulation software, has learned that its assembly lines can accommodate diverse products of similar size, as long as IBM tweaked its conveyor belts to deliver different parts and products to different work areas. One early result: with almost no retooling, IBM expanded a plant in Charlotte, NC, that had made only banking systems to include voice-messaging systems, bar-code readers and devices to program pacemakers.

"No one wants to build a new assembly line, if they can re-use the one they have," said Frank Lerchenmuller, an IBM vice president for engineering technology solutions. Of course not. But until the advent of software that can simulate assembly lines and the movements of the people who run them, no one really knew a good way to gauge what they could salvage from an older factory. Now, with software from companies like Tecnomatix Technologies Inc. of Israel, Dassault Systemes of France and Aspen Technology in the United States, it is as feasible to design or remodel an entire plant as it is to reconfigure a car.

Production engineers in industries as diverse as chemicals, automobiles and aluminum smelting are manipulating virtual pictures of their plants and processes to see whether moving a clamp or adding a new ingredient will make existing equipment more productive, or will enable the same assembly line to skip freely from product to product. Some are even testing out a new virtual reality program that enables engineers wearing special goggles to detect problems by "walking through and around" a three-dimensional model of their factory designs. The entire relationship between product design and production engineering is being turned on its ear. No longer is it enough for designers to create products that can be made and maintained efficiently. Increasingly, management is asking them whether the products can be manufactured with a minimum of retooling or work disruption—and if not, whether it is worth giving up a particular feature to wring time and money from the manufacturing process.

Of course, there is little allure to the time-honored alternative to computer modeling—setting up an actual assembly line and trouble-shooting it piece by piece. "It's too risky and expensive to try new tools or methods if you have to build prototype hardware to test every change," said Rani Agarwal Finstad, director of manufacturing math-based systems for GM Powertrain, a manufacturer of car components that recently used simulation software to squeeze a month out of the process of programming robots to make new crankshafts.

The market for simulation software is growing, if slowly. Bernard Charles, Dassault's president, calculates that companies spent $300 million globally for software that simulates manufacturing operations last year. That is nothing compared with the $4 billion he estimates that industry spent on product-design software, but "there's a growing understanding that computer simulations can replace physical mockups of plants the same way they've replaced mockups of products," Mr. Charles said. Apparently, IBM agrees; it is marketing Dassault's factory-simulation software in the United States.

Testimonials to factory modeling are easy to find. The engineers who design the automation systems sold by Rockwell International have used simulation to design assembly lines that can handle different-sized items, enabling factories to produce small batches of products cost-effectively. "We're headed to where modeling will let us design plants that can efficiently build a single-lot size of one item," said Randall L. Freeman, vice president for global marketing at Rockwell Automation.

Computer simulations of the tread-etching process has enabled tire makers like Goodyear Tire and Rubber to switch production from one type of tire to another in about an hour—a process that previously took an entire work shift. And simulations have shown cookie companies like the Nabisco unit of RJR Nabisco Holdings how to use the same packaging machines to make five-pound bags for price clubs, one-pound bags for grocery stores and six-cookie packs for vending machines.

Various forces are driving the trend toward computer modeling. For one thing, computer technology has finally caught up with manufacturing pipe dreams. "Only recently have computers been powerful enough to quickly simulate what happens if you change something in a chemical reactor," said David E. Waite, Dow's manufacturing manager for information technology.

Economic and marketplace forces are at work, too. Companies that spent much of the 1990's paring ancillary product lines and work forces are now trimming capital investment, lest shareholders think they have lost their cost-cutting touch. Consumers, meanwhile, have grown increasingly picky and expect to be able to choose among myriad colors, sizes and shapes for almost any product. That means that manufacturers must mix and match parts as the orders come in. And that, in turn, means having tools that can respond to electronic commands to switch paint wells, move clamps or change packaging and labels.

Adapted from "Digital Polish for Factory Floors; Software Simulations Head to Better Assembly Lines," by Claudia H. Deutsch, *The New York Times*, March 22, 1999.

i. *Cellular manufacturing.* Cellular manufacturing involves workstations, which are *manufacturing cells* usually containing several machines and with a central robot, each performing a different operation on the part.

j. *Flexible manufacturing systems (FMS)* integrate manufacturing cells into a large unit, all interfaced with a central computer. Flexible manufacturing systems have the highest level of efficiency, sophistication, and productivity in manufacturing. Although costly, they are capable of producing parts randomly and changing manufacturing sequences on different parts quickly, thus they can meet rapid changes in market demand for various types of products.

k. *Expert systems*, which are basically intelligent computer programs, are being developed rapidly with capabilities to perform tasks and solve difficult real-life problems as human experts would.

l. *Artificial intelligence (AI)* involves the use of machines and computers to replace human intelligence. Computer-controlled systems are becoming capable of learning from experience and making decisions that optimize operations and minimize costs. *Artificial neural networks*, which are designed to simulate the thought processes of the human brain, have the capability of modeling and simulating production facilities, monitoring and controlling manufacturing processes, diagnosing problems in machine performance, conducting financial planning, and managing a company's manufacturing strategy.

■ 10.19 SHARED MANUFACTURING

Although large corporations can afford to implement modern technology and take risks, smaller companies generally have difficulty in doing so with their limited personnel, resources, and capital. More recently, the concept of *shared manufacturing* has been proposed. This consists of a regional or nationwide network of manufacturing facilities with state-of-the-art equipment for training, prototype development and small-scale production runs, and is available to help small companies develop products that compete in the global marketplace.

In view of these advances and their potential, some experts have envisaged the factory of the future. Although highly controversial and viewed as unrealistic by some, this is a system in which production will take place with little or no direct human intervention. The human role is expected to be confined to supervision, maintenance, and upgrading of machines, computers, and software.

■ KEY WORDS

ADAPTIVE CONTROL	GROUP TECHNOLOGY
ARTIFICIAL INTELLIGENCE	INDUSTRIAL ROBOTS
ARTIFICIAL NEURAL NETWORKS	JOINING
AUTOMATED ASSEMBLY	JUST-IN-TIME PRODUCTION
AUTOMATED MATERIALS HANDLING	LIFE CYCLE ENGINEERING
CASTING	MACHINING
CELLULAR MANUFACTURING	MANUFACTURING ENGINEERING
COMPUTER-AIDED DESIGN, ENGINEERING, AND MANUFACTURING	MICROKNIVES
	MICROROBOTS
COMPUTER-AIDED PROCESS PLANNING	NANOFABRICATION
COMPUTER-INTEGRATED MANUFACTURING	NANOTECHNOLOGY
COMPUTER NUMERICAL CONTROL	NET-SHAPE MANUFACTURING
CONCURRENT ENGINEERING	PRODUCIBILITY
DESIGN FOR MANUFACTURE, ASSEMBLY, DISASSEMBLY, AND SERVICE	PRODUCT COST
	PROTOYTPES
DIMENSIONAL ACCURACY	RAPID PROTOTYPING
EXPERT SYSTEMS	ROBOTIC ASSEMBLY
FINISHING	SCALE
FIXED GAGES	SHAPING
FLEXIBLE MANUFACTURING SYSTEMS	SHARED MANUFACTURING
FORMING	STEEL RULE
	ULTRAPRECISION MANUFACTURING

■ CHAPTER SUMMARY

- Modern manufacturing involves product design, selection of materials, and selection of processes. The process of transforming raw materials into a finished product is called the manufacturing process.

- The design process requires a clear understanding of the functions and performance expected of that product.

- Concurrent engineering integrates the design process with production to optimize the life cycle of the product.

- Computer-aided design, engineering, and manufacturing are used to construct and study models (prototypes) allowing the designer to conceptualize objects more easily and more cost efficiently.

- The selection of appropriate materials is key to successful product development.

- Manufacturing processing methods have changed dramatically over the last few decades. More cost and time efficient processes can be implemented using computer-integrated manufacturing.

■ REVIEW QUESTIONS

1. List the three important phases in the manufacturing process.

2. Define concurrent engineering and explain how it can be used to enhance the design and manufacturing process.

3. Explain the benefits of rapid prototyping.

4. List four types of materials used in manufacturing today.

5. List the five broad categories of manufacturing processing.

6. Give at least two examples of nanotechnology.

7. List four types of measuring devices.

8. Give three consequences of improper selection of materials and processes.

9. List four application of computer-integrated manufacturing.

CHAPTER 11

DIMENSIONING

OBJECTIVES

After studying the material in this chapter, you should be able to:

1. Use conventional dimensioning techniques to describe size and shape accurately on an engineering drawing.

2. Create and read a drawing at a specified scale.

3. Create drawings using metric, engineering, and architect scales.

4. Correctly place dimension lines, extension lines, angles, and notes.

5. Recognize aligned and unidirectional dimensioning systems.

6. Dimension circles, arcs, and inclined surfaces.

7. Apply finish symbols and notes to a drawing.

OVERVIEW

We have all heard of the "rule of thumb." Actually, at one time an inch was defined as the width of a thumb, and a foot was simply the length of a man's foot. In old England, an inch used to be "three barley corns, round and dry." In the time of Noah and the Ark, the *cubit* was the length of a man's forearm, or about 18".

In 1791, France adopted the *meter** (1 meter = 39.37"; 1" = 25.4 mm), from which the decimalized metric system evolved. In the meantime England was setting up a more accurate determination of the *yard*, which was legally defined in 1824 by act of Parliament. A foot was $\frac{1}{3}$ yard, and an inch was $\frac{1}{36}$ yard. From these specifications, graduated rulers, scales, and many types of measuring devices have been developed to achieve even more accuracy of measurement and inspection.

Until this century, common fractions were considered adequate for dimensions. Then, as designs became more complicated and it became necessary to have interchangeable parts to support mass production, more accurate specifications were required, and it became necessary to turn to the decimal-inch system or to the SI system (see §§11.9 and 11.10).

Graphical entities on engineering drawings describe shape and position. Dimensions and notes describe size and necessary manufacturing processes if the engineering drawing is to be a complete instruction for the shop technician. Objects in engineering drawings are created to scale. Large objects are drawn smaller than life size; small objects are drawn larger than life size. The ratio of drawn size to life size is the **scale**. Dimensions describe the size and location of features of an object. The correct placement of dimensions is strictly prescribed by an extensive list of drawing conventions. CAD can accurately measure the length of a drawn line, but it takes a person skilled in drawing practices to place dimensions correctly so that their interpretation is clear and unambiguous.

■ 11.1 INTERNATIONAL SYSTEM OF UNITS

The current rapid growth of worldwide science and commerce has fostered an international system of units (SI) suitable for measurements in the physical and biological sciences and in engineering. The seven basic units of measurement are the meter (length), kilogram (mass), second (time), ampere (electric current), kelvin (thermodynamic temperature), mole (amount of substance), and candela (luminous intensity).

The SI system is gradually coming into use in the United States, especially by the many multinational companies in the chemical, electronic, and mechanical industries. A tremendous effort is now under way to convert all standards of the American National Standards Institute (ANSI) to the SI units in conformity with the International Standards Organization (ISO) standards.

See inside the front cover of this book for a table converting fractional inches to decimal inches or millimeters. For the International System of Units and their United States equivalents, see Appendix 31.

■ 11.2 SIZE DESCRIPTION

In addition to a complete *shape description* of an object, as discussed in previous chapters, a drawing of the design must also give a complete *size description*; that is, it must be *dimensioned* (see ANSI/ASME Y14.5M–1994).

The need for *interchangeability* of parts is the basis for the development of modern methods of size description. Drawings today must be dimensioned so that production personnel in widely separated places can make mating parts that will fit properly when brought together for final assembly or when used as repair or replacement parts by the customer (§11.26).

The increasing need for precision manufacturing and the necessity of controlling sizes for interchangeability has shifted responsibility for size control to the designing engineer and the drafter. The production worker no longer exercises judgment in engineering matters, but only in the proper execution of instructions given on the drawings. Therefore, engineers and designers should be familiar with materials and methods of construction and with production requirements. Engineering students and designers should seize every opportunity to become familiar with the fundamental manufacturing processes.

A drawing submitted to production should show the object in its completed condition and should contain all necessary information to bring it to that final state. Therefore, in dimensioning a drawing, the de-

---

* In the SI system the meter is now defined as a length equal to the distance traveled by light of a certain wavelength in a vacuum during a time interval of 1/299,792,458 second.

signer and the drafter should keep in mind the finished piece, the production processes required, and, above all, the function of the part in the total assembly. Whenever possible—that is, when there is no conflict with functional dimensioning (§11.4)—dimensions should be given that are convenient for the individual worker or the production engineer. These dimensions should be given so that it will not be necessary to scale or assume any dimensions. No dimensions should be specified for points or surfaces that are not accessible to the worker.

Dimensions should not be duplicated or superfluous (§11.30). Only those needed to produce and inspect the part against the design specifications should be given. Students often mistakenly give the dimensions used *to make the drawing*. These are not necessarily the dimensions required. There is much more to the theory of dimensioning, as we will see.

■ 11.3 SCALE OF DRAWING

Drawings should be made to scale, and the scale should be indicated in the title block even though the worker is never expected to scale the drawing or print for a needed dimension. (See §3.31 for indications of scales.)

A heavy straight line should be drawn under any dimension that is not to scale (see Fig. 11.15), or the abbreviation *NTS* (not to scale) should be indicated. This procedure may be necessary when a change made in a drawing is not important enough to justify making an entirely new drawing.

When a drawing is prepared on a CAD system, agreement should be maintained between the defining dimensions, true size, location, and direction in all views of the object represented.

Many preprinted drawing borders include a note such as "Do not scale drawing for dimensions."

■ 11.4 LEARNING TO DIMENSION

Dimensions are given in the form of linear distances, angles, or notes regardless of the dimensioning units being used. The ability to dimension properly in millimeters, decimal inch, or fractional inch requires the following.

1. The student must learn the *technique of dimensioning*: the character of the lines, the spacing of dimensions, the making of arrowheads, and so forth. A typical dimensioned drawing is shown in Fig. 11.1.

The Increasing Complexity of Modern Designs Requires the Production of Precisely Dimensioned Drawings. *Courtesy of Diamond Star Motors.*

Note the strong contrast between the visible lines of the object and the thin lines used for the dimensions.

2. The student must learn the rules of *placement of dimensions* on the drawing. These practices assure a logical and practical arrangement with maximum legibility.

3. The student should learn the *choice of dimensions*. Formerly, manufacturing processes were considered the governing factor in dimensioning. Now function is considered first and the manufacturing processes second. The proper procedure is to dimension tentatively for function and then review the dimensioning to see if any improvements from the standpoint of production can be made without adversely affecting the functional dimensioning. A "geometric breakdown" (§11.20) will assist the beginner in selecting dimensions. In most cases dimensions thus determined will be functional, but this method should be accompanied by a logical analysis of the functional requirements.

■ 11.5 LINES USED IN DIMENSIONING

A **dimension line** (Fig. 11.2a) is a thin, dark, solid line terminated by arrowheads, which indicates the direction and extent of a dimension. In machine drawing, the dimension line is broken, usually near the middle, to provide an open space for the dimension figure. In structural and architectural drawing, it is customary to place the dimension figure above an unbroken dimension line.

As shown in Fig. 11.2b, the dimension line nearest the object outline should be spaced at least 10 mm ($\frac{3}{8}''$)

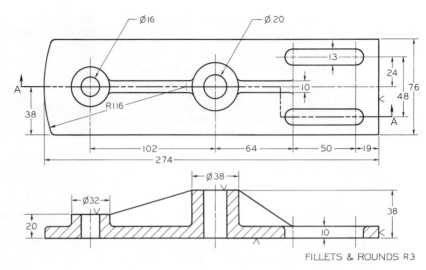

FIGURE 11.1   Dimensioning Technique. Dimensions in millimeters.

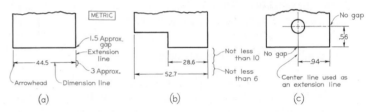

FIGURE 11.2   Dimensioning Technique.

away. All other parallel dimension lines should be at least 6 mm ($\frac{1}{4}''$) apart, and more if space is available. *The spacing of dimension lines should be uniform throughout the drawing.*

An **extension line** (Fig. 11.2a) is a thin, dark, solid line that "extends" from a point on the drawing to which a dimension refers. The dimension line meets the extension lines at right angles except in special cases (see Fig. 11.6a). A gap of about 1.5 mm ($\frac{1}{16}''$) should be left where the extension line would join the object outline. The extension line should extend about 3 mm ($\frac{1}{8}''$) beyond the outermost arrowhead (Figs. 11.2a and 11.2b).

The foregoing dimensions for lettering height, spacing, and so on should be increased approximately 50% for drawings that are to be microfilmed and then reduced to one-half size for the working print.

Otherwise the lettering and dimensioning often are not legible.

A **center line** is a thin, dark line composed of alternate long and short dashes and used to represent axes of symmetrical parts and to denote centers. As shown in Fig. 11.2c, center lines are commonly used as extension lines in locating holes and other features. When so used, the center line crosses over other lines of the drawing without gaps. A center line should always end in a long dash.

■ 11.6 PLACEMENT OF DIMENSION AND EXTENSION LINES

The correct placement of dimension lines and extension lines is shown in Fig. 11.3a. The shorter dimensions are nearest to the object outline. Dimension lines

FIGURE 11.3   Dimension and Extension Lines.

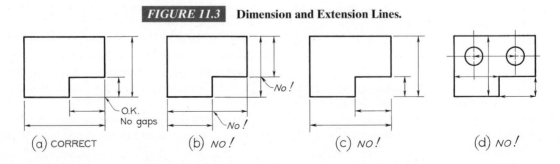

(a) CORRECT   (b) *NO!*   (c) *NO!*   (d) *NO!*

should not cross extension lines, as in Fig. 11.3b, which results from placing the shorter dimensions outside. Note that it is perfectly satisfactory to cross extension lines (Fig. 11.3a), but they should never be shortened (Fig. 11.3c). A dimension line should never coincide with, or form a continuation of, any line of the drawing (Fig. 11.3d). Avoid crossing dimension lines wherever possible.

Dimensions should be lined up and grouped together as much as possible, as in Fig. 11.4a, and not as in Fig. 11.4b.

In many cases, extension lines and center lines must cross visible lines of the object (Fig. 11.5a). When this occurs, gaps should not be left in the lines (Fig. 11.5b).

Dimension lines are normally drawn at right angles to extension lines, but an exception may be made in the interest of clarity, as in Fig. 11.6a. In crowded conditions, gaps may be left in extension lines near arrowheads so that the dimensions show clearly (Fig. 11.6b). In general, avoid dimensioning to hidden lines (Fig. 11.6c).

■ 11.7 ARROWHEADS

Arrowheads (Fig. 11.7) indicate the extent of dimensions. They should be uniform in size and style throughout the drawing and not varied according to the size of the drawing or the length of dimensions. Arrowheads should be drawn freehand, and the length and width should be in a ratio of 3:1. The length of the arrowhead should be equal to the height of the dimension whole numbers. For average use, make arrowheads about 3 mm ($\frac{1}{8}''$) long and very narrow (Fig. 11.7a). Use strokes toward the point or away from the point desired (Figs. 11.7b to 11.7d). The method in Fig. 11.7b is easier when the strokes are drawn toward the drafter. For best appearance, fill in the arrowhead, as in Fig. 11.7d.

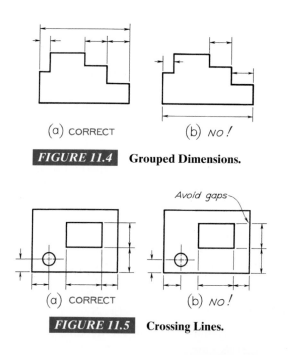

FIGURE 11.4   **Grouped Dimensions.**

FIGURE 11.5   **Crossing Lines.**

FIGURE 11.6   **Placement of Dimensions.**

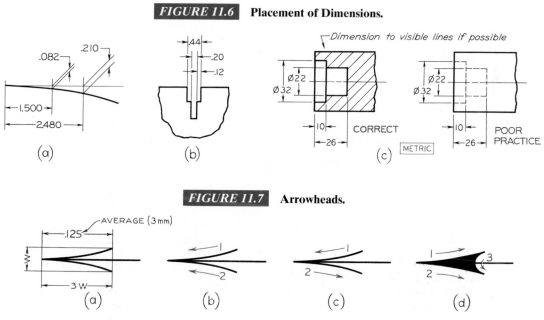

FIGURE 11.7   **Arrowheads.**

■ 11.8 LEADERS

A **leader** (Fig. 11.8) is a thin, solid line leading from a note or dimension and terminating in an arrowhead or a dot touching the part to which attention is directed. Arrowheads should always terminate on a line, such as the edge of a hole; dots should be within the outline of the object. A leader should generally be an inclined straight line, if possible, except for the short horizontal shoulder (6 mm or $\frac{1}{4}''$, approx.) extending from mid-height of the lettering *at the beginning or end of a note.*

A leader to a circle should be radial; that is, if extended, it would pass through the center. A drawing presents a more pleasing appearance if leaders near each other are drawn parallel. Leaders should cross as few lines as possible and should never cross each other. They should not be drawn parallel to nearby lines of the drawing, allowed to pass through a corner of the view, made unnecessarily long, or drawn horizontally or vertically on the sheet. A leader should be drawn at a large angle and terminate with the appropriate arrowhead, or with a dot (Fig. 11.8f).

■ 11.9 FRACTIONAL, DECIMAL, AND METRIC DIMENSIONS

In the early days of machine manufacturing in the United States, workers would scale the undimensioned design drawing to obtain any needed dimensions, and it was their responsibility to see that the parts fit together properly. Workers were skilled and very accurate, and excellent fits were obtained. Hand-built machines were often beautiful examples of precision craftsmanship.

The system of units and common fractions is still used in architectural and structural work, in which close accuracy is relatively unimportant and the steel tape or framing square is used to set off measurements. Architectural and structural drawings are therefore often dimensioned in this manner. Also, certain commercial commodities, such as pipe and lumber, are identified by standard nominal designations that are close approximations of actual dimensions.

As industry has progressed, there has been greater and greater demand for more accurate specifications of the important functional dimensions—more accurate than the $\frac{1}{64}''$ permitted by the engineers', architects', and machinists' scale. Since it was cumbersome to use still smaller fractions, such as $\frac{1}{128}$ or $\frac{1}{256}$, it became the practice to give decimal dimensions, such as 4.2340 and 3.815, for the dimensions requiring accuracy. However, some dimensions, such as standard nominal sizes of materials, punched holes, drilled holes, threads, keyways, and other features produced by tools that are so designated are still expressed in whole numbers and common fractions.

Thus, drawings may be dimensioned entirely with whole numbers and common fractions, or entirely with decimals, or with a combination of the two. However, more recent practice adopted the decimal-inch system, and current practice also utilizes the metric system as recommended by ANSI. Millimeters and inches in the decimal form can be added, subtracted, multiplied, and divided more easily than can fractions.

For inch-millimeter equivalents of decimal and common fractions, see inside the front cover of this book. For additional metric equivalents, see Appendix 31.

■ 11.10 DECIMAL SYSTEMS

A decimal system based on the decimal inch or the millimeter as a linear unit of measure has many advantages and is compatible with most measuring devices and machine tools. Metric measurement is based on the meter as a linear unit of measure, but the millimeter is used on most engineering drawings. To facilitate the changeover to metric dimensions, many drawings are dual-dimensioned in millimeters and decimal inches (see §11.11).

Complete decimal dimensioning employs decimals for all dimensions and designations except where cer-

FIGURE 11.8 Leaders.

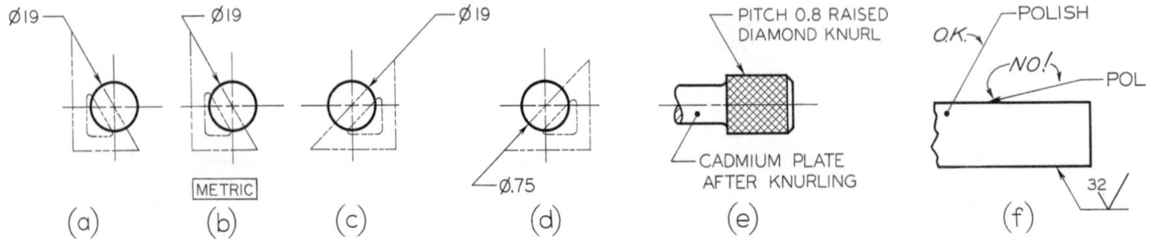

tain commercial commodities, such as pipe and lumber, are identified by standardized nominal designations. *Combination dimensioning* employs decimals for all dimensions except the designations of nominal sizes of parts or features, such as bolts, screw threads, keyseats, or other standardized fractional designations (ANSI/ASME Y14.5M–1994).

In these systems, two-place inch or one-place millimeter decimals are used when a common fraction has been regarded as sufficiently accurate.

In the combination dimensioning system, common fractions may be used to indicate nominal sizes of materials, drilled holes, punched holes, threads, keyways, and other standard features.

One-place millimeter decimals are used when tolerance limits of ±0.1 mm or more can be permitted. Two (or more)-place millimeter decimals are used for tolerance limits less than ±0.1 mm. Fractions are considered to have the same tolerance as two-place decimal-inch dimensions when determining the number of places to retain in the conversion to millimeters. Keep in mind that 0.1 mm is approximately equal to .004 inch.

Two-place inch decimals are used when tolerance limits of ±.01″ or more can be permitted. Three or more decimal places are used for tolerance limits less than ±.01″. In two-place decimals, the second place preferably should be an even digit (for example, .02, .04, and .06 are preferred to .01, .03, or .05) so that

when the dimension is divided by 2, as is necessary in determining the radius from a diameter, the result will be a decimal of two places. However, odd two-place decimals are used when required for design purposes, such as in dimensioning points on a smooth curve or when strength or clearance is a factor.

A typical example of the use of the complete decimal-inch system is shown in Fig. 11.9. The use of the preferred decimal-millimeter system is shown in Fig. 11.10.

When a decimal value is to be rounded off to fewer places than the calculated number, regardless of the unit of measurement involved, the method prescribed is as follows.

The last figure to be retained should not be changed when the figure beyond the last figure to be retained is less than 5.

EXAMPLE 3.46325, if rounded off to three places, should be 3.463.

The last figure to be retained should be increased by 1 when the figure beyond the last figure to be retained is greater than 5 (or two figures are greater than 50).

EXAMPLE 8.37652, if rounded off to three places, should be 8.377.

The last figure to be retained should be unchanged if it is even, or increased by 1 if odd, when followed by exactly 5.

FIGURE 11.9 **Complete Decimal Dimensioning.**

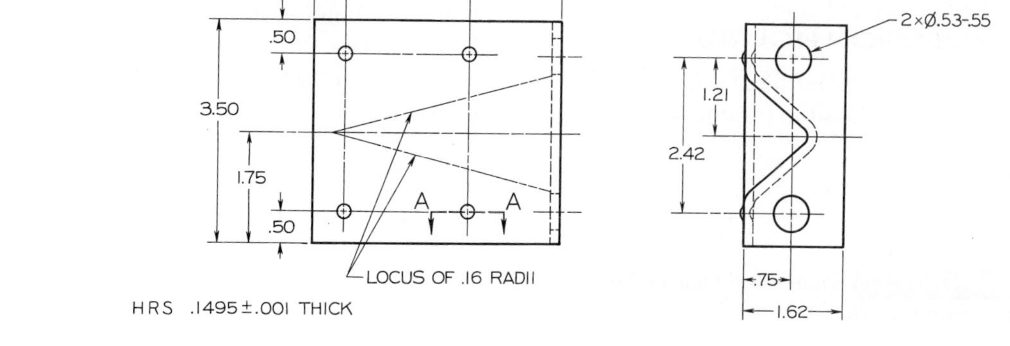

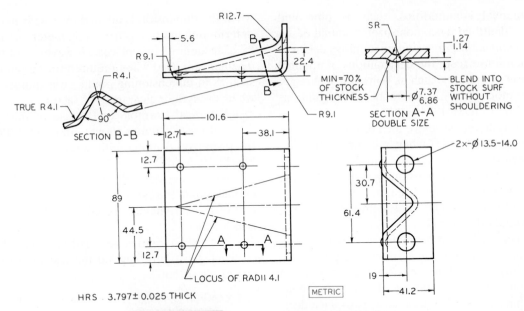

HRS 3.797± 0.025 THICK

FIGURE 11.10   **Complete Metric Dimensioning.**

EXAMPLE 4.365 becomes 4.36 when rounded off to two places. Also, 4.355 becomes 4.36 when cut off to two places.

The use of the metric system means not only a changeover of measuring equipment but also a changeover in thinking on the part of drafters and designers. They must stop thinking in terms of inches and common fractions and think in terms of millimeters and other SI units. Dimensioning practices remain essentially the same; only the units are changed (compare Figs. 11.9 and 11.10).

Shop scales and drafting scales for use in the decimal-inch and metric systems are available in a variety of forms. Refer to the inside of the front cover for a two-, three-, and four-place decimal equivalent table.

Once the metric system is installed, the advantages in computation, in checking, and in simplified dimensioning techniques are considerable.

■ 11.11 DIMENSION FIGURES

The importance of good lettering of dimension figures cannot be overstated. The shop produces according to

the directions on the drawing, and to save time and prevent costly mistakes, all lettering should be perfectly legible. A complete discussion of numerals is given in §§4.18–4.20.

Legibility should never be sacrificed by crowding dimension figures into limited spaces. For every such case there is a practical and effective method, as shown in Fig. 11.11. In Fig. 11.11a, there is only enough room for the figure, and the arrowheads are placed outside. In Fig. 11.11b, both the arrowheads and the figure are placed outside. Other methods are shown in Figs. 11.11c and 11.11d.

If necessary, a removed partial view may be drawn to an enlarged scale to provide the space needed for clear dimensioning (Fig. 7.22).

Make all decimal points bold, allowing ample space. Where the metric dimension is a whole number, neither a decimal point nor a zero is given (Figs. 11.11a and 11.11d). Where the metric dimension is less than 1 millimeter, a zero precedes the decimal point (Fig. 11.11b). Where the dimension exceeds a whole number by a fraction of 1 millimeter, the last digit to the right of the decimal point is not followed by a zero

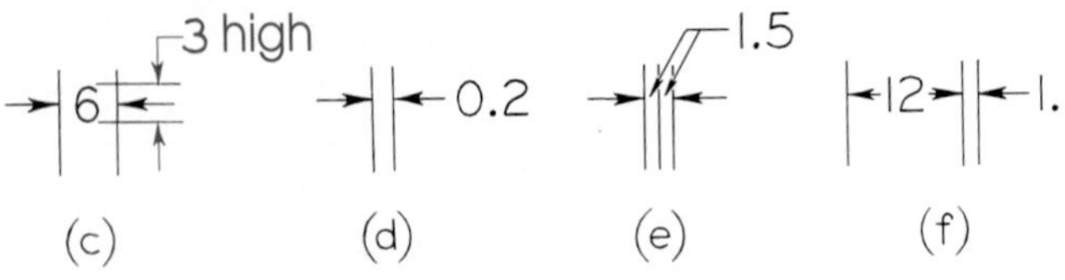

FIGURE 11.11 **Dimension Figures. Metric dimensions (c)–(f).**

(c)   (d)   (e)   (f)

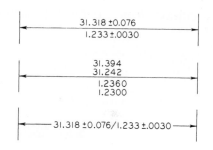

FIGURE 11.12 Decimal Dimension figures. Metric dimensions (a)–(d).

except when expressing tolerances (Figs. 11.12c and 11.12d).

Methods of lettering and displaying decimal dimension figures are shown in Fig. 11.11. Where the decimal-inch dimension is used on drawings, a zero is not used before the decimal point of values less than 1 inch (Figs. 11.12f to 11.12j). The decimal-inch dimension is expressed to the same number of decimal places as its tolerance. Thus, zeros are added to the right of the decimal point as necessary (Fig. 11.12e).

Never letter a dimension figure over any line on the drawing, but break the line if necessary. Place dimension figures outside a sectioned area if possible (Fig. 11.13a). When a dimension must be placed on a sectioned area, leave an opening in the section lining for the dimension figure (Fig. 11.13b).

In a group of parallel dimension lines, the numerals should be staggered, as in Fig. 11.14a, and not stacked up one above the other, as in Fig. 11.14b.

DUAL DIMENSIONING

Dual dimensioning is used to show metric and decimal-inch dimensions on the same drawing. Two methods of displaying the dual dimensions are as follows.

POSITION METHOD The millimeter value is placed above the inch dimension and is separated by a dimension line or an added line for some dimensions when the unidirectional system of dimensioning is used. An alternative arrangement in a single line places the millimeter dimension to the left of the inch dimension, separated by a slash line (virgule). Each drawing should illustrate the dimension identification as $\frac{\text{MILLIMETER}}{\text{INCH}}$ or MILLIMETER/INCH. (Placement of the inch dimension above or to the left of the millimeter is also acceptable.)

EXAMPLES

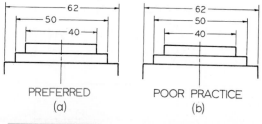

FIGURE 11.13 Dimensions and Section Lines. Metric.

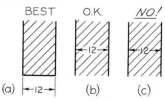

FIGURE 11.14 Staggered Numerals. Metric.

BRACKET METHOD   The millimeter dimension is enclosed in square brackets, [ ]. The location of this dimension is optional but should be uniform on any drawing—that is, above or below or to the left or right of the inch dimension. Each drawing should include a note to identify the dimension values as DIMENSIONS IN [ ] ARE MILLIMETERS.

EXAMPLES

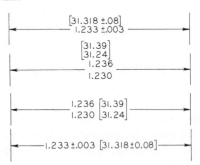

When converting a decimal-inch dimension to millimeters, multiply the inch dimension by 25.4 and round off to one less digit to the right of the decimal point than for the inch value (see §11.10). When converting a millimeter dimension to inches, divide the millimeter dimension by 25.4 and round off to one more digit to the right of the decimal point than for the millimeter value.

■ 11.12 DIRECTION OF DIMENSION FIGURES

Two systems of reading direction for dimension figures are available. In the preferred *unidirectional system*, approved by ANSI (Fig. 11.15a), all dimension figures and notes are lettered horizontally on the sheet and are read from the bottom of the drawing. The unidirectional system has been extensively adopted in the aircraft, automotive, and other industries because it is easier to use and read, especially on large drawings. In the *aligned system* (Fig. 11.15b), all dimension figures are aligned with the dimension lines so that they may be read from the right side of the sheet. Dimension lines in this system should not run in the directions included in the shaded area of Fig. 11.16, if avoidable.

In both systems, dimensions and notes shown with leaders are aligned with the bottom of the drawing. Notes without leaders should also be aligned with the bottom of the drawing.

■ 11.13 MILLIMETERS AND INCHES

Millimeters are indicated by the lowercase letters mm placed one space to the right of the numeral; thus, 12.5 mm. *Meters* are indicated by the lowercase m placed similarly; thus, 50.6 m. *Inches* are indicated by the symbol ″ placed slightly above and to the right of the numeral; thus, $2\frac{1}{2}$″. *Feet* are indicated by the symbol ′ similarly placed; thus, $3′-0, 5′-6, 10′-0\frac{1}{4}$. It is customary in such expressions to omit the inch marks.

It is standard practice to omit mm designations and inch marks on a drawing except when there is a possibility of misunderstanding. For example, 1 VALVE should be 1″ VALVE, and 1 DRILL should be 1″ DRILL or 1 mm DRILL. Where some inch dimensions are shown on a millimeter-dimensioned drawing, the abbreviation IN. follows the inch values.

In some industries, all dimensions, regardless of size, are given in inches; in others, dimensions up to 72″ inclusive are given in inches, and those greater are given in feet and inches. In structural and architectural

FIGURE 11.15   **Directions of Dimension Figures.**

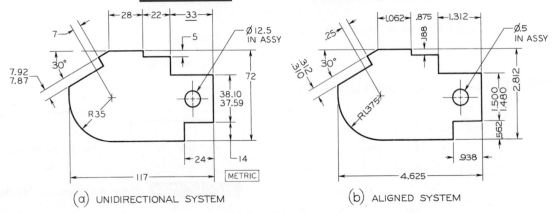

(a) UNIDIRECTIONAL SYSTEM

(b) ALIGNED SYSTEM

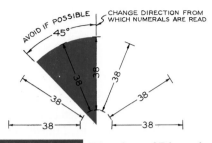

FIGURE 11.16 Directions of Dimensions.

drafting, all dimensions of 1′ or over are usually expressed in feet and inches.

If suitable, the drawing should contain a note stating "Unless otherwise specified, all dimensions are in millimeters" (or in inches, as applicable).

■ 11.14 DIMENSIONING ANGLES

Angles are dimensioned preferably by means of an angle in degrees and a linear dimension (Fig. 11.17a), or by means of coordinate dimensions of the two legs of a right triangle (Fig. 11.17b). The coordinate method is more suitable for work requiring a high degree of accuracy. Variations of angle (in degrees) are hard to control because the amount of variation increases with the distance from the vertex of the angle. Methods of indicating various angles are shown in Figs. 11.17c to 11.17f. Tolerances of angles are discussed in §12.18.

When degrees alone are indicated, the symbol ° is used. When minutes alone are given, the number should be preceded by 0°; thus 0°23′. In all cases,

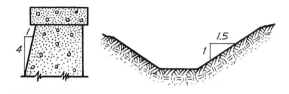

FIGURE 11.18 Angles in Civil Engineering Projects.

whether in the unidirectional system or in the aligned system, the dimension figures for angles are lettered on horizontal guide lines. (For a general discussion of angles, see §4.2.)

In civil engineering drawings, **slope** represents the angle with the horizontal, whereas **batter** is the angle referred to the vertical. Both are expressed by making one member of the ratio equal to 1, as shown in Fig. 11.18. **Grade**, as of a highway, is similar to slope but is expressed in percentage of rise per 100′ of run. Thus a 20′ rise in a 100′ run is a grade of 20%.

In structural drawings, angular measurements are made by giving the ratio of "run" to "rise," with the larger size being 12″. These right triangles are referred to as **bevels**.

■ 11.15 DIMENSIONING ARCS

A circular arc is dimensioned in the view in which its true shape is shown by giving the numerical value of its radius preceded by the abbreviation R (Fig. 11.19). The centers may be indicated by small crosses to clarify the drawing but not for small or unimportant radii.

FIGURE 11.17 Angles.

FIGURE 11.19 Dimensioning Arcs.

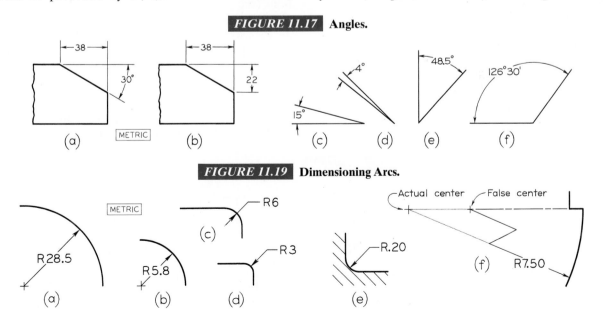

Crosses should not be shown for undimensioned arcs. As shown in Figs. 11.19a and 11.19b, when there is room enough, both the numeral and the arrowhead are placed inside the arc. In Fig. 11.19c, the arrowhead is left inside, but the numeral had to be moved outside. In Fig. 11.19d, both the arrowhead and the numeral had to be moved outside. Figure 11.19e shows an alternate method that can be used when section lines or other lines are in the way. Note that in the unidirectional system, all of these numerals are lettered horizontally on the sheet.

For a long radius (Fig. 11.19f), when the center falls outside the available space, the dimension line is drawn toward the actual center; but a false center may be indicated and dimension line "jogged" to it, as shown.

■ 11.16 FILLETS AND ROUNDS

Individual fillets and rounds are dimensioned as any arc (Figs. 11.19c to 11.19e). If there are only a few and they are obviously the same size, as in Fig. 11.43e, one typical radius is sufficient. However, fillets and rounds are often quite numerous on a drawing, and most of them are likely to be some standard size, as R3 and R6 when dimensioning in metric or R125 and R250 when using the decimal-inch system.

In such cases it is customary to give a note in the lower portion of the drawing to cover all uniform fillets and rounds; thus,

FILLETS R6 AND ROUNDS R3 UNLESS
OTHERWISE SPECIFIED

or

ALL CASTING RADII R6 UNLESS NOTED

or simply

ALL FILLETS AND ROUNDS R6

For a discussion of fillets and rounds in the shop, see §11.5.

■ 11.17 FINISH MARKS

A **finish mark** is used to indicate that a surface is to be machined, or finished, as on a rough casting or forging. To the patternmaker or diemaker, a finish mark means that allowance of extra metal in the rough workpiece must be provided for the machining (see §11.4). On drawings of parts to be machined from rolled stock, finish marks are generally unnecessary, for it is obvious that the surfaces are finished. Similarly, it is not necessary to show finish marks when an operation is specified in a note that indicates machining, such as drilling, reaming, boring, countersinking, counterboring, and broaching, or when the dimension implies a finished surface, such as Ø6.22–6.35 (metric) or Ø2.45–2.50 (decimal-inch).

Three styles of finish marks, the general ∨ symbol, the new basic ✓ symbol, and the traditional ✗ symbol, are used to indicate an ordinary smooth machined surface. The ∨ symbol is like a capital V, made about 3 mm ($\frac{1}{8}$″) high in conformity with the height of dimensioning lettering (Fig. 11.20a). The extended ✓ symbol, preferred by ANSI, is like a larger capital with the right leg extended (Fig. 11.20b). The short leg is made about 5 mm ($\frac{3}{16}$″) high and the height of the long leg is about 10 mm ($\frac{3}{8}$″). The basic symbol may be altered for more elaborate surface texture specifications (see §12.22).

For best results, all finished marks should be drawn with the aid of a template or the 30° × 60° triangle. The point of the ∨ symbol should be directed inward toward the body of metal in a manner similar to that of a tool bit. The ✓ symbol is not shown upside down (see Figs. 11.35 and 12.46).

The preferred form and placement for the ✗ symbol are shown in Fig. 11.20e. The ✗ symbol is in

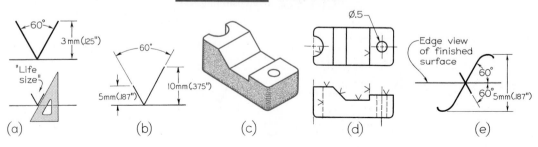

FIGURE 11.20   Finish Marks.

(a)   (b)   (c)   (d)   (e)

limited use and found mainly on drawings made in accordance with earlier drafting standards.

Figure 11.20c shows a simple casting having several finished surfaces; in Fig. 11.20d, two views of the same casting show how the finish marks are indicated on a drawing. *The finish mark is shown only on the edge view of a finished surface and is repeated in any other view in which the surface appears as a line, even if the line is a hidden line.*

The several kinds of finishes are detailed in machine shop practice manuals. The following terms are among the most commonly used: *finish all over, rough finish, file finish, sand blast, pickle, scrape, lap, hone, grind, polish, burnish, buff, chip, spotface, countersink, counterbore, core, drill, ream, bore, tap, broach,* and *knurl.* When it is necessary to control the surface texture of finished surfaces beyond that of an ordinary machine finish, the $\checkmark$ symbol is used as a base for the more elaborate surface quality symbols discussed in §12.22.

If a part is to be finished all over, finish marks should be omitted, and a general note, such as FINISH ALL OVER or FAO, should be lettered on the lower portion of the sheet.

■ 11.18 DIMENSIONS ON OR OFF VIEWS

Dimensions should not be placed on a view unless doing so promotes the clearness of the drawing. The ideal form is shown in Fig. 11.21a, in which all dimensions are placed outside the view. Compare this with the evidently poor practice in Fig. 11.21b. This is not to say that a dimension should never be placed on a view, for in many cases (particularly in complicated drawings) this is necessary (Fig. 11.21c). Certain radii and other dimensions are given on the views, but in each case investigation will reveal a good reason for placing the di-

mension on the view. *Place dimensions outside of views, except where directness of application and clarity are gained by placing them on the views, where they will be closer to the features dimensioned.* When a dimension must be placed in a sectioned area or on the view, leave an opening in the sectioned area or a break in the lines for the dimension figures (see Figs. 11.13b and 11.21c).

■ 11.19 CONTOUR DIMENSIONING

Views are drawn to describe the shapes of the various features of the object, and dimensions are given to define exact sizes and locations of those shapes. It follows that *dimensions should be given where the shapes are shown*—that is, in the views where the contours are delineated, as in Fig. 11.22a. Incorrect placement of the dimensions is shown in Fig. 11.22b.

Attachment of individual dimensions directly to the contours that show the shapes being dimensioned automatically prevents the attachment of dimensions to hidden lines, as shown for the depth 10 of the slot in Fig. 11.22b. It also prevents the attachment of dimensions to a visible line, the meaning of which is not clear in a particular view, such as dimension 20 for the height of the base in Fig. 11.22b.

Although the placement of notes for holes follows the contour rule wherever possible (Fig. 11.22a), the diameter of an external cylindrical shape is preferably given in the rectangular view, where it can be readily found near the dimension for the length of the cylinder (see Figs. 11.23b, 11.26, and 11.27).

■ 11.20 GEOMETRIC BREAKDOWN

Engineering structures are composed largely of simple geometric shapes, such as the prism, cylinder, pyramid,

FIGURE 11.21 Dimensions On or Off the Views.

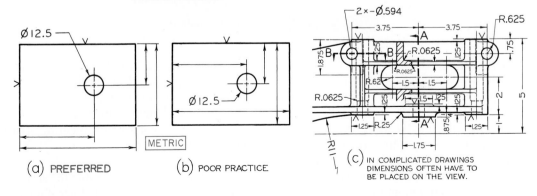

(a) PREFERRED   (b) POOR PRACTICE   (c) IN COMPLICATED DRAWINGS DIMENSIONS OFTEN HAVE TO BE PLACED ON THE VIEW.

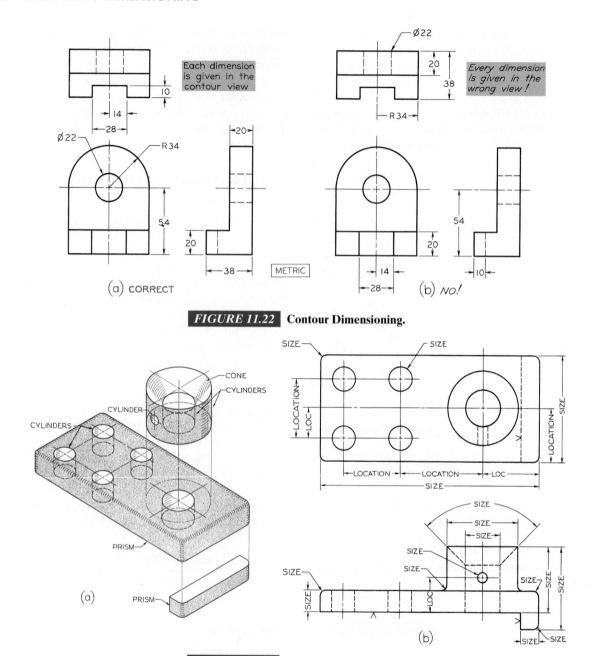

FIGURE 11.22 Contour Dimensioning.

FIGURE 11.23 Geometric Breakdown.

cone, and sphere (Fig. 11.23a). They may be exterior (positive) or interior (negative) forms. For example, a steel shaft is a positive cylinder, and a round hole is a negative cylinder.

These shapes result directly from the design necessity of keeping forms as simple as possible and from the requirements of the fundamental manufacturing operations. Forms having plane surfaces are produced by planing, shaping, milling, and so forth,

while forms having cylindrical, conical, or spherical surfaces are produced by turning, drilling, reaming, boring, countersinking, and other rotary operations (see Chapter 10, Manufacturing Design and Processes).

The dimensioning of engineering structures involves two basic steps:

1. Give the dimensions showing the sizes of the simple geometric shapes, called **size dimensions**.

2. Give the dimensions locating these elements with respect to each other, called **location dimensions**.

The process of geometric analysis is very helpful in dimensioning any object, but it must be modified when there is a conflict either with the function of the part in the assembly or with the manufacturing requirements in the shop.

Figure 11.23b is a multiview drawing of the object shown in isometric in Fig. 11.23a. Here it will be seen that each geometric shape is dimensioned with size dimensions and that these shapes are then located with respect to each other with location dimensions. Note that a *location dimension locates a three-dimensional geometric element* and not just a surface; otherwise, all dimensions would have to be classified as location dimensions.

USING CAD TO DIMENSION

Dimensions can be created somewhat automatically using CAD. The size, appearance, and dimension value can all be generated automatically. Alignment of the dimension values and whether to show metric or inch units or both can all be accomplished using CAD. Because of the complexity and intelligence that must be used in selecting which dimensions to place in the drawing, most CAD systems require you to pick what you will dimension and where to place the dimension. Some systems will generate the dimensions totally automatically, but you must review them carefully making sure that the dimensions given describe the intent for your design and the permissible tolerances.

CAD Systems Make Dimensioning Easy and Often Automatic. *Courtesy of SDRC, Milford, OH.*

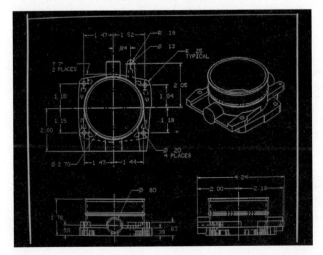

■ 11.21 SIZE DIMENSIONS: PRISMS

The right rectangular prism is probably the most common geometric shape (see Fig. 4.7). Front and top views are dimensioned as shown in Figs. 11.24a and 11.24b. The height and width are given in the front view and the depth in the top view. The vertical dimensions can be placed on the left or right, provided both of them are placed in line. The horizontal dimension applies to both the front and top views and should be placed between them, as shown, and not above the top or below the front view.

Front and side views should be dimensioned as in Figs. 11.24c and 11.24d. The horizontal dimensions can be placed above or below the views, provided both are placed in line. The dimension between views applies to both views and should not be placed elsewhere without a special reason.

An application of size dimensions to a machine part composed entirely of rectangular prisms is shown in Fig. 11.25.

■ 11.22 SIZE DIMENSIONS: CYLINDERS

The right circular cylinder is the next most common geometric shape and is commonly seen as a shaft or a hole. The general method of dimensioning a cylinder is to give both its diameter and its length in the rectangular view (Fig. 11.26). If the cylinder is drawn in a vertical position, its length or altitude may be given at the right, as in Fig. 11.26a, or on the left, as in Fig. 11.26b. If the cylinder is drawn in a horizontal position, the length may be given above the rectangular view, as in Fig. 11.26c, or below, as in Fig.11.26d. An application showing the dimensioning of cylindrical shapes is shown in Fig. 11.27. The use of a diagonal diameter in

FIGURE 11.24 Dimensioning Rectangular Prisms.

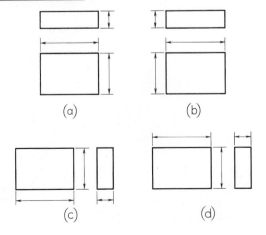

(a)

(b)

(c)

(d)

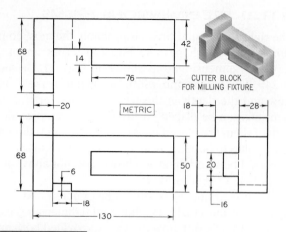

FIGURE 11.25 **Dimensioning a Machine Part Composed of Prismatic Shapes.**

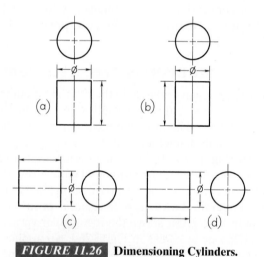

FIGURE 11.26 **Dimensioning Cylinders.**

the circular view, in addition to the method shown in Fig. 11.26, is not recommended except in special cases when clarity is improved. The use of several diagonal diameters on the same center is definitely to be discouraged, because the result is usually confusing.

The radius of a cylinder should never be given because measuring tools, such as the micrometer caliper, are designed to check diameters.

Small cylindrical holes, such as drilled, reamed, or bored holes, are usually dimensioned by means of notes specifying the diameter and the depth, with or without manufacturing operations (Figs. 11.27 and 11.32).

The diameter symbol Ø should be given before all diametral dimensions (Fig. 11.28a) (ANSI/ASME Y14.5M–1994). In some cases, the symbol Ø may be used to eliminate the circular view (Fig. 11.28b). The abbreviation DIA following the numerical value will be found on older decimal-inch drawings.

■ 11.23 SYMBOLS AND SIZE DIMENSIONS: MISCELLANEOUS SHAPES

Traditional terms and abbreviations used to describe various shapes and manufacturing processes, in addition to size specifications, are employed throughout this text. A variety of dimensioning symbols were introduced by ANSI/ASME (Y14.5M–1994) to replace traditional terms or abbreviations. These symbols are given with construction details in Fig. 11.29. Traditional terms and abbreviations are suitable for use where the symbols are not desired. Typical applications of some of these symbols are given in Fig. 11.30.

FIGURE 11.27 **Dimensioning a Machine Part That Is Composed of Cylindrical Shapes.**

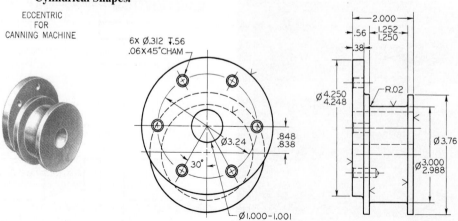

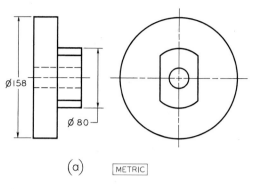

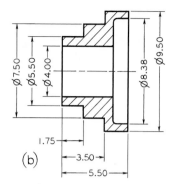

(a)   METRIC

USE OF " Ø " TO INDICATE CIRCULAR SHAPE

(b)

USE OF " Ø " TO OMIT CIRCULAR VIEW

FIGURE 11.28   Use of Ø or DIA in Dimensioning Cylinders.

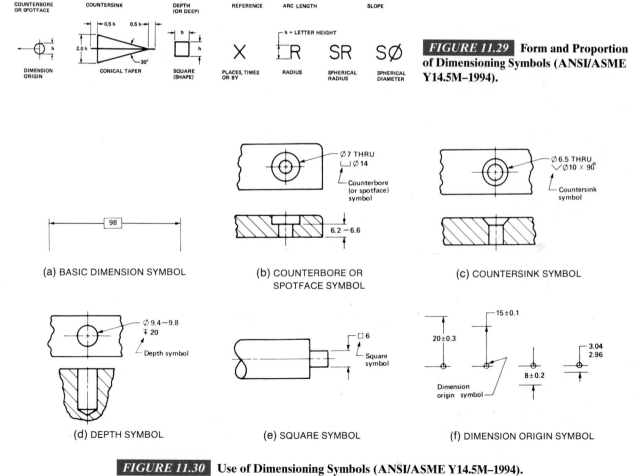

FIGURE 11.29   Form and Proportion of Dimensioning Symbols (ANSI/ASME Y14.5M–1994).

(a) BASIC DIMENSION SYMBOL

(b) COUNTERBORE OR SPOTFACE SYMBOL

(c) COUNTERSINK SYMBOL

(d) DEPTH SYMBOL

(e) SQUARE SYMBOL

(f) DIMENSION ORIGIN SYMBOL

FIGURE 11.30   Use of Dimensioning Symbols (ANSI/ASME Y14.5M–1994).

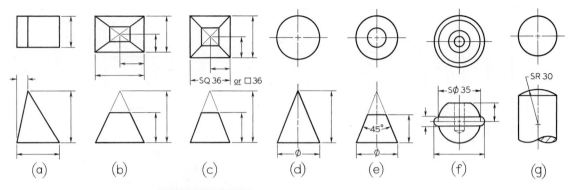

FIGURE 11.31 Dimensioning Various Shapes.

A triangular prism is dimensioned by giving the height, width, and displacement of the top edge in the front view and the depth in the top view (Fig. 11.31a).

A rectangular pyramid is dimensioned by giving the heights in the front view, and the dimensions of the base and the centering of the vertex in the top view (Fig. 11.31b). If the base is square, it is necessary to give the dimensions for only one side of the base, provided it is labeled SQ as shown or preceded by the square symbol □ (Fig. 11.31c).

A cone is dimensioned by giving its altitude and diameter of the base in the triangular view (Fig. 11.31d). A frustum of a cone may be dimensioned by giving the vertical angle and the diameter of one of the bases (Fig. 11.31e). Another method is to give the length and the diameters of both ends in the front view. Still another is to give the diameter at one end and the amount of taper per foot in a note (see §11.33).

Figure 11.31f shows a two-view drawing of a plastic knob. The main body is spherical and is dimensioned by giving its diameter preceded by the abbreviation and symbol for spherical diameter SØ or followed by the abbreviation SPHER. A bead around the knob is in the shape of a torus (see Fig. 4.7) and it

is dimensioned by giving the thickness of the ring and the outside diameter, as shown. In Fig. 11.31g, a spherical end is dimensioned by a radius preceded by the abbreviation SR.

Internal shapes corresponding to the external shapes in Fig. 11.31 would be dimensioned in a similar manner.

■ 11.24 SIZE DIMENSIONING OF HOLES

Holes that are to be drilled, bored, reamed, punched, cored, and so on are usually specified by symbols or standard notes, as shown in Figs. 11.29, 11.32, and 11.44. The order of items in a note corresponds to the order of procedure in the shop in producing the hole. Two or more holes are dimensioned by a single note, the leader pointing to one of the holes, as shown at the top of Fig. 11.32.

As illustrated in Figs. 6.40 and 11.32, the leader of a note should, as a rule, point to the circular view of the hole. It should point to the rectangular view only when clarity is thereby improved. When the circular view of the hole has two or more concentric circles, as for counterbored, countersunk, or tapped holes, the arrowhead should touch the outer circle (see Figs. 11.44b, 11.44c, and 11.44e to 11.44j).

FIGURE 11.32 Dimensioning Holes.

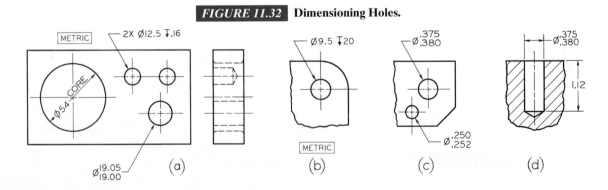

Notes should always be lettered horizontally on the drawing paper, and guide lines should always be used.

The use of decimal fractions to designate metric or inch drill sizes has gained wide acceptance [*] (Fig. 11.32b). For numbered or letter-size drills (Appendix 16), it is recommended that the decimal size be given in this manner, or given in parentheses; thus, #28 (.1405) DRILL, or "P" (.3230) DRILL. Metric drills are all decimal size and are not designated by number or letter.

On drawings of parts to be produced in large quantity for interchangeable assembly, dimensions and notes may be given without specification of the manufacturing process to be used. Only the dimensions of the holes are given, without reference to whether the holes are to be drilled, reamed, or punched (Figs. 11.32c and 11.32d). It should be realized that even though manufacturing operations are omitted from a note, the tolerances indicated would tend to dictate the manufacturing processes required.

■ 11.25   LOCATION DIMENSIONS

After the geometric shapes composing a structure have been dimensioned for size, as discussed, location dimensions must be given to show the relative positions of these geometric shapes, as shown in Fig. 11.32. Figure 11.33a shows that rectangular shapes, whether in the form of solids or of recesses, are located with reference to their faces. In Fig. 11.33b, cylindrical or

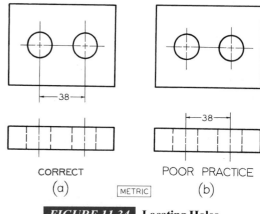

FIGURE 11.34   Locating Holes.

conical holes or bosses, or other symmetrical shapes, are located with reference to their center lines.

As shown in Fig. 11.34, location dimensions for holes are preferably given in the circular view of the holes.

Location dimensions should lead to finished surfaces wherever possible because rough castings and forgings vary in size, and unfinished surfaces cannot be relied on for accurate measurements (Fig. 11.35). Of course, the **starting dimension**, used in locating the first machined surface on a rough casting or forging, must necessarily lead from a rough surface, or from a center or a center line of the rough piece (see Figs. 13.90 and 13.91).

In general, location dimensions should be built from a finished surface as a datum plane, or from an important center or center line.

When several cylindrical surfaces have the same center line, as in Fig. 11.28b, it is not necessary to locate them with respect to each other.

FIGURE 11.33   Location Dimensions.

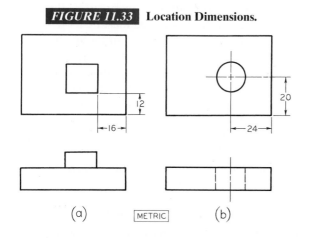

FIGURE 11.35   Dimensions to Finished Surfaces.

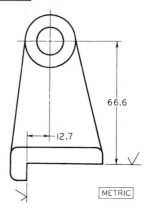

[*] Although drills are still listed fractionally in manufacturers' catalogs, many companies have supplemented drill and wire sizes with a decimal value. In many cases the number, letter, or common fraction has been replaced by the decimal-inch size. Metric drills are usually listed separately with a decimal-millimeter value.

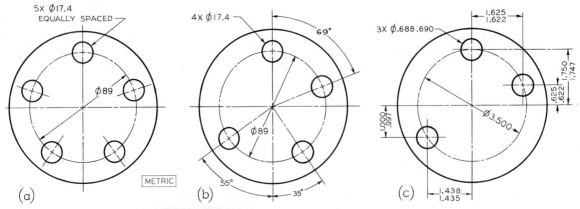

FIGURE 11.36 **Locating Holes about a Center.**

Holes equally spaced about a common center may be dimensioned by giving the diameter (diagonally) of the *circle of centers*, or *bolt circle* (Fig. 11.36a). Repetitive features or dimensions may be specified by the use of an X preceded with a numeral to indicate the number of times or places the feature is required. Allow a space between the letter X and the dimension as shown.

Unequally spaced holes are located by means of the bolt circle diameter plus angular measurements with reference to *only one* of the center lines (Fig. 11.36b).

Where greater accuracy is required, coordinate dimensions should be given (Fig. 11.36c). In this case, the diameter of the bolt circle is enclosed in parentheses to indicate that it is to be used only as a **reference dimension**. Reference dimensions are given for information only. They are not intended to be measured and do not govern the manufacturing operations. They represent calculated dimensions and are often useful in showing the intended design sizes (Fig. 11.36c).

When several nonprecision holes are located on a common arc, they are dimensioned by giving the radius and the angular measurements from a *baseline* (Fig. 11.37a). In this case, the baseline is the horizontal center line.

In Fig. 11.37b, the three holes are on a common center line. One dimension locates one small hole from the center; the other gives the distances between the small holes. Note the omission of a dimension at X. This method is used when (as is usually the case) the distance between the small holes is the important consideration. If the relation between the center hole and each of the small holes is more important, then include the distance at X, and denote the overall dimension as a reference dimension with parentheses.

Figure 11.37c shows another example of coordinate dimensioning. The three small holes are on a bolt circle whose diameter is given in parentheses for reference purposes only. From the main center, the small holes are located in two mutually perpendicular directions.

Another example of locating holes by means of linear measurements is shown in Fig. 11.37d. In this case, one such measurement is made at an angle to the coordinate dimensions because of the direct functional relationship of the two holes.

In Fig. 11.37e, the holes are located from two *baselines*, or *datums*. When all holes are located from a common datum, the sequence of measuring and machining operations is controlled, overall tolerance accumulations are avoided, and proper functioning of the finished part is assured, as intended by the designer. The datum surfaces selected must be more accurate than any measurement made from them, must be accessible during manufacture, and must be arranged to facilitate tool and fixture design. Thus, it may be necessary to specify accuracy of the datum surfaces in terms of straightness, roundness, flatness, and so forth (see §12.16).

Figure 11.37f shows a method of giving, in a single line, all the dimensions from a common datum. Each dimension except the first has a single arrowhead and is accumulative in value. The final and longest dimension is separate and complete.

These methods of locating holes are equally applicable to locating pins or other symmetrical features.

■ 11.26 MATING DIMENSIONS

In dimensioning a single part, its relation to mating parts, must be taken into consideration. For example, in Fig. 11.38a, a guide block fits into a slot in a base.

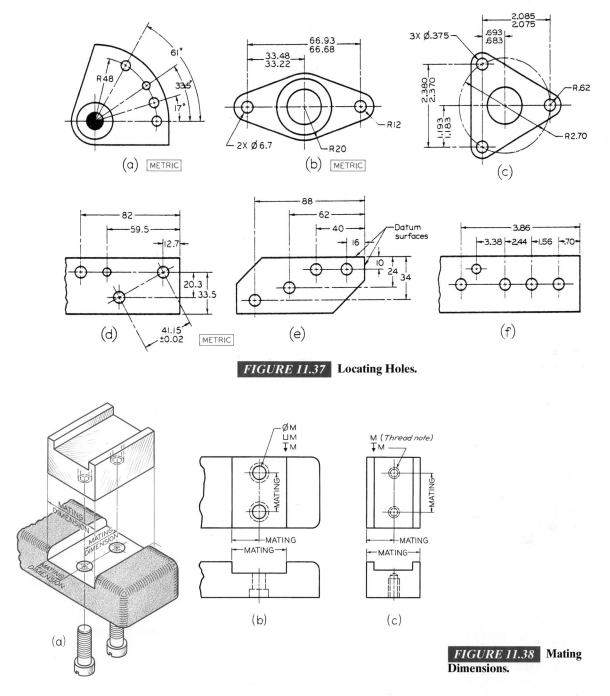

FIGURE 11.37 Locating Holes.

FIGURE 11.38 Mating Dimensions.

Those dimensions common to both parts are **mating dimensions**, as indicated.

These mating dimensions should be given on the multiview drawings in the corresponding locations (Figs. 11.38b and 11.38c). Other dimensions are not mating dimensions since they do not control the accurate fitting together of two parts. The actual *values* of two corresponding mating dimensions may not be ex-

actly the same. For example, the width of the slot in Fig. 11.38b may be dimensioned $\frac{1}{32}''$ (0.8 mm) or several thousandths of an inch larger than the width of the block in Fig. 11.38c, but these are mating dimensions figured from a single basic width. It will be seen that the mating dimensions shown might have been arrived at from a geometric breakdown (see §11.20). However, the mating dimensions need to be identified so that

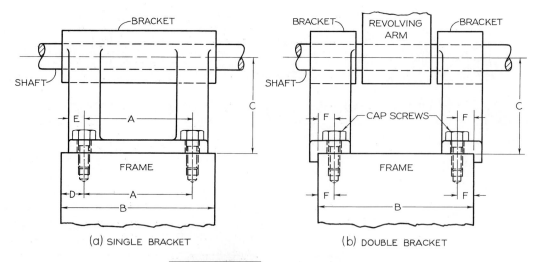

(a) SINGLE BRACKET (b) DOUBLE BRACKET

FIGURE 11.39 **Bracket Assembly.**

they can be specified in the corresponding locations on the two parts and so that they can be given with the degree of accuracy commensurate with the proper fitting of the parts.

In Fig. 11.39a the dimension A should appear on both the drawings of the bracket and of the frame and, therefore, is a necessary mating dimension. In Fig. 11.39b, which shows a redesign of the bracket into two parts, dimension A is not used on either part because it is not necessary to control closely the distance between the cap screws. But dimensions F are now essential mating dimensions and should appear correspondingly on the drawings of both parts. The remaining dimensions E, D, B, and C are not considered to be mating dimensions since they do not directly affect the mating of the parts.

■ 11.27 MACHINE, PATTERN, AND FORGING DIMENSIONS

In Fig. 11.38a, the base is machined from a rough casting; the patternmaker needs certain dimensions to make the pattern, and the machinist needs certain dimensions for the machining. In some cases one dimension will be used by both. Again, in most cases, these dimensions will be the same as those resulting from a geometric breakdown, but it is important to identify them to assign values to them.

Figure 11.40 shows the same part as in Fig. 11.38, with the machine dimensions and pattern dimensions identified by the letters M and P. The patternmaker is interested only in the dimensions required to make the pattern, and the machinist, in general, is concerned on-

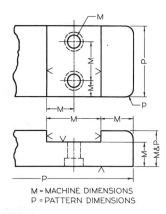

M = MACHINE DIMENSIONS
P = PATTERN DIMENSIONS

FIGURE 11.40 **Machine and Pattern Dimensions.**

ly with the dimensions needed to machine the part. Frequently, a dimension that is convenient for the machinist is not convenient for the patternmaker, or vice versa. Since the patternmaker uses the drawing only once, while making the pattern, and the machinist refers to it continuously, the dimensions should be given primarily for the convenience of the machinist.

If the part is large and complicated, two separate drawings are sometimes made, one showing the pattern dimensions and the other the machine dimensions. The usual practice, however, is to prepare one drawing for both the patternmaker and the machinist (see §11.4).

For forgings, it is common practice to make separate forging drawings and machining drawings. A forging drawing of a connecting rod, showing only the dimensions needed in the forge shop, is shown in

Fig. 13.18. A machining drawing of the same part, but containing only the dimensions needed in the machine shop, is shown in Fig. 13.19.

Unless a decimal system is used (see §11.10), the pattern dimensions are nominal, usually to the nearest $\frac{1}{16}''$, and given in whole numbers and common fractions. If a machine dimension is given in whole numbers and common fractions, the machinist is usually allowed a tolerance (permissible in variation in size) of $\pm \frac{1}{64}''$. Some companies specify a tolerance of $\pm .010''$ on all common fractions. If greater accuracy is required, the dimensions are given in decimal form. Metric dimensions are given to one or more places, and decimal-inch dimensions are given to three or more places (see §§11.10 and 12.1). Remember that 0.1 mm is approximately .004 inch.

■ 11.28 DIMENSIONING OF CURVES

Curved shapes may be dimensioned by giving a group of radii, as shown in Fig. 11.41a. Note that in dimensioning the R126 arc whose center is inaccessible, the center may be moved inward along a center line and a jog made in the dimension line (see also Fig. 11.19f). Another method is to dimension the outline envelope of a curved shape so that the various radii are self-locating from "floating centers" (Fig. 11.41b). Either a circular or a noncircular curve may be dimensioned by means of coordinate dimensions, or datums (Fig. 11.41c; see also Fig. 11.6a).

■ 11.29 DIMENSIONING OF ROUNDED-END SHAPES

The method used for dimensioning rounded-end shapes depends on the degree of accuracy required (Fig. 11.42). When precision is not necessary, the methods used are those that are convenient for manufacturing, as in Figs. 11.42a to 11.42c.

In Fig. 11.42a, the link to be cast or to be cut from sheet metal or plate is dimensioned as it would be laid out for manufacture, by giving the center-to-center distance and the radii of the ends. Note that only one such radius dimension is necessary, but that the number of places may be included with the size dimension.

In Fig. 11.42b, the pad on a casting, with a milled slot, is dimensioned from center to center for the convenience of both the patternmaker and the machinist in layout. An additional reason for the center-to-center distance is that it gives the total travel of the milling cutter, which can be easily controlled by the machinist. The width dimension indicates the diameter of the milling cutter; hence, it is incorrect to give the radius of a machined slot. On the other hand, a cored slot (see §11.3) should be dimensioned by radius in conformity with the patternmaker's layout procedure.

In Fig. 11.42c, the semicircular pad is laid out in a similar manner to the pad in Fig. 11.42b, except that angular dimensions are used. Angular tolerances can be used if necessary (see §12.18).

When accuracy is required, the methods shown in Figs. 11.42d to 11.42g are recommended. Overall lengths of rounded-end shapes are given in each case, and radii are indicated, but without specific values. In the example in Fig. 12.42f, the center-to-center distance is required for accurate location of the holes.

In Fig. 11.42g, the hole location is more critical than the location of the radius; hence, the two are located independently, as shown.

■ 11.30 SUPERFLUOUS DIMENSIONS

All necessary dimensions must be shown, but the designer should avoid giving unnecessary or superfluous dimensions (Fig. 11.43a). Dimensions should not be repeated on the same view or on different views; nor should the same information be given in two different ways.

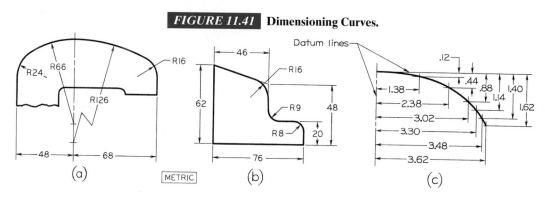

FIGURE 11.41 Dimensioning Curves.

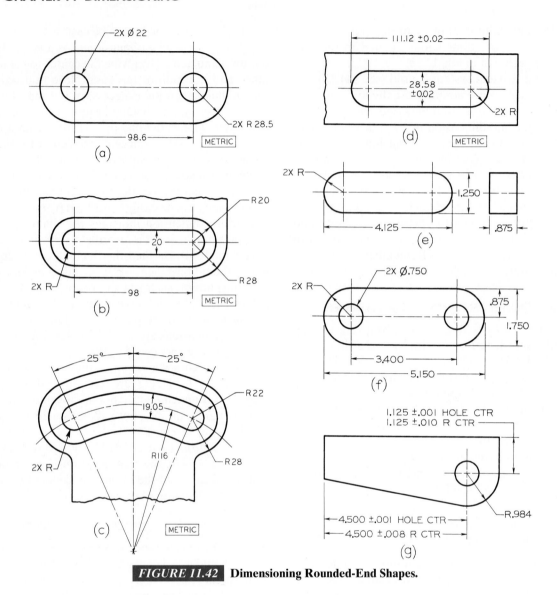

FIGURE 11.42   Dimensioning Rounded-End Shapes.

Figure 11.43b illustrates a type of superfluous dimensioning that should generally be avoided, especially in machine drawing, where accuracy is important. Production personnel should not be allowed a choice between two dimensions. *Avoid "chain" dimensioning*, in which a complete series of detail dimensions is given together with an overall dimension. In such cases, one dimension of the chain should be omitted, as shown, so that the machinist is obliged to work from one surface only. This is particularly important in tol-

erance dimensioning (see §12.1), where an accumulation of tolerances can cause serious difficulties (see also §12.9).

Some inexperienced detailers have the habit of omitting both dimensions, such as those at the right in Fig. 11.43b, on the theory that the holes are symmetrically located and will be understood to be centered. One of the two location dimensions should be given.

As shown in Fig. 11.43e, when one dimension clearly applies to several identical features, it need not

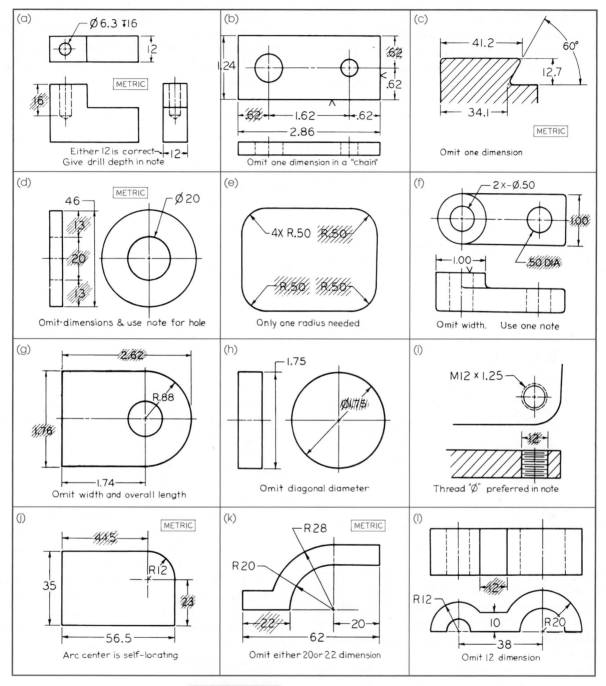

FIGURE 11.43 Superfluous Dimensions.

be repeated, but the number of places should be indicated. Dimensions for fillets and rounds and other noncritical features need not be repeated nor number of places specified.

For example, the radii of the rounded ends in Figs. 11.42a to 11.42f need not be repeated, and in Fig. 11.1 both ribs are obviously the same thickness so it is unnecessary to repeat the 10-mm dimension.

■ 11.31 NOTES

It is usually necessary to supplement the direct dimensions with notes. Notes should be brief and carefully worded to allow only one interpretation. *Notes should always be lettered horizontally on the sheet, with guide lines, and arranged in a systematic manner.* They should not be lettered in crowded places, and they should not be placed between views, if possible. They should not be lettered so close together as to confuse the reader or so close to another view or detail as to suggest application to the wrong view. Leaders should be as short as possible and cross as few lines as possible. Leaders should never run through a corner of a view or through any specific points or intersections.

Notes are classified as **general notes** when they apply to an entire drawing and as **local notes** when they apply to specific items.

GENERAL NOTES   General notes should be lettered in the lower right-hand corner of the drawing, above or to the left of the title block, or in a central position below the view to which they apply.

EXAMPLES

FINISH ALL OVER (FAO)
BREAK SHARP EDGES TO R0.8
G33106 ALLOY STEEL–BRINELL 340–380
ALL DRAFT ANGLES 3° UNLESS OTHERWISE SPECIFIED
DIMENSIONS APPLY AFTER PLATING

In machine drawings, the title strip or title block will carry many general notes, including material, general tolerances, heat treatment, and pattern information (see Fig. 13.14).

LOCAL NOTES   Local notes apply to specific operations only and are connected by a leader to the point at which such operations are performed (Fig. 11.44). The leader should be attached at the front of the first word of a note, or just after the last word, and not at any intermediate place.

For information on notes applied to holes, see §11.24.

Certain commonly used abbreviations may be used freely in notes, such as THD, DIA, MAX. The less common abbreviations should be avoided as much as possible. All abbreviations should conform to ANSI Y1.1–1989. See Appendix 4 for American National Standard abbreviations. See Figs. 11.29 and 11.30 for form and use of alternative dimensioning symbols.

In general, leaders and notes should not be placed on the drawing until the dimensioning is substantially completed. If notes are lettered first, they will almost invariably be in the way of necessary dimensions and will have to be moved.

■ 11.32 DIMENSIONING OF THREADS

Local notes are used to specify dimensions of threads. For tapped holes the notes should, if possible, be attached to the circular views of the holes, as shown in Fig. 11.44g. For external threads, the notes are usually placed in the longitudinal views, where the threads are more easily recognized, as in Figs. 11.44v and 11.44w. For a detailed discussion of thread notes, see §13.21.

■ 11.33 DIMENSIONING OF TAPERS

A **taper** is a conical surface on a shaft or in a hole. The usual method of dimensioning a taper is to give the amount of taper in a note, such as TAPER 0.167 ON DIA (often TO GAGE added), and then give the diameter at one end, plus the length, or give the diameter at both ends and omit the length. *Taper on diameter means the difference in diameter per unit of length.*

Standard machine tapers are used on machine spindles, shanks of tools, or pins, for example, and are described in "Machine Tapers" (ANSI/ASME B5.10–1994). Such standard tapers are dimensioned on a drawing by giving the diameter, usually at the large end, the length, and a note, such as NO. 4 AMERICAN NATIONAL STANDARD TAPER (Fig. 11.45a).

For not-too-critical requirements, a taper may be dimensioned by giving the diameter at the large end, the length, and the included angle, all with proper tolerances (Fig. 11.45b). Or the diameters of both ends, plus the length, may be given with necessary tolerances.

For close-fitting tapers, the amount of *taper per unit on diameter* is indicated as shown in Figs. 11.45c and 11.45d. A gage line is selected and located by a comparatively generous tolerance, while other dimensions are given appropriate tolerances as required.

■ 11.34 DIMENSIONING OF CHAMFERS

A **chamfer** is a beveled or sloping edge, and it is dimensioned by giving the length of the offset and the angle (Fig. 11.46a). A 45° chamfer also may be dimensioned in a manner similar to that shown in Fig. 11.46a, but usually it is dimensioned by note without or with the word CHAM, as in Fig. 11.46b.

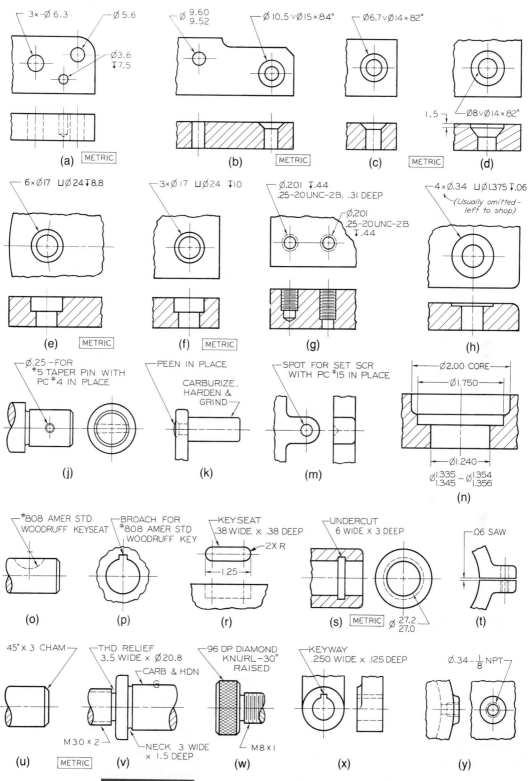

FIGURE 11.44 Local Notes. See also Figs. 7.40 and 13.32.

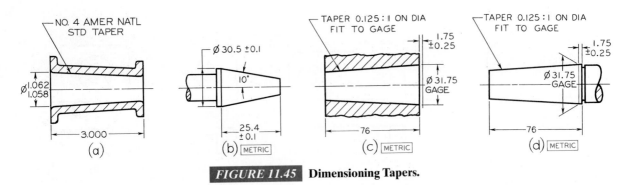

FIGURE 11.45 **Dimensioning Tapers.**

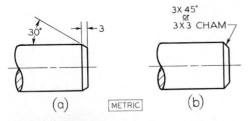

FIGURE 11.46 **Dimensioning Chamfers.**

■ 11.35 SHAFT CENTERS

Shaft centers are required on shafts, spindles, and other conical or cylindrical parts for turning, grinding, and other operations. Such a center may be dimensioned, as shown in Fig. 11.47. Normally the centers are produced by a combined drill and countersink.

■ 11.36 DIMENSIONING KEYWAYS

Methods of dimensioning keyways for Woodruff keys and stock keys are shown in Fig. 11.48. Note, in both

cases, the use of a dimension to center the keyway in the shaft or collar. The preferred method of dimensioning the depth of a keyway is to give the dimension from the bottom of the keyway to the opposite side of the shaft or hole, as shown. The method of computing such a dimension is shown in Fig. 11.48d. Values for A may be found in machinists' handbooks.

For general information about keys and keyways, see §13.34.

■ 11.37 DIMENSIONING OF KNURLS

A **knurl** is a roughened surface to provide a better handgrip or to be used for a press fit between two parts. For handgripping purposes, it is necessary only to give the pitch of the knurl, the type of knurling, and the length of the knurled area (Fig. 11.49a and 11.49b). To dimension a knurl for a press fit, the toleranced diameter before knurling should be given (Fig. 11.49c). A note should be added giving the pitch and type of knurl and the minimum diameter after knurling [see ANSI/ASME B94.6–1984 (R1995)].

■ 11.38 DIMENSIONING ALONG CURVED SURFACES

When angular measurements are unsatisfactory, chordal dimensions (Fig. 11.50a) or linear dimensions on the curved surfaces (Fig. 11.50b) may be given.

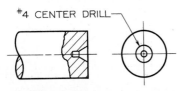

FIGURE 11.47 **Shaft Center.**

FIGURE 11.48 **Dimensioning Keyways.**

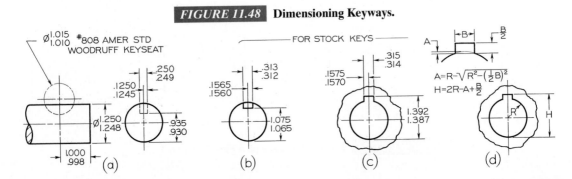

Semiautomatic Dimensioning Using CAD

DIMENSIONING CONTROLS PART

Dimensioning is an important skill because the dimensions given in the drawing control how the part will be constructed and how tolerance values will be applied. Even when a drawing or model database is exported for direct machining, the machinist must know which fits and dimensions are critical, and where the part can vary. No parts are created exactly to the same size as the dimensions specified, so the designer must make it clear what allowances are possible. Dimensioning CAD drawings is accomplished using a suite of dimensioning tools provided by the software. Programs like AutoCAD 2000 call their tools semiautomatic dimensioning because the dimension lines, values, arrowheads, and extension lines are created for you automatically, but you must still choose where you will place dimensions in the drawing.

DIMENSION STYLES

AutoCAD 2000 lets you create different families of dimension appearances, called dimension styles. You use this to change the appearance of dimensions for different types of drawings. For example, architectural drawings have a different standard for their appearance than mechanical drawings, and civil drawings may still have yet another appearance. To create dimension styles and set their appearance in AutoCAD R. 14, you use the Dimension Styles dialog box. You can quickly pick it from the Dimensioning Toolbar shown in Figure A.

PARENT & CHILD STYLES

AutoCAD 2000 uses child styles to let you change the appearance of dimension types within the style; for example radial dimensions can have a different appearance than linear dimensions, or ordinate dimensions. You can have a different appearance for each of these types of dimensions: linear, radial, angular, diameter, ordinate, and leader. You can think of child styles like this. If you have a child, they generally resemble you; have brown eyes if you do, etc. But the child may decide to dye their hair. After

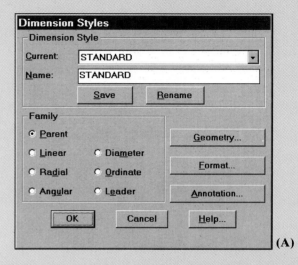

(A)

that, no amount of you dying your hair will change the appearance of the child's hair. This is essentially how child dimension styles work. You can set the child style for a type of dimension so that it looks different from the parent style. Once a characteristic of a child dimension style is set differently than the parent style, changing the parent no longer changes the child. You can use these styles to manage the appearance of the dimensions in your drawing so that you do not have to tweak individual dimensions. Dimension styles also allow you to have a consistent approach to controlling the appearance of the dimensions in the drawing so that you know how the dimensions will update if you make a change. Figure B shows the dialogue box you can use.

(B)

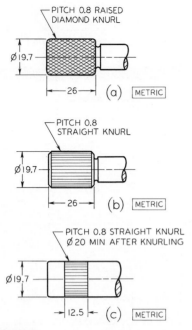

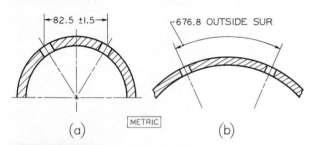

FIGURE 11.49   Dimensioning Knurls.

FIGURE 11.50   Dimensioning along Curved Surfaces.

■ 11.39   SHEET-METAL BENDS

In sheet-metal dimensioning, allowance must be made for bends. The intersection of the plane surfaces adjacent to a bend is called the *mold line*, and this line, rather than the center of the arc, is used to determine dimensions (Fig. 11.51). The following procedure for calculating bends is typical. If the two inner plane surfaces of an angle are extended, their line of intersection is called the IML or *inside mold line* (Figs. 11.52a to 11.52c). Similarly, if the two outer plane surfaces are extended, they produce the OML or *outside mold line*. The *center line of bend* (℄ B) refers primarily to the machine on which the bend is made and is at the center of the bend radius.

The length, or *stretchout*, of the pattern equals the sum of the flat sides of the angle plus the distance around the bend measured along the *neutral axis*. The

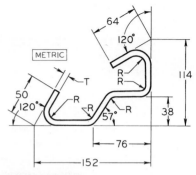

FIGURE 11.51   Profile Dimensioning.

distance around the bend is called the *bend allowance*. When metal bends, it compresses on the inside and stretches on the outside. At a certain zone in between, the metal is neither compressed not stretched, and this is called the neutral axis (Fig. 11.52d). The neutral axis is usually assumed to be 0.44 of the thickness from the inside surface of the metal.

The developed length of material, or bend allowance (BA), to make the bend is computed from the empirical formula

$$BA = (0.017453R + 0.0078T)N,$$

where R = radius of bend, T = metal thickness, and N = number of degrees of bend (Fig. 11.52c).

■ 11.40   TABULAR DIMENSIONS

A series of objects having like features but varying in dimensions may be represented by one drawing (Fig. 11.53). Letters are substituted for dimension figures on the drawing, and the varying dimensions are given in tabular form. The dimensions of many standard parts are given in this manner in catalogs and handbooks.

■ 11.41   STANDARDS

Dimensions should be given, wherever possible, to make use of readily available materials, tools, parts, and gages. The dimensions for many commonly used machine elements, such as bolts, screws, nails, keys, tapers, wire, pipes, sheet metal, chains, belts, ropes, pins, and rolled metal shapes, have been standardized, and the drafter must obtain these sizes from company standards manuals, from published handbooks, from American National Standards, or from manufacturers' catalogs. Tables of some of the more common items are given in the Appendix of this text.

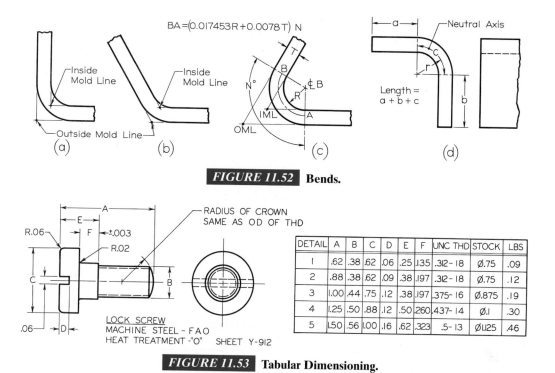

$$BA = (0.017453R + 0.0078T) N$$

FIGURE 11.52 Bends.

FIGURE 11.53 Tabular Dimensioning.

DETAIL	A	B	C	D	E	F	UNC THD	STOCK	LBS
1	.62	.38	.62	.06	.25	.135	.312-18	Ø.75	.09
2	.88	.38	.62	.09	.38	.197	.312-18	Ø.75	.12
3	1.00	.44	.75	.12	.38	.197	.375-16	Ø.875	.19
4	1.25	.50	.88	.12	.50	.260	.437-14	Ø.I	.30
5	1.50	.56	1.00	.16	.62	.323	.5-13	Ø1.125	.46

Such standard parts are not delineated on detail drawings unless they are to be altered for use, but are drawn conventionally on assembly drawings and are listed in parts lists (see §13.14). Common fractions are often used to indicate the nominal sizes of standard parts or tools. If the complete decimal-inch system is used, all such sizes ordinarily are expressed by decimals; for example, .250 DRILL instead of $\frac{1}{4}$ DRILL. If the all-metric system of dimensioning is used, then the *preferred* metric drill of the approximate same size (.2480″) will be indicated as a 6.30 DRILL.

■ 11.42 COORDINATE DIMENSIONING

In general, the basic coordinate dimensioning practices are compatible with the data requirements for tape or computer-controlled automatic production machines. However, to design for automated production, the designer and/or drafter should first consult the manufacturing machine manuals before making the drawings for production. Certain considerations should be noted.

1. A set of three mutually perpendicular datum or reference planes is usually required for coordinate dimensioning. These planes either must be obvious or clearly identified (Fig. 11.54).

2. The designer selects as origins for dimensions those surfaces or other features most important to the functioning of the part. Enough of these features are selected to position the part in relation to the set of mutually perpendicular planes. All related dimensions on the part are then made from these planes. An example of rectangular coordinate dimensioning without dimension lines is shown in Fig. 11.55.

3. All dimensions should be in decimals.

4. Angles should be given, where possible, in degrees and decimal parts of degrees.

5. Standard tools, such as drills, reamers, and taps, should be specified when required.

6. All tolerances should be determined by the design requirements of the part, not by the capability of the manufacturing machine.

■ 11.43 DO'S AND DON'TS OF DIMENSIONING

The following checklist summarizes briefly most of the situations in which a beginning designer is likely to make a mistake in dimensioning. Students should check the drawing by this list before submitting it to the instructor.

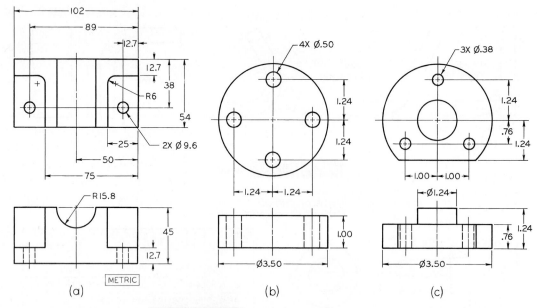

FIGURE 11.54   Coordinate Dimensioning.

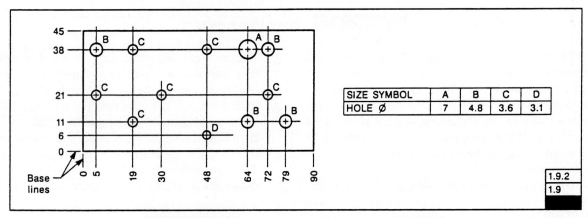

FIGURE 11.55   Rectangular Coordinate Dimensioning
Without Dimension Lines (ANSI/ASME Y14.5M–1994).

1. Each dimension should be given clearly so that it can be interpreted in only one way.

2. Dimensions should not be duplicated or the same information given in two different ways—dual dimensioning excluded—and no dimensions should be given except those needed to produce or inspect the part.

3. Dimensions should be given between points or surfaces that have a functional relation to each other or that control the location of mating parts.

4. Dimensions should be given to finished surfaces or important center lines, in preference to rough surfaces, wherever possible.

5. Dimensions should be so given that it will not be necessary for the machinist to calculate, scale, or assume any dimension.

6. Dimensions should be attached to the view where the shape is best shown (contour rule).

7. Dimensions should be placed in the views where the features dimensioned are shown true shape.

8. Dimensioning to hidden lines should be avoided wherever possible.

9. Dimensions should not be placed on a view unless clarity is promoted and long extension lines are avoided.

10. Dimensions applying to two adjacent views should be placed between views, unless clarity is promoted by placing some of them outside.

11. The longer dimensions should be placed outside all intermediate dimensions so that dimension lines will not cross extension lines.

12. In machine drawing, all unit marks should be omitted, except when necessary for clarity; for example, 1″ VALVE or 1 mm DRILL.

13. Production personnel should not be expected to assume that a feature is centered (as a hole on a plate), but a location dimension should be given from one side. However, if a hole is to be centered on a symmetrical rough casting, mark the center line and omit the locating dimension from the center line.

14. A dimension should be attached to only one view, not to extension lines connecting two views.

15. Detail dimensions should "line up" in chain fashion.

16. A complete chain of detail dimensions should be avoided; it is better to omit one; otherwise reference should be added to one detail dimension or the overall dimension by enclosing within parentheses.

17. A dimension line should never be drawn through a dimension figure. A figure should never be lettered over any line of the drawing. The line can be broken if necessary.

18. Dimension lines should be spaced uniformly throughout the drawing. They should be at least 10 mm (.38″) from the object outline and 6 mm (.25″) apart.

19. No line of the drawing should be used as a dimension line or coincide with a dimension line.

20. A dimension line should never be joined end to end (chain fashion) with any line of the drawing.

21. Dimension lines should not cross, if avoidable.

22. Dimension lines and extension lines should not cross, if avoidable. (Extension lines may cross each other.)

23. When extension lines cross extension lines or visible lines, no break in either line should be made.

24. A center line may be extended and used as an extension line, in which case it is still drawn like a center line.

25. Center lines should generally not extend from view to view.

26. Leaders for notes should be straight, not curved, and pointing to the center of circular views of holes wherever possible.

27. Leaders should slope at 45°, or 30°, or 60° with horizontal but may be made at any convenient angle except vertical or horizontal.

28. Leaders should extend from the beginning or from the end of a note, the horizontal "shoulder" extending from midheight of the lettering.

29. Dimension figures should be approximately centered between the arrowheads, except that in a "stack" of dimensions, the figures should be "staggered."

30. Dimension figures should be about 3 mm (.13″) high for whole numbers and 6 mm (.25″) high for fractions.

31. Dimension figures should never be crowded or in any way made difficult to read.

32. Dimension figures should not be lettered over lines or sectioned areas unless necessary, in which case a clear space should be reserved for the dimension figures.

33. Dimension figures for angles should generally be lettered horizontally.

34. Fraction bars should never be inclined except in confined areas, such as in tables.

35. The numerator and denominator of a fraction should never touch the fraction bar.

36. Notes should always be lettered horizontally on the sheet.

37. Notes should be brief and clear, and the wording should be standard in form.

38. Finish marks should be placed on the edge views of all finished surfaces, including hidden edges and the contour and circular views of cylindrical surfaces.

39. Finish marks should be omitted on holes or other features where a note specifies a machining operation.

40. Finish marks should be omitted on parts made from rolled stock.

41. If a part is finished all over, all finish marks should be omitted, and the general note FINISH ALL OVER or FAO should be used.

42. A cylinder is dimensioned by giving both its diameter and length in the rectangular view, except when notes are used for holes. A diagonal diameter in the circular view may be used in cases where clarity is gained thereby.

43. Holes to be bored, drilled, reamed, and so on are size-dimensioned by notes in which the leaders preferably point toward the center of the circular views of the holes. Indications of manufacturing processes may be omitted from notes.

44. Drill sizes are preferably expressed in decimals. For drills designated by number or letter, the decimal size must also be given.

45. In general, a circle is dimensioned by its diameter, an arc by its radius.

46. Diagonal diameters should be avoided, except for very large holes and for circles of centers. They may be used on positive cylinders when clarity is gained thereby.

47. A diameter dimension value should always be preceded by the symbol ∅.

48. A radius dimension should always be preceded by the letter R. The radial dimension line should have only one arrowhead, and it should pass through or point through the arc center and touch the arc.

49. Cylinders should be located by their center lines.

50. Cylinders should be located in the circular views, if possible.

51. Cylinders should be located by coordinate dimensions in preference to angular dimensions where accuracy is important.

52. When there are several rough, noncritical features obviously the same size (fillets, rounds, ribs, etc.), it is necessary to give only typical (abbreviation TYP) dimensions or to use a note.

53. When a dimension is not to scale, it should be underscored with a heavy straight line or marked NTS or NOT TO SCALE.

54. Mating dimensions should be given correspondingly on drawings of mating parts.

55. Pattern dimensions should be given in two-place decimals or in common whole numbers and fractions to the nearest $\frac{1}{16}''$.

56. Decimal dimensions should be used for all machining dimensions.

57. Cumulative tolerances should be avoided, especially in limit dimensioning, described in §11.9.

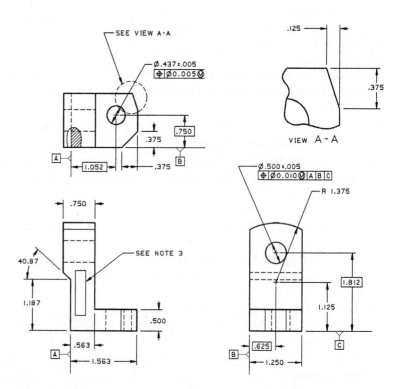

FIGURE 11.56 **Dimensioned Detail Drawing Produced by the Computervision CADDS 4X Production Drafting System.** *Courtesy of Computervision Corporation, a subsidiary of Prime Computer, Inc.*

■ KEY WORDS

METRIC SCALE	**ENGINEERING SCALE**
ARCHITECT'S SCALE	**EXTENSION LINES**
DIMENSION LINES	**LEADERS**
UNIDIRECTIONAL SYSTEM	**ALIGNED SYSTEM**
FINISH MARKS	**SIZE DIMENSIONS**
LOCATION DIMENSIONS	**ANSI STANDARDS**
NOTES	**COORDINATE DIMENSIONS**
ABBREVIATIONS	**SUPERFLUOUS**

■ CHAPTER SUMMARY

- To increase clarity, dimensions and notes are added to a drawing to precisely describe size, location, and manufacturing process.

- Drawings are scaled to fit on a standard sheet of paper. Drawings created by hand are drawn to scale. CAD drawings are drawn full size and scaled when they are printed.

- The three types of scales are metric, engineer's, and architect's.

- Dimensions and notes are placed on drawings according to prescribed standards.

- Dimensions that are incorrectly placed on a drawing are considered just as wrong as if the numbers in the dimension were incorrect.

- Special dimensioning techniques are used for surfaces that have been machined by one of the manufacturing processes.

■ REVIEW QUESTIONS

1. What are the different units used when a drawing is created with a metric scale? With an engineering scale? With an architect's scale?
2. Explain the concept of contour dimensioning.
3. Which type of line is *never* crossed by any other line when dimensioning an object?
4. How is geometric analysis used in dimensioning?
5. What is the difference between a size dimension and a location dimension?
6. Which dimension system allows dimensions to be read from the bottom and from the right? When can a dimension be read from the left?
7. Draw an example of dimensioning an angle.
8. When are finish marks used? Draw two types.
9. How are negative and positive cylinders dimensioned? Draw examples.
10. How are holes and arcs dimensioned? Draw examples.
11. What are notes and leaders used for?
12. Why is it important to avoid superfluous dimensions?

■ DIMENSIONING PROBLEMS

Most of a student's practice in dimensioning will be in connection with working drawings assigned from other chapters. However, a limited number of special dimensioning problems are available here in Figs. 11.57 and 11.58. The problems are designed for Layout A–3 (8.5″ × 11.0″) and are to be drawn with instruments and dimensioned to a full-size sale. Layout A4–3

(297 mm × 420 mm) may be used with appropriate adjustments in the title strip layout.

Since many of the problems in this and other chapters are of a general nature, they can also be solved on most computer graphics systems. If a system is available, the instructor may choose to assign specific problems to be completed by this method.

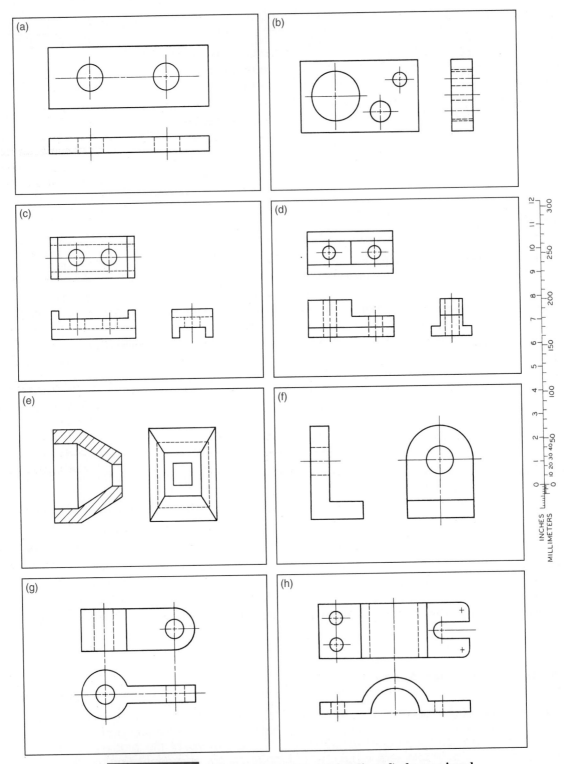

FIGURE 11.57 Using Layout A–3 or A4–3 (adjusted), draw assigned problem with instruments. To obtain sizes, place bow dividers on the views on this page and transfer to scale at the side to obtain values. Dimension drawing completely in one-place millimeters or two-place inches as assigned, full size. See inside back cover for decimal-inch and millimeter equivalents.

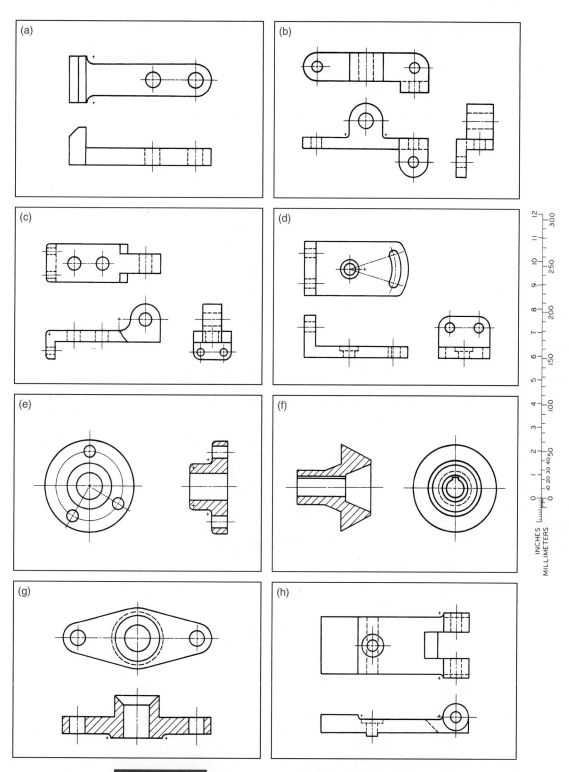

FIGURE 11.58   Using Layout A–3 or A4–3 (adjusted), draw assigned problem with instruments. To obtain sizes, place bow dividers on the views on this page and transfer to scale at the side to obtain values. Dimension drawing completely in one-place millimeters or two-place inches as assigned, full size. See inside back cover for decimal-inch and millimeter equivalents.

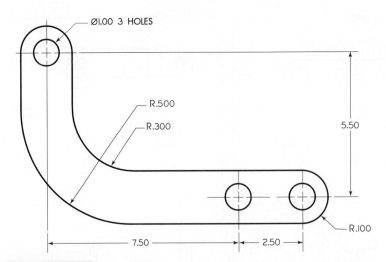

Ø1.00 3 HOLES

R.500

R.300

5.50

R.100

7.50

2.50

FIGURE 11.59   **Using Sheet Layout A–3 of A4–3 (adjusted), draw the OML and IML. Calculate the bend allowance using the formula on p. 336.**

C H A P T E R   1 2

TOLERANCING

OBJECTIVES

After studying the material in this chapter, you should be able to:

1. Read and create limit dimensions.
2. Describe the nominal size, tolerance, limits, and allowance of two mating parts.
3. Identify a clearance fit, interference fit, and transition fit.
4. Describe the basic hole and basic shaft systems.
5. Dimension two mating parts using limit dimensions, unilateral tolerances, and bilateral tolerances.
6. Describe the classes of fit and give examples of each.
7. Draw geometric tolerancing symbols.
8. Specify position and geometric tolerances.

OVERVIEW

Interchangeable manufacturing allows parts made in widely separated localities to be brought together for assembly. That the parts all fit together properly is an essential element of mass production. Without interchangeable manufacturing, modern industry could not exist, and without effective size control by the engineer, interchangeable manufacturing could not be achieved.

For example, an automobile manufacturer not only subcontracts the manufacture of many parts of a design to other companies but must also provide the parts for replacement. All parts in each category must be nearly enough alike so that any one of them will fit properly in any assembly. Unfortunately, *it is impossible to make anything to exact size.* Parts can be made to very close dimensions, even to a few millionths of an inch or thousandths of a millimeter (e.g., gage blocks), but such accuracy is extremely expensive.

Fortunately, *exact* sizes are not needed. The need is for varying degrees of accuracy according to functional requirements. A manufacturer of children's tricycles would soon go out of business if the parts were made with jet-engine accuracy—no one would be willing to pay the price. So what is wanted is a means of specifying dimensions with whatever degree of accuracy is required. The answer to the problem is the specification of a *tolerance* on each dimension.

The concept of quality in manufacturing is primarily a factor of machining tolerances. Products with small variations in shape and size are considered high quality and can command higher prices. Waste results when the manufacturing process cannot maintain shape and size within prescribed limits. By monitoring the manufacturing processes and reducing waste, a company can improve profits. This direct benefit to profits is why tolerancing is critical to manufacturing success. Tolerancing is an extension of dimensioning. It provides additional information about the shape, size, and position of every feature of a product. It communicates exacting directions about how to manufacture a product. Basic CAD program functionality is often supplemented with advanced dimensioning and tolerancing software that can assist with the tolerancing process. It takes a highly skilled designer to be able to apply tolerance dimensions correctly.

■ 12.1 TOLERANCE DIMENSIONING

Tolerance is the total amount that a specific dimension is permitted to vary; it is the difference between the maximum and the minimum limits for the dimension (ANSI/ASME Y14.5M–1994). For example, a dimension given as 1.625 ± .002 means that the manufactured part may be 1.627″ or 1.623″, or anywhere between these **limit dimensions.** The tolerance, or total amount of variation "tolerated," is .004″. Thus, it becomes the function of the detailer or designer to specify the allowance error that may be tolerated for a given dimension and still permit the satisfactory functioning of the part. Since greater accuracy costs more money, the detailer or designer will not specify the closest tolerance, but instead will specify as generous a tolerance as possible.

To control the dimensions of quantities of the two parts so that any two mating parts will be interchangeable, it is necessary to assign tolerance to the dimensions of the parts, as shown in Fig. 12.1a. The diameter of the hole may be machined not less than 1.250″ and not more than 1.251″; these two figures represent the limits and the difference between them (.001″ is the tolerance). Likewise, the shaft must be produced between the limits of 1.248″ and 1.247″; the tolerance on the shaft is the difference between these, or .001″. The metric versions for these limit dimensions for the hole and shaft are shown in Fig. 12.1b. The difference in the dimensions for either the hole or shaft is 0.03 mm, the total tolerance.

A pictorial illustration of the dimensions in Fig. 12.1a is shown in Fig. 12.2a. The maximum shaft is shown solid, and the minimum shaft is shown in phantom. The difference in diameters, .001″, is the tolerance on the shaft. Similarly, the tolerance on the hole is the difference between the two limits shown, or .001″. The loosest fit, or maximum clearance, occurs when the smallest shaft is in the largest hole (Fig. 12.2b). The tightest fit, or minimum clearance occurs when the largest shaft is in the smallest hole (Fig. 12.2c). The difference between these, .002″, is the **allowance**. The average clearance is .003″, which is the same difference as allowed in the example in Fig. 12.1a; thus, any shaft will fit any hole interchangeably.

When expressed in metric dimensions, the limits for the hole are 31.75 mm and 31.78 mm; the difference between them, 0.03 mm, is the tolerance. Similarly, the limits for the shaft are 31.70 mm and 31.67 mm;

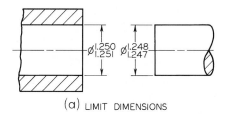
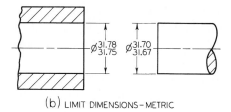

(a) LIMIT DIMENSIONS   (b) LIMIT DIMENSIONS–METRIC

FIGURE 12.1   Fits between Mating Parts.

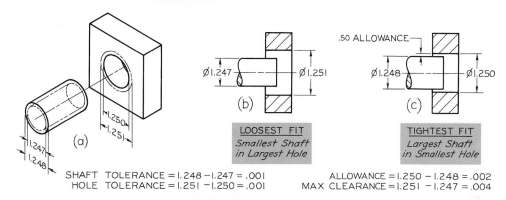

SHAFT TOLERANCE = 1.248 − 1.247 = .001   ALLOWANCE = 1.250 − 1.248 = .002
HOLE TOLERANCE = 1.251 − 1.250 = .001   MAX CLEARANCE = 1.251 − 1.247 = .004

FIGURE 12.2   Limit Dimensions.

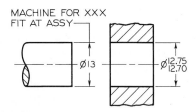

FIGURE 12.3   Noninterchangeable Fit.

the tolerance on the shaft is the difference between them, or 0.03 mm.

When parts are required to fit properly in assembly but not to be interchangeable, the size of one part need not be toleranced, but is indicated to be made to fit at assembly (Fig. 12.3).

■ 12.2 SIZE DESIGNATIONS

At this point, it is well to fix in mind the definitions of certain terms (ANSI/ASME Y14.5M–1994).

Nominal size is the designation used for general identification and is usually expressed in common fractions. In Fig. 12.1, the nominal size of both hole and shaft, which is $1\frac{1}{4}''$, would be 1.25″ or 31.75 mm in a decimal system of dimensioning.

Basic size, **or basic dimension**, is the theoretical size from which limits of size are derived by the application of allowances and tolerances. It is the size from which limits are determined for the size, shape, or location of a feature. In Fig. 12.1a, the basic size is the decimal equivalent of the nominal size $1\frac{1}{4}''$, or 1.250″ or 31.75 mm in Fig. 12.1b.

Actual size is the measured size of the finished part.

Allowance is the minimum clearance space (or maximum interference) intended between the maximum material condition (MMC) of mating parts. In Fig. 12.2c, the allowance is the difference between the smallest hole, 1.250″, and the largest shaft, 1.248″, or .002″. Allowance, then, represents the tightest permissible fit and is simply the smallest hole minus the largest shaft. For clearance fits, this difference will be positive, while for interference fits it will be negative.

■ 12.3 FITS BETWEEN MATING PARTS

"Fit is the general term used to signify the range of tightness or looseness that may result from the application of a specific combination of allowances and tolerances in mating parts." [ANSI B4.1–1967 (R1994) and ANSI B4.2–1978 (R1994)]. There are four general types of fits between parts.

1. In a **clearance fit**, an internal member fits in an external member (as a shaft in a hole) and always leaves a space or clearance between the parts. In Fig. 12.2c, the largest shaft is 1.248″ and the smallest hole is 1.250″, which permits a minimum air space of .002″ between the parts. This space is the allowance, and in a clearance fit it is always positive.

2. In an **interference fit**, the internal member is larger than the external member such that there is always an actual interference of metal. In Fig. 12.4a, the smallest shaft is 1.2513″, and the largest hole is 1.2506″, so that there is an actual interference of metal amounting to at least .0007″. Under maximum material conditions the interference would be .0019″. This interference is the allowance, and in an interference fit it is always negative.

3. A **transition fit** may result in either a clearance or interference condition. In Fig. 12.4b, the smallest shaft, 1.2503″, will fit in the largest hole, 1.2506″, with .003″ to spare. But the largest shaft, 1.2509″, will have to be forced into the smallest hole, 1.2500″, with an interference of metal (negative allowance) of .0009″.

4. In a **line fit**, the limits of size are so specified that a clearance or surface contact may result when mating parts are assembled.

■ 12.4 SELECTIVE ASSEMBLY

If allowances and tolerances are properly given, mating parts can be completely interchangeable. But for close fits, it is necessary to specify very small allowances and tolerances, and the cost may be very high. To avoid this expense, either manual or computer-controlled selective assembly is often used. In **selective assembly**, all parts are inspected and classified into several grades according to actual sizes, so that "small" shafts can be matched with "small" holes, "medium" shafts with "medium" holes, and so on. In this way, very satisfactory fits may be obtained at much less expense than by machining all mating parts to very accurate dimensions. Since a transition fit may or may not represent an interference of metal, interchangeable as-

sembly generally is not as satisfactory as selective assembly.

■ 12.5 BASIC HOLE SYSTEM

Standard reamers, broaches, and other standard tools are often used to produce holes, and standard plug gages are used to check the actual sizes. On the other hand, shafting can easily be machined to any size desired. Therefore, toleranced dimensions are commonly figured on the so-called **basic hole system**, in which the *minimum hole is taken as the basic size*, an allowance is assigned, and tolerances are applied on both sides of, and away from, this allowance.

In Fig. 12.5a, the minimum size of the hole, .500″, is taken as the basic size. An allowance of .002″ is decided on and subtracted from the basic hole size, making the maximum shaft .498″. Tolerances of .002″ and .003″, respectively, are applied to the hole and shaft to obtain the maximum hole of .502″ and the minimum shaft of .495″. Thus, the minimum clearance between the parts becomes .500″ − .498″ = .002″ (smallest hole minus largest shaft), and the maximum clearance is .502″ − .495″ = .007″ (largest hole minus smallest shaft).

In the case of an interference fit, the maximum shaft size would be found by *adding the desired allowance* (maximum interference) to the basic hole size. In Fig. 12.4a, the basic size is 1.2500″. The maximum interference decided on was .0019″, which added to the basic size gives 1.2519″, the largest shaft size.

The basic hole size can be changed to the basic shaft size by subtracting the allowance for a clearance fit, or adding it for an interference fit. The result is the largest shaft size, which is the new basic size.

■ 12.6 BASIC SHAFT SYSTEM

In some branches of industry, such as textile machinery manufacturing, in which use is made of a great deal of cold-finished shafting, the **basic shaft system** is often

FIGURE 12.4 Fits between Parts.

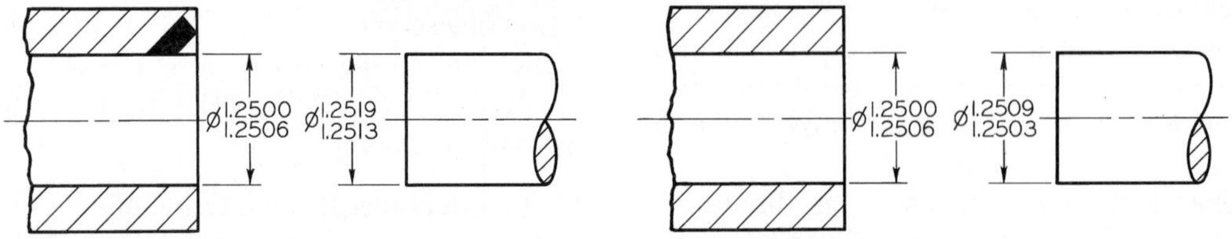

(a) INTERFERENCE FIT

(b) TRANSITION FIT

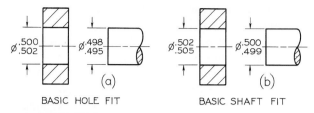

FIGURE 12.5 Basic-Hole and Basic-Shaft Systems.

used. This system should be used only when there is a reason for it. For example, it is advantageous when several parts having different fits, but one nominal size, are required on a single shaft. In this system, *the maximum shaft is taken as the basic size*, an allowance for each mating part is assigned, and tolerances are applied on both sides of, and away from, this allowance.

In Fig. 12.5b, the maximum size of the shaft, .500″, is taken as the basic size. An allowance of .002″ is decided on and added to the basic shaft size, giving the minimum hole, .502″. Tolerances of .003″ and .001″, respectively, are applied to the hole and shaft to obtain the maximum hole, .505″, and the minimum shaft, .499″. Thus, the minimum clearance between the parts is .502″ − .500″ = .002″ (smallest hole minus largest shaft), and the maximum clearance is .505″ − .499″ = .006″ (largest hole minus smallest shaft).

In the case of an interference fit, the minimum hole size would be found by *subtracting the desired allowance from the basic shaft size.*

The basic shaft size may be changed to the basic hole size by adding the allowance for a clearance fit or by subtracting it for an interference fit. The result is the smallest hole size, which is the new basic size.

■ 12.7 SPECIFICATIONS OF TOLERANCES

A tolerance of a decimal dimension must be given in decimal form to the same number of places (see Fig. 12.8).

General tolerances on decimal dimensions in which tolerances are not given may also be covered in a printed note, such as

DECIMAL DIMENSION TO BE HELD TO ± .001.

Thus, if a dimension 3.250 is given, the worker machines between the limits 3.249 and 3.251 (see Fig. 12.9).

Tolerances for metric dimensions may be covered in a note, such as the commonly used

METRIC DIMENSIONS TO BE HELD TO ± 0.08.

Thus, when the given dimension of 3.250″ is converted to millimeters, the worker machines between the limits of 82.63 mm and 82.74 mm.

Every dimension on a drawing should have a tolerance, either direct or by general tolerance note, except that commercial material is often assumed to have the tolerances set by commercial standards.

It is customary to indicate an overall general tolerance for all common fraction dimensions by means of a printed note in or just above the title block (see Fig. 12.9).

EXAMPLE

ALL FRACTIONAL DIMENSIONS ± $\frac{1}{64}$″
UNLESS OTHERWISE SPECIFIED.

General angular tolerances also may be given as

ANGULAR TOLERANCE ± 1°.

Several methods of expressing tolerances in dimensions are approved by ANSI (ANSI/ASME Y14.5M–1994) as follows.

1. Limit dimensioning. In this preferred method, the maximum and minimum limits of size and location are specified, as shown in Fig. 12.6. The high limit (maximum value) is placed above the low limit (minimum value) (Fig. 12.6a). In single-line note form, the low limit precedes the high limit separated by a dash (Fig. 12.6b).

2. Plus-or-minus dimensioning. In this method the basic size is followed by a plus-or-minus expression of tolerance resulting in either a unilateral or bilateral tolerance (Fig. 12.7). If two unequal tolerance numbers are given, one plus and one minus, the plus is placed above the minus. One of the numbers may be zero, if desired. If a single tolerance value is given, it is preceded by the plus-or-minus symbol (±) (Fig. 12.8). This method should be used when the plus and minus values are equal.

FIGURE 12.6 Method of Giving Limits.

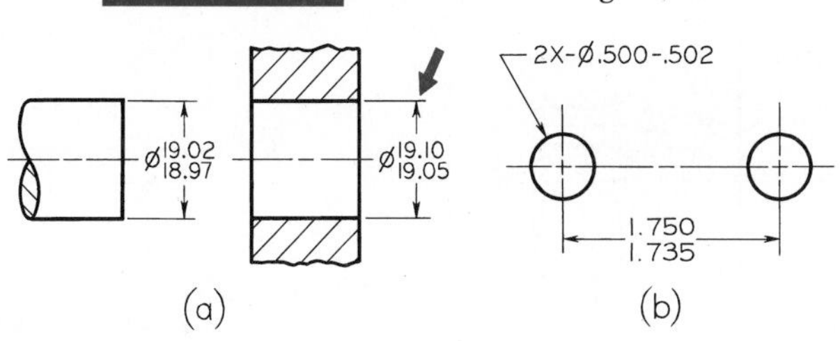

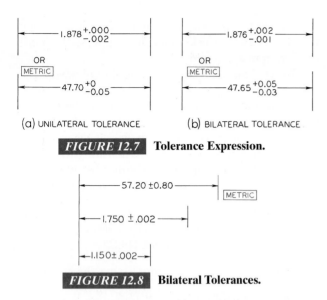

(a) UNILATERAL TOLERANCE   (b) BILATERAL TOLERANCE

FIGURE 12.7 **Tolerance Expression.**

57.20 ±0.80 — METRIC

1.750 ±.002

1.150±.002

FIGURE 12.8 **Bilateral Tolerances.**

The **unilateral system of tolerances** allows variations in only one direction from the basic size. This method is advantageous when a critical size is approached as material is removed during manufacture, as in the case of close-fitting holes and shafts. In Fig. 12.7a the basic size is 1.878″ (47.70 mm). The tolerance .002″ (0.05 mm) is all in one direction—toward the smaller size. If this is a shaft diameter, the basic size 1.878″ (47.70 mm) is the size nearest the critical size because it is nearest to the tolerance zone; hence the tolerance is taken *away* from the critical size. A unilateral tolerance is always all plus or all minus; that is, either the plus or the minus value must be zero. However, the zeros should be given as shown in Fig. 12.7a.

The **bilateral system of tolerances** allows variations in both directions from the basic size. Bilateral tolerances are usually given with location dimensions or with any dimensions that can be allowed to vary in either direction. In Fig. 12.7b, the basic size is 1.876″ (47.65 mm), and the actual size may be larger by .002″ (0.05 mm) or smaller by .001″ (0.03 mm). If it is desired to specify an equal variation in both directions, the combined plus-or-minus symbol (±) is used with a single value, as shown in Fig. 12.8

A typical example of limit dimensioning is given in Fig. 12.9.

FIGURE 12.9 **Limit Dimensions.**

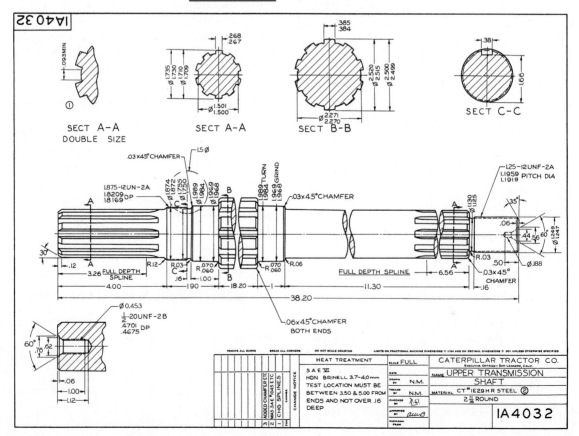

3. Single-limit dimensioning. It is not always necessary to specify both limits. MIN or MAX is often placed after a number to indicate minimum or maximum dimensions desired where other elements of design determine the other unspecified limit. For example, a thread length may be dimensioned as: $|\leftarrow 1.500 \rightarrow|$ MIN FULL THD or a radius dimensioned as .05 R MAX $\diagdown$. Other applications include depths of holes, chamfers, and so on

4. Angular tolerances are usually bilateral and in terms of degrees, minutes, and seconds.

> **EXAMPLES** $25° \pm 1°$, $25° 0' \pm 0° 15'$, or $25° \pm 0.25°$ (see also §12.18).

■ 12.8 AMERICAN NATIONAL STANDARD LIMITS AND FITS

The American National Standards Institute has issued the ANSI B4.1–1967 (R1994), "Preferred Limits and Fits for Cylindrical Parts," defining terms and recommending preferred standard sizes, allowances, tolerances, and fits in terms of the decimal inch. This standard gives a series of standard types and classes of fits on a unilateral hole basis such that the fit produced by mating parts in any one class will produce approximately similar performance throughout the range of sizes. These tables prescribe the fit for any given size or type of fit; they also prescribe the standard limits for the mating parts which will produce the fit.

The tables are designed for the basic hole system (§12.5; see Appendixes 5–9). For coverage of the metric system of tolerances and fits, see §§12.11–12.13 and Appendixes 11–14.

Letter symbols to identify the five types of fits are as follows:

RC	Running or Sliding Clearance Fits
LC	Locational Clearance Fits
LT	Transition Clearance or Interference Fits
LN	Locational Interference Fits
FN	Force or Shrink Fits

These letter symbols, plus a number indicating the class of fit within each type, are used to indicate a complete fit. Thus, FN 4 means a Class 4 Force Fit. The fits are described [ANSI B4.1–1967 (R1994)] as follows.

RUNNING AND SLIDING FITS

Running and sliding fits, for which description of classes of fits and limits of clearance are given (Appendix 5), are intended to provide a similar running performance, with suitable lubrication allowance, throughout the range of sizes. The clearances for the first two classes, used chiefly as slide fits, increase more slowly with diameter than the other classes, so that accurate location is maintained even at the expense of free relative motion.

LOCATIONAL FITS

Locational fits (Appendixes 6–8) are fits intended to determine only the location of the mating parts; they may provide rigid or accurate location, as with interference fits, or provide some freedom of location, as with clearance fits. Accordingly, they are divided into three groups: clearance fits, transition fits, and interference fits.

FORCE FITS

Force or shrink fits (Appendix 9) constitute a special type of interference fit, normally characterized by maintenance of constant bore pressures throughout the range of sizes. The interference therefore varies almost directly with diameter, and the difference between its minimum and maximum value is small to maintain the resulting pressures within reasonable limits.

In the tables for each class of fit, the range of nominal sizes of shafts or holes is given in inches. To simplify the tables and reduce the space required to present them, the other values are given in thousandths of an inch. Minimum and maximum limits of clearance are given; the top number is the least clearance, or the allowance, and the lower number the maximum clearance, or the greatest looseness of fit. Then, under the heading "Standard Limits" are given the limits for the hole and for the shaft that are to be applied algebraically to the basic size to obtain the limits of size for the parts, using the basic hole system.

For example, take a 2.0000″ basic diameter with a Class RC 1 fit. This fit is given in Appendix 5. In the column headed "Nominal Size Range, Inches," find 1.97–3.15, which embraces the 2.0000″ basic size. Reading to the right, we find under "Limits of Clearance" the values 0.4 and 1.2, representing the maximum clearance between the parts *in thousandths of an inch*. To get these values in inches, simply multiply by one thousandth; thus, $\frac{4}{10} \times \frac{1}{1000} = .0004″$. To convert 0.4 thousandths to inches, simple move the decimal point three places to the left; thus, .0004″. Therefore, for this 2.0000″ diameter, with a Class RC 1 fit, the minimum clearance, or allowance, is .0004″, and the maximum clearance, representing the greatest looseness, is .0012″.

Reading farther to the right, we find under "Standard Limits" the value + 0.5, which when converted to inches is .0005″. Add this to the basic size thus: 2.0000″ + .0005″ = 2.0005″, the upper limit of the hole. Since the other value given for the hole is zero, the lower limit of the hole is the basic size of the hole, or 2.0000″. The hole would then be dimensioned as

$$\begin{matrix} 2.0005 \\ 2.0000 \end{matrix} \text{ or } 2.0000 \begin{matrix} +.0005 \\ -.0000. \end{matrix}$$

The limits for the shaft are read as −.0004″ and −.0007″. To get the limits of the shaft, subtract these values from the basic size; thus,

$$2.000″ − .0004″ = 1.9996″ \text{ (upper limit)}$$

$$2.000″ − .0007″ = 1.9993″ \text{ (lower limit)}.$$

The shaft would then be dimensioned in inches as follows:

$$\begin{matrix} 1.9996 \\ 1.9993 \end{matrix} \text{ or } 1.9996 \begin{matrix} +.0000 \\ -.0003. \end{matrix}$$

■ 12.9 ACCUMULATION OF TOLERANCES

In tolerance dimensioning, it is very important to consider the effect of one tolerance on another. When the location of a surface in a given direction is affected by more than one tolerance figure, the tolerances are *cumulative*. For example, in Fig. 12.10a, if dimension Z is omitted, surface A will be controlled by both dimensions X and Y, and there can be a total variation of .010″ instead of the variation of .005″ permitted by dimension Y, which is the dimension directly applied to

surface A. Furthermore, if the part is made to all the minimum tolerances of X, Y, and Z, the total variation in the length of the part will be .015″, and the part can be as short as 2.985″. However, the tolerance on the overall dimension W is only .005″, permitting the part to be only as short as 2.995″. The part is superfluously dimensioned.

In some cases, for functional reasons, it may be desired to hold all three small dimensions X, Y, and Z closely without regard to the overall length. In such a case the overall dimension is just a **reference dimension** and should be denoted with parentheses. In other cases it may be desired to hold two small dimensions X and Y and the overall closely without regard to dimension Z. In that case, dimension Z should be omitted, or denoted as a reference with parentheses.

As a rule, it is best to dimension each surface so that it is affected by only one dimension. This can be done by referring all dimensions to a single datum surface, such as B, as shown in Fig. 12.10b (see also Figs. 11.37d to 11.37f).

■ 12.10 TOLERANCES AND MACHINING PROCESSES

As has been repeatedly stated in this chapter, tolerances should be as coarse as possible and still permit satisfactory use of the part. If this is done, great savings can be effected from the use of less expensive tools, lower labor and inspection costs, and reduced scrapping of material.

Figure 12.11 shows a chart of tolerance grades obtainable in relation to the accuracy of machining processes that may be used as a guide by the designer. Metric values may be ascertained by multiplying the given decimal-inch values by 25.4 and rounding off the product to one less place to the right of the decimal point than given for the decimal-inch value (see

FIGURE 12.10 **Cumulative Tolerances.**

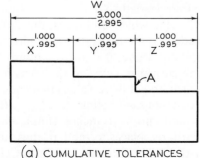

(a) CUMULATIVE TOLERANCES

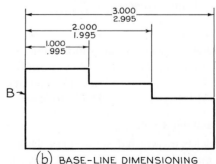

(b) BASE-LINE DIMENSIONING

Range of Sizes		Tolerances								
From	To and Including									
.000	.599	.00015	.0002	.0003	.0005	.0008	.0012	.002	.003	.005
.600	.999	.00015	.00025	.0004	.0006	.001	.0015	.0025	.004	.006
1.000	1.499	.0002	.0003	.0005	.0008	.0012	.002	.003	.005	.008
1.500	2.799	.00025	.0004	.0006	.001	.0015	.0025	.004	.006	.010
2.800	4.499	.0003	.0005	.0008	.0012	.002	.003	.005	.008	.012
4.500	7.799	.0004	.0006	.001	.0015	.0025	.004	.006	.010	.015
7.800	13.599	.0005	.0008	.0012	.002	.003	.005	.008	.012•	.020
13.600	20.999	.0006	.001	.0015	.0025	.004	.006	.010	.015	.025

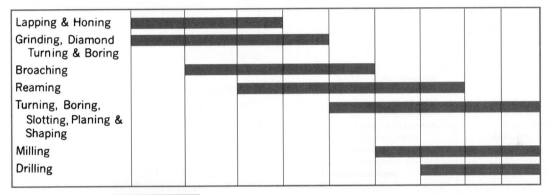

FIGURE 12.11 Tolerances Related to Machining Processes.

§11.10). For detailed information on manufacturing processes and measuring devices, see Chapter 10.

■ 12.11 METRIC SYSTEM OF TOLERANCES AND FITS

The preceding material on limits and fits between mating parts is suitable, without need of conversion, for the decimal-inch system of measurement. A system of preferred metric limits and fits by the International Organization for Standardization (ISO) is in the ANSI B4.2 standard. The system is specified for holes, cylinders, and shafts, but it is also adaptable to fits between parallel surfaces of such features as keys and slots. The following terms for metric fits, although somewhat similar to those for decimal-inch fits, are illustrated in Fig. 12.12.

1. Basic size is the size from which limits or deviations are assigned. Basic sizes, usually diameters, should be selected from a table of preferred sizes (see Fig. 12.17).

2. Deviation is the difference between the basic size

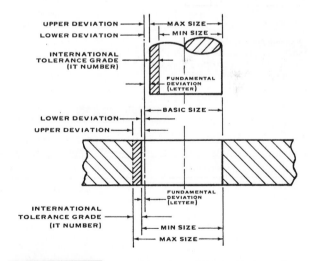

FIGURE 12.12 Terms Related to Metric Limits and Fits [ANSI B4.2– 1978 (R1994)].

and the hole or shaft size. This is equivalent to the tolerance in the decimal-inch system.

3. Upper deviation is the difference between the basic size and the permitted maximum size of the part.

This is comparable to maximum tolerance in the decimal-inch system.

4.   Lower deviation is the difference between the basic size and the minimum permitted size of the part. This is comparable to minimum tolerance in the decimal-inch system.

5.   Fundamental deviation is the deviation closest to the basic size. This is comparable to minimum allowance in the decimal-inch system.

6.   Tolerance is the difference between the permitted minimum and maximum sizes of a part.

7.   International tolerance grade (IT) is a set of tolerances that varies according to the basic size and

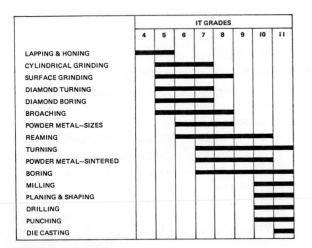

FIGURE 12.13   International Tolerance Grades Related to Machining Processes [ANSI B4.2–1978 (R1994)].

provides a uniform level of accuracy within the grade. For example, in the dimension 50H8 for a close-running fit, the IT grade is indicated by the numeral 8. (The letter H indicates that the tolerance is on the hole for the 50-mm dimension.) In all, there are 18 IT grades—IT01, IT0, and IT1 through IT16 (Figs. 12.13 and 12.14)—for IT grades related to machining processes and for the practical use of the IT grades (see also Appendix 10).

8.   Tolerance zone refers to the relationship of the tolerance to basic size. It is established by a combination of the fundamental deviation indicated by a letter and the IT grade number. In the dimension 50H8, for the close-running fit, the H8 specifies the tolerance zone (see Fig. 12.14).

9.   The hole-basis system of preferred fits is a system in which the basic diameter is the minimum size. For the generally preferred hole-basis system, the fundamental deviation is specified by the uppercase letter H (Fig. 12.15a).

10.   The shaft-basis system of preferred fits is a system in which the basic diameter is the maximum size of the shaft. The fundamental deviation is given by the lowercase letter f (Fig. 12.15b).

11.   An interference fit results in an interference between two mating parts under *all* tolerance conditions.

12.   A transition fit results in either a clearance or an interference condition between two assembled parts.

13.   Tolerance symbols are used to specify the tolerances and fits for mating parts (Fig. 12.15c). For the hole-basis system, the 50 indicates the diameter in millimeters; the fundamental deviation for the hole is in-

FIGURE 12.14   Practical Use of International Tolerance Grades.

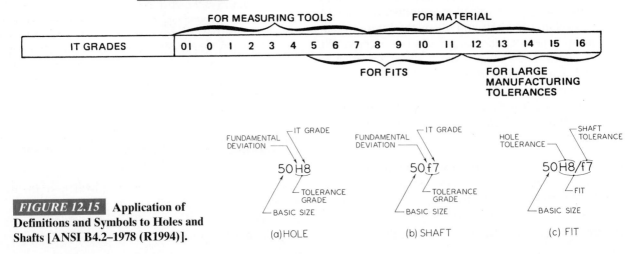

FIGURE 12.15   Application of Definitions and Symbols to Holes and Shafts [ANSI B4.2–1978 (R1994)].

50 H8   50H8$\binom{50.039}{50.000}$ $\frac{50.039}{50.000}$(50H8)

(a) PREFERRED   (b)   (c)

FIGURE 12.16 Acceptable Methods of Giving Tolerance Symbols (ANSI/ASME Y14.5M–1994).

dicated by the capital letter H, and for the shaft it is indicated by the lowercase letter f. The numbers following the letters indicate this IT grade. Note that the symbols for the hole and shaft are separated by the slash. Tolerance symbols for a 50-mm-diameter hole may be given in several acceptable forms (Fig. 12.16). The values in parentheses are for reference only and may be omitted. The upper and lower limit values may be found in Appendix 11.

■ 12.12 PREFERRED SIZES

The preferred basic sizes for computing tolerances are given in Table 12.1. Basic diameters should be selected

Basic Size, mm		Basic Size, mm		Basic Size mm	
First Choice	Second Choice	First Choice	Second Choice	First Choice	Second Choice
1		10		100	
	1.1		11		110
1.2		12		120	
	1.4		14		140
1.6		16		160	
	1.8		18		180
2		20		200	
	2.2		22		220
2.5		25		250	
	2.8		28		280
3		30		300	
	3.5		35		350
4		40		400	
	4.5		45		450
5		50		500	
	5.5		55		550
6		60		600	
	7		70		700
8		80		800	
	9		90		900
				1000	

TABLE 12.1 Preferred Sizes [ANSI B4.2–1978 (R1994)]

from the first choice column since these are readily available stock sizes for round, square, and hexagonal products.

■ 12.13 PREFERRED FITS

The symbols for either the hole-basis or shaft-basis preferred fits (clearance, transition, and interference) are given in Table 12.2. Fits should be selected from this table for mating parts where possible.

The values corresponding to the fits are found in Appendixes 11–14. Although second and third choice basic size diameters are possible, they must be calculated from tables not included in this text. For the generally preferred hole-basis system, note that the ISO symbols range from H11/c11 (loose running) to H7/u6 (force fit). For the shaft-basis system, the preferred symbols range from C11/h11 (loose fit) to U7/h6 (force fit).

Suppose that you want to use the symbols to specify the dimensions for a free-running fit (hole basis) for a proposed diameter of 48 mm. Since 48 mm is not listed as a preferred size in Table 12.1, the design is altered to use the acceptable 50-mm diameter. From the preferred fits descriptions in Table 12.2, the free-running fit (hole-basis) is H9/d9. To determine the upper and lower deviation limits of the hole as given in the preferred hole-basis table (Appendix 11), follow across from the basic size of 50 to H9 under "Free running." The limits for the hole are 50.000 and 50.062 mm. Then the upper and lower limits of deviation for the shaft are found in the d9 column under "Free running." They are 49.920 and 49.858 mm, respectively. Limits for other fits are established in a similar manner.

Limits for the shaft-basis dimensioning are determined similarly from the preferred shaft basis table in Appendix 12. See Figs. 12.16 and 12.17 for acceptable methods of specifying tolerances by symbols on drawings. A single note for the mating parts (free-running fit, hole basis) would be Ø50 H9/d9 (Fig. 12.17).

■ 12.14 GEOMETRIC TOLERANCING

Geometric tolerances state the maximum allowable variations of a form or its position from the perfect geometry implied on the drawing. The term "geometric" refers to various forms, such as a plane, a cylinder, a cone, a square, or a hexagon. Theoretically these are perfect forms, but, because it is impossible to produce perfect forms, it may be necessary to specify the

| ISO Symbol | | Description | |
Hole Basis	Shaft[a] Basis		
Clearance Fits			
H11/c11	C11/h11	***Loose-running*** fit for wide commercial tolerances or allowances on external members.	↑
H9/d9	D9/h9	***Free-running*** fit not for use where accuracy is essential, but good for large temperature variations, high running speeds, or heavy journal pressures.	More clearance
H8/f7	F8/h7	***Close-running*** fit for running on accurate machines and for accurate location at moderate speeds and journal pressures.	
H7/g6	G7/h6	***Sliding*** fit not intended to run freely, but to move and turn freely and locate accurately.	
Transition Fits			
H7/h6	H7/h6	***Locational clearance*** fit provides snug fit for locating stationary parts; but can be freely assembled and disassembled.	
H7/k6	K7/h6	***Locational transition*** fit for accurate location, a compromise between clearance and interference.	
H7/n6	N7/h6	***Locational transition*** fit for more accurate location where greater interference is permissible.	More interference
Interference Fits			
H7/p6	P7/h6	***Locational interference*** fit for parts requiring rigidity and alignment with prime accuracy of location but without special bore pressure requirements.	
H7/s6	S7/h6	***Medium drive*** fit for ordinary steel parts or shrink fits on light sections, the tightest fit usable with cast iron.	
H7/u6	U7/h6	***Force*** fit suitable for parts which can be highly stressed or for shrink fits where the heavy pressing forces required are impractical.	↓

[a] The transition and interference shaft-basis fits shown do not convert to exactly the same hole-basis fit conditions for basic sizes in the range from Q through 3 mm. Interference fit P7/h6 converts to a transition fit H7/p6 in the above size range.

TABLE 12.2   Preferred Fits *[ANSI B4.2–1978 (R1994)]*

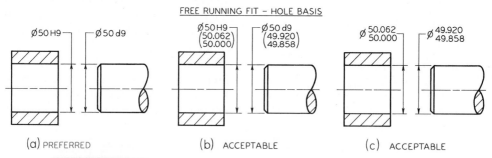

FREE RUNNING FIT – HOLE BASIS

(a) PREFERRED   (b) ACCEPTABLE   (c) ACCEPTABLE

FIGURE 12.17 Methods of Specifying Tolerances with Symbols for Mating Parts.

amount of variation permitted. These tolerances specify either the diameter or the width of a tolerance zone within which a surface or the axis of a cylinder or a hole must be if the part is to meet the required accuracy for proper function and fit. When tolerances of form are not given on a drawing, it is customary to assume that, regardless of form variations, the part will fit and function satisfactorily.

Tolerances of form and position or location control such characteristics as straightness, flatness, parallelism, perpendicularity (squareness), concentricity, roundness, angular displacement, and so on.

Methods of indicating geometric tolerances by means of *geometric characteristic symbols* are recommended rather than by traditional notes and are discussed and illustrated subsequently. See the latest Dimensioning and Tolerancing Standard (ANSI/ASME Y14.5M–1994) for more complete coverage.

■ 12.15 SYMBOLS FOR TOLERANCES OF POSITION AND FORM

Since traditional narrative notes for specifying tolerances of *position* (location) and *form* (shape) may be confusing or unclear, may require too much space, and may not be understood internationally, most multinational companies have adopted symbols for such specifications (ANSI/ASME Y14.5M–1994). These ANSI symbols provide an accurate and concise means of specifying geometric characteristics and tolerances in a

minimum of space (Table 12.3). The symbols may be supplemented by notes if the precise geometric requirements cannot be conveyed by the symbols. For construction details of the geometric tolerancing symbols, see Appendix 37.

Combinations of the various symbols and their meanings are given in Fig. 12.18. Application of the symbols to drawing are illustrated in Fig. 12.43. The geometric characteristic symbols plus the supplementary symbols are further explained and illustrated with material adapted from ANSI/ASME Y14.5M–1994, as follows.

1. The **basic dimension symbol** is identified by the enclosing frame symbol (Fig. 12.18a). The basic dimension (size) is the value used to describe the theoretically exact size, shape, or location of a feature. It is the basis from which permissible variations are

Geometric characteristic symbols				Modifying symbols	
	Type of Tolerance	Characteristic	Symbol	Term	Symbol
For individual features	Form	Straightness	—	At maximum material condition	Ⓜ
		Flatness	⟋	At least material condition	Ⓛ
		Circularity (roundness)	○	Projected tolerance zone	Ⓟ
		Cylindricity	⌭	Free state	Ⓕ
For individual or related features	Profile	Profile of a line	⌒	Tangent plane	Ⓣ
		Profile of a surface	⌓	Diameter	⌀
				Spherical diameter	S⌀
For related features	Orientation	Angularity	∠	Radius	R
		Perpendicularity	⊥	Spherical radius	SR
		Parallelism	//	Controlled radius	CR
	Location	Position	⊕	Reference	( )
		Concentricity	◎	Arc length	⌒
		Symmetry	⩬	Statistical tolerance	⟨ST⟩
	Runout	Circular runout	↗*	Between	↔
		Total runout	↗↗*		

* ARROWHEADS MAY BE FILLED OR NOT FILLED

TABLE 12.3 Geometric Characteristic and Modifying Symbols *(ASME Y14.5M–1994)*

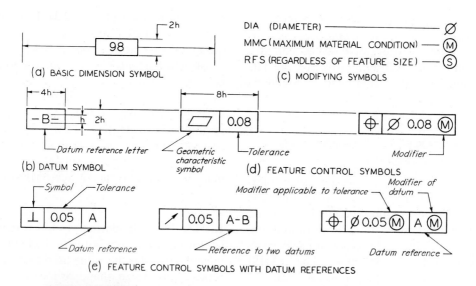

FIGURE 12.18 Use of Symbols for Tolerance of Position and Form (ASME Y14.5M–1994).

established by tolerances on other dimensions in notes, or in feature control frames.

2. The **datum identifying symbol** consists of a capital letter in a square frame and a leader line extending from the frame to the concerned feature, terminating with a triangle. The triangle may be filled or not filled (see Fig. 12.18b). Letters of the alphabet (except I, O, and Q) are used as datum-identifying letters. A point, line, plane, cylinder, or other geometric form assumed to be exact for purposes of computation may serve as

a datum from which the location or geometric relationship of features of a part may be established, as shown in Fig. 12.19.

3. **Supplementary symbols** include the symbols for MMC (maximum material condition, i.e., minimum hole diameter, maximum shaft diameter) and LMC (least material condition, i.e., maximum hole diameter, minimum shaft diameter) (Fig. 12.18c). The abbreviations MMC and LMC are also used in notes (see also Table 12.3).

FIGURE 12.19 Placement of Datum Feature Symbol (ASME Y14.5M–1994).

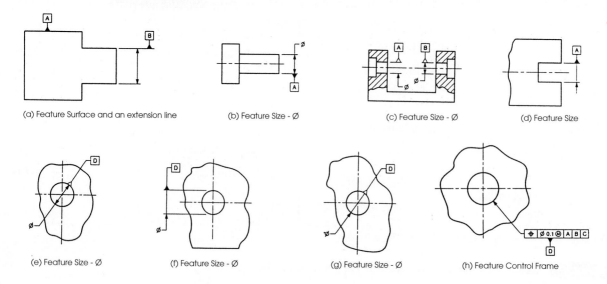

The symbol for diameter is used instead of the abbreviation DIA to indicate a diameter, and it precedes the specified tolerance in a feature control symbol (Fig. 12.18d). This symbol for diameter instead of the abbreviation DIA is used on a drawing, and it should precede the dimension. For narrative notes, the abbreviation DIA is preferred.

4. Combined symbols are found when individual symbols, datum reference letters, needed tolerances, and so on are found in a single frame (Fig. 12.18e).

A position of form tolerance is given by a feature control symbol made up of a frame about the appropriate geometric characteristic symbol plus the allowable tolerance. A vertical line separates the symbol and the tolerance (Fig. 12.18d). Where needed, the tolerance should be preceded by the symbol for the diameter and followed by the symbol for MMC or LMC.

A tolerance of position or form related to a datum is so indicated in the feature control symbol by placing the datum reference letter following either the geometric characteristic symbol or the tolerance. Vertical lines separate the entries, and, where applicable, the datum reference letter entry includes the symbol for MMC or LMC (Fig. 12.18).

■ 12.16 POSITIONAL TOLERANCES

In §12.25 there are a number of examples of the traditional methods of locating holes—that is, by means of rectangular coordinates or angular dimensions. Each dimension has a tolerance, either given directly or indicated on the completed drawing by a general note.

Figure 12.20a shows a hole located from two surfaces at right angles to each other. In Fig. 12.20b, the center may lie anywhere within a square tolerance zone, the sides of which are equal to the tolerances.

Thus, the total variations along either diagonal of the square by the coordinate method of dimensioning will be 1.4 times greater than the indicated tolerance. Hence, a .014-diameter tolerance zone would increase the square tolerance zone area 57% without exceeding the tolerance permitted along the diagonal of the square tolerance zone.

Features located by toleranced angular and radial dimensions will have a wedge-shaped tolerance zone. (see Fig. 12.29).

If four holes are dimensioned with rectangular coordinates as in Fig. 12.21a, acceptable patterns for the square tolerance zones for the holes are shown in Figs. 12.21b and 12.21c. The locational tolerances are actually greater than indicated by the dimensions.

Feature control symbols are related to the feature by one of several methods illustrated in Fig. 12.44. The following methods are preferred:

1. Adding the symbol to a note or dimension pertaining to the feature.

2. Running a leader from the symbol to the feature.

3. Attaching the side, end, or corner of the symbol frame to an extension line from the feature.

4. Attaching a side or end of the symbol frame to the dimension line pertaining to the feature.

In Fig. 12.21a, hole A is selected as a datum, and the other three are located from it. The square tolerance zone for hole A results from the tolerances on the two rectangular coordinate dimensions locating hole A. The sizes of the tolerance zones for the other three holes result from the tolerances between the holes, while their locations will vary according to the actual location of the datum hole A. Two of the many possible zone patterns are shown in Figs. 12.21b and 12.21c.

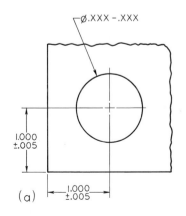

(a)

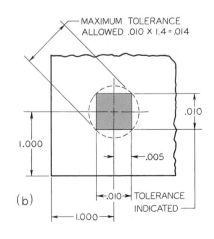

(b)

FIGURE 12.20 Tolerance Zones.

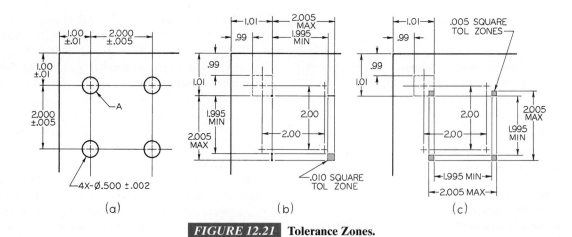

FIGURE 12.21 Tolerance Zones.

Thus, with the dimensions shown in Fig. 12.21a, it is difficult to say whether the resulting parts will actually fit the mating parts satisfactorily even though they conform to the tolerances shown on the drawing.

These disadvantages are overcome by giving exact theoretical locations by untoleranced dimensions and then specifying by a note how far actual positions may be displaced from these locations. This is called **true-position dimensioning**. It will be seen that the tolerance zone for each hole will be a circle, with the size of the circle depending on the amount of variation permitted from true position.

A true-position dimension denotes the theoretically exact position of a feature. The location of each feature, such as a hole, slot, or stud, is given by untoleranced basic dimensions identified by the enclosing frame or symbol. To prevent misunderstandings, true position should be established with respect to a datum.

In simple arrangements, the choice of a datum may be obvious and not require identification.

Positional tolerancing is identified by a characteristic symbol directed to a feature, which establishes a circular tolerance zone (Fig. 12.22).

Actually, the "circular tolerance zone" is a cylindrical tolerance zone (the diameter of which is equal to the positional tolerance while its length is equal to the length of the feature unless otherwise specified), and its axis must be within this cylinder (Fig. 12.23).

The center line of the hole may coincide with the center line of the cylindrical tolerance zone (Fig. 12.23a); it may be parallel to it but displaced so as to remain within the tolerance cylinder (Fig. 12.23b); or it may be inclined while remaining within the tolerance cylinder (Fig. 12.23c). In this last case we see that the positional tolerance also defines the limits of squareness variation.

FIGURE 12.22 True-Position Dimensioning [ANSI Y14.5M–1982 (R1988)].

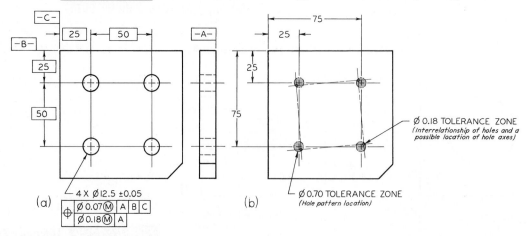

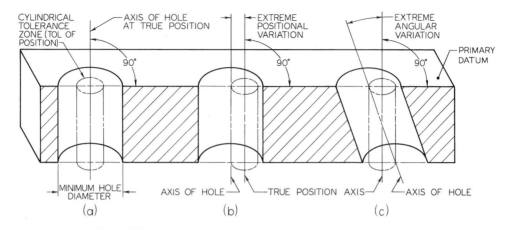

FIGURE 12.23   Cylindrical Tolerance Zone (ASME Y14.5M–1994).

In terms of the cylindrical surface of the hole, the positional tolerance specification indicates that all elements on the hole surface must be on or outside a cylinder whose diameter is equal to the minimum diameter (MMC; §12.17) or the maximum diameter of the hole minus the positional tolerance (diameter, or twice the radius), with the center line of the cylinder located at true position (Fig. 12.24).

The use of basic untoleranced dimensions to locate features at true position avoids one of the chief difficulties in tolerancing—the accumulation of tolerances (§12.9), even in a chain of dimensions (Fig. 12.25).

While features, such as holes and bosses, may vary in any direction from the true-position axis, other features, such as slots, may vary on either side of a true-position plane (Fig. 12.26).

FIGURE 12.24   True Position Interpretation (ASME Y14.5M–1994).

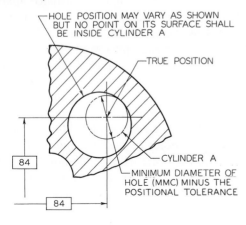

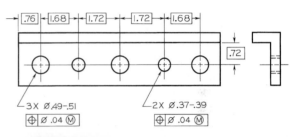

FIGURE 12.25   No Tolerance Accumulation.

Since the exact locations of the true positions are given by untoleranced dimensions, it is important to prevent the application of general tolerances to these. A note should be added to the drawing, such as

GENERAL TOLERANCES DO NOT APPLY TO BASIC TRUE-POSITION DIMENSIONS.

■ 12.17 MAXIMUM MATERIAL CONDITION

Maximum material condition, usually abbreviated MMC, means that a feature of a finished product contains the maximum amount of material permitted by the toleranced size dimensions shown for that feature. Thus, we have MMC when holes, slots, or other internal features are at minimum size, or when shafts, pads, bosses, and other external features are at their maximum size. We have MMC for both mating parts when the largest shaft is in the smallest hole and there is the least clearance between the parts.

In assigning positional tolerance to a hole, it is necessary to consider the size limits of the hole. If the hole is at MMC (smallest size), the positional

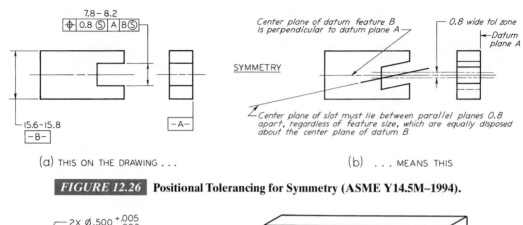

(a) THIS ON THE DRAWING . . .   (b)   . . . MEANS THIS

FIGURE 12.26 Positional Tolerancing for Symmetry (ASME Y14.5M–1994).

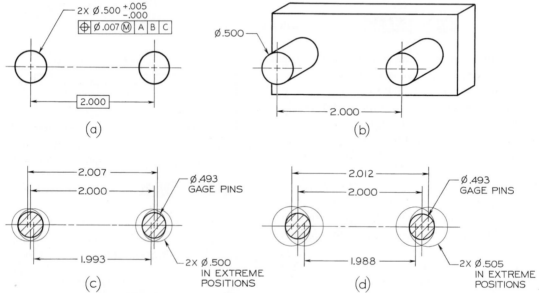

FIGURE 12.27 Maximum and Minimum Material Conditions—Two-Hole Pattern (ASME Y14.5M–1994).

tolerance is not affected, but if the hole is larger, the available positional tolerance is greater. In Fig. 12.27a, two half-inch holes are shown. If they are exactly .500″ in diameter (MMC, or smallest size) and are exactly 2.000″ apart, they should receive a gage made of two round pins .500″ in diameter fixed in a plate 2.000″ apart (Fig. 12.27b). However, the center-to-center distance between the holes may vary from 1.993″ to 2.007″.

If the .500″ diameter holes are at their extreme positions (Fig. 12.27c), the pins in the gage would have to be .007″ smaller, or .493″ diameter, to enter the holes. Thus, if the .500″ diameter holes are located at the maximum distance apart, the .493″ diameter gage pins

would contact the inner sides of the holes; and if the holes are located at the minimum distance apart, the .493″ diameter pins would contact the outer surfaces of the holes, as shown. If gagemakers' tolerances are not considered, the gage pins would have to be .493″ diameter and exactly 2.000 apart if the holes are .500 diameter, or MMC.

If the holes are .505″ diameter—that is, at maximum size (Fig. 12.27d)—they will be accepted by the same .493″ diameter gage pins at 2.000″ apart if the inner sides of the holes contact the inner sides of the gage pins and the outer sides of the holes contact the outer sides of the gage pins, as shown. Thus the holes may be 2.012″ apart, which is beyond the tolerance permitted for the center-to-center distance between the holes. Similarly, the holes may be as close together as 1.988″ from center to center, which again is outside the specified positional tolerance.

Thus, when the holes are at maximum size, a greater positional tolerance becomes available. Since all features may vary in size, it is necessary to make clear on the drawing at what basic dimension the true position applies. In all but a few exceptional cases, the additional positional tolerance available when holes are larger than minimum size is acceptable and desirable. Parts thus accepted can be freely assembled whether or not the holes or other features are within the specified positional tolerance. This practice has been recognized and used in manufacturing for years, as is evident from the use of fixed-pin gages, which have been commonly used to inspect parts and control the least favorable condition of assembly. Thus it has become common practice for both manufacturing and inspection to assume that positional tolerance applies to MMC and that greater positional tolerance becomes permissible when the part is not at MMC.

To avoid possible misinterpretation as to whether maximum material condition (MMC) applies, it should be clearly stated on the drawing by the addition of MMC symbols to each applicable tolerance or by suitable coverage in a document referenced on the drawing.

When MMC is not specified on the drawing with respect to an individual tolerance, datum reference, or both, the following rules apply:

1. True-position tolerances and related datum references apply at MMC. For a tolerance of position, RFS (regardless of feature size) may be specified on the drawing with respect to the individual tolerance, datum reference, or both, as applicable.

2. All applicable geometric tolerances, such as angularity, parallelism, perpendicularity, concentricity, and symmetry tolerances, including related datum references, or both, apply at RFS, where no modifying symbol is specified. Circular runout, total runout, concentricity, and symmetry are applicable only on an RFS basis and cannot be modified to MMC or LMC. No element of the actual feature will extend beyond the envelope of the perfect form at MMC. MMC or LMC must be specified on the drawing where it is required.

■ 12.18 TOLERANCES OF ANGLES

Bilateral tolerances have traditionally been given on angles (Fig. 12.28). Consequently, the wedge-shaped tolerance zone increases as the distance from the vertex of the angle increases. Thus, the tolerance had to be figured after considering the total displacement at the point farthest from the vertex of the angle before a tolerance could be specified that would not exceed the allowable displacement. The use of angular tolerances may be avoided by using gages. Taper turning is often handled by machining to fit a gage or by fitting to the mating part.

If an angular surface is located by a linear and an angular dimension (Fig. 12.29a), the surface must lie within a tolerance zone (Fig. 12.29b). The angular zone will be wider as the distance from the vertex increases. To avoid the accumulation of tolerances—that is, to decrease the tolerance zone—the *basic angle* tolerancing method (Fig. 12.29c) is recommended (ASME Y14.5M–1994). The angle is indicated as basic with the proper symbol and no angular tolerance is specified. The tolerance zone is now defined by two parallel planes, resulting in improved angular control (Fig. 12.29d).

■ 12.19 FORM TOLERANCES FOR SINGLE FEATURES

Straightness, flatness, roundness, cylindricity, and, in some instances, profile are form tolerances applicable to single features regardless of feature size (RFS).

1. A **straightness tolerance** specifies a tolerance zone within which an axis or all points of the considered

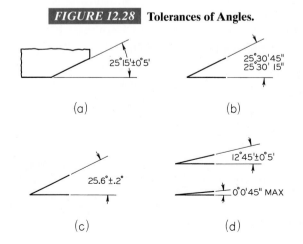

FIGURE 12.28   **Tolerances of Angles.**

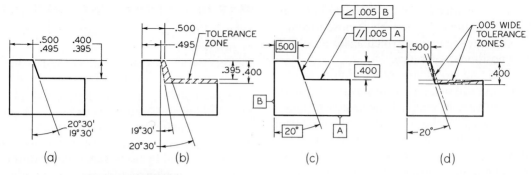

FIGURE 12.29 Angular Tolerance Zones (ASME Y14.5M–1994).

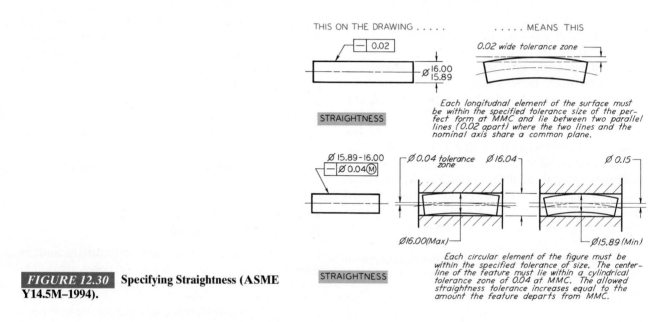

FIGURE 12.30 Specifying Straightness (ASME Y14.5M–1994).

element must lie (Fig. 12.30). Straightness is a condition in which an element of a surface or an axis is a straight line.

2. A **flatness tolerance** specifies a tolerance zone defined by two parallel planes within which the surface must lie (Fig. 12.31). Flatness is the condition of a surface having all elements in one plane.

3. A **roundness (circularity) tolerance** specifies a tolerance zone bounded by two concentric circles within which each circular element of the surface must lie (Fig. 12.32). Roundness is a condition of a surface of revolution in which, for a cone or cylinder, all points of the surface intersected by any plane perpendicular to a common axis are equidistant from that axis. For a sphere, all points of the surface intersected by any

plane passing through a common center are equidistant from that center.

4. A **cylindricity tolerance** specifies a tolerance zone bounded by two concentric cylinders within which the surface must lie (Fig. 12.33). This tolerance applies to both circular and longitudinal elements of the entire surface. Cylindricity is a condition of a surface of revolution in which all points of the surface are equidistant from a common axis. When no tolerance of form is given, many possible shapes may exist within a tolerance zone, as illustrated in Fig. 12.34.

5. A **profile tolerance** specifies a uniform boundary or zone along the true profile within which all elements of the surface must lie (Figs. 12.35 and 12.36). A profile is the outline of an object in a given plane (two-

THIS ON THE DRAWING . . . . .

FLATNESS

. . . . . MEANS THIS

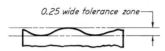

0.25 wide tolerance zone

The surface must be within the
specified tolerance of size and must lie
between two parallel planes 0.25 apart.

FIGURE 12.31 **Specifying Flatness (ASME Y14.5M–1994).**

dimensional) figure. Profiles are formed by projecting a three-dimensional figure onto a plane or by taking cross sections through the figure, with the resulting profile composed of such elements as straight lines, arcs, or other curved lines.

■ 12.20 FORM TOLERANCES FOR RELATED FEATURES

Angularity, parallelism, perpendicularity, and in some instances, profile are form tolerances applicable to related features. These tolerances control the attitude of features to one another (ASME Y14.5M–1994).

1. An **angularity tolerance** specifies a tolerance zone defined by two parallel planes at the specified basic angle (other than 90°) from a datum plane or axis

THIS ON THE DRAWING . . . . MEANS THIS

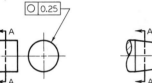

0.25 wide
tolerance zone

CYLINDER   CONE   SECTION A–A

Each circular element of the surface in any plane
perpendicular to a common axis must be within
the specified tolerance of size and must lie between two concentric circles — one having a radius
0.25 larger than the other.

ROUNDNESS

FIGURE 12.32 **Specifying Roundness for a Cylinder or Cone (ASME Y14.5M–1994).**

THIS ON THE DRAWING . . . . MEANS THIS

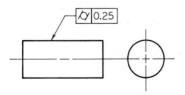

0.25 wide tolerance zone

CYLINDRICITY

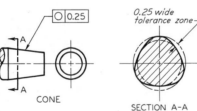

The cylindrical surface must be within
the specified tolerance of size and must
lie between two concentric cylinders — one
having a radius 0.25 larger than the other.

FIGURE 12.33 **Specifying Cylindricity (ASME Y14.5M–1994).**

THIS ON THE DRAWING . . . . MEANS THIS

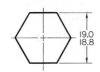

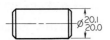

Tolerance zone or boundary within which forms may vary when no tolerance of form is given.

FIGURE 12.34 **Acceptable Variations of Form—No Specified Tolerance of Form.**

THIS ON THE DRAWING . . . . MEANS THIS

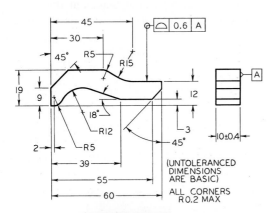

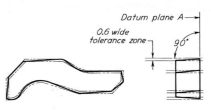

Surfaces all around must lie within two parallel boundaries 0.6 apart equally disposed about the true profile which are perpendicular to datum plane A. Radii of part corners must not exceed 0.2.

FIGURE 12.35 **Specifying Profile of a Surface All Around (ASME Y14.5M–1994).**

THIS ON THE DRAWING . . . . MEANS THIS

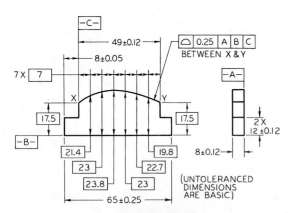

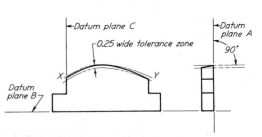

The surface between points X & Y must lie between the two profile boundaries 0.25 apart, equally disposed about the true profile, which are perpendicular to datum plane A and positioned with respect to datum planes B & C.

FIGURE 12.36 **Specifying Profile of a Surface between Points (ASME Y14.5M–1994).**

THIS ON THE DRAWING . . . . .

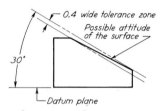

ANGULARITY

. . . . . MEANS THIS

0.4 wide tolerance zone

Possible attitude
of the surface

30°

Datum plane

The surface must be within the
specified tolerance of size and must lie
between two parallel planes 0.4 apart which
are inclined at 30° to the datum plane A.

FIGURE 12.37 **Specifying Angularity for a Plane Surface (ASME Y14.5M–1994).**

within which the surface or the axis of the feature must lie (Fig. 12.37).

2. A **parallelism tolerance** specifies a tolerance zone defined by two parallel planes or lines parallel to a datum plane or axis, respectively, within which the surface or axis of the feature must lie, or the parallelism tolerance may specify a cylindrical tolerance zone parallel to a datum axis within which the axis of the feature must lie (Figs. 12.38–12.40).

3. **Perpendicularity tolerance.** Perpendicularity is a condition of a surface, median plane, or axis at 90° to a datum plane or axis. A perpendicularity tolerance specifies one of the following:

a. A tolerance zone is defined by two parallel planes perpendicular to a datum plane, datum axis, or axis within which the surface of the feature must lie (Fig. 12.41).

b. A cylindrical tolerance zone perpendicular to a datum plane within which the axis of the feature must lie (Fig. 12.42).

4. **Concentricity tolerance.** Concentricity is the condition in which the axes of all cross-sectional elements of a feature's surface of revolution are common to the axis of a datum feature. A concentricity tolerance specifies a cylindrical tolerance zone whose axis coincides with a datum axis and within which all cross-sectional axes of the feature being controlled must lie (Fig. 12.43).

■ 12.21 APPLICATION OF GEOMETRIC TOLERANCING

The use of various feature control symbols in lieu of notes for position and form tolerance dimensions as abstracted from ASME Y14.5M–1994 is illustrated in Fig. 12.44. For a more detailed treatment of geometric tolerancing, consult the latest ASME Y14.5M dimensioning and tolerancing standard.

■ 12.22 SURFACE ROUGHNESS, WAVINESS, AND LAY

The modern demands of the automobile, the airplane, and other modern machines that can stand heavier loads and higher speeds with less friction and wear have increased the need for accurate control of surface quality by the designer regardless of the size of the

THIS ON THE DRAWING . . . . MEANS THIS

0.12 wide tolerance zone

Possible orientation of
the surface

Datum plane A

PARALLELISM

The surface must be within the specified
tolerance of size and must lie between two
planes 0.12 apart which are parallel to the
datum plane A.

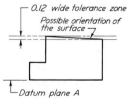

FIGURE 12.38 **Specifying Parallelism for a Plane Surface (ASME Y14.5M–1994).**

THIS ON THE DRAWING . . . . .

. . . . . MEANS THIS

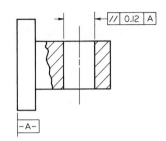

PARALLELISM

Datum plane A

0.12 wide tolerance zone

Possible orientation of the feature axis

The feature axis must be within the specified tolerance of location and must lie between two planes 0.12 apart which are parallel to the datum plane, regardless of feature size.

FIGURE 12.39 **Specifying Parallelism for an Axis Feature RFS (ASME Y14.5M–1994).**

THIS ON THE DRAWING . . . . .

. . . . . MEANS THIS

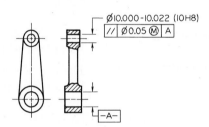

PARALLELISM

Possible orientation of feature axis

0.05 diameter tolerance zone

Datum axis A

The feature axis must be within the specified tolerance of location. Where the feature is at maximum material condition (10.00), the maximum parallelism tolerance is 0.05 diameter. Where the feature departs from its MMC size, an increase in the parallelism tolerance is allowed which is equal to the amount of such departure.

FIGURE 12.40 **Specifying Parallelism for an Axis Feature at MMC (ASME Y14.5M–1994).**

feature. Simple finish marks are not adequate to specify surface finish on such parts.

Surface finish is intimately related to the functioning of a surface, and proper specification of finish of such surfaces as bearings and seals is necessary. Surface quality specifications should be used only where needed, since the cost of producing a finished surface becomes greater as the quality of the surface called for is increased. Generally, the ideal surface finish is the roughest one that will do the job satisfactorily.

The system of surface texture symbols recommended by ANSI/ASME (Y14.36M–1996) for use on drawings, regardless of the system of measurement used, is now broadly accepted by American industry. These symbols are used to define *surface texture, roughness,* and *lay.* See Fig. 12.45 for the meaning and construction of these symbols. The basic surface texture symbol in Fig. 12.46a indicates a finished or machined surface by any method, just as does the general V symbol (Fig. 12.20a). Modifications to the basic surface texture symbol (Figs. 12.44b to 12.44d), define restrictions on material removal for the finished surface. Where surface texture values other than roughness average (R_a) are specified, the symbol must be drawn with the horizontal extension, as shown in Fig. 12.45e. Construction details for the symbols are given in Fig. 12.45f.

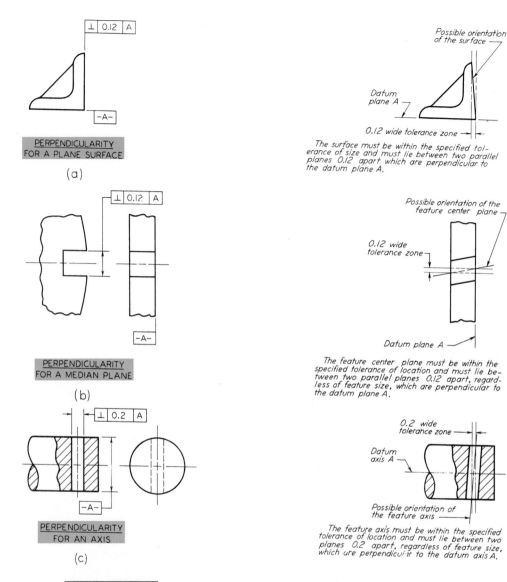

FIGURE 12.41 Specifying Perpendicularity (ASME Y14.5M–1994).

Applications of the surface texture symbols are given in Fig. 12.46a. Note that the symbols read from the bottom and/or the right side of the drawing and that they are not drawn at any angle or upside down.

Measurements for roughness and waviness, unless otherwise specified, apply in the direction that gives the maximum reading, usually across the lay (Fig. 12.46b). The recommended roughness height values are given in Table 12.4.

When it is necessary to indicate the roughness-width cutoff values, the standard values to be used are listed in Table 12.5. If no value is specified, the 0.80 value is assumed.

THIS ON THE DRAWING . . . . MEANS THIS

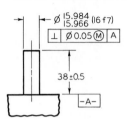

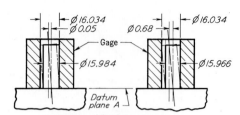

The feature axis must be within the specified
tolerance of location. Where the feature is at
MMC (15.984) the maximum perpendicularity
tolerance is 0.05 diameter. Where the feature
departs from its MMC size, an increase in the
perpendicularity tolerance is allowed which is
equal to the amount of such departure.

FIGURE 12.42 **Specifying
Perpendicularity for an Axis, Pin, or Boss
(ASME Y14.5M–1994).**

PERPENDICULARITY

THIS ON THE DRAWING . . . . MEANS THIS

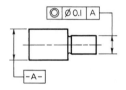

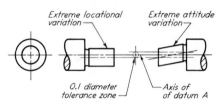

The feature axis must be within a cylindrical
zone of 0.1 diameter, regardless of feature size,
and whose axis coincides with the datum axis.

FIGURE 12.43 **Specifying
Concentricity (ASME Y14.5M–1994).**

CONCENTRICITY

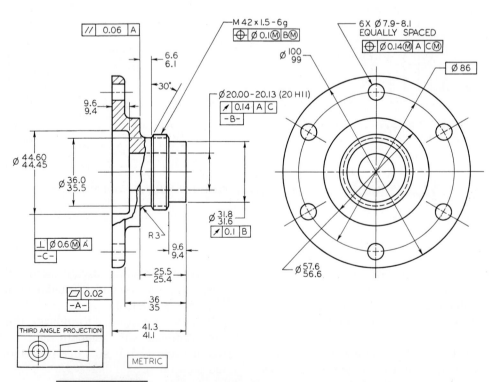

FIGURE 12.44 **Application of Symbols to Position and Form Tolerance
Dimensions (ASME Y14.5M–1994).**

Symbol	Meaning
(a)	Basic Surface Texture Symbol. Surface may be produced by any method except when the bar or circle, (b) or (d), is specified.
(b)	Material Removal By Machining Is Required. The horizontal bar indicates that material removal by machining is required to produce the surface and that material must be provided for that purpose.
(c) 3.5	Material Removal Allowance. The number indicates the amount of stock to be removed by machining in millimeters (or inches). Tolerances may be added to the basic value shown or in a general note.
(d)	Material Removal Prohibited. The circle in the vee indicates that the surface must be produced by processes such as casting, forging, hot finishing, cold finishing, die casting, powder metallurgy or injection molding without subsequent removal of material.
(e)	Surface Texture Symbol. To be used when any surface characteristics are specified above the horizontal line or to the right of the symbol. Surface may be produced by any method except when the bar or circle, (b) or (d), is specified.

(f)

MINIMUM*

3 X
1.5 X
60°
60°
3 X APPROX
3 X
0.00
1.5 X

LETTER HEIGHT = X

* This dimension is adjusted by +1 for each line of values beyond the two lines shown below the horizontal line.

FIGURE 12.45 Surface Texture Symbols and Construction (ANSI/ASME Y14.36M–1996).

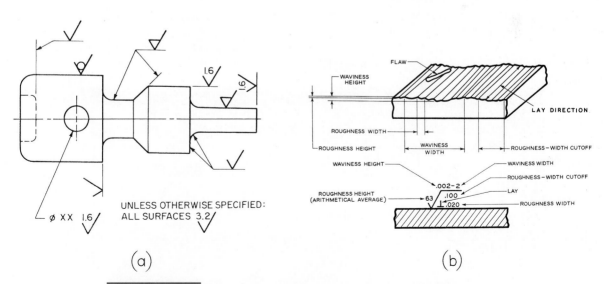

(a)

UNLESS OTHERWISE SPECIFIED:
ALL SURFACES 3.2

(b)

FIGURE 12.46 Application of Surface Texture Symbols and Surface Characteristics (ANSI/ASME Y14.36M–1996).

G R A P H I C S   S P O T L I G H T

Geometric Tolerances
With AutoCAD 2000

AutoCAD 2000 has built in dialog boxes which allow you to create feature control frames for geometric dimensioning and tolerancing. You can create a feature control frame by picking the Tolerance icon from AutoCAD's Dimensioning toolbar. When you do so the dialog box shown in Figure A will appear on your screen. It shows the standard tolerance symbols.

To begin creating a feature control frame all you need to do is double click on the symbol you want to use. For example, you could double-click on the posi-

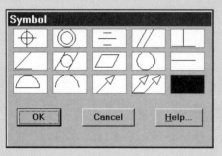

(A)

tional tolerance symbol shown in the upper left hand corner of the dialog box. When you have selected the symbol a new dialog box shown in Figure B will appear on the screen. You use it to create a single feature control frame, or a stacked feature control frame. You can also add diameter symbols, modifiers, datum references, or datum identifiers. The diameter symbol

shown in the Geometric Tolerance dialog box was added just by picking in the empty box below the heading Dia. A diameter symbol appears automatically. The area to the right of the diameter symbol is a text entry box that you use to type in the value you want to show for the tolerance.

To add a modifier symbol pick in the next empty box to the right. The Material Condition dialog box shown in Figure C pops up on the screen. You can quickly pick the modifier you want to add from the dialog box.

Datum references can be created just as quickly by picking in the appropriate box and typing in the letter you want to use. Datum references can also have a modifier when used with certain types of tolerances. Again, just pick in the empty box below the modifier and use the dialog box that appears to add a symbol.

If you want stacked tolerances or a datum identifier, you continue on with the same basic procedure. When you pick OK at the end of the process, you will be prompted to pick a location to place the tolerance in the drawing. You can also use the Leader command and select the options to place a tolerance at the end of a leader line. Using these dialog boxes you can quickly add geometric tolerance symbols as shown in Figure D to your drawings. Creating the symbols is easy, but you must give careful consideration to what the placement of the symbols in the

(B)

drawing means and make sure to reflect the intent of the design and the tolerances that are required for the part to function correctly in the assembly. Specify-ing needlessly restrictive tolerances just increases the cost of the part, without adding more functionality to the design.

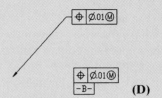

(C)

(D)

Micro-meters[a] (μm)	Micro-inches (μin.)	Micro-meters[a] (μm)	Micro-inches (μin.)
0.012	0.5	1.25	50
0.025	1	1.60	63
0.050	2	2.0	80
0.075	3	2.5	100
0.10	4	3.2	125
0.125	5	4.0	180
0.15	6	5.0	200
0.20	8	6.3	250
0.25	10	8.0	320
0.32	13	10.0	400
0.40	16	12.5	500
0.50	20	15	600
0.63	25	20	800
0.80	32	25	1000
1.00	40		

[a] Micrometers are the same as thousandths of a millimeter(1μm = 0.001 mm)

TABLE 12.4   **Preferred Series Roughness Average Values** *(R_a) (ANSI/ASME Y14.36–1996). Recommended values are in color.*

Millimeters (mm)	Inches (in.)	Millimeters (mm)	Inches (in.)
0.08	.003	2.5	.1
0.25	.010	8.0	.3
0.80	.030	25.0	1.0

TABLE 12.5   **Standard Roughness Sampling Length (Cutoff) Values** *(ANSI/ASME Y14.36–1996).*

Millimeters (mm)	Inches (in.)	Millimeters (mm)	Inches (in.)
0.0005	.00002	0.025	.001
0.0008	.00003	0.05	.002
0.0012	.00005	0.08	.003
0.0020	.00008	0.12	.005
0.0025	.0001	0.20	.008
0.005	.0002	0.25	.010
0.008	.0003	0.38	.015
0.012	.0005	0.50	.020
0.020	.0008	0.80	.030

TABLE 12.6   **Preferred Series Maximum Waviness Height Values** *(ANSI/ASME Y14.36–1996).*

When maximum waviness height values are required, the recommended values to be used are as given in Table 12.6.

When it is desired to indicate lay, the lay symbols in Fig. 12.48 are added to the surface texture symbols as per the examples given. Selected applications of the surface texture values to the symbols are given and explained in Fig. 12.49.

A typical range of surface roughness values that may be obtained from various production methods is shown in Fig. 12.50. Preferred roughness-height values are shown at the top of the chart.

■ 12.23 USING GEOMETRIC DIMENSIONING AND TOLERANCING

Geometric dimensioning and tolerancing (GDT) has evolved over the last forty years to become an indispensable tool for defining parts and features more accurately. GDT not only considers an individual part and its dimensions and tolerances, but views that part in relations to its related parts. This allows the designer more latitude in defining the parts feature's more accurately by not only considering the part's dimensions, but its tolerances at the initial design stage. GDT also simplifies the inspection process.This is accomplished through the use of ASME standards (ASME--Y14.5M), as we have discussed previously.

Individually manufactured parts and components must eventually be assembled in to products. We take for granted that each part of a lawnmower, for example, will mate properly with its other components when assembled. The wheels will slip into their axles, the pistons will fit properly into their cylinders, etc. Nothing should be too tight or too loose.

Geometric dimensioning and tolerancing, therefore, is important to both the design and manufacturing processes.

Applying GDT principles to the design process requires five steps:

Step 1: Define the part's functions. It is best to break the part down to its simplest functions. Be as specific as possible. For example, a lawnmower wheel's functions is to: (a). Give the product mobility: (b). Lift the mowing deck off the ground; (c) Add rigidity to the body, etc.

Step 2: List the functions by priority. Only one function should have top priority This step can be difficult since many parts are designed to incorporate multiple functions. In our lawnmower wheels example, the function with top priority would be to give the product mobility.

Step 3: Define the datum reference frame. This step should be based on your list of priorities This may mean creating several reference frames, each based on a priority on your list. The frame should be set up in either one, two, or three planes.

Step 4: Control selection. (See §§12.19, 12.21, 12.22). In most cases, several controls will be needed (e.g.,runout, position, concentricity, roughness, etc.). Begin with the simplest control. By "simplest" we mean least restrictive. Work from the least restrictive to the most restrictive set of controls.

Step 5: Calculate tolerances. Most tolerances are mathematically based. This step should be the easiest. Apply MMC, RFS, or LMC where indicated. (See §§12.15, 12.17, 12.19.) Avoid completing this step first, it should always be your last. See Fig. 12.47 for a worksheet outlining these five steps.

CAD programs generally allow the user to add tolerances to dimension values in the drawings. Geometric dimensioning and tolerancing symbols, finish marks, and other standard symbols are typically available as a part of the CAD program or as a symbol library.

Geometric dimensioning and tolerancing has become an essential part of today's manufacturing industry. To compete in today's marketplace, companies are required to develop and produce products of the highest quality, at lowest cost, and guarantee on-time delivery. Although considered by most to be a design specification language, GDT is a manufacturing and inspection language as well, providing a means for uniform interpretation and understanding by these various groups. It provides both a national and international contract base for customers and suppliers. See Chapter 10 for a more in-depth discussion of how GDT can be used in both the manufacturing and inspection processes.

1. DEFINE THE PART'S (FEATURE'S) FUNCTION:

Basic function _____

_____

Additional function(s) _____

_____

2. LIST THE FUNCTIONS IN ORDER OF PRIORITY

Function# _____ Function# _____ Function# _____

3. DEFINE DATUM REFERENCE FRAME

Part # _____   Function _____

Primary Datum Feature _____   Secondary Datum Feature _____

Part # _____   Function _____

Primary Datum Feature _____   Secondary Datum Feature _____

Part # _____   Function _____

Primary Datum Feature _____   Secondary Datum Feature _____

4. SELECT CONTROL TYPE

Part# _____   Part # _____

Control _____   Control _____

5. CALCULATE THE TOLERANCES

Part # _____   Part# _____   Part# _____

_____

FIGURE 12.47   **Geometric Dimensioning and Tolerancing Design Worksheet.**

LAY SYMBOLS

SYM	DESIGNATION	EXAMPLE	SYM	DESIGNATION	EXAMPLE
—	Lay parallel to the line representing the surface to which the symbol is applied.	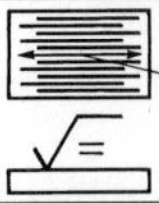 DIRECTION OF TOOL MARKS	X	Lay angular in both directions to line representing the surface to which symbol is applied.	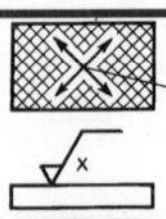 DIRECTION OF TOOL MARKS
⊥	Lay perpendicular to the line representing the surface to which the symbol is applied.	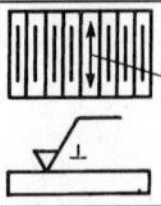 DIRECTION OF TOOL MARKS	M	Lay multidirectional	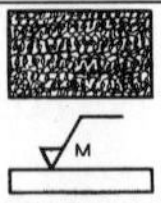
C	Lay approximately circular relative to the center of the surface to which the symbol is applied.	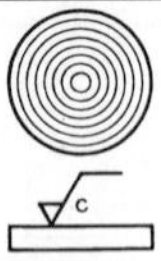	R	Lay approximately radial relative to the center of the surface to which the symbol is applied.	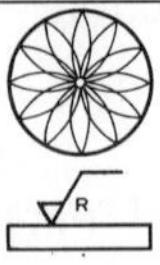

FIGURE 12.48 Lay Symbols (ANSI/ASME Y14.36M–1996).

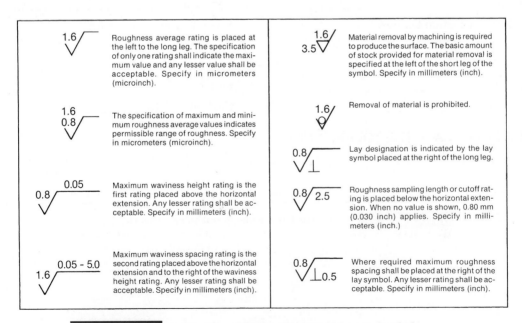

1.6 — Roughness average rating is placed at the left to the long leg. The specification of only one rating shall indicate the maximum value and any lesser value shall be acceptable. Specify in micrometers (microinch).

1.6 / 0.8 — The specification of maximum and minimum roughness average values indicates permissible range of roughness. Specify in micrometers (microinch).

0.8 / 0.05 — Maximum waviness height rating is the first rating placed above the horizontal extension. Any lesser rating shall be acceptable. Specify in millimeters (inch).

1.6 / 0.05 - 5.0 — Maximum waviness spacing rating is the second rating placed above the horizontal extension and to the right of the waviness height rating. Any lesser rating shall be acceptable. Specify in millimeters (inch).

1.6 / 3.5 — Material removal by machining is required to produce the surface. The basic amount of stock provided for material removal is specified at the left of the short leg of the symbol. Specify in millimeters (inch).

1.6 — Removal of material is prohibited.

0.8 / ⊥ — Lay designation is indicated by the lay symbol placed at the right of the long leg.

0.8 / 2.5 — Roughness sampling length or cutoff rating is placed below the horizontal extension. When no value is shown, 0.80 mm (0.030 inch) applies. Specify in millimeters (inch.)

0.8 / ⊥0.5 — Where required maximum roughness spacing shall be placed at the right of the lay symbol. Any lesser rating shall be acceptable. Specify in millimeters (inch).

FIGURE 12.49 Application of Surface Texture Values to Symbol (ANSI/ASME Y14.36M–1996).

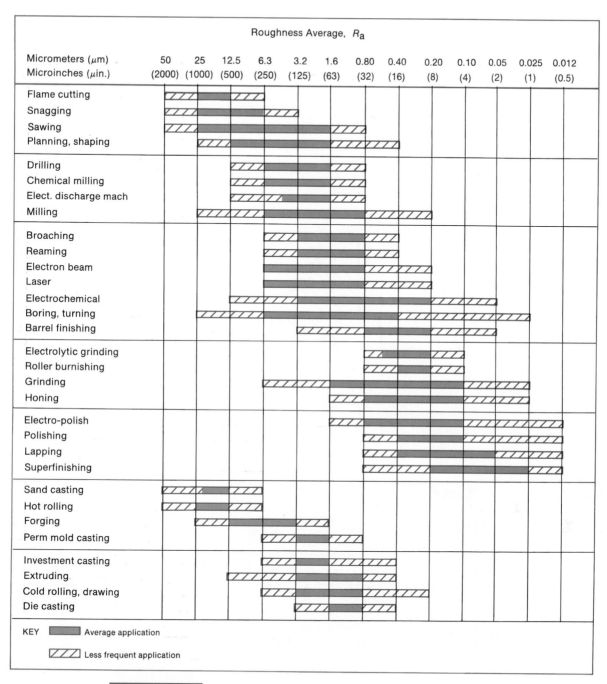

FIGURE 12.50 Surface Roughness Produced by Common Production Methods (ANSI/ASME B46.1–1985). The ranges shown are typical of the processes listed. Higher or lower values may be obtained under special conditions.

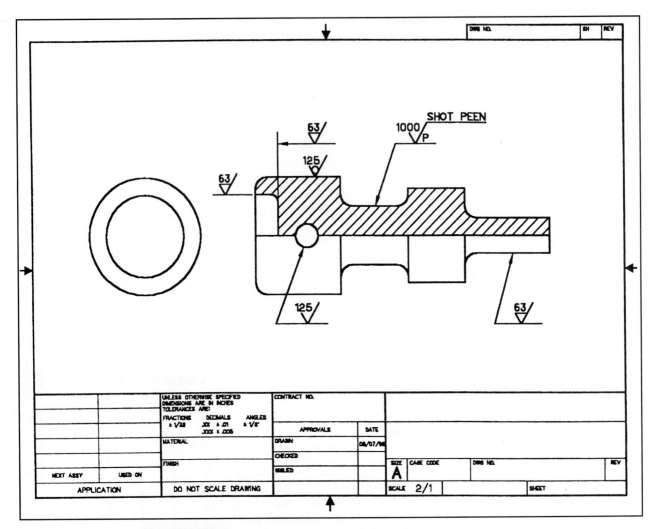

FIGURE 12.51 Application of Surface Finish Symbols with Computervision CADDS 4X Production Drafting System. *Courtesy of Design Pacifica.*

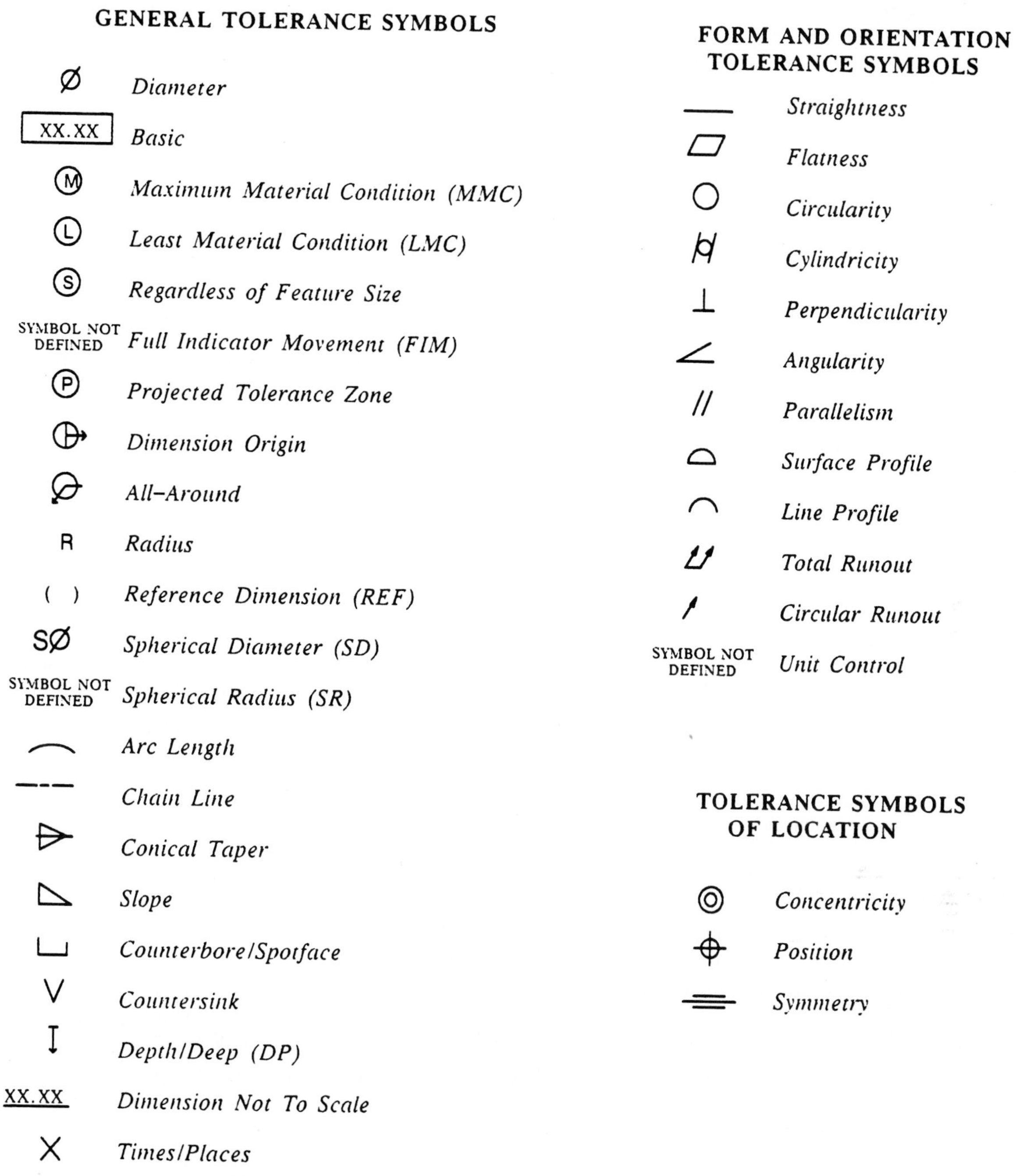

GENERAL TOLERANCE SYMBOLS

⌀   *Diameter*

XX.XX   *Basic*

Ⓜ   *Maximum Material Condition (MMC)*

Ⓛ   *Least Material Condition (LMC)*

Ⓢ   *Regardless of Feature Size*

SYMBOL NOT DEFINED   *Full Indicator Movement (FIM)*

Ⓟ   *Projected Tolerance Zone*

⊕→   *Dimension Origin*

⌀   *All–Around*

R   *Radius*

( )   *Reference Dimension (REF)*

S⌀   *Spherical Diameter (SD)*

SYMBOL NOT DEFINED   *Spherical Radius (SR)*

⌒   *Arc Length*

- - -   *Chain Line*

▷   *Conical Taper*

◿   *Slope*

⊔   *Counterbore/Spotface*

∨   *Countersink*

↧   *Depth/Deep (DP)*

XX.XX   *Dimension Not To Scale*

✗   *Times/Places*

FORM AND ORIENTATION TOLERANCE SYMBOLS

—   *Straightness*

▱   *Flatness*

○   *Circularity*

⌭   *Cylindricity*

⊥   *Perpendicularity*

∠   *Angularity*

//   *Parallelism*

⌒   *Surface Profile*

⌒   *Line Profile*

⌰   *Total Runout*

↗   *Circular Runout*

SYMBOL NOT DEFINED   *Unit Control*

TOLERANCE SYMBOLS OF LOCATION

◎   *Concentricity*

⊕   *Position*

⟳   *Symmetry*

FIGURE 12.52   **Geometric Dimensioning and Tolerancing Symbols Available in Computervision's CADDS 4X Production Drafting System.** *Courtesy of Computervision Corporation, a subsidiary of Prime Computer, Inc.*

■ KEY WORDS

TOLERANCE	LIMIT DIMENSION
ALLOWANCE	NOMINAL SIZE
BASIC-HOLE SYSTEM	LIMITS
CLEARANCE FIT	INTERFERENCE FIT
TRANSITION FIT	BASIC-SHAFT SYSTEM
UNILATERAL TOLERANCE	BILATERAL TOLERANCE
GEOMETRIC TOLERANCING	MAXIMUM MATERIAL CONDITION
PERPENDICULAR	PARALLEL
RUNOUT	CONCENTRIC
CYLINDRICITY	CIRCULARITY
CONTROL	GEOMETRIC DIMENSIONING AND TOLERANCING (GDT)
FUNCTION	

■ CHAPTER SUMMARY

- Tolerance dimensioning describes the minimum and maximum limits for a size or location of a feature.
- There are several ways of dimensioning tolerances, including limit dimensions, unilateral tolerances, bilateral tolerances, and geometric tolerancing.
- Basic-hole tolerance systems are the most commonly used tolerance system because they assume the hole is nominal size and adjust the shaft to accommodate the tolerance.
- The amount of space between two mating parts at maximum material condition is called the allowance.
- Mating parts with large allowances are classified as having a clearance fit or running and sliding fit.

- Mating parts with negative allowances are classified as having an interference fit or force fit.
- Mating parts are designed around a nominal size and class of fit. All other tolerances are calculated from these two values.
- High-quality parts are often dimensioned with geometric tolerancing to ensure that the size, shape, and relative geometric characteristics are properly defined.
- GDT has become an essential part of today's manufacturing industry. GDT is not only a design language, but an inspection language as well.

■ REVIEW QUESTIONS

1. What do the two numbers of a limit dimension mean?
2. Draw three different geometric tolerances that reference a datum. Label the information in each box.
3. Why is the basic-hole system more common than the basic-shaft system?
4. Give five examples of nominal sizes in everyday life. What is the purpose of a nominal size?
5. Give an example of two parts that would require a running and sliding fit. A force fit.
6. List five classes of fit.
7. Can one part have an allowance? Why?
8. Can two parts have a tolerance? Why?
9. Give an example of how GDT could be used as both a design and inspection tool.
10. List the five steps required to apply GDT to the design process.

C H A P T E R 1 3

THREADS, FASTENERS, AND SPRINGS

After studying the material in this chapter, you should be able to:

1. Define and label the parts of a screw thread.
2. Identify various screw thread forms.
3. Draw detailed, schematic, and simplified threads in section and elevation.
4. Define typical thread specifications.
5. Identify various fasteners and describe their use.
6. Draw various screw head types.
7. Draw springs in elevation using break conventions.

OVERVIEW

The concept of the screw thread seems to have occurred first to Archimedes, the third-century B.C. mathematician who wrote briefly on spirals and invented or designed several simple devices applying the screw principle. By the first century B.C., the screw was a familiar element but was crudely cut from wood or filed by hand on a metal shaft. Nothing more was heard of the screw thread until the fifteenth century.

Leonardo da Vinci understood the screw principle, and he created sketches showing how to cut screw threads by machine. In the sixteenth century, screws appeared in German watches and were used to fasten suits of armor. In 1569, the screw-cutting lathe was invented by the Frenchman Besson, but this method of screw production did not take hold for another century and a half; nuts and bolts continued to be made largely by hand. In the eighteenth century, screw manufacturing started in England during the Industrial Revolution.

Threads and fasteners are the principal fastening devices used for assembling component parts. The shape of the helical thread is called the **thread form**. The metric thread form is the international standard, although the unified thread form is common in the United States. Other thread forms are used in specific applications. CAD drawing programs often use software that automatically depicts threads. The thread specification is a special leader note that defines the type of thread or fastener. This is an instruction for the shop technician so the correct type of thread is created during the manufacturing process. To speed production time and reduce costs, many new types of fasteners are created every year. Existing fasteners are also modified to improve their insertion in mass production.

■ 13.1 STANDARDIZED SCREW THREADS

In early times, there was no such thing as standardization. Nuts made by one manufacturer would not fit the bolts of another. In 1841 Sir Joseph Whitworth started crusading for a standard screw thread, and soon the Whitworth thread was accepted throughout England.

In 1864 the United States adopted a thread proposed by William Sellers of Philadelphia, but the Sellers nuts would not screw onto a Whitworth bolt, or vice versa. In 1935 the American Standard thread, with the same 60° V form of the old Sellers thread, was adopted in the United States. Still there was no stan-

| (a) | (b) |

(a) Springs. (b) Screws and fasteners. *From* **Machine Design: An Integrated Approach** *by Robert Norton,* © *1996. Reprinted by permission of Prentice-Hall, Inc.*

dardization among countries. In peacetime it was a nuisance; in World War I it was a serious inconvenience; and in World War II the obstacle was so great that the Allies decided to do something about it. Talks began among the Americans, British, and Canadians, and in 1948 an agreement was reached on the unification of American and British screw threads. The new thread was called the *Unified screw thread*, and it represents a compromise between the American Standard and Whitworth systems, allowing complete interchangeability of threads in three countries.

In 1946 an International Organization for Standardization (ISO) committee was formed to establish a single international system of metric screw threads. Consequently, through the cooperative efforts of the Industrial Fasteners Institute (IFI), several committees of the American National Standards Institute, and the ISO representatives, a metric fastener standard was prepared.[*]

Today screw threads are vital to our industrial life. They are designed for hundreds of different purposes; the three basic applications are (1) to *hold parts* together, (2) to *adjust parts* with reference to each other, and (3) to *transmit power*.

■ 13.2 SCREW THREAD TERMS

The following definitions apply to screw threads in general (Fig. 13.1). For additional information regarding specific Unified and metric screw thread terms and definitions, refer to the following standards:

ANSI/ASME B1.1–1989
ANSI/ASME B1.7M–1984 (R1992)

---

[*] For a listing of ANSI standards for threads, fasteners, and springs, see Appendix 1.

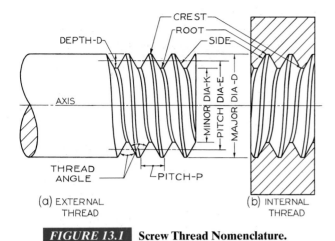

FIGURE 13.1 **Screw Thread Nomenclature.**

ANSI/ASME B1.13M–1983 (R1989)

ANSI/ASME Y14.6–1978 (R1993)

ANSI/ASME Y14.6aM–1981 (R1993).

SCREW THREAD A ridge of uniform section in the form of a helix (§4.60) on the external or internal surface of a cylinder.

EXTERNAL THREAD A thread on the outside of a member, as on a shaft.

INTERNAL THREAD A thread on the inside of a member, as in a hole.

MAJOR DIAMETER The largest diameter of a screw thread (applies to both internal and external threads).

MINOR DIAMETER The smallest diameter of a screw thread (applies to both internal and external threads).

PITCH The distance from a point on a screw thread to a corresponding point on the next thread measured parallel to the axis. The pitch P is equal to 1 divided by the number of threads per inch.

PITCH DIAMETER The diameter of an imaginary cylinder passing through the threads to make equal the widths of the threads and the widths of the spaces cut by the cylinder.

LEAD The distance a screw thread advances axially in one turn.

ANGLE OF THREAD The angle included between the sides of the thread measured in a plane through the axis of the screw.

CREST The top surface joining the two sides of a thread.

ROOT The bottom surface joining the sides of two adjacent threads.

SIDE The surface of the thread that connects the crest with the root.

AXIS OF SCREW The longitudinal center line through the screw.

DEPTH OF THREAD The distance between the crest and the root of the thread measured normal to the axis.

FORM OF THREAD The cross section of thread cut by a plane containing the axis.

SERIES OF THREAD Standard number of threads per inch for various diameters.

■ 13.3 SCREW THREAD FORMS

Various forms of threads are used to hold parts together, to adjust parts with reference to each other, or to transmit power (Fig. 13.2). The 60° *Sharp-V thread* was originally called the United States Standard thread, or the Sellers thread. For purposes of certain adjustments, the Sharp-V thread is useful with the increased friction resulting from the full thread face. It is also used on brass pipe work.

The *American National thread* with flattened roots and crests is a stronger thread. This form replaced the Sharp-V thread for general use.

The *Unified thread* is the standard thread agreed upon by the United States, Canada, and Great Britain in 1948, and has replaced the American National form. The crest of the external thread may be flat or rounded, and the root is rounded; otherwise, the thread form is essentially the same as the American National.

The *metric thread* is the standard screw thread agreed upon for international screw thread fasteners. The crest and root are flat, but the external thread is often rounded if formed by a rolling process. The form is similar to the American National and Unified threads but with less depth of thread.

The *square thread* is theoretically the ideal thread for power transmission, since its face is nearly at right angles to the axis, but due to the difficulty of cutting it with dies and because of other inherent disadvantages (such as the fact that split nuts will not readily disengage), the square thread has been displaced to a large extent by the Acme thread. The square thread is not standardized.

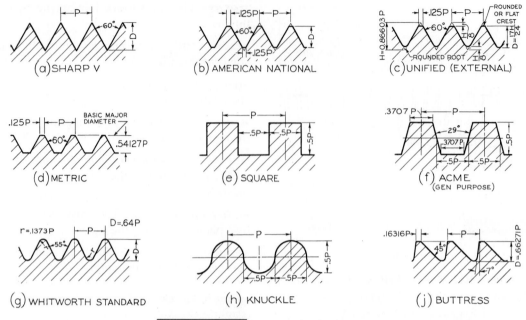

FIGURE 13.2   Screw Thread Forms.

The *Acme thread* is a modification of the square thread and has largely replaced it. It is stronger than the square thread, is easier to cut, and has the advantage of easy disengagement from a split nut, as on the lead screw of a lathe.

The *standard worm thread* (not shown) is similar to the Acme thread but is deeper. It is used on shafts to carry power to worm wheels.

The *Whitworth thread* was the British standard and has been replaced by the Unified thread. Its uses correspond to those of the American National thread.

The *knuckle thread* is usually rolled from sheet metal but is sometimes cast; in modified forms it is used in electric bulbs and sockets, bottle tops, etc.

The *buttress thread* is designed to transmit power in one direction only and is used in large guns, in jacks, and in other mechanisms having similar high-strength requirements.

■ 13.4 THREAD PITCH

The **pitch** of any thread form is the distance parallel to the axis between corresponding points on adjacent threads (Figs. 13.1a–13.3). For metric threads, this distance is specified in millimeters.

The pitch for a metric thread that is included with the major diameter in the thread designation determines the size of the thread—for example, M10 × 1.5, as shown in Fig. 13.3b. See also §13.21 or Appendix 15 for more information.

FIGURE 13.3   Pitch of Threads.

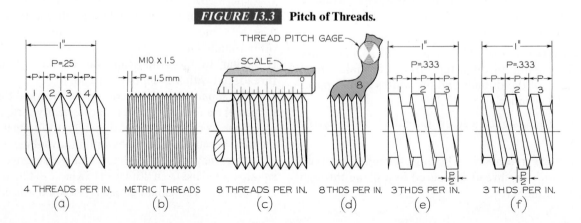

For threads dimensioned in inches, the pitch is equal to 1 divided by the number of threads per inch. The thread tables give the number of threads per inch for each standard diameter. Thus, a Unified coarse thread (Appendix 15) of 1″ diameter has eight threads per inch, and the pitch P equals $\frac{1}{8}$″ (.125).

If a thread has only four threads per inch, the pitch and the threads themselves are quite large (Fig. 13.3a). If there are, say, 16 threads per inch, the pitch is only $\frac{1}{16}$″ (.063), and the threads are relatively small, similar to those in Fig. 13.3b.

The pitch or the number of threads per inch can be measured with a scale (Fig. 13.3c) or with a *thread-pitch gage* (Fig. 13.3d).

■ 13.5 RIGHT-HAND AND LEFT-HAND THREADS

A *right-hand thread* is one that advances into a nut when turned clockwise, and a left-hand thread is one that advances into a nut when turned counterclockwise (Fig. 13.4). A thread is always considered to be right-handed (RH) unless otherwise specified. A *left-hand thread* is always labeled LH on a drawing (see Fig. 13.19a).

■ 13.6 SINGLE AND MULTIPLE THREADS

A *single thread*, as the name implies, is composed of one ridge, and the lead is therefore equal to the pitch. *Multiple threads* are composed of two or more ridges running side by side. As shown in Figs. 13.5a to 13.5c, the *slope line* is the hypotenuse of a right triangle whose short side equals .5P for single threads, P for double threads, 1.5P for triple threads, and so on. This applies to all forms of threads. In *double threads*, the lead is twice the pitch; in *triple threads*, the lead is three times the pitch, and so on. On a drawing of a single or triple thread, a root is opposite a crest; in the case of a double or quadruple thread, a root is drawn opposite a root. Therefore, in one turn, a double thread advances twice as far as a single thread, and a triple thread advances three times as far. RH double square and RH triple Acme threads are shown in Figs. 13.5d and 13.5e, respectively.

Multiple threads are used wherever quick motion, but not great power, is desired, as on fountain pens, toothpaste caps, valve stems, etc. The threads on a valve stem are frequently multiple threads, to impart quick action in opening and closing the valve. Multiple threads on a shaft can be recognized and counted by observing the number of thread endings on the end of the screw.

■ 13.7 THREAD SYMBOLS

There are three methods of representing screw threads on drawings—the *schematic*, *simplified*, and *detailed* methods. For clarity of representation and where good judgment dictates, schematic, simplified, and detailed thread symbols may be combined on a single drawing.

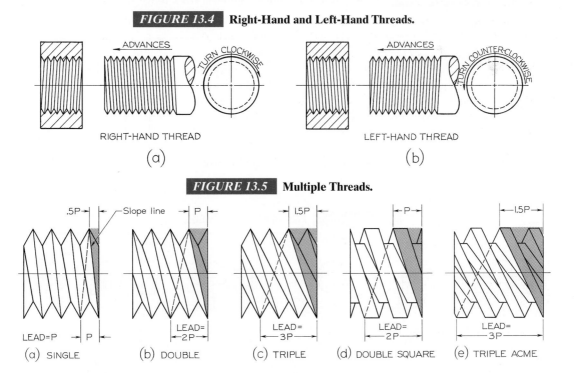

FIGURE 13.4   Right-Hand and Left-Hand Threads.

RIGHT-HAND THREAD
(a)

LEFT-HAND THREAD
(b)

FIGURE 13.5   Multiple Threads.

(a) SINGLE   (b) DOUBLE   (c) TRIPLE   (d) DOUBLE SQUARE   (e) TRIPLE ACME

Two sets of thread symbols, the schematic and the more common simplified, are used to represent the threads of small diameter, under approximately 1″ or 25-mm diameter on the drawing. The symbols are the same for all forms of threads, such as metric, Unified, square, and Acme.

The detailed representation is a close approximation of the exact appearance of a screw thread. The true projection of the helical curves of a screw thread (see Fig. 13.1) presents a pleasing appearance, but this does not compensate for the laborious task of plotting the helices. Consequently, the true projection is rarely used in practice.

When the diameter of the thread on the drawing is more than 1″ or 25 mm, a pleasing drawing may be made by the *detailed* representation method, in which the true profiles of the threads (any form of thread) are drawn; but the helical curves are replaced by straight lines (Fig. 13.6).* Whether the crests or roots are flat or rounded, they are represented by single

* A thread 42 mm or $1\frac{5}{8}$″ diameter, if drawn half size, would be less than 25 mm or 1″ diameter on the drawing and hence would be too small for this method of representation.

lines and not double lines, as in Fig. 13.1; consequently, American National and Unified threads are drawn in exactly the same way.

◼ 13.8 EXTERNAL THREAD SYMBOLS

Simplified external thread symbols are shown in Figs. 13.7a and 13.7b. The threaded portions are indicated by hidden lines parallel to the axis at the approximate depth of the thread, whether in section or in elevation. [Use the schematic depth of thread in the table in Fig. 13.9a to draw these lines (Fig. 13.7d)].

When the schematic form is shown in section (Fig. 13.7c), it is necessary to show the Vs; otherwise no threads would be evident. However, it is not necessary to show the Vs to scale or according to the actual slope of the crest lines. To draw the Vs, use the schematic thread depth (Fig. 13.9a), and let the pitch be determined by the 60° Vs.

Schematic threads in elevation (Fig. 13.7d) are indicated by alternate long and short lines at right angles to the center line, with the root lines being preferably thicker than the crest lines. Although the crest lines should theoretically be spaced according to actual pitch, the lines would often be very crowded

---

FIGURE 13.6 **Detailed Metric, American National, and Unified Threads.**

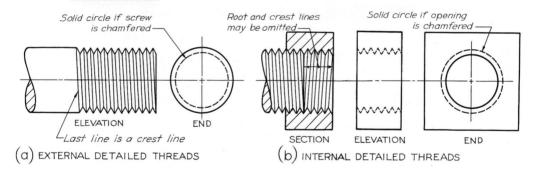

(a) EXTERNAL DETAILED THREADS

(b) INTERNAL DETAILED THREADS

FIGURE 13.7 **External Thread Symbols.**

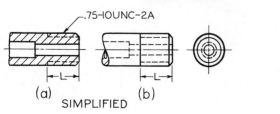

(a) (b) SIMPLIFIED

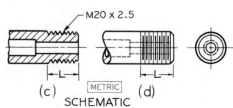

(c) (d) SCHEMATIC

and tedious to draw, thus defeating the purpose of the symbol, which is to save drafting time. In practice, the experienced drafter spaces the crest lines carefully by eye, and then adds the heavy root lines spaced by eye halfway between the crest lines. In general, the spacing should be proportionate for all diameters. For convenience in drawing, proportions for the schematic symbol are given in Fig. 13.9.

■ 13.9 INTERNAL THREAD SYMBOLS

Internal thread symbols are shown in Fig. 13.8. Note that the only differences between the schematic and simplified internal thread symbols occur in the sectional views. The representation of the schematic thread in section in Figs. 13.8m, 13.8o, and 13.8p is exactly the same as the external symbol in Fig. 13.7d. Hidden threads, by either method, are represented by pairs of hidden lines. The hidden dashes should be staggered, as shown.

In the case of blind tapped holes, the drill depth normally is drawn at least three schematic pitches beyond the thread length, as shown in Figs. 13.8d, 13.8e, 13.8n, and 13.8o. The symbols in Figs. 13.8f and 13.8p represent the use of a bottoming tap, when the length of thread is the same as the depth of drill (see also §13.24).

■ 13.10 DRAWING THREAD SYMBOLS

Figure 13.9a shows a table of values of depth and pitch to use in drawing thread symbols. These values are selected to produce a well-proportioned symbol and to be convenient to set off with the scale. An experienced drafter can carefully space the lines by eye, but a student should use the scale. Note that values of D and P are for the diameter *on the drawing*. Thus, a $1\frac{1}{2}''$ diameter thread at half-scale would be $\frac{3}{4}''$ diameter on the drawing, and values of D and P for a $\frac{3}{4}''$ major diameter would be used. Nominal diameters for decimal-inch or metric threads are treated in a similar manner.

SIMPLIFIED SYMBOLS The steps for drawing the simplified symbols for an external thread in elevation and in section are shown in Figs. 13.9b to 13.9e. The thread depth from the table is used for establishing the pairs of hidden lines that represent the threads in elevation and in section. No pitch measurement is needed. The completed symbols are shown in Figs. 13.9d and 13.9e.

The steps for drawing the simplified symbol for an internal thread in section are shown in Figs. 13.9f to 13.9h. The simplified representation for the internal thread in elevation is identical to that used for schematic representation, as the threads are indicated by pairs of hidden lines (Fig. 13.9g). The simplified

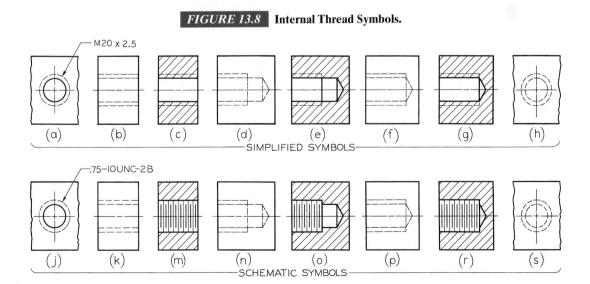

FIGURE 13.8 Internal Thread Symbols.

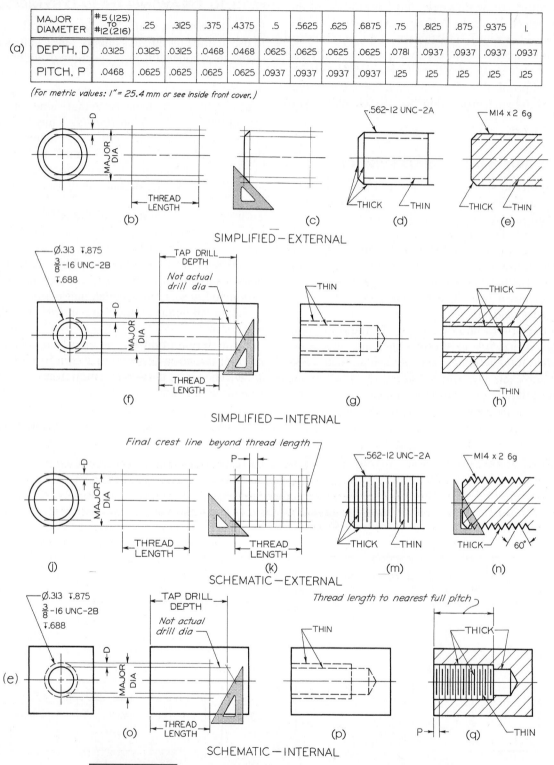

(a)

MAJOR DIAMETER	#5 (.125) TO #12 (.216)	.25	.3125	.375	.4375	.5	.5625	.625	.6875	.75	.8125	.875	.9375	I.
DEPTH, D	.03125	.03125	.03125	.0468	.0468	.0625	.0625	.0625	.0625	.0781	.0937	.0937	.0937	.0937
PITCH, P	.0468	.0625	.0625	.0625	.0625	.0937	.0937	.0937	.0937	.125	.125	.125	.125	.125

(For metric values: 1" = 25.4 mm or see inside front cover.)

SIMPLIFIED—EXTERNAL

SIMPLIFIED—INTERNAL

SCHEMATIC—EXTERNAL

SCHEMATIC—INTERNAL

FIGURE 13.9   **To Draw Thread Symbols—Simplified and Schematic.**

symbol for an internal thread in section is shown in Fig. 13.9h. The major diameter of the thread is represented by hidden lines across the sectioned area.

SCHEMATIC SYMBOLS The steps for drawing the schematic symbols for an external thread in elevation and in section are shown in Figs. 13.9j to 13.9n. Note that when the pitches P are set off in Fig. 13.9k, the final crest line for a full pitch may fall beyond the actual thread length, as shown. The completed schematic symbol for an external thread in elevation is shown in Fig. 13.9m. The completed schematic symbol for an external thread in section is shown in Fig. 13.9n. The schematic thread depth is used for drawing the Vs, and the pitch is established by the 60° Vs.

The steps for drawing the schematic symbols for an internal thread in elevation and in section are shown in Figs. 13.9o to 13.9q. Here again the symbol thread length may be slightly longer than the actual given thread length. If the tap drill depth is known or given, the drill is drawn to that depth, as shown. If the thread note omits this information, as is often done in practice, the drafter merely draws the hole three thread pitches (schematic) beyond the thread length. The tap drill diameter is represented approximately, as shown, and not to actual size. The completed schematic symbol for an internal thread in elevation is shown (Fig. 13.9p). Pairs of hidden lines represent the threads, and the hidden-line dashes are staggered. The completed schematic symbol for an internal thread in section is shown in Fig. 13.9q. The schematic internal thread in section is represented in the same manner as for the schematic external thread.

■ 13.11 DETAILED REPRESENTATION: METRIC, UNIFIED, AND AMERICAN NATIONAL THREADS

The detailed representation for metric, Unified, and American National threads is the same, since the flats, if any, are disregarded. The steps in drawing these threads are shown in Fig. 13.10.

1. Draw center line and lay out length and major diameter (Fig. 13.10a).

2. Find the number of threads per inch in Appendix 15 for American National and Unified threads (Fig. 13.10b). This number depends on the major diameter of the thread, whether the thread is internal or external. Find P (pitch) by dividing 1 by the number of

threads per inch (§13.4). The pitch for metric threads is given directly in the thread designation. For example, the M14 × 2 thread has a pitch of 2 mm (see Appendix 15). Establish the slope of the thread by offsetting the slope line $\frac{1}{2}$P for single threads, P for double threads, $1\frac{1}{2}$P for triple threads, and so on. (These offsets are the same in terms of P for any form of thread.) For right-hand external threads, the slope line slopes upward to the left; for left-hand external threads, the slope line slopes upward to the right. If the number of threads per inch conforms to the scale, the pitch can be set off directly. For example, eight threads per inch can easily be set off with the architects' scale, and ten threads per inch with the engineers' scale. Otherwise, use the bow dividers or use the parallel-line method shown in Fig. 13.10b.

3. From the pitch points, draw crest lines parallel to the slope line (Fig. 13.10c). These should be dark, thin lines. Slide your triangle along a T-square (or another triangle) to make parallel lines. Draw two Vs to establish the depth of thread, and draw guide lines for the root of thread, as shown.

4. Draw 60° Vs finished weight (Fig. 13.10d). These Vs should stand vertically; that is, they should not "lean" with the thread.

5. Draw root lines dark at once (Fig. 13.10e). Root lines will *not* be parallel to crest lines. Slide the triangle on the straightedge to make root lines parallel.

6. When the end is chamfered (usually 45° with end of shaft, sometimes 30°), the chamfer extends to the thread depth (Fig. 13.10f). The chamfer creates a new crest line, which is then drawn between the two new crest points. It is not parallel to the other crest lines. In the final drawing, all thread lines should be approximately the same weight—thin, but dark.

The corresponding internal detailed threads, in section, are drawn as shown in Fig. 13.11. Notice that for LH threads the lines slope upward to the left (Figs. 13.11a to 13.11c), while for RH threads the lines slope upward to the right (Figs. 13.11d to 13.11f). Make all final thread lines medium-thin but dark.

■ 13.12 DETAILED REPRESENTATION OF SQUARE THREADS

The steps in drawing the detailed representation of an external square thread when the major diameter is over 1″ or 25 mm (approx.) on the drawing are shown in Fig. 13.12.

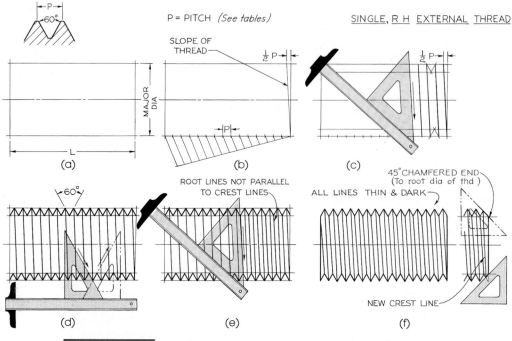

FIGURE 13.10 Detailed Representation—External Metric, Unified, and American National Threads.

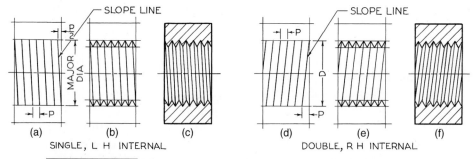

FIGURE 13.11 Detailed Representation—Internal Metric, Unified, and American National Threads.

1. Draw the center line, and lay out the length and major diameter of the thread (Fig. 13.12a). Determine P by dividing 1 by the number of threads per inch (see Appendix 22). For a single RH thread, the lines slope upward to the left, and the slope line is offset *as for all single threads of any form.* On the upper line, set off spaces equal to $\frac{P}{2}$, as shown, using a scale if possible; otherwise, use the bow dividers or the parallel-line method to space the points.

2. From the $\frac{P}{2}$ points on the upper line, draw fairly thin lines (Fig. 13.12b). Draw guide lines for root of thread, making the depth $\frac{P}{2}$ as shown.

3. Draw parallel visible back edges of threads (Fig. 13.12c).

4. Draw parallel visible root lines (Fig. 13.12d). Note enlarged detail in Fig. 13.12g.

5. Accent the lines (Fig. 13.12e). All lines should be thin and dark.

Note the end view of the shaft in Fig. 13.12f. The root circle is hidden; no attempt is made to show the true projection. If the end is chamfered, a solid circle would be drawn instead of the hidden circle.

Figure 13.13 is an assembly drawing showing an external square thread partly screwed into a nut. The detail of the square thread at A is the same as shown in Fig. 13.12. But when the external and internal threads are assembled, the thread in the nut overlaps and covers up half of the V, as shown at B.

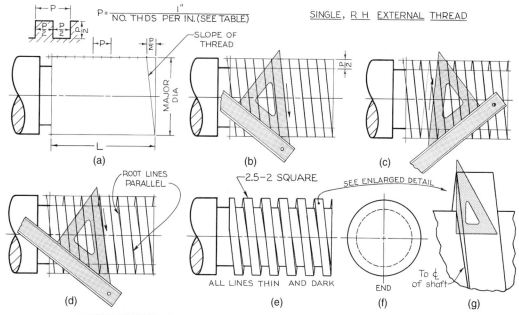

FIGURE 13.12 Detailed Representation—External Square Threads.

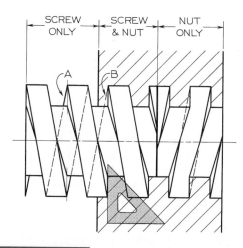

FIGURE 13.13 Square Threads in Assembly.

The internal thread construction is the same as in Fig. 13.14. Note that the thread lines representing the back half of the internal threads (since the thread is in section) slope in the opposite direction from those on the front side of the screw.

Steps in drawing a single internal square thread in section are shown in Fig. 13.14. Note in Fig. 13.14b that a crest is drawn opposite a root. This is the case for both single and triple threads. For double or quadruple threads, a crest is opposite a crest. Thus, the construction in Figs. 13.14a and 13.14b is the same for any multiple of thread. The differences appear in Fig. 13.14c,

where the threads and spaces are distinguished and outlined.

The same internal thread is shown in elevation (external view) in Fig. 13.14e. The profiles of the threads are drawn in their normal position, but with hidden lines, and the sloping lines are omitted for simplicity. The end view of the same internal thread is shown in Fig. 13.14f. Note that the hidden and solid circles are opposite those for the end view of the shaft (see Fig. 13.12f).

■ 13.13 DETAILED REPRESENTATION OF ACME THREAD

The steps in drawing the detailed representation of Acme threads when the major diameter is larger than 1″ or 25 mm (approx.) on the drawing are shown in Fig. 13.15.

1. Draw the center line, and lay out the length and major diameter of the thread (Fig. 13.15a). Determine P by dividing 1 by the number of threads per inch (see Appendix 22). Draw construction lines for the root diameter, making the thread depth $\frac{P}{2}$. Draw construction lines halfway between crest and root guide lines.

2. On the intermediate construction lines, lay off $\frac{P}{2}$ spaces (Fig. 13.15b). Setting off spaces directly with a scale is possible (for example, if $\frac{P}{2} = \frac{1}{10}″$, use the

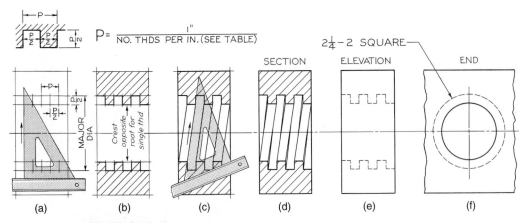

FIGURE 13.14 Detailed Representation—Internal Square Threads.

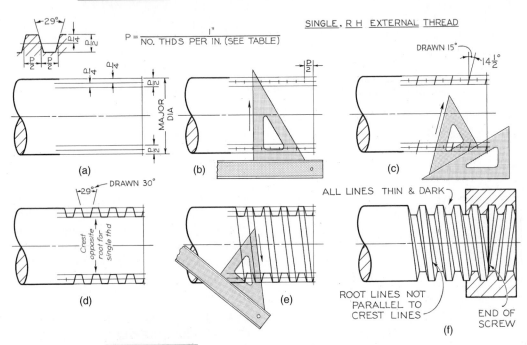

FIGURE 13.15 Detailed Representation—Acme Threads.

engineers' scale); otherwise, use bow dividers or the parallel-line method.

3. Through alternate points, draw construction lines for sides of threads (draw 15° instead of $14\frac{1}{2}°$) (Fig. 13.15c).

4. Draw construction lines for other sides of threads (Fig. 13.15d). Note that for single and triple threads, a crest is opposite a root, while for double and quadruple threads, a crest is opposite a crest. Heavy in tops and bottoms of threads.

5. Draw parallel crest lines, final weight at once (Fig. 13.15e).

6. Draw parallel root lines, final weight at once, and heavy in the thread profiles (Fig. 13.15f). All lines should be thin and dark. Note that the internal threads in the back of the nut slope in the opposite direction to the external threads on the front side of the screw.

End views of Acme threaded shafts and holes are drawn exactly like those for the square thread (Figs. 13.12 and 13.14).

■ 13.14 USE OF PHANTOM LINES

In representing objects having a series of identical features, phantom lines may be used to save time

FIGURE 13.16 Use of Phantom Lines.

(Fig. 13.16). Threaded shafts and springs thus represented may be shortened without the use of conventional breaks, but must be correctly dimensioned. The use of phantom lines is limited almost entirely to detailed drawings.

13.15 THREADS IN SECTION

Detailed representations of large threads in section are shown in Figs. 13.6 and 13.10–13.15. As indicated by the note in Fig. 13.6b, the root lines and crest lines may be omitted in internal sectional views, if desired.

External thread symbols are shown in section in Fig. 13.7. Note that in the schematic symbol, the Vs must be drawn. Internal thread symbols in section are shown in Fig. 13.8.

Threads in an assembly drawing are shown in Fig. 13.17. It is customary not to section a stud or a nut, or any solid part, unless necessary to show some internal shapes (see §13.23). Note that when external and internal threads are sectioned in assembly, the Vs are required to show the threaded connection.

13.16 AMERICAN NATIONAL THREAD

The old American National thread was adopted in 1935. The *form*, or profile (Fig. 15.2b), is the same as the old Sellers' profile, or U.S. Standard, and is known as the *National form*. The methods of representation are the same as for the Unified and metric threads. As noted earlier, American National threads have been replaced by the Unified and metric threads. However, they may still be found on earlier drawings. Five series of threads were embraced in the old ANSI standards.

1. *Coarse thread*—A general-purpose thread for holding purposes. Designated NC (National Coarse).

2. *Fine thread*—A greater number of threads per inch; used extensively in automotive and aircraft construction. Designated NF (National Fine).

3. *8-pitch thread*—All diameters have 8 threads per inch. Used on bolts for high-pressure pipe flanges, cylinder-head studs, and similar fasteners. Designated 8N (National form, 8 threads per inch).

4. *12-pitch thread*—All diameters have 12 threads per inch; used in boiler work and for thin nuts on shafts and sleeves in machine construction. Designated 12N (National form, 12 threads per inch).

5. *16-pitch thread*—All diameters have 16 threads per inch; used where necessary to have a fine thread regardless of diameter, as on adjusting collars and bearing retaining nuts. Designated 16N (National form, 16 threads per inch).

13.17 UNIFIED EXTRA FINE THREADS

The *Unified extra fine thread* series has many more threads per inch for given diameters than any series of the American National or Unified. The form of thread is the same as the American National. These small threads are used in thin metal where the length of thread engagement is small, in cases where close adjustment is required, and where vibration is great. They are designated UNEF (extra fine).

13.18 AMERICAN NATIONAL THREAD FITS

For general use, three classes of screw thread fits between mating threads (as between bolt and nut) have been established by ANSI.

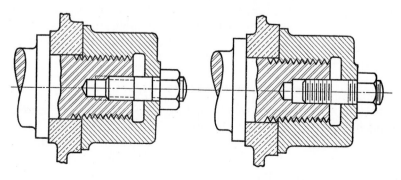

(a) SIMPLIFIED   (b) SCHEMATIC   **FIGURE 13.17** Threads in Assembly.

These fits are produced by the application of tolerances listed in the standard and are described as follows.

1. *Class 1 fit*—Recommended only for screw thread work where clearance between mating parts is essential for rapid assembly and where shake or play is not objectionable.
2. *Class 2 fit*—Represents a high quality of commercial thread product and is recommended for the great bulk of interchangeable screw thread work.
3. *Class 3 fit*—Represents an exceptionally high quality of commercially threaded product and is recommended only in cases where the high cost of precision tools and continual checking are warranted.

The class of fit desired on a thread is indicated in the thread note, as shown in §13.21.

■ 13.19 METRIC AND UNIFIED THREADS

The preferred metric thread for commercial purposes conforms to the International Organization for Standardization (ISO) publication basic profile M for metric threads. This M profile design is comparable to the Unified inch profile, but the two are not interchangeable. For commercial purposes, two series of metric threads are preferred—coarse (general purpose) and fine—thus drastically reducing the number of previously used thread series (see Appendix 15).

The Unified thread constitutes the present American National standard. Some earlier American National threads are still included in the new standard. The standard lists 11 different series of numbers of threads per inch for the various standard diameters, together with the selected combinations of special diameters and pitches.

The 11 series are the *coarse thread series* (UNC or NC), recommended for general use corresponding to the old National coarse thread; the *fine thread* series (UNF or NF), for general use in automotive and aircraft work and in applications where a finer thread is required; the *extra fine* series (UNF or NF), which is the same as the SAE extra fine series, used particularly in aircraft and aeronautical equipment and generally for threads in thin walls; and the eight series of 4, 6, 8, 12, 16, 20, 28, and 32 threads with constant pitch. The 8UN or 8N, 12UN or 12N, and 16UN or 16N series are recommended for the uses corresponding to the old 8-, 12-, and 16-pitch American National threads. In addition, there are three special thread series—UNS, NS, and UN—that involve special combinations of diameter, pitch, and length of engagement.

■ 13.20 METRIC AND UNIFIED THREAD FITS

For some specialized metric thread applications, the tolerances and deviations are specified by tolerance grade, tolerance position, class, and length of engagement. Two classes of metric thread fits are generally recognized. The first class of fits is for general-purpose applications and has a tolerance class of 6H for internal threads and a class of 6g for external threads. The second class of fits is used where closer fits are necessary and has a tolerance class of 6H for internal threads and a class of 5g6g for external threads. Metric thread tolerance classes of 6H/6g are generally assumed if not otherwise designated and are used in applications comparable to the 2A/2B inch classes of fits.

The single-tolerance designation of 6H refers to both the tolerance grade and position for the pitch diameter and the minor diameter for an internal thread. The single-tolerance designation of 6g refers to both the tolerance grade and position for the pitch diameter and the major diameter of the external thread. A double designation of 5g6g indicates separate tolerance grades for the pitch diameter and for the major diameter of the external thread.

The standard for Unified screw threads specifies tolerances and allowances defining the several classes of fit (degree of looseness or tightness) between mating threads. In the symbols for fit, the letter A refers to the external threads and B to internal threads. There are three classes of fit each for external threads (1A, 2A, 3A) and internal threads (1B, 2B, 3B). Classes 1A and 1B have generous tolerances, facilitating rapid assembly and disassembly. Classes 2A and 2B are used in the normal production of screws, bolts, and nuts, as well as a variety of general applications. Classes 3A and 3B provide for applications needing highly accurate and close-fitting threads.

■ 13.21 THREAD NOTES

Thread notes for metric, Unified, and American National screw threads are shown in Fig. 13.18. These same notes or symbols are used in correspondence, on shop and storeroom records, and in specifications for parts, taps, dies, tools, and gages.

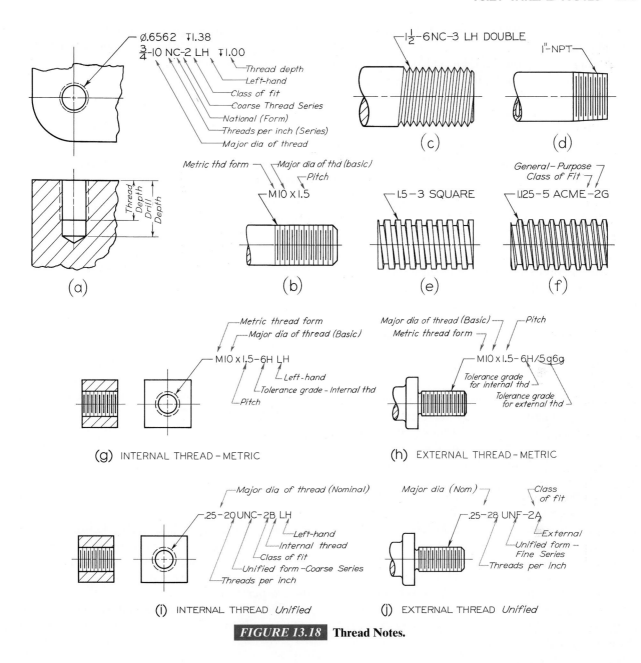

FIGURE 13.18 Thread Notes.

Metric screw threads are designated basically by the letter M for metric profile followed by the nominal size (basic major diameter) and the pitch, both in millimeters and separated by the symbol ×. For example, the basic thread note M10 × 1.5 is adequate for most commercial purposes (Fig. 13.18b).

If the generally understood tolerances need to be specified, the tolerance class of 6H for the internal thread or the tolerance class of 6g for the external thread is added to the basic note (Fig. 13.18g). Where closer mating threads are desired, a tolerance of 6H for the internal thread and tolerance classes of 5g6g for the external thread are added to the basic note.

When the thread note refers to mating parts, a single basic note is adequate with the addition of the internal and external thread tolerance classes separated by the slash. The basic note for mating threads now becomes M10 × 1.5–6H/6g for the general-purpose thread or M10 × 1.5–6H/5g6g for the close-fitting thread (Fig. 13.18h).

For a left-hand thread, LH is added to the thread note. (Absence of LH indicates an RH thread.)

If it is necessary to indicate the length of the thread engagement, the letter S (short), N (normal), or L (long) is added to the thread note. For example, the single note M10 × 1.5–6H/6g–N–LH combines the specifications for internal and external mating left-hand metric threads of 10-mm diameter and 1.5-mm pitch with general-purpose tolerances and normal length of engagement.

A thread note for a blind tapped hole is shown in Fig. 13.18a. In a complete note, the tap drill and depth should be given, though in practice they are often omitted and left to the shop. For tap drill sizes, see Appendix 15. If the LH symbol is omitted, the thread is understood to be RH. If the thread is a multiple thread, the word DOUBLE, TRIPLE, or QUADRUPLE should precede the thread depth; otherwise, the thread is understood to be single. Thread notes for holes are preferably attached to the circular views of the holes, as shown.

Thread notes for external threads are preferably given in the longitudinal view of the threaded shaft, as shown in Figs. 13.18b to 13.18f. Examples of 8-, 12-, and 16-pitch threads, not shown in the figure, are 2–8N–2, 2–12N–2, and 2–16N–2. A sample special thread designation is $1\frac{1}{2}$–7N–LH.

General-purpose Acme threads are indicated by the letter G, and centralizing Acme threads by the letter C. Typical thread notes are $1\frac{3}{4}$–4 ACME–2G or $1\frac{3}{4}$–6 ACME–4C.

Thread notes for Unified threads are shown in Figs. 13.18j and 13.18k. Unified thread notes are distinguished from American National threads by the insertion of the letter U before the series letters, and by the letters A and B (for external or internal, respectively)

after the numeral designating the class of fit. If the letters LH are omitted, the thread is understood to be RH. Some typical thread notes are:

$\frac{1}{4}$–20 UNC–2A TRIPLE

$\frac{9}{16}$–18 UNF–2B

$1\frac{3}{4}$–16 UN–2A

■ 13.22 AMERICAN NATIONAL STANDARD PIPE THREADS

The American National Standard for pipe threads, originally known as the Briggs standard, was formulated by Robert Briggs in 1882. Two general types of pipe threads have been approved as American National Standard: *taper pipe threads* and *straight pipe threads*.

The profile of the taper pipe thread is illustrated in Fig. 13.19. The taper of the standard tapered pipe thread is 1 in 16 or .75″ per foot measured on the diameter and along the axis. The angle between the sides of the thread is 60°. The depth of the sharp V is .8660p, and the basic maximum depth of the truncated thread is .800p, where p = pitch. The basic pitch diameters, E_0 and E_1, and the basic length of the effective external taper thread, L_2, are determined by the formulas

$$E_0 = D - (.050D + 1.1)\frac{1}{n}$$

$$E_1 = E + .0625L_1$$

$$L_2 = (.80D + 6.8)\frac{1}{n},$$

were D = basic outer diameter (O.D.) of pipe, E_0 = pitch diameter of thread at end of pipe, E_1 = pitch diameters of thread at large end of internal thread, L_1 = normal engagement by hand, and n = number of threads per inch.

The ANSI also recommended two modified taper pipe threads for (1) dryseal pressure-tight joints (.88″

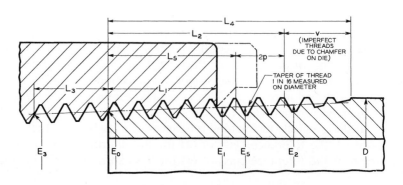

FIGURE 13.19 American National Standard Taper Pipe Thread [ANSI/ASME B1.20.1–1983 (R1992)].

per foot taper) and (2) rail fitting joints. The former is used to provide a metal-to-metal joint, eliminating the need for a sealer, and is used in refrigeration, marine, automotive, aircraft, and ordnance work. The latter is used to provide a rigid mechanical thread joint, as required in rail fitting joints.

While taper pipe threads are recommended for general use, there are certain types of joints in which straight pipe threads are used to advantage. The number of threads per inch, the angle, and the depth of thread are the same as on the taper pipe thread, but the threads are cut parallel to the axis. Straight pipe threads are used for pressure-tight joints for pipe couplings, fuel and oil line fittings, drain plugs, free-fitting mechanical joints for fixtures, loose-fitting mechanical joints for locknuts, and loose-fitting mechanical joints for hose couplings.

Pipe threads are represented by detailed or symbolic methods in a manner similar to the representation of Unified and American National threads. The symbolic representation (schematic or simplified) is recommended for general use regardless of diameter (Fig. 13.20). The detailed method is recommended only when the threads are large and when it is desired to show the profile of the thread, as, for example, in a sectional view of an assembly.

As shown in Fig. 13.20, it is not necessary to draw the taper on the threads unless there is some reason to emphasize it, since the thread note indicates whether the thread is straight or tapered. If it is desired to show the taper, it should be exaggerated, as shown in Fig. 13.21, where the taper is drawn $\frac{1}{16}''$ per $1''$ *on radius* (or $6.75''$ per $1'$ *on diameter*) instead of the actual taper of $\frac{1}{16}''$ *on diameter*. American National Standard taper pipe threads are indicated by a note giving the nominal diameter followed by the letters NPT (National pipe taper), as shown in Fig. 13.21. When straight pipe threads are specified, the letters NPS (National pipe straight) are used. In practice, the tap drill size is normally not given in the thread note.

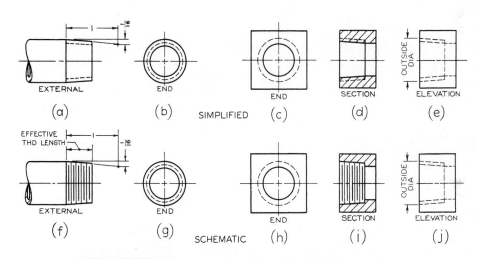

FIGURE 13.20 **Conventional Pipe Thread Representation.**

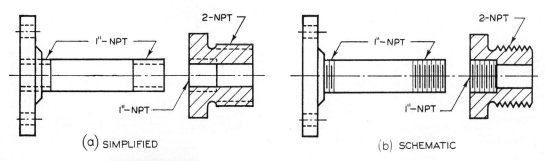

FIGURE 13.21 **Conventional Representation of Pipe Threads.**

G R A P H I C S   S P O T L I G H T

Fastener Libraries

ENGINEERS SPEND TWENTY HOURS PER MONTH REDRAWING PARTS

Many engineers spend up to twenty hours each month redrawing standard parts or parts they purchase from outside vendors. They need to show how the parts fit in assembly or in order to specify which part to use. Using a library of these standard parts can save considerable time in creating engineering drawings and specifications. Some vendors are willing to provide free drawings for their parts in standard drawing format. Many resources are available on the World Wide Web. You can also purchase libraries of standard symbols.

THOUSANDS OF PARTS AVAILABLE IN PARTSPEC

One place you can go for standard fasteners and vendor part drawings is Autodesk's PartSpec software. PartSpec is an application that runs with AutoCAD

Manufacturer: FASTENERS
Product: (Selection Available)
Model:

(Selection Available)
BINDING HEAD SCREW
BUTTON HEAD CAP SCREW
DOWEL PIN
FILLISTER HEAD SCREW
FLAT HEAD 100
FLAT HEAD 82
FLAT SOCKET
FLAT WASHER (A NARROW)
FLAT WASHER (A WIDE)
FLAT WASHER (B NARROW)
FLAT WASHER (B REGULAR)
FLAT WASHER (B WIDE)
HEX BOLT
HEX NUT
HEX NUT (HEAVY JAM)
HEX NUT (HEAVY)
HEX NUT (JAM)
HEX NUT (THICK)
HEX SCREW
HEX WASHER SCREW
LOCK WASHER (EXTRA DUTY)

(A)

Manufacturer: FASTENERS
Product: FILLISTER HEAD SCREW
Model: ANSI-FRACTION

○ Drawing
○ Picture
○ Text

Sch   T
Back L F R
Sect   B

Insert

Front View

SIZE: 1/4-20

Part Number:
Part Description: 1/4-20 FILLISTER HEAD SCREW

Previous

(B)

Release 12 or 13. You can use it to search through a part database containing thousands of parts on two CD-ROMs. The PartSpec window with standard fasteners selected is shown in Fig. A.

As you see in the figure above, you can select from various standard libraries and manufacturer databases. Once you have a library or manufacturer selected, you can choose from the list of products and the desired model for which drawings are available in the PartSpec database. Figure B shows a fillister head screw, model ANSI-fraction, size 1/4-20, is shown as the selection. The front view drawing of the fillister head screw of that size is shown at the right of the dialog box. If you want to insert that view into your current AutoCAD drawing you can pick on the icon labeled Insert in the upper right. To select from other available views, you can pick on the T, F, B, L, R, Back, Sch, and Sect. buttons for the view desired. Drawings provided with PartSpec follow a set of standards to ensure their usability.

MANUFACTURER DRAWINGS AND ORDERING INFORMATION

Over twenty different manufactures are represented in the PartSpec database. You can select a manufac-

turer from the list available, and then pick a particular product and model or type in a part number or description used to search the database. You can also qualify the search by detailed information appropriate to the part. Once you have selected a part, you can get ordering information or manufacturer's specifications in text format. Figure C shows Penn Engineering's Self-Clinching Flush Head Stud.

SAVE TIME SEARCHING FOR MATERIAL DATA WITH MATERIALSPEC

PartSpec has a counterpart for specifying materials, called MaterialSpec. It is a searchable text database on CD-ROM containing materials from five categories: plastics, metals, composites, ceramics, and military specifications (MIL5). You can choose materials by type, manufacturer, part name or number, description, property, or application. Add-on applications can provide a valuable resource for engineering. One of the biggest advantages of using CAD is that drawings can be re-used, re-scaled, or re-oriented for different purposes resulting in a valuable time savings. Remember the World Wide Web is also a valuable engineering resource for drawings and material information.

(C)

■ 13.23 BOLTS, STUDS, AND SCREWS

The term **bolt** is generally used to denote a "through bolt" that has a head on one end, is passed through clearance holes in two or more aligned parts, and is threaded on the other end to receive a nut to tighten and hold the parts together (Fig. 13.22a; see also §§13.25 and 13.26).

A hexagon head *cap screw* (Fig. 13.22b) is similar to a bolt except that it generally has greater length of thread for when it is used without a nut; in such cases, one of the members being held together is threaded to act as a nut. The cap screw is screwed on with a wrench. Cap screws are not screwed into thin materials if strength is desired (see §13.29).

A *stud* (Fig. 13.22c) is a steel rod threaded on both ends. It is screwed into place with a pipe wrench or, preferably, with a stud driver. As a rule, a stud is

passed through a clearance hole in one member and screwed into another member; a nut is used on the free end, as shown.

A *machine screw* (Fig. 13.32) is similar to the slotted-head cap screws but, in general, is smaller. It may be used with or without a nut.

A *set screw* (Fig. 13.33) is a screw with or without a head that is screwed through one member and whose special point is forced against another member to prevent relative motion between the two parts.

It is not customary to section bolts, nuts, screws, and similar parts when drawn in assembly, as shown in Figs. 13.22 and 13.31, because they do not themselves require sectioning for clearness (see §13.23).

■ 13.24 TAPPED HOLES

The bottom of a drilled hole is conical in shape, as formed by the point of the twist drill (Figs. 13.23a and 13.23b). When an ordinary drill is used in connection with tapping, it is referred to as a **tap drill**. On drawings, an angle of 30° is used to approximate the actual 31°.

The thread length is the length of full or perfect threads. The tap drill depth is the depth of the cylindrical portion of the hole and does not include the cone point. The portion A of the drill depth shown beyond the threads in Figs. 13.23c and 13.23d includes the several imperfect threads produced by the chamfered end of the tap. This distance A varies according to drill size and whether a plug tap (see Fig. 11.19h) or a bottom-

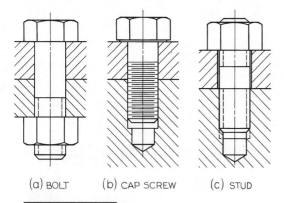

(a) BOLT (b) CAP SCREW (c) STUD

FIGURE 13.22 **Bolt, Cap Screw, and Stud.**

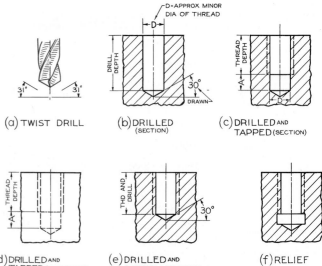

(a) TWIST DRILL (b) DRILLED (SECTION) (c) DRILLED AND TAPPED (SECTION)

(d) DRILLED AND TAPPED (ELEVATION) (e) DRILLED AND BOTTOM TAPPED (SECTION) (f) RELIEF

FIGURE 13.23 **Drilled and Tapped Holes.**

ing tap is used to finish the hole. For drawing purposes, when the tap drill depth is not specified, the distance A may be drawn equal to three schematic thread pitches (see Fig. 13.9).

A tapped hole finished with a bottoming tap is drawn as shown in Fig. 13.23e. Blind bottoming holes should be avoided wherever possible. A better procedure is to cut a relief with its diameter slightly greater than the major diameter of the thread (Fig. 13.23f).

One of the chief causes of tap breakage is insufficient tap drill depth, in which the tap is forced against a bed of chips in the bottom of the hole. Therefore, the drafter should never draw a blind hole when a through hole of not much greater length can be used. When a blind hole is necessary, however, the tap drill depth should be generous. Tap drill sizes for Unified, American National, and metric threads are given in Appendix 15. It is good practice to give the tap drill size in the thread note (see §13.21).

The thread length in a tapped hole depends on the major diameter and the material being tapped. In Fig. 13.24, the minimum engagement length X, when both parts are steel, is equal to the diameter D of the thread. When a steel screw is screwed into cast iron, brass, or bronze, $X = 1\frac{1}{2}D$; when it is screwed into aluminum, zinc, or plastic, $X = 2D$.

Since the tapped thread length contains only full threads, it is necessary to make this length only one or two pitches beyond the end of the engaging screw. In simplified or schematic representation, the threads are omitted in the bottoms of tapped holes to show the ends of the screws clearly (Fig. 13.24).

When a bolt or a screw is passed through a clearance hole in one member, the hole may be drilled 0.8

mm ($\frac{1}{32}''$) larger than the screw up to $\frac{3}{8}''$ or 10 mm diameter and 1.5 mm ($\frac{1}{16}''$) larger for larger diameters. For more precise work, the clearance hole may be only 0.4 mm ($\frac{1}{64}''$) larger than the screw up to $\frac{3}{8}''$ or 10 mm diameter and 0.8 mm ($\frac{1}{32}''$) larger for larger diameters. Closer fits may be specified for special conditions.

The clearance spaces on each side of a screw or bolt need not be shown on a drawing unless it is necessary to show that there is no thread engagement, in which case the clearance spaces are drawn about 1.2 mm ($\frac{3}{64}''$) wide for clarity.

■ 13.25 STANDARD BOLTS AND NUTS

American National Standard bolts and nuts,[*] metric and inch series, are produced in hexagon form, while the square form is only produced in the inch series (Fig. 13.25). Square heads and nuts are chamfered at 30°, and hexagon heads and nuts are chamfered at 15–30°. Both are drawn at 30° for simplicity.

BOLT TYPES Bolts are grouped according to use: *regular* bolts for general service, and *heavy* bolts for heavier service or easier wrenching. Square bolts come only in the regular type; hexagon bolts, screws, and nuts and square nuts are standard in both types.

Metric hexagon bolts are grouped according to use: regular and heavy bolts and nuts for general service and high-strength bolts and nuts for structural bolting.

[*] The ANSI standards cover several bolts and nuts. For complete details, see the standards.

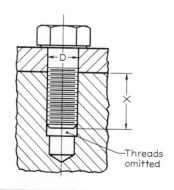

FIGURE 13.24 **Tapped Holes.**

HEXAGON BOLT AND NUT
(a)

SQUARE BOLT AND NUT
(b)

FIGURE 13.25 **Standard Bolts and Nuts.** *Courtesy of Cordova Bolt Inc., Buena Park, CA.*

FINISH Square bolts and nuts, hexagon bolts, and hexagon flat nuts are unfinished. *Unfinished* bolts and nuts are not machined on any surface except for the threads. Traditionally, hexagon bolts and hexagon nuts have been available as unfinished, semifinished, or finished. According to the latest standards, hexagon cap screws and finished hexagon bolts have been consolidated into a single product—*hex cap screws*—thus eliminating the regular semifinished hexagon bolt classification. Heavy semifinished hexagon bolts and heavy finished hexagon bolts also have been combined into a single product called *heavy hex screws*. Hexagon cap screws, heavy hexagon screws, and all hexagon nuts, except hexagon flat nuts, are considered *finished* to some degree and are characterized by a "washer face" machined or otherwise formed on the bearing surface. The washer face is $\frac{1}{64}''$ thick (drawn $\frac{1}{32}''$), and its diameter is equal to $1\frac{1}{2}$ times the body diameter D for the inch series.

For nuts the bearing surface also may be a circular surface produced by chamfering. Hexagon screws and hexagon nuts have closer tolerances and may have a more finished appearance but are not completely machined. There is no difference in the drawing for the degree of finish on finished screws and nuts.

Metric bolts, cap screws, and nuts are produced in the hexagon form (Fig. 13.25a). The hexagon heads and nuts are chamfered at 15–30°. Both are drawn at 30° for simplicity.

PROPORTIONS Sizes based on diameter D of the bolt body (including metric) (Fig. 13.26), which are either exact formula proportions or close approximations for drawing purposes, are as follows.
Regular hexagon and square bolts and nuts:

$$W = 1\tfrac{1}{2}D \qquad H = \tfrac{2}{3}D \qquad T = \tfrac{7}{8}D,$$

where W = width across flats, H = head height, and T = nut height.
Heavy hexagon bolts and nuts and square nuts:

$$W = 1\tfrac{1}{2}D + \tfrac{1}{8}'' \text{ (or } + 3 \text{ mm)}$$

$$H = \tfrac{2}{3}D \qquad T = D$$

The washer face is always included in the head or nut height for finished hexagon screw heads and nuts.

THREADS Square and hex bolts, hex cap screws, and finished nuts in the inch series are usually Class 2 and may have coarse, fine, or 8-pitch threads. Unfinished nuts have coarse threads, Class 2B. For diameter and pitch specifications for metric threads, see Appendix 18.

THREAD LENGTHS

For bolts or screws up to 6″ (150 mm) in length:

$$\text{Thread length} = 2D + \tfrac{1}{4}'' \text{ (or } + 6 \text{ mm)}$$

For bolts or screws over 6″ in length,

$$\text{Thread length} = 2D + \tfrac{1}{2}'' \text{ (or } + 12 \text{ mm)}$$

Fasteners too short for these formulas are threaded as close to the head as practicable. For drawing purposes, this may be taken as approximately three pitches. The threaded end may be rounded or chamfered, but is usually drawn with a 45° chamfer from the thread depth (Fig. 13.26).

BOLT LENGTHS Lengths of bolts have not been standardized because of the endless variety required by industry. Short bolts are typically available in standard length increments of $\frac{1}{4}''$ (6 mm), while long bolts come in increments of $\frac{1}{2}$ to 1″ (12 to 25 mm). For dimensions of standard bolts and nuts, see Appendix 18.

■ 13.26 DRAWING STANDARD BOLTS

In practice, standard bolts and nuts are not shown in detail drawings unless they are to be altered, but they appear so frequently on assembly drawings that a suitable but rapid method of drawing them must be used. Time-saving templates are available, or they may be drawn from exact dimensions taken from tables (see Appendix 18) if accuracy is important, as in figuring clearances. However, in the great majority of cases the conventional representation, in which proportions based on the body diameter are used, will be sufficient, and a considerable amount of time may be saved. Three typical bolts illustrating the use of these proportions for the regular bolts are shown in Fig. 13.26.

Although the curves produced by the chamfer on the bolt heads and nuts are hyperbolas, in practice these curves are always represented approximately by means of circular arcs (Fig. 13.26).

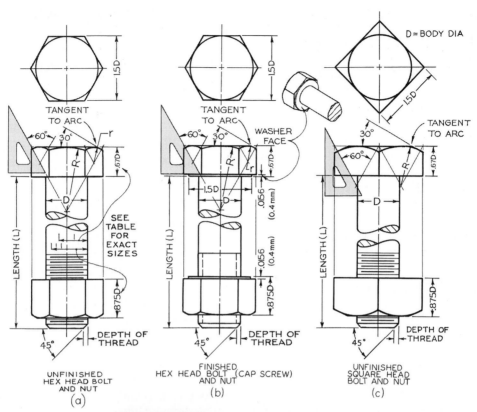

FIGURE 13.26 Bolt Proportions (Regular).

Generally, bolt heads and nuts should be drawn "across corners" in all views, regardless of projection. This conventional violation of projection is used to prevent confusion between the square and hexagon heads and nuts and to show actual clearances. Only when there is a special reason should bolt heads and nuts be drawn across flats. In such cases, the conventional proportions are used (Fig. 13.27).

Steps in drawing hexagon bolts, cap screws, and nuts are illustrated in Fig. 13.28, and those for square bolts and nuts in Fig. 13.29. Before drawing a bolt, the diameter of the bolt, the length (from the underside of the bearing surface to the tip), the style of head (square or hexagon), and the type (regular or heavy), as well as the finish, must be known.

If only the longitudinal view of a bolt is needed, it is necessary to draw only the lower half of the top views in Figs. 13.28 and 13.29 with light construction lines to project the corners of the hexagon or square to

the front view. These construction lines may then be erased if desired.

The head and nut heights can be spaced off with the dividers on the shaft diameter and then transferred as shown in both figures, or the scale may be used as in Fig. 5.15. The heights should not be determined by arithmetic.

The $\frac{1}{64}''$ (0.4 mm) washer face has a diameter equal to the distance across flats of the bolt head or nut. It appears only on the metric and finished hexagon screws or nuts, and the washer face thickness is drawn at $\frac{1}{32}''$ (1 mm) for clearness. The $\frac{1}{32}''$ (1 mm) is included in the head or nut height.

Threads should be drawn in simplified or schematic form for body diameters of 1″ (25 mm) or less on the drawing (Fig. 13.9b or 13.9d) and by detailed representation for larger diameters (§§13.7 and 13.8). The threaded end of the screw should be chamfered at 45° from the schematic thread depth (Fig. 13.9a).

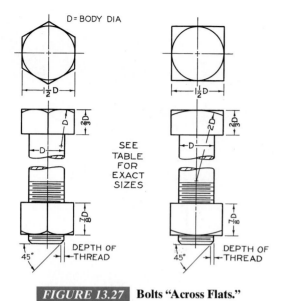

FIGURE 13.27 **Bolts "Across Flats."**

■ 13.27 SPECIFICATIONS FOR BOLTS AND NUTS

In specifying bolts in parts lists, in correspondence, or elsewhere, the following information must be covered in order.

1. Nominal size of bolt body.
2. Thread specification or thread note (see §13.21).
3. Length of bolt.
4. Finish of bolt.
5. Style of head.
6. Name.

   EXAMPLE (Complete Decimal-Inch)

   .75–10 UNC-2A × 2.50 HEXAGON CAP SCREW

   EXAMPLE (Abbreviated Decimal-Inch)

   .75 × 2.50 HEX CAP SCR

   EXAMPLE (Metric)

   M8 X 1.25–40, HEX CAP SCR

Nuts may be specified as follows.

   EXAMPLE (Complete)

   $\frac{5}{8}$–11 UNC–2B SQUARE NUT

On drawings of small bolts or nuts under approximately $\frac{1}{2}''$ diameter (12 mm), where the chamfer is hardly noticeable, the chamfer on the head or nut may be omitted in the longitudinal view.

Many styles of templates are available for saving time in drawing bolt heads and nuts.

FIGURE 13.28 **Steps in Drawing (a) Finished Hexagon Head Bolt (Cap Screw) and Hexagon Nut, and (b) Square-Itead Bolt and Square Nut.**

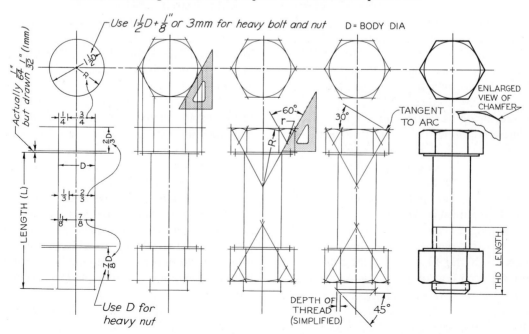

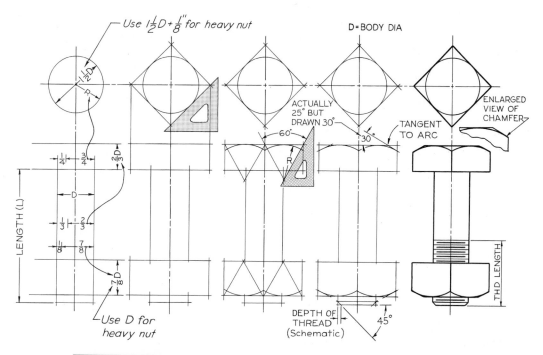

FIGURE 13.29 Steps in Drawing Square-Head Bolt and Square Nut.

EXAMPLE (Abbreviated)

$\frac{5}{8}$ SQ NUT

EXAMPLE (Metric)

M8 × 1.25 HEX NUT

For either bolts or nuts, the words REGULAR or GENERAL PURPOSE are assumed if omitted from the specification. If the heavy series is intended, the word HEAVY should appear as the first word in the name of the fastener. Likewise, HIGH STRENGTH STRUCTURAL should be indicated for such metric fasteners. However, the number of the specific ISO standard is often included in the metric specifications—for example, HEXAGON NUT ISO 4032 M12 × 1.75. Similarly, finish need not be mentioned if the fastener or nut is correctly named (see §13.25).

■ 13.28 LOCKNUTS AND LOCKING DEVICES

Many types of special nuts and devices to prevent nuts from unscrewing are available, some of the most common of which are illustrated in Fig. 13.30. The Ameri-

can National Standard *jam nuts* (Figs. 13.30a and 13.30b) are the same as the hexagon or hexagon flat nuts, except that they are thinner. The application shown in Fig. 13.30b, where the larger nut is on top and is screwed on more tightly, is recommended. They are the same distance across flats as the corresponding hexagon nuts ($1\frac{1}{2}D$ or $1\frac{1}{2}D + \frac{1}{8}''$). They are slightly over $\frac{1}{2}D$ in thickness but are drawn $\frac{1}{2}D$ for simplicity. They are available with or without the washer face in the regular and heavy types. The tops of all are flat and chamfered at 30°, and the finished forms have either a washer face or a chamfered bearing surface.

The lock washer, shown in Fig. 13.30c, and the cotter pin, shown in Figs. 13.30e, 13.30g, and 13.30h, are very common (see Appendixes 27 and 30). The set screw (Fig. 13.30f) is often made to press against a plug of softer material, such as brass, which in turn presses against the threads without deforming them. For use with cotter pins (see Appendix 30), a hex slotted nut (Fig. 13.30g) and a hex castle nut (Fig. 13.30h), as well as a hex thick slotted nut and a heavy hex thick slotted nut, are recommended.

Similar metric locknuts and locking devices are available. See fastener catalogs for details.

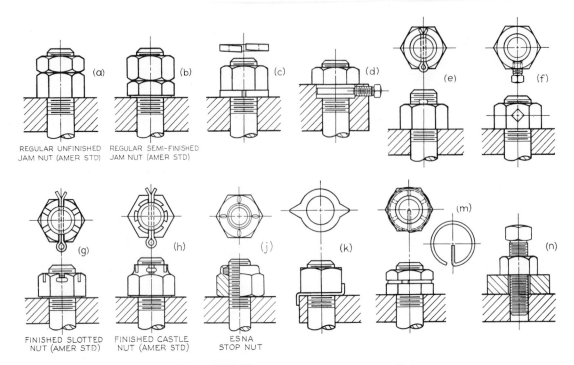

FIGURE 13.30 Locknuts and Locking Devices.

■ 13.29 STANDARD CAP SCREWS

Five types of American National Standard cap screws are shown in Fig. 13.31. The first four of these have standard heads, while the socket head cap screws (Fig. 13.31e) have several different shapes of round heads and sockets. Cap screws are regularly produced in finished form and are used on machine tools and other machines for which accuracy and appearance are important. The ranges of sizes and exact dimensions are given in Appendixes 18 and 19. The hexagon head cap screw and hex socket head cap screw in several forms are available in metric (see Appendix 18).

Cap screws ordinarily pass through a clearance hole in one member and screw into another (§13.24). The clearance hole need not be shown on the drawing when the presence of the unthreaded clearance hole is obvious.

Cap screws are inferior to studs if frequent removal is necessary; hence, they are used on machines requiring few adjustments. The slotted or socket-type heads are best under crowded conditions.

The actual standard dimensions may be used in drawing the cap screws whenever exact sizes are necessary, but this is seldom the case. In Fig. 13.31 the dimensions are given in terms of body diameter D, and they closely conform to the actual dimensions. The resulting drawings are almost exact reproductions and

are easy to draw. The hexagonal head cap screw is drawn in the manner shown in Fig. 13.28. The points are drawn chamfered at 45° from the schematic thread depth.

For correct representation of tapped holes, see §13.24. For information on drilled, countersunk, or counterbored holes, see §6.33.

In an assembly section, it is customary not to section screws, bolts, shafts, or other solid parts whose center lines lie in the cutting plane. Such parts in themselves do not require sectioning and are, therefore, shown "in the round" (Fig. 13.31).

Note that screwdriver slots are drawn at 45° in the circular views of the heads, without regard to true projection, and that threads in the bottom of the tapped holes are omitted so that the ends of the screws may be clearly seen. A typical cap screw note is as follows:

EXAMPLE (Complete)

$\frac{3}{8}$–16 UNC-2A × $2\frac{1}{2}$ HEXAGON HEAD CAP SCREW

EXAMPLE (Abbreviated)

$\frac{3}{8}$ × $2\frac{1}{2}$ HEX HD CAP SCR

EXAMPLE (Metric)

M20 × 2.5 × 80 HEX HD CAP SCR

a

A fully rendered and assembled turbine display.
(All courtesy of Autodesk, Inc.)

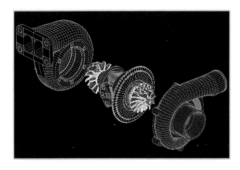

b

An exploded wireframe display.

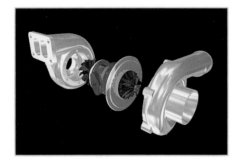

c

A surface-rendered exploded display.

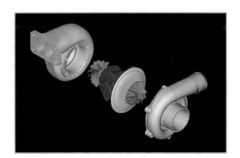

d

A surface-rendered, color-coded display.

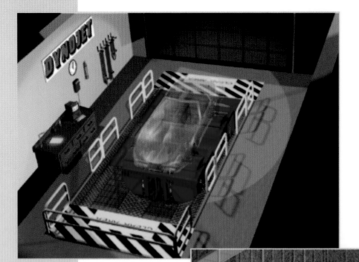

a, b, c

VRML images found at the Dynojet Research Incorporated web site. When viewed with a VRML browser, these images have a 3D quality. *(All courtesy of Dynojet Research, Inc.)*

a, b

Use of CAD speeds the design process and reduces costs. These models of a robot welder and a Zip drive let manufacturers simulate their products before they were even built. *(Reproduced by permission of SDRC, Inc.)*

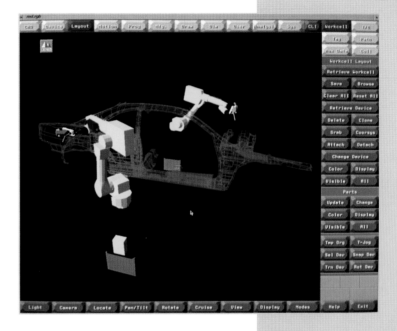

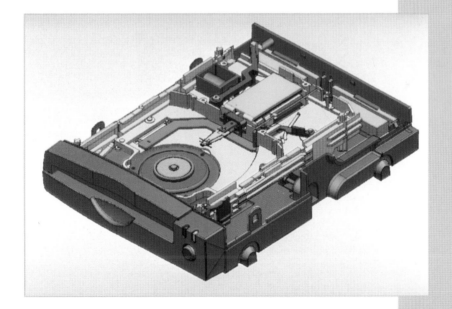

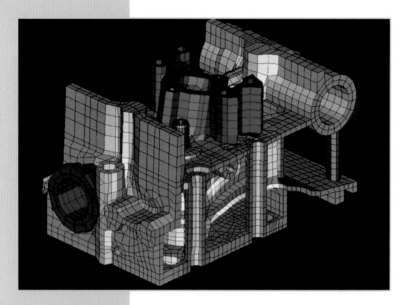

a

A Finite Element Model of diesel engine components created with I-DEAS software. *(Courtesy of SDRC, Inc.)*

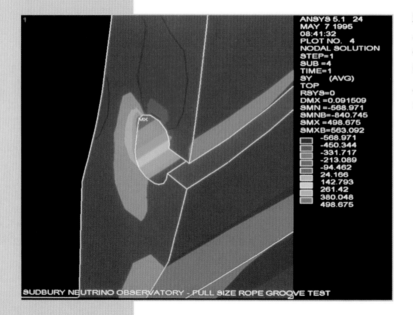

b

A structural stress analysis of the Sudbury Neutrino Observatory created with Auto FEA by ANSYS. *(Courtesy of ANSYS)*

PLATE 5

a

An engineer using virtual reality to help in autmobile design. A special helmet, goggles, and glove allow the engineer to simulate the driving experience as seen below. *(Courtesy of Ford Motor Company)*

b

CAD software and virtual reality equipment allow users to experience "virtual rooms" and decide whether they like the design before construction. *(Courtesy of Truevision)*

Peter Greene Residence   Aspen Colorado

James Biebl   Gaviota Graphics   303 963−3309

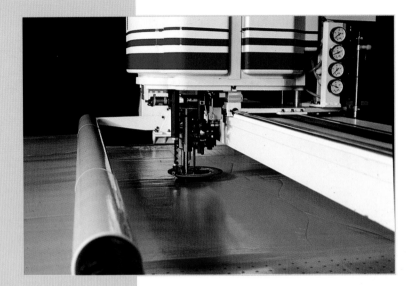

a

An automated cutting system uses patterns designed with CAD software to produce precise designs. *(Courtesy of Gerber Scientific, Inc.)*

b

Modern manufacturing depends heavily on computer controlled robots. At this Chrysler plant, 66 robots apply spot welds. *(Courtesy of Chrysler Motor Corp.)*

PLATE 7

a, b, c

Examples of solid models
created with I-DEAS soft-
ware. Solid modelling can
help bring designs to life.
(All courtesy of SDRC, Inc.)

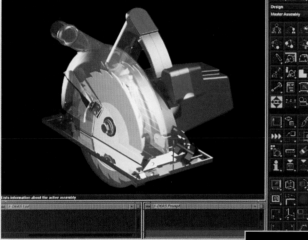

a, b, c

CAD software is very useful as a tool to visualize architectural designs. All these images were created in AutoCAD. *(All courtesy of Autodesk, Inc.)*

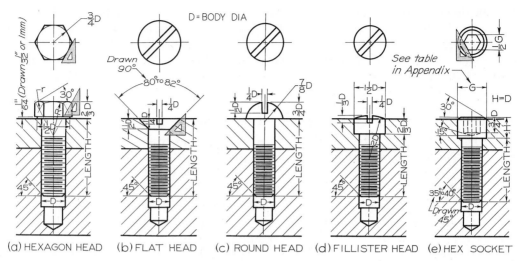

(a) HEXAGON HEAD   (b) FLAT HEAD   (c) ROUND HEAD   (d) FILLISTER HEAD   (e) HEX SOCKET

Hexagon Head Screws   Coarse, Fine, or 8-Thread Series, 2A. Thread length = $2D + \frac{1}{4}''$ up to 6″ long and $2D + \frac{1}{2}''$ if over 6″ long. For screws too short for formula, threads extend to within $2\frac{1}{2}$ threads of the head for diameters up to 1″. Screw lengths not standardized. For suggested lengths for metric Hexagon Head Screws, see Appendix 15.

Slotted Head Screws   Coarse, Fine, or 8-Thread Series, 2A. Thread length = $2D + \frac{1}{4}''$. Screw lengths not standardized. For screws too short for formula, threads extend to within $2\frac{1}{2}$ threads of the head.

Hexagon Socket Screws   Coarse or Fine Threads, 3A. Coarse thread length = $2D + \frac{1}{2}''$ where this would be over $\frac{1}{2}L$; otherwise thread length = $\frac{1}{2}L$. Fine thread length = $1\frac{1}{2}D + \frac{1}{2}''$ where this would be over $\frac{3}{8}L$; otherwise thread length = $\frac{3}{8}L$. Increments in screw lengths = $\frac{1}{8}''$ for screws $\frac{1}{4}''$ to 1″ long, $\frac{1}{4}''$ for screws 1″ to 3″ long, and $\frac{1}{2}''$ for screws $3\frac{1}{2}''$ to 6″ long.

FIGURE 13.31   **Standard Cap Screws. See Appendixes 18 and 19.**

■ 13.30 STANDARD MACHINE SCREWS

Machine screws are similar to cap screws but are in general smaller (.060″ to .750″ dia). There are eight ANSI-approved forms of heads shown in Appendix 20. The hexagonal head may be slotted if desired. All others are available in either slotted or recessed-head forms. Standard machine screws are regularly produced with a naturally bright finish, not heat treated, and are regularly supplied with plain-sheared ends, not chamfered. For similar metric machine screw forms and specifications, see Appendix 20.

Machine screws are particularly adapted to screwing into thin materials, and all the smaller-numbered screws are threaded nearly to the head. They are used extensively in firearms, jigs, fixtures, and dies. Machine screw nuts are used mainly on the round head, pan head, and flat head types and are usually hexagonal in form.

Exact dimensions of machine screws are given in Appendix 20, but they are seldom needed for drawing purposes. The four most common types of machine screws are shown in Fig. 13.32, in which proportions based on diameter D conform closely to the actual dimensions and produce almost exact drawings. Clearance holes and counterbores should be made slightly larger than the screws, as explained in §13.24.

FIGURE 13.32   **Standard Machine Screws. See Appendix 20.**

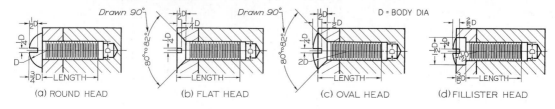

(a) ROUND HEAD   (b) FLAT HEAD   (c) OVAL HEAD   (d) FILLISTER HEAD

Note that the threads in the bottom of the tapped holes are omitted so that the ends of the screws will be clearly seen. Observe also that it is conventional practice to draw the screwdriver slots at 45° in the circular view without regard to true projection.

A typical machine screw note is as follows:

EXAMPLE (Complete)

No. 10 (.1900)–32 NF-3 × $\frac{5}{8}$ FILLISTER HEAD MACHINE SCREW

EXAMPLE (Abbreviated)

No. 10 (.1900) × $\frac{5}{8}$ FILL HD MACH SCR

EXAMPLE (Metric)

M8 × 1.25 × 30 SLOTTED PAN HEAD MACHINE SCREW

■ 13.31 STANDARD SET SCREWS

The function of set screws (Fig. 13.33) is to prevent relative motion, usually rotary, between two parts, such as the movement of the hub of a pulley on a shaft. A set screw is screwed into one part so that its point bears firmly against another part. If the point of the set screw is cupped (Fig. 13.33e), or if a flat is milled on the shaft (Fig. 13.33a), the screw will hold much more firmly. Obviously, set screws are not efficient when the load is heavy or is suddenly applied. Usually they are manufactured of steel and case hardened.

The American National Standard square head set screw and slotted headless set screw are shown in Figs. 13.33a and 13.33b. Two American National Standard socket set screws are illustrated in Figs. 13.33c and 13.33d. American National Standard set screw points are shown in Figs. 13.33e to 13.33k. Headless set screws have come into greater use because the projecting head of headed set screws has caused many industrial casualties; this has resulted in legislation prohibiting their use in many states.

Most of the dimensions in Fig. 13.33 are American National Standard formula dimensions, and the resulting drawings are almost exact representations.

Metric hexagon socket headless set screws with the full range of points are available and are represented in the same manner as shown in Fig. 13.33. Nominal diameters of metric hex socket set screws are 1.6, 2, 2.5, 3, 4, 5, 6, 8, 10, 12, 16, 20, and 24 mm.

Square head set screws have coarse, fine, or 8-pitch threads, Class 2A, but are usually furnished with coarse threads since the square head set screw is generally used on the rougher grades of work. Slotted headless and socket set screws have coarse or fine threads, Class 3A.

Nominal diameters of set screws range from number 0 up through 2″; set screw lengths are standardized in increments of $\frac{1}{32}$″ to 1″ depending on the overall length of the set screw.

Metric set screw length increments range from 0.5 to 4 mm, again depending on overall screw length.

FIGURE 13.33 American National Standard Set Screws.

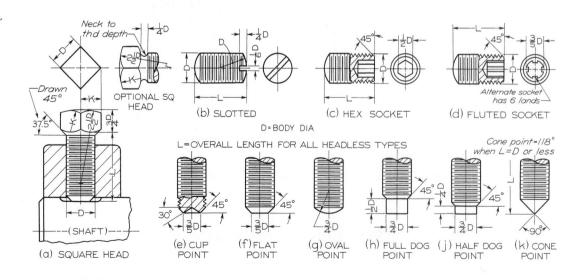

Set screws are specified as follows.

EXAMPLE (Complete)

$$\tfrac{3}{8}\text{–16 UNC-2A} \times \tfrac{3}{4}\text{SQUARE HEAD}$$
FLAT POINT SET SCREW

EXAMPLE (Abbreviated)

$$\tfrac{3}{8}\text{–} \times 1\tfrac{1}{4}\text{ SQ HD FL PT SS}$$
$$\tfrac{7}{16} \times \tfrac{3}{4}\text{ HEX SOC CUP PT SS}$$
$$\tfrac{1}{4}\text{–20 UNC 2A} \times \tfrac{4}{8}\text{ SLOT. HDLS CONE PT SS}$$

EXAMPLE (Metric)

M10 × 1.5 × 12 HEX SOCKET HEAD SET SCREW

■ 13.32 AMERICAN NATIONAL STANDARD WOOD SCREWS

Wood screws with three types of heads—flat, round, and oval—have been standardized (Fig. 13.34). The dimensions shown closely approximate the actual dimensions and are more than sufficiently accurate for use on drawings.

The Phillips-style recessed head is also available on several types of fasteners as well as wood screws. Three styles of cross recesses have been standardized by the ANSI. A special screwdriver is used, as shown in Fig. 13.35q, and results in rapid assembly without damage to the head.

■ 13.33 MISCELLANEOUS FASTENERS

Many other types of fasteners have been devised for specialized uses. Some of the more common types are shown in Fig. 13.35. A number of these are American National Standard round-head bolts, including carriage, button head, step, and countersunk bolts.

Helical coil threaded inserts (Fig. 13.35p) are shaped like a spring except that the cross section of the wire conforms to threads on the screw and in the hole. These are made of phosphor bronze or stainless steel, and they provide a hard, smooth protective lining for tapped threads in soft metals and plastics.

■ 13.34 KEYS

Keys are used to prevent relative movement between shafts and wheels, couplings, cranks, and similar

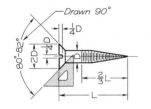
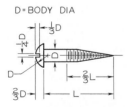
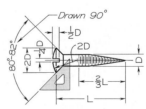

FIGURE 13.34 American National Standard Wood Screws.

FIGURE 13.35 Miscellaneous Bolts and Screws.

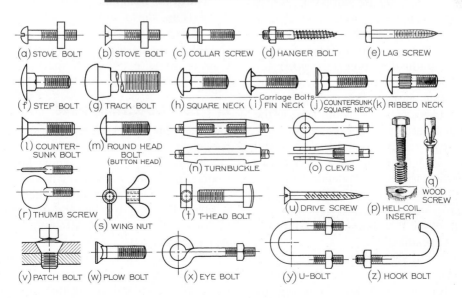

(a) STOVE BOLT (b) STOVE BOLT (c) COLLAR SCREW (d) HANGER BOLT (e) LAG SCREW
(f) STEP BOLT (g) TRACK BOLT (h) SQUARE NECK (i) FIN NECK (j) COUNTERSUNK SQUARE NECK (k) RIBBED NECK
Carriage Bolts
(l) COUNTERSUNK BOLT (m) ROUND HEAD BOLT (BUTTON HEAD) (n) TURNBUCKLE (o) CLEVIS (q) WOOD SCREW
(r) THUMB SCREW (s) WING NUT (t) T-HEAD BOLT (u) DRIVE SCREW (p) HELI-COIL INSERT
(v) PATCH BOLT (w) PLOW BOLT (x) EYE BOLT (y) U-BOLT (z) HOOK BOLT

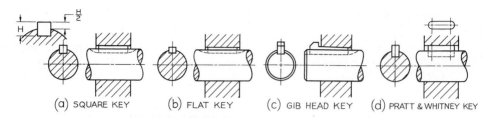

(a) SQUARE KEY   (b) FLAT KEY   (c) GIB HEAD KEY   (d) PRATT & WHITNEY KEY

FIGURE 13.36   **Square and Flat Keys.**

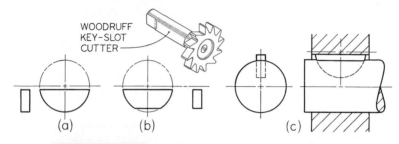

(a)   (b)   (c)

FIGURE 13.37   **Woodruff Keys and Key-Slot Cutter.**

machine parts attached to or supported by shafts (Fig. 13.36).

For heavy-duty functions, rectangular keys (flat or square) are suitable, and sometimes two rectangular keys are necessary for one connection. For even stronger connections, interlocking *splines* may be machined on the shaft and in the hole (see Fig. 13.9).

A *square key* is shown in Fig. 13.36a and a *flat key* in Fig. 13.36b. The widths of keys generally used are about one fourth the shaft diameter. In either case, one half the key is sunk into the shaft. The depth of the keyway or the keyseat is measured on the side—not the center (Fig. 13.36a). Square and flat keys may have the top surface tapered $\frac{1}{8}''$ per foot, in which case they become square taper or flat taper keys.

A rectangular key that prevents rotary motion but permits relative longitudinal motion is a *feather key* and is usually provided with *gib heads*, or otherwise fastened so it cannot slip out of the keyway. A *gib head key* is shown in Fig. 13.36c. It is exactly the same as the square taper or flat taper key except that a gib head, which provides for easy removal, is added. Square and flat keys are made from cold-finished stock and are not machined. For dimensions, see Appendix 21.

The *Pratt & Whitney key* (P & W) (Fig. 13.36d) is rectangular in shape, with semicylindrical ends. Two thirds of the height of the P & W key is sunk into the shaft keyseat (see Appendix 25).

The *Woodruff key* is semicircular in shape (Fig. 13.37). The key fits into a semicircular key slot cut with a Woodruff cutter, as shown, and the top of the key fits into a plain rectangular keyway. Sizes of keys for given shaft diameters are not standardized, but for average conditions it will be found satisfactory to select a key whose diameter is approximately equal to the shaft diameter. For dimensions, see Appendix 23.

A **keyseat** is in a shaft; a **keyway** is in the hub or surrounding part.

Typical specifications for keys are as follows:

$\frac{1}{4} \times 1\frac{1}{2}$SQ KEY
No. 204 WOODRUFF KEY
$\frac{1}{4} \times \frac{1}{16} \times 1\frac{1}{2}$FLAT KEY
No. 10 P & W KEY

Notes for nominal specifications of keyways and keyseats are shown in Figs. 11.44o, 11.44p, 11.44r, and 11.44x. For production work, keyways and keyseats should be dimensioned as shown in Fig. 11.48. See manufacturers' catalogs for specifications for metric counterparts.

■ 13.35 MACHINE PINS

Machine pins include taper pins, straight pins, dowel pins, clevis pins, and cotter pins. For light work, the taper pin is effective for fastening hubs or collars to shafts (Fig. 13.38) in which the hole through the collar and shaft is drilled and reamed when the parts are assembled. For slightly heavier duty, the taper pin may be used parallel to the shaft as for square keys (see Appendix 29).

Dowel pins are cylindrical or conical in shape and are used for a variety of purposes, chief of which is to keep two parts in a fixed position or to preserve alignment. The dowel pin is most commonly used and is

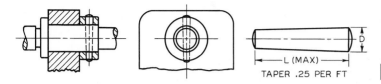

FIGURE 13.38   Taper Pin.

recommended where accurate alignment is essential. Dowel pins are usually made of steel and are hardened and ground in a centerless grinder.

The clevis pin is used in a clevis and is held in place by a cotter pin. For the latter, see Appendix 30.

■ 13.36 RIVETS

Rivets are regarded as permanent fastenings as distinguished from removable fastenings, such as bolts and screws. Rivets are generally used to hold sheet metal or rolled steel shapes together and are made of wrought iron, carbon steel, copper, or occasionally other metals.

To fasten two pieces of metal together, holes are punched, drilled, or punched and then reamed, slightly larger in diameter than the shank of the rivet. Rivet diameters in practice are made from $d = 1.2 \sqrt{t}$ to $d = 1.4 \sqrt{t}$, where d is the rivet diameter and t is the metal thickness. The larger size is used for steel and single-riveted joints, and the smaller may be used for multiple-riveted joints. In structural work it is common practice to make the hole 1.6 mm ($\frac{1}{16}''$) larger than the rivet.

When the red-hot rivet is inserted, a "dolly bar," having a depression the shape of the driven head, is held against the head. A riveting machine is then used to drive the rivet and to form the head on the plain end. This action causes the rivet to swell and fill the hole tightly.

Large rivets or heavy hex structural bolts were commonly used in structural work of bridges and buildings and in ship and boiler construction. Although not used in modern construction, an engineer could be involved in updating existing structures using rivets. They are shown in their exact formula proportions in Fig. 13.39. The button head (Fig. 13.39a) and countersunk head (Fig. 13.39e) types are the rivets most commonly used in structural work. The button head and cone head are commonly used in tank and boiler construction.

Typical riveted joints are illustrated in Fig. 13.40. Notice that the longitudinal view of each rivet shows the shank of the rivet with both heads made with circular arcs, and the circular view of each rivet is represented by only the outside circle of the head. In structural drafting, where there may be many such circles to draw, the drop spring bow (Fig. 3.53) is a convenient instrument.

Since many engineering structures are too large to be built in the shop, they are built in the largest units possible and then are transported to the desired location. Trusses are common examples of this. The rivets driven in the shop are called *shop rivets*, and those driven on the job are called *field rivets*. However, heavy steel bolts are commonly used on the job for structural work. Solid black circles are used to represent field rivets, and other standard symbols are used to show other features, as shown in Fig. 13.41.

For light work, small rivets are used. American National Standard small solid rivets are illustrated with dimensions showing their standard proportions in Fig. 13.42 [ANSI/ASME B18.1.1–1972 (R1995)]. Included in the same standard are tinners', coppers', and belt rivets. Metric rivets are also available. Dimensions for large rivets can be found in ANSI/ASME B18.1.2–1972 (R1995). See manufacturers' catalogs for additional details.

FIGURE 13.39   Standard Large Rivets.

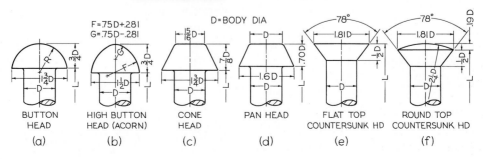

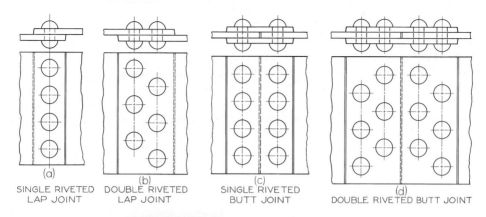

SINGLE RIVETED
LAP JOINT

DOUBLE RIVETED
LAP JOINT

SINGLE RIVETED
BUTT JOINT

DOUBLE RIVETED BUTT JOINT

FIGURE 13.40   **Common Riveted Joints.**

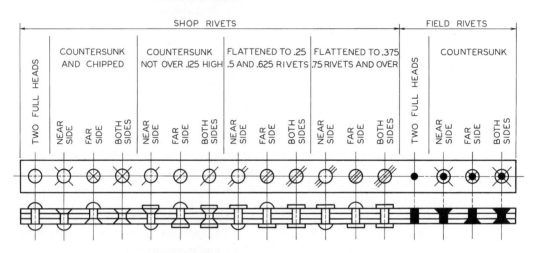

FIGURE 13.41   **Conventional Rivet Symbols.**

FIGURE 13.42   **American National Standard Small Solid Rivet Proportions.**

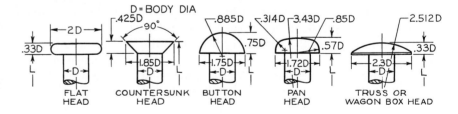

Blind rivets, commonly known as Pop Rivets™ (Fig. 13.43), are often used for fastening together thin sheet-metal assemblies. Blind rivets are hollow and are installed with manual or power-operated rivet guns which grip a center pin or mandrel, pulling the head into the body and expanding the rivet against the sheet metal. They are available in aluminum, steel, stainless steel, and plastic. As with any fastener, the designer should be careful to choose an appropriate material to avoid corrosive action between dissimilar metals.

■ 13.37 SPRINGS

A **spring** is a mechanical device designed to store energy when deflected and to return the equivalent amount of energy when released [ANSI Y13.13M–1981 (R1992)]. Springs are commonly made of *spring steel*, which may be music wire, hard-drawn wire, or oil-tempered wire. Other materials used for compression springs include stainless steel, beryllium copper, and phosphor bronze. In addition, compression springs made of urethane plastic are used in applications where conventional springs would be

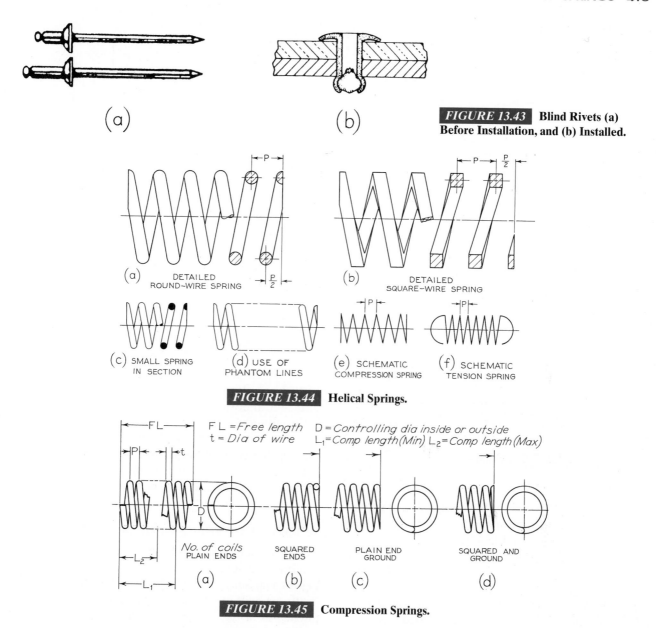

FIGURE 13.43 Blind Rivets (a) Before Installation, and (b) Installed.

FIGURE 13.44 Helical Springs.

FIGURE 13.45 Compression Springs.

affected by corrosion, vibration, or acoustic or magnetic forces. Springs are classified as *helical springs* (Fig. 13.44) or *flat springs* (Fig. 13.48). Helical springs may be cylindrical or conical but are usually the former.

There are three types of helical springs: *compression springs*, which offer resistance to a compressive force (Fig. 13.45); *extension springs*, which offer resistance to a pulling force (Fig. 13.46); and *torsion springs*, which offer resistance to a torque or twisting force (Fig. 13.47).

On working drawings, true projections of helical springs are never drawn because of the labor involved.

Instead, as in the drawing of screw threads, the detailed and schematic methods are used, where straight lines replace helical curves (Fig. 13.44).

The elevation view of the square-wire spring is similar to the square thread with the core of the shaft removed (Fig. 13.12). Standard section lining is used if the areas in section are large, as in Figs. 13.44a and 13.44b. If these areas are small, the sectioned areas may be made solid black (Fig. 13.44c). In cases where a complete picture of the spring is not necessary, phantom lines may be used to save time in drawing the coils (Fig. 13.44d). If the drawing of the spring is too small

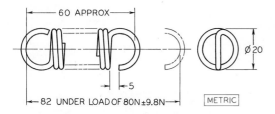

MATERIAL: 2.00 OIL TEMPERED SPRING STEEL WIRE
                     14.5 COILS RIGHT HAND
                     MACHINE LOOP AND HOOK IN LINE
                     SPRING MUST EXTEND TO 110 WITHOUT SET
FINISH: BLACK JAPAN

FIGURE 13.46   Extension Spring Drawing.

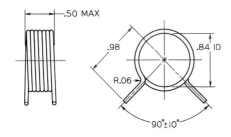

MATERIAL : .059 MUSIC WIRE
                    6.75 COILS RIGHT HAND NO INITIAL TENSION
TORQUE : 2.50 INCH LB AT 155° DEFLECTION SPRING MUST
                   DEFLECT 180° WITHOUT PERMANENT SET AND
                   MUST OPERATE FREELY ON .75 DIAMETER SHAFT
FINISH : CADMIUM OR ZINC PLATE

FIGURE 13.47   Torsion Spring Drawing.

to be represented by the outlines of the wire, it may be drawn by the schematic method, in which single lines are used (Figs. 13.44e and 13.44f).

Compression springs have *plain ends* (Fig. 13.45a) or *squared (closed) ends* (Fig. 13.45b). The ends may be *ground* (Fig. 13.45c) or both *squared and ground* (Fig. 13.45d). Required dimensions are indicated in the figure. When required, RH or LH is specified.

Many companies use standard printed spring drawings with a printed form to be filled in by the drafter, providing the necessary information, plus load at a specified deflected length, the load rate, finish, type of service, and other data.

An extension spring may have any one of many types of ends, and it is therefore necessary to draw the spring or at least the ends and a few adjacent coils (Fig. 13.46). Note the use of phantom lines to show the continuity of coils. Printed forms are used when a given form of spring is produced with differences in verbal specification only.

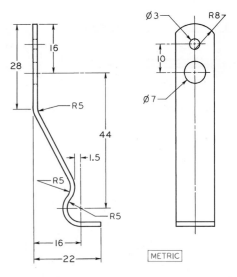

MATERIAL : 1.20 X 14.0 SPRING STEEL
HEAT TREAT : 44–48 C ROCKWELL
FINISH : BLACK OXIDE AND OIL

FIGURE 13.48   Flat Spring.

A typical torsion spring drawing is shown in Fig. 13.47. Here also printed forms are used when there is sufficient uniformity in product to permit a common representation.

A typical flat spring drawing is shown in Fig. 13.48. Other types of flat springs are *power springs* (or flat coil springs), *Belleville springs* (like spring washers), and *leaf springs* (commonly used in automobiles).

■ 13.38 DRAWING HELICAL SPRINGS

The construction for a schematic elevation view of a compression spring having six total coils is shown in Fig. 13.49a. Since the ends are closed, or squared, two of the six coils are "dead" coils, leaving only four full pitches to be set off along the top of the spring, as shown.

If there are $6\frac{1}{2}$ total coils (Fig. 13.49b), the $\frac{P}{2}$ spacings will be on opposite sides of the spring. The construction of an extension spring with six active coils and loop ends is shown in Fig. 13.49c.

Figure 13.50 shows the steps in drawing a sectional view and an elevation view of a compression spring by detailed representation. The given spring is shown pictorially in Fig. 13.50a. In Fig. 13.50b, a cutting plane has passed through the center line of the spring, and the front portion of the spring has been removed. In Fig. 13.50c, the cutting plane has been removed. Steps in

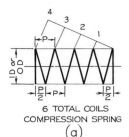

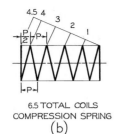

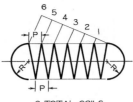

6 TOTAL COILS
COMPRESSION SPRING
(a)

6.5 TOTAL COILS
COMPRESSION SPRING
(b)

6 TOTAL COILS
EXTENSION SPRING
(c)

FIGURE 13.49 Schematic Spring Representation. *Courtesy of SDRC, Milford, OH.*

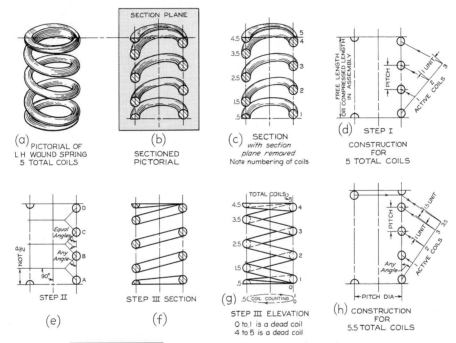

(a) PICTORIAL OF L H WOUND SPRING 5 TOTAL COILS

(b) SECTIONED PICTORIAL

(c) SECTION *with section plane removed* Note numbering of coils

(d) STEP I CONSTRUCTION FOR 5 TOTAL COILS

(e) STEP II

(f) STEP III SECTION

(g) STEP III ELEVATION 0 to 1 is a dead coil 4 to 5 is a dead coil

(h) CONSTRUCTION FOR 5.5 TOTAL COILS

FIGURE 13.50 Steps in Detailed Representation of Spring.

constructing the spring through several stages to obtain the sectional view are shown in Figs. 13.50d to 13.50f. The corresponding elevation view is shown in Fig. 13.50g.

If there is a fractional number of coils, such as $5\frac{1}{2}$ coils in Fig. 13.50h, note that the half-rounds of sectional wire are placed on opposite sides of the spring.

■ 13.39 COMPUTER GRAPHICS

Standard representations of threaded fasteners and springs, in both detailed and schematic forms, are available in CAD symbol libraries. Use of computer graphics frees the drafter from the need to draw time-consuming repetitive features by hand and also make it easy to modify drawings if required.

In 3D modeling, thread is not usually represented because it can be difficult to create and computer intensive to view and edit. Instead the nominal diameter of a threaded shaft or hole is usually created along with notation calling out the thread. Sometimes the depth of the thread is shown in the 3D drawing, to call attention to the thread and to help in determining fits and clearances.

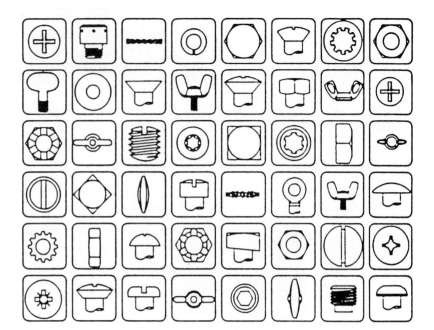

FIGURE 13.51 **Examples of Fastener Symbols Available for Use with AutoCAD in SPOCAD's Autofasteners Library.** (*Courtesy SPOCAD*).

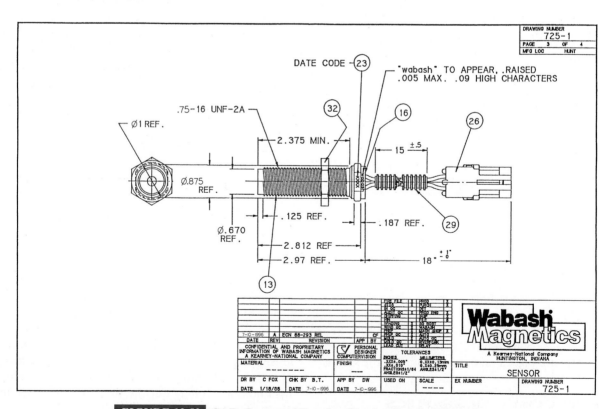

FIGURE 13.52 **CAD-Generated Drawing Showing Detailed Thread Representation.** *Courtesy of Computervision Corporation, a Subsidiary of Prime Computer, Inc.*

■ KEY WORDS

THREAD FORM	MAJOR DIAMETER
MINOR DIAMETER	INTERNAL THREAD
EXTERNAL THREAD	PITCH
LEAD	THREAD ANGLE
CREST	ROOT
CLASS OF FIT	UNIFIED THREAD FORM
METRIC THREAD FORM	SCHEMATIC THREAD REPRESENTATION
MULTIPLE LEAD	SIMPLIFIED THREAD REPRESENTATION
PHANTOM LINES	CAP SCREWS
MACHINE SCREWS	SET SCREWS
KEYS	PINS
HELIX	BOLTS
STUDS	SPRINGS
RIVETS	NUTS

■ CHAPTER SUMMARY

- There are many types of thread forms; however, metric and Unified are the most common.

- The method of showing threads on a drawing is called the thread representation. The three types of thread representation are detailed, schematic, and simplified.

- The major diameter, pitch, and form are the most important parts of a thread specification.

- Thread specifications are a special type of leader note. The thread specification tells the manufacturing technician what kind of thread needs to be created.

- The nut and bolt is still the most common type of fastener. Many new types of fasteners are being created to streamline the production process.

- Keys and pins are special fasteners that attach a pulley to a shaft.

- The screw head determines what kind of tool will be necessary to install the fastener.

■ REVIEW QUESTIONS

1. Draw a typical screw thread and label the parts of the thread.

2. Sketch a long spring and show how phantom lines are used to represent the middle part of the spring.

3. Draw several types of screw heads.

4. List five types of screws.

5. Why is the simplified thread representation the most commonly used drawing style?

6. List five fasteners that do *not* have any threads.

7. Write out a metric thread specification and a Unified thread specification and label each part of the specification.

8. Which type of thread form is used on a light bulb?

■ THREAD AND FASTENER PROBLEMS

Students are expected to make use of the information in this chapter and in various manufacturers' catalogs in connection with the working drawings at the end of the next chapter, where many different kinds of threads and fasteners are required. However, several problems are included here for specific assignment in this area (Figs. 13.53 to 13.56). All are to be drawn on tracing paper or detail paper, size B or A3 sheet (see inside front cover).

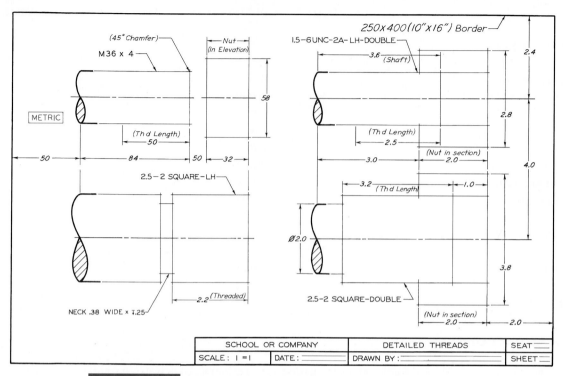

FIGURE 13.53 Draw specified detailed threads arranged as shown. Using Layout B–3 or A3–3. Omit all dimensions and notes given in inclined letters. Letter only the thread notes and the title strip.

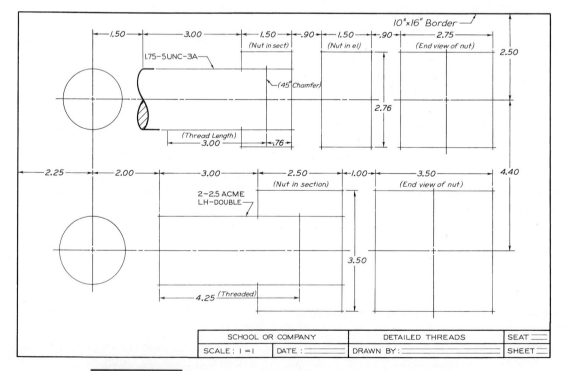

FIGURE 13.54 Draw specified detailed threads, arranged as shown. using Layout B–3 or A3–3. Omit all dimensions and notes given in inclined letters. Letter only the thread notes and the title strip.

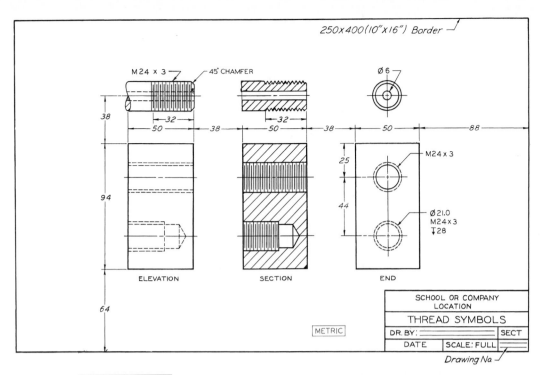

FIGURE 13.55 Draw specified thread symbols, arranged as shown. Draw simplified or schematic symbols, as assigned by instructor. Using Layout B–5 or A3–5. Omit all dimensions and notes given in inclined letters. Letter only the drill and thread notes, the titles of the views, and the title strip.

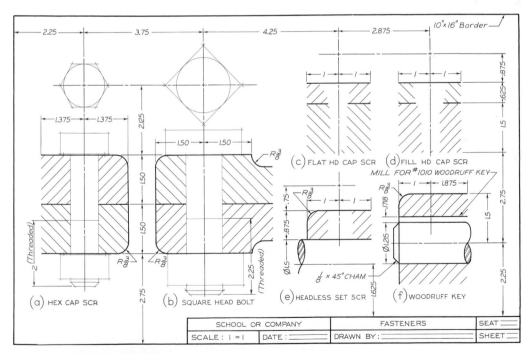

FIGURE 13.56 Draw fasteners, arranged as shown. Using Layout B–3 or A3–3. At (a) draw $\frac{7}{8}$–9 UNC-2A × 4 Hex Cap Screw. At (b) draw $1\frac{1}{8}$–7 UNC-2A × $4\frac{1}{4}$ Sq Hd Bolt. At (c) draw $\frac{3}{8}$–16 UNC-2A × $1\frac{1}{2}$ Flat Hd Cap Screw. At (d) draw $\frac{7}{16}$–14 UNC-2A × 1 Fill Hd Cap Screw. At (e) draw $\frac{1}{2}$ × 1 Headless Slotted Set Screw. At (f) draw front view of No. 1010 Woodruff Key. Draw simplified or schematic thread symbols as assigned. Letter titles under each figure as show.

419

C H A P T E R   1 4

DESIGN
AND WORKING
DRAWINGS

OBJECTIVES

After studying the material in this chapter, you should be able to:

1. Define the different types of design.
2. Define engineering design and examine how the proper objectives and motivation can turn any one into a good designer.
3. Discuss the different sources for design ideas for both the individual designer as well as a design team.
4. Describe the stages of the design process.
5. Discuss solid modeling, parametric modeling, prototyping and rapid prototyping.
6. Use a case study to illustrate of each stage of the design process.
7. Identify the elements of a detail drawing and create a simple detail drawing.
8. List the common elements of a title block and record strip.
9. Create a typical drawing sequence of numbers.
10. Describe the process for revising drawings.
11. List the parts of an assembly drawing.
12. Describe the special requirements of a patent drawing.

OVERVIEW

The many products, systems, and services that enrich our standard of living are largely the result of the design activities of engineers. It is principally this design activity that distinguishes engineering from science and research; the engineer is a designer, a creator, or a "builder."

The design process is an exciting and challenging effort, and the engineer-designer relies heavily on graphics as a means to create, record, analyze, and communicate to other design concepts or ideas. The ability to communicate verbally, symbolically, and graphically is essential.

The design team progresses through five stages in the design process. To be a successful member of the design team, every member must understand the process and know how to fulfill his or her role. Various types of drawings are required at each stage of the design process. CAD can help the drafter create drawings, but it takes a skilled drafter to know which drawings are required at each design stage. Much of the design process is refinement of existing products. Refinement creates improved products, lowers cost, and increases profit for the company that manufacturers the product. Models are an important part of the design process. Some models are created to scale in a model shop. Other models are generated by the computer and printed or displayed in virtual reality. All models allow interaction with the design to further improve the design. Revising a drawing is an important part of the design process. Revisions must be tracked, identified, logged, and saved for future reference. Both paper and electronic storage is an important part of the drafter's responsibility on the design team. Assembly and working drawings show how multiple parts fit together. They describe the end result of creating individual pieces that must fit together to work.

FIGURE 14.1   Aesthetic and functional design combine to give this sports car not only a look of elegance, but of speed. Although pleasing to the eye, this car is the superb product of aerodynamic and mechanical engineering. *Reprinted with permission from Bavarian Autosport.*

thetic design while the enhancement of product development is considered *functional design*. Aesthetics and function can work hand in hand to create a product that is not only appealing to the senses, but fulfills specific product demands. A well-designed automobile is a good example of how aesthetics and function can work together. (See Fig. 14.1.)

There are two general types of design: *empirical design*, sometimes referred to as conceptual design, and *scientific design*. In scientific design, use is made of the principles of physics, mathematics, chemistry, mechanics, and other sciences in the new or revised design of devices, structures, or systems intended to function under specific conditions. In empirical design, much use is made of the information in handbooks, which in turn has been learned by experience. Nearly all technical design is a combination of scientific and empirical design. Therefore, a competent designer has both adequate engineering and scientific knowledge and access to the many handbooks related to the field.

■ 14.1 "DESIGN" DEFINED

Design is a process, a series of linked steps with stated objectives. It is a way of conceiving and creating new ideas and then communicating those ideas to others in a way that can be easily understood. This is accomplished most efficiently through the use of graphics. Design can be used to reflect personal expressions or to enhance product development. This reflection of personal expression is most often referred to as *aes-*

■ 14.2 "ENGINEERING DESIGN" DEFINED

Engineering design is also a process. This process is used to solve society's needs, desires, and problems through the application of scientific principles, experience, and creativity. Some people are creative and are naturally gifted at design, but everyone can become a designer if they learn to use the proper tools and techniques involved with the design process. Becoming a designer is much like learning to play a musical instrument; some people are better at it then others, but everyone can learn to play if they learn the steps involved.

Two key elements to any successful design plan is gaining the proper *motivation* and stating the *objectives* to the plan. Design is the single most important activity practiced by engineers. Design separates engineering from the rest of the sciences in that it is the application of scientific principles to create solutions. Your motivation in any design plan should be to create the most efficient solution to any given problem. The objective statement will provide a framework within which any engineering design problem can be addressed in a methodical manner. Proper planning and scheduling are also key to successful designs plans. Setting a deadline for the completion of each design phase is imperative. We will the discuss the steps in the design process at length in §14.4

■ 14.3 DESIGN CONCEPTS— SOURCES FOR NEW IDEAS

INDIVIDUAL CREATIVITY TECHNIQUES   New ideas or design concepts usually begin in the mind of a single individual—the designer. But how does one go about developing new ideas? There is an old saying in the engineering industry:

> "Good design is to borrow. Genius design is to steal."

Unlike your other courses of study where plagiarism is considered bad and should be subject to punishment, copying good ideas is highly advisable in design. Students are urged to copy not only from existing products and classmates, but they should study catalogs, manufacturers' patents, and nature.

Look through industry catalogs and handbooks for existing designs. Think of ways in which these existing designs can be used or modified to work in your design plan. Manipulate them through freehand sketches (see Chapter 3) or through the use of computer software (CAD).

STUDY PATENT DRAWINGS (See Fig. 14.2)   A patent is issued by the U.S. government granting the holder the "right to exclude others from making, using or selling" a specific product. The patent process was first developed as a way of disclosing technical advances by granting a period of protection for a limited

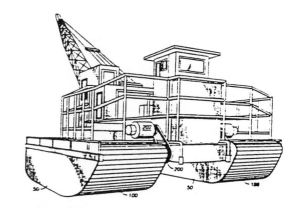

FIGURE 14.2   **Pictorial Patent Drawing.**

amount of time (a patent is issued for 17 years). The U.S. Patent and Trademark Office has extremely strict regulations as to the presentation of materials (i.e., no freehand sketches are accepted and all patent drawings must be made with drafting instruments or by a process which will make them easily reproducible). You cannot use or copy any existing patents, but they can be a great source of ideas. The U.S. Patent Office, therefore, can be a valuable resource in quest for design ideas. A discussion of how to apply for a patent for a new design appears in §14.26

EXAMINE MANUFACTURED PRODUCTS. Dismantle them, evaluate them, and study how their parts are designed to work together. This is referred to a *reverse engineering*. Sophisticated reverse engineering involves evaluating a product using a machine called a *coordinate measuring machine* (CMM). (See Fig. 14.3.) The machine is a electromechanical devices containing a probe on one end. The probe measures the object and then places all of the pertinent information into a CAD database where it can be manipulated. Although you may not have access to such a complicated machine, examine manufactured products that are available to you. Think of ways to improve or change these existing designs. Where could they be improved? What would you do differently?

Study a product that is no longer performing to the manufacturer's existing or upgraded standards, referred to as *functional decomposition*. How could you expand/change the design to guarantee better performance? What could you do to expand the life of the product? How could you make it more efficient, more cost effective, etc.?

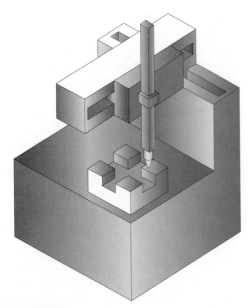

FIGURE 14.3 Coordinate Measuring Machine (CMM). Used to Accurately Measure a Part for Reverse Engineering or Quality Control.

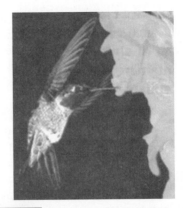

FIGURE 14.5 The wings of a hummingbird are aerodynamic wonders. Study of such designs in nature can be enlightening to any designer. *Courtesy of Photo Researchers, Inc.*

the resources to invest in numerous new software programs, be sure to make use of those readily available at your school. Even programs such as Adobe Illustrator® or Photoshop® can help in your search for new ways of viewing and conceiving objects.

Excellent resources for engineering and design are available on the World Wide Web. Search for terms like design, engineering, technology, or for more specific terms, depending on your interests. The following Web sites are useful for engineering design:

- http://www.yahoo.com/headlines/
  Yahoo's site for the latest technology news and a one-week archive

- http://www.techweb.com/
  TechWeb site from CMP media

- http://www.uspto.gov/
  U.S. Patent Office on-line search site

Using the techniques discussed above and listed in Fig. 14.6, the designer should have a few ideas as to where to begin. In order to capture, preserve, and develop these ideas, the designer makes liberal use of freehand sketches of views and pictorials (see Chapter 3). These sketches are revised or redrawn as the concept is developed. All sketches should be preserved for reference and dated as a record of the development of the design.

STUDYING THE NATURAL WORLD. Noting how other creatures interact with their surroundings can provide a wealth of information and creativity. Such things as bee hives or spiders' webs are masterpieces of structural design. A hummingbird's wings are aerodynamic wonders. (See Figs. 14.4 and 14.5.) Study them and expand on their designs.

Use all computer hardware and software available to you. Animation and CAD programs have become numerous in recent years. Although you may not have

FIGURE 14.4 Studying designs in nature, such as this spider's web, can provide the designer with new information and ideas. *Courtesy of Peter Arnold, Inc.*

GROUP CREATIVITY TECHNIQUES At some point in the development of the idea, you will probably find it to your advantage to pool your ideas with those of others and begin working in a team effort; such a team may include others familiar with problems of materials, production, marketing, and so on. In industry, the project becomes a team effort long before the product is produced and marketed. Obviously, the design process is not a haphazard operation of an inventor working in a garage or basement, although it might well begin in that manner. Industry could not long survive if its products were determined in a haphazard manner. Hence, nearly all successful companies support a well-organized design effort, and the vitality of the company depends, to a large extent, on the *planned* output of its designers and design teams. Groups play an important role in the creative design process.

The two most commonly used group creativity techniques are brainstorming and storyboarding. *Brainstorming* occurs when a group of individuals come to together to discuss new ideas. A brainstorming session can stimulate, enlighten, and motivate designers to look at their product designs in a different light based upon input from other team members. The most important rule to follow during a brainstorming session is that no criticism of others' work should be tolerated. Criticism only serves to stymie the creative process. A second rule for any good brainstorming session is to come prepared with as many new ideas as possible. This is not the time to be conservative; be open to presenting new ideas.

Storyboarding is a technique often used by designers to graphically illustrate the progression of their designs, as well as the manufacturing process required to create a final product. Storyboards are rough sketches, usually created freehand by the designer. Storyboards are a valuable tool in any brainstorming session as they can be used as a base to be built on creatively by group members during the session. These freehand sketches should be rough so that they can be updated, revised, or modified based on input from team members.

Reintegration of the ideas generated during the brainstorming session into an individual's design is of utmost importance. Once the design team has settled on a specific design, it is imperative that the individual designer incorporate this input into his/her design. (See Fig. 14.6.)

Since it is important for you to be able to work effectively with others in a group or team, you must be able to express yourself clearly and concisely. Do not underestimate the importance of your communication skills, your ability to express your ideas verbally (written and spoken), symbolically (equations, formulas, etc.), and *graphically*.

INDIVIDUAL CREATIVITY TECHNIQUES INCLUDE:

Studying Industry Catalogs/Handbooks

Examining Manufactured Items

Studying Patent Drawings

Conducting Reverse Engineering

Examining Functional Decomposition

Studying the Natural World

Using Software Products

Utilizing Design, Engineering, and Technology Web Sites

GROUP CREATIVITY TECHNIQUES INCLUDE:

Brainstorming

Storyboarding

Reintegration

FIGURE 14.6 **Individual and group creativity techniques.**

These graphical skills include the ability to present information and ideas clearly and effectively in the form of sketches, drawings, graphs, and so on. This textbook is dedicated to helping you develop your communication skills in graphics.

■ 14.4 THE DESIGN PROCESS

Design is the ability to combine ideas, scientific principles, resources, and often existing products into a solution of a problem. This ability to solve problems in design is the result of an organized and orderly approach to the problem known as the *design process*.

The design process leading to manufacturing, assembly, marketing, service, and the many activities necessary for a successful product is composed of several easily recognized phases. Although many industrial groups may identify them in their own particular way, a convenient procedure for the design of a new or improved product is in five stages as follows:

1. Identification of problem, need, or "customer."

2. Concepts and ideas.

3. Compromise solutions.

4. Models and/or prototypes.

5. Production and/or working drawings.

Ideally, the design moves through the stages as shown in Fig. 14.7, but if a particular stage proves

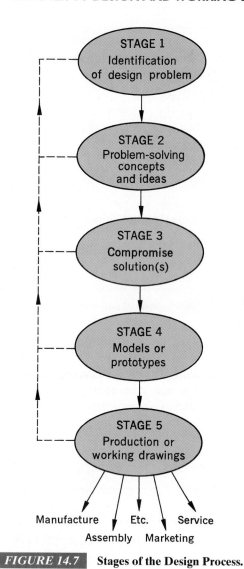

FIGURE 14.7   **Stages of the Design Process.**

sign and the related design process ultimately should be driven by its end users. Determine if the design should be geared toward a single, very specific user, a specific purchaser or purchasers, a manufacturer or group of manufacturers, or to the general public. A part to be used in the space shuttle, for example, would not need to be designed or manufactured for operation by the general public. It has a limited market and customer base. But a design for a new home gym, which requires the user to complete the final assembly, should take into account a wide range of users and mechanical abilities. It is important that the designer identify the end user before beginning the design process.

It is also important to determine if the product to be designed must meet with any government standards/regulations or adhere to any professional organizations' standards or codes before starting the design process.

Any design process will involve compromises such that all of the designer's original requirements cannot be met. Government standards may limit the use of certain materials, for example, and the design may have to take a different form than that originally conceived by the designer. Materials or manufacturing processes may become too costly, resources may become unavailable, etc. It is important for the designer to approach the design process knowing that compromises may have to be made. Prioritize the design requirements. It is a good idea to break the *design requirements* into four categories: essential, important, desirable, or beneficial. If the designer has the requirements broken into these categories before implementing the design, decision making as to what to change and when will become easier when the need arises.

Engineering design problems may range from the simple and inexpensive container opener such as the pull tab (Fig. 14.8) commonly used on beverage cans to the more complex problems associated with the needs of air and ground travel, space exploration, environmental control, and so forth. Although the product may be very simple, such as the pull tab on a beverage can, the production tools and dies require considerable engineering and design effort. The airport automated transit system design, Fig. 14.9, meets the need of moving people efficiently between the terminal areas. The system is capable of moving 3300 people every 10 minutes.

The Lunar Roving Vehicle, Fig. 14.10, is a solution to a need in the space program to explore larger areas

unsatisfactory, it may be necessary to return to a previous stage and repeat the procedure as indicated by the dashed-line paths. This repetitive procedure is often referred to as *looping*.

■ 14.5 STAGE 1—IDENTIFICATION OF THE PROBLEM AND THE CUSTOMER

The design activity begins with the recognition of a *problem* and/or the determination of a *need* or want for a product, service, or system and the economic feasibility of fulfilling this need.

The designer not only must identify the problem or need but also the *customer*. Who will be affected or influenced by the design? The creation of any new de-

FIGURE 14.8   **Pull-Tab Can Opener.** *John Schultz—PAR/NYC.*

FIGURE 14.9   **Airport Transit System.** *Courtesy of Westinghouse Electric Corp.*

FIGURE 14.10   **Lunar Roving Vehicle.** *Courtesy of NASA.*

of the lunar surface. This vehicle is the end result of a great deal of design work associated with the support systems and the related hardware.

At the problem identification stage, either the designer recognizes that there does exist a need requiring a design solution or, perhaps more often, a directive is received to that effect from management. No attempt is made at this time to set goals or criteria for the solution.

Information concerning the identified problem becomes the basis for a problem proposal, which may be a paragraph or a multipage report presented for formal consideration. A proposal is a plan for action that will be followed to solve the problem. The proposal, if approved, becomes an agreement to follow the plan. In the classroom, the agreement is made between you and your instructor on the identification of the problem and your proposed plan of action.

Following approval of the proposal, further aspects of the problem are explored. Available information related to the problem is collected, and parameters or guidelines for time, cost, function, and so on are defined within which you will work. For example: What is the design expected to do? What is the estimated cost limit? What is the market potential? What can it be sold for? When will the prototype be ready for testing? When must production drawings be ready? When will production begin? When will the product be available on the market?

The parameters of a design problem, including the time schedule, are established at this stage. Nearly all designs represent a compromise, and the amount of time budgeted to a project is no exception.

■ 14.6 STAGE 2—CONCEPTS AND IDEAS

At this stage, many ideas are collected—reasonable and otherwise—for possible solutions to the problem. The ideas are broad and unrestricted to permit the possibility of new and unique solutions. The ideas may be from individuals, or they may come from group or team brainstorming sessions where one suggestion often generates many more ideas from the group. As the ideas are elicited, they are recorded for future consideration and refinement. No attempt is made to evaluate ideas at this stage. All notes and sketches are signed, dated, and retained for possible patent proof.

The larger the collection of ideas, the greater are the chances of finding one or more ideas suitable for further refinement. All sources of ideas, such as technical literature, reports, design and trade journals, patents, and existing products are explored. Ideas can come from such sources as the Greenfield Village Museum in Dearborn, Michigan, the Museum of Science and Industry in Chicago, trade exhibitions, the World Wide Web, large hardware and supply stores, mail order catalogs. Even the user of an existing product is an excellent source, because that person often has suggestions for improvement. The potential user may be helpful with specific reactions to the proposed solution.

No attempt is made to evaluate ideas at this stage. All notes and sketches are signed, dated, and retained for possible patent proof.

■ 14.7 STAGE 3—COMPROMISE SOLUTIONS

Various features of the many conceptual ideas generated in the preceding stages are selected after careful consideration and combined into one or more promising compromise solutions. At this point the best solution is evaluated in detail, and attempts are made to simplify it so that it performs efficiently and is easy to manufacture, repair, and even dispose of when its lifetime is over.

Refined design sketches are often followed by a study of suitable materials and of motion problems that may be involved. What source of power is to be used—manual, electric motor, or what? What type of motion is needed? Is it necessary to translate rotary motion into linear motion or vice versa? Many of these problems are solved graphically using schematic drawings in which various parts are shown in skeleton form. For example, pulleys and gears are represented by circles, an arm by a single line, and a path of motion by centerlines. Certain basic calculations, such as those related to velocity and acceleration, may also be made at this time.

Preliminary studies are followed by a design layout—usually an accurate CAD drawing, showing actual sizes so that proportions and fits can be clearly visualized—or by a clearly dimensioned layout sketch. An example is shown in Fig. 14.11. At this time all parts are carefully designed for strength and function. Costs are constantly kept in mind, because no matter how well the device performs, it must sell at a profit; otherwise the time and development costs will have been a loss.

During the layout process, experience provides a sense of proportion, size, and fit that permits noncritical features to be designed by eye or with the aid of empirical data. Stress analysis and detailed computation may be necessary in connection with high speeds, heavy loads, or special requirements or conditions.

Figure 14.12 shows the layout of basic proportions of parts and how they fit together in an assembly drawing. Special attention is given to clearances of moving parts, ease of assembly, and serviceability. Standard parts are used wherever possible, because they are less expensive than custom parts. Most companies maintain some form of an engineering standards manual, which contains much of the empirical data and detailed information that is regarded as "company standard." Materials and costs are carefully considered. Although functional considerations must come first, manufacturing problems must be kept constantly in mind.

■ 14.8 STAGE 4—MODELS AND PROTOTYPES

A model to scale is often constructed to study, analyze, and refine a design. To instruct the model-shop craftsperson in the construction of the prototype or model, dimensioned sketches or three dimensional computer models are required. A full-size working model made to final specifications, except possibly for materials, is known as a prototype. The prototype is tested and modified where necessary, and the results are noted in the revision of the sketches and working drawings. Figure 14.13 shows a prototype of the magnetic levitation train.

If the prototype is unsatisfactory, it may be necessary to return to a previous stage in the design process and repeat the procedures. It must be remembered that time and expenses always limit the duration of this looping. Eventually a decision must be reached for the production model.

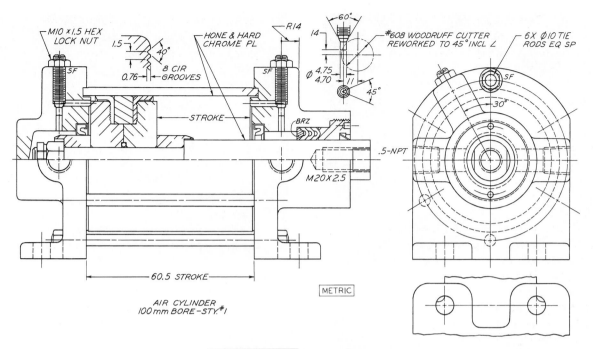

FIGURE 14.11   Design Layout.

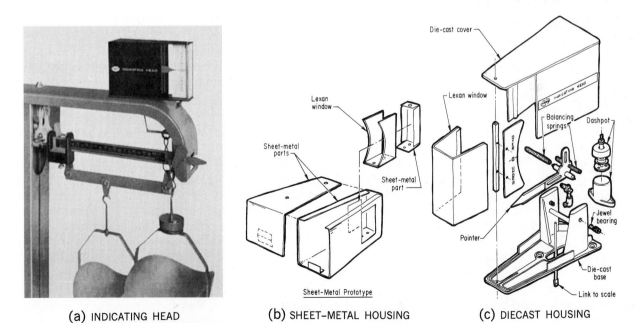

(a) INDICATING HEAD   (b) SHEET-METAL HOUSING   (c) DIECAST HOUSING

FIGURE 14.12   Improved Design of Indicating Head. *Courtesy of Ohaus Scale Corp. and Machine Design.*

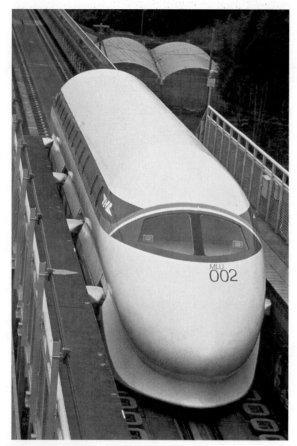

FIGURE 14.13 **A Prototype of the Magnetic Levitation Train Car During a Test Run.**

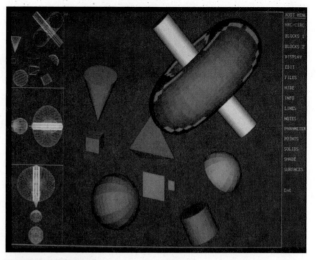

FIGURE 14.14 **Examples of Solids Created with CAD.** *Courtesy of American Small Business Computers.*

SOLID OR 3D MODELING CAD systems that offer solid modeling usually have options or commands for creating complex solids using primitive shapes, including boxes, prisms, cylinders, spheres, cones, tori, and sometimes wedges and pyramids. Figure 14.14 shows an example of some solids generated using CAD. Many CAD systems also have a primitive command which creates any regular solid. If there is not a specific command to create the solid you want, you usually have the option of creating new solid objects through processes called extrusion and revolution (both generally referred to as sweeping solids).

Extrusion is named for the manufacturing process which forms material by forcing it through a shaped opening. You can also think of extrusion as taking the cross-sectional shape of the object and sweeping it along a path to enclose a solid volume. Features which have a continuous cross section along a straight axis can be created using extrusion. Some CAD software

can only extrude shapes along a straight-line path, while others can do straight or curved paths. Most have the option to taper the extruded feature as it is created. An example of an extruded solid model is shown in Fig. 14.15.

Revolution is the process of forming a solid by revolving the cross-sectional shape of the object along a circular path to enclose a solid volume. Many objects that cannot be created by extrusion can be created by revolution. A solid object created by revolving a shape by 270° is shown in Fig. 14.16.

Complex solid models can be formed by joining primitives and solids formed by extrusion and revolution using *Boolean operators*. Boolean operators are named for 18th century mathematician and logician, Charles Boole. Most CAD programs that allow solid modeling support three Boolean operators: *union* (sometimes called addition), *difference* (sometimes called subtraction), and *intersection*. Venn diagrams are often used to show how sets are joined together using Boolean operators. Figure 14.17 shows Venn diagrams for union, difference, and intersection. Boolean operators can also be used to join solids you create by extruding or revolving other solids or solid primitives.

The union of solid A and solid B forms a single new solid that is their combined volume without any duplication where they have overlapped. Solid A difference solid B is similar to subtracting B from A. The order of the operation does make a difference in the result (unlike union). For A difference B, any volume from solid B that overlaps solid A is eliminated and

Venn diagrams depicting Boolean operators

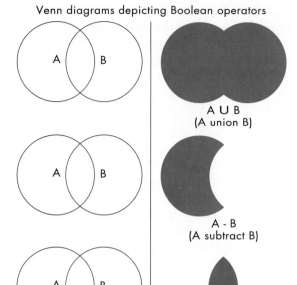

A ∪ B
(A union B)

A - B
(A subtract B)

A ∩ B
(A intersect B)

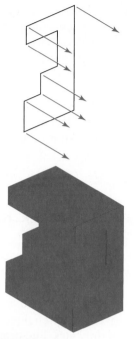

FIGURE 14.15 Extruded Solid.

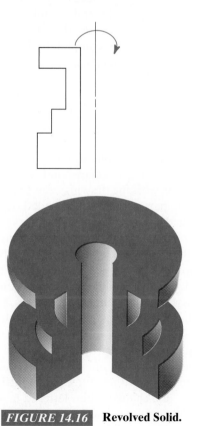

FIGURE 14.16 Revolved Solid.

Results of Boolean operators on solids

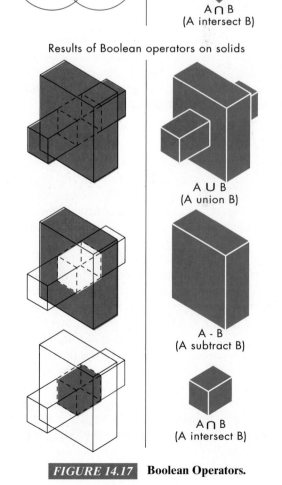

A ∪ B
(A union B)

A - B
(A subtract B)

A ∩ B
(A intersect B)

FIGURE 14.17 Boolean Operators.

the result forms a new solid. A intersect B results in a new solid where only the volume common both to solid A and solid B is retained. You can use these primitives, extrusion, revolution, and Boolean operators to create solid models for a variety of objects. Objects that cannot be created in this way are warped surfaces, such as those on the exteriors of automobiles and airplanes.

PARAMETRIC SOLID MODELING CAD systems, especially those allowing parametric solid modeling (where design parameters control the model geometry), provide many benefits for shortening the design cycle time. (In parametric design, the curve paths are controlled by mathematical functions rather than a set of coordinates.) In this way, parametric solid modeling allows concurrent design, where members of the design team along with members from the company's manufacturing and marketing divisions can work together at the same time to provide a total design solution. In parametric models, constraints and parametric dimensions control the model geometry. As the design changes, so can the constraints and dimensions; the model and drawings update automatically. Three-dimensional models can also be exported to rapid prototyping equipment and to direct manufacturing to allow quick progress from design to product.

RAPID PROTOTYPING While refining the design ideas, engineers often work concurrently with manufacturing to determine the best ways to make an assemble the necessary parts. After several cycles of refining, analyzing, and synthesizing the best ideas, the final design is ready to go into production. *Rapid prototyping* systems allow parts to quickly be generated directly from 3D models for mockup and testing. Rapid prototyping can also be used in situations were prototypes are still deemed necessary for various reasons (i.e., customer request). Many companies are designing their products with 3D design packages an then feeding the CAD data into separate software programs that not only evaluate the design but also generate a separate set of data, which is then usable in a variety of ways. The new data can be forwarded to a CNC mill and a sample model can then be cut from the data. Data can also be forwarded to rapid prototyping equipment using such technologies as *stereolithography* (SLA), *selective laser sintering* (SLS), *ballistic particle manufacturing* (BPM), and *laminated object manufacturing* (LOM).

SLA builds part from layers of laser-cured photopolymer. The SLS process builds parts layer by layer with a laser from powdered materials such as nylon, polycarbonate, or a composite glass-nylon material. Some rapid prototyping machines build objects by spraying molten particles of a thermoplastic. The LOM process builds layer by layer from rolls of sheet goods similar to paper. (Refer to the Graphics Spotlight on pages 436 to 437 for a good example of rapid prototyping in action.)

■ 14.9 STAGE 5—PRODUCTION OR WORKING DRAWINGS

To produce or manufacture a product, a final set of production or working drawings is made, checked, and approved.

In industry the approved production design layouts are turned over to the engineering department for the production drawings. The drafter, or detailers, "pick off" the details from the layouts with the aid of the scale or dividers. The necessary views are drawn for each part to be made, and complete dimensions and notes, Chapter 11, are added so that the drawings will describe these parts completely. These working drawings of the individual parts are also known as *detail drawings*, §14.11.

Unaltered standard parts do not require a detail drawing but are shown conventionally on the assembly drawing and listed with specifications in the parts list, §14.15.

A detail drawing of one of the parts from the design layout of Fig. 14.11 is shown in Fig. 14.19. For details concerning working drawings, see §§14.11–14.19.

After the parts have been detailed, an *assembly drawing* is made, showing how all the parts go together in the complete product. The assembly may be made by tracing the various details in place directly from the detail drawings, or the assembly may be traced from the original design layout, but if either is done, the value of the assembly for checking purposes, §14.25, will be largely lost. The various types of assemblies are discussed in §§14.20–14.25.

Finally, in order to protect the manufacturer, a *patent drawing*, which is often a form of assembly, is prepared and filed with the u.s. patent office. patent drawings are line shaded, often lettered in script, and otherwise follow rules of the patent office, §14.25.

FIGURE 14.18 Solid modeling allows you to quickly get from design to product. *Courtesy of Solid Concepts, Inc.*

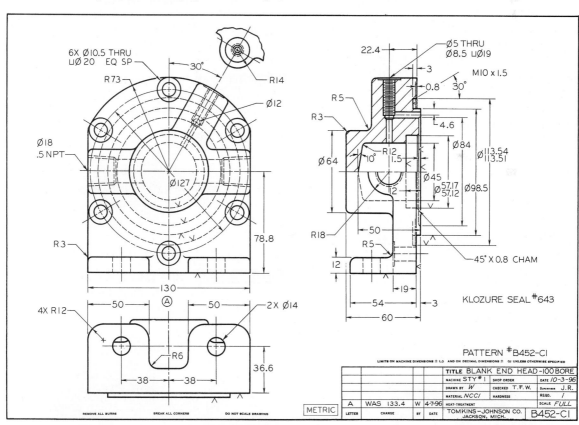

FIGURE 14.19 A Detail Drawing.

433

■ 14.10 DESIGN OF A NEW PRODUCT

An example of the design and development of a new product is that of IBM's ThinkPad 701C subnotebook computer, shown in Fig. 14.20.

STAGE 1   Problem Identification. When a company wants to determine the feasibility of a new product, it solicits opinions and ideas from many sources, including engineers, designers, drafters, managers, and potential consumers. Price ranges and estimated sales are also carefully explored.

In the case of the ThinkPad, IBM wanted to produce a subnotebook-size computer with a full-size keyboard—that is, a keyboard with the same size keys and the same spacing between keys as in a desktop computer. They also wanted the computer to have the largest available display, but the end product had to be thin, lightweight, and inexpensive enough to be competitive.

STAGE 2   Concepts and Ideas. In order to fit a full-size keyboard into a subnotebook case, a mechanical engineer working on the project came up with the idea of splitting the keyboard into two pieces that would interlock when the computer case was open. The two pieces could hang over the sides of the open case, thereby giving a little extra space for the relatively large keyboard. To close the computer case, the two pieces of the keyboard would need to separate and slide inward into new positions.

Once the idea of the split keyboard was accepted, the team moved from rough sketches and mockups to solid modeling.

STAGE 3   Compromise Solution. Using IBM's CATIA CAD/CAM software and an IBM RISC System/6000 workstation, the development team created a variety of possible keyboard designs for the ThinkPad. This system enabled them to produce virtual prototypes without building and rebuilding a series of actual 3D models. For example, they had to design a system to move the two parts of the keyboard when the computer case was opened and closed. However, the keyboard halves could not simply be moved by the case itself because if the system jammed, forcing the case open would break the keyboard. One feature of the CATIA software is a system that can identify areas of possible physical interference between parts of a model. Thus, the team could see on a computer screen where solid parts of the new computer might hit each other.

When the ThinkPad case is opened, the Track-Write keyboard is driven into position by a spring-loaded mechanism that moves the two halves of the keyboard asymmetrically. Closing the computer case moves the two halves of the keyboard back into their storage position by way of an axial cam and resets the spring (see Figs. 14.21 and 14.22).

FIGURE 14.21   IBM ThinkPad. *Courtesy of International Business Machines Corporation. Unauthorized use not permitted.*

FIGURE 14.20   IBM ThinkPad 701 C Subnotebook. *Courtesy of International Business Machines Corporation. Unauthorized use not permitted.*

FIGURE 14.22   IBM ThinkPad. *Courtesy of International Business Machines Corporation. Unauthorized use not permitted.*

STAGE 4 Prototypes Although in the past prototypes referred to actual working models, in the production of the ThinkPad computer, many of the prototypes existed only as 3D computer images. For example, early prototype designs for the keyboard—in the form of CATIA files—were given to the IBM unit that was developing other elements of the computer so that all the parts could be integrated into a functional whole.

STAGE 5 Production In the final stages of the design process, completed CATIA models were sent to outside vendors who used them to program numerically control (NC) tools that created the molds for parts of the computer. The same models created by the design process functioned throughout the various stages of the project. (See the Graphics Spotlight for a real-life example of a team using the design process.)

■ 14.11 WORKING DRAWINGS

Working drawings, which normally include assembly and details, are the specifications for the manufacture of a design. Therefore, they must be neatly made and carefully checked. The working drawings of the individual parts are also referred to as *detail drawings*, §§14.12–14.19.

■ 14.12 NUMBER OF DETAILS PER SHEET

Two general methods are followed in industry regarding the grouping of details on sheets. If the machine or structure is small or composed of few parts, all the details may be shown on one large sheet, Fig. 14.23.

When larger or more complicated mechanisms are represented, the details may be drawn on several large sheets, several details to the sheet, and the assembly is drawn on a separate sheet. Most companies have now adopted the practice of drawing only one detail per sheet, however simple or small. The basic 8.50 × 11.00 or 210 mm × 297 mm sheet is most commonly used for details, multiples of these sizes being used for larger details or the assembly. For standard sheet sizes.

When several details are drawn on one sheet, careful consideration must be given to spacing. The drafter should determine the necessary views for each detail and *block in all views lightly before beginning to draw any view*, as shown in Fig. 14.23. Ample space

FIGURE 14.23 Blocking in the Views.

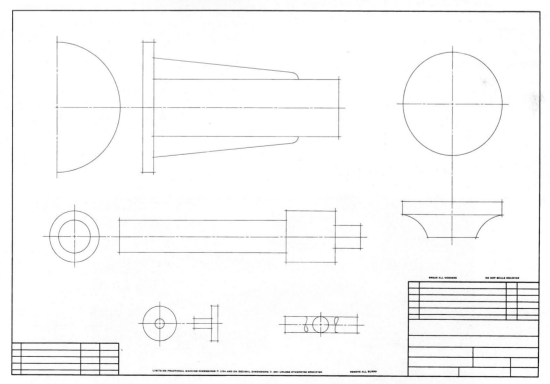

A Day at Ideo U

The rules for Cannonball Run, the final competition at Ideo University, are simple: To win, a team needs to:

1. Build a device that will propel a steel "cannonball" farther than contraptions built by the three opposing teams.

2. Make the ball signal the end of its journey by setting off a buzzer in the designated "cannonball catcher," a round target with the approximate circumference of a coffee cup.

3. Use the provided material, which consists of a bundle of wooden dowels, a chunk of cushiony foam, a roll of electrical tape, a stack of index cards, some paint sticks, a fistful of rubber bands, an extrememly long piece of black string, a mousetrap, and the tubes all this came in.

4. Complete a cannonball launcher within 90 minutes.

Not the simplest of tasks, but then again, no one ever said that learning creativity was easy. In fact, the competiton at Ideo U. consisted of one journalist and 15 engineers from the networking giant Cisco Systems. At the beginning of the competition, they were all skeptical. Creativity couldn't possibliy be taught, let alone at a one-day innovation and design workshop.

Ideo U. Design Workshops

Ideo, the Silicon Valley firm that's famous for the designing products like Apple's first mouse, a no-mess toothpaste tube for Procter & Gamble's Crest, and, most recently, the Palm V, the silvery, wafer-thin, light-as-a feather addition to 3Com's blockbuster line of hand held organizers.

Founded in 1978 by Standford design professor, David Kelley, Ideo has designed more than 3,000 products for a roster of clients, which include a sizeable portion of the FORTUNE 500. Ideo began offering design workshops for its customers and potential customers after many began clamouring to know exactly how they came up with there designs. Since then, companies ranging from NEC to Kodak to Steelcase have been sending employees to Ideo U.

Can Creativity be Taught?

"They say that genius is 99% perspiration and 1% inspiration," says Dennis Boyle, the Ideo principal leading the Cannonball Run Workshop. "Most companies have that 99%. It's the 1% that's really hard, and that's why our clients are asking us to work with their people and not just their product."

To home in on that elusive 1%, Ideo's employees explain the techniques they use to design products and then force participants to put them into practice. That's where Cannonball Run comes in.

Brainstorming

The first step: brainstorming. Earlier that afternoon, Brendan Boyle, head of Ideo's toy-invention studio had explained brainstorming the Ideo way. Ideo takes its rules for brainstorming so seriously that they are printed on a large banner that runs across the top of the classroom's whiteboards They are:

- Defer judgment (otherwise you'll interrupt the flow of ideas);
- Build on the ideas of others (it's far more productive than hogging the glory of your own insights);
- Stay focused on the topic (no tangents);

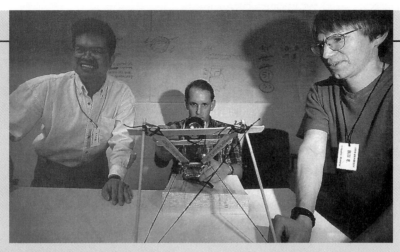

- One person at a time (so you don't drown out that quiet, brilliant mumbler in the corner);
- Go for quantity (when Ideo staffers brainstorm, they shoot for 150 ideas in 30 to 45 minutes);
- Encourage wild ideas (to paraphrase Einstein, "If the first idea doesn't sound absurd, then there's no hope for it); and
- Be visual (sketch ideas to help people understand them).

Team 2 stuck to these rules and within a few minutes came up with 15-20 ideas for the launcher.

Rapid Prototyping

To narrow their ideas down, Team 2 used rapid prototyping, another Ideo technique. The idea behind rapid prototyping is that it's easier to discuss a model of something, no matter how primitive, than to talk about a bunch of ideas. "If a picture is worth a thousand words," says Ideo's Steve Vassallo, "a prototype is worth ten thousand."

Rapid prototyping consists of three Rs: Rough, Rapid, and Right. The first two Rs are fairly self-explanatory—make your models rough and make them rapidly. In the early stages perfecting a model is a waste of time. "You learn just as much from a models that's wrong as you do from one that's right," says Vassalo. Even the final R (Right), doesn't mean that your model has to work. Instead it refers to building lots of small models that focus on specific problems. For Team 2's Cannonball Run project, the three Rs worked.

After making a few rough models with the mousetrap, they were certain that they could never fire the cannonball through the air with any precision. That nudged them towards a safer option—building a ramp that would guide the ball from a table top to its target on the floor via two guide rails made out of the wooden dowels. They got busy building their device. Thirty minutes before deadline, they realized that one of the other teams were doing the exact same thing, with one crucial difference: They were taping togeth-

er the index cards (the most useless looking of all the provided materials) to create a 30-foot long track that would guide the cannonball from their ramp to their cannonball catcher at the far end of the room. What's more they were carefully folding up the sides of each card to make sure the ball didn't get derailed on its way to the destination. Brilliant!

Teams 3 and 4 had gone with different designs that don't look nearly as impressive. So what did Team 2 do? They stole the idea. Suddenly, the competition was reduced to which team, Team 1 or Team 2, would be able to tape the index cards together faster.

Using Brainstorming and Rapid Prototyping Techniques in Real Life

Of course, in real life, you can't always glance across the room and steal your competitor's ideas. But the Cannonball Run project drove home Ideo's philosophy about brainstorming and rapid prototyping methods: They get you to stop dithering and start doing. What you can come up with on the fly won't be nearly as bad as you think—in fact it'll often be better that what you come up with working slowly and deliberately. Besides, coming up with something—anything— is often half the battle.

All 15 participants walked away from the experience with changed ideas of how to approach even the simplest of projects. "I like how it showed that you don't have to spend tons of money to prototype. You can do a lot of trial-and-error modeling on your own without paying a lot of money to go outside just to find something that doesn't work," stated one Team 2 participant. Another teammate added, "When I went to work the next day, I called an emergency brainstorming session and set a goal of 100 ideas in an hour. I thought maybe we'd get 50. We got 103." Would he recommend the workshop to his colleagues? "Oh man, are you kidding me? Absolutely!" But they could be biased. After all, Team 2 won.

Adapted from "Staying Smart: A Day at Innovation U.," by Ed Brown, *Fortune*, April 12, 1999, pp. 163-165.

should be allowed for dimensions and notes. A simple method to space the views is to cut out rectangular scraps of paper roughly equal to the sizes of the views and to move these around on the sheet until a suitable spacing is determined. The corner locations are then marked on the sheet, and the scraps of paper are discarded.

The same scale should be used for all details on a single sheet, if possible. When this is not possible, the scales for the dissimilar details should be clearly noted under each.

■ 14.13 TITLE AND RECORD STRIPS

The function of the title and record strip is to show, in an organized manner, all necessary information not given directly on the drawing with its dimensions and notes. Obviously, the type of title used depends on the filing system in use, the processes of manufacture, and the requirements of the product. The following information should generally be given in the title form:

1. Name of the object represented.
2. Name and address of the manufacturer.
3. Name and address of the purchasing company, if any.
4. Signature of the drafter who made the drawing and the date of completion.
5. Signature of the checker and the date of completion.
6. Signature of the chief drafter, chief engineer, or other official, and the date of approval.

7. Scale of the drawing.
8. Number of the drawing.

Other information may be given, such as material, quantity, heat treatment, finish, hardness, pattern number, estimated weight, superseding and superseded drawing numbers, symbol of machine, and many other items, depending on the plant organization and the peculiarities of the product. Some typical commercial titles are shown in Figs. 14.24, 14.25, and 14.26. See the inside back cover for traditional title forms and ANSI-approved sheet sizes.

The title form is usually placed along the bottom of the sheet, Fig. 14.24, or in the lower right-hand corner of the sheet, Fig. 14.26, because drawings are often filed in flat, horizontal drawers, and the title must be easily found. However, many filing systems are in use, and the location of the title form is governed by the system employed.

Lettering should be single-stroke vertical or inclined Gothic capitals. The items in the title form should be lettered in accordance with their relative importance. The drawing number should receive the greatest emphasis, closely followed by the name of the object and the name of the company. The date, scale, and drafter's and checker's names are important, but they do not deserve prominence. Greater importance of items is indicated by heavier lettering, larger lettering, wider spacing of letters, or by a combination of these methods. See Table 14.1 for recommended letter heights.

FIGURE 14.24   Title Strip.

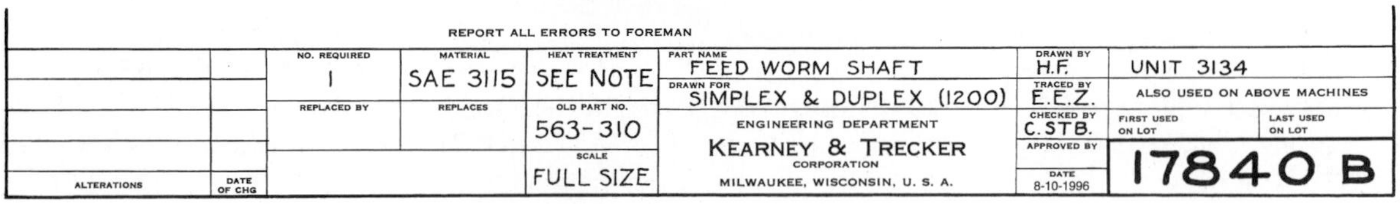

FIGURE 14.25   Title Strip.

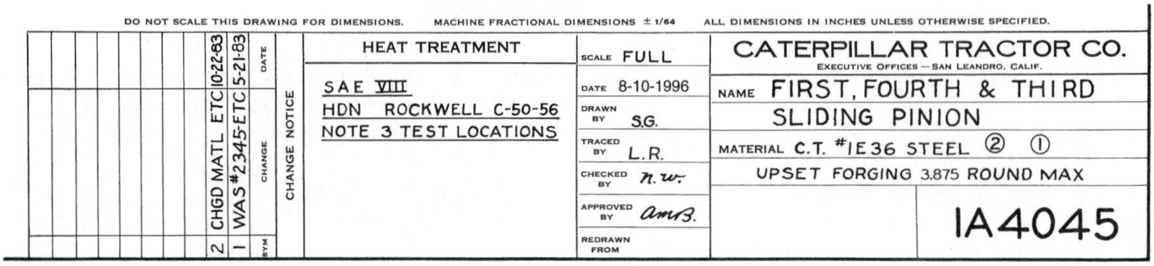

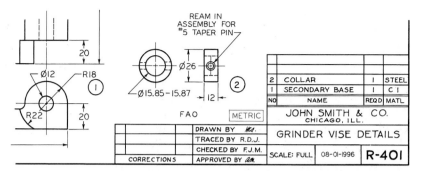

FIGURE 14.26 Identification of Details with Parts List.

	Minimum Letter Heights		Drawing Size
Use	**Freehand**	**Instrumental**	
Drawing number in title block	.312″ ($\frac{5}{16}$) 7 mm	.290″ 7 mm	Larger than 17″ × 22″
	.250″ ($\frac{1}{4}$) 7 mm	.240″ 7 mm	Up to and including 17″ × 22″
Drawing title	.250″ ($\frac{1}{4}$) 7 mm	.240″ 7 mm	
Section and tabulation letters	.250″ ($\frac{1}{4}$) 7 mm	.240″ 7 mm	All
Zone letters and numerals in borders	.188″ ($\frac{3}{16}$) 5 mm	.175″ 5 mm	
Dimensions, tolerances, limits, notes, subtitles for special views, tables, revisions, and zone letters for the body of the drawing	.125″ ($\frac{1}{8}$) 3.5 mm	.120″ 3.5 mm	Up to and including 17″ × 22″
	.156″ ($\frac{5}{32}$) 5 mm	.140″ 5 mm	Larger than 17″ × 22″

[a] ANSI Y14.2M–1979 (R1987).

TABLE 14.1 Recommended[a] Minimum Letter Heights.

Many companies have adopted their own title forms or those preferred by ANSI and have them printed on standard-size sheets, so that the drafters need merely fill in the blank spaces.

Drawings constitute important and valuable information regarding the products of a manufacturer. Hence, carefully designed, well-kept, systematic files are generally maintained for the filing of drawings.

■ 14.14 DRAWING NUMBERS

Every drawing should be numbered. Some companies use serial numbers, such as 60412, or a number with a prefix or suffix letter to indicate the sheet size, as A60412 or 60412-A. The size A sheet would probably be the standard 8.50 × 11.00 or 9.00 × 12.00, and the B size a multiple thereof. Many different numbering schemes are in use in which various parts of the drawing number indicate different things, such as model number of the machine and the general nature or use of the part. In general, it is best to use a simple numbering system and not to load the number with too many indications.

The drawing number should be lettered 7 mm (.2500) high in the lower-right and upper-left corners of the sheet.

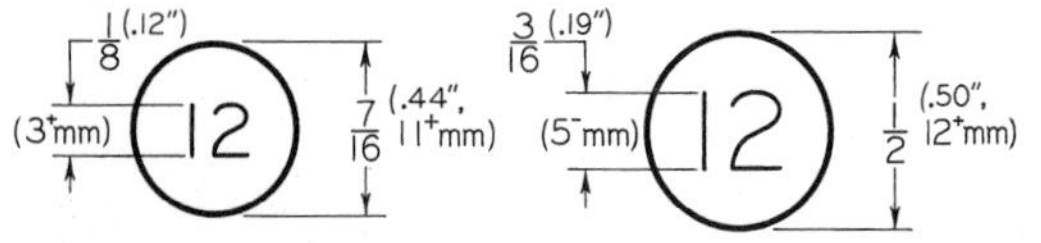

FIGURE 14.27 **Identification Numbers.**

■ 14.15 PARTS LISTS

A bill of material, or *parts list*, consists of an itemized list of the several parts of a structure shown on a detail drawing or an assembly drawing [ANSI Y14.34M–1982 (R1988)]. This list is often given on a separate sheet, but is frequently lettered directly on the drawing. The title strip alone is sufficient on detail drawings of only one part, Fig. 14.25, but a parts list is necessary on detail drawings of several parts, Fig. 14.26.

Parts lists on machine drawings contain the part numbers or symbols, a descriptive title of each part, the number required, the material specified, and frequently other information, such as pattern numbers, stock sizes of materials, and weights of parts.

Parts are listed in general order of size or importance. The main castings or forgings are listed first, parts cut from cold-rolled stock second, and standard parts such as fasteners, bushings, and roller bearings third. If the parts list rests on top of the title box or strip, the order of the items should be from the bottom upward, Figs. 14.26 and 14.32, so that new items can be added later, if necessary. If the parts list is placed in the upper-right corner, the items should read downward.

Each detail on the drawing may be identified with the parts list by the use of a small circle containing the part number, placed adjacent to the detail, as in Fig. 14.26. One of the sizes in Fig. 14.27 will be found suitable, depending on the size of the drawing.

Standard parts whether purchased or company produced, are not drawn but are included in the parts list. Bolts, screws, bearings, pins, keys, and so on are identified by the part number from the assembly drawing and are specified by name and size or number.

■ 14.16 ZONING

To facilitate locating an item on a large or complex drawing, regular ruled intervals are labeled along the margins, often in the right and lower margins only. The intervals on the horizontal margin are labeled from right to left with numerals, and the intervals on the vertical margin are labeled from bottom to top with letters.

■ 14.17 CHECKING

The importance of accuracy in technical drawing cannot be overestimated. In commercial offices, errors sometimes cause tremendous unnecessary expenditures. *The drafter's signature on a drawing identifies who is responsible for the accuracy of the work.*

In small offices, checking is usually done by the designer or by one of the drafters. In large offices, experienced engineers are employed who devote a major part of their time to checking drawings.

The pencil drawing, upon completion, is carefully checked and signed by the drafter who made it. The drawing is then checked by the designer for function, economy, practicability, and so on. Corrections, if any, are then made by the original drafter.

The final checker should be able to discover all remaining errors, and, to be effective, the work must be done in a systematic way. The checker should study the drawing with particular attention to the following points.

1. Soundness of design, with reference to function, strength, materials, economy, manufacturability, serviceability, ease of assembly and repair, lubrication, and so on.
2. Choice of views, partial views, auxiliary views, section line work, lettering, and so on.
3. Dimensions, with special reference to repetition, ambiguity, legibility, omissions, errors, and finish marks. Special attention should be given to tolerances.
4. Standard parts. In the interest of economy, as many parts as possible should be standard.
5. Notes, with special reference to clear wording and legibility.
6. Clearances. Moving parts should be checked in all possible positions to assure freedom of movement.
7. Title form information.

■ 14.18 DRAWING REVISIONS

Changes on drawings are necessitated by changes in design, changes in tools, desires of customers, or errors in design or in production. In order that the sources of all changes of information on drawings may be understood, verified, and accessible, an accurate record of all changes should be made on the drawings. The record should show the character of the change, by whom, when, and why made.

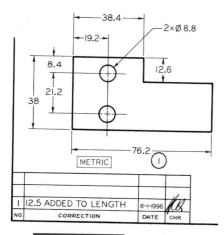

FIGURE 14.28 Revisions.

The changes are made by erasures directly on the original drawing or by means of erasure fluid on a reproduction print. Additions are simply drawn in on the original. The removal of information by crossing out is not recommended. If a dimension is not noticeably affected by a change, it may be underlined with a heavy line to indicate that it is not to scale. In any case, prints of each issue or microfilms are kept on file to show how the drawing appeared before the revision. New prints are issued to supersede old ones each time a change is made.

If considerable change on a drawing is necessary, a new drawing may be made and the old one then stamped OBSOLETE and placed in the "obsolete" file. In the title block of the old drawing, the words "SUPERSEDED BY ..." or "REPLACED BY ..." are entered followed by the number of the new drawing. On the new drawing, under "SUPERSEDES ..." or "REPLACES ...," the number of the old drawing is entered.

Various methods are used to reference the area on a drawing where the change is made, with the entry in the revision block. The most common is to place numbers or letters in small circles near the places where the changes were made and to use the same numbers or letters in the revision block, Fig. 14.28. On zoned drawings, §14.16, the zone of the correction would be shown in the revision block. In addition, the change should be described briefly, and the date and the initials of the person making the change should be given.

■ 14.19 SIMPLIFIED DRAFTING

Drafting time is a considerable element of the total cost of a product. Consequently, industry attempts to reduce drawing costs by simplifying its drafting practices, but without loss of clarity to the user.

The American National Standard Drafting Manual, published by the American National Standards Institute, incorporates the best and the most representative practices in this country, and the authors are in full accord with them. These standards advocate simplification in many ways, for example, partial views, half views, thread symbols, piping symbols, and single-line spring drawings. Any line or lettering on a drawing that is not needed for clarity should be omitted.

A summary of practices to simplify drafting is as follows.

1. Use word description in place of drawing wherever practicable.

2. Never draw an unnecessary view. Often a view can be eliminated by using abbreviations or symbols such as HEX, SQ, DIA, Ø, , and cL.

3. Draw partial views instead of full views wherever possible. Draw half views of symmetrical parts.

4. Avoid elaborate, pictorial, or repetitive detail as much as possible. Use phantom lines to avoid drawing repeated features.

5. List rather than draw, when possible, standard parts such as bolts, nuts, keys, and pins.

6. Omit unnecessary hidden lines.

7. Use outline section lining in large sectioned areas wherever it can be done without loss of clarity.

8. Omit unnecessary duplication of notes and lettering.

9. Use symbolic representation wherever possible, such as piping symbols and thread symbols.

10. Draw freehand, or mechanically plus freehand, wherever practicable.

11. Avoid hand lettering as much as possible. For example, parts lists should be typed on a separate sheet.

12. Use labor saving devices wherever feasible, such as templates and plastic overlays.

13. Use electronic devices or computer graphics systems wherever feasible for design, drawing, and repetitive work.

Some industries have attempted to simplify their drafting practices even more. Until these practices are accepted generally by industry and in time find their way into ANSI standards, the students should follow the ANSI standards as exemplified throughout this book. Fundamentals should come first—shortcuts perhaps later.

■ 14.20 ASSEMBLY DRAWINGS

An assembly drawing shows the assembled machine or structure, with all detail parts in their functional positions. Assembly drawings vary in character according to use, as follows: (1) design assemblies, or layouts, discussed in §14.7, (2) general assemblies, (3) working drawing assemblies, (4) outline or installation assemblies, and (5) check assemblies.

■ 14.21 GENERAL ASSEMBLIES

A set of working drawings includes the *detail drawings* of the individual parts and the *assembly drawing* of the assembled unit. The detail drawings of an automobile

connecting rod are shown in Figs. 14.29 and 14.30, and the corresponding assembly drawing is shown in Fig. 14.31. Such an assembly, showing only one unit of a larger machine, is often referred to as a *subassembly*.

An example of a complete general assembly appears in Fig. 14.32, which shows the assembly of a hand grinder. Another example of a subassembly is shown in Fig. 14.33.

1. Views. In selecting the views for an assembly drawing, the purpose of the drawing must be kept in mind: to show how the parts fit together in the assembly and to suggest the function of the entire unit, not to describe the shapes of the individual parts. The assembly worker receives the actual finished parts. If more information is needed about a part that cannot be obtained form the part itself, the detail drawing must be checked. Thus, the assembly drawing purports to show

FIGURE 14.29 **Forging Drawing of Connecting Rod.** *Courtesy of Cadillac Motor Car Division.*

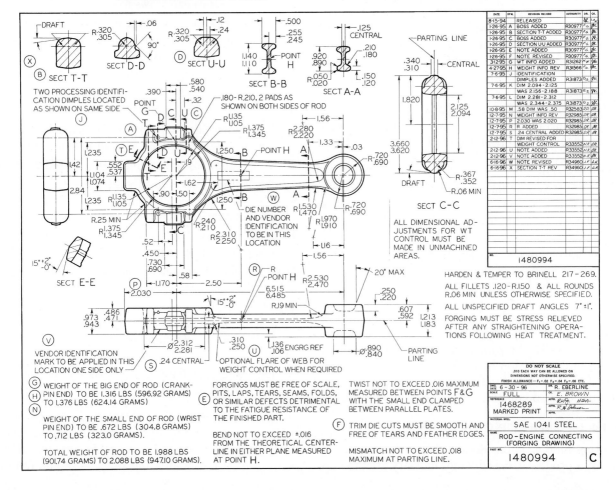

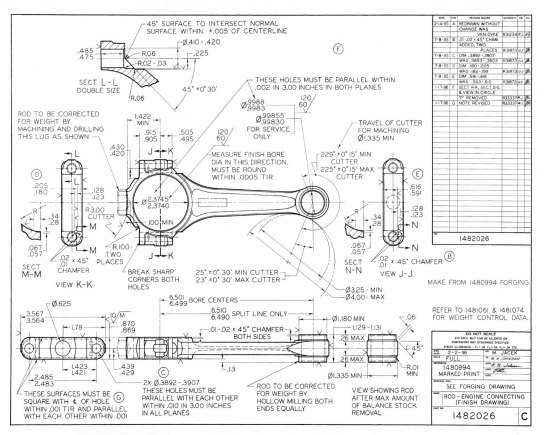

FIGURE 14.30 Detail Drawing of Connecting Rod. *Courtesy of Cadillac Motor Car Division.*

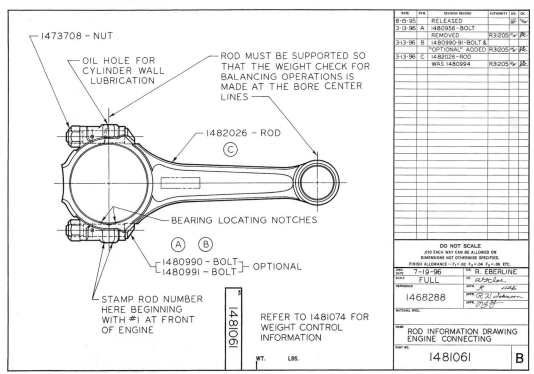

FIGURE 14.31 Assembly Drawing of Connecting Rod. *Courtesy of Cadillac Motor Car Division.*

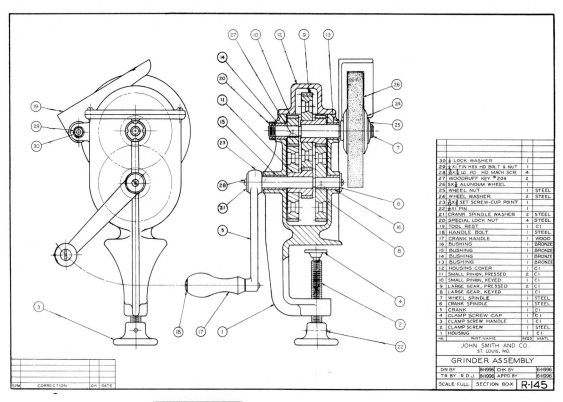

FIGURE 14.32 Assembly Drawing of Grinder.

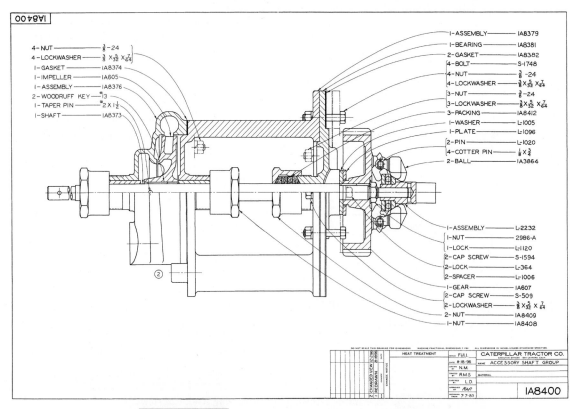

FIGURE 14.33 Subassembly of Accessory Shaft Group.

relationships of parts, *not shapes.* The view or views selected should be the minimum views or partial views that will show how the parts fit together. In Fig. 14.31, only one view is needed, while in Fig. 14.32 only two views are necessary.

2. Sections. Since assemblies often have parts fitting into or overlapping other parts, hidden-line delineation is usually out of the question. Hence, in assemblies, sectioning can be used to great advantage. For example, in Fig. 14.32, try to imagine the right-side view drawn in elevation with interior parts represented by hidden lines. The result would be completely unintelligible.

Any kind of section may be used as needed. A broken-cut section is shown in Fig. 14.32, a half section in Fig. 14.33, and several removed sections are shown in Fig. 14.29. For general information on assembly sectioning, see §14.22. For methods of drawing threads in sections, see §14.16.

3. Hidden lines. As a result of the extensive use of sectioning in assemblies, hidden lines are often not needed. However, they should be used wherever necessary for clearness.

4. Dimensions. As a rule, dimensions are not given on assembly drawings, since they are given completely on the detail drawings. If dimensions are given, they are limited to some function of the object as a whole, such as the maximum height of a jack, or the maximum opening between the jaws of a vise. Or when machining is required in the assembly operation, the necessary dimensions and notes may be given on the assembly drawing.

5. Identification. The methods of identification of parts in an assembly are similar to those used in detail drawings where several details are shown on one sheet, as in Fig. 14.26. Circles containing the part numbers are placed adjacent to the parts, with leaders terminated by arrowheads touching the parts as in Fig. 14.32. The circles shown in Fig. 14.27 are, with the addition of radial leaders, satisfactory for assembly drawings. Note, in Fig. 14.32, that these circles are placed in orderly horizontal or vertical rows and not scattered over the sheet. Leaders are never allowed to cross, and adjacent leaders are parallel or nearly so.

The parts list includes the part numbers or symbols, a descriptive title of each part, the number required per machine or unit, the material specified, and frequently other information, such as pattern numbers, stock sizes, weights, and so on. Frequently the parts list is lettered or typed on a separate sheet.

Another method of identification is to letter the part names, numbers required, and part numbers, at the end of leaders as shown in Fig. 14.33. More commonly, however, only the part numbers are given, together with ANSI-approved straight-line leaders.

6. Drawing revisions. Methods of recording changes are the same as those for detail drawings, Fig. 14.29, for example. See §14.18.

■ 14.22 ASSEMBLY SECTIONING

In assembly sections it is necessary not only to show the cut surfaces but also to distinguish between adjacent parts. This is done by drawing the section lines in opposing directions, as shown in Fig. 14.34. The first

FIGURE 14.34   Section Lining (Full Size).

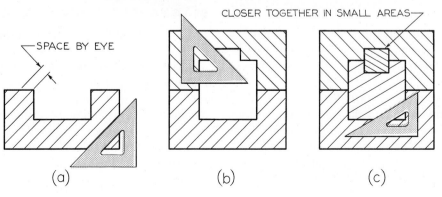

(a)   (b)   (c)

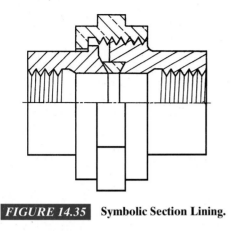

FIGURE 14.35 Symbolic Section Lining.

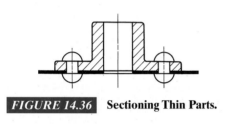

FIGURE 14.36 Sectioning Thin Parts.

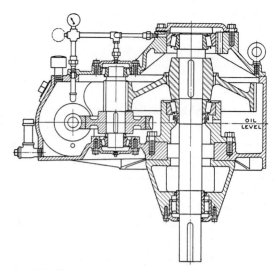

FIGURE 14.37 Assembly Section. *Courtesy of Hewitt-Robins, Inc.*

large area, (a), is section-lined at 458. The next large area, (b), is section-lined at 458 in the opposite direction. Additional areas are then section-lined at other angles, such as 308 or 608 with horizontal, as shown at (c). If necessary, "odd" angles may be used. Note at (c) that in small areas it is necessary to space the section lines closer together. The section lines in adjacent areas should not meet at the visible lines separating the areas.

For general use, the cast-iron general-purpose section lining is recommended for assemblies. Wherever it is desired to give a general indication of the materials used, symbolic section lining may be used, as in Fig. 14.35.

In sectioning relatively thin parts in assembly, such as gaskets and sheet-metal parts, section lining is ineffective, and such parts should be shown in solid black, Fig. 14.36.

Often solid objects, or parts that themselves do not require sectioning, lie in the path of the cutting plane. It is customary and standard practice to show such parts unsectioned, or "in the round." These include bolts, nuts, shafts, keys, screws, pins, ball or roller

bearings, gear teeth, spokes, and ribs among others. Many are shown in Fig. 14.37, and similar examples are shown in Figs. 14.32 and 14.33.

■ 14.23 WORKING DRAWING ASSEMBLY

A working drawing assembly, Fig. 14.38, is a combined detail and assembly drawing. Such drawings are often used in place of separate detail and assembly drawings when the assembly is simple enough for all its parts to be shown clearly in the single drawing. In some cases, all but one or two parts can be drawn and dimensioned clearly in the assembly drawing, in which event these parts are detailed separately on the same sheet. This type of drawing is common in valve drawings, locomotive subassemblies, aircraft subassemblies, and drawings of jigs and fixtures.

■ 14.24 INSTALLATION ASSEMBLIES

An assembly made specifically to show how to install or erect a machine or structure is an *installation assembly*. This type of drawing is also often called an *outline assembly*, because it shows only the outlines and the relationships of exterior surfaces. A typical installation assembly is shown in Fig. 14.39. In aircraft drafting, an installation drawing (assembly) gives complete information for placing details or subassemblies in their final positions in the airplane.

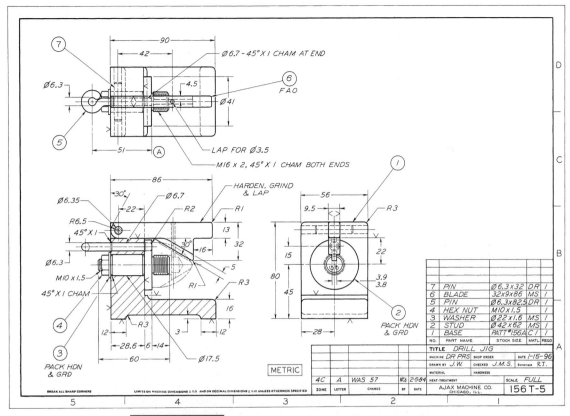

FIGURE 14.38 Working Drawing Assembly of Drill Jig.

■ 14.25 CHECK ASSEMBLIES

After all detail drawings of a unit have been made, it may be necessary to make a *check assembly*, especially if a number of changes were made in the details. Such an assembly is drawn accurately to scale in order to check graphically the correctness of the details and their relationship in assembly. After the check assembly has served its purpose, it may be converted into a general assembly drawing.

■ 14.26 PATENT DRAWINGS

The patent application for a machine or device must include drawings to illustrate and explain the invention. It is essential that all patent drawings be mechanically correct and constitute complete illustrations of every feature of the invention claimed. The strict re-

quirements of the U.S. Patent Office in this respect serve to facilitate the examination of applications and the interpretation of patents issued thereon. A typical patent drawing is shown in Fig. 14.40.

The drawings for patent applications are pictorial and explanatory in nature; hence, they are not detailed as are working drawings for production purposes. Center lines, dimensions, notes, and so forth are omitted. Views, features, and parts, for example, are identified by numbers that refer to the descriptions and explanations given in the specification section of the patent application.

Patent drawings are made with India ink on heavy, smooth, white paper, exactly 10.00×15.00 with 1.00 borders on all sides. A space of not less than 1.250 from the shorter border, which is the top of the drawing, is left blank for the heading of title, name, number, and other data to be added by the Patent Office.

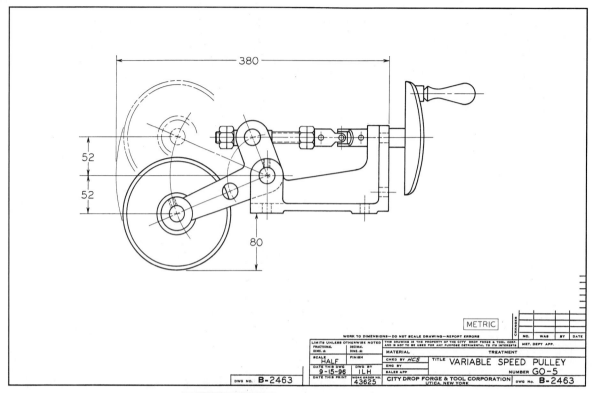

FIGURE 14.39   Installation Assembly.

5,511,508
PONTOON RUNNER SYSTEM
John M. Wilson, Sr., and Dean R. Wilson, both of Marrero,
La., assignors to Wilco Marsh Buggies & Draglines, Inc., La.
Filed Apr. 21, 1994, Ser. No. 230,618
Int. Cl.^{6} B63B 3/00

U.S. Cl. 114—356   10 Claims

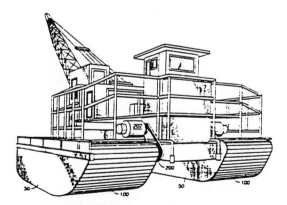

FIGURE 14.40   Pictorial Patent Drawing.

All lines must be solid black and suitable for reproduction at a smaller size. Line shading is used whenever it improves readability.

The drawings must contain as many figures as necessary to show the invention clearly. There is no restriction on the number of sheets. The figures may be plan, elevation, section, pictorial, and detail views of portions or elements, and they may be drawn to an enlarged scale if necessary. The required signatures must be placed in the lower right-hand corner of the drawing, either inside or outside the border line.

Because of the strict requirements of the Patent Office, applicants are advised to employ competent drafters to make their drawings. To aid drafters in the preparation of drawings for submission in patent applications, the *Guide for Patent Draftsmen* has been prepared by and can be obtained from the Superintendent of Documents, U.S. Government Printing Office, Washington, D.C. 20402.

■ KEY WORDS

STAGES OF DESIGN	CONCEPTS
DESIGN REQUIREMENTS	REFINEMENT
MODELS	PARAMETRIC SOLID MODELING
VIRTUAL REALITY	COMPUTER SIMULATION
DETAIL DRAWING	WORKING DRAWING
3D MODELING	PROTOTYPE
SOLID MODELING	RAPID PROTOTYPING
ASSEMBLY DRAWING	TITLE BLOCKS
PRODUCTION DRAWINGS	DRAWING NUMBERS
DIFFERENCE INTERSECTION	PATENT DRAWINGS
RECORD STRIPS	REVERSE ENGINEERING
EXTRUSION REVOLUTION	COORDINATE MEASURING MACHINE
CHECKING AND PROOFING	FUNCTIONAL DECOMPOSITION
FUNCTIONAL DESIGN	BRAINSTORMING
AESTHETIC DESIGN	STORYBOARDING
MOTIVATION	REINTEGRATION
OBJECTIVES	BOOLEAN OPERATORS
CREATIVITY TECHNIQUES	REVOLVED SOLID
EXTRUDED SOLID	VENN DIAGRAMS
CUSTOMER	

■ CHAPTER SUMMARY

- The design team moves through five stages during the design process. Each stage helps the team refine the design until it meets all product requirements.
- The final drawings created during the design process include assembly drawings, working drawings, design drawings, and patent drawings.
- There are many revisions to drawings during the design process. The drafter must keep track of each version and what changes were made.

- Models are an important way of testing the way parts are assembled. Both scale models created in a model shop and computer-generated virtual reality models are used by the design team to test their design.
- During the design process, all members of the team must understand their specific roles and how they relate and interact with the rest of the team. Effective teamwork is an essential part of the design process.

■ REVIEW QUESTIONS

1. What are the special requirements of a patent drawing?

2. What kinds of information are included in an assembly drawing?

3. Name three individual creativity techniques. Name two group creativity techniques.

4. Define SLA, SLS, BPM, and LOM and give an example of each used in a manufacturing setting.

5. What are the five stages of the design process? Describe each stage.

6. How is a detail drawing different from an assembly drawing?

7. Why are drawings numbered? Why is this numbering so important?

8. Describe the drawing revision process. Why is it so important to keep track of revisions?

9. How are revised paper drawings stored? How are revised CAD drawings stored?

10. What are the advantages of computer modeling? What are the disadvantages?

■ DESIGN AND WORKING DRAWING PROBLEMS

DESIGN PROBLEMS

The following suggestions for project assignments are of a general and very broad nature, and it is expected that they will help generate many ideas for specific design projects. Much design work is undertaken to improve an existing product or system by utilization of new materials, new techniques, or new systems or procedures. In addition to the design of the product itself, another large amount of design work is essential for the tooling, production, and handling of the product. You are encouraged to discuss with your instructor any ideas you may have for a project.

Prob. 14.1 Design new or improved playground, recreational, or sporting equipment. For example, a new child's toy could be both recreational and educational.

Prob. 14.2 Design new or improved health equipment. For example, physically handicapped people need special equipment.

Prob. 14.3 Design security or safety devices. Fire, theft, or poisonous gases are a threat to life and property.

Prob. 14.4 Design devices and/or systems for waste handling. Home and factory waste disposal needs serious consideration.

Prob. 14.5 Design new or improved educational equipment. Both teacher and student would welcome more efficient educational aids.

Prob. 14.6 Design improvements in our land, sea, and air transportation systems. Vehicles, controls, highways, and airports need further refinement.

Prob. 14.7 Design new or improved devices or material handling. A dispensing device for a powdered product is an example.

Prob. 14.8 Improve the design of an existing device or system.

Prob. 14.9 Design or redesign devices for improved portability.

Prob. 14.10 Design an airport luggage handling system that will reduce damage to all types of baggage.

Prob. 14.11 Break up into design teams. See how many different ideas each team can come up with for a new layout of your classroom. Time limit is 20 minutes.

Prob. 14.12 Design a new or improved bike safety lock and chain. Integrate the locking devices into the bike's frame, if possible.

Each solution to a design problem, whether prepared by an individual student or formulated by a group, should be in the form of a *report*, which should be typed or carefully lettered, assembled, and bound. It is suggested that the report contain the following (or variations of the following, as specified by your instructor).

1. A title sheet. The title of the design project should be placed in approximately the center of the sheet, and your name or the names of those in the group in the lower right-hand corner. The symbol PL should follow the name of the project leader.

2. Table of contents with page numbers.

3. Statement of the purpose of the project with appropriate comments.

4. Preliminary design sketches, with comments on advantages and disadvantages of each, leading to the final selection of the *best* solution. All work should be signed and dated.

5. An accurately made pictorial and/or assembly drawing(s), using traditional drawing methods or CAD as assigned, if more than one part is involved in the design.

6. Detail working drawings, freehand, mechanical, or CAD-produced as assigned. The 8.5″ × 11.0″ sheet size is preferred for convenient insertion in the report. Larger sizes may be bound in the report with appropriate folding.

7. A bibliography or credit for important sources of information, if applicable.

WORKING DRAWING PROBLEMS

The problems in Figs. 14.41–14.105 are presented to give you practice in making regular working drawings of the type used in industry. Many problems, especially those of the assemblies, offer an excellent opportunity for you to exercise your ability to redesign or improve on the existing design. Due to the variations in sizes and in scales that may be used, you are required to select the sheet sizes and scales, when these are not specified, subject to the approval of the instructor. Standard sheet layouts are shown inside the front cover of this book.

The statements for each problem are intentionally brief, so that the instructor may amplify or vary the requirements when making assignments. Use the preferred metric system or the acceptable complete decimal-inch system, as assigned. Either the preferred unidirectional or acceptable aligned dimensioning may be assigned.

It should be clearly understood that in problems presented in pictorial form, the placement of dimensions and finish marks cannot always be followed in the drawing. *The dimensions given are in most cases those needed to make the parts, but due to the limitations of pictorial drawings they are not in all cases the dimensions that should be shown on the working drawing.* In the pictorial problems the rough and finished surfaces are shown, but finish marks are usually omitted. You should add all necessary finish marks and place all dimensions in the preferred places in the final drawings.

Each problem should be preceded by a sketch, fully dimensioned. Any of the title blocks shown inside the back cover of this book may be used, with modification if de-

sired, or you may design the title block if so assigned by the instructor.

Since many of the problems in this chapter are of a general nature, they can also be solved on most CAD systems. If a system is available, the instructor may choose to assign specific problems to be completed by this method.

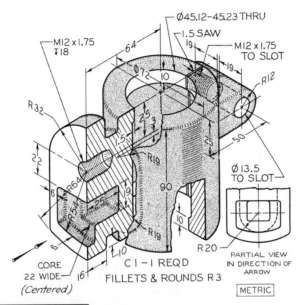

FIGURE 14.41   **Table Bracket.**
Prob. 14.13: Make detail drawing using size B or A3 sheet.

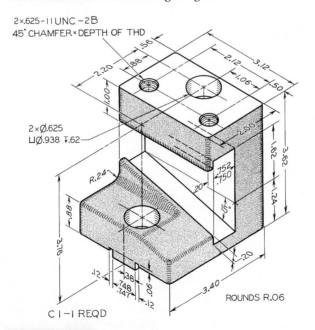

FIGURE 14.42   **RH Tool Post.**
Prob. 14.14: Make detail drawing using size B or A3 sheet.
If assigned, convert dimensions to metric system.

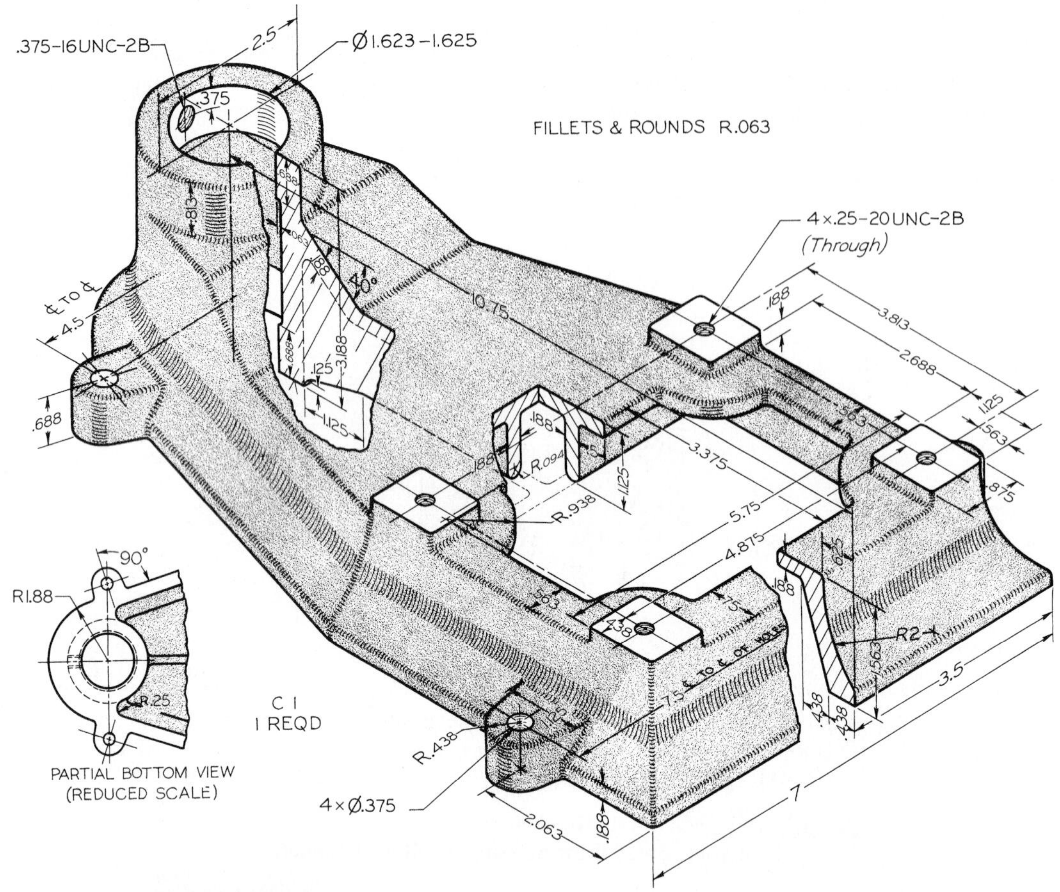

.375–16UNC–2B   2.5   Ø1.623–1.625

FILLETS & ROUNDS R.063

.375

4×.25–20UNC–2B
(Through)

.188

3.813

2.688

1.25

.563

.875

10.75

40°

.125
3.188

1.125

.188

R.094

3.375

R.938

5.75

4.875

R2

3.5

7.5 & TO ℄ OF HOLES

.188

.438
.38

R.438

.188

7

4×Ø.375

2.063

℄ TO ℄
4.5

.688

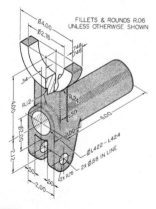

90°

R1.88

R.25

C 1
1 REQD

PARTIAL BOTTOM VIEW
(REDUCED SCALE)

FIGURE 14.43   Drill Press Base.
Prob. 14.15: Make detail drawing using size C or A2 sheet. Use unidirectional
metric or decimal-inch dimensions.

FILLETS & ROUNDS R.06
UNLESS OTHERWISE SHOWN

Ø4.00
Ø2.76

.748
.746

.34

4.00

R.12
2.00

2.00

.50

6.00

2×Ø76

Ø1.422–1.424
2×Ø.68 IN LINE

.50   .50

2.12

2.00

2×R.76

FIGURE 14.44   Shifter Fork.
Prob. 14.16: Make detail drawing using size B or A3 sheet.
If assigned, convert dimensions to metric system.

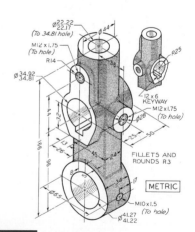

Ø22.22
22.17
(To 34.81 hole)

Ø44

M12×1.75
(To hole)
R14

R25

Ø34.92
34.81

12×6
KEYWAY

M12×1.75
(To hole)

Ø28

FILLETS AND
ROUNDS R3

METRIC

M10×1.5
(To hole)

Ø41.27
41.22

FIGURE 14.45   Idler Arm.
Prob. 14.17: Make detail drawing using size B or A3 sheet.

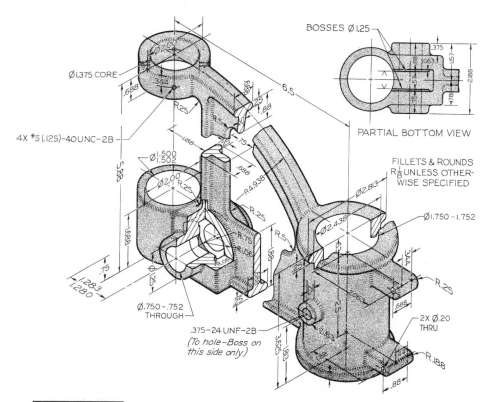

FIGURE 14.46 Drill Press Bracket.
Prob. 14.18: Make detail drawing using size C or A2 sheet. If assigned, convert dimensions to decimal inches or redesign the part with metric dimensions.

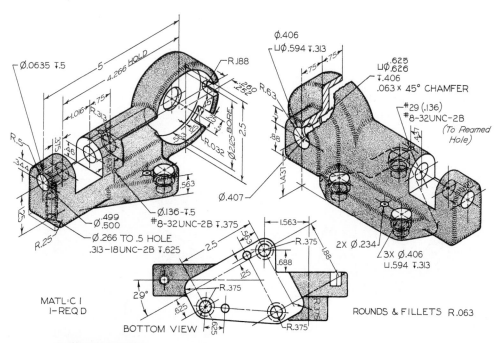

FIGURE 14.47 Dial Holder.
Prob. 14.19: Make detail drawing using size C or A2 sheet. If assigned, convert dimensions to decimal inches or redesign the part with metric dimensions.

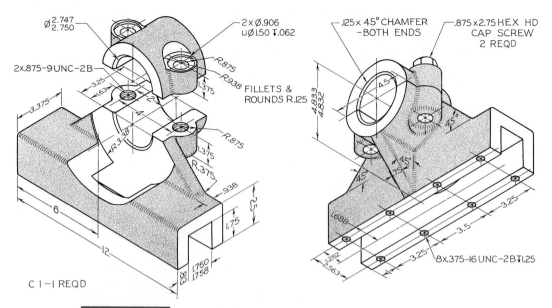

FIGURE 14.48   **Rack Slide.**
Prob. 14.20: Make detail drawings half size on size B or A3 sheet. If assigned, convert dimensions to decimal inches or redesign the part with metric dimensions.

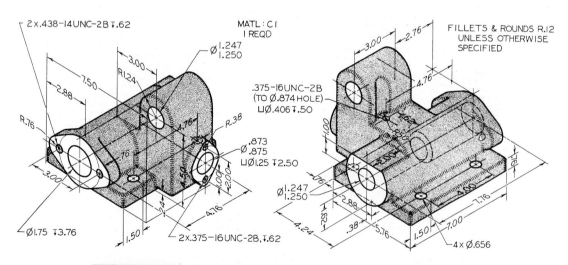

FIGURE 14.49   **Automatic Stop Box.**
Prob. 14.21: Make detail drawing half size on size B or A3 sheet. If assigned, redesign the part with metric dimensions.

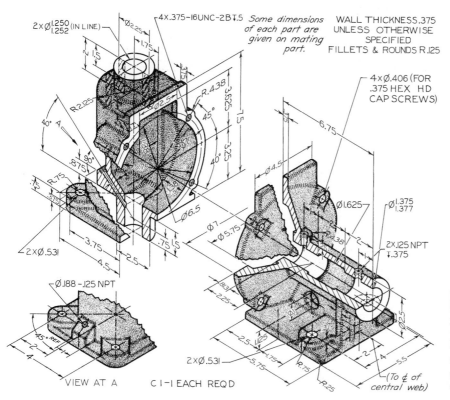

WALL THICKNESS .375
UNLESS OTHERWISE
SPECIFIED
FILLETS & ROUNDS R.125

Some dimensions of each part are given on mating part.

VIEW AT A   C I-I EACH REQ D

FIGURE 14.50 Conveyor Housing.
Prob. 14.22: Make detail drawings half size on size C or A2 sheets. If assigned, convert dimensions to decimal inches or redesign the parts with metric dimensions.

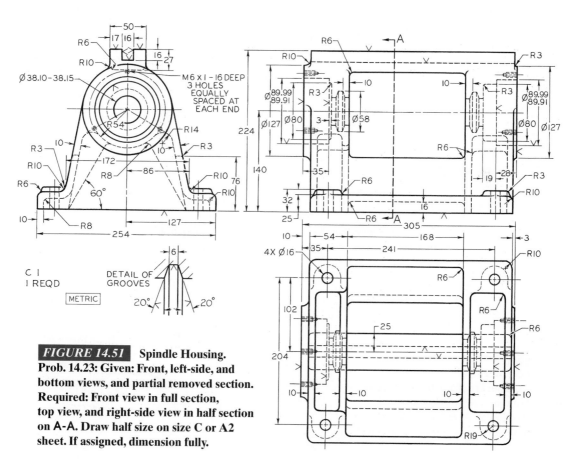

FIGURE 14.51 Spindle Housing.
Prob. 14.23: Given: Front, left-side, and bottom views, and partial removed section.
Required: Front view in full section, top view, and right-side view in half section on A-A. Draw half size on size C or A2 sheet. If assigned, dimension fully.

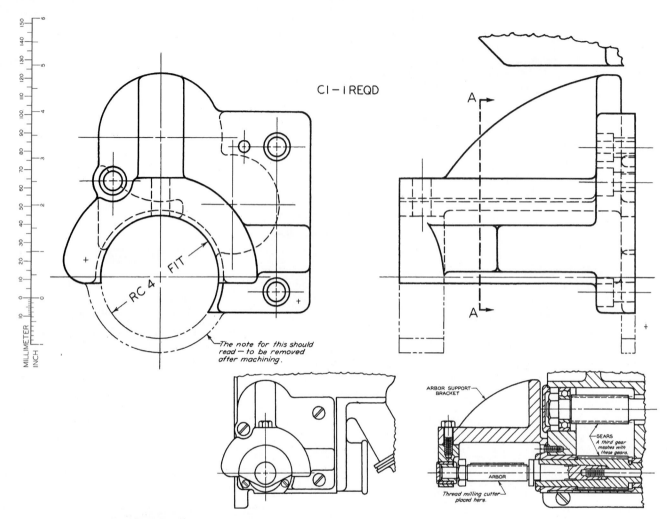

CI – I REQD

The note for this should read — to be removed after machining.

FIGURE 14.52 **Arbor Support Bracket.**
Prob. 14.24: Given: Front and right-side views.
Required: Front, left-side, and bottom views, and a detail section A-A. Use American National Standard tables for indicated fits and if required convert to metric values (see Appendixes 5–14). If assigned, dimension in the metric or decimal-inch system.

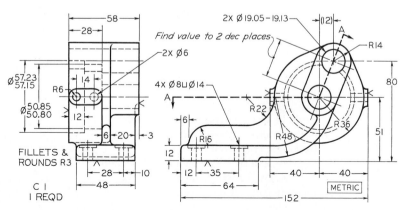

FIGURE 14.53 **Pump Bracket for a Thread Milling Machine.**
Prob. 14.25: Given: Front and left-side views. Required: Front and right-side views, and top view in section on A-A. Draw full size on size B or A3 sheet. If assigned, dimension fully.

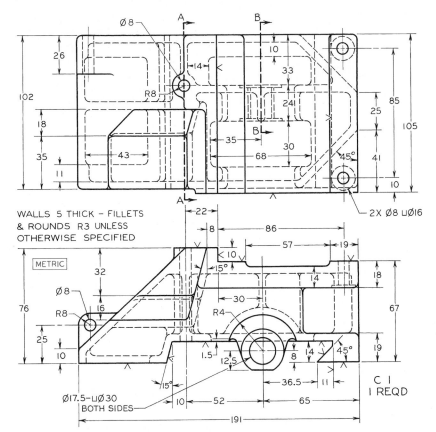

WALLS 5 THICK – FILLETS
& ROUNDS R3 UNLESS
OTHERWISE SPECIFIED

METRIC

Ø17.5–⌴Ø30
BOTH SIDES

C 1
1 REQD

FIGURE 14.54 Support Base for Planer.
Prob. 14.26: Given: Front and top views.
Required: Front and top views, left-side view in full section A-A, and removed section B-B. Draw full size on size C or A2 sheet. If assigned, dimension fully.

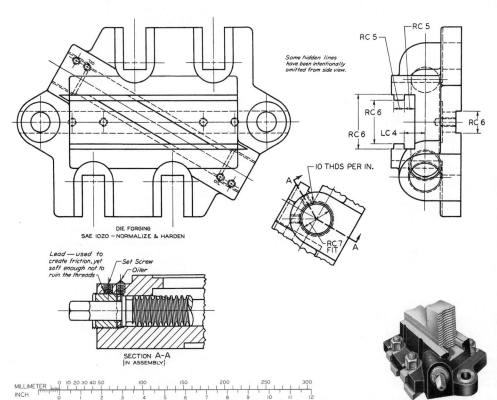

FIGURE 14.55 Jaw Base for Chuck Jaw.
Prob. 14.27: Given: Top, right-side, and partial auxiliary views.
Required: Top, left-side (beside top), front, and partial auxiliary views complete with dimensions, if assigned. Use metric or decimal-inch dimensions. Use American National Standard tables for indicated fits or convert for metric values. See Appendixes 5–14.

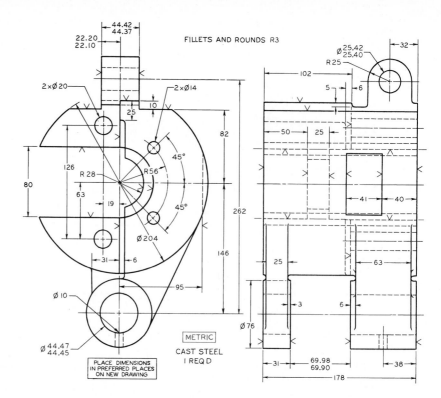

FIGURE 14.56 **Fixture Base for 60-Ton Vertical Press.**

Prob. 14.28: Given: Front and right-side views.

Required: Revolve front view 90° clockwise; then add top and left-side views. Draw half size on size C or A2 sheet. If assigned, complete with dimensions.

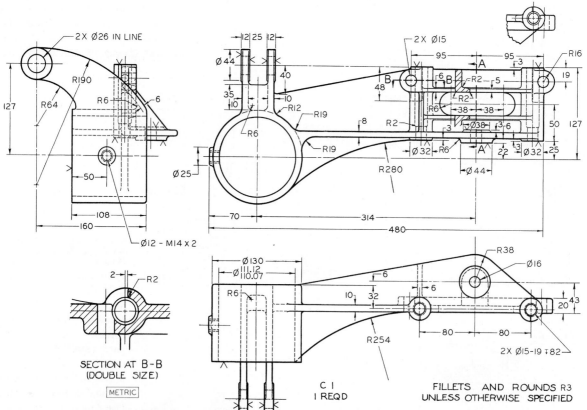

FIGURE 14.57 **Bracket.**

Prob. 14.29: Given: Front, left-side, and bottom views, and partial removed section.
Required: Make detail drawing. Draw front, top, and right-side views, and removed sections A-A and B-B. Draw half size on size C or A2 sheet. Draw section B-B full size. If assigned, complete with dimensions.

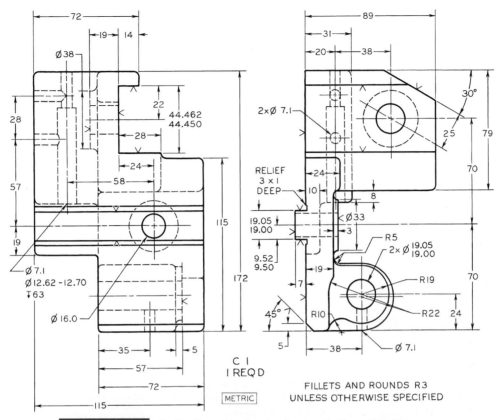

FIGURE 14.58   Roller Rest Bracket for Automatic Screw Machine.
Prob. 14.30: Given: Front and left-side views.
Required: Revolve front view 90° clockwise; then add top and left-side views.
Draw half size on size C or A2 sheet. If assigned, complete with dimensions.

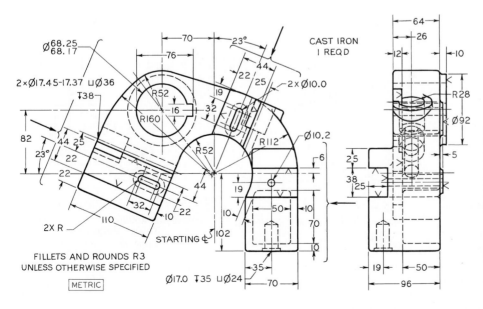

FIGURE 14.59   Guide
Bracket for Gear Shaper.
Prob. 14.31: Given: Front and
right-side views.
Required: Front view, a partial
right-side view, and two partial
auxiliary views taken in
direction of arrows. Draw half
size on size C or A2 sheet. If
assigned, complete with
unidirectional dimensions.

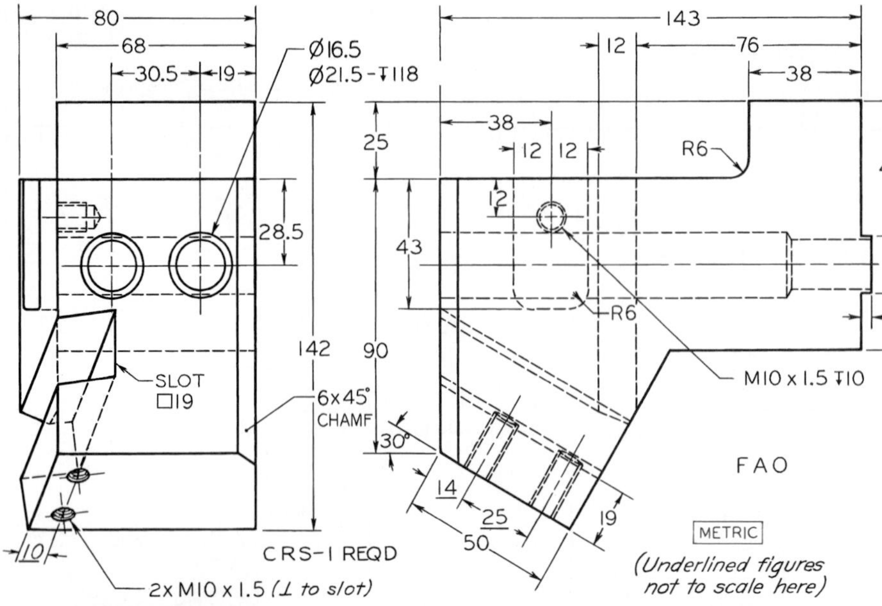

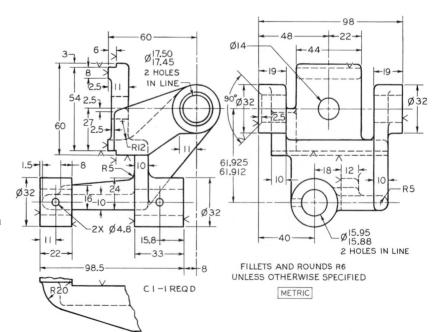

FIGURE 14.60 **Rear Tool Post.**
Prob. 14.32: Given: Front and left-side views.
Required: Take left-side view as new top view; add front and left-side views,
approx. 215 mm apart, a primary auxiliary view, then a secondary view taken
so as to show true end view of 19-mm slot. Complete all views, except show
only necessary hidden lines in auxiliary views. Draw full size on size C or A2
sheet. If assigned, complete with dimensions.

FIGURE 14.61 **Bearing for a Worm Gear.**
Prob. 14.33: Given: Front and right-side views.
Required: Front, top, and left-side views. Draw full size on size C or A2 sheet. If assigned, complete with dimensions.

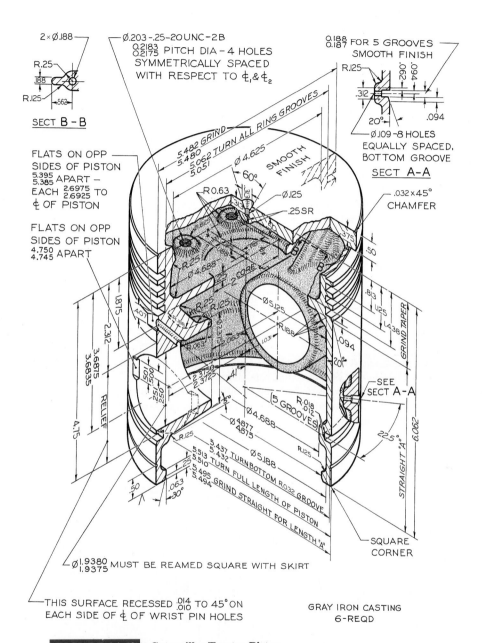

FIGURE 14.62 Caterpillar Tractor Piston.
Prob. 14.34: Make detail drawing full size on size C or A2 sheet. If assigned, use unidirectional decimal-inch system, converting all fractions to two-place decimal dimensions, or convert all dimensions to metric.

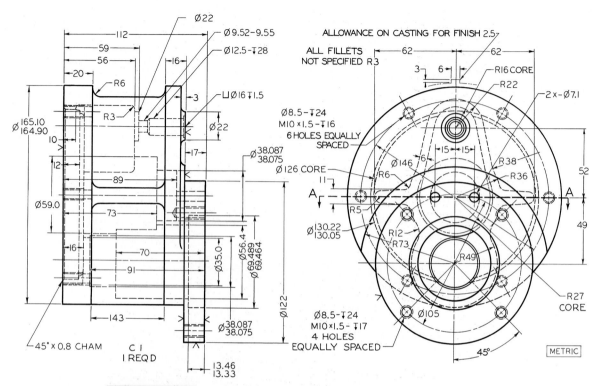

FIGURE 14.63 **Generator Drive Housing.**
Prob. 14.35: Given: Front and left-side views.
Required: Front view, right-side view in full section, and top view in full section on **A-A**. Draw full size on size C or A2 sheet. If assigned, complete with dimensions.

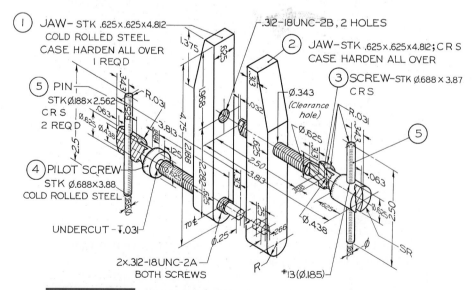

FIGURE 14.64 **Machinist's Clamp.**
Prob. 14.36: Draw details and assembly. If assigned, use unidirectional two-place decimal-inch dimensions or redesign for metric dimensions.

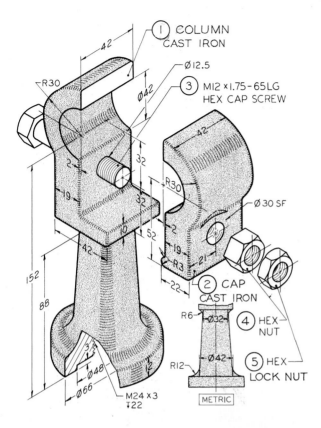

① COLUMN
CAST IRON

42

R30

Ø12.5

Ø42

③ M12×1.75−65LG
HEX CAP SCREW

42

32

R30

32

Ø30 SF

52

2

R3

22

② CAP
CAST IRON

R6

Ø32

④ HEX
NUT

Ø42

⑤ HEX
LOCK NUT

R12

152

88

48

66

M24×3
↧22

Ø48

METRIC

FIGURE 14.65 Hand Rail Column.
Prob. 14.37: (1) Draw details. If assigned, complete with dimensions. (2) Draw assembly.

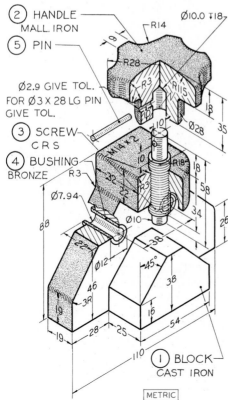

② HANDLE
MALL. IRON

⑤ PIN

Ø2.9 GIVE TOL.
FOR Ø3×28 LG PIN
GIVE TOL.

③ SCREW
CRS

④ BUSHING
BRONZE

Ø7.94

Ø10

R14

Ø10.0 ↧18

R28

R3

R115

18

35

Ø28

M14×2

R18

18

58

34

26

88

Ø12

R3

22

46

3R

19

19

38

45°

38

16

28

25

54

110

① BLOCK
CAST IRON

METRIC

FIGURE 14.66 Drill Jig.
Prob. 14.38: (1) Draw details. If assigned, complete with dimensions. (2) Draw assembly.

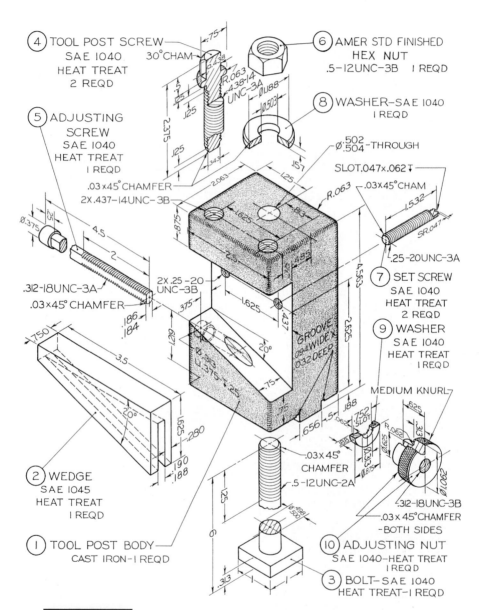

FIGURE 14.67 Tool Post.
Prob. 14.39: (1) Draw details. (2) Draw assembly. If assigned, use
unidirectional two-place decimals for all fractional dimensions or redesign
for all metric dimensions.

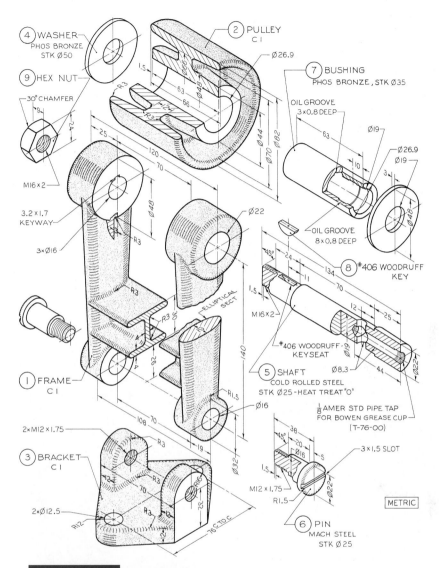

FIGURE 14.68   Belt Tightener.

Prob. 14.40: (1) Draw details. (2) Draw assembly. It is assumed that the parts are to be made in quantity and they are to be dimensioned for interchangeability on the detail drawings. Use tables in Appendixes 11–14 for limit values. Design as follows.

a. Bushing fit in pulley: Locational interference fit.
b. Shaft fit in bushing; Free running fit.
c. Shaft fits in frame: Sliding fit.
d. Pin fit in frame: Free running fit.
e. Pulley hub length plus washers fit in frame: Allowance 0.13 and tolerances 0.10.
f. Make bushing 0.25 mm shorter than pulley hub.
g. Bracket fit in frame: Same as e above.

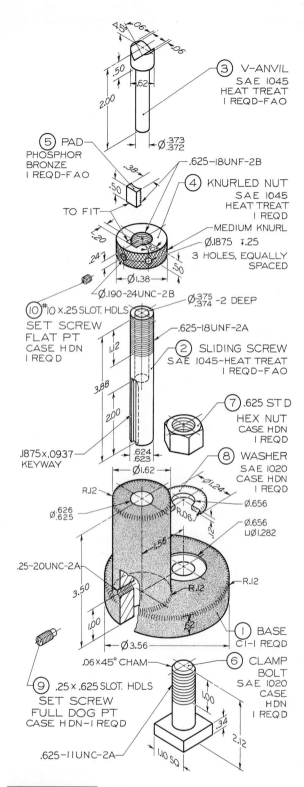

FIGURE 14.69 Milling Jack.
Prob. 14.41: (1) Draw details. (2) Draw assembly. If assigned, convert dimensions to metric or decimal-inch system.

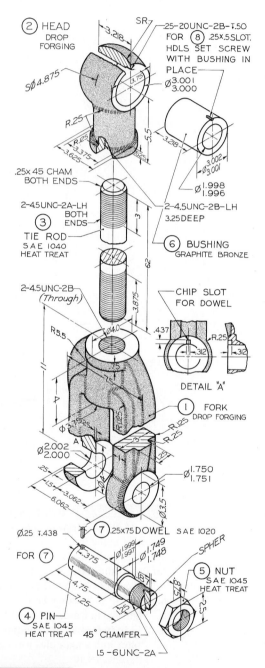

FIGURE 14.70 Connecting Bar.
Prob. 14.42: (1) Draw details. (2) Draw assembly. If assigned, convert dimensions to metric or decimal-inch system.

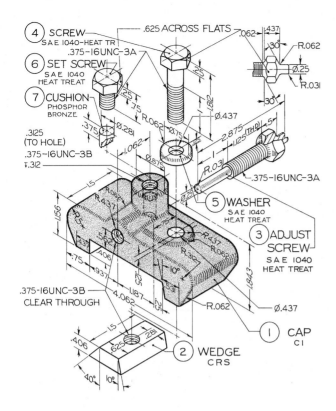

④ SCREW
SAE 1040-HEAT TR
.375-16UNC-3A

⑥ SET SCREW
SAE 1040
HEAT TREAT

⑦ CUSHION
PHOSPHOR
BRONZE

.625 ACROSS FLATS
.062 .437
30 R.062
Ø.25
R.031
.812
Ø.437
30

.3125
(TO HOLE)
.375-16UNC-3B
.312

.281
R.062
Ø.75
Ø.875
2.875 (THD)
1.125
.375-16UNC-3A
.5

⑤ WASHER
SAE 1040
HEAT TREAT

③ ADJUST
SCREW
SAE 1040
HEAT TREAT

.375-16UNC-3B
CLEAR THROUGH
4.062
1.187
R.062
10°
Ø.437

① CAP
C1

② WEDGE
CRS
.406
.625
.28
40° 10°
1.5

FIGURE 14.71 Clamp Stop.
Prob. 14.43: (1) Draw details. (2) Draw assembly. If
assigned, convert dimensions to decimal-inch system or
redesign for metric dimensions.

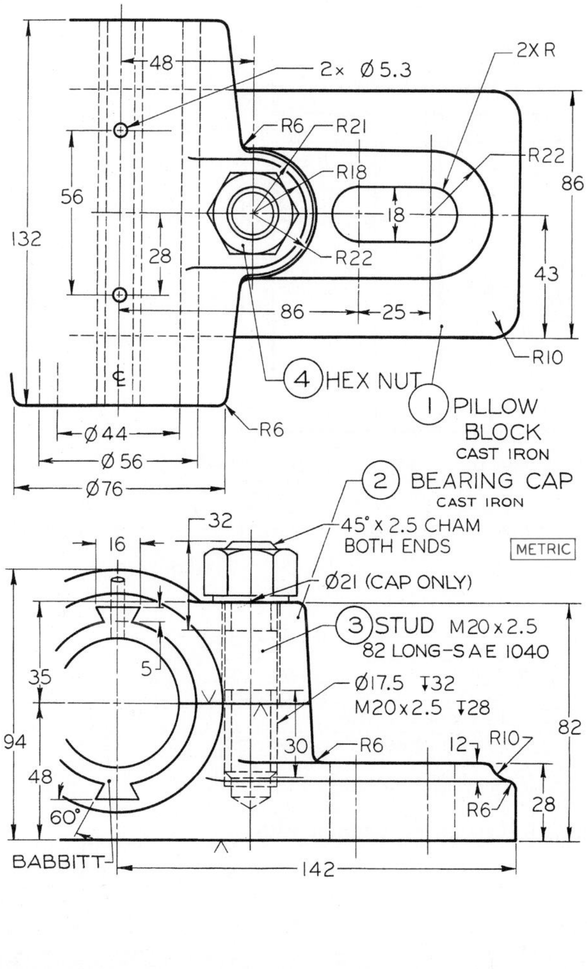

48
2× Ø 5.3
2X R

R6 R21
R18
R22
86
18
56
28
R22
43
86 25
132
R10

⑥ HEX NUT

① PILLOW
BLOCK
CAST IRON

② BEARING CAP
CAST IRON

Ø 44
Ø 56
Ø 76
R6

45° × 2.5 CHAM
BOTH ENDS
Ø21 (CAP ONLY)

METRIC

32
16
③ STUD M20×2.5
82 LONG-SAE 1040
Ø17.5 ⌀32
M20×2.5 ⌀28
35
5
94
48
60°
30 R6 12 R10
R6
82
28
142
BABBITT

FIGURE 14.72 Pillow Block Bearing.
Prob. 14.44: (1) Draw details. (2) Draw assembly. If
assigned, complete with dimensions.

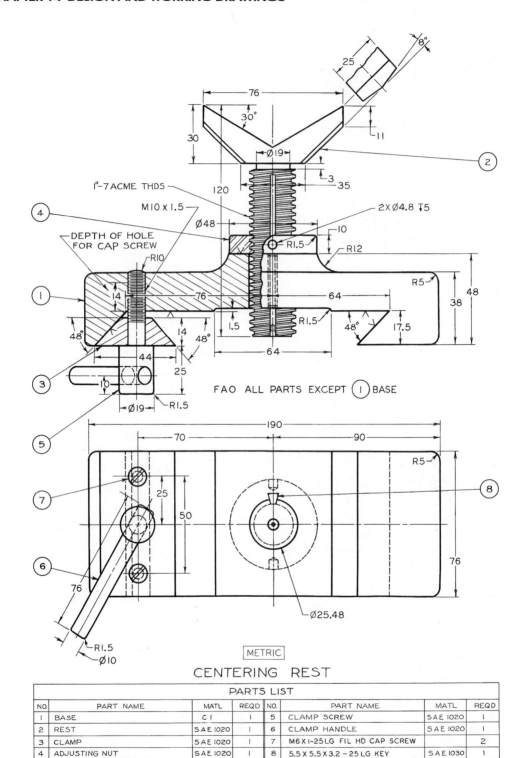

FAO ALL PARTS EXCEPT ①BASE

METRIC

CENTERING REST

NO.	PART NAME	MATL	REQD	NO.	PART NAME	MATL	REQD
				PARTS LIST			
1	BASE	C I	1	5	CLAMP SCREW	SAE 1020	1
2	REST	SAE 1020	1	6	CLAMP HANDLE	SAE 1020	1
3	CLAMP	SAE 1020	1	7	M6 X 1-25 LG FIL HD CAP SCREW		2
4	ADJUSTING NUT	SAE 1020	1	8	5.5 X 5.5 X 3.2 – 25 LG KEY	SAE 1030	1

FIGURE 14.73 **Centering Rest.**
Prob. 14.45: (1) Draw details. (2) Draw assembly. If assigned, complete with dimensions.

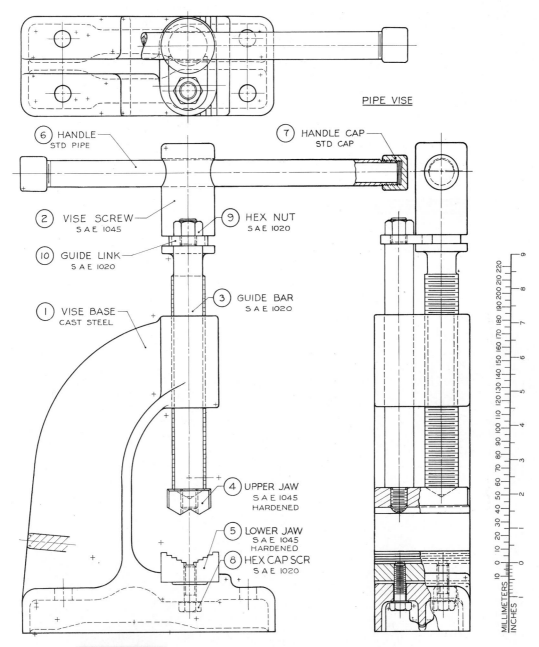

PIPE VISE

6 HANDLE
STD PIPE

7 HANDLE CAP
STD CAP

2 VISE SCREW
S.A.E. 1045

9 HEX NUT
S.A.E. 1020

10 GUIDE LINK
S.A.E. 1020

1 VISE BASE
CAST STEEL

3 GUIDE BAR
S.A.E. 1020

4 UPPER JAW
S.A.E. 1045
HARDENED

5 LOWER JAW
S.A.E. 1045
HARDENED

8 HEX CAP SCR
S.A.E. 1020

MILLIMETERS
INCHES

FIGURE 14.74   Pipe Vise.

Prob. 14.46: (1) Draw details. (2) Draw assembly. To obtain dimensions, take distances directly from figure with dividers; then set dividers on printed scale and read measurements in millimeters or decimal inches as assigned. All threads are general-purpose metric threads (see Appendix 15) or Unified coarse threads except the American National Standard pipe threads on handle and handle caps.

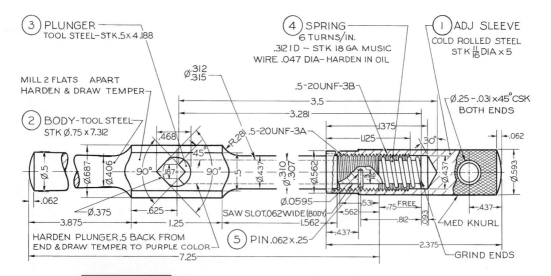

FIGURE 14.75 Tap Wrench.
Prob. 14.47: (1) Draw details. (2) Draw assembly. If assigned, use
unidirectional two-place decimals for all fractional dimensions or redesign
for metric dimensions.

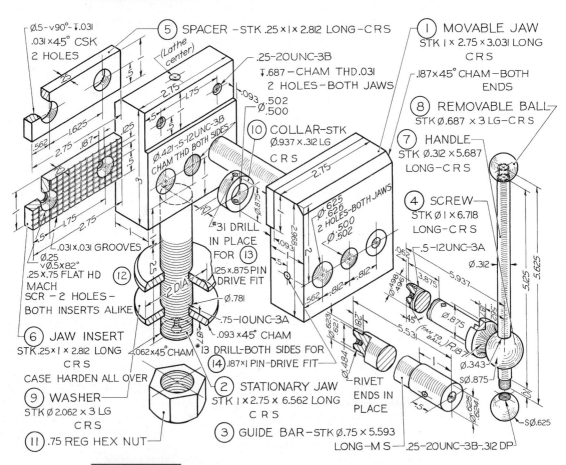

FIGURE 14.76 Machinist's Vise.
Prob. 14.48: (1) Draw details. (2) Draw assembly. If assigned, use
unidirectional two-place decimals for all fractional dimensions or redesign
for metric dimensions.

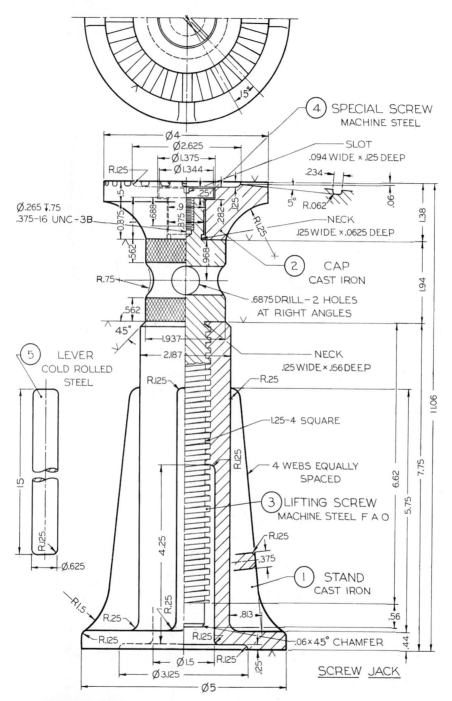

FIGURE 14.77   Screw Jack.

Prob. 14.49: (1) Draw details. See Fig. 14.21, showing "boxed-in" views on sheet layout C–678 or A2–678 (see inside front cover). (2) Draw assembly. If assigned, convert dimensions to decimal inches or redesign for metric dimensions.

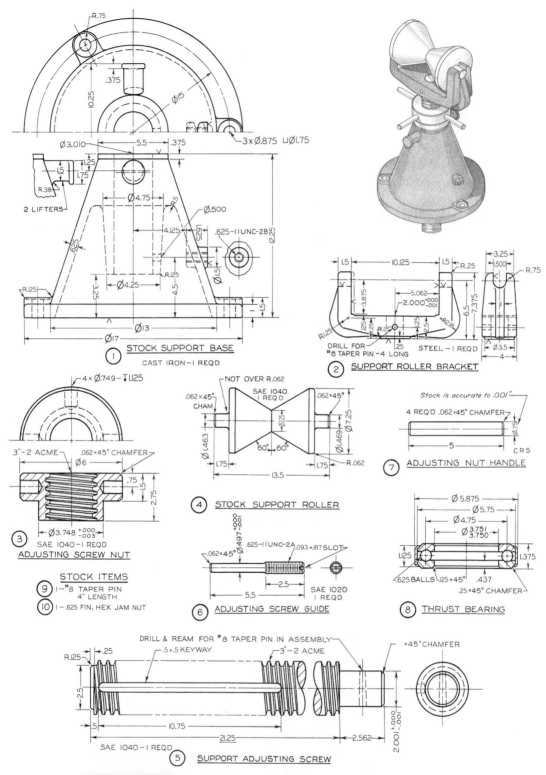

FIGURE 14.78   **Stock Bracket for Cold Saw Machine.**
Prob. 14.50: (1) Draw details. (2) Draw assembly. If assigned, use
unidirectional decimal dimensions or redesign for metric dimensions.

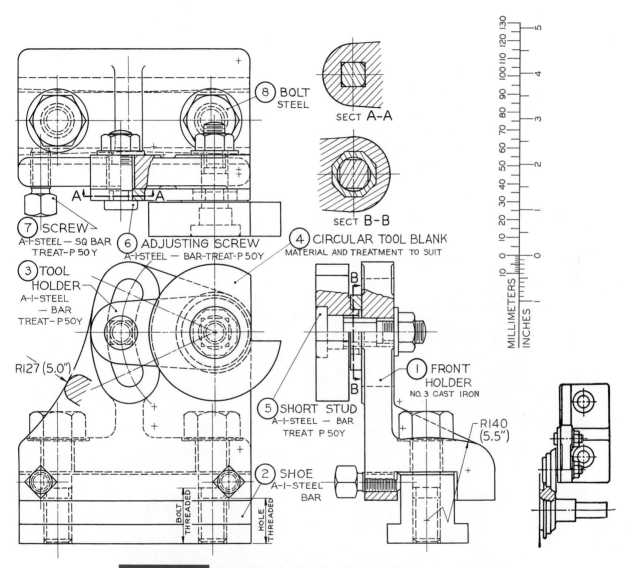

FIGURE 14.79   **Front Circular Forming Cutter Holder.**
Prob. 14.51: (1) Draw details. (2) Draw assembly. To obtain dimensions, take
distances directly from figure with dividers and set dividers on printed scale.
Use metric or decimal-inch dimensions as assigned.

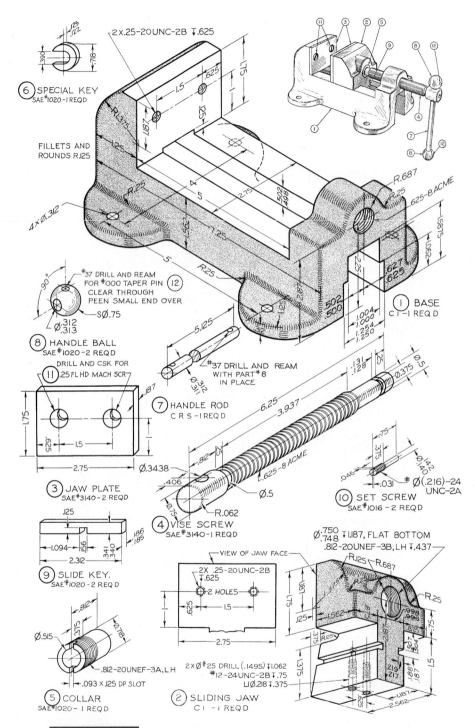

FIGURE 14.80 **Machine Vise.**
**Prob. 14.52: (1) Draw details. (2) Draw assembly. If assigned, convert
dimensions to the decimal-inch system or redesign with metric dimensions.**

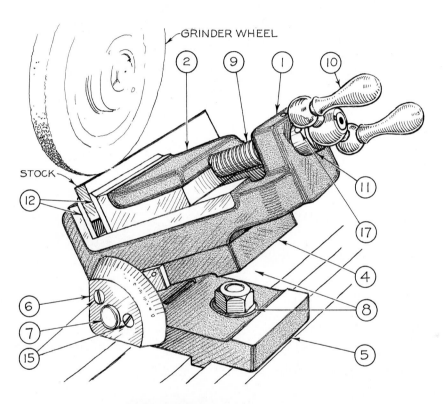

GRINDER WHEEL

STOCK

FIGURE 14.81   Grinder Vise.
Prob. 14.53: See Figs. 14.82 and 14.83.

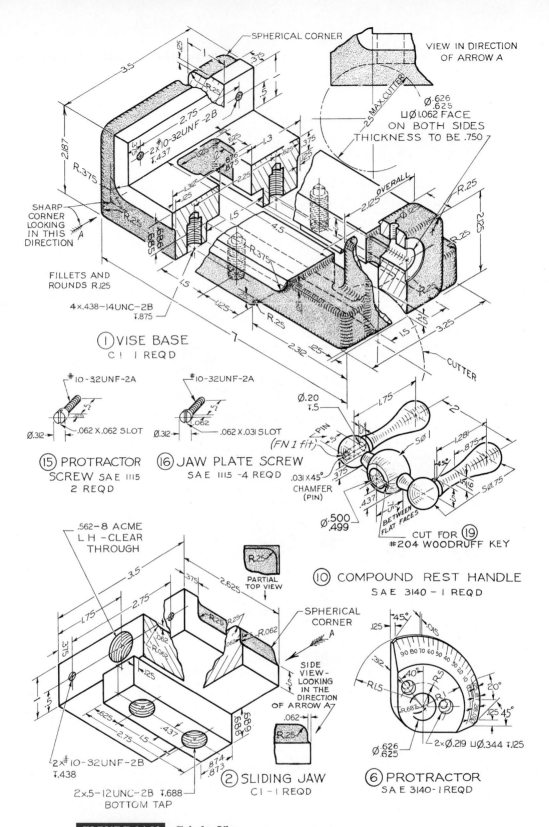

SPHERICAL CORNER

VIEW IN DIRECTION
OF ARROW A

Ø.626 / .625
⊔Ø1.062 FACE
ON BOTH SIDES
THICKNESS TO BE .750

2×#10-32UNF-2B
T.437

SHARP
CORNER
LOOKING
IN THIS
DIRECTION

FILLETS AND
ROUNDS R.125

4×.438-14UNC-2B
T.875

① VISE BASE
C I I REQD

#10-32UNF-2A
.5
Ø.312   .062 X .062 SLOT

#10-32UNF-2A
.5
Ø.312   .062 X .031 SLOT

⑮ PROTRACTOR
SCREW SAE 1115
2 REQD

⑯ JAW PLATE SCREW
SAE 1115 -4 REQD

Ø.20
T.5

(FN 1 fit)

.031 X 45°
CHAMFER
(PIN)

Ø.500 / .499

CUT FOR ⑲
#204 WOODRUFF KEY

⑩ COMPOUND REST HANDLE
SAE 3140 – I REQD

.562-8 ACME
L H - CLEAR
THROUGH

PARTIAL
TOP VIEW

SPHERICAL
CORNER

SIDE
VIEW-
LOOKING
IN THE
DIRECTION
OF ARROW A

2×#10-32UNF-2B
T.438

2×.5-12UNC-2B T.688
BOTTOM TAP

② SLIDING JAW
C I - I REQD

Ø.626 / .625

2×Ø.219 ⊔Ø.344 T.125

⑥ PROTRACTOR
SAE 3140-I REQD

FIGURE 14.82 Grinder Vise.
**Prob. 14.53, continued: (1) Draw details. (2) Draw assembly. See Figs. 14.81
and 14.83. If assigned, convert dimensions to decimal inches or redesign with
metric dimensions.**

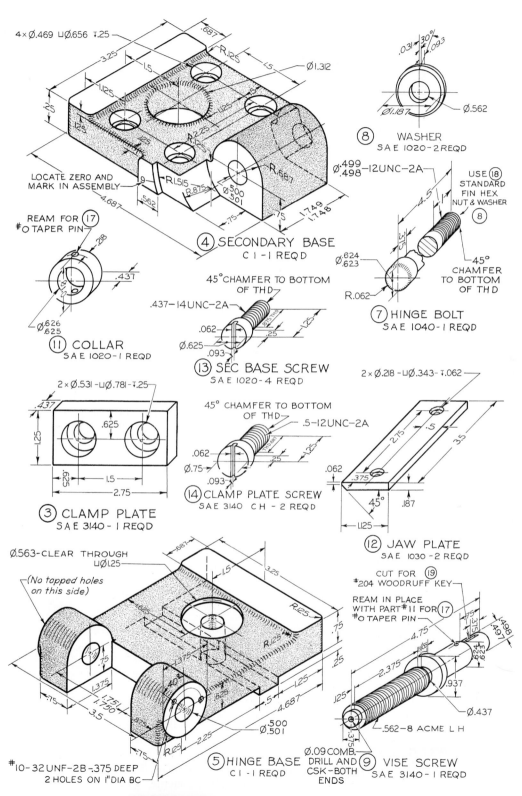

4 × Ø.469 ⊔Ø.656 ⬇.25
.687
3.25
1.5
1.5   Ø1.312
R.125
.25
1.125
.125
R2.25
R.515
.562
4.687
9
.562
R.875
Ø.500
.501
.75
.75
1.749
1.748

LOCATE ZERO AND
MARK IN ASSEMBLY

④ SECONDARY BASE
C I - 1 REQD

.031   30°   .093
Ø1.187   Ø.562

⑧ WASHER
S A E 1020 - 2 REQD

Ø.499
.498 -12UNC-2A
4.5
.375
Ø.624
.623
R.062

USE ⑱
STANDARD
FIN HEX
NUT & WASHER
⑧
45°
CHAMFER
TO BOTTOM
OF THD

⑦ HINGE BOLT
S A E 1040 - 1 REQD

REAM FOR ⑰
#O TAPER PIN
.28
.437
Ø.626
.625

⑪ COLLAR
S A E 1020 - 1 REQD

45° CHAMFER TO BOTTOM
OF THD
.437 - 14UNC - 2A
.062
Ø.625
.093
1.25
.25

⑬ SEC BASE SCREW
S A E 1020 - 4 REQD

45° CHAMFER TO BOTTOM
OF THD
.5 - 12UNC - 2A
.062
Ø.75
.093
1.25
.25

⑭ CLAMP PLATE SCREW
SAE 3140   CH - 2 REQD

2 × Ø.531 ⊔Ø.781 ⬇.25
.437
1.25
.625
.625
1.5
2.75

③ CLAMP PLATE
S A E 3140 - 1 REQD

2 × Ø.218 - ⊔Ø.343 - ⬇.062
2.75   .5
3.5
.062
.375
45°   .187
1.125

⑫ JAW PLATE
S A E 1030 - 2 REQD

Ø.563 - CLEAR THROUGH
⊔1.25
(No tapped holes
on this side)
.687
1.5
3.25
R.125
.75
R.125
.25
1.375
1.375
.75
.40
1.751
1.750
3.5
.625
1.25
.5
4.687
Ø.500
.501
2.25
R.125
R.125
.75

#10-32UNF-2B -.375 DEEP
2 HOLES ON 1"DIA BC

⑤ HINGE BASE
C I - 1 REQD

Ø.09 COMB.
DRILL AND
CSK-BOTH
ENDS

CUT FOR ⑲
#204 WOODRUFF KEY
REAM IN PLACE
WITH PART #11 FOR ⑰
#O TAPER PIN
4.75
.75
.375
Ø.629
.498
.497
2.375
.937
.125
Ø.437
.562-8 ACME L H
.375

⑨ VISE SCREW
S A E 3140 - 1 REQD

FIGURE 14.83   Grinder Vise.
Prob. 14.53, continued: See Fig. 14.82 for instructions.

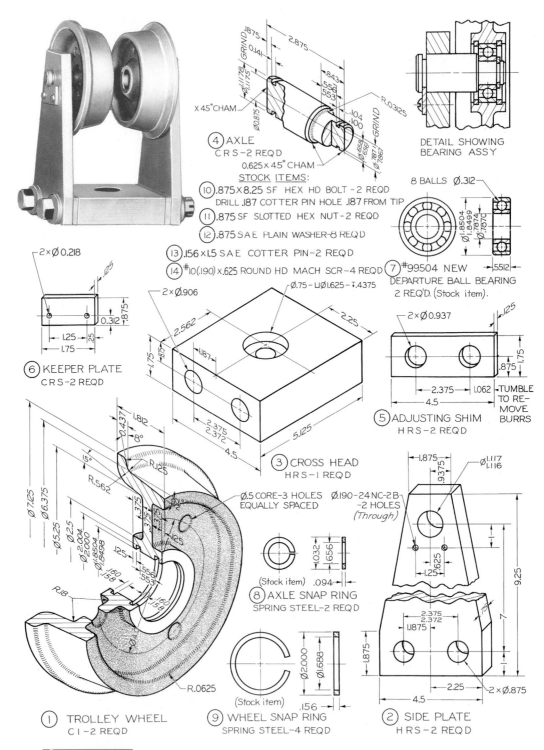

FIGURE 14.84 Trolley.
Prob. 14.54: (1) Draw details, omitting parts 7–14. (2) Draw assembly. If assigned, convert dimensions to decimal inches or redesign for metric dimensions.

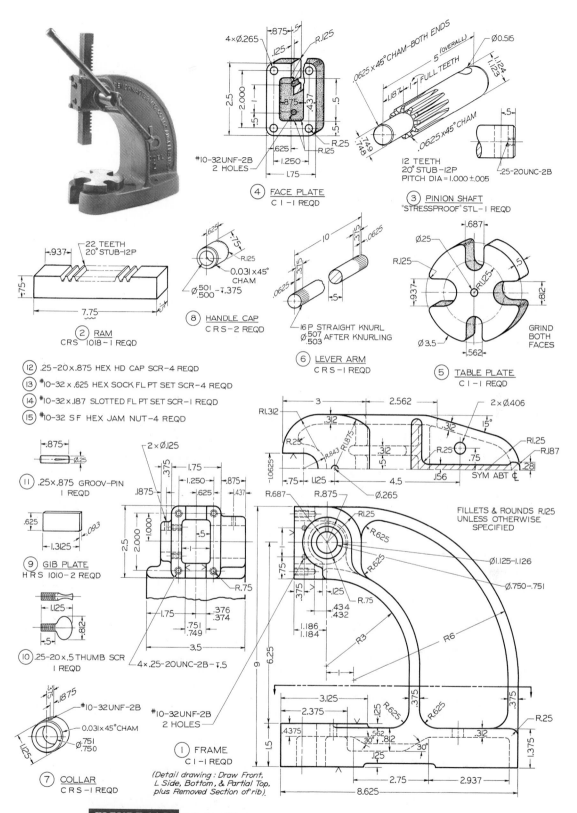

FIGURE 14.85 **Arbor Press.**

Prob. 14.55: (1) Draw details. (2) Draw assembly. If assigned, convert dimensions to decimal inches or redesign for metric dimensions.

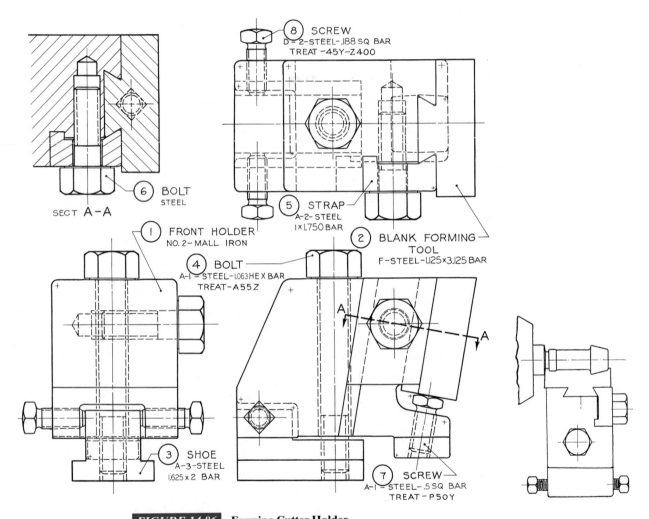

FIGURE 14.86 Forming Cutter Holder.
Prob. 14.56: (1) Draw details using decimal or metric dimensions. (2) Draw assembly. Above layout is half size. To obtain dimensions, take distances directly from figure with dividers and double them. At left is shown the top view of the forming cutter holder in use on the lathe.

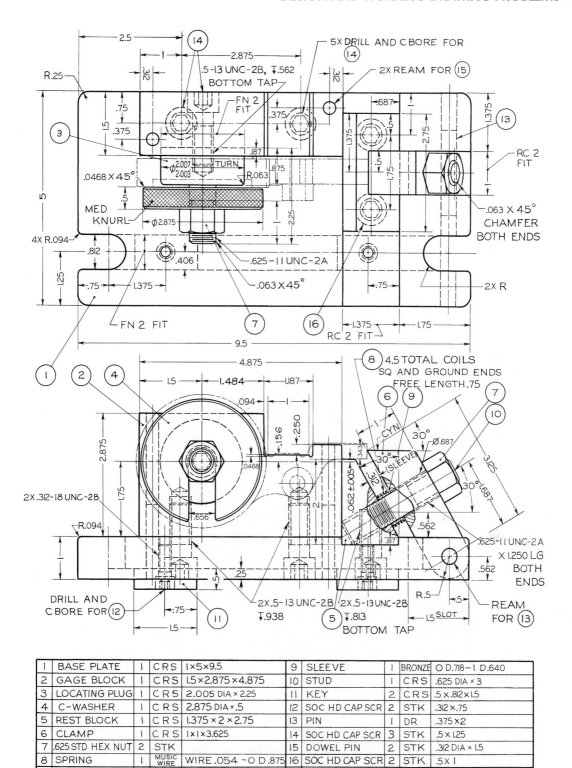

Item	NAME	Amt	MATL	REMARKS	Item	NAME	Amt	MATL	REMARKS
1	BASE PLATE	1	CRS	1×5×9.5	9	SLEEVE	1	BRONZE	O D.718−1 D.640
2	GAGE BLOCK	1	CRS	1.5×2.875×4.875	10	STUD	1	CRS	.625 DIA × 3
3	LOCATING PLUG	1	CRS	2.005 DIA × 2.25	11	KEY	2	CRS	.5 ×.812×1.5
4	C-WASHER	1	CRS	2.875 DIA×.5	12	SOC HD CAP SCR	2	STK	.312 ×.75
5	REST BLOCK	1	CRS	1.375 × 2 × 2.75	13	PIN	1	DR	.375 ×2
6	CLAMP	1	CRS	1×1×3.625	14	SOC HD CAP SCR	3	STK	.5 × 1.25
7	.625 STD HEX NUT	2	STK		15	DOWEL PIN	2	STK	.312 DIA × 1.5
8	SPRING	1	MUSIC WIRE	WIRE .054 − O D.875	16	SOC HD CAP SCR	2	STK	.5×1
Item	NAME	Amt	MATL	REMARKS	Item	NAME	Amt	MATL	REMARKS

FIGURE 14.87   Milling Fixture for Clutch Arm.
Prob. 14.57: (1) Draw details using the decimal-inch system or redesign for
metric dimensions, if assigned. (2) Draw assembly.

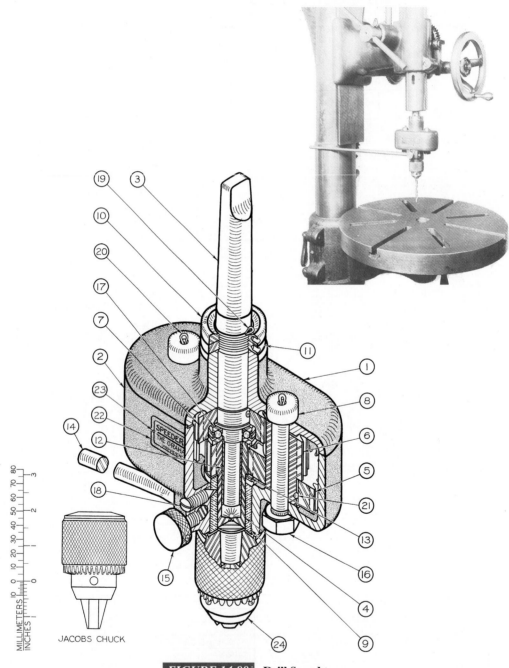

JACOBS CHUCK

FIGURE 14.88 Drill Speeder.
Prob. 14.58: See Figs. 14.89 and 14.90.

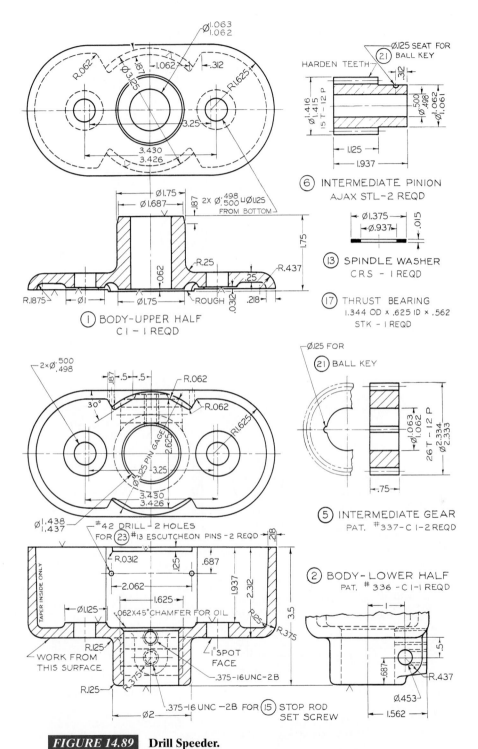

FIGURE 14.89   **Drill Speeder.**
Prob. 14.58, continued: (1) Draw details. (2) Draw assembly. See Fig. 14.88. If assigned, convert dimensions to decimal inches or redesign with metric dimensions.

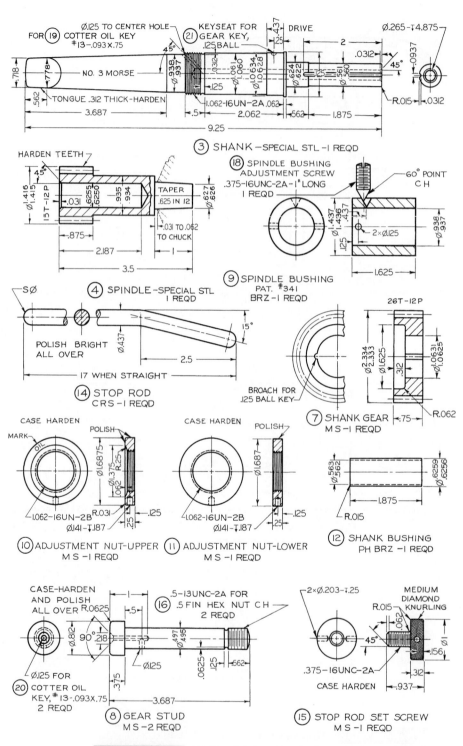

FIGURE 14.90 Drill Speeder.
Prob. 14.58, continued: See Fig. 14.89 for instructions.

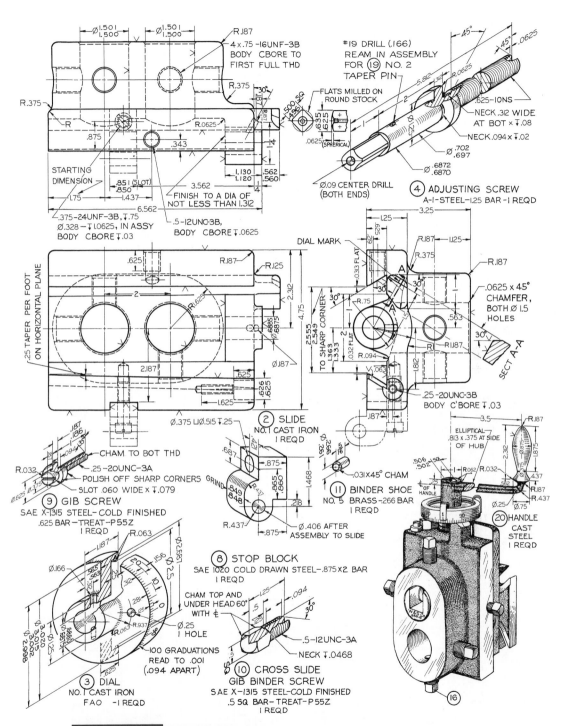

FIGURE 14.91 **Vertical Slide Tool.**
Prob. 14.59: (1) Draw details. If assigned, convert dimensions to decimal inches or redesign for metric system. (2) Draw assembly. For part 2: Take given top view as front view in the new drawing; then add top and right-side views. See also Fig. 14.92. If assigned, use unidirectional dimensions.

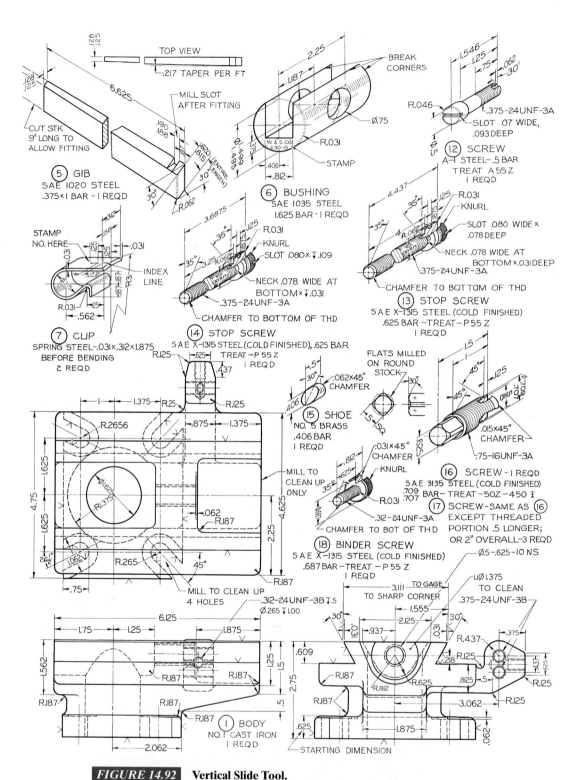

FIGURE 14.92 Vertical Slide Tool.
Prob. 14.59, continued: See Fig. 14.91 for instructions. For part 1: Take top
view as front view in the new drawing; then add top and right-side views.

FIGURE 14.93 Slide Tool.
Prob. 14.60: Make assembly drawing. See Figs. 14.95–14.97.

TOTAL ON MACH.	**NO. PCS.**	**NAME OF PART**	**PART NO.**	**CAST FROM PART NO.**	**TRACING NO.**	**MATERIAL**	**ROUGH WEIGHT PER PC.**	**DIA.**	**LENGTH**	**MILL**	**PART USED ON**	**NO. REQ. FINISH**
1		Body	219-12		D-17417	A-3-S D F						
	1	Slide	219-6		D-19255	A-3-S D F					219-12	
	1	Nut	219-9		E-19256	#10 BZ					219-6	
	1	Gib	219-1001		C-11129	S A E 1020					219-6	
	1	Slide Screw	219-1002		C-11129	A-3-S					219-12	
	1	Dial Bush.	219-1003		C-11129	A-1-S					219-1002	
	1	Dial Nut	219-1004		C-11129	A-1-S					219-1002	
	1	Handle	219-1011		E-18270	(Buy from Cincinnati Ball Crank Co.)					219-1002	
	1	Stop Screw (Short)	219-1012		E-51950	A-1-S					219-6	
	1	Stop Screw (Long)	219-1013		E-51951	A-1-S					219-6	
	1	Binder Shoe	219-1015		E-51952	#5 Brass					219-6	
	1	Handle Screw	219-1016		E-62322	X-1315 C.F.					219-1011	
	1	Binder Screw	219-1017		E-63927	A-1-S					219-6	
	1	Dial	219-1018		E-39461	A-1-S					219-1002	
	2	Gib Screw	219-1019		E-52777	A-1-S		$\frac{1}{4}$-20	1		219-6	
	1	Binder Screw	280-1010		E-24962	A-1-S					219-1018	
	2	Tool Clamp Screws	683-F-1002		E-19110	D-2-S					219-6	
	1	Fill Hd Cap Scr	1-A			A-1-S		$\frac{3}{8}$	1 $\frac{3}{8}$		219-6 219-9	
	1	Key	No.404 Woodruff								219-1002	

PARTS LIST NO. OF SHEETS __2__ SHEET NO. __1__ MACHINE NO. _M-219_

NAME __NO. 4 SLIDE TOOL (SPECIFY SIZE OF SHANK REQ'D.)__ LOT NUMBER

NO. OF PIECES

FIGURE 14.94 **Prob. 14.61: Slide Tool Parts List.**

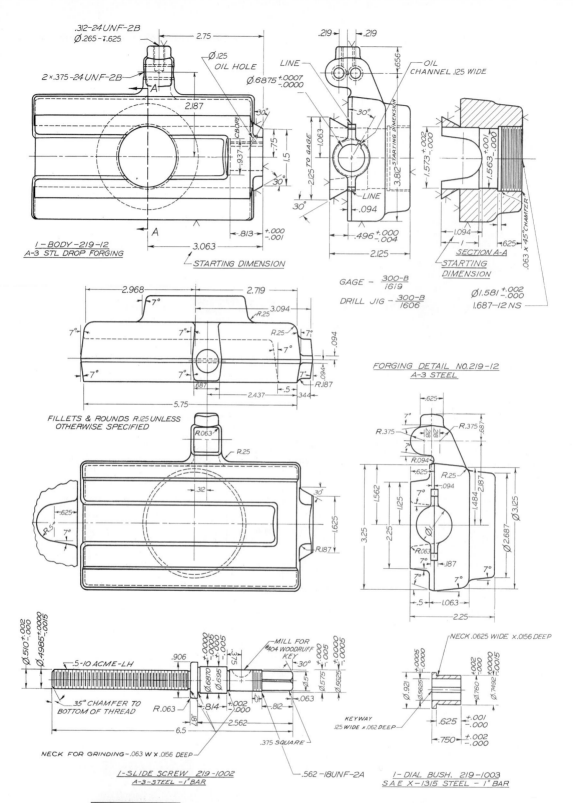

FIGURE 14.95 Slide Tool.
Prob. 14.61, continued: (1) Draw details using decimal-inch dimensions or
redesign with metric dimensions, if assigned. (2) Draw assembly. See
Fig. 14.93.

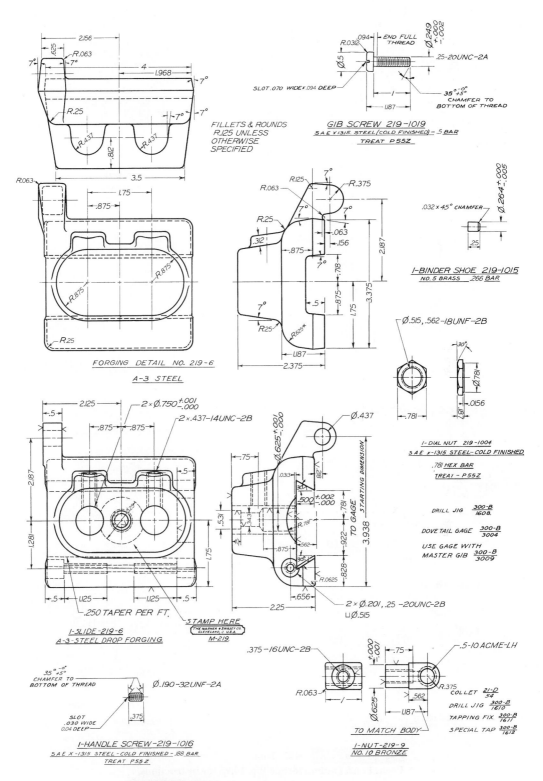

FIGURE 14.96 Slide Tool.
Prob. 14.61, continued: See. Fig. 14.95 for instructions.

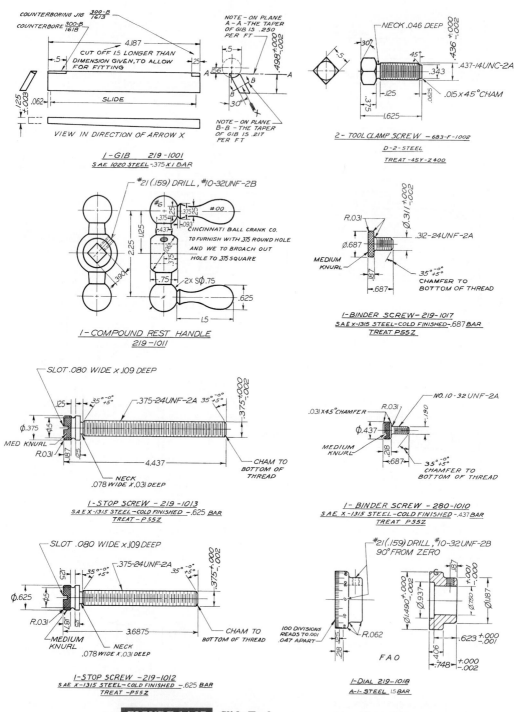

FIGURE 14.97 Slide Tool.
Prob. 14.61, continued: See Fig. 14.95 for instructions.

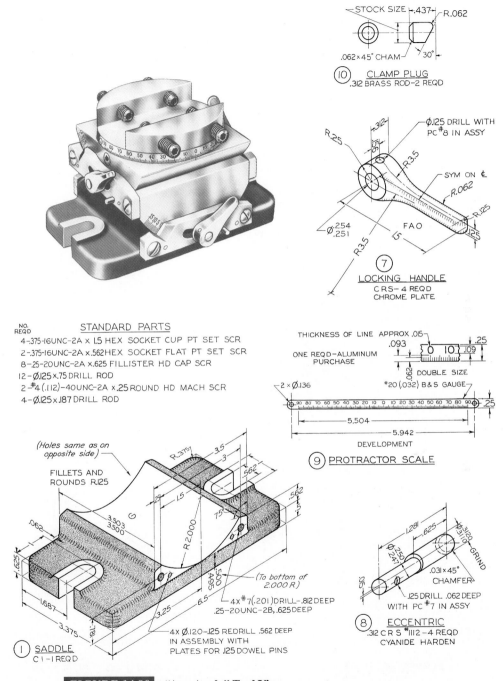

STOCK SIZE — .437 — R.062

.062 × 45° CHAM — 30°

⑩   CLAMP PLUG
.312 BRASS ROD–2 REQD

Ø.125 DRILL WITH
PC #8 IN ASSY

R.25
.312
.156
R.3.5
SYM ON ₵
R.062
R.125
Ø.254
.251
FAO   .5
R.3.5

⑦
LOCKING HANDLE
C RS– 4 REQD
CHROME PLATE

NO.
REQD   STANDARD PARTS
4–.375–16UNC–2A × 1.5 HEX SOCKET CUP PT SET SCR
2–.375–16UNC–2A ×.562 HEX SOCKET FLAT PT SET SCR
8–.25–20UNC–2A ×.625 FILLISTER HD CAP SCR
12–Ø.125 ×.75 DRILL ROD
2–#4 (.112)–40UNC–2A ×.25 ROUND HD MACH SCR
4–Ø.125 ×.187 DRILL ROD

THICKNESS OF LINE APPROX .015
.25
.093   .109
ONE REQD–ALUMINUM
PURCHASE
.062
DOUBLE SIZE
2 × Ø.136   #20 (.032) B&S GAUGE
.25
5.504
5.942
DEVELOPMENT

⑨ PROTRACTOR SCALE

(Holes same as on
opposite side)

FILLETS AND
ROUNDS R.125

R.375
3.5
3
.562
.062
.25
1.5
.562
.5
.5
3.503
3.500
R.2.000
75°
.062
.625
.500
.495
1.687
3.25
6.5
3.375
.781
4× #7 (.201) DRILL–.812 DEEP
.25–20UNC–2B, .625 DEEP
(To bottom of
2.000 R)
4× Ø.120–.125 REDRILL .562 DEEP
IN ASSEMBLY WITH
PLATES FOR .125 DOWEL PINS

① SADDLE
C I –1 REQD

1.281
.625
Ø.3120 GRIND
Ø.250
.247
.031 × 45°
CHAMFER
.125 DRILL .062 DEEP
WITH PC #7 IN ASSY
.25

⑧   ECCENTRIC
.312 C R S #1112 –4 REQD
CYANIDE HARDEN

FIGURE 14.98   "Any Angle" Tool Vise.
**Prob. 14.62: (1) Draw details using decimal-inch dimensions or redesign with
metric dimensions, if assigned. (2) Draw assembly. See also Fig. 14.99.**

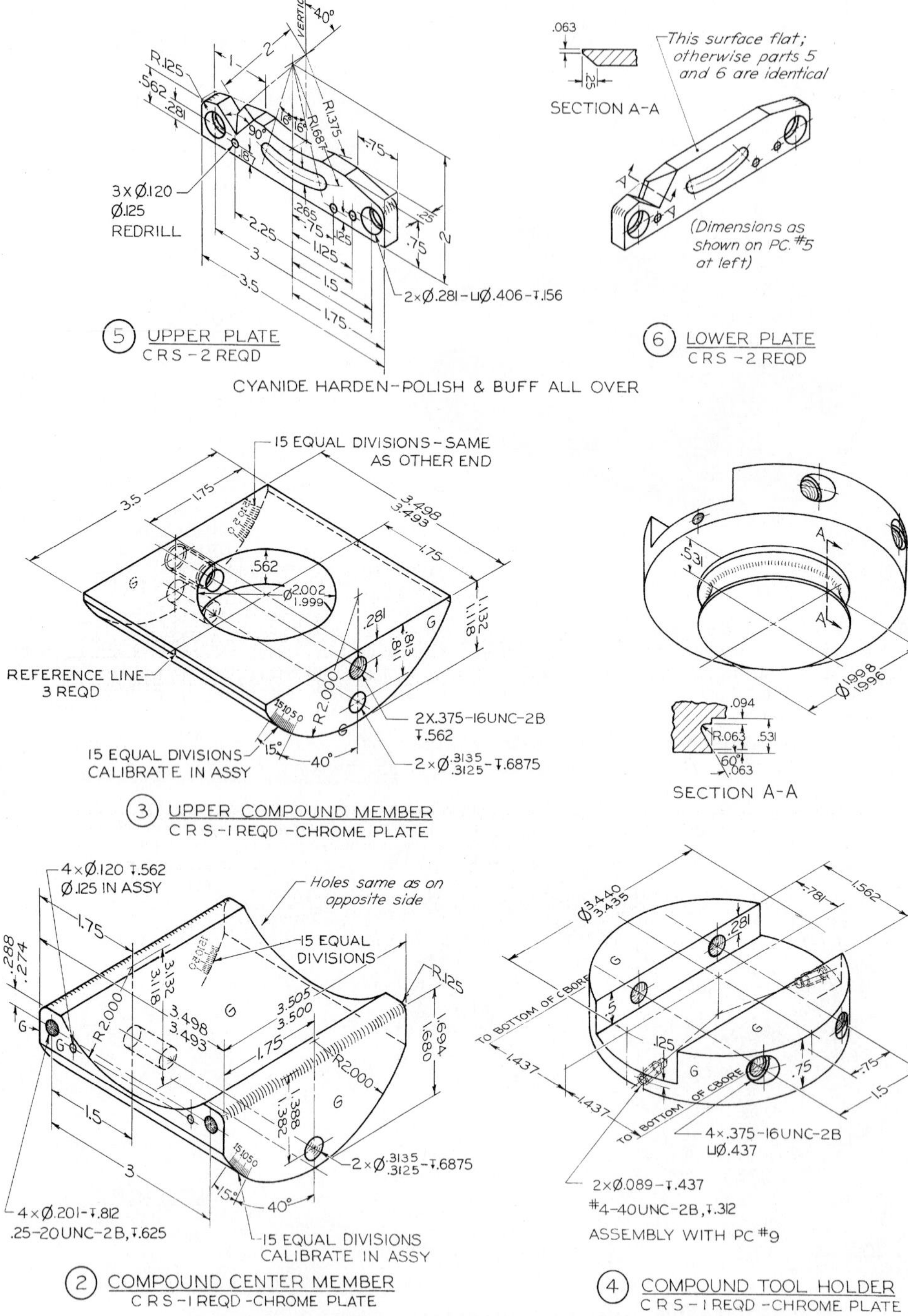

5 UPPER PLATE
CRS-2 REQD
3 x Ø.120
Ø.125
REDRILL
2xØ.281-⌴Ø.406-↧.156
6 LOWER PLATE
CRS-2 REQD
This surface flat; otherwise parts 5 and 6 are identical
SECTION A-A
(Dimensions as shown on PC. #5 at left)
CYANIDE HARDEN-POLISH & BUFF ALL OVER
15 EQUAL DIVISIONS-SAME AS OTHER END
REFERENCE LINE-3 REQD
15 EQUAL DIVISIONS-CALIBRATE IN ASSY
2X.375-16UNC-2B ↧.562
3 UPPER COMPOUND MEMBER
CRS-1 REQD-CHROME PLATE
SECTION A-A
4xØ.120 ↧.562
Ø.125 IN ASSY
Holes same as on opposite side
15 EQUAL DIVISIONS
4xØ.201-↧.812
.25-20UNC-2B,↧.625
15 EQUAL DIVISIONS
CALIBRATE IN ASSY
2 COMPOUND CENTER MEMBER
CRS-1 REQD-CHROME PLATE
TO BOTTOM OF CBORE
4x.375-16UNC-2B
⌴Ø.437
2xØ.089-↧.437
#4-40UNC-2B,↧.312
ASSEMBLY WITH PC #9
4 COMPOUND TOOL HOLDER
CRS-1 REQD-CHROME PLATE

FIGURE 14.99 **"Any-Angle" Tool Vise.**
Prob. 14.62, continued: See Fig. 14.98 for instructions.

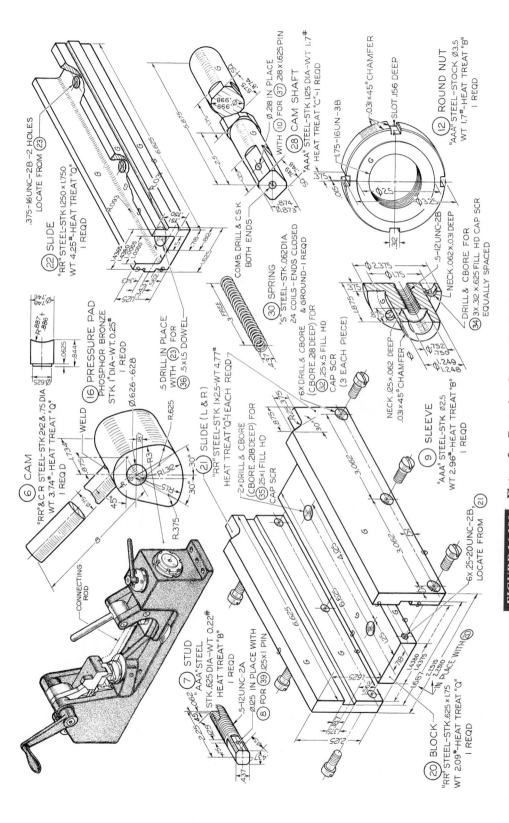

FIGURE 14.100 **Fixture for Centering Connecting Rod.**
Prob. 14.63: (1) Draw details using decimal-inch dimensions or redesign with metric dimensions, if assigned. (2) Draw assembly. See also Figs. 14.101 and 14.102.

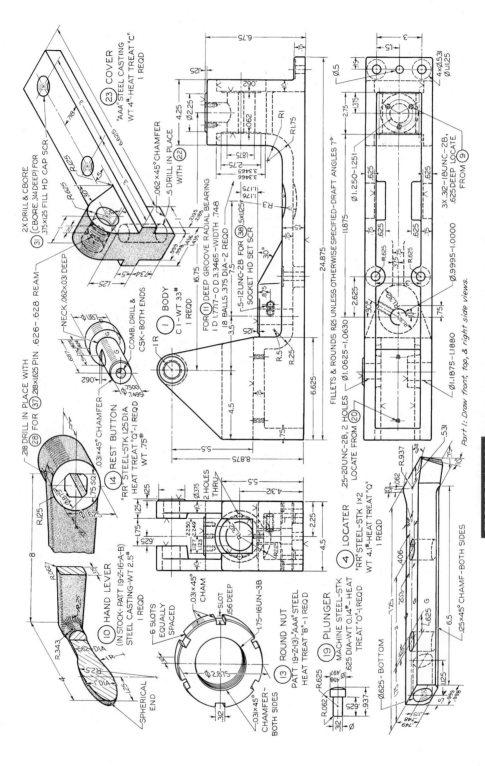

FIGURE 14.101 **Fixture for Centering Connecting Rod.**
Prob. 14.63, continued: See Fig. 14.100 for instructions.

Part 1: Draw front, top, & right side views.

494

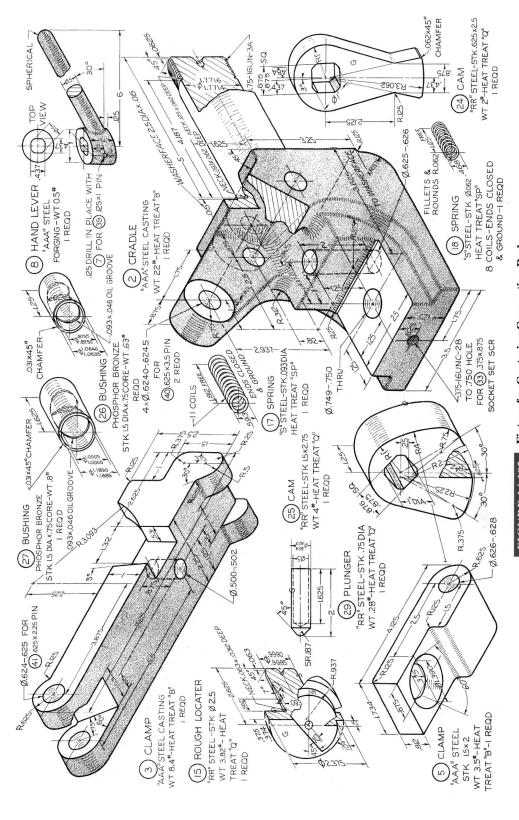

TOP VIEW — SPHERICAL

⑧ HAND LEVER
"AAA" STEEL
FORGING—WT 0.5#
I REQD

.125 DRILL IN PLACE WITH
⑦ FOR ㉟ .125×1 PIN
.093×.046 OIL GROOVE

㉖ BUSHING
PHOSPHOR BRONZE
STK 1.5 DIA×.75CORE—WT .63#
REQD

② CRADLE
"AAA" STEEL CASTING
WT 22#—HEAT TREAT "B"
I REQD

⑳ BUSHING
PHOSPHOR BRONZE
STK 1.5 DIA×.75CORE—WT .8#
I REQD
.093×.046 OIL GROOVE

㉗ CLAMP
"AAA" STEEL CASTING
WT 8.4#—HEAT TREAT "B"
I REQD

⑮ ROUGH LOCATER
"RR" STEEL—STK Ø 2.5
WT 3.82#—HEAT
TREAT "Q"
I REQD

⑰ SPRING
"S" STEEL—STK.093DIA
HEAT TREAT "SP"
I REQD

㊵ .625×3.5 PIN
2 REQD

㉕ CAM
"RR" STEEL—STK 1.5×2.75
WT 4#—HEAT TREAT "Q"
I REQD

㉙ PLUNGER
"RR" STEEL—STK .75 DIA
WT .28#—HEAT TREAT "Q"
I REQD

㉔ CAM
"RR" STEEL—STK.625×2.5
WT 2#—HEAT TREAT "Q"
I REQD

⑱ SPRING
"S" STEEL—STK .062
HEAT TREAT "SP"
8 COILS—ENDS CLOSED
& GROUND—I REQD

.375-16UNC-2B
TO .750 HOLE
FOR ㉝ .375×.875
SOCKET SET SCR

⑤ CLAMP
"AAA" STEEL
STK 3.5#
WT 3.5#—HEAT
TREAT "B"—I REQD

FIGURE 14.102 Fixture for Centering Connecting Rod.
Prob. 14.63, continued: See Fig. 14.100 for instructions.

495

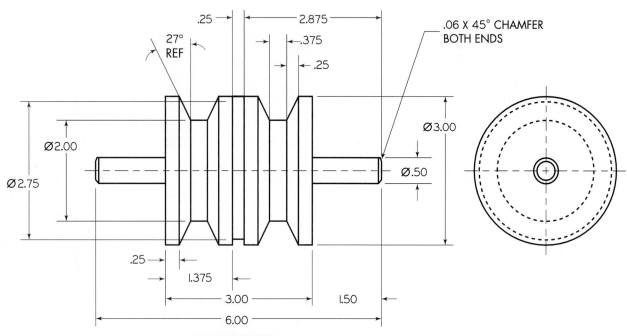

FIGURE 14.103   Alignment Wheel.
Prob. 14.64. (1) Draw top and left-side views.
(2) Redesign with metric dimensions.

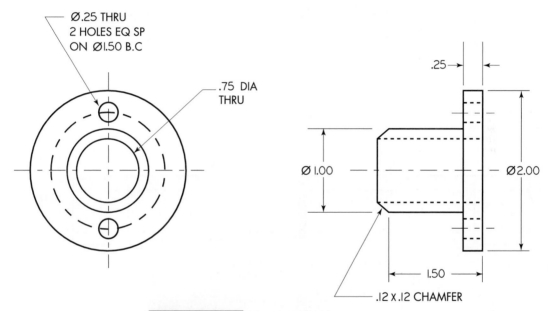

Ø.25 THRU
2 HOLES EQ SP
ON Ø1.50 B.C

.75 DIA
THRU

.25

Ø 1.00

Ø2.00

1.50

.12 X .12 CHAMFER

FIGURE 14.104   **Rubber Bushing.**
Prob. 14.65. (1) Draw top and left-side views.
(2) Redesign with metric dimensions.

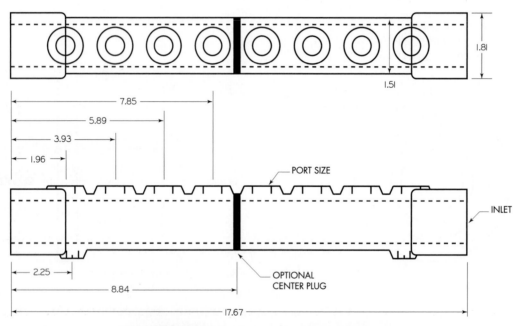

FIGURE 14.105 Problem 14.66. 8-Port Nylon
Manifold. Draw bottom and left-side view. Redesign
with metric dimensions.

CHAPTER 15

REPRODUCTION AND CONTROL OF DRAWINGS

OBJECTIVES

After studying the material in this chapter, you should be able to:

1. Describe blueprinting and digital engineering printing.

2. Discuss the differences between Diazo-Moist and Diazo-Dry Processes.

3. Understand the differences between CD-ROM and Optical Disk Storage Systems.

4. Discuss how the processes of Xerography and fax technology work.

5. Discuss how documents may be handled in a "paperless office."

OVERVIEW

An essential part of designers' or drafters' education is a thorough knowledge of reproduction techniques and processes and drawing storage. Specifically, they should be familiar with the various processes available for the reproduction of drawings: *blueprint, diazo, microfilm, microfiche,* etc.

Because the average engineering drawing requires a considerable economic investment, adequate control and protection of the original drawing are mandatory. Such items as drawing numbers, §15.13, methods of filing, microfilming, security files, print making and distribution, drawing changes, and retrieval of drawings are all important. A proper drawing-control system will enable those in charge of drawings (1) to know the location and status of the drawing at all times; (2) to minimize the damage to original drawings from the handling required for revisions, printing, and so on; and (3) to provide distribution of prints to proper persons.

Those organizations with computer-aided design (CAD) systems often use computer storage of finished drawings. In addition, digitized drawing information about frequently required components and elements, such as standard bolts, nuts, screws, pins, and piping valves, is stored in the computer for recall and replacement on drawings as needed. CAD systems of this type have obvious advantages for production, storage, control, and recall of drawings. However, design documentation requirements generally dictate that all of the approved drawing revisions be stored in a permanent record. Many companies fail to properly store the design history of their drawings in a suitable fashion. Because CAD drawings stored on a hard drive or tape can be edited, they are not considered a permanent record. WORM (write once read many) CD-ROM drives make a good permanent storage medium for CAD drawing. Most small and medium-size companies and a substantial number of large firms may not be able to justify the acquisition of high-capacity permanent storage systems for their CAD documents and thus will continue to use the conventional methods of reproduction, storage, retrieval, and control of drawings described in this chapter. But technologies such as WORM CD-ROMs are becoming more compact and less expensive, and as advancing technology makes such systems more affordable, it is expected that many companies will make the change to storing their permanent drawings on a computer storage system.

■ 15.1 STORAGE OF DRAWINGS

Drawings may be stored flat in large flat-drawer files or hung vertically in cabinets especially designed for the purpose. Exceptionally large drawings are often rolled and stored in tubes in racks or cabinets. Prints are often folded and stored in standard office file cases. Proper control procedures will enable the user of the drawing to find it in the file, to return it to its proper place, and to know where the drawing is when not in the file.

■ 15.2 REPRODUCTION OF DRAWINGS

After the drawings of a machine or structure have been completed, it is usually necessary to supply copies to many different persons and firms. Obviously, therefore, some means of exact, rapid, and economical reproduction must be used. Even when a drawing is plotted from a CAD system, it is frequently faster to use a reproduction process to distribute the drawing. Plots can be created on mylar, or on specialty papers that allow prints to be made directly from the plot. Large format copy machines are also popular for creating distribution drawings of a print. Networks, groupware, internet and intranet, and modems are allowing many workplaces to go to a paperless office, where drawings are distributed electronically (Fig. 15.1). However, when there is a need to take drawings into the field, printing is still frequently required.

■ 15.3 BLUEPRINT PROCESS AND DIGITAL ENGINEERING PRINTING/COPYING

Of the several processes in use for reproduction, the *blueprint* is the oldest process used for making prints of large drawings. It is essentially a photographic process in which the original drawing is the negative.

The blueprint process, which was the common method used for the reproduction of drawings for many years, has now been replaced to a large extent by other more convenient and efficient processes.

High-speed digital engineering printers, such as the one shown in Fig. 15.2, are most widely used in the engineering and manufacturing industry. Such printers can produce output from either CAD files, hard copy, or a mixture of both. These printers can also produce multi-media products and digital sorting for automated set production.

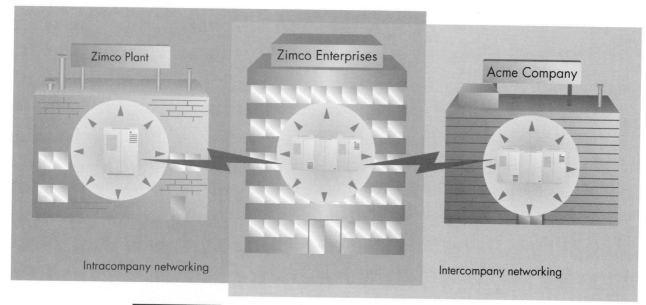

FIGURE 15.1   **Intracompany and Intercompany Networking.** *From Computers, 4/E by Long/Long. © Copyright 1996. Reprinted by permission of Prentice-Hall, Inc.*

FIGURE 15.2   **Digital Engineering Printer.** *Courtesy of Océ-USA, Inc., Chicago, IL.*

■ 15.4 DIAZO-MOIST PRINTS

A black-and-white print, composed of nearly black lines on a white background, may be made from ordinary pencil or ink tracings by exposure in the same manner as for blueprints, directly on special blackprint paper, cloth, or film.

Exposure may be made in a blueprint machine or any machine using light in a similar way. However, the prints are not washed as in blueprinting, but must be fed through a special developer that dampens the coated side of the paper with a developing solution.

Colored-line prints in red, brown, or blue lines on white backgrounds may be made on the same machine simply by using the appropriate paper in each case. These prints, together with diazo-dry prints, §15.5, are still used today, put recent improvements in digital printing has made that the most common form of outputting today.

■ 15.5 THE DIAZO-DRY PROCESS

The *diazo-dry* process is based on the sensitivity to light of certain dyestuff intermediates that have the characteristic of decomposing into colorless substances if exposed to ultraviolet light, and of reacting with coupling components to form an azo dyestuff upon exposure to ammonia vapors. It is a contact method of reproduction, and depends on the transmission of light through the original for the reproduction of positive prints. The subject matter may be pen or pencil lines, typewritten or printed matter, or any opaque image. There is no negative step involved; positives are used to obtain positive prints. Sensitized materials can be handled under normal illumination.

The diazo whiteprint method of reproduction consists of two simple steps—exposure and dry development by means of ammonia vapors. Exposure is made in a printer equipped with a source of ultraviolet light, a mercury vapor lamp, fluorescent lamp, or carbon arc. The light emitted by these light sources brings about a

FIGURE 15.3 **Diazit Omnitrac EFP Model Machine.** *Courtesy of Diazit Company, Inc.*

photochemical decomposition of the light-sensitive yellow coating of the paper except in those places where the surface is protected by the opaque lines of the original. The exposed print is developed dry in a few seconds in a dry-developing machine by the alkaline medium produced by ammonia vapors.

A popular combination printer (exposer) and developer, Diazit Omnitrac printer, is shown in Fig. 15.3. The tracing and the sensitized paper are fed into the machine, and when they emerge, the print is practically dry and ready for use.

Another exposer and developer combined in one machine is the Ozalid Whiteprinter. Two operations are involved: (1) the tracing and the sensitized paper are fed into the printer slot for exposure to light, as shown in Fig. 15.4 (a), and (2) the paper is then fed through the developer slot for exposure to ammonia vapors, as shown at (b). If it is desired to remove the ammonia odor completely, the print is then fed through the printer with the back of the sheet next to the warm glass surrounding the light.

■ 15.6 XEROGRAPHY

Xerox prints are positive prints with black lines on a white background (Fig. 15.5). A selenium-coated and electrostatically charged plate is used. A special camera is used to project the original onto the plate; hence, reduced or enlarged reproductions are possible. A negatively charged plastic powder is spread across the

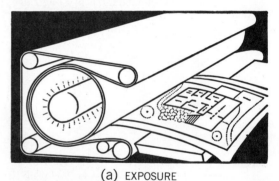

(a) EXPOSURE

(b) AMMONIA DEVELOPMENT

FIGURE 15.4 **Exposure and Development.** *Courtesy of Ozalid Corporation.*

plate and adheres to the positively charged areas of the image. The powder is then transferred to paper by means of a positive electric charge and is baked onto the surface to produce the final print. Full-sized prints

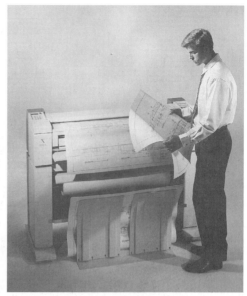

FIGURE 15.5 Xerox Engineering Document Systems. *Courtesy of the Xerox Corporation.*

or reductions can be made inexpensively and quickly in the fully automated Xerox or other similar copy machines. The process is dry and sensitive to all types of copy. The Xerox process is used also to produce mats for the Multilith offset duplicating method, §15.12.

Recent application of xerography includes volume print making from original drawings or microfilms, §15.16.

■ 15.7 FAX TECHNOLOGY

Telecopiers, also called facsimile or fax machines, Fig. 15.10, can receive or send documents (usually $8\frac{1}{2}'' \times 11''$) over standard telephone lines in the office or in the field. Most fax machines can transmit data at 9,600 bits per second (bps). After the telephone circuit is established, the document is fed into the sending machine. The copy is read and translated into signals for the receiving machine, which reproduces the document. A computer can send and receive documents directly, generally as a raster or bitmap type image, avoiding the need for a paper copy of the drawing.

■ 15.8 DIGITAL IMAGE PROCESSING

Modern digital techniques have made possible the direct production of drawings on a laser printer from a variety of input sources, including computers and electronic video equipment.

In addition to drawings produced with the assistance of a CAD program, Chapter 2, conventional hand-produced drawings may be stored in computer memory through the use of digital scanning techniques or by manually digitizing the drawing. Also, color laser copiers can reproduce drawings in four colors, with black lettering (Fig. 15.6), and some are available with optional built-in computer processing unit and monitor, video player, and film projector to permit convenient viewing and editing of drawings.

■ 15.9 RECORDABLE CD-ROM STORAGE SYSTEMS

Recordable CD-ROM systems (Fig. 15.7) let you store digital information such as CAD drawings, digital audio and video, data, multimedia projects, and other digitally stored records. Write-once-read-many (WORM) CD-ROM storage devices provide excellent storage for CAD documentation. Once the CD has been written it cannot be erased or re-written. The shelf life for storage of the media is at least 100 years, so it qualifies as an archival media for permanent storage. CDs are compact and easy to store, and CD-ROM players are standard equipment on many CAD systems. Another advantage of CD-ROM systems is that they are random access storage systems, so that you can go directly to the document you wish to retrieve, unlike tape systems, which must wind through all of the previous tape. Systems that can automatically retrieve from a selection of multiple CDs, called juke

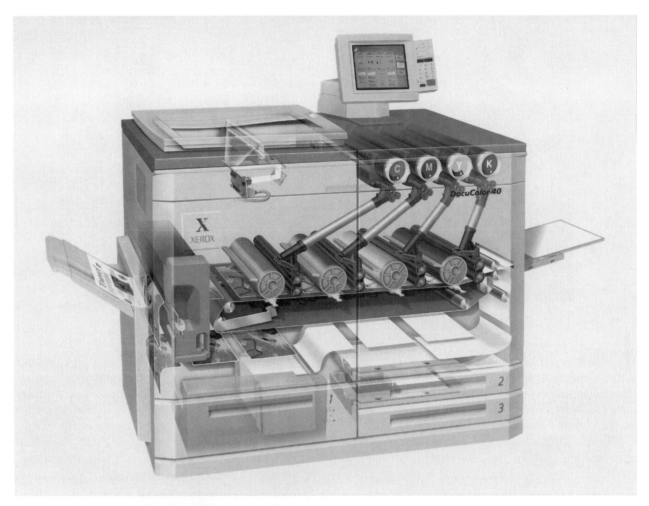

FIGURE 15.6   **Internal Components of a Xerox Digital Color Production System.** *Courtesy of the Xerox Corporation.*

box systems, are available for quickly retrieving documents in a network storage situation.

■ 15.10 OPTICAL DISK STORAGE SYSTEMS

Optical disk storage systems, Fig. 15.8, use optical magnetic media to store capacities up to 4.6 GB (gigabytes) on a single removable disk. They are rewritable media. This means that they are not suitable for archival storage of permanent records. However, the fast retrieval speeds of up to 6 MB/sec, which is as fast as most hard drives, low cost per megabyte of storage, thirty year shelf life, and high capacities make them a useful storage system for CAD documentation.

■ 15.11 DOCUMENT MANAGEMENT SOFTWARE

Specialized software is available to help manage document revision history, approval, storage, file naming, and other issues of managing digital documentation, such as that produced by CAD systems. Automanager Workflow™® is one such software package. In order to have an effective document management system, a lot of planning and setup needs to be done to ensure success. A software package alone will not provide instant success in managing the large number of files and meeting legal requirements for document storage. It requires setup time, training, and on-going effort to make it effective. If you are unsure how long different

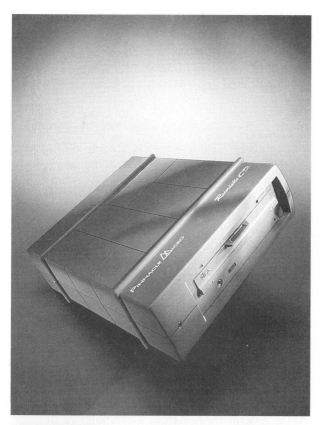

FIGURE 15.7   **This Pinnacle Recordable CD-Rom System is Priced for Market Sales to both the Consumer and Commercial.** *Courtesy of Pinnacle Micro, Inc.*

FIGURE 15.8   **An Affordable Pinnacle Optical Hard Drive and Disk.** *Courtesy of Pinnacle Micro, Inc.*

documents must be retained, ARMA (the American Records Management Association) is a good place to go for information.

■ 15.12 OFFSET PRINTING

The Photolith, Multilith, and Planograph methods are generally known as *offset printing*. A camera is used to reproduce the original, enlarged or reduced if necessary, upon an aluminum or zinc sheet. This master plate is then mounted on a rotary drum that revolves in contact with a rubber roller that picks up the ink from the image and transfers it to the paper. The prints are excellent positive reproductions.

■ 15.13 PHOTOGRAPHIC CONTACT PRINTS

Either transparent or opaque drawings may be reproduced the same size as the original by means of *contact printing*. The original is pressed tightly against a sheet of special photographic paper, either by mechanical spring pressure or by means of suction as in the "vacuum printer." The paper is exposed by the action of the transmitted or reflected light, and the print is developed in the manner of a photograph in a dark or semi-dark room.

Excellent duplicate tracings can be made on paper, on film, and on either opaque map cloth or transparent tracing cloth through this process. Poor pencil drawings or tracings can be duplicated and improved by intensifying the lines so as to be much better than the original. Pencil drawings can be transformed into "ink-line" tracings. Also, by this process, a reproduction can be made directly from an existing print.

■ 15.14 PHOTOSTATS

The Visual Graphics Total Camera II, Fig. 15.9, is essentially a highly specialized camera. It is a completely modular daylight graphic reproduction system that includes a precision black-and-white stat camera with self-contained lighting, automatic focusing and processing, and programmable memory and microprocessor control. Various plug-in modular components are also available for special-purpose photo reproductions, such as enlargements or reductions, line or halftone stats, offset plates, reverses, dropouts, color prints and film transparencies, and other special effects for a multitude of applications.

A photostat print may be the same size or larger or smaller than the original, while photographic

FIGURE 15.9 **Visual Graphics Total Camera II.**
Courtesy of Visual Graphics Corporation.

contact prints, §15.13, must be the same size. The original may be transparent or opaque. It is simply fastened in place, the camera is adjusted to obtain the desired type and size of print, and the print is made, developed, and dried in the machine (no darkroom is required). For black-and-white stats, the result is a negative print with white lines on a near-black background. A positive print having near-black lines on a white background is made by photostating the negative print.

■ 15.15 LINE ETCHING

Line etching is a photographic method of reproduction. The drawing, in black lines on white paper or on tracing cloth or film, is placed in a frame behind a glass and photographed. This photographic negative is then mounted on a pane of glass and is printed on a sheet of planished zinc or copper. After the print has been

specially treated to render the lines acid resistant, the plate is washed in a nitric acid solution, which eats away the metal between the lines, leaving them standing above the surface of the plate, like type. The plate is then mounted upon a hardwood base, which can be used in any printing press, as are blocks of type.

■ 15.16 MICROFILM AND MICROFICHE

A microfilm is a photographic image of information, records, or drawings that is stored on film at a greatly reduced scale.

Microfilm has come into considerable use, especially where large numbers of drawings are involved, to furnish complete duplication of all drawings in a small filing space. Microfilm makes a good permanent storage media because of its long archival period and its relatively small size. Microfilm also provides duplicate copies for security reasons. Enlargement copies can be made from the film through photographic, electrostatic, or photocopy methods. The films may be 16, 35, or 70 mm and may be produced in rolls, jacketed film, or individually mounted on aperture cards, Fig. 15.10.

The individual cards may be viewed in a reader and, if desired, a full-size print may be had from a reader-printer unit. Many readers use automated retrieval systems connected to a computer so that the operator only has to search for the desired record on the computer screen. Marks are created on the film as it is photographed to allow the computer system to index the exact spot on the film for retrieval.

Since the aperture equipment is essentially a standard data card used for electronic computers, mechanized equipment is capable of sorting, filing, and retrieving the cards. Duplicate aperture cards and reproductions of the drawings to the size desired may be

FIGURE 15.10 **A Standard Aperture Card. Courtesy of C. S. Barber.**

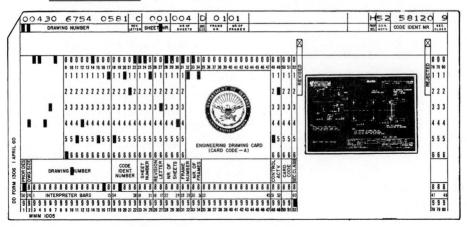

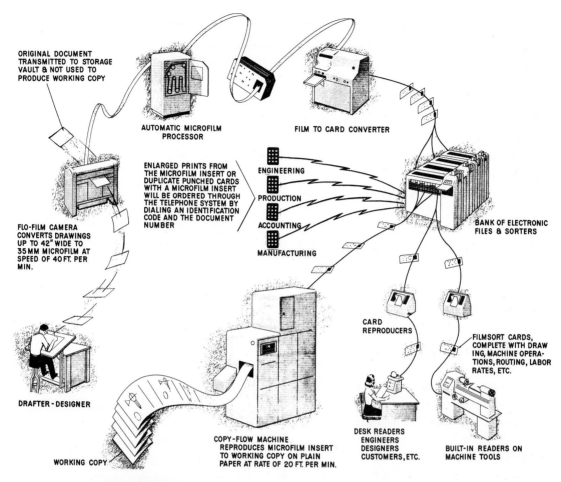

ORIGINAL DOCUMENT TRANSMITTED TO STORAGE VAULT & NOT USED TO PRODUCE WORKING COPY

AUTOMATIC MICROFILM PROCESSOR

FILM TO CARD CONVERTER

ENLARGED PRINTS FROM THE MICROFILM INSERT OR DUPLICATE PUNCHED CARDS WITH A MICROFILM INSERT WILL BE ORDERED THROUGH THE TELEPHONE SYSTEM BY DIALING AN IDENTIFICATION CODE AND THE DOCUMENT NUMBER

ENGINEERING

PRODUCTION

ACCOUNTING

MANUFACTURING

FLO-FILM CAMERA CONVERTS DRAWINGS UP TO 42" WIDE TO 35 MM MICROFILM AT SPEED OF 40 FT. PER MIN.

BANK OF ELECTRONIC FILES & SORTERS

CARD REPRODUCERS

FILMSORT CARDS, COMPLETE WITH DRAWING, MACHINE OPERATIONS, ROUTING, LABOR RATES, ETC.

DRAFTER-DESIGNER

WORKING COPY

COPY-FLOW MACHINE REPRODUCES MICROFILM INSERT TO WORKING COPY ON PLAIN PAPER AT RATE OF 20 FT. PER MIN.

DESK READERS ENGINEERS DESIGNERS CUSTOMERS, ETC.

BUILT-IN READERS ON MACHINE TOOLS

FIGURE 15.11 A Mechanized Engineering Data Handling System. *Courtesy of United Technology Center, Sunnyvale, CA.*

produced on specialized equipment. A typical mechanized engineering data handling system is shown in Fig. 15.11.

A *microfiche* is a cardlike film containing many rows of images or records or drawings. Card sizes used for storage are 3″ × 5″, 4″ × 6″, and 5″ × 8″. A typical 4″ × 6″ microfiche will contain the equivalent of 270 pages of information. The individual cards may be viewed on a reader and, if desired, a full-size copy may be made by using a reader-printer (Fig. 15.12).

Computer-output microfilm (COM) refers to a process used to produce drawings and records on microfilm, with the aid of a computer. A COM unit will produce a microfilm from database information converted to an image on a high-resolution screen that is then photographed. The main advantages of COM are storage capability and speed.

FIGURE 15.12 Microfilm Reader-Printer. *Courtesy of 3M Co.*

■ KEY WORDS

BLUEPRINT	DIGITAL IMAGING
DIAZO	OFFSET PRINTING
OPTICAL DISK STORAGE	MICROFILM
WORM	MICROFICHE
XEROGRAPHY	PHOTOSTAT

■ CHAPTER SUMMARY

• Technical drawings are detailed and often very expensive to create. It is important to undertand how to reproduce and manage them safely.

• Past management systems involved physically filing drawings. Most new systems now archive and distribute electronically.

• The blueprint, Diazo-Moist and Diazo-Dry processes of reproduction are older but still used in many companies.

• Xerography, Fax Technology and Digital Image Processing are processes now commonly used which allow great flexibility in control of the drawing.

• Microfilm and Microfiche methods store information on a greatly reduced scale and allow for large numbers of drawings to be stored in a small space.

• Digital printers have become the most widely used output devices in the engineering industry.

■ REVIEW QUESTIONS

1. How do the Diazo-Dry and the Diazo-Moist processes differ?

2. What are some of the benefits and the dangers of using an electronic system to transfer and archive drawings?

3. Explain how CD-ROMs are used in the field of Technical Drawing.

4. Explain how the Xerography process works.

5. Describe how your class archives and manages its drawings?

6. What are some advantages to using digital printers versus other output devices?

C H A P T E R   1 6

AXONOMETRIC PROJECTION

OBJECTIVES

After studying the material in this chapter, you should be able to:

1. Describe the differences between multiview projection, axonometric projection, oblique projection, and perspective.

2. Sketch examples of an isometric cube, a dimetric cube, and a trimetric cube.

3. List the advantages of multiview projection, axonometric projection, oblique projection, and perspective.

4. Create an isometric drawing given a multiview drawing.

5. Measure along each isometric axis.

6. Draw inclined and oblique surfaces in isometric.

7. Draw angles, ellipses, and irregular curves in isometric.

OVERVIEW

As described in Chapter 6, multiview drawing makes it possible to represent accurately the most complex forms of a design by showing a series of exterior views and sections. This type of representation has two limitations, however: Its execution requires a thorough understanding of the principles of multiview projection, and its reading requires a definite exercise of the constructive imagination. Although multiview drawings are common in most technical drawings, they do not show length, width, and height in a silngle view.

For communication of design ideas, it is often necessary to prepare accurate and scientifically correct than can be easily understood by persons without technical training. **Axonometric projection** rotates the object with respect to the observer so that all three dimensions can be seen in one view, approximately as they appear to an observer. These projections are often called *pictorial drawings* [ANSI/ASME Y14.4M–1989 (R1994)] because they look more like a picture than multiview drawings. Since pictorial drawing shows only the appearance of an object, it is not satisfactory for completely describing complex or detailed forms.

Various types of pictorial drawing are used extensively in catalogs, sales literature, and technical work to supplement and amplify multiview drawings. For example, pictorial drawing is used in Patent Office drawings; in piping diagrams; in machine, structural, and architectural designs; and in furniture design.

The most common axonometric projection is *isometric*, which means "equal measure." When a cube is drawn in isometric, the axes are equally spaced (120° apart). Though not as photorealistic as perspective drawings, isometric drawings are much easier to draw. CAD software is particularly effective at creating isometric drawings, which are often called 3D models. Dimetric and trimetric are more realistic than isometric but are much harder to draw and are therefore used less frequently in engineering work.

■ 16.1 METHODS OF PROJECTION

The four principal types of projection are illustrated in Fig. 16.1, and all except the regular multiview projection (Fig. 16.1a) are pictorial types since they show several sides of the object in a single view. In all cases, the views, or projections, are formed by the piercing points in the plane of projection of an infinite number of visual rays of projectors.

In both multiview projection and *axonometric projection* (Fig. 16.1b), the observer is considered to be at infinity, and the visual rays are parallel to each other and perpendicular to the plane of projection. Therefore, both are classified as *orthographic projections* (§1.10).

In *oblique projection* (Fig. 16.1c), the observer is considered to be at infinity, and the visual rays are parallel to each other but oblique to the plane of projection (see Chapter 17).

In *perspective* (Fig. 16.1d), the observer is considered to be at a finite distance from the object, and the visual rays extend form the observer's eye, or station point (SP), to all points of the object to form a "cone of rays" (see Chapter 18).

■ 16.2 TYPES OF AXONOMETRIC PROJECTION

The distinguishing feature of axonometric projection, as compared to multiview projection, is the inclined position of the object with respect to the plane of projection. Since the principal edges and surfaces of the object are inclined to the plane of projection, the lengths of the lines, the sizes of the angles, and the general proportions of the object vary with the infinite number of possible positions in which the object may be placed with respect to the plane of projection. Three of these are shown in Fig. 16.2.

In these cases the edges of the cube are inclined to the plane of projection and are therefore foreshortened (see Fig. 6.20c). The degree of foreshortening of any line depends on its angle with the plane of projection; the greater the angle, the greater the foreshortening. If the degree of foreshortening is determined for each of the three edges of the cube that meet at one corner, scales can be easily constructed for measuring along these edges or any other edges parallel to them (see Figs. 16.41a and 16.46).

It is customary to consider three edges of the cube that meet at the corner nearest the observer as the *axonometric axes*. In Fig. 16.1b, the axonometric axes, or simply the *axes*, are OA, OB, and OC. As shown in Fig. 16.2, axonometric projections are classified as (a) isometric projection, (b) dimetric projection, and (c) trimetric projection, depending on the number of scales of reduction required.

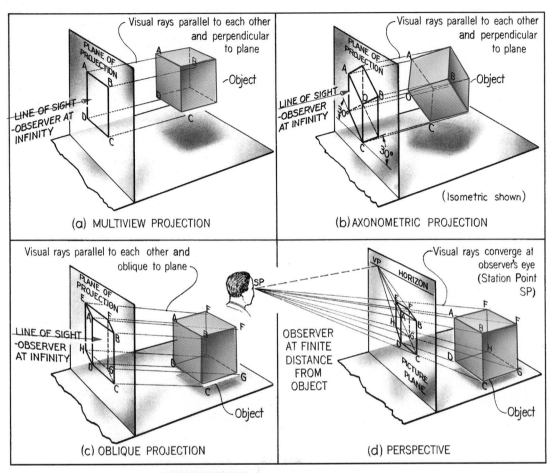

FIGURE 16.1   Four Types of Projection.

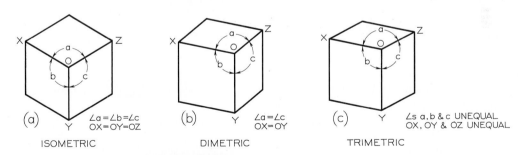

FIGURE 16.2   Axonometric Projections.

ISOMETRIC PROJECTION

■ 16.3 THE ISOMETRIC METHOD OF PROJECTION

An **isometric projection** as one in which all angles between the axonometric axes are equal. To produce an isometric projection (again, isometric means "equal measure"), it is necessary to place the object so that its principal edges, or axes, make equal angles with the plane of projection and are therefore foreshortened equally (see Fig. 5.21). In this position the edges of a cube would be projected equally and would

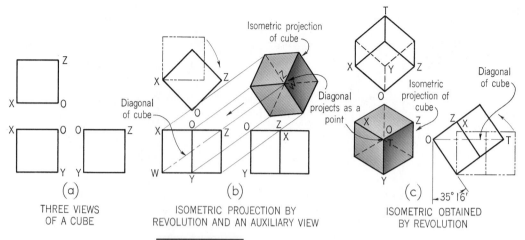

FIGURE 16.3 **Isometric Projection.**

make equal angles with each other (120°), as shown in Fig. 16.2a.

Figure 16.3a shows a multiview drawing of a cube. In Fig. 16.3b, the cube is shown revolved through 45° about an imaginary vertical axis. Now an auxiliary view in the direction of the arrow will show the cube diagonal ZW as a point, and the cube appears as a true isometric projection. The cube may be further revolved (Fig. 16.3c). This time the cube is titled forward about an imaginary horizontal axis until the three edges OX, OY, and OZ make equal angles with the frontal plane of projection and are therefore foreshortened equally. Here again, a diagonal of the cube, in this case OT, appears as a point in the isometric view. The front view thus obtained is a true isometric projection. In this projection the 12 edges of the cube make angles of about 35° 16′ with the frontal plane of projection. The lengths of their projections are equal to the lengths of the edges multiplied by $\sqrt{\frac{2}{3}}$, or by 0.816, approximately. Thus the projected lengths are about 80% of the true lengths, or still more roughly, about three fourths of the true lengths. The projections of the axes OX, OY, and OZ make angles of 120° with each other and are called the *isometric axes*. Any line parallel to one of these is called a *nonisometric line*. The angles in the isometric projection of the cube are either 120° or 60°, and all are projections of 90° angles. In an isometric projection of a cube, the faces of the cube, and any planes parallel to them, are called *isometric planes*.

■ 16.4 THE ISOMETRIC SCALE

A correct isometric projection may be drawn with the use of a special isometric scale, prepared on a strip of paper or cardboard (Fig. 16.4). All distances in the isometric scale are $\sqrt{\frac{2}{3}}$ times true size, or approximately 80% of true size. The use of the isometric scale is illustrated in Fig. 16.5a. A scale of 9″ = 1′-0, or $\frac{3}{4}$-size scale (or metric equivalent), could be used to approximate the isometric scale.

■ 16.5 ISOMETRIC DRAWING

When a drawing is prepared with an isometric scale, or when the object is actually *projected* on a plane of projection, it is an *isometric projection* (Fig. 16.5a). When a drawing is prepared with an ordinary scale, it is an *isometric drawing* (Fig. 16.5b). The isometric drawing is

FIGURE 16.4 **Isometric Scale.**

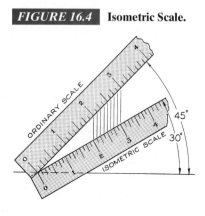

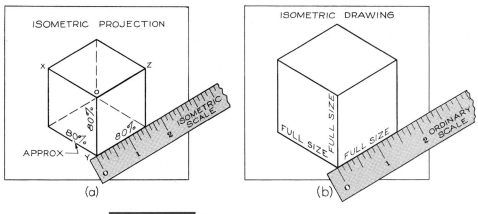

FIGURE 16.5   Isometric and Ordinary Scales.

about 25% larger than the isometric projection, but the pictorial value is obviously the same in both.

Since the isometric projection is foreshortened and an isometric drawing is full-scale size, it is usually advantageous to make an isometric drawing rather than an isometric projection. The drawing is much easier to execute and, for all practical purposes, is just as satisfactory as the isometric projection.

■ 16.6 MAKING AN ISOMETRIC DRAWING

The construction of an isometric drawing of an object composed only of normal surfaces (§6.19) is illustrated in Fig. 16.6. Notice that all measurements are made parallel to the main edges of the enclosing box—that is, parallel to the isometric axes. No measurement along a diagonal (nonisometric line) on any surface or through the object can be set off directly with the scale. The object may be drawn in the same position by beginning at the corner Y, or any other corner, instead of at the corner X.

The construction of an isometric drawing of an object composed partly of inclined surfaces (and oblique edges) is shown in Fig. 16.7. Notice that inclined surfaces are located by *offset*, or *coordinate*, *measurements* along the isometric lines. For example, dimensions E and F are set off to locate the inclined surface M, and dimensions A and B are used to locate surface N.

Sketching in isometric is discussed in §§5.11–5.13.

■ 16.7 OBLIQUE SURFACES IN ISOMETRIC

Oblique surfaces in isometric may be drawn by establishing the intersections of the oblique surface with the isometric planes. For example, in Fig. 16.8a, the oblique plane is known to contain points A, B, and C. To establish the plane (Fig. 16.8b), line AB is extended to X and Y, points that are in the same isometric planes as C. Lines XC and YC locate points E and F. Finally, AD and ED are drawn, using the rule of parallelism of lines. The completed drawing is shown in Fig. 16.8c.

■ 16.8 OTHER POSITIONS OF THE ISOMETRIC AXES

The isometric axes may be placed in any desired position according to the requirements of the problem (Fig. 16.9), but the angle between the axes must remain 120°. The choice of the directions of the axes is determined by the position from which the object is usually viewed (Fig. 16.10), or by the position that best describes the shape of the object. If possible, both requirements should be met.

If the object is characterized by considerable length, the long axis may be placed horizontally for best effect (Fig. 16.11).

■ 16.9 OFFSET LOCATION MEASUREMENTS

The method of locating one point with respect to another is illustrated in Figs. 16.12 and 16.13. In each

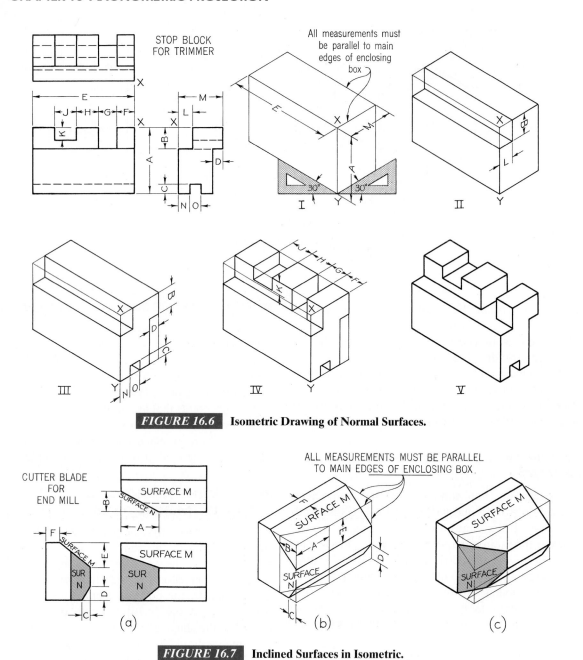

<u>FIGURE 16.6</u> **Isometric Drawing of Normal Surfaces.**

<u>FIGURE 16.7</u> **Inclined Surfaces in Isometric.**

case, after the main block has been drawn, the offset lines CA and BA in the multiview drawing are drawn full size in the isometric drawing, thus locating corner A of the small block or rectangular recess. These measurements are called *offset measurements*, and since they are parallel to certain edges of the main block in the multiview drawings, they will be parallel, respec-

tively, to the same edges in the isometric drawings (see §6.25).

■ 16.10 HIDDEN LINES

The use of hidden lines in isometric drawing is governed by the same rule as in all other types of projec-

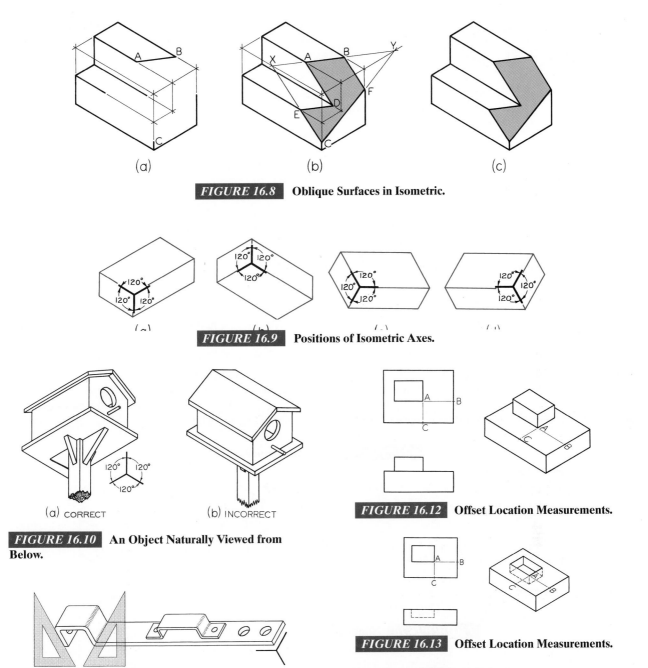

FIGURE 16.8   Oblique Surfaces in Isometric.

FIGURE 16.9   Positions of Isometric Axes.

(a) CORRECT   (b) INCORRECT

FIGURE 16.10   An Object Naturally Viewed from Below.

FIGURE 16.11   Long Axis Horizontal.

FIGURE 16.12   Offset Location Measurements.

FIGURE 16.13   Offset Location Measurements.

■ 16.11 CENTER LINES

The use of center lines in isometric drawing is governed by the same rules as in the multiview drawing: *Center lines are drawn if they are needed to indicate symmetry or if they are needed for dimensioning* (Fig. 16.14). In general, center lines should be used sparingly and omitted in cases of doubt. The use of too many

tion: *Hidden lines are omitted unless they are needed to make the drawing clear.* Figure 16.14 shows a case in which hidden lines are needed because a projecting part cannot be clearly shown without them.

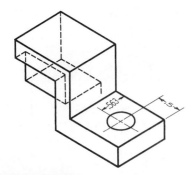

FIGURE 16.14 **Use of Hidden Lines.**

center lines may produce a confusion of lines. Examples in which center lines are not needed are shown in Figs. 16.10 and 16.11. Examples in which they are needed are seen in Figs. 16.14 and 16.39a.

■ 16.12 BOX CONSTRUCTION

Objects of rectangular shape may be more easily drawn by means of *box construction*, which consists simply in imagining the object to be enclosed in a rectangular box whose sides coincide with the main faces of the object. For example, the object shown in two views (Fig. 16.15) is imagined to be enclosed in a construction box. This box is then drawn lightly with con-

struction lines (Fig. 16.15I), the irregular features are then constructed (Fig. 16.15II), and finally, the required lines are made heavy (Fig. 16.15III).

■ 16.13 NONISOMETRIC LINES

Since the only lines of an object that are drawn true length in an isometric drawing are the isometric axes or lines parallel to them, nonisometric lines cannot be set off directly with the scale. For example, in Fig. 16.16a, the inclined lines BA and CA are shown in their true lengths (54 mm) in the top view, but since they are not parallel to the isometric axes, they are not true length in the isometric. Such lines are drawn in isometric by means of box construction and offset measurements. First, as shown in Fig. 16.16b, the measurements 44 mm, 18 mm, and 22 mm can be set off directly since they are made along isometric lines. The nonisometric 54 mm dimension cannot be set off directly, but if one half of the given top view is constructed full size to scale (Fig. 16.16I), the dimension X can be determined. This dimension is parallel to an isometric axis and can be transferred with dividers to the isometric (Fig. 16.16II). The dimensions 24 mm and 9 mm are parallel to isometric lines and can be set off directly (Fig. 16.16III).

FIGURE 16.15 **Box Construction.**

FIGURE 16.16 **Nonisometric Lines (metric dimensions).**

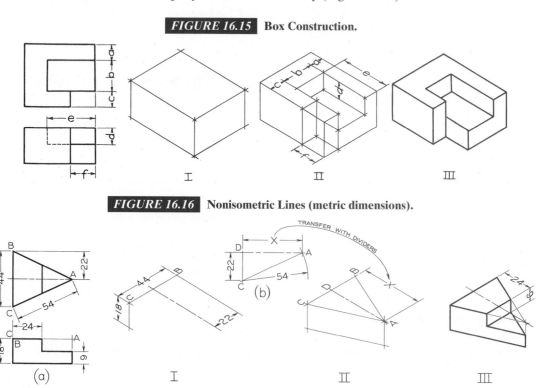

To realize the fact that nonisometric lines will not be true length in the isometric drawing, set your dividers on BA in (II) and then compare with BA on the given top view in Fig. 16.16a. Do the same for line CA. It will be seen that BA is shorter and CA is longer in the isometric than the corresponding lines in the given views.

■ 16.14 ANGLES IN ISOMETRIC

As shown in §6.26, angles project true size only when the plane of the angle is parallel to the plane of projection. An angle may project larger or smaller than true size, depending on its position. Since in isometric the various surfaces of the object are usually inclined to the plane of projection, it follows that angles generally will not be projected true size. For example, in the multiview drawing in Fig. 16.17a, none of the three 60° angles will be 60° in the isometric drawing. To realize this fact, measure each angle in the isometric in Fig. 16.17II with the protractor and note the number of degrees compared to the true 60°. No two angles are the same; two are smaller and one is larger than 60°.

As shown in Fig. 16.17I, the enclosing box can be drawn from the given dimensions, except for dimen-

sion X, which is not given. To find dimension X, draw triangle BDA from the top view full size, as shown in Fig. 16.17b. Transfer dimension X to the isometric in (I) to complete the enclosing box.

To locate point A in Fig. 16.17II, dimension Y must be used, but this is not given in the top view (Fig. 16.17a). Dimension Y is found by the same construction (Fig. 16.17b) and then transferred to the isometric, as shown. The completed isometric is shown in Fig. 16.17III, where point E is located by using dimension K, as shown.

A regular protractor cannot be used to set off angles in isometric.* *Angular measurements must be converted to linear measurements along isometric lines.*

Figure 16.18a shows two views of an object to be drawn in isometric. Point A can easily be located in the isometric by measuring .88″ down from point O (Fig. 16.18I). However, in the given drawing, the location of point B depends on the 30° angle, and to locate B in the isometric linear dimension X must be known. This distance can be found graphically by drawing the right triangle BOA attached to the isometric, as shown. The

---

* Isometric protractors for setting off angles on isometric surfaces are available from drafting supplies dealers.

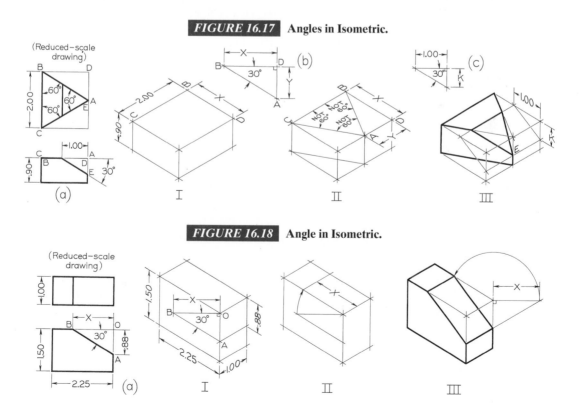

FIGURE 16.17 Angles in Isometric.

FIGURE 16.18 Angle in Isometric.

distance X is then transferred to the isometric with the compass or dividers (Fig. 16.18II). Actually, the triangle could be attached in several different positions. One of these is shown in Fig. 16.18III.

When angles are given in degrees, it is necessary to convert the angular measurements into linear measurements. This is best done by drawing a right triangle separately, as in Fig. 16.17b, or attached to the isometric (Fig. 16.18).

■ 16.15 IRREGULAR OBJECTS

If the general shape of an object does not conform somewhat to a rectangular pattern (Fig. 16.19), it may be drawn using the box construction discussed previously (Fig. 16.19a). Various points of the triangular base are located by means of offsets a and b along the edges of the bottom of the construction box. The vertex is located by means of offsets OA and OB on the top of the construction box.

However, it is not necessary to draw the complete construction box. If only the bottom of the box is drawn (Fig. 16.19b), the triangular base can be constructed as before. The orthographic projection of the vertex O′ on the base can then be located by offsets O′A and O′B, as shown, and from this point, the vertical line O′O can be erected, using measurement C.

An irregular object may be drawn by means of a series of sections. The edge views of a series of imaginary cutting planes are shown in the top and front views of the multiview drawing in Fig. 16.20a. In Fig. 16.20I, the various sections are constructed in isometric, and in Fig. 16.20II, the object is completed by drawing lines through the corners of the sections. In the isometric in Fig. 16.20I, all height dimensions are taken from the front view in Fig. 16.20a, and all depth dimensions from the top view.

■ 16.16 CURVES IN ISOMETRIC

Curves may be drawn in isometric by means of a series of offset measurements similar to those discussed in §16.9. In Fig. 16.21 any desired number of points, such as A, B, and C, are selected at random along the curve in the given top view in Fig. 16.21a. Enough points should be chosen to fix accurately the path of the curve; the more points used, the greater the accuracy. Offset grid lines are then drawn from each point parallel to the isometric axes.

Offset measurements a and b are laid off in the isometric to locate point A on the curve (Fig. 16.21I). Points B, C, and D are located in a similar manner (Fig. 16.21II). A light freehand curve is sketched smoothly through the points (Fig. 16.21III). Points A′, B′, C′,

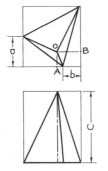
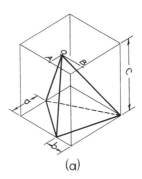
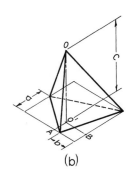

FIGURE 16.19 Irregular Object in Isometric.

(a) (b)

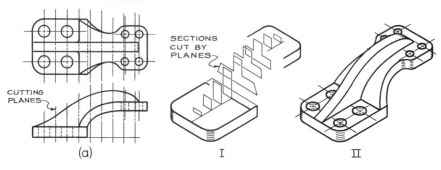

FIGURE 16.20 Use of Sections in Isometric.

(a) I II

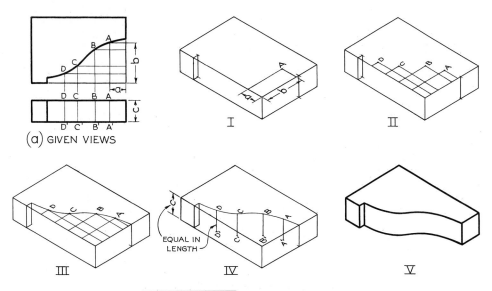

FIGURE 16.21 Curves in Isometric.

and D′ are located directly under points A, B, C, and D by drawing vertical lines downward, making all equal to dimension c, the height of the block (Fig. 16.21IV). A light freehand curve is then drawn through the points. The final curve is heavied in with the aid of the irregular curve, and all straight lines are darkened to complete the isometric (Fig. 16.21V).

■ 16.17 TRUE ELLIPSES IN ISOMETRIC

As shown in §§4.48, 5.12, and 6.30, if a circle lies in a plane that is not parallel to the plane of projection, the circle will be projected as a true ellipse. The ellipse can be constructed by the method of offsets (§16.16). As shown in Fig. 16.22a, draw parallel lines spaced at random across the circle; then transfer these lines to the isometric with the aid of the dividers (Fig. 16.22b). To locate points in the lower ellipse, transfer points of the upper ellipse down a distance equal to the height d of the block and draw the ellipse, part of which will be hidden, through these points. Draw the final ellipses with the aid of the irregular curve.

A variation of the method of offsets, which provides eight points on the ellipse, is illustrated in Figs. 16.22c and 16.22d. If more points are desired, parallel lines, as in Fig. 16.22a, can be added. As shown in Fig. 16.22c, circumscribe a square around the given circle, and draw diagonals. Through the points of intersection of the diagonals and the circle, draw another square, as shown. Draw this construction in the isometric, as shown in Fig. 16.22d, transferring distances a and b with the dividers.

A similar method that provides 12 points on the ellipse is shown in Fig. 16.22e. The given circle is divided

FIGURE 16.22 True Isometric Ellipse Construction.

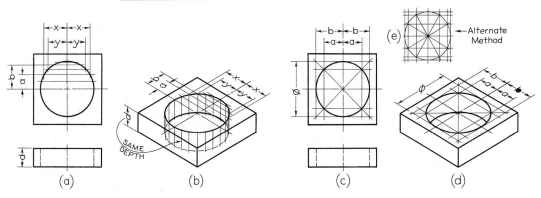

into 12 equal parts, using the 30° × 60° triangle. Lines parallel to the sides of the square are drawn through these points. The entire construction is then drawn in isometric, and the ellipse is drawn through the points of intersection.

When the center lines shown in the top view in Fig. 16.22a are drawn in isometric (Fig. 16.22b), they become the conjugate diameters of the ellipse. The ellipse can then be constructed on the conjugate diameters by the methods of Figs. 4.51 and 4.52b.

When the 45° diagonals (Fig. 16.22c) are drawn in isometric (Fig. 16.22d), they coincide with the major and minor axes of the ellipse, respectively. Note that the minor axis is equal in length to the sides of the inscribed square in Fig. 16.22c. The ellipse can be constructed on the major and minor axes by any of the methods in §§4.45 and 4.48.

Remember the rule: *The major axis of the ellipse is always at right angles to the center line of the cylinder, and the minor axis is at right angles to the major axis and coincides with the center line.*

Accurate ellipses may be drawn with the aid of ellipse guides (§§4.53 and 16.21) or with a special *ellipsograph.*

If the curve lies in a nonisometric plane, not all offset measurements can be applied directly. For example, in Fig. 16.23a, the elliptical face shown in the auxiliary view lies in an inclined nonisometric plane. The cylinder is enclosed in a construction box, and the box is then drawn in isometric (Fig. 16.23I). The base is drawn by the method of offsets, as shown in Fig. 16.22. The inclined ellipse is constructed by locating a number of points on the ellipse in the isometric and drawing the final curve by means of the irregular curve.

Measurements a, b, c, and so on are parallel to an isometric axis and can be set off in the isometric on each side of the center line X–X (Fig. 16.23I). Measurements e, f, g, and so on are not parallel to any isometric axis and cannot be set off directly in isometric. However, when these measurements are projected to the front view and down to the base (Fig. 16.23a), they can then be set off along the lower edge of the construction box, as shown in (I). The completed isometric is shown in (II).

The ellipse may also be drawn with the aid of an appropriate ellipse template selected to fit the major and minor axes established along X–X and Y–Y, respectively.

■ 16.18 APPROXIMATE FOUR-CENTER ELLIPSE

An approximate ellipse is sufficiently accurate for nearly all isometric drawings. The method commonly used, called the *four-center ellipse,* is illustrated in Figs. 16.24 to 16.26. It can be used only for ellipses in isometric planes.

To apply this method (Fig. 16.24), draw, or conceive to be drawn, a square around the given circle in the multiview drawing; then do the following:

I. Draw the isometric of the square, which is an equilateral parallelogram whose sides are equal to the diameter of the circle (Fig. 16.24I).

FIGURE 16.23 **Ellipse in Inclined Plane.**

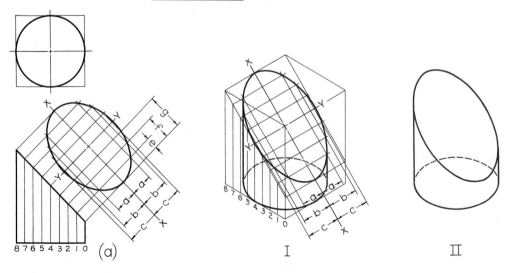

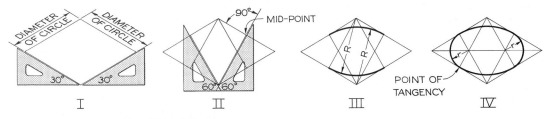

FIGURE 16.24 Steps in Drawing Four-Center Ellipse.

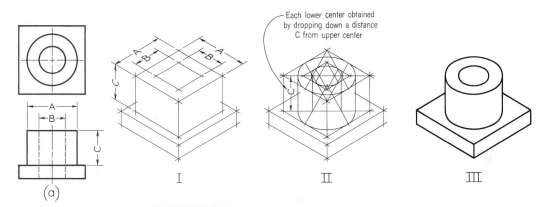

FIGURE 16.25 Isometric Drawing of a Bearing.

II. Erect perpendicular bisectors to each side, using the 30° × 60° triangle (Fig. 16.24II). These perpendiculars will intersect at four points, which will be centers for the four circular arcs.

III. Draw the two large arcs, with radius R, from the intersections of the perpendiculars in the two closest corners of the parallelogram (Fig. 16.24III).

IV. Draw the two small arcs, with radius r, from the intersections of the perpendiculars within the parallelogram, to complete the ellipse (Fig. 16.24IV). As a check on the accurate location of these centers, a long diagonal of the parallelogram may be drawn, as shown. The midpoints of the sides of the parallelogram are points of tangency for the four arcs.

A typical drawing with cylindrical shapes is illustrated in Fig. 16.25. Note that the centers of the larger ellipse cannot be used for the smaller ellipse, though the ellipses represent concentric circles. Each ellipse has its own parallelogram and its own centers. Observe also that the centers of the lower ellipse are obtained by projecting the centers of the upper large

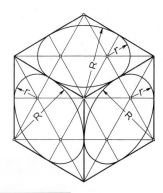

FIGURE 16.26 Four-Center Ellipses.

ellipse down a distance equal to the height of the cylinder.

The construction of the four-center ellipse on the three visible faces of a cube is shown in Fig. 16.26, a study of which shows that all diagonals are horizontal or 60° with horizontal; hence, the entire construction is made with the T-square and 30° × 60° triangle.

Actually, the four-center ellipse deviates considerably from the true ellipse. As shown in Fig. 16.27a, the

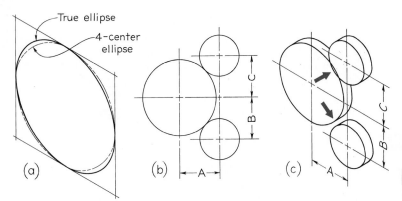

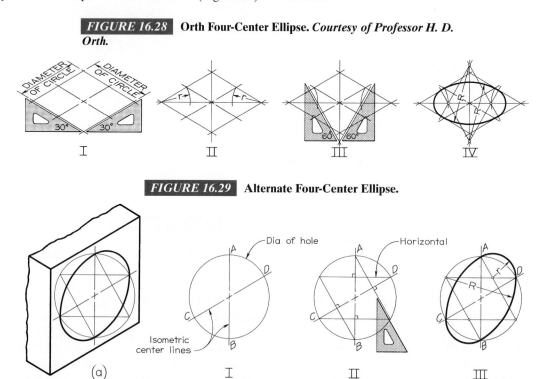

FIGURE 16.27 Faults of Four-Center Ellipse.

four-center ellipse is somewhat shorter and "fatter" than the true ellipse. In constructions where tangencies or intersections with the four-center ellipse occur in the zones of error, the four-center ellipse is unsatisfactory (Figs. 16.27b and 16.27c).

For a much closer approximation to the true ellipse, the Orth four-center ellipse (Fig. 16.28), which requires only one more step than the regular four-center ellipse, will be found sufficiently accurate for almost any problem.

When it is more convenient to start with the isometric center lines of a hole or cylinder in drawing the ellipse, rather than the enclosing parallelogram, the *alternate four-center ellipse* is recommended (Fig. 16.29).

A completely constructed ellipse is shown in Fig. 16.29a, and the steps followed are shown in Figs. 16.29b to 16.29d.

1. Draw the isometric center lines (Fig. 16.29I). From the center, draw a construction circle equal to the actual diameter of the hole or cylinder. The circle will intersect the center lines at four points A, B, C, and D.

2. From the two intersection points on one center line, erect perpendiculars to the other center line (Fig. 16.29II); then from the two intersection points on the other center line, erect perpendiculars to the first center line.

FIGURE 16.28 Orth Four-Center Ellipse. *Courtesy of Professor H. D. Orth.*

FIGURE 16.29 Alternate Four-Center Ellipse.

3. With the intersections of the perpendiculars as centers, draw two small arcs and two large arcs (Fig. 16.29III).

NOTE The preceding steps are exactly the same as for the regular four-center ellipse of Fig. 16.24 except for the use of the isometric center lines instead of the enclosing parallelogram.

■ 16.19 SCREW THREADS IN ISOMETRIC

Parallel partial ellipses spaced equal to the symbolic thread pitch (Fig. 13.9a) are used to represent the crests only of a screw thread in isometric (Fig. 16.30). The ellipses may be drawn by the four-center method of §16.18, or with the ellipse template, which is much more convenient.

■ 16.20 ARCS IN ISOMETRICS

The four-center ellipse construction is used in drawing circular arcs in isometric. Figure 16.31a shows the complete construction. However, it is not necessary to draw the complete constructions for arcs, as shown in Figs. 16.31b and 16.31c. In each case the radius R is set off from the construction corner; then at each point, perpendiculars to the lines are erected, and their intersection is the center of the arc. Note that the R distances are equal in Figs. 16.31b and 16.31c, but that the actual radii used are quite different.

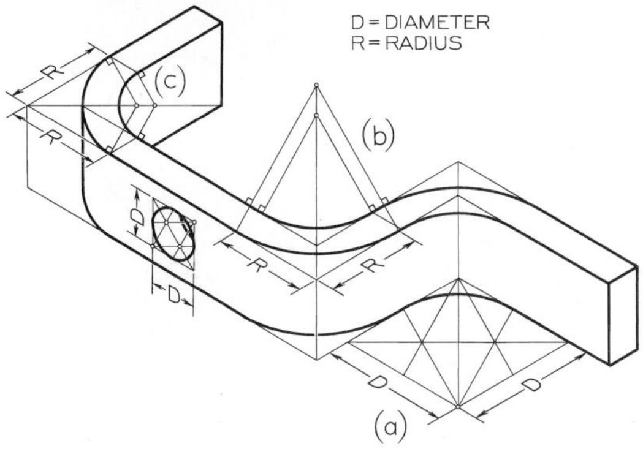

If a truer elliptic arc is required, the Orth construction (Fig. 16.28) can be used, or a true elliptic arc may be drawn by the method of offsets (§16.17) or with the aid of an ellipse template (§16.21).

■ 16.21 ELLIPSE TEMPLATES

One of the principal time-consuming elements in pictorial drawing is the construction of ellipses. A wide variety of ellipse templates is available for ellipses of various sizes and proportions. They are not available in every possible size, of course, and it may be necessary to "use the fudge factor," such as leaning the pencil or pen when inscribing the ellipse, or shifting the template slightly for drawing each quadrant of the ellipse.

The design of the ellipse template in Fig. 16.32 combines the angles, scales, and ellipses on the same instrument. The ellipses are provided with markings to coincide with the isometric center lines of the holes—a convenient feature in isometric drawing.

■ 16.22 INTERSECTIONS

To draw the elliptical intersection of a cylindrical hole in an oblique plane in isometric (Fig. 16.33a), draw the ellipse in the isometric plane on top of the construction box (Fig. 16.33b); then project points down to the oblique plane as shown. It will be seen that the construction for each point forms a trapezoid, which is produced by a slicing plane parallel to a lateral surface of the block.

FIGURE 16.30 Screw Threads in Isometric.

FIGURE 16.31 Arcs in Isometric.

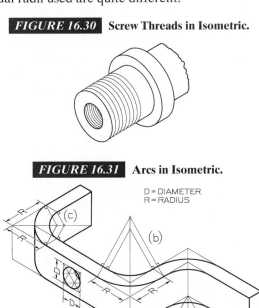

D = DIAMETER
R = RADIUS

FIGURE 16.32 Instrumaster Isometric Template.

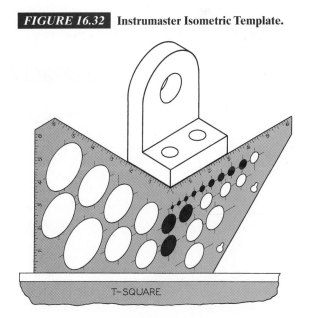

T–SQUARE

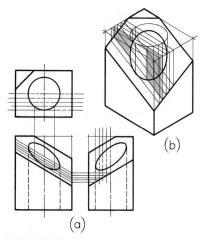

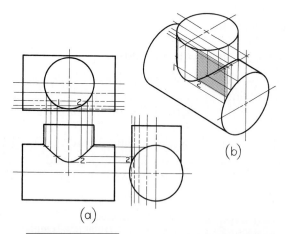

FIGURE 16.33   **Oblique Plane and Cylinder.**

FIGURE 16.34   **Intersection of Cylinders.**

To draw the curve of intersection between two cylinders (Fig. 16.34), pass a series of imaginary cutting planes through the cylinders parallel to their axes, as shown. Each plane will cut elements on both cylinders that intersect at points on the curve of intersection (Fig. 16.33b). As many points should be plotted as necessary to assure a smooth curve. For most accurate work, the ends of the cylinders should be drawn by the Orth construction, with ellipse guides, or by one of the true-ellipse constructions.

■ 16.23 THE SPHERE IN ISOMETRIC

The isometric drawing of any curved surface if evidently the envelope of all lines that can be drawn on that surface. For the sphere, the great circles (circles cut by any plane through the center) may be selected as the lines on the surface. Since all great circles, except those that are perpendicular or parallel to the plane of projection, are shown as ellipses having equal major axes, it follows that their envelope is a circle whose diameter is the major axis of the ellipses.

In Fig. 16.35a, two views of a sphere enclosed in a construction cube are shown. The cube is drawn in (I) together with the isometric of a great circle that lies in a plane parallel to one face of the cube. Actually, the ellipse need not be drawn, for only the points on the diagonal located by measurements a are needed. These points establish the ends of the major axis from which the radius R of the sphere is determined. The resulting drawing shown in (II) is an *isometric drawing*, and its diameter is, therefore, $\sqrt{\frac{3}{2}}$ times the actual diameter of the sphere. The *isometric projection* of the

FIGURE 16.35   **Isometric of a Sphere.**

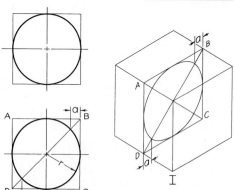

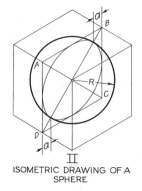

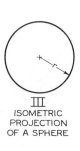

sphere is simply a circle whose diameter is equal to the true diameter of the sphere, as shown in (III).

■ 16.24 ISOMETRIC SECTIONING

In drawing objects characterized by open or irregular interior shapes, isometric sectioning is as appropriate as in multiview drawing. An *isometric full section* is shown in Fig. 16.36. In such cases it is usually best to draw the cut surface first and then to draw the portion of the object that lies behind the cutting plane. Other examples of isometric full sections are shown in Figs. 6.92, 7.11b, and 7.12d.

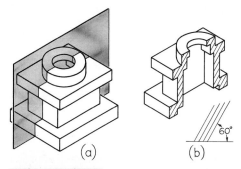

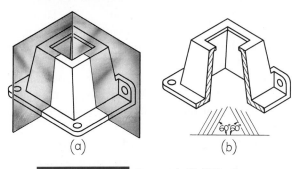

FIGURE 16.36 **Isometric Full Section.**

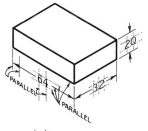

FIGURE 16.37 **Isometric Half Section.**

An *isometric half section* is shown in Fig. 16.37. The simplest procedure in this case is to make an isometric drawing of the entire object and then the cut surfaces. Since only a quarter of the object is removed in a half section, the resulting pictorial drawing is more useful than full sections in describing both exterior and interior shapes together. Other typical isometric half sections are shown in Figs. 7.13, 7.41, and 7.42.

Isometric broken-out sections are also sometimes used. Examples are shown in Figs. 6.99 and 7.51.

Section lining in isometric drawing is similar to that in multiview drawing. Section lining at an angle of 60° with horizontal (Figs. 16.36 and 16.37) is recommended, but the direction should be changed if at this angle the lines would be parallel to a prominent visible line bounding the cut surface, or to other adjacent lines of the drawing.

■ 16.25 ISOMETRIC DIMENSIONING

Isometric dimensions are similar to ordinary dimensions used on multiview drawings but are expressed in pictorial form. Two methods of dimensioning are approved by ANSI—namely, the pictorial plane (aligned) system and the unidirectional system (Fig. 16.38). Note that vertical lettering is used for either system of dimensioning. Inclined lettering is not recommended for pictorial dimensioning. The method of drawing numerals and arrowheads for the two systems is shown in Figs. 16.38a and 16.38b. For the 64-mm dimension in the aligned system in Fig. 16.38a, the extension lines, dimension lines, and lettering are all drawn in the isometric plane of one face of the object. The "horizontal" guide lines for the lettering are drawn parallel to the dimension line, and the "vertical" guide lines are drawn parallel to the extension lines. The

FIGURE 16.38 **Numerals and Arrowheads in Isometric (metric dimensions).**

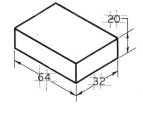

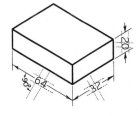

(a) ALIGNED   (b) UNIDIRECTIONAL   (c) INCORRECT

barbs of the arrowheads should line up parallel to the extension lines.

For the 64-mm dimension in the unidirectional system (Fig. 16.38b), the extension lines and dimension lines are all drawn in the isometric plane of one face of the object, and the barbs of the arrowheads should line up parallel to the extension lines, as in Fig. 16.38a. However, the lettering for the dimensions is vertical and reads from the bottom of the drawing. This simpler system of dimensioning is often used on pictorials for production purposes.

As shown in Fig. 16.38c, the vertical guide lines for the letters should not be perpendicular to the dimension lines. The example in Fig. 16.38c is incorrect because the 64-mm and 32-mm dimensions are lettered neither in the plane of corresponding dimension and extension lines nor in a vertical position to read from the bottom of the drawing. The 20-mm dimension is awkward to read because of its position.

Correct practice in isometric dimensioning using the aligned system of dimensioning is shown in Fig. 16.39a. Incorrect practice is shown in Fig. 16.39b, where the $3\frac{1}{8}''$ dimension runs to a wrong extension line at the right, and consequently the dimension does not lie in an isometric plane. Near the left side, a number of lines cross one another unnecessarily and terminate on the wrong lines. The upper $\frac{1}{2}''$ drill hole is located from the edge of the cylinder when it should be dimensioned from its center line. Study these two drawings carefully to discover additional mistakes in Fig. 16.39b.

The dimensioning methods described apply equally to fractional, decimal, and metric dimensions.

Many examples of isometric dimensioning are given in the problems at the ends of Chapters 6, 7, 8, and 14, and you should study these to find samples of almost any special case you may encounter.

■ 16.26 EXPLODED ASSEMBLIES

Exploded assemblies are often used in design presentations, catalogs, sales literature, and in the shop to show all the parts of an assembly and how they fit together. They may be drawn by any of the pictorial methods, including isometric (Fig. 16.40). Other isometric exploded assemblies are shown in Chapter 14.

■ 16.27 PIPING DIAGRAMS

Isometric and oblique drawings are well suited for representation of piping layouts, as well as for all other structural work to be represented pictorially.

DIMETRIC PROJECTION
■ 16.28 THE DIMETRIC METHOD OF PROJECTION

A **dimetric projection** is an axonometric projection of an object so placed that two of its axes make equal angles with the plane of projection and the third axis makes either a smaller or a greater angle. Hence, the two axes making equal angles with the plane of projection are foreshortened equally, while the third axis is foreshortened in a different ratio.

Generally, the object is so placed that one axis will be projected in a vertical position. However, if the relative positions of the axes have been determined, the

FIGURE 16.39   **Correct and Incorrect Isometric Dimensioning (Aligned System).**

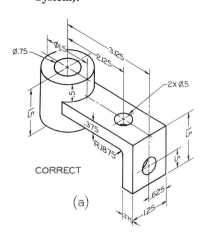

(a)

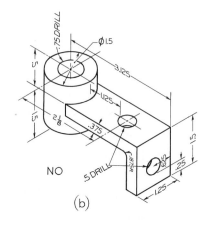

(b)

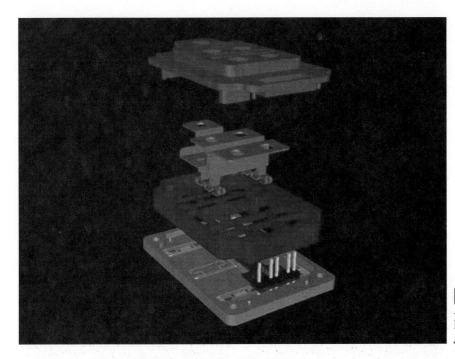

FIGURE 16.40 **Exploded Assembly of Power Module for Zero-Emission Automobiles.** *Courtesy of SDRC, Milford, OH.*

projection may be drawn in any revolved position, as in isometric drawing (see §16.8).

The angles between the projection of the axes must not be confused with the angles which the axes themselves make with the plane projection.

The positions of the axes may be assumed such that any two angles between the axes are equal and over 90°, and the scales are determined graphically, as shown in Fig. 16.41a, in which OP, OL, and OS are the projections of the axes or converging edges of a cube. In this case, angle POS = angle LOS. Lines PL, LS, and SP are the lines of intersection of the plane of projection with the three visible faces of the cube. From descriptive geometry we know that since line LO is perpendicular to the plane POS, in space, its projection LO is perpendicular to PS, the intersection of the plane POS and the plane of projection. Similarly, OP is perpendicular to SL, and OS is perpendicular to PL.

FIGURE 16.41 **Dimetric Projection.**

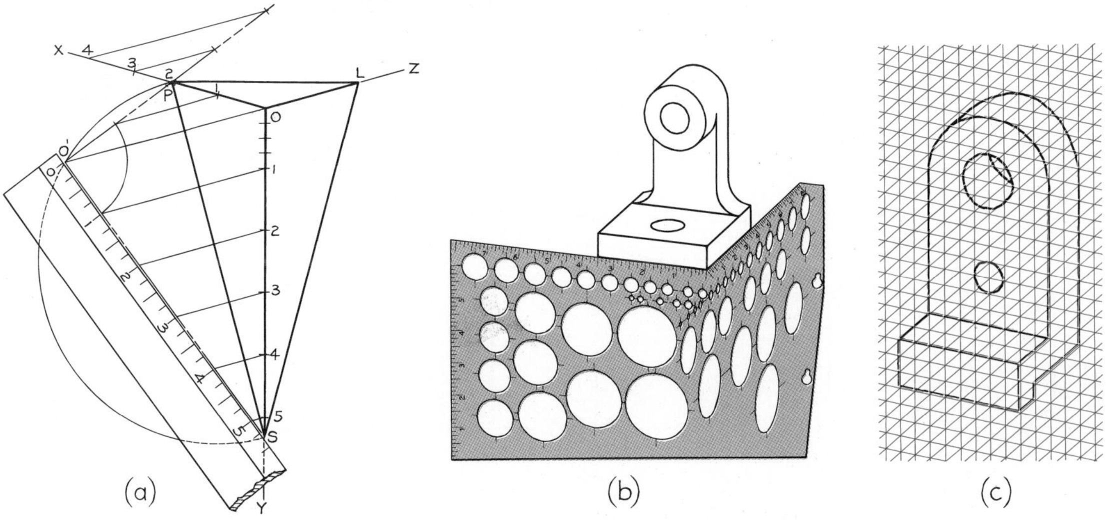

(a) (b) (c)

If the triangle POS is revolved about the line PS as an axis into the plane of projection, it will be shown in its true size and shape as PO'S. If regular full-size scales are marked along the lines O'P and O'S, and the triangle is counterrevolved to its original position, the dimetric scales may be laid off on the axes OP and OS, as shown.

To avoid the preparation of special scales, use can be made of available scales on the architects' scale by assuming the scales and calculating the positions of the axes, as follows:

$$\cos a = -\frac{\sqrt{2h^2v^2 - v^4}}{2hv},$$

where a is one of the two equal angles between the projections of the axes, h is one of the two equal scales, and v is the third scale. Examples are shown in the upper row of Fig. 16.42, in which the assumed scales, shown encircled, are taken from the architects' scale. One of these three positions of the axes will be found suitable for almost any practical drawing.

The Instrumaster Dimetric Template (Fig. 16.41b) has angles of approximately 11° and 39° with horizontal, which provides a picture similar to that in Fig. 16.42III. In addition, the template has ellipses corre-

sponding to the axes and accurate scales along the edges. For other information on drawing of ellipses, see §16.32.

The Instrumaster Dimetric Graph paper (Fig. 16.41c) can be used to sketch in dimetrics as easily as to sketch isometrics on isometric paper. The grid lines slope in conformity to the angles on the Dimetric Template in Fig. 16.41b, and when printed on vellum, the grid lines do not reproduce on prints.

■ 16.29 APPROXIMATE DIMETRIC DRAWING

Approximate dimetric drawings, which closely resemble true dimetrics, can be constructed by substituting for the true angles shown in the upper half of Fig. 16.42 angles that can be obtained with the ordinary triangles and compass, as shown in the lower half of the figure. The resulting drawings will be sufficiently accurate for all practical purposes.

The procedure in preparing an approximate dimetric drawing, using the position in Fig. 16.42VI, is shown in Fig. 16.43. The offset method of drawing a curve is shown in the figure. Other methods for drawing ellipses are the same as in trimetric drawing (§16.32).

FIGURE 16.42   **Angles of Axes Determined by Assumed Scales.**

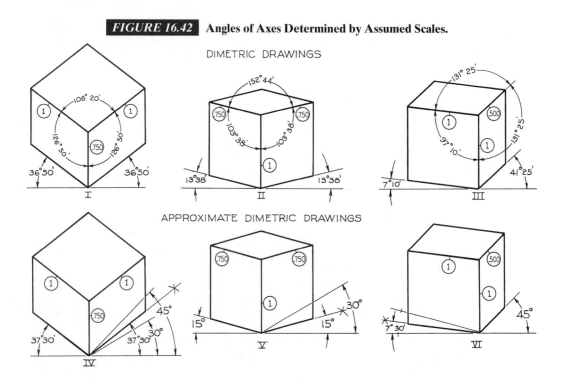

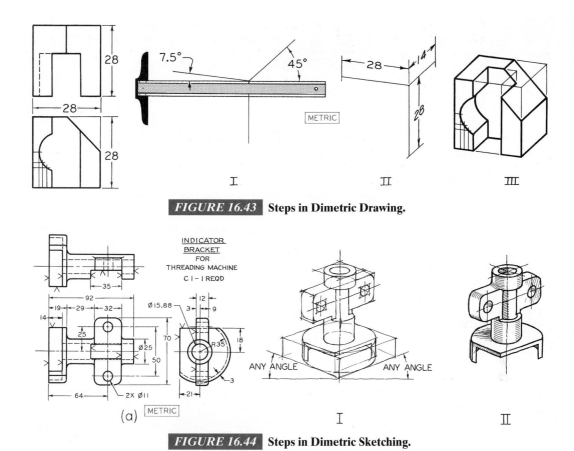

FIGURE 16.43 Steps in Dimetric Drawing.

FIGURE 16.44 Steps in Dimetric Sketching.

The steps in making a dimetric sketch, using a position similar to that in Fig. 16.42V, are shown in Fig. 16.44. The two angles are equal. Angles of about 20° with the horizontal will produce the most pleasing effect.

An exploded approximate dimetric drawing of an adding machine is shown in Fig. 16.45. The dimetric axes used are those in Fig. 16.42IV. Pictorials such as this are often used in service manuals.

TRIMETRIC PROJECTION

■ 16.30 THE TRIMETRIC METHOD OF PROJECTION

A **trimetric projection** is an axonometric projection of an object so placed that no two axes make equal angles with the plane of projection. In other words, each of the three axes and the lines parallel to them, respectively, have different ratios of foreshortening when projected to the plane of projection. If the three axes are assumed in any position on paper such that none

of the angles is less than 90°, and if neither an isometric nor a dimetric position is deliberately arranged, the result will be a trimetric projection.

■ 16.31 TRIMETRIC SCALES

Since the three axes are foreshortened differently, three different trimetric scales must be prepared and used. The scales are determined as shown in Fig. 16.46a, the method being the same as explained for the dimetric scales in §16.28. As shown, any two of the three triangular faces can be revolved into the plane of projection to show the true lengths of the three axes. In the revolved position, the regular scale is used to set off inches or fractions thereof. When the axes have been counterrevolved to their original positions, the scales will be correctly foreshortened, as shown. These dimensions should be transferred to the edges of three thin cards and marked OX, OZ, and OY for easy reference.

A special trimetric angle may be prepared from Bristol Board or plastic, as shown in Fig. 17.46b.

GRAPHICS SPOTLIGHT

3D Pictorials Aid Designers of Future Electric Cars

They rode on rockets to the moon and proved they can generate enough electricity to power an automobile. But fuel cells are still years away from widespread use in the automobile industry.

Standing in their path are high costs and the problems of a complex technology. Together, these forces ensure large numbers of fuel-cell-powered vehicles will not hit showroom floors for a decade or more.

"There's a lot of serious work going into it, and there's a lot of potential there," says Bernard Robertson, DaimlerChrysler's vice president of engineering technologies. "It's just that the challenges are pretty formidable."

FUEL CELLS POSE THORNY PROBLEMS FOR DESIGN TEAM

A fuel cell uses sophisticated membranes to strip electrons from hydrogen atoms creating a charge imbalance and electrical current. The cell recombines the hydrogen with oxygen to form water vapor. DaimlerChrysler is trying to lead the industry in fuel cells by forming joint ventures with Ford Motor Co. and Ballard Powers Systems, Inc. of Vancouver, British Columbia, a supplier of fuels cell stacks. The goal is to be the first to manufacture complete fuel-cell powertrains for sale in the world market.

The task is daunting. At DaimlerChrysler's fuel-cell development center near Stuttgart, Germany, 900

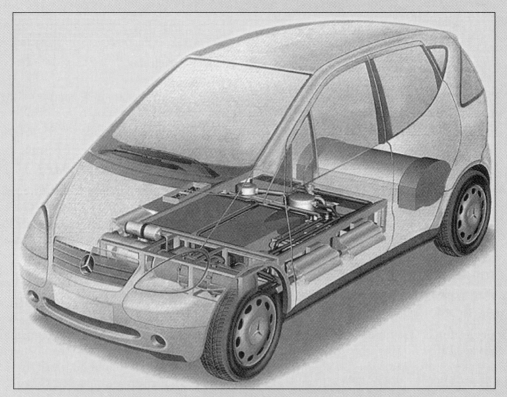

3D pictorial of Necar 4 fuel-cell-powered car. Such 3D images aid designers in overcoming current design problems.

technical people are devoted exclusively to fuel-cell research. Necar 4, the fourth generation of the center's New Electric Car series of fuel-cell concepts, was unveiled to the public in March 1999. Necar is technically impressive because it crams the entire fuel-cell system in a 6-inch-deep space under the floor, but also illustrated the shortcomings of current fuel-cell technology.

DaimlerChrysler officials admit the vehicle is overweight by more than 600 pounds and is astronomically expensive. A mass-produced fuel cell system would cost $30,000 using today's technology, although Necar 4's hand-built engine is estimated to have cost $350,000. Gasoline engines typically cost $3,000. Specialized hardware is what drives the fuel cell's costs. The largest expense is the row of electricity-conducting bipolar plates in the fuel cell stack. These plates, made from ultra hard carbon-graphite, have dozens of intricate channels that must be individually cut by computer-controlled machine tools. For maximum efficiency, the channels must be machined to the high tolerance usually reserved for jet engine turbines.

CAD HELPS DESIGNER SEE POSSIBILITIES

With all of this precision machining involved, the engineers at DaimlerChrsyler have been assigned two separate tasks: (1) develop a fuel cell that is small and light enough to meet size and weight restrictions; and (2) develop manufacturing equipment capable of mass producing such a product at a reasonable cost.

Since making even one model of such a cell or vehicle is cost restrictive, the design teams have turned to computer-design and computer-manufacturing software programs for help. 3D pictorials let them see not only how they can better design the fuel cell for each vehicle, but what is needed in the way of retooling for mass production of such cells. By keeping costs down, engineers are hopeful that they will be able to design and develop an efficient electric car by the year 2004. Retooling of assembly lines could take a few years longer, but DaimlerChrysler is confident that they will have an efficient, reliable, fuel-cell powered vehicle on the world market within the next decade.

Adapted from "Fuel Cells Still Pose Thorny Problems," by Aaron Robinson, "Automotive News," March 29, 1999.

Perhaps six or seven such guides, using angles for a variety of positions of the axes, would be sufficient for all practical requirements. [*]

■ 16.32 TRIMETRIC ELLIPSES

The trimetric center lines of a hole, or on the end of a cylinder, become the conjugate diameters of the ellipse when drawn in trimetric. The ellipse may be drawn on the conjugate diameters (Figs. 4.51 or 4.52b), or the major and minor axes may be determined from the conjugate diameters (Fig. 4.53c) and the ellipse constructed on them by any of the methods of Figs. 4.48–4.50 and 4.52a, or with the aid of an ellipse template (Fig. 4.55).

[*] Plastic templates of this type are available from drafting supplies dealers.

One advantage of trimetric projection is the infinite number of positions of the object available. The angles and scales can be handled without too much difficulty, as shown in §16.31. However, the infinite variety of ellipses has been a discouraging factor.

In drawing any axonometric ellipse, keep the following in mind:

1. On the drawing, the major axis is always perpendicular to the center line, or axis, of the cylinder.
2. The minor axis is always perpendicular to the major axis; that is, on the paper it coincides with the axis of the cylinder.
3. The length of the major axis is equal to the actual diameter of the cylinder.

Thus we know at once the directions of both the major and minor axes, and the length of the major axis.

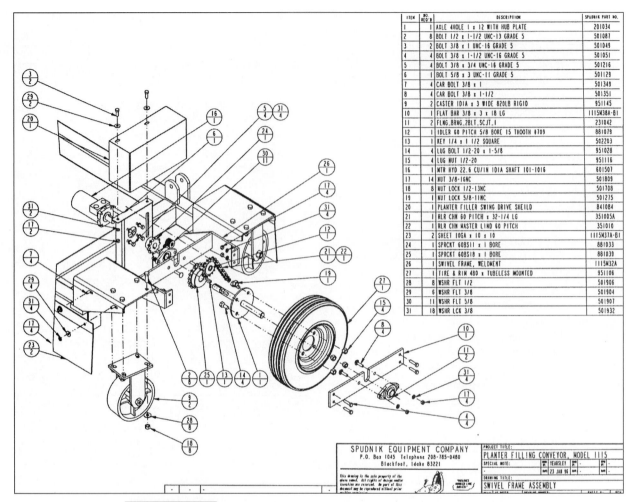

FIGURE 16.45 Exploded Diametric of Planter Filling Conveyor.
Courtesy of Spudnik Equipment Company, Inc.

We do not know the length of the minor axis. If we can find it, we can easily construct the ellipse with the aid of an ellipse template or any of a number of ellipse constructions mentioned earlier.

In Fig. 16.47a, center O is located as desired, and horizontal and vertical construction lines that will contain the major and minor axes are drawn through O. Note that the major axis will be on the horizontal line perpendicular to the axis of the hole, and the minor axis will be perpendicular to it, or vertical.

Set the compass for the actual radius of the hole and draw the semicircle, as shown, to establish the ends A and B of the major axis. Draw AF and BF parallel to the axonometric edges WX and YX, respectively, to locate F, which lies on the ellipse. Draw a vertical line through F to intersect the semicircle at F′ and join

F′ to B′, as shown. From D′, where the minor axis, extended, intersects the semicircle, draw D′E and ED parallel to F′B and BF, respectively. Point D is one end of the minor axis. From center O, strike arc DC to locate C, the other end of the minor axis. On these axes, a true ellipse can be constructed, or drawn with the aid of an ellipse template. A simple method for finding the "angle" of ellipse template to use is shown in Fig. 4.55c. If an ellipse template is not available, an approximate four-center ellipse (Fig. 4.56) will be found satisfactory in most cases.

In constructions where the enclosing parallelogram for an ellipse is available or easily constructed, the major and minor axes can be readily determined, as shown in Fig. 16.47b. The directions of both axes and the length of the major axis are known. Extend

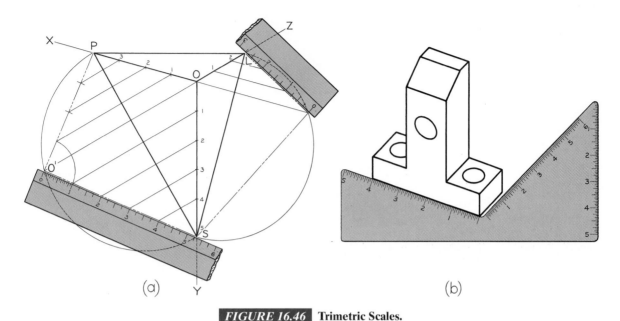

FIGURE 16.46   Trimetric Scales.

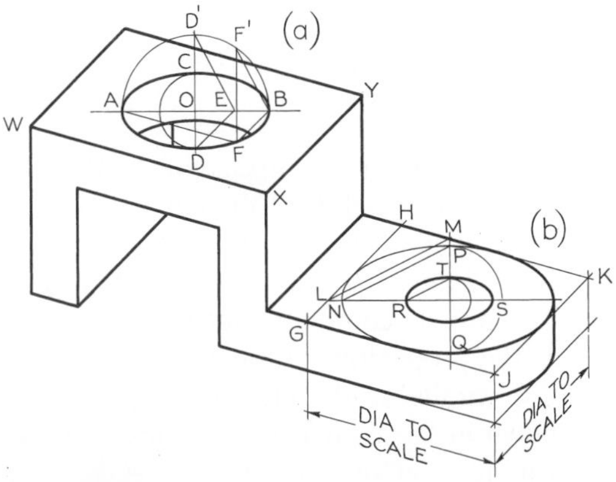

FIGURE 16.47   Ellipses in Trimetric. *Method (b) courtesy of Professor H. E. Grant.*

the axes to intersect the sides of the parallelogram at L and M, and join the points with a straight line. From one end N of the major axis, draw a line NP parallel to LM. The point P is one end of the minor axis. To find one end T of the minor axis of the smaller ellipse, it is only necessary to draw RT parallel to LM or NP.

The method of constructing an ellipse on an oblique plane in trimetric is similar to that shown for isometric in Fig. 16.33.

■ 16.33 AXONOMETRIC PROJECTION BY THE METHOD OF INTERSECTIONS

Instead of constructing axonometric projections with the aid of specially prepared scales, as explained in the preceding paragraphs, an axonometric projection can be obtained directly by projection from two orthographic views of the object. This method, called the *method of intersections*, was developed by Profs. L. Eckhart and T. Schmid of the Vienna College of Engineering and was published in 1937.

To understand this method, let us assume that the axonometric projection of a rectangular object is given, and it is necessary to find the three orthographic projections: the top view, front view, and side view (Fig. 16.48).

Assume that the object is placed so that its principal edges coincide with the coordinate axes, and assume that the plane of projection (the plane on which the axonometric projection is drawn) intersects the three coordinate planes in the triangle ABC. From descriptive geometry, we know that lines BC, CA, and AB will be perpendicular, respectively, to axes OX, OY, and OZ. Any one of the three points A, B, or C may be assumed anywhere on one of the axes, and the triangle ABC may be drawn.

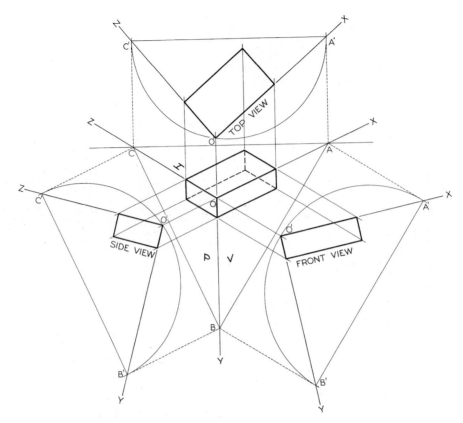

FIGURE 16.48 Views from an Axonometric Projection.

To find the true size and shape of the top view, revolve the triangular portion of the horizontal plane AOC, which is in front of the plane of projection, about its base CA, into the plane of projection. In this case, the triangle is revolved *inward* to the plane of projection through the smallest angle made with it. The triangle will then be shown in its true size and shape, and the top view of the object can be drawn in the triangle by projection from the axonometric projection, as shown, since all width dimensions remain the same. In the figure, the base CA of the triangle has been moved upward to C'A' so that the revolved position of the triangle will not overlap its projection.

In the same manner, the true sizes and shapes of the front view and side view can be found, as shown.

It is evident that if the three orthographic projections, or in most cases any two of them, are given in their relative positions, as shown in Fig. 16.48, the directions of the projections could be reversed so that the intersections of the projecting lines would determine the required axonometric projection.

To draw an axonometric projection by the method of intersections, it is helpful to make a sketch of the desired general appearance of the projection (Fig. 16.49). Even if the object is a complicated one, this sketch need not be complete, but only an enclosing box. Draw the projections of the coordinate axes OX, OY, and OZ parallel to the principal edges of the object, as shown in the sketch, and the three coordinate planes with the plane of projection.

Revolve the triangle ABO about its base AB as the axis into the plane of projection. Line OA will revolve to O'A, and this line, or one parallel to it, must be used as the baseline of the front view of the object. The projecting lines from the front view to the axonometric must be drawn parallel to the projection of the unrevolved Z-axis, as indicated in the figure.

Similarly, revolve the triangle COB about its base CB as the axis into the plane of projection. Line CO will revolve to CO", and this line, or one parallel to it, must be used as the baseline of the side view. The direction of the projecting lines must

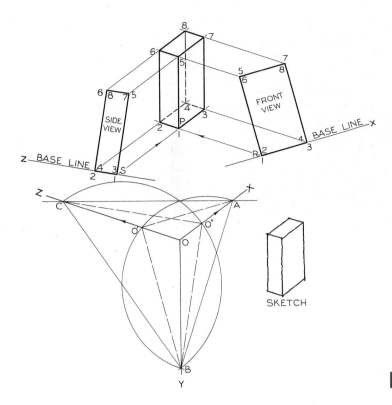

FIGURE 16.49   Axonometric Projection.

be parallel to the projection of the unrevolved X-axis, as shown.

Draw the front-view baseline at a convenient location, but parallel to O'X, and with it as the base, draw the front view of the object. Draw the side-view baseline also at a convenient location, but parallel to O"C, and with it as the base, draw the side view of the object, as shown. From the corners of the front view, draw projecting lines parallel to OZ, and from the corners of the side view, draw projecting lines parallel to OX. The intersections of these two sets of projecting lines determine the desired axonometric projection. It will be an isometric, a dimetric, or a trimetric projection, depending on the form of the sketch used as the basis for the projections (§16.2). If the sketch is drawn so that the three angles formed by the three coordinate axes are equal, the resulting projection will be an isometric projection; if two of the three angles are equal, the resulting projection will be a dimetric projection; and if no two of the three angles are equal, the resulting projection will be a trimetric projection.

To place the desired projection on a specific location on the drawing (Fig. 16.49), select the desired projection P of the point 1, for example, and draw two projecting lines PR ands PS to intersect the two baselines and thereby to determine the locations of the two views on their baselines.

Another example of this method of axonometric projection is shown in Fig. 16.50. In this case, it was deemed necessary only to draw a sketch of the plan or base of the object in the desired position, as shown. The axes are then drawn with OX and OZ parallel, respectively, to the sides of the sketch plan, and the remaining axis OY is assumed in a vertical position. The triangles COB and AOB are revolved, and the two baselines are drawn parallel to O"C and O'A, as shown. Point P, the lower front corner of the axonometric drawing, is then chosen at a convenient place, and projecting lines are drawn toward the baselines parallel to axes OX and OZ to locate the positions of the views on the baselines. The views are drawn on the baselines or cut apart from another drawing and fastened in place with drafting tape.

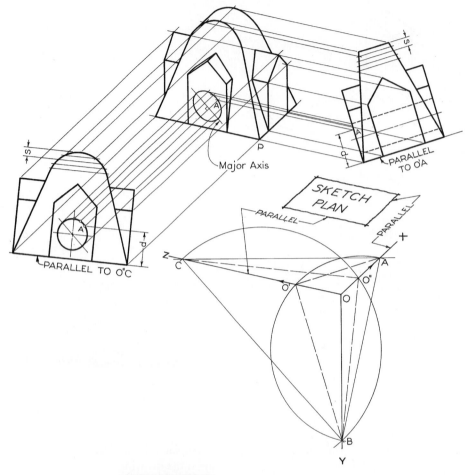

FIGURE 16.50 Axonometric Projection.

To draw the elliptical projection of the circle, assume any points, such as A, on the circle in both front and side views. Note that point A is the same altitude d above the baseline in both views. The axonometric projection of point A is found simply by drawing the projecting lines from the two views. The major and minor axes may be easily found by projecting in this manner or by methods shown in Fig. 16.47; the true ellipse may be drawn by any of the methods of Figs. 4.48–4.50 and 4.52a, or with the aid of an ellipse template. An approximate ellipse is satisfactory for most drawings and may be used.

■ 16.34 USING CAD

Pictorial drawings of all sorts can be created using 3D CAD. To create pictorials using 2D CAD, you would use similar projection techniques to those presented in this chapter. The advantage of 3D CAD is that once you make a 3D model of a part or assembly you can change the viewing direction at any time for orthographic, isometric, or perspective views. You can also apply different materials to the drawing objects and shade them to produce a high degree of realism in the pictorial view.

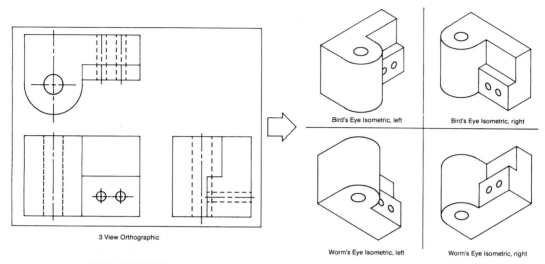

3 View Orthographic

Bird's Eye Isometric, left

Bird's Eye Isometric, right

Worm's Eye Isometric, left

Worm's Eye Isometric, right

FIGURE 16.51 **Orthographic to Isometric Conversion. The Auto-trol Orthographic to Axonometric Package (OTAP) system can be used to convert an orthographic drawing to axonometric.** *Courtesy of Auto-trol Technology Corporation.*

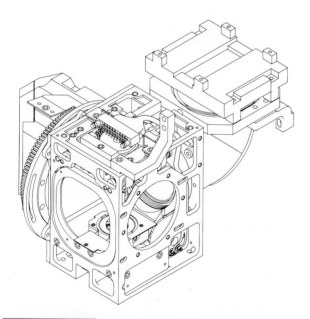

FIGURE 16.52 **Isometric Assembly Drawing Produced by Using the Computervision Designer System.** *Courtesy of Computervision Corporation, a subsidiary of Prime Computer, Inc.*

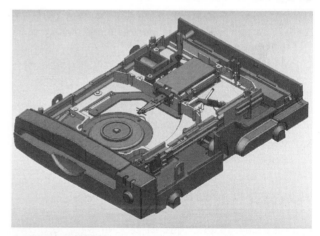

FIGURE 16.53 **3D Pictorial Drawing of a Zip Drive.** *Courtesy of SDRC, Milford, OH.*

■ KEY WORDS

AXONOMETRIC	ISOMETRIC
DIMETRIC	TRIMETRIC
PERSPECTIVE	INFINITY
LINE OF SIGHT	VISUAL RAYS
PLANE OF PROJECTION	FINITE
INCLINED SURFACE	OBLIQUE
ISOMETRIC AXES	BOX CONSTRUCTION
FOUR-CENTER ELLIPSE	

■ CHAPTER SUMMARY

- Axonometric projection is a method of creating a pictorial representation of an object. It shows all three dimensions of length, width, and height in one view.
- Isometric is the easiest of the axonometric projections to draw and is therefore the most common pictorial drawing. Isometric drawings created with CAD are often called 3D models.
- The spaces between the axes of an isometric drawing each are 120°. Isometric axes are drawn at 30° to the horizontal and vertical.
- The only lines on an isometric drawing that are to scale are parallel to the three isometric axes.
- An axonometric drawing is created by rotating an object about imaginary vertical and horizontal axes until three adjacent views, usually the top, front, and right side view, can all be seen at the same time.
- Inclined surfaces and oblique surfaces must determined by plotting the endpoints of each edge of the surface.
- Angles, irregular curves, and ellipses require special construction techniques for accurate representation.
- A common method of drawing an object in isometric is by creating as isometric box and drawing the features of the object within the box.
- Unlike perspective drawing, in which parallel lines converge on a vanishing point, parallel lines are drawn parallel in axonometric drawings.

■ REVIEW QUESTIONS

1. Why is isometric drawing more common than perspective drawing in engineering work?
2. What are the differences between axonometric projection and perspective?
3. What type of projection is used when creating a 3D model with CAD?
4. At what angles are the isometric axes drawn?
5. What are the three views that are typically shown in an isometric drawing?
6. Which type of projection places the observer at a finite distance from the object? Which types place the observer at an infinite distance?
7. Why is isometric easier to draw than dimetric or trimetric?
8. Is the four-circle ellipse a true ellipse or an approximation?
9. Is an ellipse in CAD a four-circle ellipse or a true conic section?

■ AXONOMETRIC PROBLEMS

Figures 16.54–16.58 consist of problems to be drawn axonometrically. The earlier isometric sketches may be drawn on isometric paper (§5.13), and later sketches should be made on plain drawing paper. On drawings to be executed with instruments, show all construction lines required in the solutions.

For additional problems, see Figs. 6.52–6.54, 17.23–17.26, and 18.34.

Since many of the problems in this chapter are of a general nature, they can also be solved on most computer graphics systems. If a system is available, the instructor may choose to assign specific problems to be completed by this method.

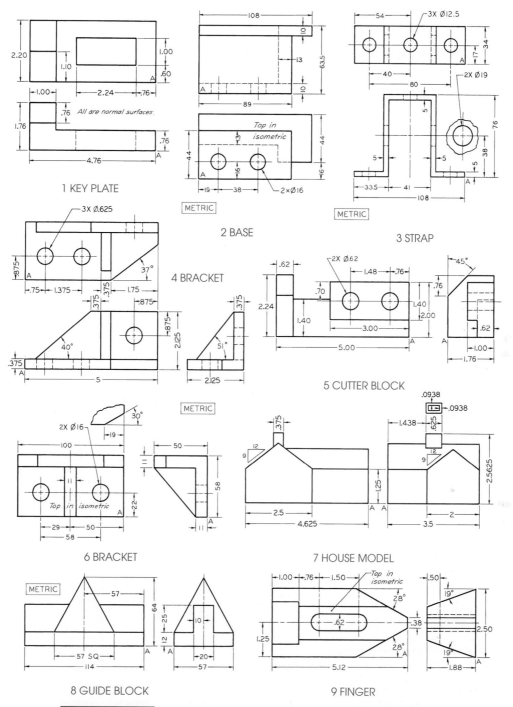

1 KEY PLATE

2 BASE

3 STRAP

4 BRACKET

5 CUTTER BLOCK

6 BRACKET

7 HOUSE MODEL

8 GUIDE BLOCK

9 FINGER

FIGURE 16.54 (1) Make freehand isometric sketches. (2) Make isometric drawings with instruments on Layout A–2 or A4–2 (adjusted). (3) Make dimetric drawings with instruments, using Layout A–2 or A4–2 (adjusted), and position assigned from Fig. 16.42. (d) Make trimetric drawings, using instruments, with axes chosen to show the objects to best advantage. If dimensions are required, study §16.25.

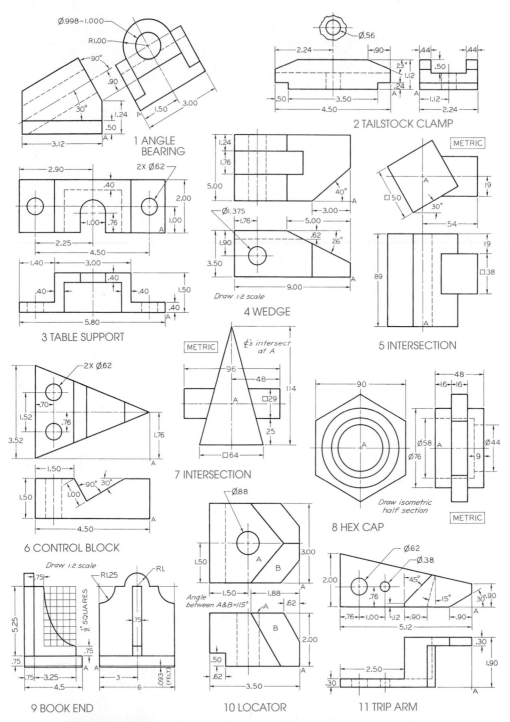

FIGURE 16.55 (1) Make freehand isometric sketches. (2) Make isometric drawings with instruments on Layout A–2 or A4–2 (adjusted). (3) Make dimetric drawings with instruments, using Layout A–2 or A4–2 (adjusted), and position assigned from Fig. 16.42. (4) Make trimetric drawings, using instruments, with axes chosen to show the objects to best advantage. If dimensions are required, study §16.25.

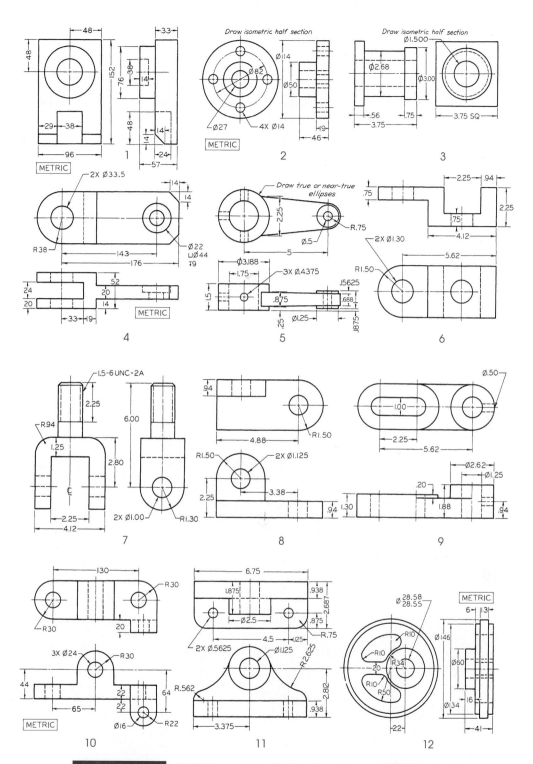

FIGURE 16.56 (1) Make isometric freehand sketches. (2) Make isometric drawings with instruments, using Size A or A4 sheet or Size B or A3 sheet, as assigned. (3) Make dimetric drawings with instruments, using Size A or A4 sheet or Size B or A3 sheet, as assigned, and position assigned from Fig. 16.42. (4) Make trimetric drawings, using instruments, with axes chosen to show the objects to best advantage. If dimensions are required, study §16.25.

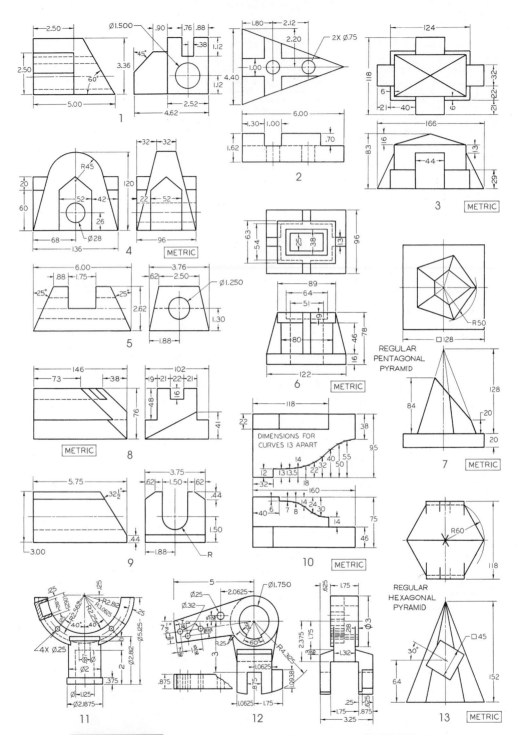

FIGURE 16.57 (1) Make isometric freehand sketches. (2) Make isometric drawings with instruments, using Size A or A4 sheet or Size B or A3 sheet, as assigned. (3) Make dimetric drawings with instruments, using Size A or A4 sheet or Size B or A3 sheet, as assigned, and position assigned from Fig. 16.42. (4) Make trimetric drawings, using instruments, with axes chosen to show the objects to best advantage. If dimensions are required, study §16.25.

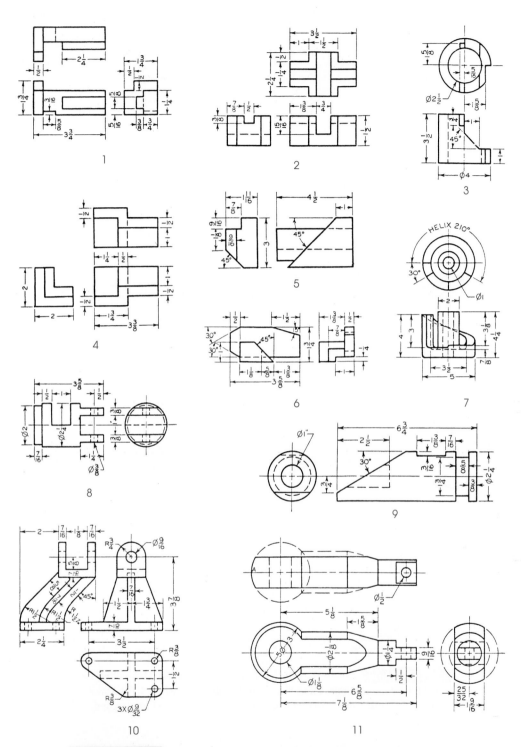

FIGURE 16.58 (1) Make isometric freehand sketches. (2) Make isometric drawings with instruments, using Size A or A4 sheet or Size B or A3 sheet, as assigned. (3) Make dimetric drawings with instruments, using Size A or A4 sheet or Size B or A3 sheet, as assigned, and position assigned from Fig. 16.42. (4) Make trimetric drawings, using instruments, with axes chosen to show the objects to best advantage. If dimensions are required, study §16.25.

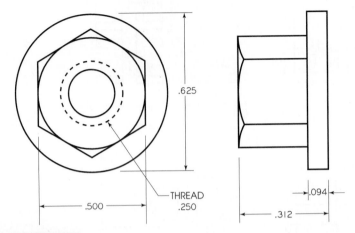

FIGURE 16.59 *Nylon Collar Nut.* **(1) Make isometric freehand sketch. (2) Make isometric drawing with instruments, using Size A or A4 sheet or Size B or A3 sheet, as assigned.**

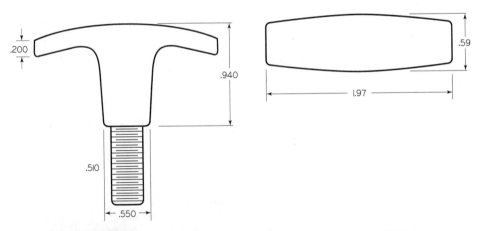

FIGURE 16.60 *Plastic T-Handle Plated Steel Stud.* **(1) Make dimetric drawing with instruments, using Size A or A4 sheet as assigned. (2) Make trimetric drawing, using instruments. Use Size A or A4 sheet as assigned.**

C H A P T E R   1 7

OBLIQUE
PROJECTION

OBJECTIVES

After studying the material in this chapter, you should be able to:

1. Describe how an oblique projection is created.
2. List the advantages of oblique projection.
3. Draw cavalier and cabinet oblique drawings.
4. Know how to place circles when creating an oblique drawing.
5. Describe why CAD software does not automatically create oblique drawings.

OVERVIEW

If an observer is considered to be stationed at an infinite distance from an object and looking toward the object so that the projectors are parallel to each other and oblique to the plane of projection, the resulting drawing is an **oblique projection** (see Fig. 16.1c). As a rule, the object is placed with one of its principal faces parallel to the plane of projection. This is equivalent to holding the object in your hand and viewing it approximately as shown in Fig. 5.27.

Oblique projection provides an easy method for drawing circular features that are parallel to the plane of projection. Axonometric circles are tedious constructions using paper and pencil. With oblique projection, the front view is the same as the front view in a multiview drawing. Circles and angles parallel to the projection plane are true size and shape and are therefore easy to construct. Oblique projection is not as photorealistic as axonometric projection. Depth is distorted and must be compensated for to approximate a realistic appearance for the object. There is little reason to create oblique drawing with CAD since isometric drawings are created automatically once a 3D model is created. While circles in the front view are easy to draw, they are very difficult to create in the top or side views. Therefore, oblique is primarily used on manually constructed drawings, where the majority of circles appear in the front view. Sometimes an object is rotated to position circles in the front view.

■ 17.1 OBLIQUE AND OTHER PROJECTIONS COMPARED

A comparison of oblique projection and orthographic projection is shown in Fig. 17.1. In the oblique projection, the front face A′B′C′D′ is identical with the front view, or orthographic projection, $A^V B^V C^V D^V$. Thus, if an object is placed with one of its faces parallel to the plane of projection, that face will be projected true size and shape in oblique projection as well as in orthographic or multiview projection. This is why oblique is preferable to axonometric projection in the pictorial representation of certain objects. Note that surfaces of the object that are not parallel to the plane of projection will not project in true size and shape. For example, surface ABFE on the object (a square) projects as a parallelogram A′B′F′E′ in the oblique projection.

In axonometric projection, circles on the object nearly always lie in the surfaces inclined to the plane of projection and project as ellipses. In oblique projection, the object may be positioned so that those surfaces are parallel to the plane of projection, in which case the circles will project as true circles and can be easily drawn with the compass.

A comparison of the oblique and orthographic projections of a cylindrical object is shown in Fig. 17.2. In both cases, the circular shapes project as true circles. Note that although an observer looking in the direction of the oblique arrow sees these shapes as ellipses, the drawing, or projection, represents not what is seen but what is projected on the plane of projection. This curious situation is peculiar to oblique projection.

Observe that the axis AB of the cylinder projects as a point $A^V B^V$ in the orthographic projection, since the line of sight is parallel to AB. But in the oblique projection, the axis projects as a line A′B′. The more nearly the direction of sight approaches the perpendicular with respect to the plane of projection—that is, the larger the angle between the projectors and the plane—the closer the oblique projection moves toward the orthographic projection, and the shorter A′B′ becomes.

■ 17.2 DIRECTIONS OF PROJECTORS

In Fig. 17.3, the projectors make an angle of 45° with the plane of projection; hence, the line CD′, which is perpendicular to the plane, projects true length at C′D′. If the projectors make a greater angle with the plane of projection, the oblique projection is shorter, and if the projectors make a smaller angle with the plane of projection, the oblique projection is longer. Theoretically, CD′ could project in any length from zero to infinity. However, the line AB is parallel to the plane and will project in true length regardless of the angle the projectors make with the plane of projection.

In Fig. 17.1, the lines AE, BF, CG, and DH are perpendicular to the plane of projection and project as parallel inclined lines A′E, B′F, C′G, and D′H′ in the oblique projection. These lines on the drawing are called the **receding lines**. As we have seen, they may be any length, from zero to infinity, depending on the direction of the line of sight. Our next concern is this: What angle do these lines make on paper with respect to horizontal?

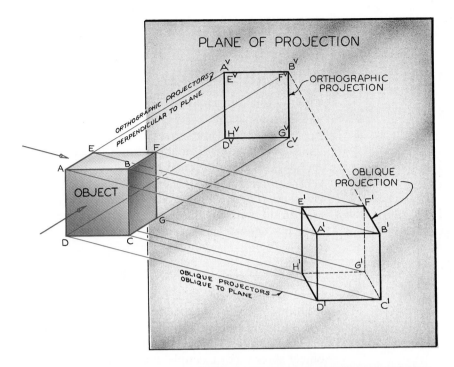

FIGURE 17.1 Comparison of Oblique and Orthographic Projections.

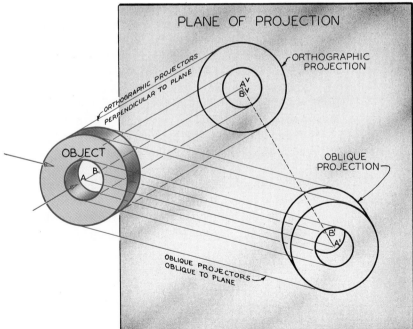

FIGURE 17.2 Circles Parallel to Plane of Projection.

In Fig. 17.4, the line AO is perpendicular to the plane of projection, and all the projectors make angles of 45° with it; therefore, all the oblique projections BO, CO, DO, and so on are equal in length to the line AO. It can be seen from the figure that the projectors may be selected in any one of an infinite number of direc-

tions and yet maintain any desired angle with the plane of projection. It is also evident that the directions of the projections BO, CO, DO, and so on are independent of the angles the projectors make with the plane of projection. Ordinarily, this inclination of the projection is 45° (CO in the figure), 30°, or 60° with

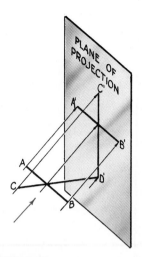

FIGURE 17.3   **Lengths of Projections.**

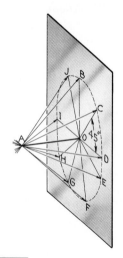

FIGURE 17.4   **Directions and Projections.**

horizontal since these angles may be easily drawn with the triangles.

■ 17.3 ANGLES OF RECEDING LINES

The receding lines may be drawn at any convenient angle. Some typical drawings with the receding lines in various directions are shown in Fig. 17.5. The angle that should be used in an oblique drawing depends on the shape of the object and the location of its significant features. For example, in Fig. 17.6a a large angle was used to obtain a better view of the rectangular recess on the top, while in Fig. 17.6b a small angle was chosen to show a similar feature on the side.

■ 17.4 LENGTH OF RECEDING LINES

Since the eye is accustomed to seeing objects with all receding parallel lines appearing to converge, an oblique projection presents an unnatural appearance, with more or less serious distortion depending on the object shown. For example, the object shown in Fig. 17.7a is a cube, and the receding lines are full length, but the receding lines appear to be too long and to diverge toward the rear of the block. A striking example of the unnatural appearance of an oblique drawing when compared with the natural appearance of a perspective is shown in Fig. 17.8. This example points up one of the chief limitations of oblique projection: Objects characterized by great length should not be drawn in oblique with the long dimension perpendicular to the plane of projection.

The appearance of distortion may be reduced by decreasing the length of the receding lines (remember, we established in §17.2 that they could be any length). In Fig. 17.7 a cube is shown in five oblique drawings with varying degrees of foreshortening of the receding

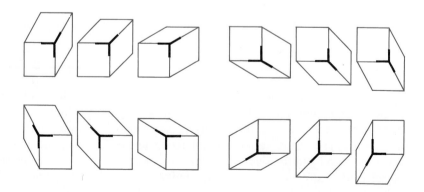

FIGURE 17.5   **Variation in Direction of Receding Axis.**

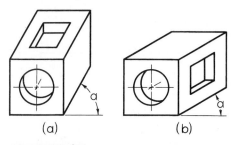

FIGURE 17.6 Angle of Receding Axis.

lines. The range of scales chosen is sufficient for almost all problems, and most of the scales are available on the architects', engineers', or metric scales.

When the receding lines are true length—that is, when the projectors make an angle of 45° with the plane of projection—the oblique drawing is called a **cavalier projection** (Fig. 17.7a). Cavalier projections originated in the drawing of medieval fortifications and were made on horizontal planes of projection. On

these fortifications the central portion was higher than the rest, and it was called *cavalier* because of its dominating and commanding position.

When the receding lines are drawn to half size (Fig. 17.7d), the drawing is commonly known as a **cabinet projection**. The term is attributed to the early use of this type of oblique drawing in the furniture industries. A comparison of cavalier projection and cabinet projection is shown in Fig. 17.9.

■ 17.5 CHOICE OF POSITION

The face of an object showing the essential contours should generally be placed parallel to the plane of projection (Fig. 17.10). If this is done, distortion will be kept to a minimum and labor reduced. For example, in Figs. 17.10a and 17.10c, the circles and circular arcs are shown in their true shapes and may be quickly drawn with the compass, while in Figs. 17.10b and 17.10d these curves are not shown in their true shapes and must be plotted as free curves or in the form of ellipses.

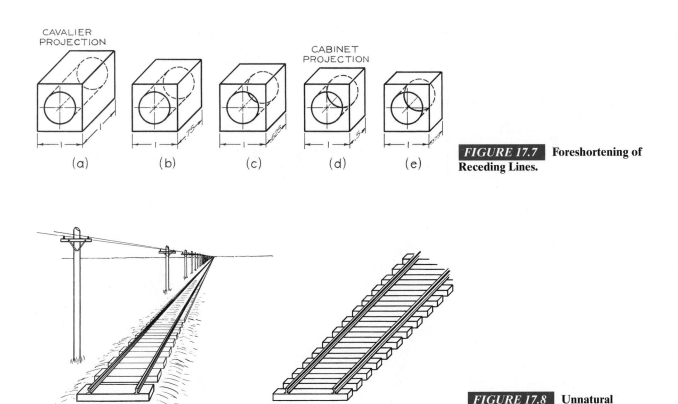

FIGURE 17.7 Foreshortening of Receding Lines.

FIGURE 17.8 Unnatural Appearance of Oblique Drawing.

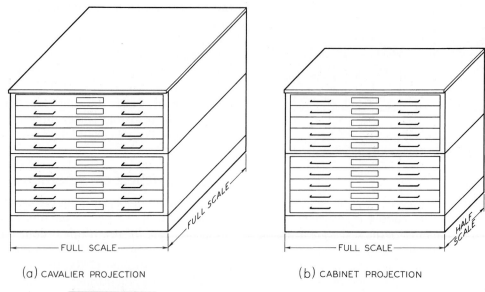

(a) CAVALIER PROJECTION   (b) CABINET PROJECTION

FIGURE 17.9 **Comparison of Cavalier and Cabinet Projections.**

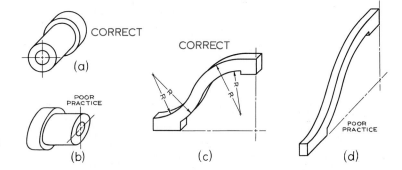

FIGURE 17.10 **Essential Contours Parallel to Plane of Projection.**

The longest dimension of an object should generally be placed parallel to the plane of projection, as shown in Fig. 17.11.

FIGURE 17.11 **Long Axis Parallel to Plane of Projection.**

■ 17.6 STEPS IN OBLIQUE DRAWING

The steps in drawing a cavalier drawing of a rectangular object are shown in Fig. 17.12. Being by drawing the axes OX and OY perpendicular to each other and the receding axis OZ at any desired angle with horizontal (Fig. 17.12I). On these axes, construct an enclosing box, using the overall dimension of the object. Block in the various shapes in detail (II), and heavy in all final lines (III).

Many objects most adaptable to oblique representation are composed of cylindrical shapes built on axes

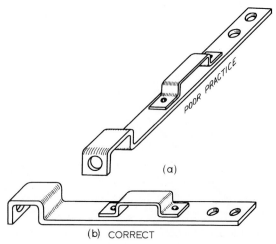

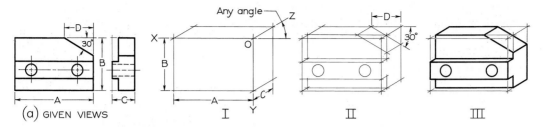

FIGURE 17.12 Steps in Oblique Drawing—Box Construction.

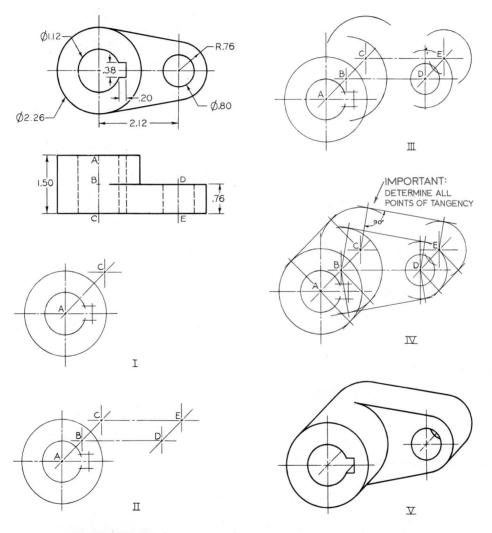

FIGURE 17.13 Steps in Oblique Drawing—Skeleton Construction.

or center lines. In such cases, the oblique drawing is best constructed on the projected center lines, as shown in Fig. 17.13. The object is positioned so that the circles shown in the given top view are parallel to the plane of projection and, hence, can be readily

drawn with the compass to their true shapes. The general procedure is to draw the center-line skeleton, as shown in (I) and (II), and then to build the drawing on these center lines. It is very important to construct all points of tangency, as shown in (IV), especially if the drawing is to be inked. For a review of tangencies, see §§4.30–4.38. The final cavalier drawing is shown in (V).

■ 17.7 FOUR-CENTER ELLIPSE

It is not always possible to place an object so that all its significant contours are parallel to the plane of

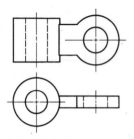

FIGURE 17.14 Circles and Arcs Not Parallel to Plane of Projection.

(a) OBJECT WITH CIRCLES IN DIFFERENT PLANES

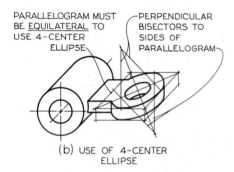

PARALLELOGRAM MUST BE EQUILATERAL TO USE 4-CENTER ELLIPSE

PERPENDICULAR BISECTORS TO SIDES OF PARALLELOGRAM

(b) USE OF 4-CENTER ELLIPSE

projection. For example, the object shown in Fig. 17.14a has two sets of circular contours in different planes, and both cannot be placed parallel to the plane of projection.

In the oblique drawing in Fig. 17.14b, the regular four-center method of Fig. 16.24 was used to construct ellipses representing circular curves not parallel to the plane of projection. This method can be used only in cavalier drawing, in which case the enclosing parallelogram is equilateral—that is, the receding axis is drawn to full scale. The method is the same as in the isometric: Erect perpendicular bisectors to the four sides of the parallelogram; their intersections will be centers for the four circular arcs. If the angle of the receding lines is other than 30° with horizontal, as in this case, the centers of the two large arcs will not fall in the corners of the parallelogram.

The regular four-center method is not convenient in oblique drawing unless the receding lines make 30° with horizontal so that the perpendicular bisectors may be drawn easily with the 30° × 60° triangle and the T-square, parallel rule, or drafting machine without the necessity of first finding the midpoints of the sides. A more convenient method is the alternate four-center ellipse drawn on the two center lines, as shown in Fig. 17.15. This is the same method as used in isometric (Fig. 16.29), but in oblique drawing it varies slightly in appearance according to the different angles of the receding lines.

First, draw the two center lines. Then, from the center, draw a construction circle equal in diameter to the actual hole or cylinder. The circle will intersect each center line at two points. From the two points on

FIGURE 17.15 Alternate Four-Center Ellipse.

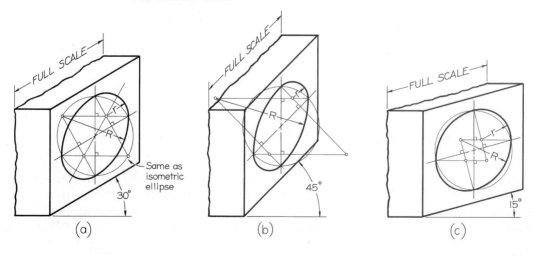

(a)   (b)   (c)

one center line, erect perpendiculars to the other center line. Then, from the two points on the other center line, erect perpendiculars to the first center line. From the intersections of the perpendiculars, draw four circular arcs, as shown.

It must be remembered that the four-center ellipse can be inscribed only in an *equilateral* parallelogram; hence, it cannot be used in any oblique drawing in which the receding axis is foreshortened. Its use is limited, therefore, to cavalier drawing.

■ 17.8 OFFSET MEASUREMENTS

Circles, circular arcs, and other curved or irregular lines may be drawn by means of offset measurements, as shown in Fig. 17.16. The offsets are first drawn on the multiview drawing of the curve (Fig. 17.16a), and these are transferred to the oblique drawing (Fig. 17.16b). In this case, the receding axis is full scale, and therefore all offsets can be drawn full scale. The four-center ellipse could be used, but the method here is more accurate. The final curve is drawn with the aid of the irregular curve.

If the oblique drawing is a cabinet drawing (Fig. 17.16c) or any oblique drawing in which the receding axis is drawn to a reduced scale, the offset measurements parallel to the receding axis must be drawn to the same reduced scale. In this case, there is no choice of methods since the four-center ellipse cannot be used. A method of drawing ellipses in a cabinet drawing of a cube is shown in Fig. 17.16d.

As shown in Fig. 17.17, a free curve may be drawn in oblique by means of offset measurements. This figure also illustrates a case in which hidden lines are used to make the drawing clearer.

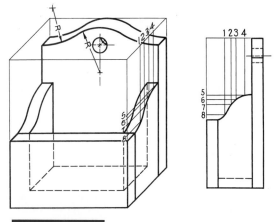

FIGURE 17.17 Use of Offset Measurements.

The use of offset measurements in drawing an ellipse in a plane inclined to the plane of projection is shown in Fig. 17.18. In Fig. 17.18a, a number of parallel lines are drawn to represent imaginary cutting planes. Each plane will cut a rectangular surface between the front end of the cylinder and the inclined surface. These rectangles are drawn in oblique, as shown in Fig. 17.18b, and the curve is drawn through corner points, as indicated. The final cavalier drawing is shown in Fig. 17.18c.

■ 17.9 ANGLES IN OBLIQUE PROJECTION

If an angle that is specified in degrees lies in a receding plane, it is necessary to convert the angle into linear measurements to draw the angle in oblique. For example, in Fig. 17.19a, an angle of 30° is given. To draw the angle in oblique, we need to know dimensions AB and

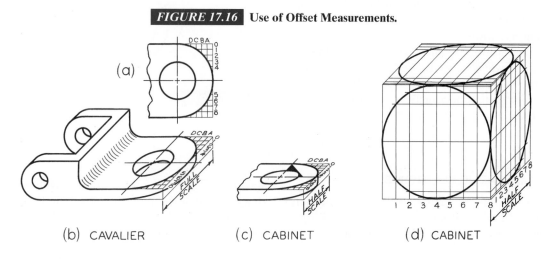

FIGURE 17.16 Use of Offset Measurements.

(a)

(b) CAVALIER   (c) CABINET   (d) CABINET

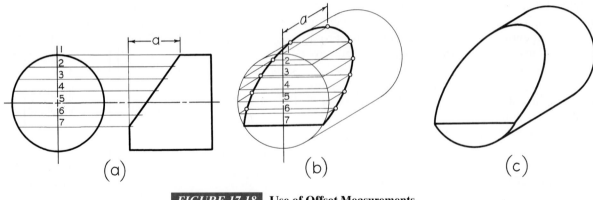

FIGURE 17.18   Use of Offset Measurements.

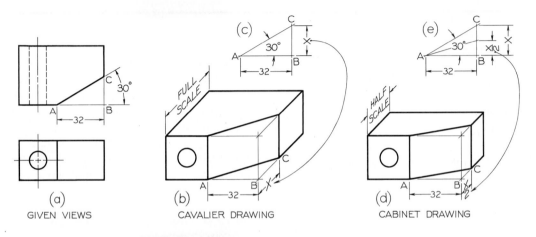

GIVEN VIEWS   CAVALIER DRAWING   CABINET DRAWING

FIGURE 17.19   Angles in Oblique Projection.

BC. The distance AB is given as 32 mm and can be set off directly in the cavalier drawing (Fig. 17.19b). Distance BC is not known, but can easily be found by constructing the right triangle ABC (Fig. 17.19c) from the given dimensions in the top view in Fig. 17.19a. The length BC is then transferred with dividers to the cavalier drawing, as shown.

In cabinet drawing, it must be remembered that *all receding dimensions* must be reduced to half size. Thus, in the cabinet drawing in Fig. 17.19d, the distance BC must be half the side BC of the right triangle in Fig. 17.19e.

FIGURE 17.20   Oblique Half Section.

■ 17.10 OBLIQUE SECTIONS

Sections are often useful in oblique drawing, especially in the representation of interior shapes. An **oblique half section** is shown in Fig. 17.20. Other examples are shown in Figs. 7.41, 7.42, and 7.45. **Oblique full sections**, in which the plane passes completely through the object, are seldom used because they do not show enough of the exterior shapes. In general, all the types of sections discussed in §16.24 for isometric drawing may be applied equally to oblique drawing.

17.11 SCREW THREADS IN OBLIQUE

Parallel partial circles spaced equal to the symbolic thread pitch (see Fig. 13.9a) are used to represent the crests only of a screw thread in a cavalier oblique (Fig. 17.21). For cabinet oblique the space would be one half of the symbolic pitch. If the thread is so positioned to require ellipses, they may be drawn by the four-center method of §17.7.

17.12 OBLIQUE DIMENSIONING

An oblique drawing may be dimensioned in a manner similar to that described in §16.25 for isometric drawing (Fig. 17.22). The general principles of dimensioning, as outlined in Chapter 11, must be followed. As shown in Fig. 17.22, all dimension lines, extension lines, and arrowheads must lie in the planes of the object to which they apply. The dimension figures also will lie in the plane when the aligned dimensioning system is used (Fig. 17.22a). For the unidirectional system of dimensioning, all dimension figures are set horizontal and read form the bottom of the drawing (Fig. 17.22b). This simpler system is often used on pictorials for production purposes. Vertical lettering should be used for all pictorial dimensioning.

Dimensions should be placed outside the outlines of the drawing except when greater clarity or directness of application results from placing the dimensions directly on the view. The dimensioning methods described apply equally to fractional, decimal, and metric dimensions. For many other examples of oblique dimensioning, see Figs. 6.61, 6.65, 6.66, and others on following pages.

17.13 OBLIQUE SKETCHING

Methods of sketching in oblique on plain paper are illustrated in Fig. 5.27. Ordinary graph paper is very useful in oblique sketching (see Fig. 5.28). The height and width proportions can be easily controlled by simply counting the squares. A very pleasing depth proportion can be obtained by sketching the receding lines at 45° diagonally through the squares and through half as many squares as the actual depth would indicate.

17.14 COMPUTER GRAPHICS

Using computer graphics, the drafter can easily create an oblique drawing that will provide the desired amount of foreshortening along the receding axis as well as the preferred direction of the axis. CAD programs also permit curves and circular features, which are not parallel to the frontal plane, to be readily shown on the drawing. Oblique sections (§17.10) and repetitive features such as screw threads (§17.11) may be quickly and accurately depicted.

FIGURE 17.21   Screw Threads in Oblique.

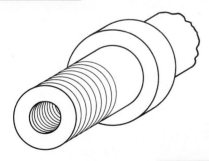

FIGURE 17.22   Oblique Dimensioning.

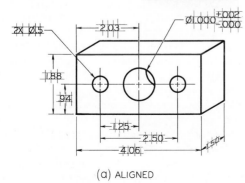

(a) ALIGNED

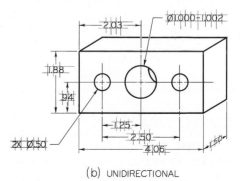

(b) UNIDIRECTIONAL

High Technology Is First Mate In the Race for America's Cup[*]

A klaxon sounds and the sleek hull of the racing yacht begins knifing through cool, calm waters at increasing speed. The bow heaves as it pushes through the flat water, raising a wave as it moves faster and sending ripples fanning from the waterline. People riding along feel the breeze in their faces as the hull glides at speeds of up to 14 knots, but the motion and sensations last only a few seconds. This dash by a 25-foot-long scale model covers only about 800 feet and takes place in a huge water tank, not the open ocean. But it nonetheless signals a beginning of the next race for the America's Cup, the world's most prestigious sailing competition.

Challengers hoping to capture the America's Cup from New Zealand in 2000 have already begun extensive research and testing directed toward designing and buiding the fastest sailboats in their class. Months, even years, before construction begins on the boats, teams of naval architects, designers, computer scientists, model builders and engineers engage in a technological competition to create machines that can complete a two-hour race a few minutes or seconds ahead of rivals. It is this competition that brings yacht builders to the David Taylor Model Basin at the Naval Surface Warfare Center. Here, where the Navy tows and tests models of its future destroyers, frigates and other warships in the world's largest towing tanks, John K. Marshall, president of the New York Yacht Club, watches as fiberglass models of different racing hulls go through their paces.

"This is our surface warfare, so perharps it's appropriate that we are here," said Mr. Marshall, director of the Young America campaign, a $40 million effort by the New York Yacht Club to build the boat that wins the right to challenge Team New Zealand for the oldest trophy in international sport.

"Sailing in a sport, an athletic competition for which people must train and develop their skills," Mr. Marshall said. "But the America's Cup is also a technology competition and it's always been that way".

"If your boat is as fast, you can win sailing skill," said Bob Billingham of America One, the group organizing the entry for the St. Francis Yacht Club of San Francisco. "But you can't win with a slower boat.

So team worldwide labor to refine their designs, jealously guarding studies of hull shapes, sails and even rigging hardware to deny the smallest secret to competitors. Yachts in this class are so closely matched that no advantage is insignificant. Veterans remember when an Australian challenger in 1983 snatched the cup from the United States for the first time with help from a radical innovation, a winged keel. The Australians kept the keel secret until the last moment to prevent competitors from trying to copy it.

Getting the fastest boat requires working within the strictures of tight rules governing the overall design of this class of vessels. From 1958 to 1987, sailors competed for the America's Cup in 12-Meter Class yachts. But in 1989, a multinational group of yacht designers developed rules for a new International America's Cup Class, which first appeared in the 1992 competition.

The new America's Cup boats are lighter, faster, narrower and longer and carry more sail than their predecessors, with canoe-like bodies made of carbon fiber material instead of aluminum. The design is based on a mathematical formula that balances a boat's waterline length, sail area and displacement so that adding significantly to one dimension requires decreasing others. Generally, an America's Cup contender in about 75 feet long, supports a mast that stands 115 feet off the water, has a keel 14 feet deep and weighs 45,000 to 48,000 pounds. More than 40,000 pounds are in a lead ballast bulb at the base of the keel.

Registering to challenge Team New Zealand, who gained the prize in 1995 sailing Black Magic, 16 yacht clubs and syndicates from 10 nations, including 5 from the United States, have so far paid their $200,000 entry fees. Experts estimate that 10 or 12 of these groups will raise enough money to build at least one boat and that perhaps 4 or 5 of the competitors will muster the talent and expertise to produce first-rate vessels with a chance of winning. The competitors are to assemble in the harbor off Auckland in October 1999 and begin a series of match races to determine the best boat to be named the official challenger for the America's Cup. The winning boat will then race Team New Zealand's best new yacht in a best-of-seven series held in February and March of 2000 in the Hauraki Gulf, northeast of Auckland.

With such intense competition, any weight reduction or change in hull shape, sail design, or the placement of components, like the keel or rudder, that resluts in even a 1 percent increase in performance is significant.

Increasingly, the teams rely on computer simulation and the ability to test many design ideas in the cyberseas of a mathematical model before building and trying them in the real word. "Engineers use sophistcated software, known as *computational fluid dynamics programs*", said John Kuhn, a naval architect at Science Applications International Corporation in San Diego, a technical firm supporting the San Francisco group. The programs simulate the fluid flow around hulls and appendages, like rudders, keels and ballast bulbs, or the movement of air around masts or sails. Results from the programs, which calculate pressure and drag, give engineers the information they need to design components that are then tested in tanks or wind tunnels.

Information from these tests the *fluid dynamics programs* then go into a larger computer simulation called a *velocity prediction program*, or VPP. This program combines design specifications with environmental variables like wind, wave and temperature to make predictions on how fast a boat will sail in specific conditions. "A VPP integrates the work of different people working on different parts of a boat and predicts how an overall design will perform on water," said Mr. Kuhn, technical coordinator for the San Francisco group. "These programs are not perfect, but they help tell you how basic elements contribute to a design".

With each competition, said Duncan MacLane, technology project manager for the New York group, the designers are seeing better matches between computer predictions and actual performance. Still, advice from naval architects and other experts like Bruce Farr, the principal designer for the New York group, remains crucial.

"There is still a lot of art in the design process," Mr. MacLane said, "with many of the improvements we are considering coming from the intuition of designers. There is still a designer in front of the computer screen dealing with the nuances, making very subtle alignments in the design that produces a winner".

Tom Schnackenberg, who heads the design team for the New Zealand group, said that although computational and model testing made significant contributions to producing a boat, only full-scale testing and analysis of a real vessel could confirm its design performance. "In the real world", Mr. Scknackenberg said, "we often find full-scale results at variance with predictions". Team New Zealand plans to build at least one boat and use its older championship yachts in its preparations, said Alan Sefton, a spokesman for the group.

Many of the teams preparing new boats are paying extra attention to the sails and riggings, partly because of differences in these areas that appeared in the last cup race, said Mr. Kuhn of the San Francisco group. The mast and rigging on Black Magic, the New Zealand boat, was set farther back than those on other races, and the New Zealanders displayed sails with unusual shapes.

"Everyone in looking at sails and rigging because sail aerodynamics is one of the least understood elements of design," Mr. Kuhn said. "This is where we may be able to squeeze out some more performance". He and other experts said any group that wished to seek the America's Cup but had yet to start this kind of research and planning was probably out of luck. "The race," Mr. Kuhn said, "has already begun".

*Adapted from "High Technology is First Mate in the Race for America's Cup," by Warren E. Leary, *New York Times*, July 21, 1998.

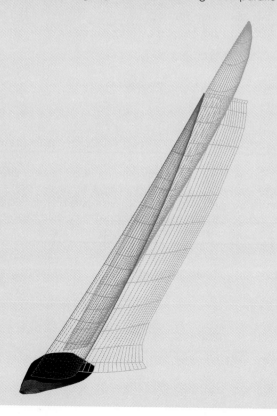

■ KEY WORDS

OBLIQUE PROJECTION	PROJECTORS
PLANE OF PROJECTION	RECEDING LINES
CAVALIER	CABINET
SKELETON CONSTRUCTION	OFFSET MEASUREMENT

■ CHAPTER SUMMARY

• Oblique projection makes drawing circles in the projection plane easier than with other pictorial projection methods.

• There are two commonly used types of oblique projection: cavalier and cabinet.

• Width and height are drawn true size and true shape in oblique projection. The depth axis for cavalier (full depth) is usually 30°, while the depth axis for cabinet (half depth) is usually 45°.

• Oblique drawings of circular features are often created by first drawing a skeleton of center lines.

• There is usually no reason for creating oblique drawings using CAD, since isometric drawings are easier to make with CAD and appear more photorealistic.

• Oblique projection is a common sketching method because the front view is true size and true shape and easier to draw.

■ REVIEW QUESTIONS

1. What is the primary advantage of an oblique projection?

2. Which is the most photorealistic: isometric, perspective, or oblique projection? Which is the least photorealistic?

3. If a hockey puck were to be drawn using oblique projection, how should it be positioned to appear as a circle?

4. Can an angle on an oblique drawing be measured in the front view? In the right side view? In the top view?

5. Why are oblique drawings seldom created with CAD software?

6. What is the first thing that should be drawn when creating an oblique drawing?

7. Describe how to plot an irregular curve in an oblique drawing.

■ OBLIQUE PROJECTION PROBLEMS

Many problems to be drawn in oblique—either cavalier or cabinet—are given in Figs. 17.23–17.26. They may be drawn freehand (§5.14) using graph paper or plain drawing paper as assigned by the instructor, or they may be drawn with instruments. In the latter case, all construction lines should be shown on the completed drawing.

Many additional problems suitable for oblique projection will be found in Figs. 6.52–6.54, 8.27–8.29, 16.54–16.57, and 18.34.

Since many of the problems in this chapter are of a general nature, they can also be solved on most computer graphics systems. If a system is available, the instructor may choose to assign specific problems to be completed by this method.

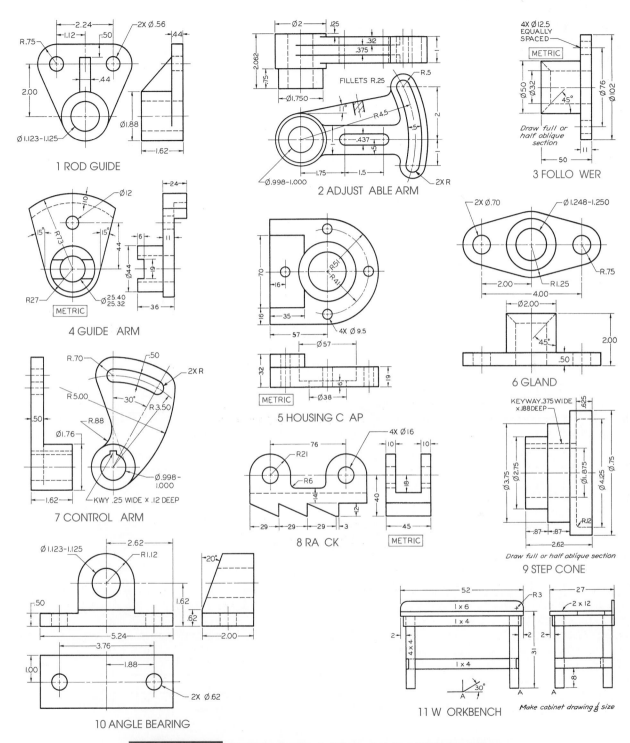

1 ROD GUIDE

2 ADJUST ABLE ARM

3 FOLLO WER

4 GUIDE ARM

5 HOUSING C AP

6 GLAND

7 CONTROL ARM

8 RA CK

9 STEP CONE

10 ANGLE BEARING

11 W ORKBENCH

Make cabinet drawing ⅛ size

FIGURE 17.23 **(1) Make freehand oblique sketches. (2) Make oblique drawings with instruments, using Size A or A4 sheet, or Size B or A3 sheet, as assigned. If dimensions are required, study §17.12.**

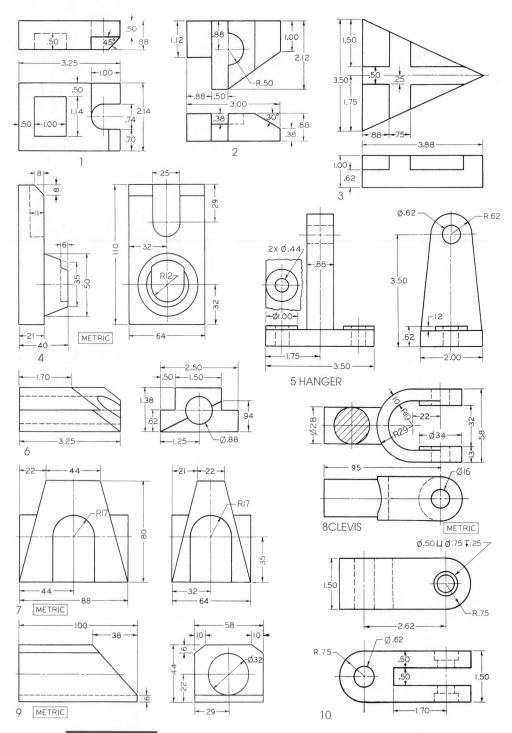

FIGURE 17.24   Make oblique drawings with instruments, using Size A or A4 sheet, or Size B or A3 sheet, as assigned. If dimensions are required, study §17.12.

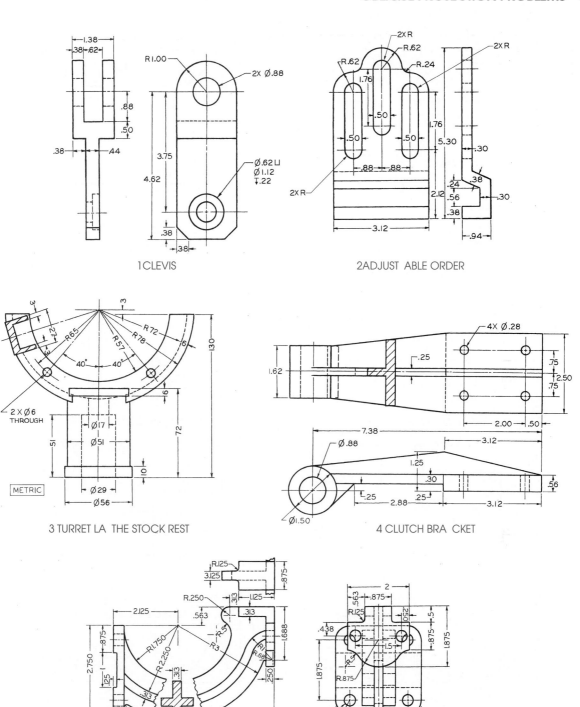

1 CLEVIS

2 ADJUST ABLE ORDER

3 TURRET LA THE STOCK REST

4 CLUTCH BRA CKET

5 RAIL SUPPORT

FIGURE 17.25 Make oblique drawings with instruments, using Size A or A4 sheet, or Size B or A3 sheet, as assigned. If dimensions are required, study §17.12.

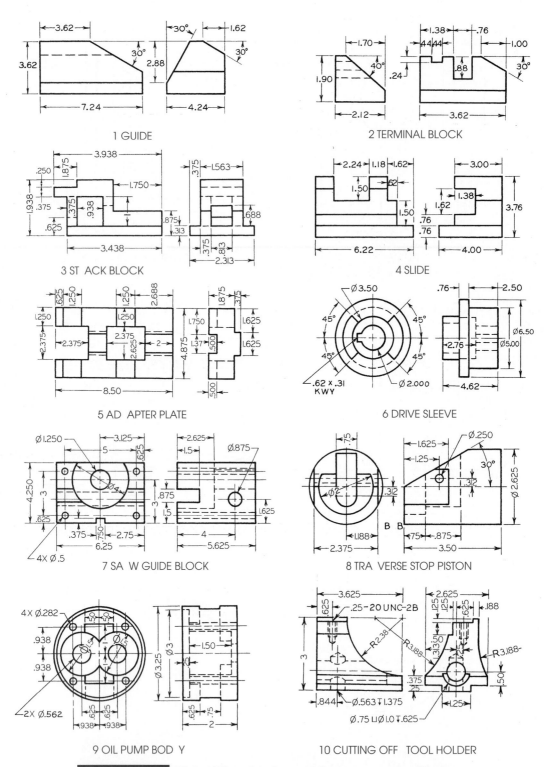

1 GUIDE

2 TERMINAL BLOCK

3 ST ACK BLOCK

4 SLIDE

5 AD APTER PLATE

6 DRIVE SLEEVE

7 SA W GUIDE BLOCK

8 TRA VERSE STOP PISTON

9 OIL PUMP BOD Y

10 CUTTING OFF TOOL HOLDER

FIGURE 17.26 Make oblique drawings with instruments, using Size A or A4 sheet, or Size B or B3 sheet, as assigned. If dimensions are required, study §17.12.

CHAPTER 18

PERSPECTIVE

OBJECTIVES

After studying the material in this chapter, you should be able to:

1. Identify a drawing created using perspective projection.
2. List the differences between perspective projection and axonometric projection.
3. Create a drawing using multiview perspective.
4. Describe three types of perspective.
5. Measure distances in perspective projection.

OVERVIEW

Perspective, or **central projection**, excels over all other types of projection in the pictorial representation of objects because it more closely approximates the view produced by the human eye (Fig. 18.1). Geometrically, an ordinary photograph is a perspective. While perspective is of major importance to architects, industrial designers, and illustrators, engineers at one time or another are also certain to be concerned with the pictorial representation of objects and should understand the basic principles of perspective (see ANSI/ASME Y14.4M–1989 (R1994)).

Unlike axonometric projection, which places the observer at an infinite distance from the object, perspective projection defines the observer at a finite distance. This projection technique causes parallel edges to converge at vanishing points, as in a photograph. This technique duplicates the image produced by the human eye of the real world. To accomplish this image accuracy, every point and edge of an object must be projected from a multiview drawing.

The three types of perspective are one-point, two-point, and three-point perspective. Perspective projection is complicated and requires both theoretical understanding of projection and proficiency in construction techniques. This complexity exceeds the capabilities of many engineering CAD software programs for personal computers. Advanced CAD software programs that run on faster workstation computers can easily render perspective projection.

◼ 18.1 GENERAL PRINCIPLES

As explained in §1.9, a perspective involves four main elements: (1) the observer's eye, (2) the object being viewed, (3) the plane of projection, and (4) the projectors from the observer's eye to all points on the object. With rare exceptions (see §18.6), the plane of projection is placed between the observer and the object, as shown in Fig. 1.9, and the collective piercing points in the plane of projection of all the projectors produce the perspective.

In Fig. 18.2, the observer is shown looking along a boulevard and through an imaginary plane of projection. This plane is called the **picture plane**, or PP. The position of the observer's eye is called the **station point**, or SP. The lines from SP to the various points in the scene are the projectors, or more properly in per-

spective, **visual rays**. The points where the visual rays pierce PP are the **perspectives** of the respective points. Collectively, these piercing points form the perspective of the object or the scene as viewed by the observer. The perspective thus obtained is shown in Fig. 18.3.

In Fig. 18.2, the perspective of lamp post 1–2 is shown at 1'–2', on the picture plane; the perspective of lamp post 3–4 is shown at 3'– 4', and so on. Each succeeding lamp post, as it is farther from the observer, will be projected smaller than the one preceding. A lamp post at an infinite distance from the observer would appear as a point on the picture plane. A lamp post in front of the picture plane would be projected taller than it is, and a lamp post in the picture plane would be projected in true length. In the perspective (Fig. 18.3), the diminishing heights of the posts are apparent.

In Fig. 18.2, the line representing the *horizon* is the edge view of the *horizon plane*, which is parallel to the ground plane and passes through SP. In Fig. 18.3, the horizon is the line of intersection of this plane with the picture plane and represents the eye level of the observer, or SP. Also, in Fig. 18.2, the *ground plane* is the edge view of the ground on which the object usually rests. In Fig. 18.3, the *ground line*, or GL, is the intersection of the ground plane with the picture plane.

In Fig. 18.3, it will be seen that lines that are parallel to each other but not parallel to the picture plane, such as curb lines, sidewalk lines, and lines along the tops and bottoms of the lamp posts, all converge toward a single point on the horizon. This point is called the **vanishing point**, or VP, of the lines. Thus, the first rule to learn in perspective is this: *All parallel lines that are not parallel to* PP *vanish at a single vanishing point*, and *if these lines are parallel to the ground, the vanishing point will be on the horizon.* Parallel lines that are also parallel to PP, such as the lamp posts, remain parallel and do not converge toward a vanishing point.

◼ 18.2 MULTIVIEW PERSPECTIVE

A perspective can be drawn by the ordinary methods of multiview projection, as shown in Fig. 18.4. The upper portion of the drawing shows the top view of the station point, the picture plane, the object, and the visual rays. At the right are the right-side view of the same station point, picture plane, object, and visual rays. In the front view, the picture plane coincides with the plane of the paper, and the perspective is drawn on it. Note the method of projecting from the top view to

FIGURE 18.1   **A CAD-produced Perspective of an Airport.** *The material has been reprinted with permission from and under the copyright of Autodesk, Inc.*

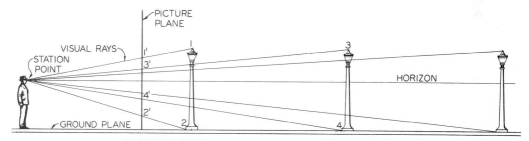

FIGURE 18.2   **Looking Through the Picture Plane.**

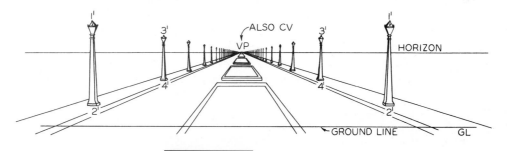

FIGURE 18.3   **A Perspective.**

the side view, which conforms to the usual multiview methods shown in Fig. 6.7.

To obtain the perspective of point 1, a visual ray is drawn in the top view from SP$_T$ to point 1 on the object. From the intersection 1′ of this ray with the picture plane, a projection line is drawn downward until it meets a similar projection line from the side view. This intersection is the perspective of point 1, and the perspectives of all other points are found in a similar manner.

Observe that all parallel lines that are also parallel to the picture plane (the vertical lines) remain parallel

and do not converge, whereas the other two sets of parallel lines converge toward vanishing points. Because the vanishing points are not needed in the multiview construction of Fig. 18.4, they are not shown. But if the converging lines were extended, it would be found that they meet at two vanishing points (one for each set of parallel lines).

The perspective of any object may be constructed in this way, but if the object is placed at an angle with the picture plane, as is usually the case, the method is a bit cumbersome because of the necessity of constructing the side view in a revolved position. The revolved side view can be dispensed with, as shown in the following section.

Parallel Railroad Tracks Seem to Converge in the Distance.
Willard Clay; Tony Stone Worldwide.

FIGURE 18.4   **Multiview Method of Drawing Perspective.**

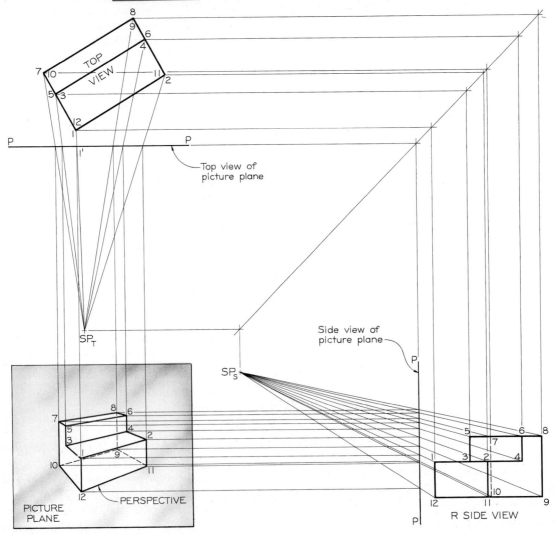

■ 18.3 THE SETUP FOR A SIMPLE PERSPECTIVE

The construction of a perspective of a simple form is shown in Fig. 18.5. The upper portion of the drawing, as in Fig. 18.4, shows the top views of SP, PP, and of the object. The lines SP–1, SP–2, SP–3, and SP–4 are the top views of the visual rays.

In the side view, a departure from Fig. 18.4 is made, in that a revolved side view is not required. All that is needed is any elevation view that will provide the necessary elevation or height measurements. If these dimensions are known, no view is required.

The perspective itself is drawn in the front-view position, the picture plane being considered as the plane of the paper on which the perspective is drawn. The ground line is the edge view of the ground plane or the intersection of the ground plane with the picture plane. The horizon is a horizontal line in the picture plane that is the line of intersection of the horizon plane with the picture plane. Since the horizon plane passes through the observer's eye, or SP, the horizon is drawn at the level of the eye—that is, at the distance above the ground line representing, to scale, the altitude of the eye above the ground.

The *center of vision*, or CV, is the orthographic projection (or front view) of SP on the picture plane, and since the horizon is at eye level, CV will always be on the horizon, except in three-point perspective (see §18.11). In Fig. 18.5, the top view of CV is CV′, found by dropping a perpendicular from SP to PP. The front view CV is found by projecting downward from CV′ to the horizon.

FIGURE 18.5 Perspective of a Prism.

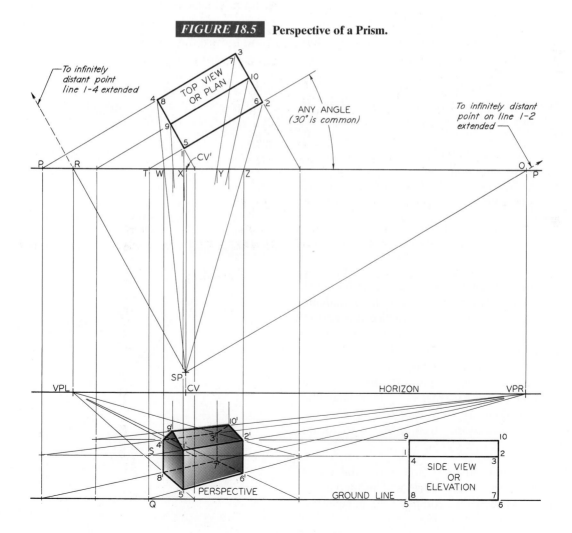

■ 18.4 DRAWING AN ANGULAR PERSPECTIVE

Since objects are defined principally by edges that are straight lines, the drawing of a perspective resolves itself into drawing the *perspective of a line*. A drafter who can draw the perspective of a line can draw the perspective of any object, no matter how complex.

To draw the perspective of any horizontal straight line not parallel to PP—for example, the line 1–2 in Fig. 18.5—proceed as follows.

I. *Find the piercing point in* PP *of the line*. In the top view, extend line 1–2 until it pierces PP at T; then project downward to the level of the line 1–2 projected horizontally from the side view. The point S is the piercing point of the line.

II. *Find the vanishing point of the line*. The vanishing point of a line is the piercing point in PP of a line drawn through SP parallel to that line. Hence, the vanishing point VPR of the line 1–2 is found by drawing a line from SP parallel to that line and finding the top view of its piercing point O, and then projecting downward to the horizon. The line SP–O is actually a visual ray drawn toward the infinitely distant point on line 1–2 of the object, extended, and the vanishing point is the intersection of this visual ray with the picture plane. The vanishing point is, then, the perspective of the infinitely distant point on the line extended

III. *Join the piercing point and the vanishing point with a straight line*. The line S–VPR is the line joining these two points, and it is the perspective of a line of infinite length containing the required perspective of the line 1–2.

IV. *Locate the endpoints of the perspective of the line.* The endpoints 1' and 2' can be found by projecting down from the piercing points of the visual rays in PP, or by simply drawing the perspectives of the remaining horizontal edges of the object. In practice, it is best to use both methods as a check on the accuracy of the construction. To locate the endpoints by projecting from the piercing points, draw visual rays from SP to the points 1 and 2 on the object in the top view. The top views of the piercing points are X and Z. Since the perspectives of points 1 and 2 must lie on the line S–VPR, project downward from X and Z to locate points 1' and 2'.

After the perspectives of the horizontal edges have been drawn, the vertical edges and inclined edges can be drawn, as shown, to complete the perspective of the object. Note that *vertical heights can be measured only in the picture plane*. If the front vertical edge 1–5 of the object was actually PP—that is, if the object was situated with the front edge in PP—the vertical height could be set off directly full size. If the vertical edge is behind PP, a plane of the object, such as surface 1–2–5–6, can be extended forward until it intersects PP in line TQ. The line TQ is called a *measuring line*, and the true height SQ of line 1–5 can be set off with a scale or projected from the side view as shown.

If a large drawing board is not available, one vanishing point, such as VPR, may fall off the board. By using one vanishing point VPL and projecting down from the piercing points in PP, vanishing point VPR may be eliminated. However, a valuable means of checking the accuracy of the construction will be lost.

■ 18.5 POSITION OF THE STATION POINT

The center line of the cone of visual rays should be directed toward the approximate center, or center of interest, of the object. In two-point perspective, the type shown in Fig. 18.5, the location of the station point (SP) in the plan view should be slightly to the left, not directly in front of the center of the object, and at such a distance that the object can be viewed at a glance without turning the head. This is accomplished if a cone of rays with its vertex at SP and a vertical angle of about 30° entirely closes the object, as shown in Fig. 18.6.

In the perspective portion of Fig. 18.5, SP does not appear because the station point is in front of the pic-

FIGURE 18.6   **Distance from Station Point to Object.**

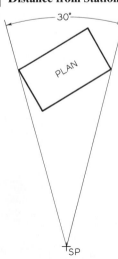

ture plane. However, the orthographic projection CV of SP in the picture plane does show the height of the station point with respect to the ground plane. Since the horizon is at eye level, it also shows the altitude of the station point. Therefore, in the perspective portion of the drawing, the horizon is drawn a distance above the ground line at which it is desired to assume the station point. For most small and medium-size objects, such as machine parts or furniture, the station point is best assumed slightly above the top of the object. Large objects, such as buildings, are usually viewed from a station point about the altitude of the eye above the ground, or about 5′–6″.

■ 18.6 LOCATION OF THE PICTURE PLANE

In general, the picture plane is placed in front of the object, as in Fig. 18.7b and Fig. 18.7c. However, it may be placed behind the object, as in Fig. 18.7a, and it may even be placed behind SP, as in Fig. 18.7d, in which event the perspective is reversed, as is the case of a camera. Of course, the usual position of the picture plane is between SP and the object. The perspectives in Fig. 18.7 differ in size but not in proportion. As in Figs. 18.7b and 18.7c, with the picture plane between SP and the object, the farther that plane is from the object, the smaller the perspective will be. This distance may be assumed, therefore, with the thought of controlling the scale of the perspective. In practice, however, the object is usually assumed with the front corner in the picture plane to facilitate vertical measurements (see Fig. 18.12).

■ 18.7 POSITION OF THE OBJECT WITH RESPECT TO THE HORIZON

To compare the elevation of the object with that of the horizon is equivalent to referring it to the level of the eye (SP) because the horizon is on a level with the eye, except in three-point perspective (see §18.11). The differences in effect produced by placing the object on, above, or below the horizon are shown in Fig. 18.8.

If the object is placed above the horizon, it is above the level of the eye, or above SP, and will appear as seen from below. Likewise, if the object is below the horizon, it will appear as seen from above.

■ 18.8 THE THREE TYPES OF PERSPECTIVES

Perspective drawings are classified according to the number of vanishing points required, which in turn de-

pends on the position of the object with respect to the picture plane.

If the object is situated with one face parallel to the plane of projection, only one vanishing point is required, and the result is a *one-point perspective*, or *parallel perspective* (§18.9).

If the object is situated at an angle with the picture plane but with vertical edges parallel to the picture plane, two vanishing points are required, and the result is a *two-point perspective*, or an *angular perspective*. This is the most common type of perspective drawing and is the one described in §18.4 (see also §18.10).

If the object is situated so that no system of parallel edges is parallel to the picture plane, three vanishing points are necessary, and the result is a *three-point perspective* (see §18.11).

■ 18.9 ONE-POINT PERSPECTIVE

In one-point perspective (see Fig. 16.1d), the object is placed so that two sets of its principal edges are parallel to PP and the third set is perpendicular to PP. This third set of parallel lines will converge toward a single vanishing point in perspective, as shown.

In Fig. 18.9, the view shows the object with one face parallel to the picture plane. If desired, this face could be placed in the picture plane. The piercing points of the eight edges perpendicular to PP are found by extending them to PP and then projecting downward to the level of the lines as projected across from the elevation view.

To find the VP of these lines, a visual ray is drawn from SP parallel to them (the same as in step II of §18.4), and it is found that the *vanishing point of all lines perpendicular to* PP *is in* CV. By connecting the eight piercing points with the vanishing point CV, the indefinite perspectives of the eight edges are obtained.

To cut off on these lines the definite lengths of the edges of the object, horizontal lines are drawn from the ends of one of the edges in the top view and at any desired angle with PP—45°, for example, as shown. The piercing points and the vanishing point VPR of these lines are found, and the perspectives of the lines are drawn. The intersections of these with the perspectives of the corresponding edges of the object determine the lengths of the receding edges. The perspective of the object may then be completed as shown.

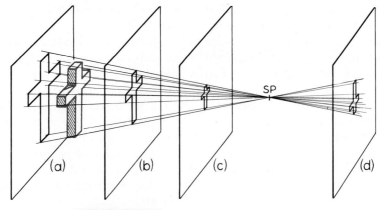

FIGURE 18.7   Location of Picture Plane.

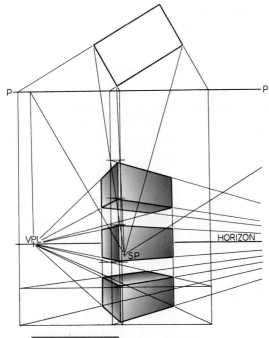

FIGURE 18.8   Object and Horizon.

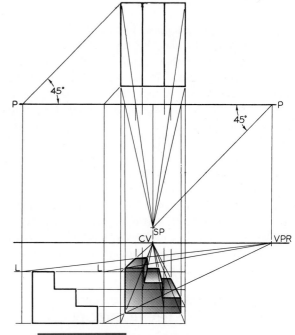

FIGURE 18.9   One-Point Perspective.

One of the most common uses of parallel perspective is in the representation of interiors of buildings, as illustrated in Fig. 18.10.

An adaptation of one-point perspective, which is simple and effective in representing machine parts, is shown in Fig. 18.11. The front surface of the cylinder is placed in PP, and all circular shapes are parallel to PP; hence, these shapes will be projected as circles and circular arcs in the perspective. SP is located in front and to one side of the object, and the horizon is placed well above the ground line. The single vanishing point is on the horizon in CV.

The two circles and the keyway in the front surface of the object will be drawn true shape because they lie in PP. The circles are drawn with the compass on center O′. To locate R′, the perspective center of the large arc, draw visual ray SP–R; then, from its intersection X with PP, project down to the center line of the large cylinder, as shown.

To find the radius T′W′ at the right end of the perspective, draw visual rays SP–T and SP–W, and from their intersections with PP, project down to T′ and W′ on the horizontal center line of the hole.

FIGURE 18.10 One-Point Perspective. *Courtesy of Autodesk, Inc.*

■ 18.10 TWO-POINT PERSPECTIVE

In two-point perspective, the object is placed so that one set of parallel edges is vertical and has no vanishing point, while the two other sets have vanishing points. This is the most common type and is the method discussed in §18.4. It is suitable especially for representing buildings in the architectural drawing, or large structures in civil engineering, such as dams or bridges.

The perspective drawing of a small building is shown in Fig. 18.12. It is common practice (1) to assume a vertical edge of an object in PP so that direct measurements may be made on it and (2) to place the object so that it faces make unequal angles with PP; for example, one angle may be 30° and the other 60°. In practical work, complete multiview drawings are

usually available, and the plane and elevation may be fastened in position, used in the construction of the perspective, and later removed.

Since the front corner AB lies in PP, its perspective A′B′ may be drawn full size by projecting downward from the plan and across from the elevation. The lengths of the receding lines from this corner are cut off by vertical lines SC′ and RE′ drawn from the intersections S and R, respectively, of the visual rays to these points of the object. The perspectives of the tops of the windows and the door are determined by the lines A′–VPR and A′–VPL, and their widths and lateral spacings are determined by projecting downward from the intersections with PP of the respective visual rays. The bottom lines of the windows are determined by the lines V′–VPR and V′–VPL.

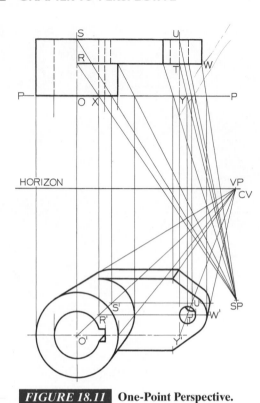

FIGURE 18.11 One-Point Perspective.

The perspective of the line containing the ridge of the roof is found by joining N′, the point where the ridge line pierces the picture plane, and VPR. The ridge ends O′ and Q′ are found by projecting downward from the intersections of the visual rays with PP, or by drawing the perspectives of any two lines intersecting at the points. The perspective of the roof is completed by joining the points O′ and Q′ to the end of the eaves.

■ 18.11 THREE-POINT PERSPECTIVE

In three-point perspective, the object is placed so that none of its principal edges is parallel to the picture plane (PP); therefore, each of the three sets of parallel edges will have a separate vanishing point (VP) (Fig. 18.13). The picture plane is assumed approximately perpendicular to the center line of the cone of rays.

In this figure, think of the paper as the picture · plane, with the object behind the paper and placed so that all its edges make an angle with the picture plane. If a point CV is chosen, it will be the orthographic projection of your eye, or the station point, on the picture plane. The vanishing points P, Q, and R are found by conceiving lines to be drawn from a station point in

space parallel to the principal axes of the object and finding their piercing points in the picture planes. It will be recalled that the basic rule for finding the vanishing point of a line in any type of perspective is to draw a visual ray, or line, from the station point parallel to the edge of the object whose vanishing point is required and finding the piercing point of this ray in the picture plane. Since the object is rectangular, these lines to the vanishing points are at right angles to each other in space exactly as the axes are in axonometric projection (see Fig. 17.48). The lines PQ, QR, and RP are perpendicular, respectively, to CV–R, CV–P, and CV–Q and are the *vanishing traces*, or horizon lines, of planes through SP parallel to the principal faces of the object.

The imaginary corner O is assumed in the picture plane and may coincide with CV; but as a rule the front corner is placed at one side near CV, thus determining how nearly the observer is assumed to be directly in front of this corner.

In this method the perspective is drawn directly from measurements and not projected from views. The dimensions of the object are given by the three views, and these will be set off on *measuring lines* GO, EO, and OF (see §18.13). The measuring lines EO and OF are drawn parallel to the vanishing trace PQ, and the measuring line GO is drawn parallel to RQ. These measuring lines are actually the lines of intersection of principal surfaces of the object, extended, with PP. Since these lines are in PP, true measurements of the object can be set off along them.

Three *measuring points* M_1, M_2, and M_3 are used in conjunction with the *measuring lines*. To find M_1, revolve triangle CV–R–Q about RQ as an axis. Since it is a right triangle, it can be easily constructed true size with the aid of a semicircle, as shown. With R as center and R–SP_1 as radius, strike arc SP_1–M_1, as shown. M_1 is the measuring point for the measuring line GO. Measuring points M_2 and M_3 are found in a similar manner.

Height dimensions, taken from the given views, are set off full size or to any desired scale, along measuring line GO, at points 3, 2, and 1. From these points, lines are drawn to M_1, and heights on the perspective are the intersections of these lines with the perspective front corner OT of the object. Similarly, the true depth of the object is set off on measuring line EO from 0 to 5, and the true width is set off on measuring line OF from 0 to 8. Intermediate points can be constructed in a similar manner.

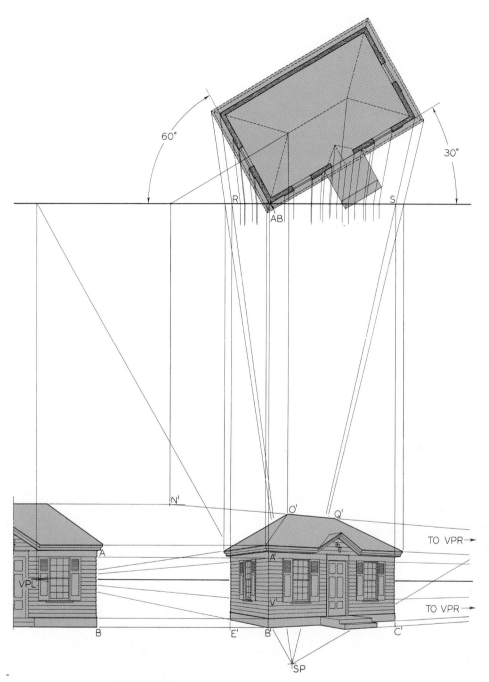

FIGURE 18.12   **Perspective Drawing of a Small Building.**

■ 18.12 THE PERSPECTIVE LINEAD AND TEMPLATE

The **perspective linead** consists of three straight-edged blades that can be clamped to each other at any desired angles (Fig. 18.14a). This instrument is conve-

nient in drawing lines toward a vanishing point outside the limits of the drawing.

Before starting such a drawing, a small-scale diagram should be made in which the relative positions of the object, PP, and SP are assumed, and the

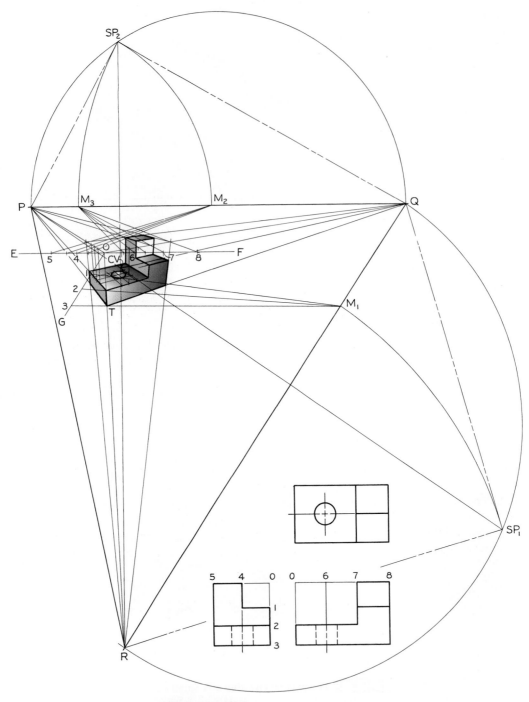

Three-Point Perspective.

distances of the vanishing points from CV determined (Fig. 18.14b). Draw any line LL through a vanishing point as shown; then on the full-size drawing, assume CV and locate LL, as shown in Fig. 18.14a.

To set the linead, clamp the blades in any convenient position; set the edge of the long blade along the horizon; and draw the lines BA and BC along the short blades. Then set the edge of the long the blade along line LL, and draw the lines DE and DF to intersect the lines first drawn at points G and H. Set pins at these points. If the linead is moved so that the short blades touch the pins, all lines drawn along the edge of the long blade will pass through VPL. This method is based on the principle that an angle inscribed in a circle is measured by half the arc it subtends.

A **perspective template** (Fig. 18.15) of thin wood or heavy cardboard, cut in the form of a circular arc, may be used instead of a perspective linead. If the template is attached to the drawing board so that the inaccessible VP is at the center of the circular arc and the T-square is moved so that the head remains in contact with the template, lines drawn along the edge of the blade will, if extended, pass through the inaccessible VP.

If the edge of the blade does not pass through the center of the head, the lines drawn will be tangent to a circle whose center is at VP and whose radius is equal to the distance from the center of the head to the edge of the blade.

■ 18.13 MEASUREMENTS IN PERSPECTIVE

As explained in §18.1, all lines in PP are shown in their true lengths, and all lines behind PP are foreshortened.

Let it be required to draw the perspective of a line of telephone poles (Fig. 18.16). Let OB be the line of intersection of PP with the vertical plane containing the poles. In this line, the height AB of a pole is set off directly to the scale desired, and the heights of the perspectives of all poles are determined by drawing lines from A and B to VPR.

To locate the bottoms of the poles along the line B–VPR, set off along PP the distances 0–1, 1–2, 2–3, . . . ,

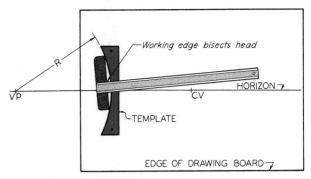

FIGURE 18.15 **Perspective Template.**

equal to the distance from pole to pole; draw the lines 1–1, 2–2, 3–3, . . . , forming a series of isosceles triangles 0–1–1, 0–2–2, 0–3–3, ... The lines 1–1, 2–2, 3–3, . . . , are parallel to each other and, therefore, have a common vanishing point MP, which is found in the usual manner by drawing from SP a line SP–T parallel to the lines 1–1, 2–2, 3–3, . . . , and finding its piercing point MP (*measuring point*) in PP.

Since the line SP–X is parallel to the line of poles 1–2–3, . . . , the triangle SP–X–T is an isosceles triangle, and T is the top view of MP. The point T may be determined by setting off the distance X–T equal to SP–X or simply by drawing the arc SP–T with center at X and radius SP–X.

Having the measuring point MP, find the piercing points in PP of the lines 1–1, 2–2, 3–3, . . . , and draw their perspectives as shown. Since these lines are horizontal lines, their piercing points fall in a horizontal line BZ in PP, at the bottom of the drawing. Along BZ the true distances between the poles are set off; hence, BZ is called a *measuring line*. The intersections 1', 2',

FIGURE 18.14 **Perspective Linead.**

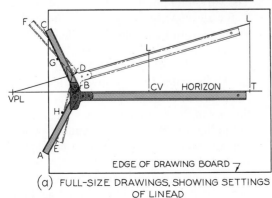

(a) FULL-SIZE DRAWINGS, SHOWING SETTINGS OF LINEAD

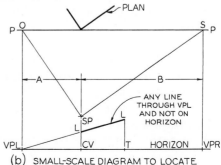

(b) SMALL-SCALE DIAGRAM TO LOCATE VANISHING POINTS AND LINE LL

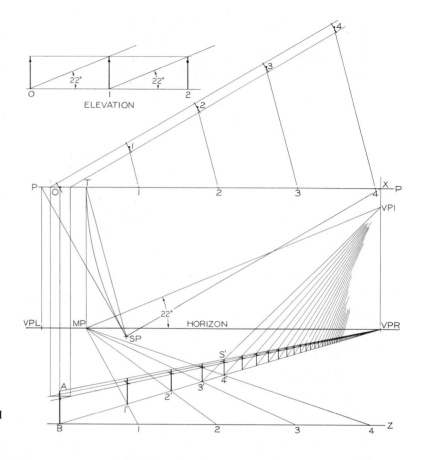

FIGURE 18.16 Measurement of Vertical and Horizontal Lines.

3′, . . ., of the perspectives of the lines 1–1, 2–2, 3–3, . . ., with the line B–VPR determine the spacing of the poles.

It will be seen that only a few measurements may be made along the measuring line BZ within the limits of the drawing. For additional measurements, the *diagonal method* of spacing may be employed, as shown. Since all diagonals from the bottom of each pole to the top of the succeeding pole are parallel, they have a common vanishing point VPI, which may be found as explained in §18.14. It is evident that the diagonal method exclusively may be used in the solution of this problem.

The method of direct measurements may also be applied to lines inclined to PP and to the ground plane, as illustrated in Fig. 18.17 for the line XE, which pierces PP at X. If the end of the house is conceived to be revolved about a vertical axis XO into PP, the line XE will be shown in its true length and inclination at XY. This line XY may be used as the measuring line for XE; it remains only to find the corresponding measuring point MP. The line YE is the horizontal base of an isosceles triangle having its vertex at X, and a line drawn parallel to it through SP will determine MP, as described for Fig. 18.16.

■ 18.14 VANISHING POINTS OF INCLINED LINES

The vanishing point of an inclined line is determined, as for all other lines, by finding the piercing point in PP of a line drawn from SP parallel to the given line.

Fig. 18.18 shows the perspective of a small building. The vanishing point of the inclined roof line C′E′ can be determined as follows: If a plane is conceived to be passed through the station point and parallel to the end of the house (plan view), it would intersect PP in the line XY, through VPL, and perpendicular to the horizon. Since the line drawn from SP parallel to C′E′ (in space) is in the plane SP-X-Y, it will pierce PP at some point T in XY. To find the point T, conceive the plane SP-X-Y revolved about the line XY as an axis in-

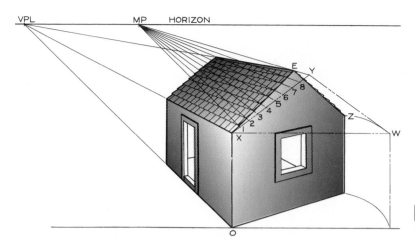

FIGURE 18.17 Measurement of Inclined Lines.

to PP. The point SP will then fall on the horizon at a point shown by O in the top view and by MR in the front view. From the point MR draw the revolved position of the line SP–T (now MR–T) making an angle of 30° with the horizon and thus determining the point T, which is the vanishing point of the line C′E′ and of all lines parallel to that line. The vanishing point S of the line D′E′ is evidently in the line XY, because D′E′ is in the same vertical plane as the line C′E′. The vanishing point S is as far below the horizon as T is above the horizon, because the line E′D′ slopes downward at the same angle at which the line C′E′ slopes upward.

The perspectives of inclined lines can generally be found without finding the vanishing points, by finding the perspectives of the endpoints and joining them. *The perspective of any point may be determined by finding the perspectives of any two lines intersecting at the point.* Obviously, it would be best to use horizontal lines, parallel, respectively, to systems of lines whose vanishing points are already available. For example, in Fig. 18.18, to find the perspective of the inclined line EC, the point E′ is the intersection of the horizontal lines R′–VPR and B′–VPL. The point C′ is already established, since it is in PP; but if it were not in PP, it could be easily found in the same manner. The perspective of the inclined line EC is, therefore, the line joining the perspectives of the endpoints E′ and C′.

■ 18.15 CURVES AND CIRCLES IN PERSPECTIVE

If a circle is parallel to PP, its perspective is a circle. If the circle is inclined to PP, its perspective may be any one of the conic sections in which the base of the cone is the given circle, the vertex is SP, and the cutting

plane is PP. But since the center line of the cone of rays should be approximately perpendicular to the picture plane, the perspective will generally be an ellipse. The ellipse may be constructed by means of lines intersecting the circle, as shown in Fig. 18.19. The radial lines in the elevation view at the left can be easily drawn with the 45° and 30° × 60° triangles. A convenient method for determining the perspective of any plane curve is shown in Fig. 18.20.

■ 18.16 THE PERSPECTIVE PLAN METHOD

A perspective may be drawn by drawing first the perspective of the plan of the object (Fig. 18.21a), then the vertical lines (Fig. 18.21b), and finally the connecting lines (Fig. 18.21c). However, in drawing complicated structures, the superimposition of the perspective on the perspective plan causes a confusion of lines. For this reason, the perspective of the plan from which the location of vertical lines is determined is drawn either above or below its normal location. A suggestion of the range of possible positions of the perspective plan is given in Fig. 18.22; use of the perspective plan below the perspective is shown in Fig. 18.23.

The chief advantages of the perspective plan method over the ordinary plan method are that the vertical lines of the perspective can be spaced more accurately and that a considerable portion of the construction can be made above or below the perspective drawing, so that a confusion of lines on the required perspective is avoided.

When the perspective plan method is used, the ordinary plan view can be omitted and measuring points used to determine distances along horizontal edges in the perspective.

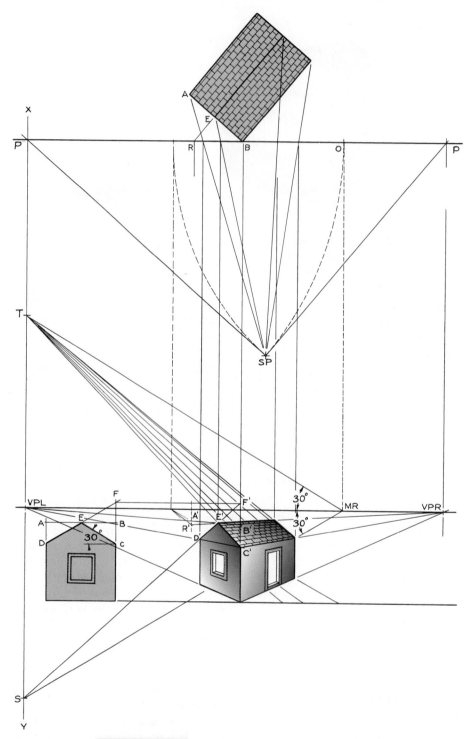

FIGURE 18.18   Vanishing Points of Inclined Lines.

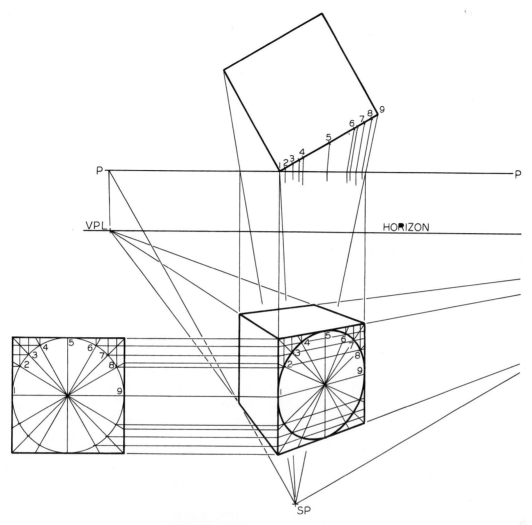

FIGURE 18.19 Circles in Perspective.

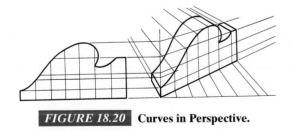

FIGURE 18.20 Curves in Perspective.

■ 18.17 PERSPECTIVE DIAGRAM

The spacing of vanishing points and measuring points may be determined graphically or may be calculated. In Fig. 18.24 a simple diagram of the plan layout shows the position of the object, the picture plane, the station point, and the constructions for finding the vanishing points and measuring points for the problem in Fig. 18.23. As indicated in the figure, the complete plan need not be drawn. The diagram should be drawn to any small convenient scale, and vanishing points and measuring points should be set off in the perspective to the larger scale desired.

In practice, structures are usually considered in one of a limited number of simple positions with reference to the picture plane, such as 30° × 60°, 45° × 45°, and 20° × 70°. Therefore, a table of measurements for locating vanishing points and measuring points may be easily prepared, to avoid the necessity of a special construction for each drawing.

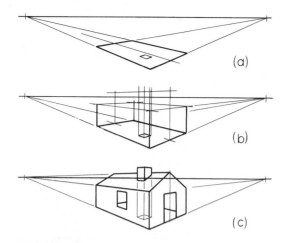

FIGURE 18.21 Building upon the Perspective Plan.

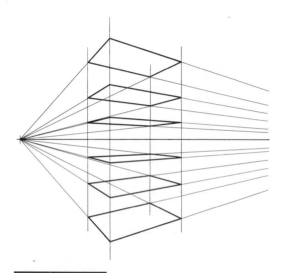

FIGURE 18.22 Positions of Perspective Plan.

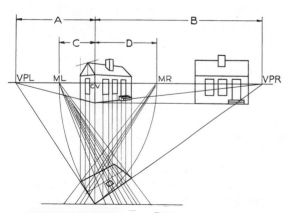

FIGURE 18.23 Perspective Plan Method.

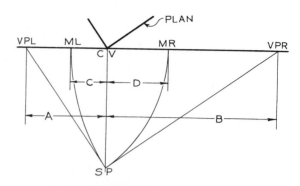

FIGURE 18.24 Perspective Diagram.

■ 18.18 SHADING

The effect of light can be utilized advantageously in describing the shapes of objects and in the finish and embellishment of such drawings as display drawings, patent drawings, and industrial pictorial drawings. Ordinary working drawings are not shaded.

Since the purpose of an industrial pictorial drawing is to show clearly the shape and not to be artistic, the shading should be simple and limited to producing a clear picture. Some of the common types of shading are shown in Fig. 18.25. Pencil or ink lines are drawn mechanically (Fig. 18.25a) or freehand (Fig. 18.25b). Two methods of shading fillets and rounds are shown in Figs. 18.25c and 18.25d. Shading produced with pen dots is shown in Fig. 18.25e, and pencil "tone" shading is shown in Fig. 18.25f. Pencil shading applied to pictorial drawings on tracing paper may be reproduced with good results by making a whiteprint or a blueprint.

Examples of line shading on pictorial drawings often used in industrial sales literature are shown in Figs. 18.25 to 18.27.

■ 18.19 COMPUTER GRAPHICS

Perspective drawings, which provide pictorials most resembling photographs, are also the most time-consuming types of pictorials to draw. CAD programs are available that will produce either wireframe (Fig. 18.28), or solid perspective representations, with user selection of viewing distance, focal point, z-axis convergence, and arc resolution scale. Historically, perspectives have seen far greater application in architectural than in engineering drawing. Now the availability of these computer graphics routines makes perspective drawing a viable alternative for the drafter wishing to employ a pictorial representation of an object.

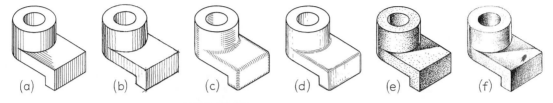

FIGURE 18.25   Methods of Shading.

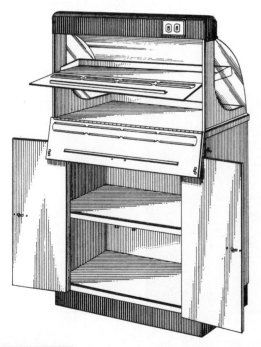

FIGURE 18.26   Surface Shading Applied to Pictorial Drawing of Display Case.

FIGURE 18.27   A Line-Shaded Drawing of an Adjustable Support for Grinding. *Courtesy of A. M. Byers Co.*

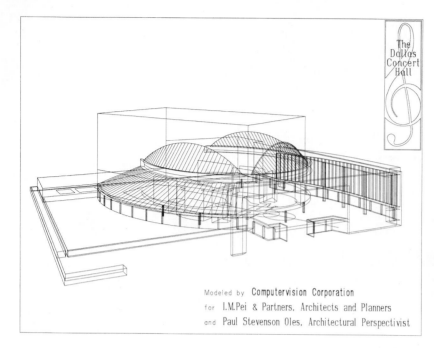

Modeled by Computervision Corporation
for I.M.Pei & Partners, Architects and Planners
and Paul Stevenson Oles, Architectural Perspectivist

FIGURE 18.28 **Perspective Drawing Produced by Using the Computervision Designer System for Building and Management (BDM).** *Courtesy of Computervision Corporation, a subsidiary of Prime Computer, Inc.*

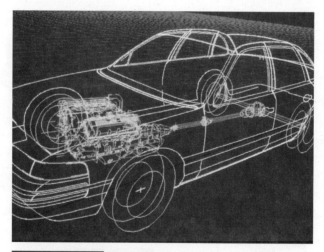

FIGURE 18.29 **Wireframe CAD Perspective of a Car.** *Courtesy of Ford Motor Company.*

■ KEY WORDS

PERSPECTIVE PROJECTION	OBSERVER'S POINT OF VIEW
CENTER OF VISION (CV)	VANISHING POINT (VP)
GROUND LINE (GL)	PICTURE PLANE (PP)
HORIZON	STATION POINT (SP)
VISUAL RAYS	ANGULAR PERSPECTIVE

■ CHAPTER SUMMARY

- The most photorealistic projection is perspective.

- There are three types of perspective projection: one-point, two-point, and three-point perspective.

- In perspective projection, parallel edges converge to one or more vanishing points, which replicate the image of objects as seen by the human eye.

- Object edges are projected onto a picture plane via a straight line to the observer's eye (station point).

- Perspective projection requires two orthographic views (usually top and right side) to construct the perspective view.

- The location and relationship between the vanishing points, the picture plane, and the object determine the appearance of the perspective view.

- In one-point perspective, the object is placed so that two of the three primary axes of the object are parallel to the picture plane.

- In two-point perspective, the object is placed so that only one of the three primary axes of the object are parallel to the picture plane.

- In three-point perspective, the object is placed so that none of the three primary axes of the object are parallel to the picture plane.

■ REVIEW QUESTIONS

1. What is the primary advantage of a perspective projection?
2. Why is perspective projection rarely used in engineering?
3. What is the purpose of the picture plane?
4. What is the station point?
5. How does the distance between the station point and the ground line affect the final perspective drawing?

6. What is the relationship between the station point and the horizon?

7. What type of perspective is often used for rendering interior spaces in architectural drawings?

8. What tools are available to assist the drafter in creating perspective drawings on paper?

■ PERSPECTIVE PROBLEMS

Layouts for perspective problems are given in Figs. 18.30–18.32. Draw on a size B or A3 sheet of paper, vellum, or film, with your name, date, class, and other information lettered below the border as specified by the instructor. Omit dimensions.

Draw problems from Fig. 18.33 on size B or A3 paper. Determine the arrangement on the sheet to produce the most effective perspective in each case. For problems in Fig. 18.34, select both sheet size and scale.

Since many of the problems in this chapter are of a general nature, they can also be solved on most computer graphics systems. If a system is available, the instructor may choose to assign specific problems to be completed by this method.

In addition to these problems, many suitable problems for perspective are found among the axonometric and oblique problems at the ends of Chapters 16 and 17.

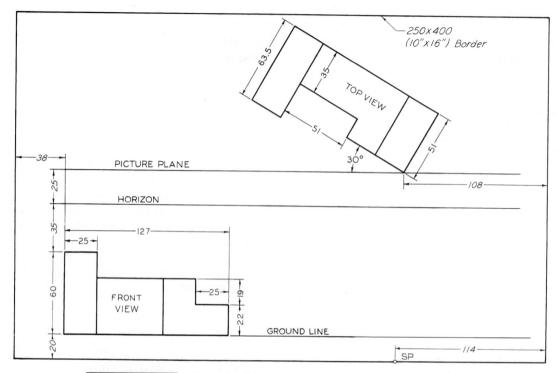

FIGURE 18.30 Draw views and perspective. Omit dimensions. Use Size B or A3 sheet.

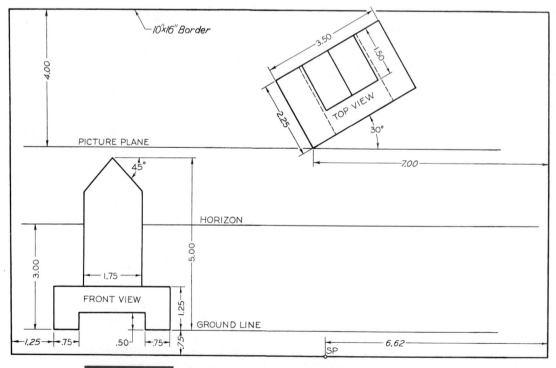

FIGURE 18.31 Draw views and perspective. Omit dimensions. Use Size B or A3 sheet.

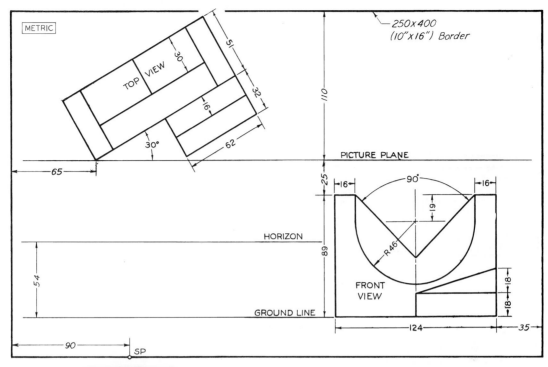

FIGURE 18.32 Draw views and perspective. Omit dimensions. Use Size B or A3 sheet.

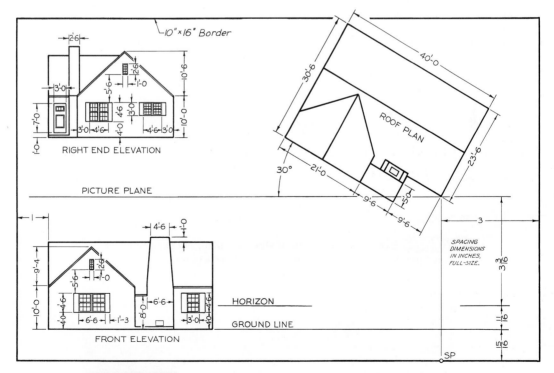

FIGURE 18.33 Draw front elevation, plan, and perspective. Omit dimensions. Scale: $\frac{1}{8}'' = 1' - 0$. Use Size B or A3 sheet.

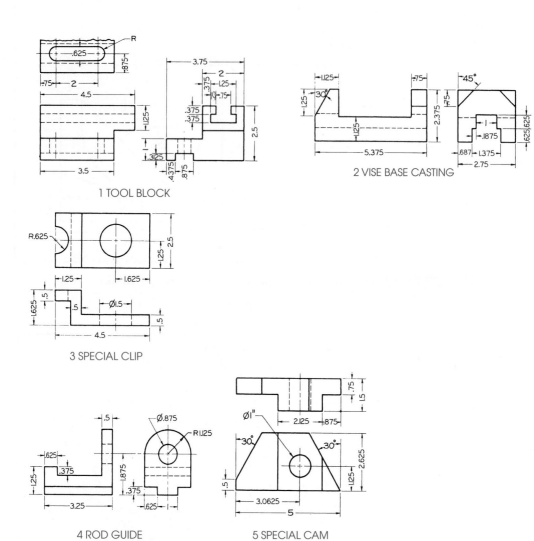

FIGURE 18.34 Draw side or front elevation, plan, and perspective of assigned problem. Omit dimensions. Use Size B or A3 sheet.

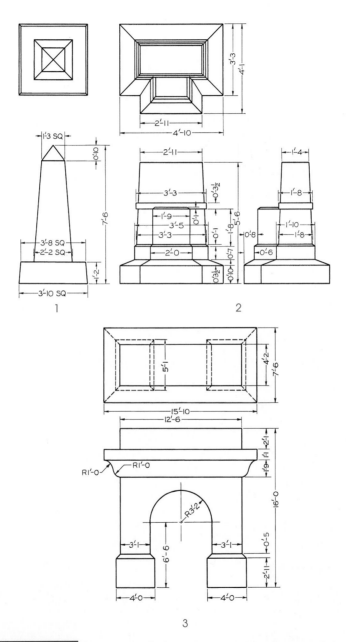

FIGURE 18.35 Draw side or front elevation, plan, and perspective of assigned problem. Omit dimensions. Select sheet size and scale

C H A P T E R   1 9

POINTS, LINES, AND PLANES

OBJECTIVES

After studying the material in this chapter, you should be able to:

1. Correctly label points, lines, and planes in space using standard descriptive geometry notation.
2. Determine whether lines in space intersect by examining two views.
3. Draw the visibility of two non-intersecting lines given two views.
4. Draw the visibility of a line intersecting a plane given two views.
5. Define and draw a frontal, horizontal, and profile line in three views.
6. Define and draw an inclined line in three views.
7. Define and draw an oblique line in three views.
8. Determine true length and true angle of a line.
9. Determine bearing and grade of a line.
10. Determine point view of a line by auxiliary view.
11. Draw the edge view of a plane by auxiliary view.
12. Find the dihedral angle between two planes.
13. Draw the true size of an oblique plane by revolution.
14. Find the piercing point of a line and a plane.
15. Find the intersection of two planes.

OVERVIEW

The science of graphical representation and the solution of spatial relationships of points, lines, and planes by means of projections are the concerns of *descriptive geometry*.

All of engineering graphics is based on the mathematical theory of descriptive geometry. Descriptive geometry uses a graphical process for describing the relation of objects in space. Using descriptive geometry techniques, the true length, true size, true angle, and intersection of lines and planes can be determined simply and with graphical precision. Descriptive geometry using special notation, labeling, and various construction techniques that assist in reading and writing solution to graphical problems. Many descriptive geometry solutions to complex spatial problems are remarkably simple, especially when compared to the alternate trigonometric solutions. The methods of representation of solid objects, wherein the planes of projection of the "glass box" and views of the object were projected upon the planes, have been discussed in §§6.1, 6.2, 6.19-6.25. The elements of the objects—points, lines, and planes—now will be discussed and explained.

During the latter part of the eighteenth century the French mathematician Gaspard Monge developed the principles of descriptive geometry to solve spatial problems related to military structures. In France and Germany, Monge's descriptive geometry soon became a part of national education. In 1816 Claude Crozet introduced descriptive geometry into the curriculum of the United States Military Academy at West Point. In 1821 Crozet published his *Treatise on Descriptive Geometry*, the first important English work on descriptive geometry published in this country. Since then descriptive geometry has been taught in many engineering colleges, and today no study of engineering graphics is considered complete without a detailed study of descriptive geometry. See also §1.5. Chapters 19–25 of this text present the basic principles and applications that every engineer should know.

■ 19.1 BASIC GEOMETRIC ELEMENTS

We start with the representation of a single point. In Fig. 19.1 (a) the front, top, and right-side views of point 1 are shown. The projections of the point are indicated by a small cross and the number 1. The folding line H/F is shown between the front and top views, and the folding line F/P is shown between the front and side views. Thus the names of the views are indicated by the letters H, F, and P (for horizontal, frontal, and pro-

file planes of projection). See §6.3. As indicated by the dimensions D, the distance of the top view to the H/F folding line is equal to the distance of the side view to the F/P folding line.

In Fig. 19.1 (b) the front and top views, or projections, of two connected points (line 1–2) are shown. Note the thin projection lines between the views. (Projection lines are usually drawn in pencil as very light construction lines.) At (c) are shown the front and top views of three connected points, or plane 1–2–3. Again, note the projection lines between the views.

In order to describe an object, lay out a mechanism, or begin the graphical solution of an engineering problem, the relative positions of two or more points must be specified. For example, in Fig. 19.2(a), the relative positions of points 1 and 2 could be described as follows: point 2 is 32 mm to the right of point 1, 12 mm below (or lower than) point 1, and 16 mm behind (or to the rear of) point 1.

When points 1 and 2 are connected, as at (b), observe that the preceding specifications have placed point 2 at a definite distance from point 1 along line 1–2 (more properly line *segment* 1–2, since line 1–2 could be extended).

When point 3 is introduced, as at (c), and is connected to point 2, line 2–3 is established. Since lines 1–2 and 2–3 have point 2 in common, they are *intersecting* lines.

If line 2–3 of Fig. 19.2 (c) is altered to position 3–4, Fig. 19.3 (a), do we still have intersecting lines? The possible point in common could only be at 5 in the top view and 5′ in the front view. The question then becomes "Are 5 and 5′ in fact views of the same point?" Since adjacent views of a point must be aligned, §6.2, a vertical projection line is added at (b), and, since this line connects 5 and 5′, it is evident that lines 1–2 and 3–4 are actually intersecting lines.

In Fig. 19.4 (a) another pair of lines, 1–2 and 3–4 is shown. In this case apparent points of intersection 5 and 5′ are not aligned with the projection lines between views and hence do not represent views of the same point. Therefore, these two lines do not intersect. Such nonintersecting, nonparallel lines are called *skew* lines. The relationship of these skew lines will now be considered in more detail.

Since the lines do not intersect, one must be above the other in the region of point 5. At (b) this region has been assigned two numbers, 5 and 6, in the top view, and it is arbitrarily decided that 5 is a point on line 1–2 and 6 is on line 3–4. These points are then projected to the front view as shown. The direction of sight for the top view is downward toward the front view of the pair of points 5

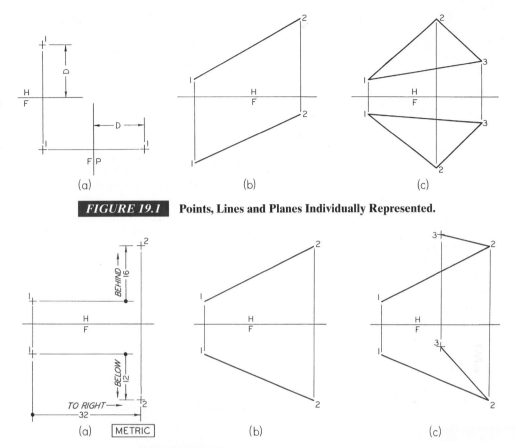

FIGURE 19.1   Points, Lines and Planes Individually Represented.

FIGURE 19.2   Views of Points and Lines.

and 6. It is observed that point 5 on line 1–2 is *higher* in space or *nearer* to the observer than is point 6 on line 3–4. Line 1–2 thus passes above line 3–4 close to point 5.

In like manner, at (c), numbers 7 and 8 are assigned to the apparent crossing point in the front view and projected to the top view, with 8 assigned to line 1–2 and 7 to line 3–4. It is now noted that the direction of sight for a front view is upward (on the paper) toward the top view. With this in mind, it is observed that point 7, and therefore line 3–4, are *nearer* to the observer (in the front view) than are point 8 and line 1–2. Line 3–4 thus passes in front of line 1–2 in the vicinity of point 7.

The foregoing is useful in determining the visibility of nonintersecting members of a structure or of pipes and tubes, Fig. 19.5. At (a) the views are incomplete because it has not been determined which of the two rods is visible at the apparent crossover in each view. Only the relative positions of the center lines need be investigated. As before, concentration is limited temporarily to the apparent crossing point in the top view, with numbers 5 and 6 assigned to the region.

Point 5 is projected to line 1–2 and point 6 to line 3–4 in the front view, where it is discovered that point 5 is above point 6. Line 1–2 therefore passes above line 3–4 and rod 1–2 is visible at the crossover in the top view, as shown at (b). Rod 3–4 is, of course, hidden where it passes below rod 1–2 and is completed accordingly, as shown.

Attention is now directed to the apparent crossing in the front view, Fig. 19.5 (b), and numbers 7 and 8 are assigned. Projected to the top view, these reveal that point 7 on line 3–4 is in front of point 8 on line 1–2. Rod 3–4 is therefore visible in the front view and rod 1–2 is hidden. The completed views are shown at (c).

This discussion has been in terms of front and top views, but it should be realized that the principle applies to *any pair of adjacent views* of the same structure. For example, study Figs. 19.4 and 19.5 with this book held upside down. Observe that the top views become the front views and vice versa but that the visibility is not altered.

Observe finally that *any* two adjacent views, Fig. 19.6, have this same fundamental relationship. The di-

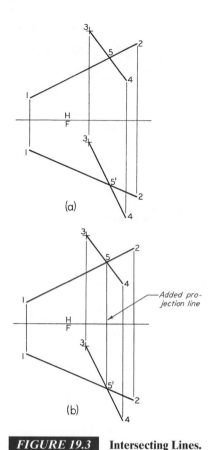

(a)

(b)

—Added pro-
jection line

FIGURE 19.3 Intersecting Lines.

rection of sight for either view is always directed *toward* the adjacent view. Hence at (a), in view B, it is observed that point 5 on edge 1–3 is the nearer of the two assigned points 5 and 6. Thus edge 1–3 is visible in view B, and it follows that edge 2–4 is hidden. Note that this

procedure reveals nothing about the visibility of the interior lines of view A of the tetrahedron. At (b), the positions of points 7 and 8 relative to the direction of sight for view A reveal that edge 1–3 is visible in view A.

■ 19.2 INCLINED LINE AND ANGLE WITH PLANE OF PROJECTION

By definition, an inclined line appears true length on the plane to which it is parallel, §6.22. For convenience or precision, inclined lines are frequently classified as *frontal, horizontal,* or *profile,* Fig. 19.7.

Observe that the true-length view of an inclined line is always in an inclined position, while the foreshortened views are in either vertical or horizontal positions.

Note that additional information is available in the true-length views. The true angle between a line and a plane may be measured when the line is true length and the plane is in edge view in a single view. For example, in Fig. 19.7 (a), the horizontal and profile surfaces of the cube appear as lines (in edge view) in the front view, where edge 1–2 is true length. Thus the angles between edge 1–2 and the horizontal plane ($\angle$H) and between edge 1–2 and the profile plane ($\angle$P) may be measured in the front view. Similarly, $\angle$F and $\angle$P for edge 1–3 are measured in the top view, Fig. 19.7 (b), while $\angle$F and $\angle$H for edge 2–3 appear in the side view at (c).

■ 19.3 TRUE LENGTH OF OBLIQUE LINE AND ANGLE WITH PLANE OF PROJECTION

By definition, an oblique line does not appear true length in any principal view—front, top, or side, §6.24. It

FIGURE 19.4 Nonintersecting Lines.

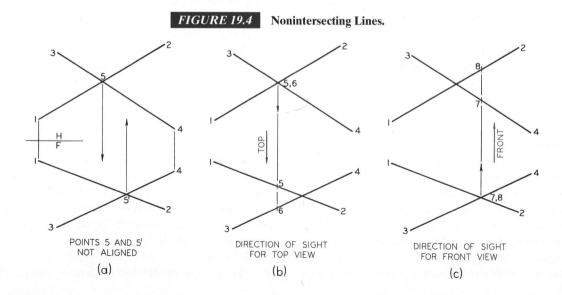

POINTS 5 AND 5'
NOT ALIGNED

(a)

DIRECTION OF SIGHT
FOR TOP VIEW

(b)

DIRECTION OF SIGHT
FOR FRONT VIEW

(c)

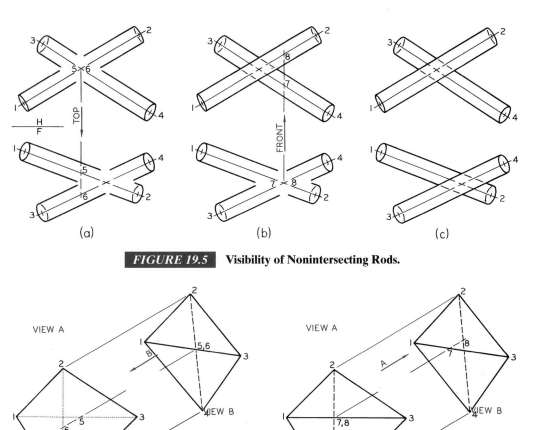

FIGURE 19.5   Visibility of Nonintersecting Rods.

FIGURE 19.6   Visibility of Nonintersecting Lines of a Tetrahedron.

follows that the angles formed with the planes of projection also cannot be measured in the principal views.

In §8.18, the true length (TL) of hip rafter 1–2 was obtained by assuming a direction of sight perpendicular to the front view of rafter 1–2 and constructing a depth auxiliary view. In Fig. 19.8 (a) this construction is repeated with the remainder of the roof omitted. At (b) a portion of the front wall of the building has been added, passing through point 2. Note that, because every point of the wall is at distance D from the folding line H/F in the top view, all points of the wall are at this same distance D from folding line F/1 in the auxiliary view. The entire wall thus appears as a line (in edge view) in these views. Since line 1–2 appears true length in auxiliary view 1, the angle between line 1–2 and the edge view of the front wall, ∠F, can be measured.

In civil engineering, mining, and geology the most important principal plane is the horizontal plane be-

cause a map (of a relatively small area) is a horizontal projection and thus corresponds to a top view. The angle between a line, such as the center line of a highway, and a horizontal plane is a very important factor in the engineering description of the highway. If the angle (∠H) is measured in degrees, it is sometimes called the *slope* of the highway. More commonly it is measured by the ratio between the horizontal and vertical displacements and is called the *grade*. See §19.5.

To measure the slope (∠H) of line 1–2 in Fig. 19.9 (a), a view must be obtained in which line 1–2 appears true length and a horizontal plane appears in edge view. Any horizontal plane appears in edge view in the front view (parallel to the H/F folding line). Thus every point of a particular horizontal plane is at distance H (height) from H/F and any height auxiliary view, §8.8, will show the horizontal plane in edge view. At (b) a direction of sight perpendicular to the top view is cho-

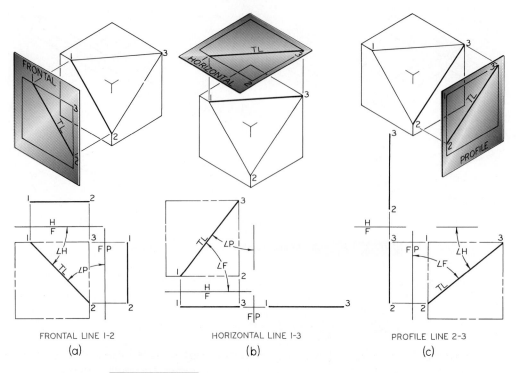

FIGURE 19.7 Frontal, Horizontal, and Profile Lines.

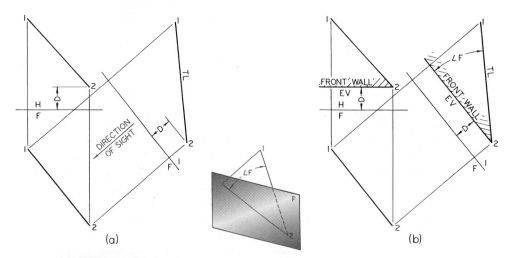

FIGURE 19.8 True Length of Line and Angle with Frontal Plane (∠ F).

sen to obtain a true-length view of line 1–2. The resulting auxiliary view then shows the slope of line 1–2.

The observant student has probably noted that the angle could just as well be measured with respect to folding line H/1 in the auxiliary view. There is actually no need to introduce a special horizontal plane, providing the working space is suitable, since the H/F and H/1 folding lines represent edge views of the horizontal plane in the front view and in any height auxiliary view.

To obtain ∠P of line 1–2 in Fig. 19.10 (a), a view must be obtained in which line 1–2 appears true length and a profile plane appears in edge view.

At (b) side view P is constructed, which shows line 1–2 as it appears projetted on a profile plane. At (c) the direction of sight is established perpendicular to the side view of line 1–2. The resulting auxiliary view shows a true-length view of line 1–2 and any profile plane in edge view and parallel to folding line P/1. The

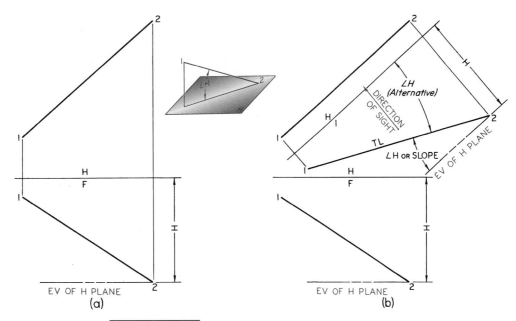

FIGURE 19.9   **True Length and Slope (∠H) of Line.**

∠P may then be measured with respect to the edge view of the profile plane, as indicated.

In summary, note that each of the angles ∠F, ∠H (slope), and ∠P is obtained by a separate auxiliary view, Figs. 19.8, 19.9, and 19.10, respectively; that is, an auxiliary view can show no more than one of these angles of an oblique line.

■ 19.4 TRUE LENGTH AND ANGLE WITH PLANE OF PROJECTION BY REVOLUTION

The true length of a line may also be obtained by revolution, §9.10. In Fig. 19.11 a vertical axis of revolution is employed to find the true length of line 1–2. The path of revolution lies in a horizontal plane seen edgewise in the front view, as indicated. As the line revolves, its angle with horizontal (∠H) remains unchanged in space. Thus in the true-length position this angle may be measured as indicated.

Note that as line 1–2 revolves about the chosen axis, its angles with the other two planes, frontal and profile, continually change. Hence this particular revolution, Fig. 19.11, cannot be used to find the angle the line forms with these planes.

The axis of revolution in Fig. 19.12 is perpendicular to a profile plane. Hence the angle revealed at the true-length position is ∠P.

To determine ∠F for line 1–2, it is necessary to establish the axis of revolution perpendicular to a frontal plane, Fig. 19.13. In practice, it is not necessary to show the axis of revolution, since the remaining construction

makes the position of the axis obvious. Note that a separate revolution is needed for each angle of an oblique line with a projection plane.

■ 19.5 BEARING AND GRADE

The position of a line in space, as is often found in geology, mining, and navigation, is described by its *bearing* and *grade* or by its bearing and slope. The bearing of a line is the direction of a line on a map or horizontal projection. Since for practical purposes a limited area of the earth's surface may be considered a horizontal plane, a map is a top view of the area. Thus the bearing of a line is measured in degrees with respect to north or south in the *top view* of the line, Fig. 19.14 (a).

It is customary to consider north as being toward the top of the map unless information to the contrary is given. Hence the small symbol showing the directions of north, east, south, and west is not usually needed. These are the directions assumed in the absence of the symbol. Note that the bearing indicated, N 45° W, is that of line 1–2 (from 1 toward 2). The conventional practice is to give either the abbreviation for north or south first, chosen so that the angle is less than 90°, and then the angle, followed by the abbreviation for east or west, as appropriate. For example: N 45° W, as shown in the figure.*

* Special cases are "Due north," "Due south," "Due east," and "Due west."

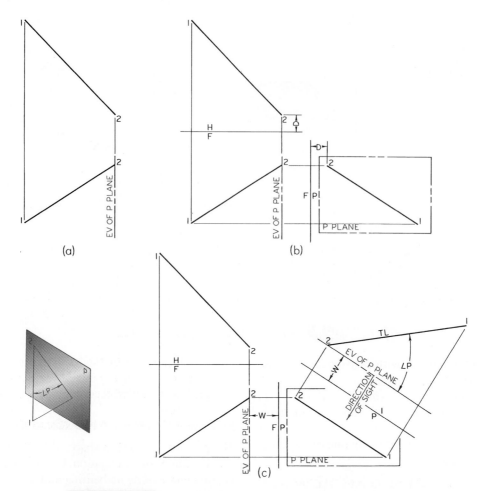

FIGURE 19.10 True Length and Angle with Profile Plane (∠ P).

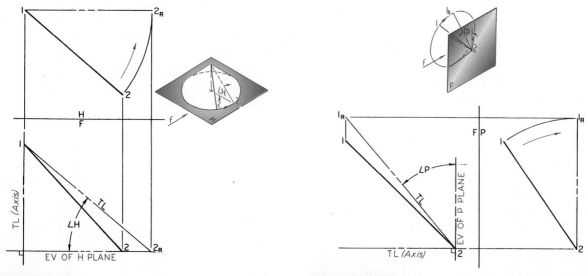

FIGURE 19.11 True Length and Angle with Horizontal Plane by Revolution.

FIGURE 19.12 True Length and Angle with Profile Plane by Revolution.

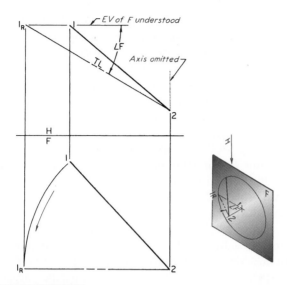

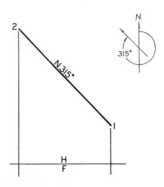

FIGURE 19.15   Azimuth Bearing.

FIGURE 19.13   True Length and Angle with Frontal Plane by Revolution.

At (b) a true-length auxiliary view is added, projected from the top view. As discussed in §20.3 this is the auxiliary view appropriate for measuring the slope of line 1–2 or ∠H. However, in this case another method, known as *grade,* is used to measure the inclination. The grade is the ratio of the vertical displacement (rise) to the horizontal displacement (run) expressed as a percentage.

A construction line horizontal in space (parallel to H/1) is drawn through a point of the line—point 1 in this example. Along this horizontal line 100 units of any appropriate scale are set off. In this instance the

1/20 scale was used. At the 100th division a line is drawn perpendicular to the folding line H/1 and extended to intersect line 1–2 as shown. The length of this line, as measured by the previously used scale, becomes a numerical description of the inclination of line 1–2 expressed as −50%,* because

$$\frac{-50 \text{ units vertically}}{100 \text{ units horizontally}} \times 100 = -50\% \ ^\dagger$$

This is the *percent grade,* or simply the *grade,* of line 1–2.

Another means of describing the bearing of a line is by its *azimuth* bearing, Fig. 19.15. Here the total *clockwise* angle from the base direction, usually north, is given. Line 1–2 here has the same direction as line

* It is common practice to designate a vertical distance as positive or negative according to whether it is measured upward or downward, respectively.

† The student familiar with trigonometry will recognize the ratio 50:100 as the *tangent* of ∠H.

FIGURE 19.14   Bearing and Percent Grade.

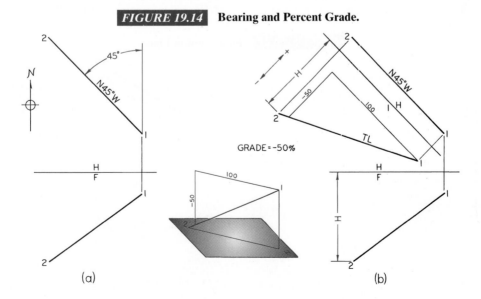

(a)   (b)

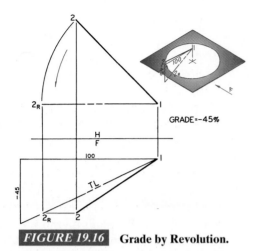

FIGURE 19.16 **Grade by Revolution.**

1–2 of Fig. 19.14 (point 1 toward 2) so that the clockwise angle is 360° − 45° or 315°. If it is understood by all concerned that north is the base direction, the N may be omitted. Thus it is common for an aircraft pilot to describe his or her flight direction as "a course of 315°." On a drawing or map, however, it is best to retain the N to avoid possible confusion.

Grade may also be obtained by revolution. In Fig. 19.16, since the top view is revolved, the axis of revolution (not shown) projects as a point coincident with the top view of point 1. Hence the axis is vertical, and point 2 moves horizontally to the front view to 2_R. In the true-length position, the grade of −45% can be measured as shown.

■ 19.6 POINT VIEW OF LINE

If a direction of sight for a view is parallel to a true-length view of a line, that line will appear as a point in the resulting view. See §6.15. In Fig. 19.17 (a) the vertical line 1–2 appears as a point in the top view, since a vertical line is true length in any height view.

At (b) line 3–4 appears true length in the front view. A direction of sight is chosen, as indicated by arrow 1, parallel to the true-length view. The resulting auxiliary view 1 is a point view, since all points of line 3–4 are the same distance D from the folding lines.

In Fig. 19.17 (c) auxiliary view 1 is necessary to show line 5–6 in true length. Direction of sight 2 is then introduced parallel to the true-length view. The resulting view 2, which is a secondary auxiliary view, shows the point view of line 5–6, §8.19.

Figure 19.17 (c) illustrates an important use of point views: finding the shortest distance from a point to a line. Since the shortest distance is measured along a perpendicular from the point to the line, the perpendicular will appear true length when the given line appears in point view. Observe point 7 in the illustration. An even more important use of the point view of a line is in obtaining an edge view of plane, §§8.11 and 19.9.

■ 19.7 REPRESENTATION OF PLANES

We have discussed planes as surfaces of objects, §6.15, bounded by straight lines or curves. Planes can be established or represented even more simply, Fig. 19.18, by intersecting lines, (a), parallel lines, (b), three points not in a straight line, (c), or a line and a point not on the line, (d). Careful study of Fig. 19.18 will reveal that the same plane 1–2–3 is represented in all four examples. One method can be converted to another by adding or deleting appropriate lines without changing the position of the plane. Most problem solutions involving planes require adding lines at one stage or an-

FIGURE 19.17 **Point View of Line.**

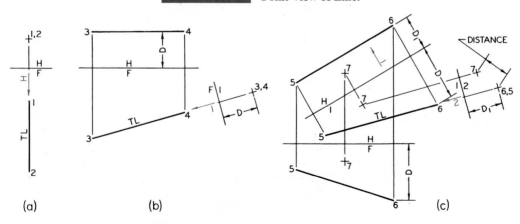

(a)   (b)   (c)

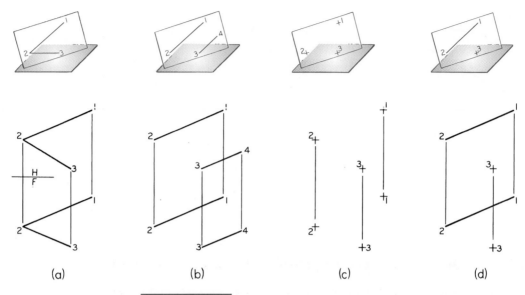

FIGURE 19.18 Representation of a Plane.

other, so that in practice the plane, regardless of its original representation, is in the end represented by intersecting lines.

19.8 POINTS AND LINES IN PLANES

One formal definition of a plane is that it is a surface such that a straight line joining any two points of the surface lies in the surface. It follows that two straight lines in the same plane must intersect, unless the lines are parallel. These concepts are used constantly in working with points and lines in planes.

Figure 19.19 (a) shows a typical elementary problem of this nature. The top view of a line 4–5 is given.

The problem is to find the front view of line 4–5 that lies in plane 1–2–3. Since lines 4–5 and 1–2 are obviously not parallel, they must intersect at point 6 as shown at (b). Point 6 is then located by projecting vertically to the front view of line 1–2. Line 4–5 extended (top view) intersects line 2–3 at 7, which is projected to line 2–3 in the front view. Line segment 4–5 in the front view then lies along a construction line through points 6 and 7, and points 4 and 5 are established by projection from the top view as shown.

A point may be placed in a plane by locating it on a line known to be in the plane. In Fig. 19.20 (a) we are given the front view of a point 4 in plane 1–2–3

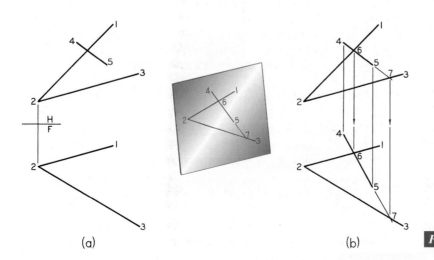

FIGURE 19.19 Straight Line in a Plane.

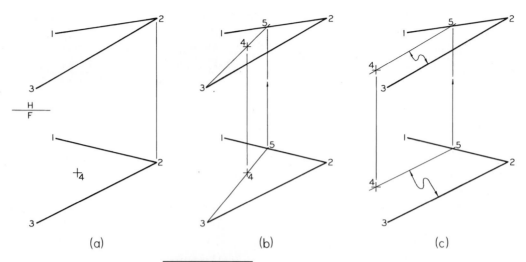

(a)   (b)   (c)

FIGURE 19.20 Point in a Plane.

and desire to find the top view. At (b) a line is introduced through points 3 and 4 and, when extended, intersects line 1–2 at point 5. Point 5 is projected to the top view, establishing line 3–5 in that view. Point 4 is then projected from the front view to the top view of line 3–5. Theoretically, any line could be drawn through point 4 to solve this problem. However, lines approaching parallel to the projection lines between views should be avoided as they may lead to significantly inaccurate results.

A different solution of a similar problem—using the principle of parallelism*—is shown in Fig. 19.20 (c). Here lines drawn through point 1 or point 3 and given point 4 lead to inconvenient intersections. If a line is drawn through point 4 in the

front view parallel to line 2–3 and intersecting line 1–2 at point 5 as shown, it will not intersect line 2–3. Therefore, according to the principles stated at the beginning of this section, the line must be parallel to line 2–3. Thus, after intersection point 5 is projected to the top view, the new line through point 5 is drawn parallel to the top view of line 2–3, and point 4 is projected to it to complete the top view. Use of this parallelism principle requires minimum construction.

Another example of locating a point on a plane is shown in Fig. 19.21. Here it is desired to locate in plane 1–2–3 a point P that is 10 mm above point 2 and 12 mm behind point 3.

At (a) a horizontal line 10 mm above (higher than) point 2 is added to the front view of plane 1–2–3. Its intersection points 4 and 5 with lines 1–2 and 2–3 are

* Parallelism is discussed in more detail in Chapter 21.

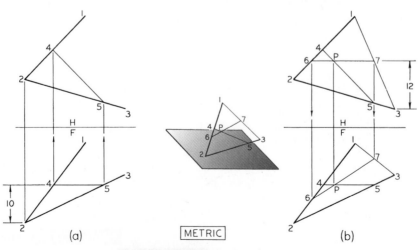

(a)   METRIC   (b)

FIGURE 19.21 Locus Problem

projected to the top view as shown. Any point along line 4–5 lies in plane 1–2–3 and is 10 mm above point 2. Line 4–5 is said to be the *locus* of such points.

At (b) a frontal line 12 mm behind (to the rear of) point 3 is added to the top view of the plane. Its front view is obtained by projection of intersection points 6 and 7. (Note the addition of line 1–3 to secure point 7.) Line 6–7 is the locus of points in plane 1–2–3 that are 12 mm behind point 3.

The intersection point of lines 4–5 and 6–7 is the required point P. The views of P at (b) are checked with a vertical projection line, as shown, to make sure that they are views of the same point.

■ 19.9 EDGE VIEWS OF PLANES

In order to get the edge view of a plane, we must get the point view of a line in the plane, §8.11. For Fig. 19.22 (a) the edge view of plane 1–2–3 could be obtained by getting the true-length view and then the point view of any one of the three given lines of the plane. Since these lines are all oblique lines, obtaining their point views would each entail two successive auxiliary views, §19.6. It is easier to add a line that appears true length in one of the principal views, thus eliminating the need of a second auxiliary view.

At (b) line 2–4 is drawn parallel to the H/F folding line (horizontal on the paper) in the top view. Thus it is a frontal line, §19.2, and its front view, obtained by projecting point 4 to the front view of line 1–3, is true length, as indicated.

Thus a true-length line in plane 1–2–3 has been established without drawing an auxiliary view, and we may now proceed as at (c) by assuming a direction of sight 1 parallel to the true-length view of 2–4. The resulting auxiliary view is the desired edge view. Note that all points of the plane, not just a minimum two

points, are actually projected to the auxiliary view. This provides a convenient check on accuracy, since obviously the points must lie on a straight line in the auxiliary view.

Edge views are useful as the first step in obtaining the true-size view of an oblique plane, §§8.21 and 19.10. They are also employed in showing dihedral angles, §8.11. In Fig. 9.11 (c) it was possible to show the true angle by drawing a primary auxiliary view, since the line of intersection of the planes appeared in true length in the top view. In Fig. 19.23 (a) the line of intersection 1–2 between surfaces A and B is not shown true length in either view. Accordingly, at (b) auxiliary view 1 is constructed, with the direction of sight 1 perpendicular to line 1–2. At (c) secondary auxiliary view 2 is then added, with direction of sight 2 parallel to the true-length view of line 1–2. As a check, all points of the planes are located in the auxiliary views. When they fall on the respective straight-line (edge) views of the surfaces A and B, confidence in accuracy is established.

■ 19.10 TRUE-SIZE VIEWS OF OBLIQUE PLANES

The procedure for obtaining the true size of an oblique surface of an object is treated in §8.20. Many problems of a more abstract nature are also solved through obtaining true-size views of oblique planes.

For example, let it be required to find the center of the circle passing through points 1, 2, and 3 of Fig. 19.24. The plane geometry construction is that of §4.31. However, the construction illustrated in that section took place in the plane of the paper, which we now recognize as a form of a true-size view. Since this is not the situation at (a), it is necessary to proceed as at (b) with the edge view and true-size views of plane

FIGURE 19.22 Edge View of Plane.

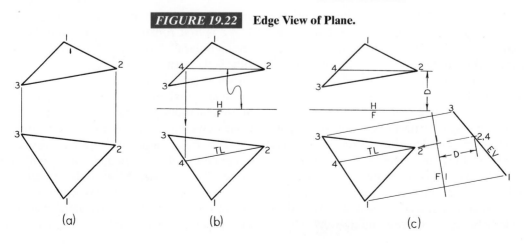

(a) (b) (c)

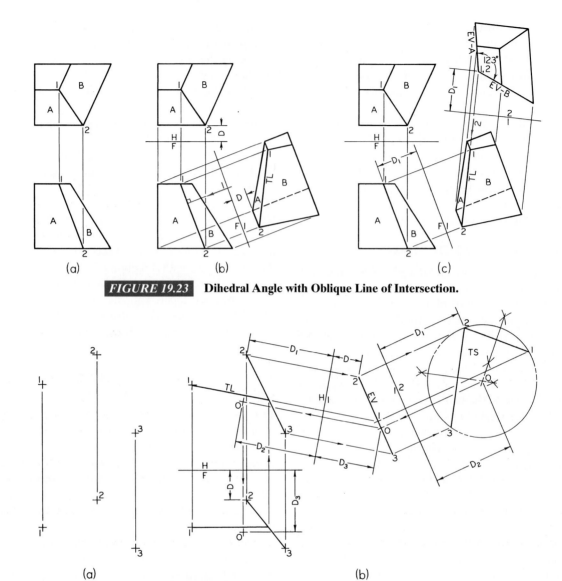

FIGURE 19.23   Dihedral Angle with Oblique Line of Intersection.

FIGURE 19.24   Center of Circle in Oblique Plane.

1–2–3. In auxiliary view 2 the construction shown in §4.31 is performed, locating center O for a circle through points 1, 2, and 3. If desired, point O can be located in the divider distances D_2 and D_3 as shown. The circle, if drawn, would appear elliptical in the front and top views, §8.24.

The revolution method explained in §10.11 may also be applied to finding the true-size view of an oblique surface for geometric construction. In Fig. 19.25, the problem is to bisect the plane angle 1–2–3. The true bisector of a plane angle lies in the same plane as the angle, and only under special circumstances when the plane of the angle is not in true size will a view of the bisector actually bisect the corresponding view of the

angle. To solve this problem, the edge view of plane 1–2–3 is first constructed at (a) and then is revolved until it is parallel to folding line F/1. The revolved front view is then true size and angle 1_R–2–3_R is bisected as shown at (b). The revolution thus takes the place of a secondary auxiliary view. This has the major advantage of compactness of construction, but the overlapping front views may in some cases be confusing.

The front and top views of the bisector are obtained by selecting an additional point on the bisector in the true-size view and reversing the whole process—*counterrevolving*. Point 4_R on the true-length line through point 3_R is particularly convenient in this case because it will counterrevolve to the true-length

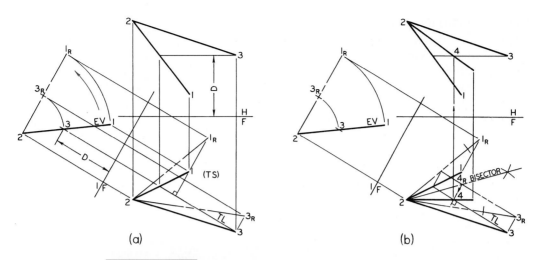

(a)   (b)

FIGURE 19.25 True of Oblique Plane—Revolution Method.

line through point 3 and can then be projected to the top view, bypassing the auxiliary view. If desired, however, the selected point can be projected to the revolved edge view, counterrevolved to the original edge view, and then returned to the front and top views by the usual methods.

■ 19.11 PIERCING POINTS

If a straight line is not parallel to a plane, it must intersect that plane in a single point called a *piercing point.* It may be necessary to extend the line, or plane, or both; but this is permissible, since the abstract terms *line* and *plane* do not imply any limits on their extent.

There are two recommended methods for finding piercing points.

EDGE-VIEW METHOD   All points of a plane are shown along its edge view. These, of course, include the piercing point of any lines that happen to be present. In Fig. 19.26 (a) the frontal line through point 3 is introduced to get a true-length line and thus the edge view of plane 1–2–3. In this case it is necessary to extend line 4–5 to find the piercing point (encircled). At (b) the piercing point is projected first to the front view and then to the top view. Note the use of divider distance D_1, to check the accuracy of location of the top view. This procedure, under some circumstances, is more accurate than direct projection.

FIGURE 19.26 Piercing Point—Edge-View Method.

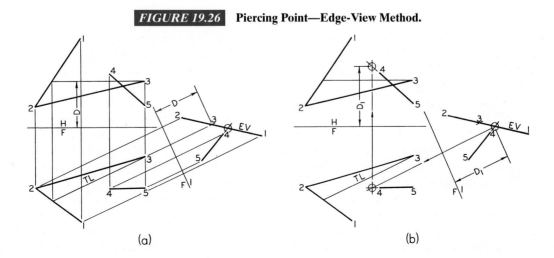

(a)   (b)

Note that a horizontal line could have been introduced into plane 1–2–3, thereby establishing a different true-length line. This would have produced a different edge view, but would not give a different piercing point, as there is only one piercing point for a particular straight line and plane. This procedure would not be considered a different method but merely an alternative approach.

CUTTING-PLANE METHOD OR GIVEN-VIEW METHOD If a cutting plane A–A containing line 4–5 is introduced, Fig. 19.27 (a), it will cut line 6–7 from given plane 1–2–3. Lines 6–7 and 4–5, being in the same plane A–A, must intersect at the piercing point (encircled). To make the method practical, an edge-view cutting plane is used, as shown at (b). To contain line 4–5, the cutting plane must coincide with a view of the line. At (b) it was chosen to have plane A–A coincide with the top view of 4–5. The line cut from plane 1–2–3 is line 6–7. Projected to the front view, line 6–7 locates the front view of the piercing point, which is then projected to the top view.

Actually, there is no need for some of the lettering shown at (b). At (c) the symbol EV adequately identifies the cutting plane, and the numbers 6 and 7 may be omitted as being of little value other than for purposes of discussion.

Note that this illustration is similar to Fig. 19.26, except that (1) the piercing point is within the line segment 4–5 and (2) the plane is *limited* or completely bounded. It is then feasible to consider the bounded area to be opaque. The line then becomes hidden after

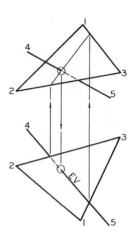

FIGURE 19.28   Piercing Point—Cutting-Plane Method (Alternative Solution).

it pierces the plane. The visibility displayed at (c) was determined by the methods of §19.1, investigating in each view any convenient point where line 4–5 crosses one of the boundary lines of the plane.

The problem in Fig. 19.27 is shown again in Fig. 19.28, this time with the edge view of the cutting plane introduced coincident with the front view of line 4–5. Of course, the same answer is obtained, and it is a matter of personal choice and convenience as to which view is chosen for introduction of the edge-view cutting plane. For the convenience of the reader, always include the letters EV as shown when the problem solutions involve such cutting planes.

FIGURE 19.27   Piercing Point—Cutting-Plane Method.

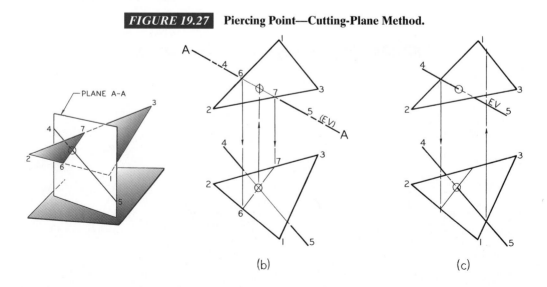

(b)

(c)

■ 19.12 INTERSECTIONS OF PLANES

The intersection of two planes is a straight line containing all points common to the two planes. Since planes are themselves represented by straight lines, §19.7 points common to two intersecting planes may be located by finding piercing points of lines of one plane with the other plane, by the use of edge-view method or the cutting-plane method of §19.11.

EDGE-VIEW METHOD   In Fig. 19.29, two planes are given: 1–2–3 and 4–5–6. If the edge view of either plane is constructed, the piercing points of the lines of the other plane will lie along the edge view. At (a) a horizontal line is introduced through point 3 of plane 1–2–3 in order to secure a true-length line in the top view. (A horizontal line in plane 4–5–6 or frontal lines in either plane would serve just as well in this problem.) Auxiliary view 1 is then constructed with its direction of sight parallel to the true-length view of the line. The completed auxiliary view 1 shows the edge view of plane 1–2–3 and the piercing points of lines 4–5 and 5–6 as indicated by the encircled points.

At (b) the piercing points are projected to the top view and then to the front view. (It is good practice to check accuracy by divider distances, as indicated by dimension D_1). Since the given planes are not completely bounded, there is no reason to restrict the drawn length of the segment of the line of intersection (LI). However, the views of the line of intersection should be compatible from view to view.

Since the LI is common to both planes, it must intersect or be parallel to each line of both planes. As a check on accuracy, observe in this case that the LI intersects line 1–2 at 7 and is parallel to line 2–3.

CUTTING-PLANE METHOD   Because it requires no additional views, the cutting-plane method is frequently used to find the intersection of two planes, Fig. 19.30. At (a) it is arbitrarily decided to introduce an edge-view cutting plane coinciding with the top view of line 5–6, with the intention of finding the piercing point of line 5–6 in plane 1–2–3–4. The student should realize that one could introduce cutting planes in either view coinciding with any of the lines of the planes. With so many possibilities it is imperative that the choice be indicated with proper use of the symbol EV, both to avoid confusion on the student's part and as a courtesy to the person who must read the drawing.

In this case the introduced plane cuts line 8–9 from plane 1–2–3–4. Point 8 is on line 1–2 and point 9 is on line 2–3. Observe this carefully to avoid mistakes in projecting to the front view. The front view of line 8–9 intersects line 5–6 at the encircled piercing point which, after projection to the top view, represents one point common to the given planes.

At (b) another piercing point is located by introducing an edge-view cutting plane along line 5–7. The line of intersection, LI, passes through the two piercing points as shown.

In this illustration the given planes are bounded and can therefore be considered limited as at (c). The

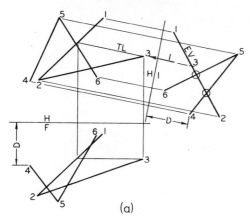 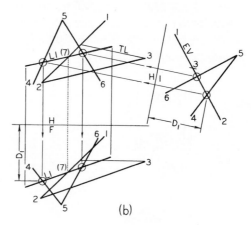

FIGURE 19.29   Intersection of Two Planes—Edge-View Method.

(a)

(b)

GRAPHICS SPOTLIGHT

Potatoes and Productivity

DOCUMENTED PRODUCTIVITY IMPROVEMENT

Spudnik Equipment Company of Blackfoot, Idaho, uses Pro/ENGINEER software to design specialty potato handling equipment. They design and manufacture trucks, conveyors, pilers, and storage equipment to automate large scale potato storage and handling. They have only used Pro/ENGINEER one year but are impressed with the capabilities to create and document equipment designs. Drex Pincock, CAD Designer, says that in the previous five years using their previous CAD package, he was only able to complete three sets of design documentation. Drex and one other person have created three entire sets of design documentation in just six months since switching to Pro/ENGINEER.

WELDMENTS AND SHEET METAL

Many of their parts are created from stock metal welded together, and from sheet metal. One of the benefits for Spudnik of using Pro/ENGINEER is its complete associativity. The part drawings are linked into the proper assembly drawings so that if a part or stock size is changed, the assembly drawing automatically updates. Also since parametric solid modeling is used to create the design for an entire part, the necessary drawings for custom machines can quickly be generated by changing the dimension values. They also use Pro/ENGINEER's ability to automatically generate the flat pattern for sheet metal layout from the 3D model to shorten the time to go from design to manufacturing.

At Spudnik Pro/ENGINEER runs on a 275 MHz DEC Alphastation with 128 MB RAM. They use it to create very large assemblies with parts such as weldments, sheet metal, bearings, and rollers. When working in a large assembly, they turn off the parts they are not currently working so that they are not loaded into RAM in order to make working in the large assemblies faster.

QUICK DESIGN OF CUSTOM MACHINES & DOCUMENTATION

Three entire machine designs have been completed at Spudnik since they purchased Pro/ENGINEER. One is for a planter-filler, which loads seed potatoes into a planter. They have also designed and built a pup conveyor. A pup conveyor is a small 8 to 10 foot device that can hang off a larger conveyor or other equipment so more than one truck can unload potatoes at the same time. The third device is a portable eliminator, which is used for either sorting out smaller potatoes right at the time of unloading or for sorting out dirt and debris from the potatoes. Spudnik creates many custom machines for their customers and needs to quickly produce documentation for the custom machines. With Pro/ENGINEER, they can save time designing the custom machinery and producing the necessary shop drawings to manufacture the equipment. Then they go one step more and create exploded isometric and isometric drawings for the part books and customer manuals directly from their design database.

To get started using Pro/ENGINEER, all of the designers went to one week of basic training. Then key personnel like Drex went to additional training on assembly modeling and advanced part modeling. Drex says that Pro/ENGINEER has a lot of features for modeling irregular surfaces that they do not really use because their parts are mainly manufactured from flat and bar stock. The key feature that they feel saves time in designing, manufacturing, and documenting their products is the associativity of the parts, drawings, and assemblies.

Courtesy of Spudnik Equipment Company, Inc.

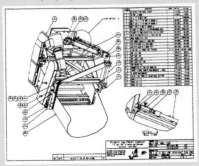

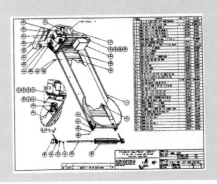

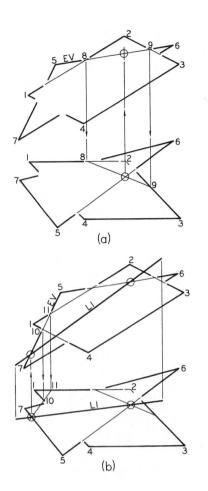

(a)

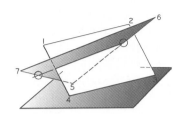

(b)

(c)

FIGURE 19.30 **Intersection of Two Planes—Cutting-Plane Method.**

piercing point of line 5–7 falls outside plane 1–2–3–4 and is therefore not on the "real" portion of the line of intersection, which is drawn as a visible line only in the area common to the views of both planes. The termination of this segment is at the point that is actually the piercing point of line 1–4 in plane 5–6–7. However, this result was not obvious at the start of the construction. Visibility was determined by the method of §19.1. Usually it is necessary to examine only one apparent crossing point in each view. After visibility is established in one such region, the spatial relations of the remaining lines of that view are evident, since each boundary line in turn can change visibility only where it meets a piercing point or a boundary line of the other given plane.

SPECIAL CUTTING-PLANE METHOD The line of intersection of two planes also may be found through the use of cutting planes that do not coincide with the views of given lines. Any plane cutting the two given

planes, Fig. 19.31 (a), cuts one line from each. Since these lines lie in the cutting plane as well as in the given planes, they will intersect at a point common to the given planes. A second cutting plane will establish a second common point, giving two points on the line of intersection. For convienence, the edge-view cutting planes employed are usually drawn parallel to a regular coordinate plane, but this is not necessary. It is suggested that the two planes be introduced in the same view for more control of the distance between the points secured.

At (b) horizontal plane EV_1 cuts lines 9–4 and 8–10 from the given planes. When these lines are projected to the top view, they intersect at point 11, which is then one point on the required line of intersection. The front view of point 11 is on line EV_1 as shown. At (c) a second horizontal plane EV_2 is introduced, cutting lines 12–13 and 14–15, which intersect in the top view at point 16. After point 16 is projected to line EV_2 the line of intersection 11–16 (LI) is drawn to any desired length. Note carefully the parallelism of the lines in the top view at (c). This affords a convenient and a very desirable check on accuracy.

This method involves more construction than did the previous methods and can be confusing when the given views occupy overlapping areas, as in Figs. 19.29 and 19.30. This particular method is therefore recommended primarily for problems in which the given views of the planes are separated.

■ 19.13 ANGLE BETWEEN LINE AND OBLIQUE PLANE

The true angle between a line and a plane of projection (frontal, horizontal, or profile) is seen in the view in which the given line is true length and the plane in ques-

tion is in edge view, §19.3. This is a general principle that applies to any plane: normal, inclined, or oblique.

In Fig. 19.32 (a) two views of a plane 1–2–3 and a line 5–6 are given. One cannot expect a primary auxiliary view to show plane 1–2–3 in edge view and also line 5–6 in true length, for generally the directions of sight for these two purposes will not be parallel. Note that in Fig. 19.8 the direction of sight for the auxiliary view is toward the front view, which shows the true-size view of all frontal planes. In Fig. 19.9 the direction of sight is toward the top view, which shows the true-size view of all horizontal planes. In summary, *any view projected from a true-size view of a plane shows an edge view of that plane.*

In Fig. 19.32 (a) frontal line 2–4 is added to given plane 1–2–3. We thus now have a true-length line in

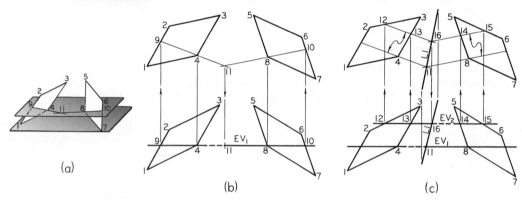

FIGURE 19.31 Intersection of Two Planes—Special Cutting-Plane Method.

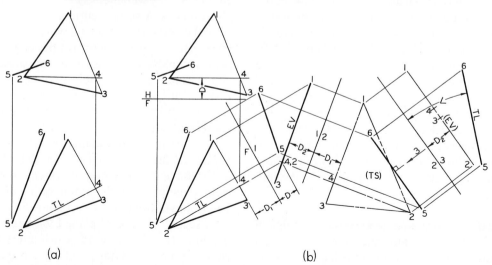

FIGURE 19.32 Angle Between Line and Plane—True-Size Method.

the front view, and edge-view auxiliary view 1 is projected from it. The true-size secondary auxiliary view 2 is then constructed in the customary manner.

Any view projected from view 2 will show plane 1–2–3 in edge view. Therefore to show the true angle between line 5–6 and the plane, direction of sight 3 is established at right angles to view 2 of line 5–6. View 3 then shows the required angle ($\angle$). Because in this chain of views the edge view of the plane in view 3 is always parallel to folding line 2/3 (note divider distance D_2), the construction can be simplified if desired by omitting plane 1–2–3 from auxiliary views 2 and 3. The required angle is then measured between line 4–5 and the folding line 2/3.

■ KEY WORDS

FRONTAL PLANE OF PROJECTION	PROFILE PLANE OF PROJECTION
DIHEDRAL ANGLE	HORIZONTAL PLANE OF PROJECTION
VISIBILITY	FRONTAL LINE
PROFILE LINE	HORIZONTAL LINE
REVOLUTION	BEARING
GRADE	LOCUS
PIERCING POINTS	OBLIQUE
EDGE VIEW	CUTTING PLANE

■ CHAPTER SUMMARY

- The basic geometric elements are represented using special notation in descriptive geometry. Correct solutions for spatial problems include the correct labeling of the basic geometric elements.

- Descriptive geometry is the basic for all engineering graphics. The fundamental principles for all engineering drawing techniques are based on the theory of descriptive geometry.

- Identifying, creating, and manipulating lines in each of the three primary planes of projection are fundamental techniques for solving descriptive geometry problems.

- Bearing, slope, and grade are easily calculated using simple descriptive geometry techniques.

- Auxiliary view and revolution are two common solution techniques for determining true length, true angle, true size, and the intersection between lines and planes.

- Finding the point view of a line and the edge view of a plane are common solution techniques for determine visibility of lines and planes.

- The intersection of a line with a plane (piercing point) like many descriptive geometry problems, has several methods for determining solutions.

■ REVIEW QUESTIONS

1. Why were military applications the first applications for descriptive geometry?

2. Why are the principles of descriptive geometry so important that they are taught at engineering schools?

3. How are the end points of lines labeled in descriptive geometry?

4. What is meant by the visibility of two lines?

5. What is the name of the intersection of a line and a plane?

6. What is the name for a line that is parallel to the frontal plane of projection?

7. What is an inclined line?

8. What is a normal line?

9. What is an oblique line?

10. How many views are required to determine true length of a line using the revolution method? Using the auxiliary view method?

11. In which view is bearing measured?

12. Grade can be measured in any view adjacent to which primary view?

13. What is a dihedral angle?

14. What do the labels EV, TL and TS mean?

■ POINT, LINE, AND PLANE PROBLEMS

The problems in Figs. 19.33–19.44 cover points and lines, intersecting and nonintersecting lines, visibility, true length and angles with principal planes, auxiliary-view method and revolution method, point views, points and lines in planes, dihedral angles, edge view and true size of planes, piercing points, intersection of planes, and angle between line and oblique plane.

Use Layout A–1 or A4–1 (adjusted) and divide the working area into four equal areas for problems to be assigned by the instructor. Some problems will require a single problem area, and others will require two problem areas (half the sheet). Data for the layout for each problem are given by a coordinate system. For example, in Fig. 19.33, Prob. 1, point 1 is located by the full-scale coordinates, 22 mm, 38 mm, 75 mm. The first coordinate locates the front view of the point from the left edge of the problem area. The second coordinate locates the front view of the point from the bottom edge of the problem area. The third coordinate locates either the top view of the point from the bottom edge of the problem area or the side view of the point from the left edge of the problem area. Inspection of the given problem layout will determine which application to use.

Since many of the problems in this chapter are of a general nature, they can also be solved on most computer graphics systems. If a system is available, the instructor may choose to assign specific problems to be completed by this method.

Additional problems, in convenient form for solution, are available in *Engineering Graphics Problems*, Series 1, by Spencer, Hill, Loving, Dygdon, and Novak, designed to accompany this text and published by Prentice Hall Publishing Company.

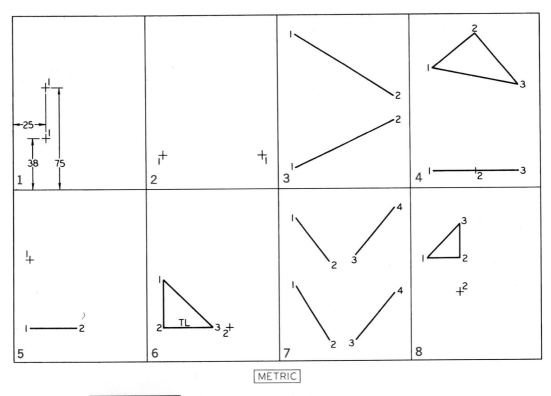

FIGURE 19.33 Lay out and solve four problems per sheet as assigned.
Use Layout A–1 or A4–1 (adjusted) divided into four equal areas.

1. Given point 1 (25, 38, 75), locate the front and top views of point 2, which is 50 mm to the right of point 1, 25 mm below point 1, and 30 mm behind point 1.
2. Given point 1 (12, 25, 90), locate the front and side views of line 1–2 such that point 2 is 38 mm to the right of point 1, 45 mm above point 1, and 25 mm in front of point 1. Add the top view of line 1–2.
3. Find the views of points 3, 4, and 5 on line 1 (12, 15, 115)–2 (90, 50, 70) that fit the following descriptions: point 3, 20 mm above point 1; point 4, 65 mm to the left of point 2; and point 5, 25 mm in front of point 1.
4. Triangle 1 (18, 12, 90)–2 (50, 12, 115)–3 (85, 12, 75) is the base of a pyramid. The vertex V is 8 mm behind point 1, 8 mm to the left of point 2, and 45 mm above point 3. Complete the front and top views of the pyramid.
5. Line 1 (12, 25, 75)–2 (48, 25, ?) is 43 mm long (2 behind 1). Line 1–3 is a 50 mm frontal line, and line 2–3 is a profile line. Find the true length of line 2–3.
6. Line 1 (12, 60, ?)–2 (12, 25, 64) is 45 mm long. The front view of line 2–3 (50, 25, ?) is true length as indicated. Complete the front and side views and add a top view of triangle 1–2–3.
7. Point 5 is on line 1 (12, 56, 106)–2 (38, 15, 74), 18 mm below point 1. Point 6 is on line 3 (58, 15, 80)–4 (90, 50, 115). Line 5–6 is frontal. Find the true length of line 5–6.
8. Line 2 (38, 50, 75)–3 (38, ?, 100) is 38 mm long. Line 3–1 (12, ?, 75) is horizontal. How long is line 1–2?

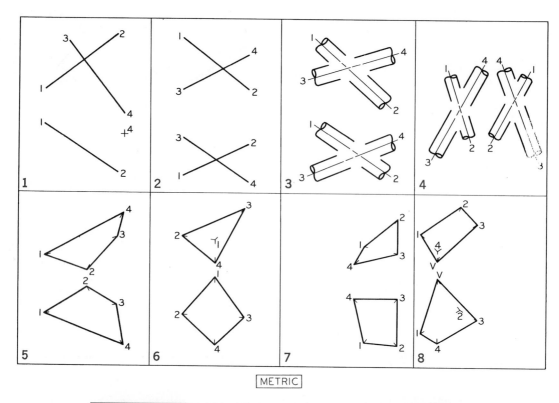

METRIC

FIGURE 19.34 Lay out and solve problems as assigned. Use Layout A-1 or A4-1 (adjusted) divided into four equal areas.

1. Lines 1 (20, 50, 75)–2 (75, 15, 117) and 3 (48, ?, 112)–4 (82, 43, 58) are intersecting lines. Complete the front view.

2. Demonstrate that lines 1 (25, 12, 114)–2 (75, 48, 75) and 3 (25, 40, 75)–4 (75, 8, 100) do not intersect. Then move point 4 *vertically* in space to a new position 4′ such that line 3–4′ intersects line 1–2.

3. Lines 1 (25, 50, 114)–2 (84, 12, 64) and 3 (20, 12, 84)–4 (90, 40, 104) are the center lines of two nonintersecting cylindrical tubes, each of which has a diameter of 200 mm. Scale: 1/20. Complete the views including correct visibility.

4. Rod 1 (12, 92, 90)–2 (43, 38, 58) has a diameter of 10 mm. Rod 3 (12, 25, 94)–4 (53, 96, 66) has a diameter of 12 mm. Complete the views including correct visibility.

5. Points 1 (20, 38, 80), 2 (53, 58, 70), 3 (75, 46, 96), and 4 (90, 15, 115) are the vertices of a tetrahedron. Complete the given views with proper visibility.

6. Points 1 (50, 65, 94), 2 (25, 38, 96), 3 (73, 35, 116), and 4 (50, 15, 75) are the vertices of a tetrahedron. Complete the given views with proper visibility.

7. Points 1 (63, 17, 88), 2 (88, 15, 118), 3 (88, 50, 83), and 4 (55, 50, 75) are the vertices of a tetrahedron. Complete the given views and add a left-side view, all with proper visibility.

8. Points 1 (5, 25, 98), 2 (35, 43, 118), 3 (48, 35, 116), and 4 (18, 18, 86) are corners of the base of a pyramid. Point V (18, 66, 78) is the vertex. Complete the views and add a right-side view, including proper visibility.

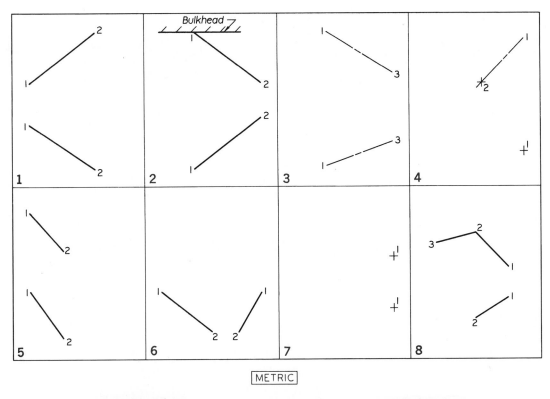

METRIC

FIGURE 19.35   Lay out and solve problems as assigned. Use Layout A-1
or A4-1 (adjusted) divided into four equal areas.

1. Find and measure the true length of line 1 (12, 46, 75)–2 (63, 12, 114) and the angle it forms with a horizontal plane ($\angle$H).

2. Find and dimension the true length of and the angle formed by control cable 1 (38, 12, 114)–2 (80, 50, 75) and the frontal bulkhead. Scale: 1/10.

3. Measure the bearing and slope ($\angle$H) of pipe center line 1 (38, 15, 114)–3 (88, 33, 83).

4. The center line of a segment of a highway runs from point 1 (88, 25, 110) through point 2 (56, ?,75) to a point 3. The line slopes downward from point 1 at an angle of 15°. The length of segment 1–3 is 350 m. Scale: 1/5000. Find the top and front views of point 3.

5. Find and measure the true length of line 1 (12, 50, 110)–2 (38, 15, 80) and the true angle it forms with a profile plane ($\angle$P).

6. Find and measure the bearing and percent grade of line 1 (12, 50, 88)–2 (50, 20, 70).

7. A tunnel bears N 40° W from point 1 (88, 38, 75) on a downgrade of 30% to point 2 at a distance of 230 m along the tunnel. Scale: 1/4000. Find the front and top views of tunnel 1–2.

8. If segments 1 (75, 45, 68)–2 (50, 30, 94) and 2–3 (20, ?, 86) of pipeline 1–2–3 have the same grade, find the front view of 2–3.

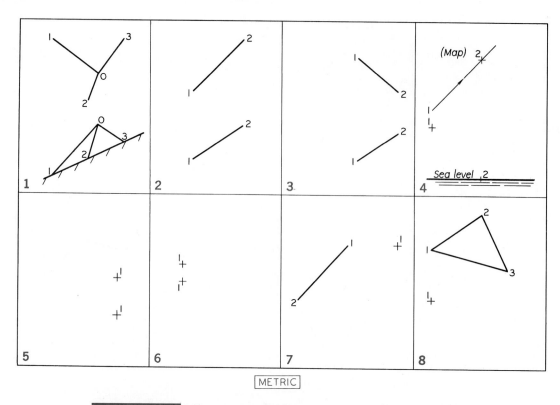

METRIC

FIGURE 19.36 **Lay out and solve four problems per sheet, as assigned.
Note: Use revolution for these problems. Use Layout A-1 or A4-1
(adjusted) divided into four equal areas.**

1. Find the true lengths of the tripod legs O $(60, 50, 88)$–1 $(25, 12, 114)$, 0–2 $(53, 25, 68)$, and 0–3 $(80, 38, 114)$. Scale: $1/20$.
2. Find and measure the true length and angles with H and F for line 1 $(33, 25, 75)$–2 $(70, 50, 114)$.
3. Find the true length and angle with P of line 1 $(58, 25, 100)$–2 $(88, 46, 75)$. Scale: $1/200$.
4. An aircraft flies from position 1 $(12, 50, 63)$ toward point 2 $(50, 12, 100)$. If the craft is losing altitude at the rate of 500 m in each 1000 m (map distance), at what altitude will it pass over point 2? Scale: $1/20\,000$. If the aircraft fails to pull out of the dive, show the front and top views of the point of impact.
5. Pipe center line 1 $(75, 38, 65)$–2 $(?, ?, ?)$ has an azimuth bearing of N 310°, a downgrade of 30%, and a true length of 240 m. Scale: $1/4000$. Find the front and top views of line 1–2.
6. Line 1 $(25, 64, 75)$–2 $(?, ?, ?)$ has a bearing of N 40° E, is 60 mm in length, makes an angle of 30° with a horizontal plane, and slopes downward. Complete the front and top views of 1–2.
7. Line 1 $(50, 90, 88)$–2 $(12, 50, ?)$ forms an angle of 35° with a frontal plane. The lines slopes forward. Complete the side view.
8. Line 1 $(12, 50, 88)$–2 $(50, ?, 114)$ has a downward slope of 40°. Line 2–3 $(70, ?, 74)$ has an upward slope of 20°. What is the slope of line 1–3?

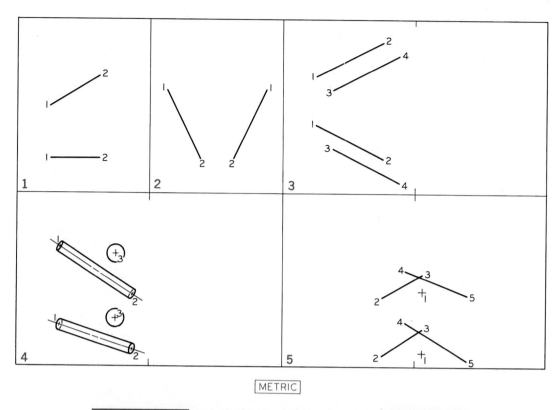

METRIC

FIGURE 19.37 Lay out and solve problems as assigned. Use Layout A-1 or A4-1 (adjusted) divided into four equal areas.

1. Find a point view of line 1 (25, 25, 63)–2 (63, 25, 86).
2. Find a point view of line 1 (12, 75, 88)–2 (38, 25, 63).
3. Find the true distance between parallel lines 1 (25, 50, 86)–2 (75, 25, 112) and 3 (38, 33, 75)–4 (88, ?, ?).
4. Find the clearance between 0.8 m diameter cylinder 1 (32, 30, 88)–2 (88, 12, 53) and 1.25 m diameter sphere 3 (75, 35, 83). Scale: 1/100.
5. Determine if point 1 (106, 10, 55) is nearer to line 2 (75, 7, 50)–3 (106, 28, 68) or to line 4 (94, 33, 70)–5 (142, 5, 53).

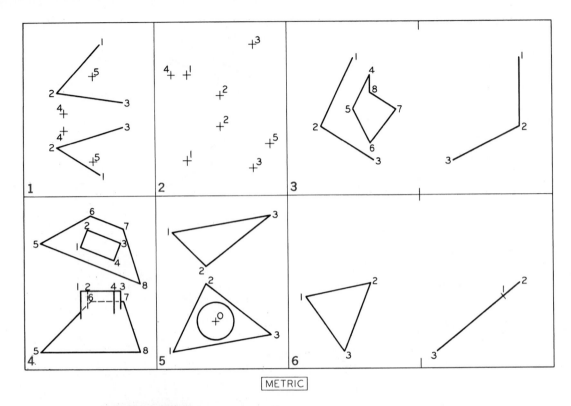

METRIC

FIGURE 19.38   **Lay out and solve problems as assigned. Use Layout
A–1 or A4–1 (adjusted) divided into four equal areas.**

1. Determine if either (or both) of points 4 (30, 48, 60) and 5 (53, 25, 88) lies in plane 1 (58, 15, 112)–2 (25, 35, 75)–3 (75, 50, 68).

2. Line 4 (12, ?, 88)–5 (88, 38, ?) lies in plane 1 (25, 25, 88)–2 (50, 50, 74)–3 (75, 20, 112). Complete the views of line 4-5.

3. Pentagon 4(63, 88, ?)–5(50, 63, ?)–6(63, 38, ?)–7(84, 63, ?)–8(63, 75, ?) lies in plane 1 (50, 102, 178)–2 (25, 50, 178)–3 (65, 25, 126). Complete the side view of the pentagon.

4. Complete the front view including the opening in roof plane 5 (12, 12, 94)–6 (50, 50, 114)–7 (75, 50, –)–8 (88, 12, 63) for vertical chimney 1 (43, 58, 90)–2 (48, 58, 104)–3 (73, 58, –)–4 (68, 58, –).

5. Plot the top view of the curve centered at 0 (46, 35, ?) and lying in plane 1 (12, 12, 100)–2 (38, 63, 75)–3 (88, 25, 114). The circular front view of the curve has a diameter of 28 mm.

6. Find the front and side views of the center of the circle inscribed in triangle 1 (12, 53, –)–2 (63, 63, 178)–3 (43, 12, 114). Also locate the views of the points of tangency of the circle with the sides of the triangle. If assigned, plot the views of the circle.

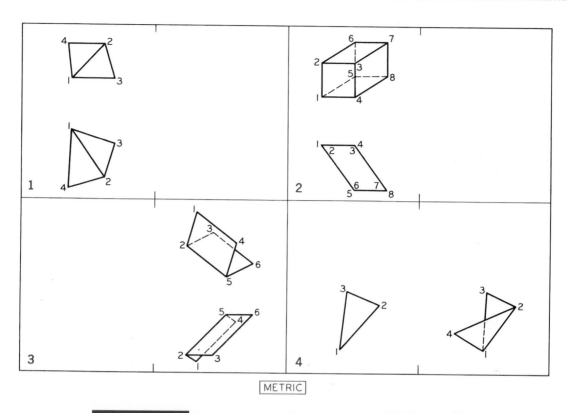

METRIC

FIGURE 19.39 **Lay out and solve problems as assigned. Use Layout A-1 or A4-1 (adjusted) divided into four equal areas.**

1. Find the true size of the dihedral angle formed by plane 1(38, 50, 88)–2(63, 15, 114)–3(70, 40, 88) and plane 1-2-4(35, 7, 114).

2. Find the dihedral angles between the lateral faces of the prism. Bases 1(25, 40, 75)–2-3-4 and 5(50, 7, 90)–6-7-8 are 25 mm square.

3. Determine the dihedral angle of clip angle 1(134, 7, 119)–2(126, 12, 94)–3(147, 12, 104)–4(165, 38, 147)–5(–, 43, –)–6 (–, 43, –).

4. Plane 1(40, 18, 152)–2(70, 50, 178)–3(45, 60, 175) and plane 1-2-4(?, 30, 130) form a dihedral angle of 60°. Complete the views.

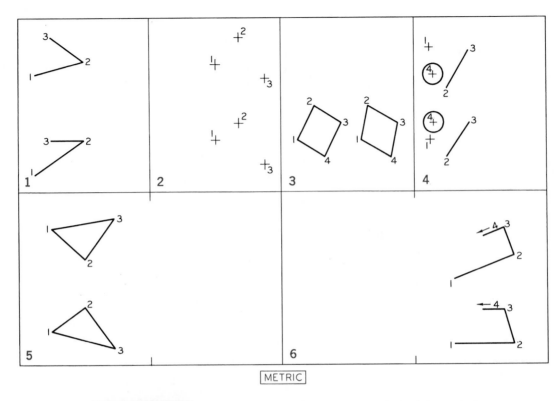

METRIC

FIGURE 19.40 **Lay out and solve problems as assigned. Use Layout A-1 or A4-1 (adjusted) divided into four equal areas.**

1. Obtain an edge view of plane $1\,(12, 12, 86)$–$2\,(50, 38, 96)$–$3\,(25, 38, 114)$.
2. Obtain an edge view of plane $1\,(50, 38, 94)$–$2\,(68, 50, 114)$–$3\,(88, 20, 84)$.
3. Obtain an edge view of plane $1\,(12, 38, 60)$–$2\,(25, 62, 66)$–$3\,(45, 50, 88)$–$4\,(-, -, -)$.
4. Find the clearance between plane $1\,(12, 38, 76)$–$2\,(25, 25, 75)$–$3\,(43, 50, 104)$ and 300 mm diameter sphere $4\,(15, 50, 86)$ by obtaining an edge view of plane 1–2–3. Scale: 1/20.
5. Obtain a true-size view of triangle $1\,(25, 25, 100)$–$2\,(50, 42, 78)$–$3\,(73, 12, 118)$ and calculate its area. Scale: 1/1.
6. Trapezoid $1\,(132, 15, 63)$–$2\,(178, 15, 80)$–$3\,(170, 40, 100)$–$4\,(?, 40, ?)$ has an area of 175 m^2. Complete the front and top views of the trapezoid. Scale: 1/400.

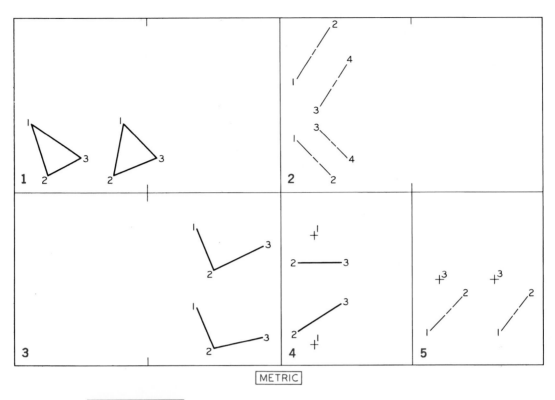

METRIC

FIGURE 19.41   Lay out and solve problems as assigned. Use Layout A-1 or A4-1 (adjusted) divided into four equal areas.

1. Find the front and side views of the center of the circle inscribed in triangle 1 (12, 50, 84)–2 (25, 12, 75)–3 (50, 25, 118). If assigned, find the views of the points of tangency. Also, if assigned, plot the views of the circle.

2. Pipe center lines 1 (12, 38, 84)–2 (38, 12, 122) and 3 (30, 45, 63)–4 (50, –, –) are to be connected with a feeder branch, using 45° lateral fittings (see Appendix 42). One fitting is to be located at the midpoint of pipe 1–2. Find the front and top views of the center line of the feeder branch.

3. Find the front and top views of the bisector of angle 1 (140, 43, 100)–2 (152, 12, 70)–3 (190, 20, 88).

4. Join point 1 (25, 15, 96) to line 2 (12, 25, 75)–3 (45, 45, 75) with a line forming an angle of 45° with line 1–2. Use revolution instead of a secondary auxiliary view.

5. Structural member 1 (12, 25, 68)–2 (38, 58, 88) is connected to point 3 (20, 63, 63) with another structural member that is 1.8 m in length. Find the front and side views of the center line of the connecting member. Use revolution instead of a secondary auxiliary view. Scale: 1/50.

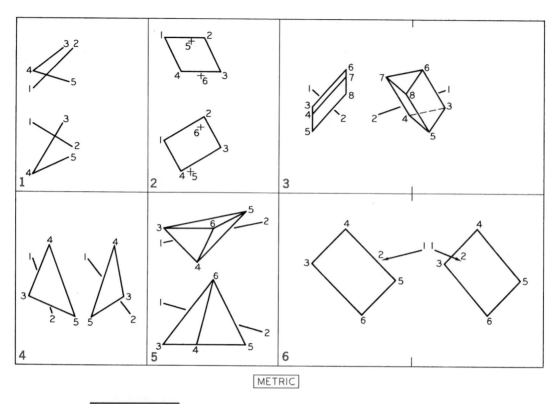

METRIC

FIGURE 19.42   **Lay out and solve problems as assigned. Use Layout A–1 or A4–1 (adjusted) divided into four equal areas.**

1. By the edge-view method, find the piercing point of line 1 (12, 50, 75)–2 (45, 32, 106) in plane 3 (35, 50, 106)–4 (12, 12, 88)–5 (40, 25, 80).

2. By the edge-view method, complete the views of bulkhead 1 (12, 38, 114)–2 (43, 55, 114)–3 (55, 33, 88)–4 (–, –, –) and intersecting cable 5 (33, 15, 112)–6 (40, 48, 86). Show visibility.

3. Find the piercing points of line 1 (25, 75, 127)–2 (46, 58, 75) with the surfaces of prism 3 (25, 63, 127)–4 (25, 58, 98)–5 (25, 45, 114)–6 (50, 90, 108)–7 (50, –, –)–8 (50, –, –). Omit that portion of line 1–2 within the prism.

4. By the cutting-plane method, find the piercing point of line 1 (12, 78, 55)–2 (28, 38, 114) with plane 3 (10, 50, 83)–4 (25, 88, 75)–5 (45, 35, 58). Show visibility.

5. By the cutting-plane method, find the piercing points of line 1 (12, 45, 91)–2 (88, 25, 106) with the surfaces of pyramid 3 (12, 15, 100)–4 (38, 15, 75)–5 (75, 15, 114)–6 (50, 63, 100). Omit that portion of line 1–2 within the pyramid, but show visibility otherwise.

6. Find the point at which light ray 1 (109, 86, 119)–2 (78, 78, 134) strikes mirror 3 (25, 75, 127)–4 (50, 100, 152)–5 (88, 63, 185)–6 (–, –, –).

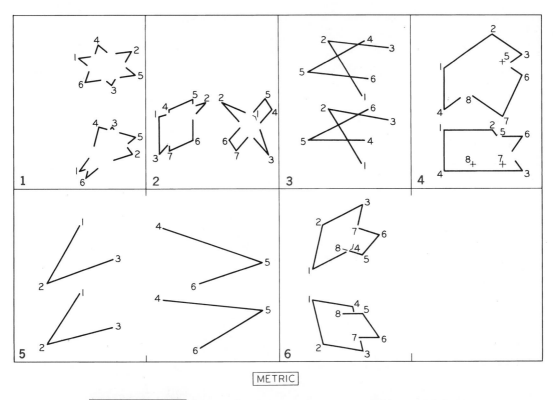

METRIC

FIGURE 19.43 Lay out and solve problems as assigned. Use Layout A-1 or A4-1 (adjusted) divided into four equal areas.

1. By the edge-view method, find the intersection of planes 1 (50, 12, 96)–2 (91, 25, 100)–3 (75, 45, 75) and 4 (66, 45, 106)–5 (94, 38, 83)–6 (55, 7, 78). Show complete visibility.

2. By the edge-view method, find the intersection of planes 1 (10, 53, 83)–2 (45, 63, 58)–3 (10, 25, 94) and 4 (17, 58, 96)–5 (35, 65, 91)–6 (35, –, –)–7 (17, 28, 68) Show complete visibility.

3. By the cutting-plane method, find the intersection of planes 1 (63, 17, 68)–2 (38, 58, 109)–3 (83, 50, 104) and 4 (66, 35, 109)–5 (23, 35, 86)–6 (68, 58, 81).

4. By the cutting-plane method, find the intersection of roof planes 1 (25, 43, 88)–2 (60, 43, 114)–3 (86, 12, –)–4 (25, 12, 58) and 5 (71, 38, –)–6 (86, 38, 84)–7 (71, 17, 53)–8 (48, 17, 68) and complete the views.

5. By the special cutting-plane method, find the intersection of planes 1 (50, 50, 100)–2 (25, 12, 58)–3 (75, 25, 75) and 4 (114, 48, 99)–5 (190, 38, 73)–6 (142, 10, 58).

6. Find and measure the dihedral angle between planes 1 (25, 46, 68)–2 (33, 12, 100)–3 (63, 8, 117)–4 (–, –, –) and 5 (63, 35, 68)–6 (75, 17, 94)–7 (58, 17, –)–8 (48, 35, 84).

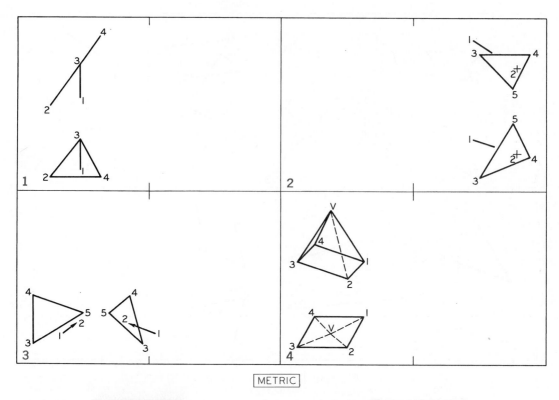

METRIC

FIGURE 19.44   Lay out and solve problems as assigned. Use Layout
A–1 or A4–1 (adjusted) divided into four equal areas.

1. Find the angle between line 1 (48, 15, 68)–3 (48, 38, –) and plane 2 (50, 10, 63)–3-4 (63, 10, 114).
2. Find the angle between cable 1 (147, 38, 112)–2 (180, 28, 91) and bulkhead 3 (152, 10, 100)–4 (190, 25, 100)–5 (177, 50, 75). Show visibility.
3. Find the angle between force vector 1 (35, 23, 106)–2 (45, 30, 86) and plane 3 (12, 15, 96)–4 (12, 50, 86)–5 (50, 38, 70).
4. Find the angle between lateral edge V (38, 23, 114)–1 (163, 35, 75) and base plane 1–2 (50, 12, 63)–3 (12, 12, 75)–4 (–, 35, –) of the pyramid.

C H A P T E R 2 0

PARALLELISM AND PERPENDICULARITY

OBJECTIVES

After studying the material in this chapter, you should be able to:

1. Identify whether lines and planes are parallel.
2. Construct a line parallel to a given line in space.
3. Construct a line parallel to a given plane.
4. Construct a line perpendicular to a given line.
5. Construct a plane perpendicular to a given line.
6. Construct a plane perpendicular to a given plane.

OVERVIEW

The parallelism of lines is a condition that is preserved in orthographic projection as demonstrated in §7.25. Parallel lines in space will be projected as parallel lines in any view. Although the parallel lines may appear as points or coincide as a single line in a view, these special cases are not regarded as exceptions to the rule.

The perpendicularity relationship of a line to a plane includes all lines in the plane through the foot of the perpendicular in addition to any other lines in the plane. If two lines are perpendicular in space, they will appear perpendicular in any view that shows at least one of the lines in true length, except if one of the lines appears as a point.

Creating lines and planes parallel to given entities are common construction techniques, especially when using CAD software. Lines that appear parallel or perpendicular in a given view may not actually be parallel or perpendicular in space. Lines and planes can be tested for parallelism and perpendicularity using descriptive geometry construction techniques. Parallel to an entity and perpendicular to an entity are common construction techniques in CAD software. While CAD can perform these constructions much faster than is possible by hand, it is important that the principles behind the construction techniques be understood to achieve desired results.

■ 20.1 PARALLEL LINES

In most situations it is valid to state that two lines drawn parallel in two adjacent views are parallel in space, Fig. 20.1 (a). However, there is an important exception: If the two adjacent views of two lines are perpendicular to the folding line between them, the lines are parallel in the views, but the lines may or may not be parallel in space, Fig. 20.1 (b), (c), and (d).

Let us study further the case of Fig. 20.1 (c). When the top view is constructed, Fig. 20.2 (a), it becomes obvious that the lines are not parallel.

If it is desired to make lines 1–2 and 3–4 parallel in space, one given view may be left incomplete, for example, the side view of line 3–4 at (b)—that is, one end point (point 4) is not immediately located. The top view is then constructed, with the top view of line 3–4 drawn parallel to line 1–2, point 4 being located by the vertical projection line from its front view. Divider distance D_1 is then used as shown at (c) to locate the side view of point 4. Thus, when two or more lines in adjacent views appear perpendicular to the folding line between the views, the test for parallelism is the construction of a third view whose direction of sight is other than parallel to the given views.

■ 20.2 PARALLEL PLANES

Planes may be established parallel to each other by drawing their edge views parallel, Fig. 20.3. Let it be assumed, as at (a), that plane 1–2–3 and point 4 are given, and it is desired to establish a plane 4–5–6 parallel to plane 1–2–3. First, the edge view of plane 1–2–3 is constructed as shown. Then the edge view of the other plane is drawn through point 4 parallel to the first edge view.

Because no further information about points 5 and 6 is given, points 5 and 6 are assumed at any random locations along the edge view of plane 4–5–6, as at (b). However, it is known that to represent a plane in space, three points must not be in a straight line, §20.7. Accordingly, projection lines are drawn from points 5 and 6 in the auxiliary view to the top view, and points 5 and 6 are placed arbitrarily in the top view along the respective projection lines. The front views are then located by projection from the top view and transfer of distances such as D_1.

A somewhat simpler and more commonly used procedure for drawing parallel planes is shown in Fig. 20.4. The method depends on this principle: *If a pair of intersecting lines in one plane is parallel to a pair of intersecting lines in a second plane, the planes are parallel.* Thus the preceding problem, Fig. 20.3 (a), is readily solved as in Fig. 20.4. Line 4–3′ is drawn parallel to line 2–3 in both views, and line 4–1′ is drawn parallel to line 1–2. Plane 3–4–1′ is parallel to plane 1–2–3. The lines 4–3′ and 4–1′ may be of any desired length.

The intersecting-line principle can also be used to check for parallelism. In Fig. 20.5, line 6–1′ was added in the top view of plane 4–5–6 and parallel to line 1–2 of plane 1–2–3. When the front view of point 1′ is located, line 6–1′ is found to be parallel to line 1–2. Continuing the investigation: In the top view of plane 4–5–6, line 4–3′ is added parallel to line 2–3. When the front view of line 4–3′ is located, however, it is seen that it *is not* parallel to the front view of line 2–3. Hence planes 1–2–3 and 4–5–6 are *not* parallel in space.

■ 20.3 LINE PARALLEL TO PLANE

A line is parallel to a plane if it is parallel to a line in the plane. To establish line 4–5′ parallel to the plane 1–2–3, Fig. 20.6, line 3–5 is arbitrarily selected and added to the plane 1–2–3. Line 4–5′ is then drawn parallel to line 3–5 in the plane. It is possible that line 4–5′ is *in* plane 1–2–3 (if extended). Even if true, this is not considered an exception to the general principle.

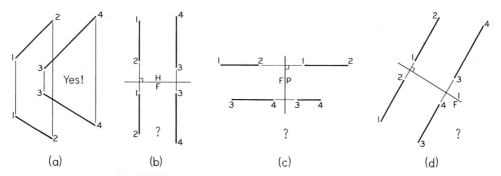

FIGURE 20.1 Parallel Views: Are the Lines Parallel in Space?

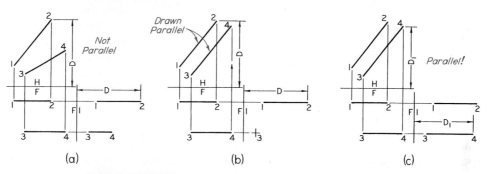

FIGURE 20.2 Construction of Parallel Horizontal Lines.

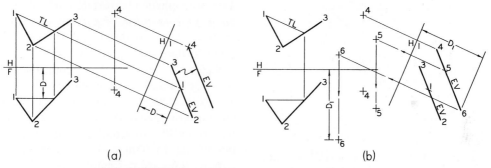

FIGURE 20.3 Parallel Planes by Parallel Edge Views.

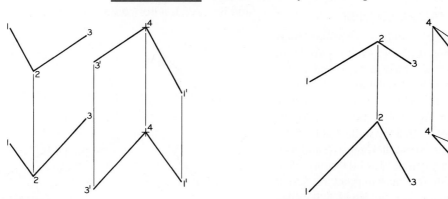

FIGURE 20.4 Parallel Planes by Parallel Lines.

FIGURE 20.5 Checking Parallelism of Planes with Intersecting Lines.

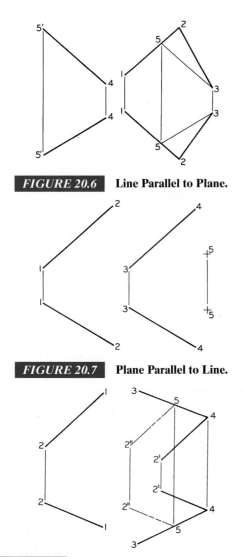

FIGURE 20.6 Line Parallel to Plane.

FIGURE 20.7 Plane Parallel to Line.

FIGURE 20.8 Plane Through One or Two Skew Lines and Parallel to the Other.

■ 20.4 PLANE PARALLEL TO LINE

A plane is parallel to a line if it contains a line that is parallel to the given line. Thus, if plane 1–2–3 is parallel to line 4–5′, Fig. 20.6, line 3–5 added to the plane 1–2–3 and parallel to line 4–5 will be parallel to line 4–5′ in all views. This is the converse of the principle of §20.3. Logically if line 4–5′ in Fig. 20.6 is parallel to plane 1–2–3, it follows that plane 1–2–3 is parallel to line 4–5′.

Suppose we are given two parallel lines 1–2 and 3–4, Fig. 20.7. How many planes can be "passed through" line 3–4 parallel to line 1–2? If any random point 5 is added (not on 3–4), a plane 3–4–5 is established, §20.7. The plane is parallel to line 1–2 because it contains line 3–4. Thus it is seen that an infinite number of planes can be

passed through one of two parallel lines and parallel to the other.

On the other hand, Fig. 20.8, let two *skew* lines 1–2 and 3–4 be given, and let it be required to pass a plane through line 3–4 and parallel to line 1–2. If a line is added, such as 4–2′, parallel to line 1–2, plane 3–4–2′ is parallel to line 1–2. Note that the added line could be made to intersect line 3–4 at any point, such as 5, resulting in plane 3–5–2″. However, this is merely a revised representation of plane 3–4–2′ It is evident, then, that through one of two skew lines only one plane can be passed parallel to the other line.

In Fig. 20.9, let skew lines 1–2 and 3–4 and point 5 be given, and let a plane be required through the point and parallel to the skew lines. If line 5–2′ is drawn parallel to line 1–2, and line 5–4′ is drawn parallel to line 3–4, plane 2′–5–4′ is parallel to both lines 1–2 and 3–4, even though nonintersecting lines 1–2 and 3–4 do not represent a plane.

■ 20.5 PERPENDICULAR LINES

In §6.26 it was observed that a 90° angle is projected in true size, even though it is in an inclined plane, provided one leg of the angle is a *normal* line, Fig. 6.29 (d). This principle can be restated in broader terms: *A 90° angle appears in true size in any view showing one leg true length,* provided the other leg does not appear as a point in the same view. Thus, in Fig. 20.10, lines 2–3, 2–4, 2–5, and 2–6 are all perpendicular to line 1–2, and they appear at 90° to the true-length front view of line 1–2. Note that the 90° angle is not observed in the top view where none of the lines is true length.

In Fig. 20.11 each of the lines 2–3 and 2–4 is perpendicular to oblique line 1–2 because their true-length views are perpendicular to the corresponding views of line 1–2. (Note that line 2–3 is a frontal line and line 2–4 is a horizontal line, §20.2.)

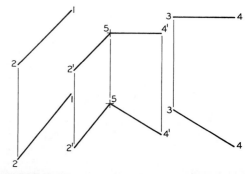

FIGURE 20.9 Plane Through Point Parallel to Skew Lines.

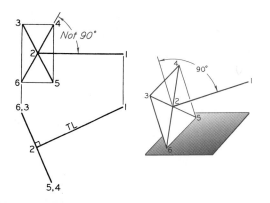

FIGURE 20.10 **Lines Perpendicular to True-Length Line.**

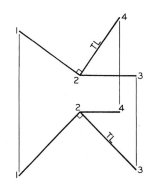

FIGURE 20.11 **True-Length Lines Perpendicular to Oblique Line.**

■ 20.6 PLANE PERPENDICULAR TO LINE

GIVEN-VIEW METHOD To establish a plane perpendicular to a line, Fig. 20.12 (a), true-length lines are drawn perpendicular to given line 1–2 in the same manner as in Fig. 20.11. Now consider plane 3–2–4 in Fig. 20.12 (a). Since it is represented by intersecting lines, each of which is perpendicular to line 1–2, the plane is perpendicular to line 1–2.

Then consider the case at (b) where a plane through given point 5 and perpendicular to line 1–2 is desired. If plane 3'–5–4' is constructed parallel to plane 3–2–4 of part (a), plane 3'–5–4' will also be perpendicular to line 1–2. However, plane 3'–5–4' could be drawn directly—without the use of plane 3–2–4—merely by drawing the true–length views of lines 3'–5 and 5–4', respectively, perpendicular to the corresponding views of line 1–2, as indicated at (b).

Note that lines 3'–5 and 5–4' do *not* intersect line 1–2. (You may prove this for yourself by extending the lines and checking vertical alignment of crossing points.) Thus, for present purposes, it is useful to regard the perpendicular true-length view *position principle* as indicating perpendicular lines, without regard to whether the lines intersect.

APPLICATION OF GIVEN-VIEW METHOD

Given: Views of a line 1–2 and a point 3, Fig. 20.13 (a).
Req'd: Find a line from point 3 perpendicular to and intersecting line 1–2. Use only the given views.
Analysis and Procedure: If a horizontal line 3–4 is drawn through point 3 with its true-length view perpendicular to the top view of 1–2 as at (b), it does not intersect line 1–2. Similarly, if a frontal line 3–5 is drawn perpendicular to line 1–2, it also does not intersect line 1–2. Neither of these lines is the required line. We conclude that the required line will not appear perpendicular in the given views. However, 4–3–5 represents a *plane* perpendicular to line 1–2, and all lines in the plane are perpendicular to line 1–2. Plane 4–3–5 is the locus

FIGURE 20.12 **Plane Perpendicular to Line—Given-View Method.**

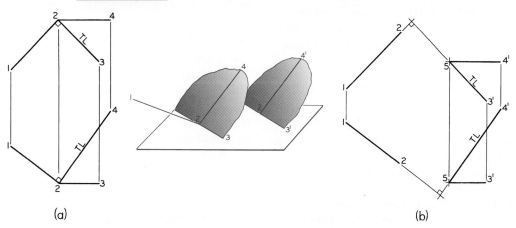

(a)   (b)

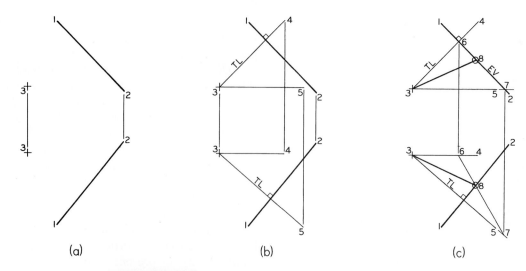

(a) (b) (c)

FIGURE 20.13 **Plane Perpendicular to Line—Application.**

of all lines through point 3 perpendicular to line 1–2. The required line belongs to this locus or family of lines.

By the cutting-plane method of §20.11, the piercing point 8 of line 1–2 in plane 4–3–5 is found, Fig. 20.13 (c). This is the only point on line 1–2 that is in the plane, and thus line 3–8 is the only possible solution to the problem.

AUXILIARY-VIEW METHOD A plane also may be constructed perpendicular to a line by drawing its edge view perpendicular to the true-length view of the line, Fig. 20.14. At (a) given line 1–2 and point 3 are projected to the true-length auxiliary view, where the edge view of the required plane is drawn through point 3 and perpendicular to the true-length view as indicated. If it is desired to represent the perpendicular plane in the top and front views, as at (b), any convenient pair of points, such as 4 and 5, is selected on the edge view. In the top view points 4 and 5 may be placed at any desired locations along the projection line from view 1. Points 4 and 5 are then located in the front view by the divider distances as indicated.

APPLICATION OF AUXILIARY-VIEW METHOD

Given: A right square prism, Fig. 4.7 (6), has its axis along line 1–2 and one corner at point 3, Fig. 20.15 (a). The other base is centered at point 4.
Req'd: Find the views of the prism.
Analysis and Procedure: Auxiliary view 1 is added showing axis 1–2 in true length. Because a *right* prism is required, the bases must be perpendicular

to axis 1–2 and appear in view 1 in edge view and perpendicular to axis 1–2. At (b) the point view of axis 1–2 is added. This view shows the true shape of the square bases, and the size of the square is established by the position of corner 3. By the method of Fig. 4.23 (c), the square is constructed.

The projection process is now reversed. The corners are projected from view 2 to view 1, which is then completed. Next, the corners are projected to the top view and located with divider distances such as D_2. Observe that a square prism is composed of three sets of parallel lines. Check the view for parallelism and correct any errors before proceeding.

Finally, the front view is projected in similar fashion. Again, the construction work should be checked for accurate parallelism before the views are completed.

■ 20.7 LINE PERPENDICULAR TO PLANE

A line perpendicular to a plane is perpendicular to all lines in the plane. In practice, it is sufficient to state that a line is perpendicular to a plane if it is perpendicular to at least two nonparallel lines in the plane. This line will also appear perpendicular (and in true length) to the edge view of the plane. Either principle may be used in the construction of a line perpendicular to a plane.

It is desired to construct by the given-view method, Fig. 20.16, a line from point 4 perpendicular to plane 1–2–3. Since the plane is oblique, the perpendicular line will be oblique and will not appear in true length. Although the required line will be perpendicu-

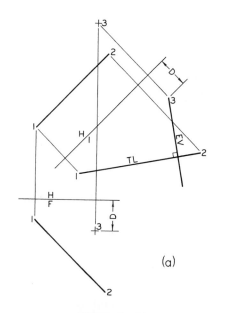

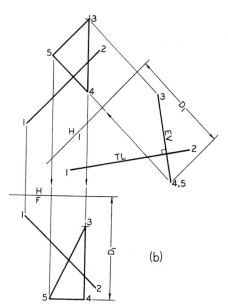

FIGURE 20.14   Plane Perpendicular to Line—Auxiliary-View Method.

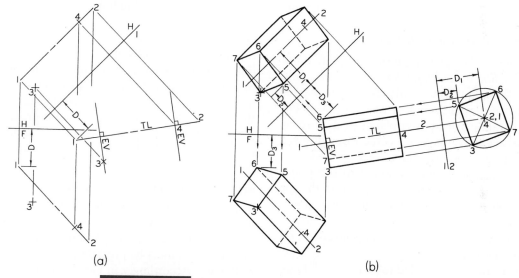

FIGURE 20.15   Plane Perpendicular to Line—Application.

lar in space to lines 1–2 and 2–3, the 90° angles between the given lines 1–2 and 2–3 and the required line will not appear in the given views because lines 1–2 and 2–3 are not shown in true length.

If a frontal line is added to plane 1–2–3, as at (b), its front view is true length, and the front view of the required line will be perpendicular to the true-length view of the frontal line as shown. It must be realized that at this stage nothing has been determined about the direction of the perpendicular in the top view. Nor is the point at which the perpendicular strikes the

plane—its piercing point—known. (Frequently the location of the piercing point is of no interest.)

At (c) a horizontal line is added. The true-length view of the horizontal line determines the direction of the top view of the required perpendicular. Again, the piercing point has not been determined, but two views have been established and thus a space description of the required perpendicular from point 4 with plane 1–2–3 has been constructed. Point 5 is arbitrarily selected as an end point of the line, not necessarily in the plane.

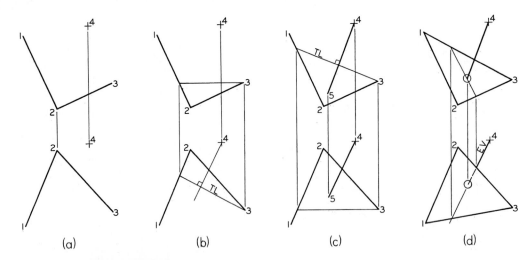

(a) (b) (c) (d)

FIGURE 20.16 Plane Perpendicular to Plane—Given-View Method.

If it is desired to terminate the perpendicular in plane 1–2–3, it will be necessary to find the piercing point by one of the methods of §20.11. At (d) the cutting-plane method was used to minimize construction.

■ 20.8 PERPENDICULAR PLANES

If a line is perpendicular to a given plane, any plane containing the line is perpendicular to the given plane. Since an infinite number of planes can be passed through such a perpendicular, it was chosen in Fig. 20.17 to illustrate perpendicular planes by a more restricted example: How to pass a plane through given line 4–5 and perpendicular to given plane 1–2–3.

At (a) a horizontal line is added to the plane. Since a horizontal line appears true length in the top view, it determines the direction of the required perpendicular in that view (but, remember, not in the front view). Point 5 is selected as being convenient for the origin of the perpendicular.

At (b) a frontal line is added to plane 1–2–3 to establish a true-length line in the front view, which is needed to determine the direction of the front view of the required perpendicular. It is assumed that the piercing point of the perpendicular is of no interest here. Accordingly, any arbitrary point, such as 6, is used to terminate the perpendicular and complete the representation of the required perpendicular plane 4–5–6.

■ 20.9 COMMON PERPENDICULAR BETWEEN SKEW LINES

The shortest distance, or clearance, between any two lines is measured along a line perpendicular to both

lines. Study Fig. 20.18. Suppose line 5–6 to be perpendicular to line 1–2, intersecting line 3–4 at 6. Now let line 5–6 move to position 5'–6', still perpendicular to line 1–2 and intersecting line 3–4 at point 6'. Continue the process of moving line 5–6 upward along line 1–2. Note that it will gradually shorten until eventually it reaches a minimum length and will begin to lengthen if moved further. At its minimum length it will be perpendicular to line 3–4 also. This is the common perpendicular representing the shortest distance between the skew lines 1–2 and 3–4. Several procedures are available for locating the views of the common perpendicular in multiview projection, two of which will be discussed here.

POINT-VIEW METHOD If a point view of any given line is constructed, a line that is perpendicular to the given line will show in true length in the view showing the given line as a point. As noted in §20.6, a point view of a line must be preceded by a true-length view of the line. Accordingly, in Fig. 20.19 (a) line 3–4 is arbitrarily chosen to be shown in true length and view 1 is projected. As shown at (b), view 2 is then constructed showing line 3–4 as a point. In this view, any line perpendicular to line 3–4 (including the shortest connector) must appear in true length. Since the shortest connector is also perpendicular to line 1–2, it must appear at 90° to line 1–2 as shown, even though line 1–2 is not true length. If only the shortest distance is required, it is measured in view 2 and the construction is complete.

If, in addition, the views of the common perpendicular are required, we proceed as shown at (b). Point

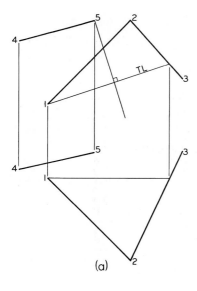

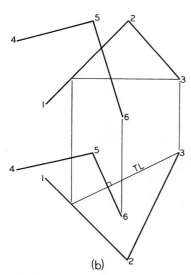

FIGURE 20.17   **Perpendicular Planes.**

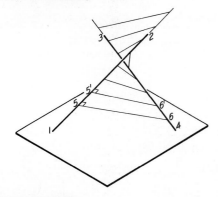

FIGURE 20.18   **Common Perpendicular.**

5 is projected to line 1–2 in view 1. In this view the common perpendicular is not true length. Line 3–4 *is* true length, however, so line 5–6 is drawn perpendicular to line 3–4 in view 1 as shown. It is then routine to project line 5–6 to the top view and then to the front view. Divider distances such as D_2 and D_3 are used to check the accuracy of the construction.

PLANE METHOD   If a plane containing one of two skew lines is parallel to the second line, the perpendicular distance from the second line to the plane is the shortest distance between the two lines.

In Fig. 20.20 (a) a line is drawn through point 2 parallel to line 3–4, thus establishing a plane containing line 1–2 and parallel to line 3–4, §20.4. The edge view of the plane is then established in auxiliary view 1. In this view the shortest distance is measured as shown. The shortest distance being obtained in only one additional view is an advantage of the plane method over the point-view method.

If, in addition, the views of the common perpendicular are required, a second auxiliary view is necessary and the total amount of construction is slightly more than in the point-view method. At (b) the second auxiliary view is constructed with its direction of sight parallel to the shortest distance (or perpendicular to the edge-view plane). In this view the shortest distance or common perpendicular must appear as a point. This can be only at the apparent crossing point 5, 6 of lines 1–2 and 3–4. This locates the common perpendicular, which is then projected back to the other views as shown.

■ 20.10 SHORTEST HORIZONTAL LINE CONNECTING SKEW LINES

Related to the preceding plane method is the problem of finding the shortest line at zero slope or grade, the shortest horizontal line connecting two skew lines.

In Fig. 20.21 (a) a plane is constructed containing line 1–2 and parallel to line 3–4. The edge view of the plane is constructed through the use of a *horizontal* line added to the plane because in the auxiliary view 1 any horizontal connecting line will be parallel to folding line H/1.

Auxiliary view 2 is then constructed at (b) with its direction of sight parallel to folding line H/1. The shortest connector 5–6 appears in point view in view 2

G R A P H I C S   S P O T L I G H T

CAD Design

CAD DESIGN OF GEARS AND CAMS

Once you know the design requirements for a particular gear or cam, a CAD software package can be used to automatically generate the drawing. It can be very time consuming to create drawings of gears and cams by hand. Using CAD frees the designer to spend more time on other aspects of the design.

AUTOMATIC DRAWING GENERATION

One program you can use to automatically generate drawings for gears and cams is Cimlogic™ Toolbox Power Transmission software. It is designed to run with AutoCAD® and AutoCAD Mechanical Desktop®. Toolbox Power Transmission software lets you type the data defining the motions for the cam or the details for the gear into user friendly dialog boxes.

The first step in creating a cam is to choose whether you will design a circular cam or a linear cam. Figure A shows a circular cam selected and follower diameter of .75.

ONE PICK SPECIFIES CAM MOTION

Once you have selected the type of cam and entered the default radius and starting point for the motion, you use a new dialog box shown in Figure B to define the various motions for the cam. You can choose constant acceleration, cycloidal, harmonic, double harmonic, dwell, modified sine, modified trapezoid, polynomial, and uniform motions. To define a motion, you just pick on the type of motion desired, type in the ending radius (if appropriate) and the number of degrees to perform this motion. Once you have the dialog box filled out to draw the portion of the cam, just pick on create. You can add additional motions for circular cams up to 360° of travel.

DISPLACEMENT DIAGRAM AND SUMMARY CREATED FOR YOU

The cam drawing shown in Figure C was created in minutes using AutoCAD Mechanical Desktop and

Toolbox Power Transmission software. You can select whether or not to show the follower path (it is shown in the figure). The displacement diagram and the cam summary were generated automatically, just by picking where to place the information. Toolbox Power Transmission software also will generate gear and sprocket drawings in a similar fashion. In addition, the regular toolbox software can be used to automatically create fasteners, keys, seals, standard symbols, and other useful items.

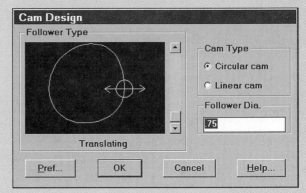

(A)

(B)

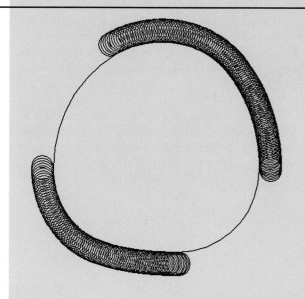

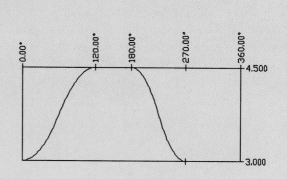

CIRCULAR CAM SUMMARY				
FOLLOWER TYPE: TRANSLATING			FOLLOWER DIAMETER: 0.7500	
STARTING RADIUS: 3.0000		STARTING ANGLE: 0.000		
NO.	MOTION TYPE	MOT. DIV.	ENDING RAD.	DEGREES MOTION
1	CONSTANT ACCELERATION	1.0000°	4.5000	120.000
2	DWELL	—	—	60.000
3	CONSTANT ACCELERATION	—	3.0000	90.000
4	DWELL	—	—	90.000

(C)

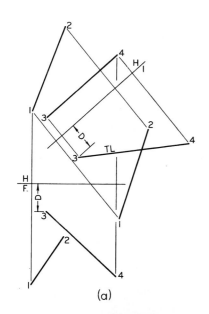

(a)

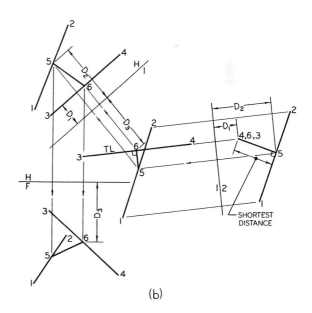

(b)

FIGURE 20.19   Common Perpendicular—Point-View Method.

at the apparent crossing point of lines 1–2 and 3–4. It is then projected back to the other views.

■ 20.11 SHORTEST LINE AT SPECIFIED SLOPE CONNECTING SKEW LINES

For a specified slope other than zero (horizontal), the method of §20.10 requires only slight modification. In this example, it is assumed that 38° is specified for the slope of the connecting line. The first portion of the construction is the same, Fig. 20.22 (a): A plane is passed through line 1–2 and parallel to line 3–4, and

an edge view of the plane is constructed, as shown, in an auxiliary view adjacent to the top view.

At (b) projection lines for view 2 are drawn at the prescribed slope angle with folding line H/1. In view 2, the apparent crossing point 5, 6 of lines 1–2 and 3–4 is the point view of the shortest connector 5–6 at the specified slope of 38°. The other views of line 5–6 are then completed by projection as before.

It should be noted that the general procedure illustrated could be readily modified for other specifications, such as the shortest connecting line at a prescribed grade or the shortest frontal line connecting two skew lines.

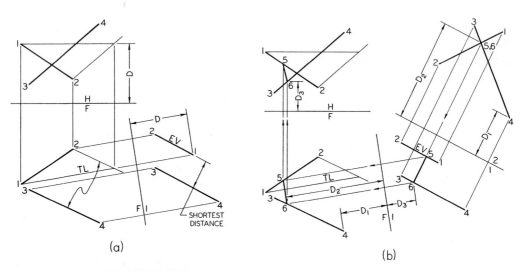

(a)   (b)

FIGURE 20.20   **Common Perpendicular—Plane Method.**

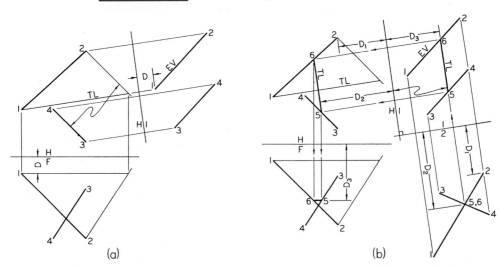

(a)   (b)

FIGURE 20.21   **Shortest Horizontal Line Connecting Two Skew Lines.**

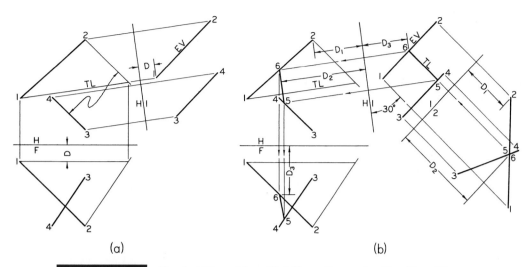

(a)   (b)

FIGURE 20.22 **Shortest Line at Specified Slope Connecting Two Skew Lines.**

■ KEY WORDS

PARALLEL	PERPENDICULAR
HORIZONTAL	VERTICAL LINES
PROFILE LINES	TRUE LENGTH
GIVEN VIEW METHOD	SKEW
COMMON PERPENDICULAR	

■ CHAPTER SUMMARY

• In all but one case, parallel lines that appear parallel in two adjacent views are parallel. The exception in the case when the lines are perpendicular to the fold lines between adjacent views.

• Two planes are parallel when their edge views appear parallel.

• A line is parallel to a plane if it is parallel to any line on the plane.

• A line is perpendicular to a plane if its measured angle is 90° with the plane when the line appears true length.

• The shortest distance between any two lines occurs on a line that is perpendicular to both lines.

■ REVIEW QUESTIONS

1. Do parallel lines always appear parallel in all views?

2. If two planes appear in edge view, are they parallel? What condition would make the planes parallel?

3. Is the point view of a line that intersects a plane perpendicular to the plane? What condition would make the line perpendicular to the plane?

4. Describe how to use intersercting lines to check for parallelism between planes.

5. How is a true length line drawn on a plane?

6. Describe the given-view method for constructing a plane perpendicular to a line.

7. When using the auxiliary view method to create a line perpendicular to a plane, does the plane need to be in edge view, or does the line need to be true length in the auxiliary view?

8. What is the name given to two lines that are not parallel and not intersecting?

9. When is the shortest distance between two skew lines NOT a line that is perpendicular to both given lines?

■ PROBLEMS IN PARALLELISM AND PERPENDICULARITY

In Figs. 20.23–20.26 are problems covering parallel lines, lines parallel to a plane, plane parallel to a line, plane parallel to a plane or skew lines, perpendicular lines, lines perpendicular to planes, common perpendicular, and shortest line at specified angle.

Use Layout A–1 or A4–1 (adjusted) and divide the working area into four equal areas for problems to be assigned by the instructor. Some problems will require a single problem area, and others will require two problem areas or one half sheet. Data for the layout for each problem are given by a coordinate system using metric dimensions. For example, in Fig. 20.23, Prob. 1, point 6 is located by the full-scale coordinates (35, 20, 50). The first coordinate locates the front view of the point from the left edge of the problem area. The second coordinate locates the front view of the point from the bottom edge of the problem area. The third coordinate locates either the top view of the point from the bottom edge of the problem area or the side view of the point from the left edge of the problem area. Inspection of the given problem layout will determine which application to use.

Since many of the problems in this chapter are of a general nature, they can also be solved on most computer graphics systems. If a system is available, the instructor may choose to assign specific problems to be completed by this method.

Additional problems, in convenient form for solution, are available in *Engineering Graphics Problems*, Series 1, by Spencer, Hill, Loving, Dygdon, and Novak designed to accompany this text and published by Prentice Hall Publishing Company.

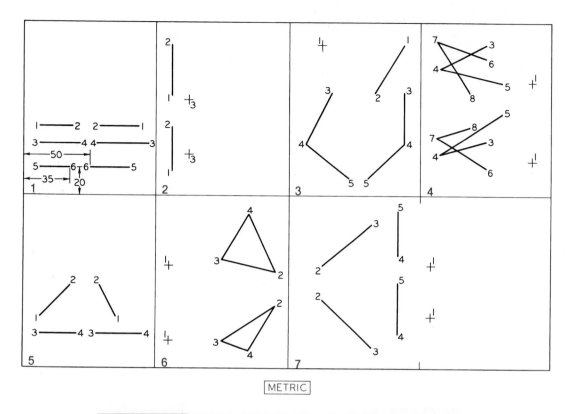

METRIC

FIGURE 20.23   Lay out and solve problems as assigned. Use Layout
A–1 or A4–1 (adjusted) divided into four equal areas.

1. Determine which, if any, of the lines 1 (12, 50, 88)–2 (38, 50, 58), 3 (12, 38, 96)–4 (43, 38, 55), or 5 (12, 20, 81)–6 (35, 20, 50) are parallel.

2. Draw a line 3 (25, 30, 71)–4, 25 mm in length and parallel to line 1 (12, 18, 73)–2 (12, 50, 112).

3. Complete the side view of line 1 (25, 112, 88)–2 (?, 75, 65) that is parallel to plane 3 (33, 75, 88)–4 (12, 38, 88)–5 (45, 12, 60).

4. Through point 1 (86, 25, 83) draw a line 1–2 parallel to planes 3 (50, 40, 112)–4 (15, 30, 94)–5 (63, 60, 84) and 6 (50, 20, 100)–7 (12, 43, 114)–8 (38, 50, 75).

5. Pass a plane through line 1 (12, 38, 70)–2 (35, 60, 58) parallel to line 3 (12, 25, 56)–4 (40, 25, 91). Add the top view.

6. By means of a horizontal line and a frontal line represent a plane containing point 1 (10, 20, 75) and parallel to plane 2 (91, 45, 70)–3 (50, 20, 81)–4 (70, 12, 114).

7. Pass a plane through point 1 (109, 38, 75) parallel to lines 2 (25, 50, 75)–3 (63, 15, 108) and 4 (85, 25, 84)–5 (85, 63, 107).

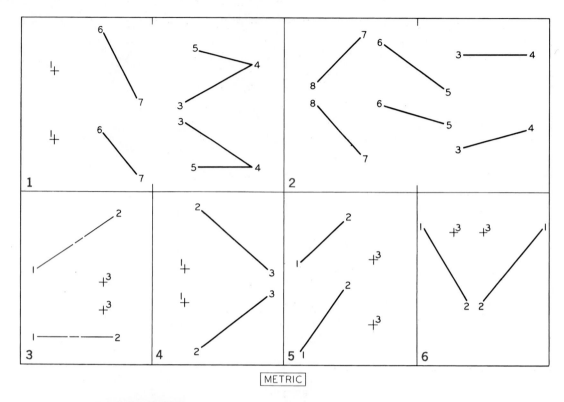

METRIC

FIGURE 20.24   Lay out and solve problems as assigned. Use Layout
A–1 or A4–1 (adjusted) divided into four equal areas.

1. Find a line 1 (25, 38, 88)–2 that is parallel to plane 3 (127, 50, 66)–4 (178, 17, 94)–5 (137, 17, 104) and intersects line 6 (63, 43, 117)–7 (88, 12, 60).

2. Establish a line 1–2 that is parallel to line 3 (137, 33, 100)–4 (185, 45, 100) and intersects lines 5 (122, 50, 75)–6 (75, 63, 109) and 7 (58, 28, 114)–8 (25, 63, 81).

3. Line 1 (12, 20, 70)–2 (70, 20, 109) is the center line of a pipe. Connect this pipe to point 3 (63, 40, 60) with a 90° elbow (pipe fitting) at the juncture on 1–2. Find the true length of the center line of the connecting pipe. Scale: 1/10.

4. Draw a 50 mm frontal line from point 1 (25, 45, 70) perpendicular to line 2 (38, 12, 115)–3 (88, 50, 70). Also draw a 50 mm horizontal line from point 1 perpendicular to line 2–3. Use only the given views. (Note that these lines do not intersect line 2–3.)

5. Using only the given views, find a line 3 (68, 30, 78)–4 that is perpendicular to line 1 (12, 10, 75)–2 (45, 55, 106) and also intersects line 1–2.

6. Find the center of the smallest sphere that has its center on line 1 (5, 100, 96)–2 (38, 48, 50) and has point 3 (28, 88, 50) on its surface. Use only the given views. If assigned, find the diameter of the sphere.

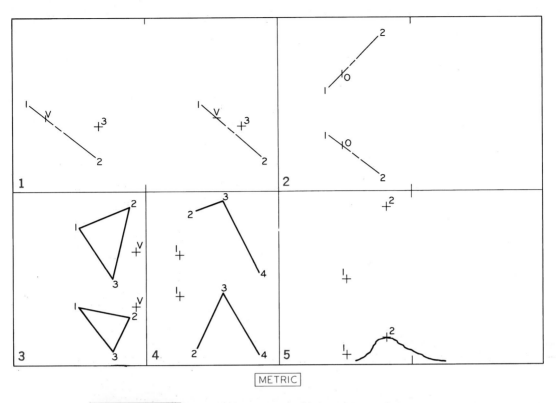

METRIC

FIGURE 20.25 **Lay out and solve problems as assigned. Use Layout
A–1 or A4–1 (adjusted) divided into four equal areas.**

1. The axis of a right square pyramid lies along center line $1(12, 63, 144)$–$2(63, 25, 190)$. One corner of the base is at point $3(66, 48, 175)$. The vertex is at point $V(25, -, -)$. Find the front and side views of the pyramid.

2. The axis of a right prism lies along center line $1(38, 40, 75)$–$2(75, 12, 114)$. One base is centered at $O(48, -, -)$. The bases are equilateral triangles inscribed in 36 mm diameter circles. The lowest side of each base is a horizontal line. The altitude of the prism is 35 mm. Complete the views.

3. If an oblique cone is drawn with its vertex at $V(94, 43, 84)$ and its base in plane $1(50, 43, 100)$–$2(88, 35, 116)$–$3(75, 10, 63)$, what is its altitude in millimeters? Use an auxiliary view to solve this problem. Show the front and top views of the altitude.

4. Find the shadow of point $1(25, 50, 80)$ on plane $2(38, 12, 114)$–$3(57, 53, 122)$–$4(84, 7, 68)$ if light rays are perpendicular to the plane.

5. An aircraft on a landing approach course of N 45° passes 300 m above point $1(50, 7, 63)$. It is losing altitude at the rate of 200 m in 1000 m. Point $2(80, 20, 116)$ represents the peak of a hill. How close to point 2 does the aircraft pass? Scale: 1/10,000.

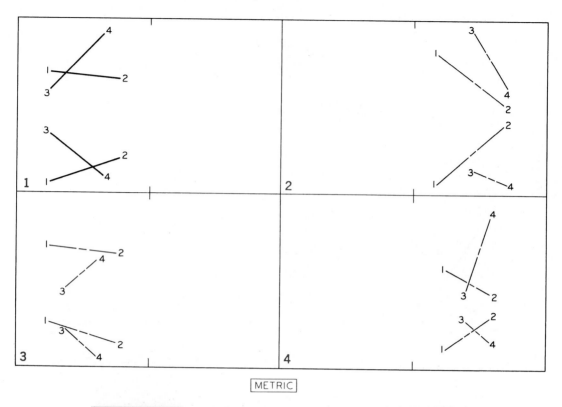

METRIC

FIGURE 20.26 Lay out and solve problems assigned. Use Layout A–1 or A4–1 (adjusted) divided into four equal areas.

1. Find the clearance between high-voltage lines $1(25, 7, 88)$–$2(78, 25, 84)$ and $3(25, 43, 75)$–$4(66, 12, 117)$. Show the views of a line representing this clearance. Scale: $1/40$.

2. Determine the bearing, grade, and length of the shortest shaft connecting tunnels $1(120, 7, 104)$–$2(170, 50, 65)$ and $3(147, 18, 120)$–$4(172, 7, 78)$. Scale: $1/80$.

3. Ski slopes represented by lines $1(25, 33, 88)$–$2(75, 18, 84)$ and $3(38, 28, 58)$–$4(61, 7, 78)$ are to be connected by the shortest possible horizontal path. Find the views and measure the length of this path. Scale: $1/4000$.

4. Tunnel $1(127, 15, 73)$–$2(160, 38, 56)$ is to be connected to tunnel $3(142, 35, 58)$–$4(162, 20, 112)$ with the shortest tunnel at a downgrade of 10%. Find the length and bearing of this connector. Scale: $1/2000$.

C H A P T E R 2 1

INTERSECTIONS

OBJECTIVES

After studying the material in this chapter, you should be able to:

1. Define the term intersection.
2. Describe examples of intersections in the real world.
3. Draw the intersection between a plane and a prism.
4. Draw the intersection between a plane and a pyramid.
5. Draw the intersection between a plane and a cylinder.
6. Draw the intersection between a plane and a cone.
7. Draw the intersection between any combination of prisms, cylinders, cones, and pyramids.

OVERVIEW

Intersections are common in building construction, sheet-metal work, machine parts and structures, and the drafter or designer must know how to construct them. The intersection of a line and a surface is a *point*. Two surfaces will intersect in a *line of intersection*. The complete intersection between two solids is called a *figure of intersection*. Although intersections may be of many different kinds, they all involve some combinations of lines, planes, and solids. Accurate representation of the intersecting surfaces therefore becomes very important since precision fit is necessary for function and appearance.

Intersections can be defined as a locus of points that the surfaces of two solids share when their elements pierce one another. Intersections are an important design consideration in metal work for aviation and automotive applications, plastics design, and general consumer and industrial products design. Intersections are one of the more complex spatial calculations that CAD software must make. Surface modeling software uses the same engineering graphics principles for intersection that drafters use on the drawing board. All but the most sophisticated CAD software require the user to understand the basic principles of intersection in order to solve these problems on the computer. Typical intersection problems involve planes that intersect with basic solid shapes, and solids that intersect with other solids.

■ 21.1 SURFACES AND SOLIDS

A *surface* is a two-dimensional geometric entity having (theoretically) no thickness. It thus has *area* but not *volume*. A surface may be generated by a moving line, straight or curved, called the *generatrix*. Any position of the generatrix is an *element* of the surface. See Fig. 6.31 (a).

A *ruled surface is* generated by a moving straight line. The surface generated may be a *plane, a single-curved surface*, or a *warped surface*.

A *plane is* a ruled surface generated by a straight line one point of which moves along another straight line (the *directrix),* while the generatrix remains parallel to its original position. Many of the geometric solids are bounded by plane surfaces, Fig. 4.7.

A *single-curved surface is* a *developable* ruled surface, §22.1; that is, it can be unrolled to coincide with a

plane. A single-curved surface is generated by a straight line one point of which moves along a plane-curve directrix, while the generatrix continues to pass through a fixed point (the vertex) as for a cone, or remains parallel to its original position as for a cylinder. See Fig. 4.7.

A *warped surface* is a ruled surface that is not developable, Fig. 21.1. These surfaces are generated by straight lines moving in various patterns, all characterized by the principle that no two adjacent positions of the generatrix lie in the same plane. Many exterior "streamlined" surfaces of automobiles or aircraft are warped surfaces.

A *double-curved surface* may be generated only by a curved line, and has no straight-line elements. Such a surface, generated by revolving a curved line about a straight line in the plane of the curve, is called a *double-curved surface of revolution*. Common examples are the *sphere, torus,* and *ellipsoid,* Fig. 4.7.

Solids bounded by plane surfaces are *polyhedra,* the most common forms of which are the *pyramid* and *prism*, Fig. 4.7. Convex solids whose faces are all congruent regular polygons are the *regular polyhedra,* which include the *tetrahedron, cube, octohedron, dodecahedron,* and *icosahedron,* known as the five *Platonic solids*.

Plane surfaces that bound polyhedra are *faces* of the solids. Lines of intersection of faces are *edges* of the solids.

A solid generated by revolving a plane figure about an axis in the plane of the figure is a *solid of revolution*.

Solids bounded by warped surfaces have no group name. A common example of such solids is the screw thread.

In practice the term *solid is* frequently used for convenience for any three-dimensional, closed form, even though it may be a hollow sheet-metal form such as a "tin" can or heating duct, and not solid at all.

INTERSECTIONS OF PLANES AND SOLIDS

■ 21.2 INTERSECTIONS OF PLANES AND POLYHEDRA

The principles involved in intersections of planes and solids have their practical application in the cutting of openings in roof surfaces for flues and stacks or in wall surfaces for pipes, chutes, and so on, and in the building of sheet-metal structures such as tanks and boilers.

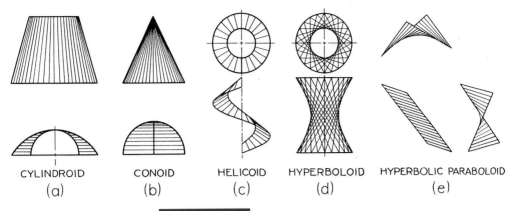

CYLINDROID	CONOID	HELICOID	HYPERBOLOID	HYPERBOLIC PARABOLOID
(a)	(b)	(c)	(d)	(e)

FIGURE 21.1   **Warped Surfaces.**

In these applications the problem is basically one of determining the views of the intersection of a plane and one of the more common geometric solids. The intersection of a plane and a solid is the locus of the points of intersection (piercing points) of the elements of the solid with the plane. For solids bounded by plane surfaces, the intersection consists of successive segments of lines of intersection of the plane and the plane surfaces of the solid §19.12.

For solids bounded by curved surfaces, it is necessary to find the piercing points of several elements of the solid with the plane, §19.11, and then to draw a smooth curve through these points. The curve of intersection of a plane and a circular cone is a *conic section,* Fig. 4.46.

INTERSECTION OF PLANE AND PRISM   *Fig. 21.2* To determine the intersection of the given plane 1–2–3 and the vertical prism, it is convenient to obtain the piercing points of the parallel lateral edges of the

prism by the edge-view method of §19.11. Since the top view of line 2–3 is true length, an edge view of plane 1–2–3 is construeted as at (b), with its direction of sight parallel to line 2–3. The piercing points are 4, 5, 6, and 7. The top views of the piercing points coincide with the point views of the edges of the prism. At (c) the piercing points are located in the front view by means of divider distances such as D_1.

Fig. 21.3   For the plane and oblique prism of Fig. 21.3, it is chosen to obtain piercing points by the cutting-plane method of §19.11. An edgeview cutting plane is introduced to coincide with a lateral edge of the prism, as shown in the top view at (b). It cuts a line 4–5 from plane 1–2–3. Line 4–5, after projection to the front view, locates the front view of point A, the piercing point of one lateral edge.

In like fashion, (c), other edge-view cutting planes are introduced to find the remaining piercing points.

FIGURE 21.2   **Intersection of Plane and Vertical Prism.**

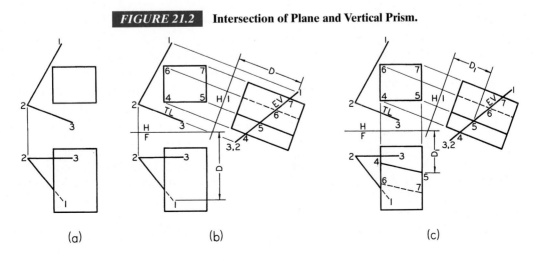

(a)	(b)	(c)

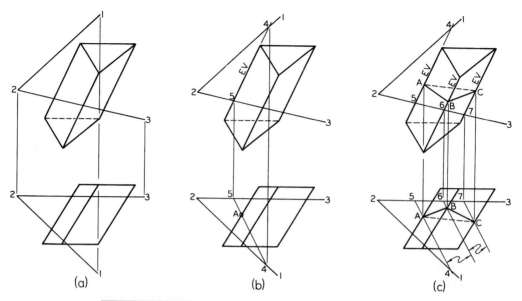

FIGURE 21.3 Intersection of Plane and Oblique Prism.

Because the lateral edges of a prism are parallel, the cutting planes in this case are parallel and cut parallel lines from plane 1–2–3. See §6.25. Hence it is only necessary to project points 6 and 7 to the front view and draw the respective lines parallel to line 4–5 to find the front views of points B and C.

Joining points A, B, and C with straight lines in the proper order and visibility completes the front view of the intersection. The top view is then completed by projecting points A, B, and C to their respective lateral edges in the top view and joining these points with proper visibility as shown.

INTERSECTION OF PLANE AND PYRAMID *Fig. 21.4* Because the plane 1–2–3–4 is limited—that is, it has definite boundaries—some details of construction for the intersection with the pyramid are different from the more general case illustrated in Fig. 21.3.

At (b) point A is found as before by the cutting-plane method of §19.11. When this method was applied to the other lateral edges of the pyramid (construction not shown), it was discovered that these piercing points are beyond the boundaries of plane 1–2–3–4 and are therefore theoretical or imaginary. Further consideration leads to the realization that we

FIGURE 21.4 Intersection of Limited Plane and Pyramid.

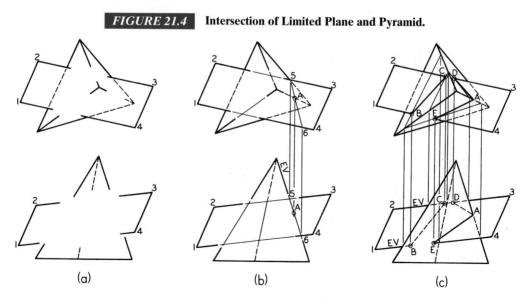

(a)   (b)   (c)

must find the piercing points of certain edges or boundary lines of the plane with the lateral surfaces of the pyramid.

Accordingly, at (c), edge-view cutting planes are introduced coinciding with the front view of lines 2–3 and 1–4. The resulting piercing points B, C, D, and E, together with point A, permit the completion of the intersection as shown.

■ 21.3 INTERSECTION OF PLANE AND CURVED SURFACE

PLANE AND RIGHT-CIRCULAR CYLINDER *Fig. 21.5* To establish the intersection of a plane and a circular cylinder, the cylinder is treated as a multisided prism through the introduction of an appropriate number of elements, as at (a). While theoretically these elements need not be equally spaced, the symmetrical arrangement shown tends both to simplify the construction of the intersection and to facilitate development, §22.4. Treated in this way, the construction becomes similar to that of Fig. 21.2. The finished solution with proper visibility is shown in Fig. 21.5 (b).

PLANE AND INCLINED CYLINDER *Fig. 21.6* The procedure illustrated here is similar to that of Fig. 21.5 except for the different position of the cylinder.[*]

---

[*] For simplicity the cylinder in Fig. 22.6 has been given circular bases. The cylinder is actually elliptical in cross section. No generality is lost.

Again the cylinder is treated as a multisided prism by dividing one of the bases into an appropriate number of equal parts to establish the location of elements in the surface of the cylinder. An auxiliary view is then constructed showing the given plane ABC in edge view, Fig. 21.6 (a). The established elements are then projected to the top and auxiliary views. In the auxiliary view, the piercing points of the elements with plane ABC coincide with the edge view.

At (b) the projections of points on elements 6 and 10 to the front and top views are illustrated. Note the use of divider distance D_1 to check the accuracy of location of the points in the top view. This procedure is repeated until all piercing points are located and smooth curves are drawn through the points to complete the views. Proper visibility is established by noting the visibility of individual elements in each view. Thus element 10 and the associated local portion of the curve of intersection are invisible in the front view, but visible in the top view. Element 6 is visible in the front view. In the top view, the point on element 6 is important in that it is the point at which the curve of intersection changes visibility and is tangent to element 6.

PLANE AND OBLIQUE CYLINDER *Fig. 21.7* To determine the curve of intersection of a plane and an oblique cylinder, the cylindrical surface is described in more detail by the introduction of a large number of equally spaced elements, as shown at (a). (For simplicity only alternate elements are shown in the figure.)

FIGURE 21.5   **Intersection of Plane and Right-Circular Cylinder.**

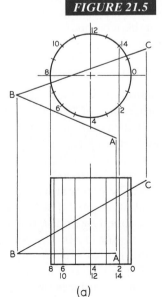

(a)

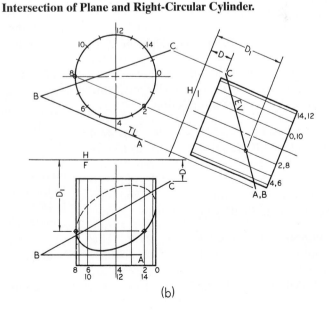

(b)

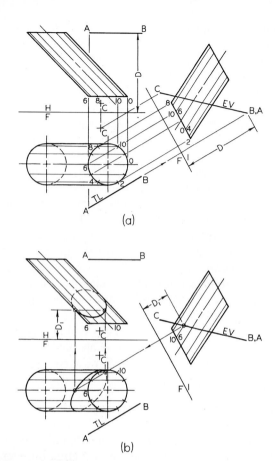

(a)

(b)

FIGURE 21.6 **Intersection of Plane and Inclined Cylinder.**

Since the elements are parallel, introduction of edge-view cutting planes, coincident with the elements in one view, establishes a system of parallel planes. These planes in turn cut a series of parallel lines from given plane ABC, as shown at (b), establishing piercing points of the elements in plane ABC. Smooth curves drawn through the points, with proper visibility, complete the views of the curve of intersection.

INCLINED PLANE AND OBLIQUE CONE *Fig. 21.8*
Equally spaced elements are introduced at (a) to represent the conical surface in more detail. Since given plane ABC is in edge view, the piercing points of the elements are apparent in the front view at (b), and their top views may be established by projection as shown.

OBLIQUE PLANE AND CONE *Fig. 21.9* This problem could be solved by constructing an auxiliary view showing plane ABC in edge view similar to the procedures shown in Figs. 21.5 and 21.6. However, a

different procedure is illustrated for this case. Note that a horizontal cutting plane, shown at (a), cuts a straight line 1–2 from plane ABC and parallel to line BC. This cutting plane also cuts a horizontal circle from the cone. Line 1–2 intersects the circle at two points (encircled) in the top view that are on the curve of intersection. The points are then projected to the front view of the cutting plane as shown.

The procedure is repeated at (b) by adding more cutting planes until a sufficient number of points is obtained to establish a smooth curve of intersection. Note the use of special cutting planes marked EV in the top view to secure critical points 3, 4, and 5 in the front view.

INTERSECTION OF SOLIDS
■ 21.4 PRINCIPLES OF INTERSECTIONS

An intersection of two solids is called a *figure of intersection*. Two plane surfaces intersect in a straight line; hence, if two solids that are composed of plane surfaces intersect, the figure of intersection will be composed of straight lines, as shown in Figs. 21.10 and 21.11. The general method consists in finding the piercing points of the edges of one solid in the surfaces of the other solid and joining these points with straight lines.

If curved surfaces intersect, or if curved surfaces and plane surfaces intersect, the figure of intersection will be composed of curves, as shown in Figs. 21.5–21.8 and Figs. 21.12–21.23. The method generally consists of finding the piercing points of elements of one solid in the surfaces of the other. A smooth curve is then drawn through these points, with the aid of the irregular curve.

■ 21.5 INTERSECTION OF TWO PRISMS

The determination of the intersection of two prisms involves the procedures of §19.12.

In Fig. 21.10, the points at which edges 1, 2, 3, and 4 of the inclined prism pierce the surfaces of the vertical prism are apparent in the top view and are projected downward to the corresponding edges 1, 2, 3, and 4 in the front view. The points at which edges B, C, and D of the vertical prism pierce the surfaces of the inclined prism are found by application of the cutting-plane method of §19.11. In this example, frontal edgeview cutting planes are introduced coincident with edges B, C, and D, respectively, in the top view. These cut lines in the surfaces of the inclined prism,

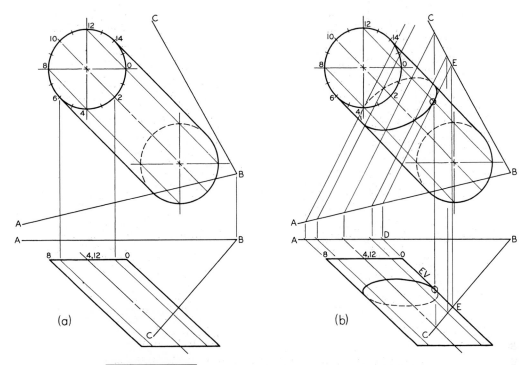

FIGURE 21.7   Intersection of Plane and Oblique Cylinder.

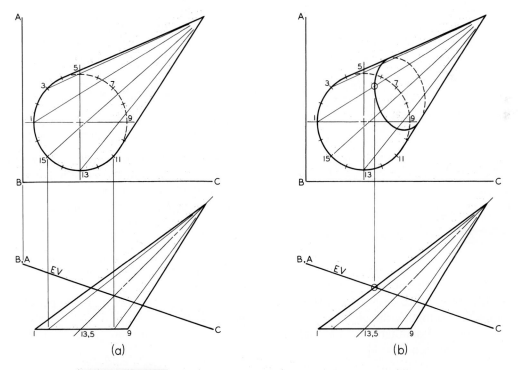

FIGURE 21.8   Intersection of Inclined Plane and Oblique Cone.

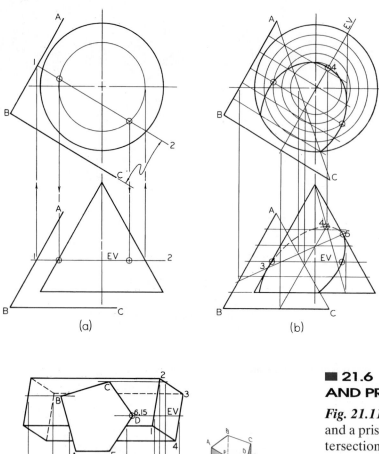

FIGURE 21.9   Intersection of Oblique Plane and Cone.

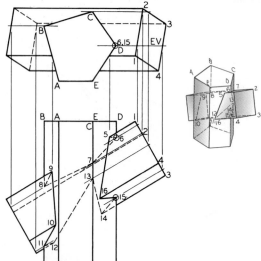

FIGURE 21.10   Intersection of Two Prisms.

which are projected to the front view, locating points 8 and 11, 7 and 13, and 6 and 15, respectively. The intersection is completed by joining the piercing points in order with straight lines. Note that in the final solution the figure of intersection is a closed path.

■ 21.6 INTERSECTION OF PYRAMID AND PRISM

Fig. 21.11   As in §21.4, the intersection of a pyramid and a prism is composed of segments of the lines of intersection of the plane surfaces, §19.12. The piercing points of the edges of each solid in turn with the surfaces of the other solid are the end points of these segments. The figure of intersection is completed by joining these points in order.

■ 21.7 INTERSECTION OF PRISM AND RIGHT CIRCULAR CONE

Fig. 21.12   Points in which the edges of the prism pierce the surface of the cone are shown in the side view at A, C, and F. Intermediate points, such as B, D, E, and G, are piercing points of lines on the lateral surface of the prism parallel to the edges of the prism. Through all the piercing points in the side view, elements of the cone are drawn and are then located in the top and front views in that order. The intersections of the elements of the cone with the edges of the prism (and lines along the prism drawn parallel thereto) are points of the intersection. The figures of intersection are drawn through these points with the aid of the irregular curve.

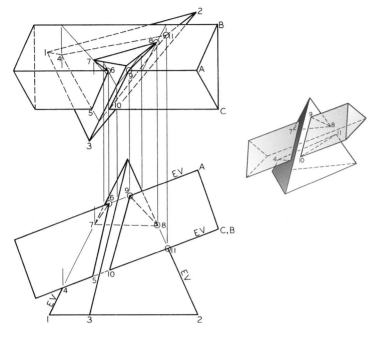

FIGURE 21.11 Intersection of Pyramid and Prism.

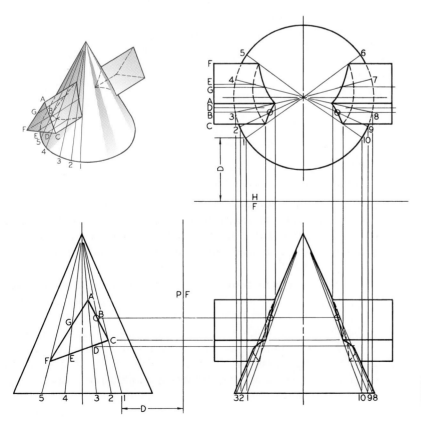

FIGURE 21.12 Intersection of Prism and Cone.

The elements 5, 4, 3, ..., in the side view of the cone may be regarded as the edge views of cutting planes that cut these elements on the cone and edges or elements on the prism. Edges or elements that lie in the same cutting plane intersect at points on the figure of intersection.

Another method of finding the figure of intersection is to pass a series of horizontal parallel planes through the solids in the manner of Fig. 21.13. Each plane cuts a circle on the cone and straight lines on the prism, and their intersections are points on the figure of intersection. Since the prism is vertical, the cut lines on its surface coincide with the edge views of the vertical surfaces of the prism in the top view. Note in this case that this portion of the figure of intersection is a segment of a hyperbola, Fig. 4.46 (e).

■ 21.8 INTERSECTIONS OF CIRCULAR CYLINDERS AND CONES

INTERSECTIONS OF CIRCULAR CYLINDERS   Fig. 21.14   To determine the intersection of two circular cylinders, a series of elements (preferably equally spaced if the surfaces are to be developed, §22.4) is assumed on the horizontal cylinder and numbered 1, 2, 3, ..., in the side view. The elements are then established in the top and front views. The points of intersection of the elements with the surface of the vertical cylinder are shown in the top view at A, B, C, ..., and

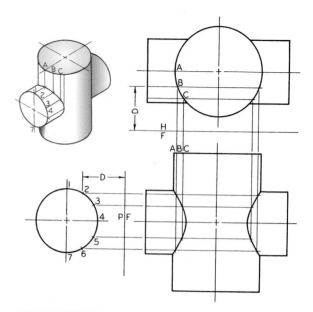

FIGURE 21.14   Intersection of Circular Cylinders.

these points may be found in the front view by projection to their intersections with the corresponding elements 1, 2, 3, ... When a sufficient number of points has been found to determine the figure of intersection, the curve is drawn through the points with the aid of the irregular curve. See also Fig. 6.38 (c).

This solution may be regarded as equivalent to passing a series of frontal cutting planes through the two cylinders and parallel to the axes. The elements cut from both cylinders intersect at points on the figure of intersection.

INTERSECTION OF CIRCULAR CYLINDERS AND CONES   Fig. 21.15   In each example, (a) and (b), points in which elements of the cylinder intersect the surface of the cone are shown in the side view at A, B, C, ... These are point views of elements of the cylinder. A series of cutting planes containing these points and the vertex of the cone will appear in edge view in the side view, and will cut corresponding elements 1, 2, 3, ..., from the cone. When the elements of both solids are located in the top and front views, they intersect at points on the figure(s) of intersection. The curves are then drawn through the points with the aid of the irregular curve.

Alternatively, a series of parallel horizontal planes could be passed through the solids that will cut circles from the cone and elements from the cylinder and intersect at points on the figure of intersection. See Fig. 21.13. This method has the disadvantage, particularly

FIGURE 21.13   Intersection of Prism and Cone.

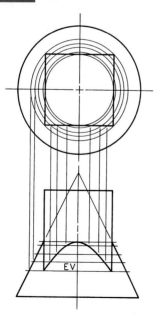

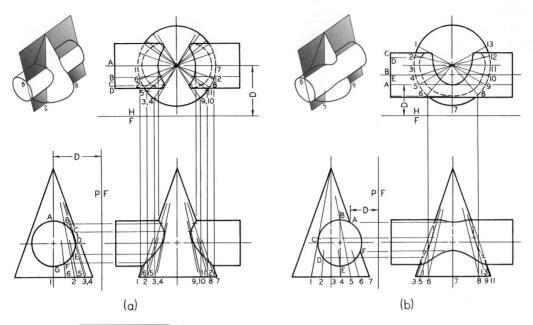

FIGURE 21.15 Intersection of Circular Cylinders and Cones, Axes Intersecting and Nonintersecting.

for Fig. 21.15 (b), that critical points where the curve is tangent to a cone or cylinder element are less precisely determined.

INTERSECTIONS OF CIRCULAR CYLINDERS AND CONES BY SPHERE METHOD If a large spherical ball rests on the open end of a circular wastebasket, either cylindrical or conical, Fig. 21.16 (a), it makes contact all around the periphery or circular lip of the basket. In geometric terms, this means that a sphere (if large enough) centered on the axis of a circular cylin-

der or cone, (b) or (c), cuts circles from the solid. Moreover, where the axis of the solid appears in true length, the circles appear in edge view—as straight lines, as shown. A limiting case is shown at (c) where, for the particular chosen sphere center, the sphere is just large enough to be tangent internally to the conical surface. The edge-view circle is thus a *circle of tangency* rather than a "cut" circle.

The preceding concept is applicable to finding the intersections of curved surfaces when all three of the following conditions are met.

FIGURE 21.16 Speres Cutting Circular Cylinders and Cones.

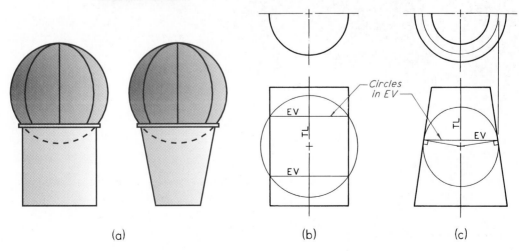

(a) (b) (c)

GRAPHICS SPOTLIGHT

Electronic Symbol Libraries

CREATE STANDARD SYMBOLS AT THE CLICK OF YOUR MOUSE

Libraries of standard electronic symbols are available from a number of different sources to help you quickly and effectively create electrical and electronic schematics. You can insert the symbols at any scale or rotation angle in your drawing. Using a library of standard symbols saves you time in creating schematics, and can have the added benefit of ensuring the appropriate standards are being followed. When each designer draws their own symbols, some may not follow the current standard.

1,300 ANSI STANDARD SYMBOLS CAN BE PURCHASED

One company that produces a set of over 1,300 ANSI Y32.2 standard electrical and electronic symbols, as well as ANSI/IEEE Standard 91 logic symbols and ANSI Y32.9 electrical wiring symbols, is the Berol® Corporation, who have long been known for their manual drafting instruments and templates. The Electrical and Electronic Drawing Symbols Library™ is available for software that can read either AutoCAD drawing format or Macintosh Pict format. The

AutoCAD version of the symbol library also contains a custom AutoCAD menu that allows you to pick the name of the symbol to insert from the screen menu. You can also use the symbols that are provided to create your own custom menus and toolbars for AutoCAD.

Some symbols from the Berol RapiDesign® Electrical and Electronic Symbol Library™ for AutoCAD are shown below as they can be used with AutoCAD Release 13. In the central part of the screen are a number of the different electronic symbols that have been inserted to illustrate some of the symbols available. The screen menu to the right shows the major categories of symbols available for use in the drawing.

OLD DRAWINGS CAN AUTOMATICALLY UPDATE

When you use standard libraries to insert symbols into CAD drawings, you save time and produce drawings with a higher level of standardization. One additional advantage is that if the symbology changes as new standards are released, most CAD drawings can automatically update the entire drawing when the new symbol is inserted using the same name as the old symbol.

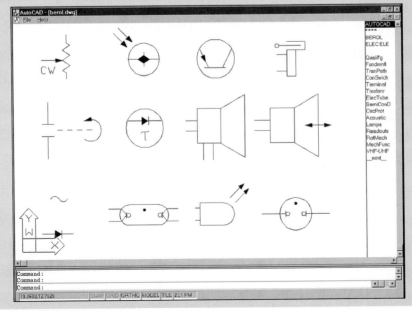

1. The forms must be circular (surfaces of revolution).

2. Their axes must intersect.

3. Both axes must appear true length in the same view.

While these conditions may appear to be excessively restrictive, they are more often met than not under practical considerations of design.

As an example, consider the circular cone and cylinder of Fig. 21.17. Only the front view is absolutely necessary, which is the advantage of the sphere method. Compare with Fig. 21.15 (a).

At Fig. 21.17 (a), a cutting sphere A is introduced with its center at the intersection of the cylinder and cone axes. The radius selected for the sphere is such that the sphere is tangent internally to the cone and thus contacts the cone in the single horizontal circle shown in edge view. The sphere cuts the cylinder in two vertical edge-view circles as shown. The circles intersect at points 1 and 2 on the figures of intersection of the cone and cylinder. Note that any sphere smaller than the one shown would not contact the cone.

At (b) a slightly larger sphere B is introduced, again with its center at the intersection of the axes. This sphere intersects both the cone and the cylinder, cutting two circles (in edge view) from each. These intersect in four points, 3, 4, 5, and 6, as shown.

This process is repeated until a sufficient number of points is obtained, after which a smooth curve is drawn through the points.

If a top view of the figure of intersection becomes necessary, the horizontal circles cut by the spheres are easily located. In Fig. 21.18 cutting sphere A is shown again. The horizontal circle cut from the cone by sphere A is projected to the top view, and the points located by this sphere are projected upward to the circle in the top view. This process is repeated for other cutting spheres until the points on the top view of the figure of intersection are established. A smooth curve is then drawn through the points.

■ 21.9 INTERSECTION OF OBLIQUE CONE AND CYLINDER

When intersecting cylinders or cones are vertical, horizontal, or inclined, cutting planes are conveniently introduced in edge-view form to establish the figure of intersection. If one or both of the intersecting solids are oblique—and in particular if the solids are not circular—the introduction of edge-view cutting planes can be awkward.

To cut elements from a cone, a cutting plane must (1) pass through the vertex and (2) cut or be tangent to the base. In Fig. 21.19 (a) such planes are represented by intersecting lines, §19.7. Any plane containing the line through the vertex of the cone, such as VP, passes through the vertex. This line intersects the extended base plane of the cone at point P Line P–1, drawn tangent to the circular base at point 1, completes the representation of a plane tangent to the cone, §23.2. The element of tangency is line V–1. A line

FIGURE 21.17 **Intersection of Circular Cylinder and Cone by Sphere Method.**

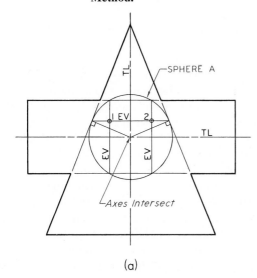

(a)

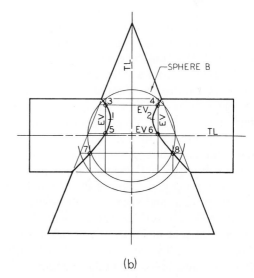

(b)

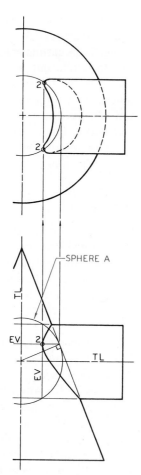

FIGURE 21.18   **Constructing the Top View of a Curve of Intersection.**

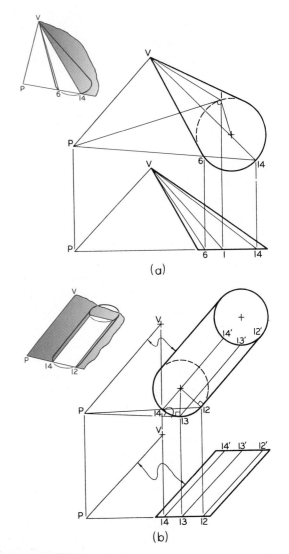

FIGURE 21.19   **Plane Cutting Elements from Cone and Cylinder.**

such as P–14, drawn *secant* to the circular base, completes representation of a plane VP–14, which cuts the base of the cone at points 6 and 14 and thus cuts elements V–6 and V–14 from the cone.

To cut elements from a cylinder, a cutting plane must (1) be parallel to the elements or axis (center line) and (2) cut or be tangent to the base.

In Fig. 21.19 (b) a line VP is drawn through a selected point V parallel to the elements of the cylinder. Any plane containing this line will be parallel to the cylinder, §21.4. From point P, the piercing point of line VP in the plane of the lower base of the cylinder, lines are drawn tangent to and secant to the lower base. These are lines P–13 and P–12, respectively. Thus are represented a tangent plane VP–13, §23.3, with element of tangency 13–13′, and a plane VP–12, cutting elements 12–12′ and 14–14′ from the cylinder.

To determine the figure of intersection of an oblique cone and cylinder, the methods shown in Fig. 21.19 are combined as illustrated in Fig. 21.20. At (a) line VP is drawn through the vertex of the cone and parallel to the cylinder. At point P, where line VP intersects the plane of the lower bases, line P–9 is drawn tangent to the base of the cone and secant to the base of the cylinder. Line P–13 is drawn tangent to the base of the cylinder and secant to the base of the cone. Planes VP–9 and VP–13 are called *limiting* planes because any plane outside of the space between them fails to cut at least one of the solids. The required figure of intersection is thus limited to the narrow wedge

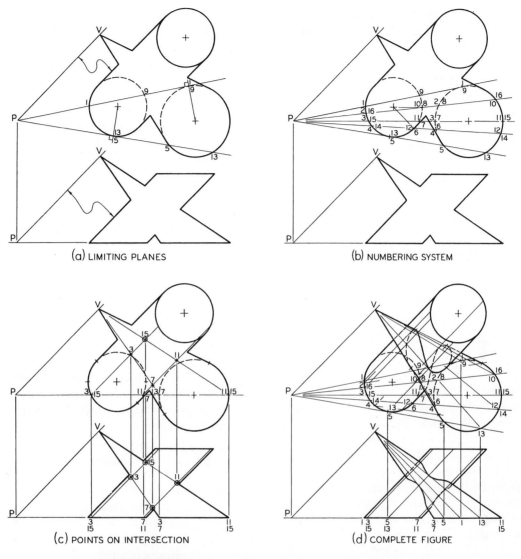

(a) LIMITING PLANES

(b) NUMBERING SYSTEM

(c) POINTS ON INTERSECTION

(d) COMPLETE FIGURE

FIGURE 21.20   **Intersection of Oblique Cone and Cylinder.**

of space between the limiting planes. Each limiting plane cuts one element from one solid and two elements from the other solid, so that each limiting plane secures two points on the figure of intersection.

At (b) several intermediate secant lines are added through point P. Each, together with line VP, represents a cutting plane secant to both solids. At (c) plane VP–15 is shown separately. Note that it cuts two elements from each solid: elements 3, 15 and 7, 11 from the cylinder, and elements 3, 7 and 11, 15 from the cone. These in turn intersect at the four points 3, 7, 11, and 15 on the figure of intersection. This system of numbering is valuable in minimizing confusion while

drawing the figure of intersection through the established points. The complete numbering system is shown in the top view at (b) and (d).

NUMBERING SYSTEM   The rules for the numbering system are as follows.

1. Either base may be selected, and any tangent or secant point in the base may be designated as point 1.

2. The points in the selected base are then numbered consecutively, either clockwise or counterclockwise, except that the progression must not be continued beyond a secant limiting plane.

3. When a secant limiting plane is reached, the direction is reversed, placing only *one* number at the secant point. This process is continued until all base points have two numbers, except for secant points of limiting planes.

4. On the other base, numbers 1 and 2 are assigned to the same lines as points 1 and 2 of the first base.

5. With the direction of progression thus established, rules (2) and (3) are followed until all base points of the second base have two numbers, except for secant points of limiting planes.

As a final check, the pattern should be inspected to make certain the same numbers appear along each secant line.

Note that the system automatically assigns a single number to each point on the figure of intersection, Fig. 21.20 (c). Moreover, in the final result, at (d), the points are connected in numerical order. In this case the curve is a single continuous curve from points 1 through 16 and back to point 1. This situation always occurs when one limiting plane is tangent to one base and the other to the second base.

VISIBILITY The visibility of the curve can change only where the curve is tangent to an extreme or *contour* element of one of the solids. Hence to determine the visibility of the entire curve, it is necessary only to check one or two points on segments between such tangency points. The visibility of a point is determined by the visibility of the individual elements that intersect to locate that point. This must be checked for each view individually, since an element may be visible in one view and invisible in an adjacent view.

A point is visible in a view only if *both* intersecting elements that determine the point are individually visible in that view. Study Fig. 21.20 (c) carefully. Only the encircled points are visible: point 15 in the top view and points 3, 7, 11, and 15 in the front view. This is determined by observing the origin of each element at the base of its solid, with respect to the adjacent contour elements.

Note that elements 3, 7 and 11, 15 are contour elements of the cone in the front view and are thus points at which the figure of intersection changes visibility, as shown at (d).

Also at (c) observe how the front views are established. The front views of the elements are located for both solids, their intersections establishing the front views of the points. It is advisable, for accuracy, to employ the vertical projection lines between the views of the points only as an accuracy check and not as a means of locating the points. [When the views fail to align properly, it usually signifies that the original construction at (a) was faulty.]

The construction is continued until all the numbered points are located. Intermediate points, if needed, can be mentally identified by the numbered points between which they lie. Thus numbering of additional base points is unnecessary.

The final curve is drawn through the points in numerical order with the aid of the irregular curve.

■ 21.10 INTERSECTION OF TWO OBLIQUE CONES

Fig. 21.21 The solution for the figure of intersection of two oblique cones differs only in detail from the procedure demonstrated in Fig. 21.20. Two observations of particular interest may be made concerning the determination of the intersection of the illustrated oblique cones.

1. The line through the two vertices is common to all cutting planes and determines, in the plane of the bases, point P from which the secant or tangent lines emanate.

2. The two limiting planes are tangent to the same cone base in this example.

Note that the numbering system on that base never reverses direction but continues until every base point has two numbers. The rules of the numbering system, §21.9, cause it to divide naturally on the second base into two separate groups of numbers, points 1 to 8 and points 9 to 16. The significance of this is that all elements of the first cone pierce the second cone, and there are two separate figures of intersection. Other examples of double and single figures of intersection are shown in Fig. 21.15.

■ 21.11 INTERSECTION OF TWO OBLIQUE CYLINDERS

Fig. 21.22 To find the figure of intersection for two oblique cylinders, cutting planes are required that cut elements in both cylindrical surfaces. To do this, as shown at (a), an arbitrary point V is selected, and a plane is passed through it and parallel to both cylinders. The plane thus established may or may not actually cut the cylinders. What is important is that it is

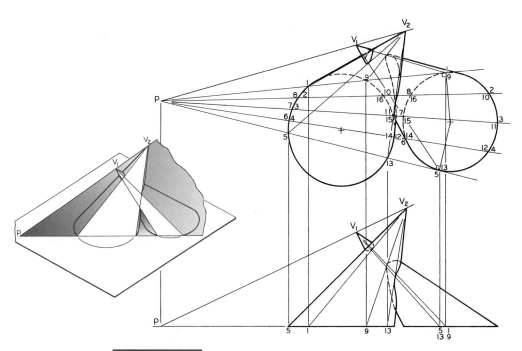

FIGURE 21.21   Intersection of Two Oblique Cones.

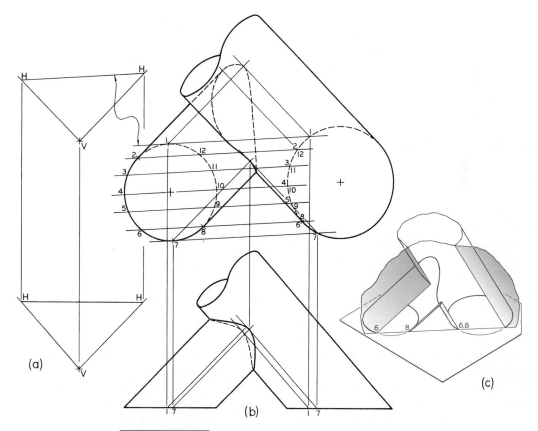

FIGURE 21.22   Intersection of Two Oblique Cylinders.

parallel to all planes that do cut elements. Since the lines that cut the bases (the secant or tangent lines) are in this case horizontal lines and lie in parallel cutting planes, they are all parallel. Their direction is established by adding any convenient horizontal line HH to the plane through V as shown at (a). The secant lines are then drawn parallel to the top view of line HH as shown at (b). The construction then proceeds as in previous examples.

Of interest here is the fact that the smaller cylinder *joins* the larger cylinder but does not extend through it, a common circumstance in duct work. Since the two limiting planes are tangent to the same base, there are theoretically two separate figures of intersection. See §21.10. In this instance, the curve for the second figure of the intersection is not plotted. In applying the numbering system, only the first group of numbers is employed.

■ 21.12 INTERSECTION OF CONE AND CYLINDER—BASES IN NONPARALLEL PLANES

Fig 21.23 In Figs. 21.20–21.22 the bases of the solids are in the same plane. If they were not, but were in parallel planes, it would be necessary to extend or shorten one solid until it had a new base in the same plane as a base of the second solid. Appropriate adjustments in lengths would be made in all views used for the determination of the intersection. In Fig. 21.23, however, the bases are in perpendicular planes. This does not alter the theoretical considerations, which are

the same as in §21.8 but does introduce certain complications to the original construction.

When a line is drawn through cone vertex V and parallel to the cylinder, it is observed that the line pierces the base planes of the solids in two separate points: X_1 for the cone and X_2 for the cylinder. Hence lines cutting the base of the cone must radiate from point X_1 in the side view, while those cutting the base of the cylinder must extend from point X_2 in the top view. For any one cutting plane, corresponding lines from points X_1 and X_2 must intersect at the line of intersection of the two base planes, which appears as point B–A in the front view and as the line through A and B in the top and side views.

After trial constructions for the determination of the limiting planes, it is found that line X_1A, drawn tangent to the base circle of the cone, locates point A such that in the top view line X_2A is secant to the base of the cylinder.

Plane X_1AX_2 is then one limiting plane for the construction. Similarly, plane X_2BX_1 is the other limiting plane. Because the limiting planes are tangent to different bases, there will be only one continuous curve for the figure of intersection, §21.9.

Intermediate cutting planes are now added. The elements of the cone, as established in the side view, are located in the top view by transfer of distances such as D_1. Their intersections with the corresponding elements of the cylinder in the top view establish points on the figure of intersection as shown. In this example, the figure of intersection is not shown in the side view.

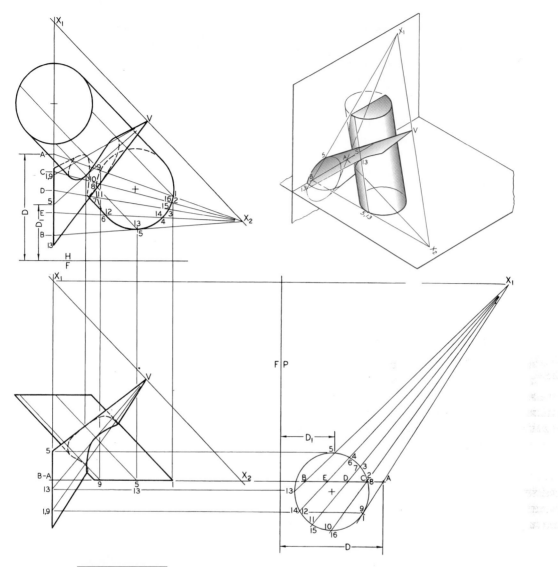

FIGURE 21.23   Intersection of Cone and Cylinder—Bases in Nonparallel Planes.

■ KEY WORDS

POLYHEDRA	OBLIQUE CYLINDER
OBLIQUE PRISM	INCLINED CONE
OBLIQUE CONE	LIMITING PLANES
INCLINED PRISM	SECANT LINES
CIRCLE OF TANGENCY	PIERCING POINTS
CUTTING PLANE	SURFACES
CUTTING PLANE ELEMENTS	SOLIDS
INTERSECTION	GENERATRIX
WARPED SURFACE	SPHERE
INCLINED CYLINDER	PYRAMID

■ CHAPTER SUMMARY

- Piercing points are the basis for finding the intersection of a plane and a solid object. The cutting plane method is a common method for graphically determining the intersection of a plane and a prism.
- Radial elements drawn on the surface of curved solids like cylinders and cones help define piercing points with a plane.
- Numbering all vertices and radial elements is essential for accurate graphical solutions.
- The intersection of a prism and another solid can be broken down into several sub-problems involving individual planes of the prism that intersect the solid.

- Solving for the edge view of a plane is a common method of identifying its intersection with a solid.
- When two curved solids intersect, their intersection can be broken down into several line intersection problems, where elements drawn on each solid pierce the other solid.
- Visibility is an important part of solving any intersection problem.

■ REVIEW QUESTIONS

1. List five warped surfaces. Which do not contain straight line elements?
2. When solving for the intersection of a plane and a prism, what is shown in edge view to solve the problem?
3. Which geometric solid has parallel elements: a cone or a cylinder?
4. Define an oblique cone.
5. Can the elements of a pyramid ever be parallel?
6. When looking down the axis of a cylinder, what is the shape of its intersection with a pyramid?

7. Why is it important to number elements when solving intersection problems?
8. Why are multiple cutting planes so useful in solving complex intersection problems?
9. Do cutting planes have to be parallel when solving intersection problems?
10. Give seven examples of real life objects that demonstrate the intersection of geometric solids.

■ INTERSECTION PROBLEMS

In Figs. 21.24–21.29 are problems covering intersection of plane with prism, pyramid, cylinder, or cone and intersections of solids.

Use Layout A–1 or A4–1 (adjusted) and divide the working area into four equal areas for problems to be assigned by the instructor. Some problems will require two problem areas or one half sheet, and others will require a full sheet. Unless otherwise indicated, the data for the layout for each problem are given by a coordinate system using metric dimensions. For example, in Fig. 21.24, Prob. 1, point 2 is located by the full-scale coordinates (45 mm, 40 mm, 83 mm). The first coordinate locates the front view of the point from the left edge of the problem area. The second coordinate locates the front view of the point from the bottom edge of the problem area. The third coordinate locates either the top view of the point from the bottom edge of the problem area or the side view of the point from the left edge of the problem area.

Inspection of the given problem layout will determine which application to use.

If development is required, Chapter 22, the development may be constructed on a second A–1 or A4–1 (adjusted) Layout or the entire problem may be drawn on one B–3 or A3–3 Layout. See development problems in Chapter 23. If assigned, dimensions may be included on the given views.

Since many of the problems in this chapter are of a general nature, they can also be solved on most computer graphics systems. If a system is available, the instructor may choose to assign specific problems to be completed by this method.

Additional problems, in convenient form for solution, are available in *Engineering Graphics Problems*, Series 1, by Spencer, Hill, Loving, Dygdon, and Novak, designed to accompany this text and published by Prentice Hall Publishing Company.

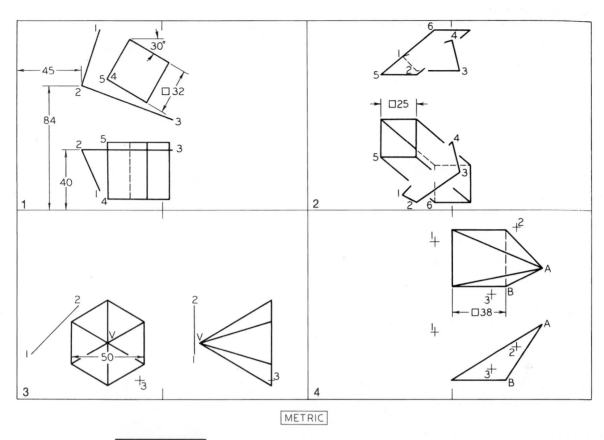

FIGURE 21.24   Lay out and solve problems as assigned. Use layout
A–1 or A4–1 (adjusted) divided into four equal areas.

1. Find the views of the intersection of unlimited plane 1 (58, 12, 122)–2 (45, 40, 83)–3 (79, 40, 60) with the square prism having edge 4 (63, 7, 88)–5 (63, 45, 88).

2. Complete the views showing the intersection of bulkhead plate 1 (65, 10, 104)–2 (75, 5, 93)–3 (106, 25, 94)–4 (100, 45, 114) and the prismatic duct with edge 5 (50, 35, 91)–6 (88, 5, 121).

3. Find the views of the intersection of unlimited plane 1 (10, 30, 124)–2 (43, 63, 124)–3 (86, 12, 175) with the pyramid having its vertex at V (63, 38, 127).

4. Find the views of the intersection of unlimited plane 1 (88, 45, 106)–2 (147, 35, 100)–3 (129, 20, 70) with the pyramid having edge A (165, 50, 88)–B (139, 12, 75).

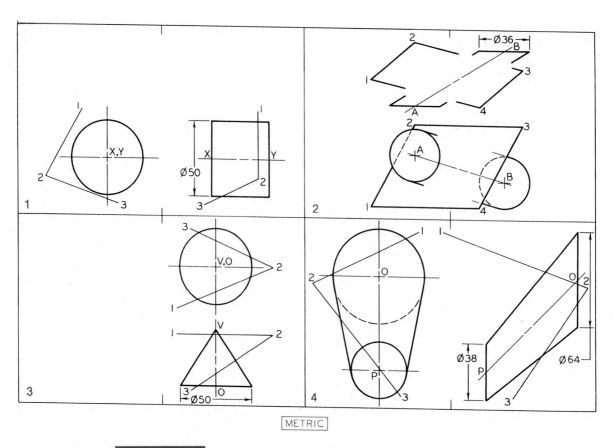

FIGURE 21.25 Lay out and solve problems as assigned. Use layout
A–1 or A4–1 (adjusted) divided into four equal areas.

1. Find the intersection of unlimited plane 1 (45, 70, 170)–2 (20, 25, 170)–3 (70, 8, 132) with the right-circular cylinder having axis X (63, 38, 134)–Y (63, 38, 177).

2. Show with complete visibility the intersection of wall section (limited plane) 1 (45, 7, 94)–2 (75, 63, 119)–3 (–, –, –)–4 (122, 7, 75) with pipe A (75, 43, 75)–B (139, 25, 114).

3. Find the intersection of unlimted plane 1 (112, 48, 66)–2 (180, 48, 44)–3 (122, 10, 119) with right-circular cone V (140, 50, 94)–O (140, 12, 94).

4. Find the intersection of unlimited plane 1 (78, 119, 96)–2 (5, 84, 198)–3 (66, 7, 145) with the conical offset transition piece O (50, 88, 190)–P (50, 25, 127).

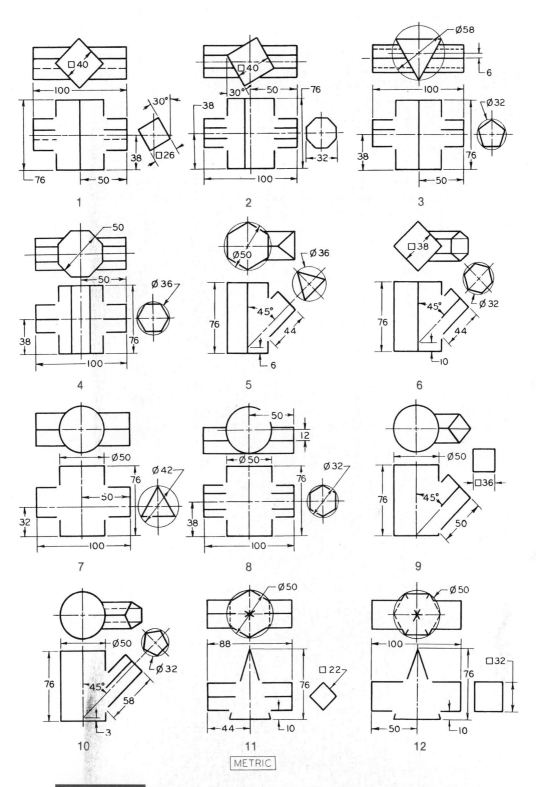

METRIC

FIGURE 21.26 Draw the given views of the assigned forms and complete the intersection. Use Layout A–1 or A4–1 (adjusted). Use Layout B–3 or A3–3 if development is required. See development problems in Chapter 22.

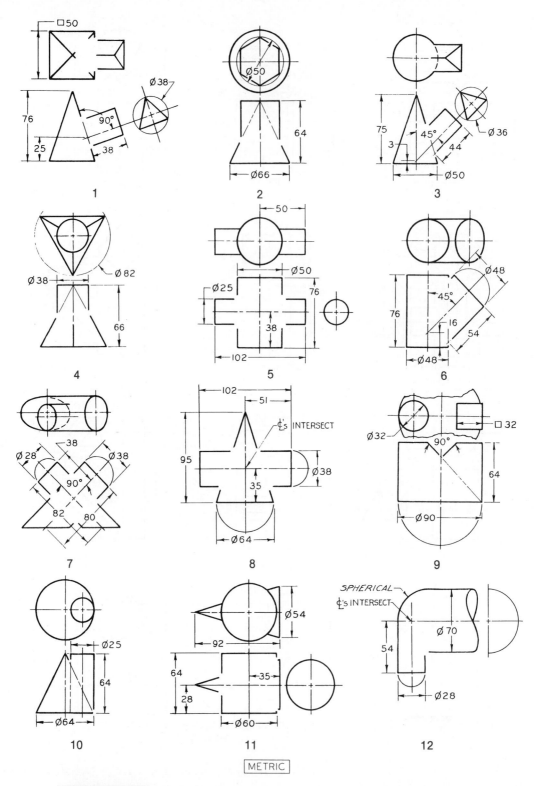

FIGURE 21.27 Draw the given views of the assigned forms and complete the intersection. Use Layout A–1 or A4–1 (adjusted). Use Layout B–3 or A3–3 if development is required. See development problems in Chapter 22.

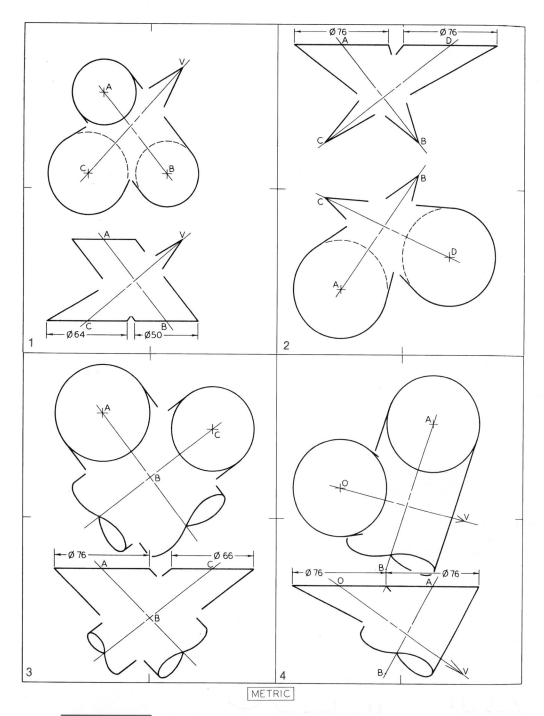

FIGURE 21.28 (opposite) Draw the given views of the assigned forms and complete the intersection. Use Layout A–1 or A4–1 (adjusted).

1. Complete the views of cone C (50, 25, 140)–V (127, 88, 224) and cylinder A (63, 88, 203)–B (114, 25, 140), including their figure of intersection.
2. Find the intersection and complete the views of cones A (50, 50, 240)–B (114, 140, 165) and C (38, 122, 165)–D (140, 75, 240).
3. Find the intersection and complete the views of cylindrical tubes A (63, 88, 210)–B (100, 50, 160) and B–C (152, 88, 198).
4. Find the intersection in which conical funnel 0 (50, 75, 152)–V (152, 7, 127) joins cylindrical duct A (127, 75, 200)–B (88, 7, 88).

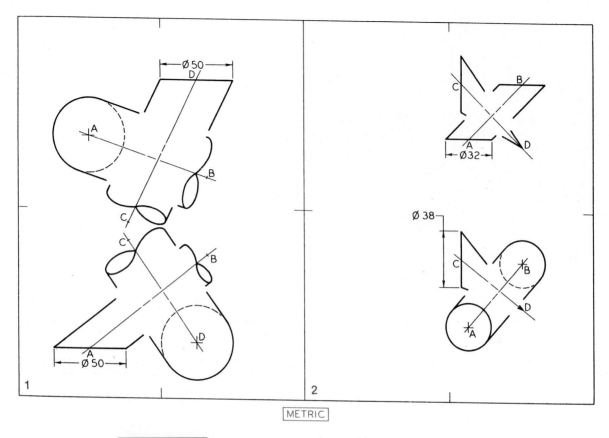

FIGURE 21.29 Draw the given views of the assigned forms and complete the intersection. Use Layout A–1 or A4–1 (adjusted).

1. Complete the views of pipes A(50, 33, 177)–B(127, 124, 150) and C(78, 106, 119)–D(127,38,215), including their figure of intersection.
2. Find the intersection of cylinder A(114, 50, 177)–B(152, 94, 215) and cone C(109, 96, 215)–D(152, 63, 172).

C H A P T E R   2 2

DEVELOPMENTS

OBJECTIVES

After studying the material in this chapter, you should be able to:

1. Define the term development.
2. List examples of surface development in the real world.
3. Solve development problems for prism and cylinders using parallel line techniques.
4. Create the development of cones and pyramids using radial line techniques.
5. Use triangulation to solve for transitions between two geometric shapes.
6. Create developments of two solids that intersect.

OVERVIEW

The development of a surface is the construction of a pattern that represents the unfolded or unrolled surface of a form. The resulting plane figure gives the true size of each area of the form, so connected that when it is folded, rolled, or fabricated, the desired form is obtained. Practical applications of developments occur in sheet-metal work, stone cutting, pattern making, consumer packaging, and package design. Containers, pipes, ducts, and other products are often made of flat stock or sheet materials and are first laid out on the flat sheets before the bending or rolling operations produce the desired form.

There are three basic types of sufaces development: parallel line, radial line, and triangulation. Each technique involves creating the true size of plane surfaces that are joined at shared edges or fold lines. When all the true size planes are joined together, the development can be cut out and folded to form the three-dimensional shape. Curved surfaces like cylinders and cones require the approximation of small planes on the surface development by using parallel or radial elements. Transition development uses small triangular planes approximating small curved surfaces that form the transition between two geometric solids. While development solution techniques are simple on the drawing board, their CAD equivalents require complex calculations. In order to effectively create a development on many CAD programs the user must understand the drawing techniques used on the traditional drawing board. Computer cartography programs use many different types of surface approximation techniques to represent the curved surface of the Earth's spheroid shape on the flat computer monitor screen.

■ 22.1 SURFACE DEVELOPMENT

A developable surface is one that may be unfolded or unrolled to coincide with a plane surface. The *development* of a surface is the *pattern* of that surface laid out on a plane surface, Fig. 22.1

Single-curved surfaces and the surfaces of polyhedra can be developed. See §22.1. Warped surfaces and double-curved surfaces can be developed only approximately, by dividing them into sections and substituting for each section a developable surface, that is, a plane or a single-curved surface. If the material used is sufficiently pliable, the flat sheets may be stretched, pressed, stamped, spun, or otherwise forced to assume the desired shape.

In sheet-metal layout, extra material must be provided for laps or seams, Fig. 22.2. If the material is heavy, the thickness may be a factor, and the compression or stretching of metal in bends must be considered in terms of a *bend allowance*, §12.39. These are considerations that involve for the most part modifications of the theoretical development. The basic developments that follow are the foundation for such work.

The designer must take into account stock sizes of materials and make the layouts for the most economical use of material and labor. In preparing developments, it is often best to locate the seam on the shortest edge (or in the center of the shortest panel). Usually, bases are not necessary, but, if needed, they should be shown on the layout and attached at a long matching edge.

Except in certain activities involving relatively thin metals, such as aircraft and heating and ventilating fabrication, it is customary to draw development layouts with the inside surfaces up. In this way, all bend lines and other markings are related directly to inside measurements, which are the important dimensions for tanks and other vessels; and the usual machines in fabricating shops are designed for working with the developments with the markings up.

PARALLEL-LINE DEVELOPMENTS
■ 22.2 PARALLEL-LINE DEVELOPMENTS

Those common solids that are characterized by parallel lateral edges or elements, such as prisms and cylinders, unroll into flat patterns in which this parallelism is retained, Fig. 22.1 (a) and (b). The methods for laying out these surfaces are based directly on this parallelism. Perpendicularity is also involved, since the perimeter of any section (cross section) of a prism or cylinder taken at right angles to the lateral edges rectifies into a straight line in the development. Such a line is called a *stretch-out line*, and all lateral edges or elements in the development are perpendicular to it.

The fundamental information needed in constructing a parallel-line development is obtained as follows.

Select or introduce a *right section* and obtain its true-size view. Its periphery is the length of the stretch-out line, and the distances between the intermediate edges or elements determine the intervals along the stretch-out line. The distances from the stretch-out line to the end points of each edge or element are mea-

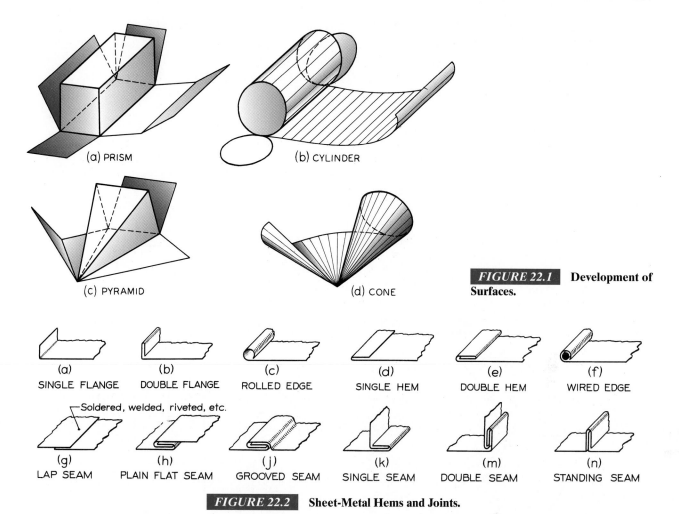

FIGURE 22.1   Development of Surfaces.

(a) PRISM

(b) CYLINDER

(c) PYRAMID

(d) CONE

(a) SINGLE FLANGE

(b) DOUBLE FLANGE

(c) ROLLED EDGE

(d) SINGLE HEM

(e) DOUBLE HEM

(f) WIRED EDGE

Soldered, welded, riveted, etc.

(g) LAP SEAM

(h) PLAIN FLAT SEAM

(j) GROOVED SEAM

(k) SINGLE SEAM

(m) DOUBLE SEAM

(n) STANDING SEAM

FIGURE 22.2   Sheet-Metal Hems and Joints.

sured perpendicular to the right section in a view showing the edges or elements in true length.

■ 22.3 DEVELOPMENT OF PRISMS

DEVELOPMENT OF RIGHT PRISM *Fig. 22.3* The lower base 1-2-3-4 is a right section, and its true size appears in the top view, (a). In the development, (b), base 1-2-3-4 unrolls into stretch-out line 1-1, its length being determined by setting off the successive widths of the faces 1-2, 2-3, . . . , taken from the true-size top view. At the interval points along the stretch-out line perpendiculars are erected and on each is set off the length of the respective lateral edge taken from the front view. In the present arrangement, it is convenient to do this by projecting across from the front view, as shown. The points thus found are joined by straight lines to complete the development of the lateral surface.

In duct work it is not usually necessary to consider bases, but, if desired, as for a model, the bases may be attached at longer edges, as shown. The result is the development of the entire surface of the frustum of the prism.

DEVELOPMENT OF OBLIQUE PRISM *Fig. 22.4* In the given front and top views at (a), the lateral edges are not true length, and the bases are not perpendicular to the lateral edges. It is thus necessary to introduce a right section, which is most convenient in a true-length view of the lateral edges. Hence auxiliary view 1 is first constructed showing the lateral edges in true length. At any convenient location in this view, cutting plane A-A is introduced perpendicular to the lateral edges. This establishes right section 1-2-3-4, which will later unroll into the stretch-out line 1-1. First, however, it is necessary to find the lengths of the

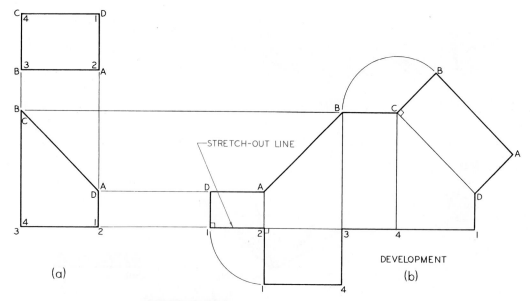

(a)

(b)

FIGURE 22.3 **Development of Right Prism.**

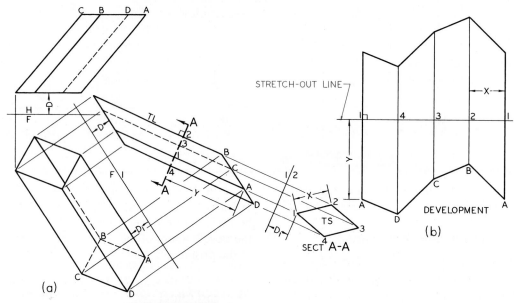

(a)

(b)

FIGURE 22.4 **Development of Oblique Prism.**

sides 1–2, 2–3, ... (periphery), of the right section. Auxiliary view 2 is now added for this purpose, showing the true size of the right section. (Note that auxiliary view 2 does *not* show the bases of the prism in true size.)

The development is now constructed at (b) in a manner similar to Fig. 22.3.

1. The segments 1–2, 2–3, ..., are set off along the selected stretch-out line 1–1, taking the lengths such as X from auxiliary view 2.

2. At these points on the stretch-out line perpendiculars are erected, and the lengths of the edges, such as Y measured from A–A in auxiliary view 1, are set off.

3. The end points of the edges are joined with straight lines to complete the development of the lateral surface.

DEVELOPMENT OF PRISMS WITH INTERSECTION
Fig. 22.5 The figure of intersection of the prisms is

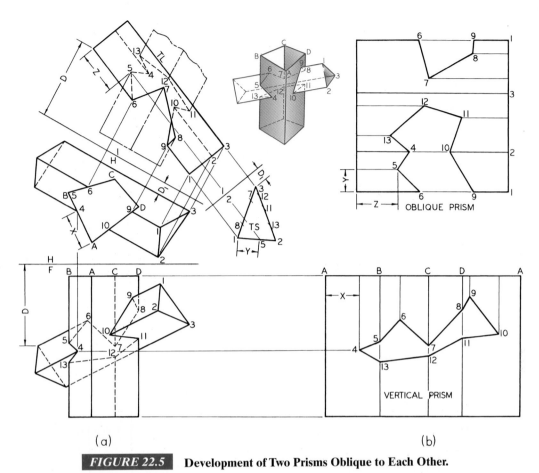

FIGURE 22.5   Development of Two Prisms Oblique to Each Other.

found as in §22.5. The developments are constructed as explained previously for Figs. 22.3 and 22.4. The true lengths of the lateral edges of the vertical prism are shown in the front view, and the true size of its right section appears in the top view. The true lengths of the lateral edges of the oblique prism appear in the primary auxiliary view, and the true size of its right section is shown in the secondary auxiliary view.

Lines parallel to the respective lateral edges are drawn through the vertices of the intersection. These are then located in the developments in the same manner as the lateral edges. Point 5, located by distances Y and Z on the oblique prism, is an example.

22.4 DEVELOPMENT OF RIGHT-CIRCULAR CYLINDER

Fig. 22.6   To construct the layout for a right-circular cylinder, an appropriate number of equally spaced elements is introduced in the top view, which in this example is a true-size half-view of the lower base or right section of the cylinder. These elements are then projected to the front view, where they appear true length.

The length of the stretch-out line 0–0 is the circumference of the complete right section. This length may be approximated by stepping off the chords 0–1, 1–2, . . . , but it is obtained more accurately by calculation, as indicated.

The stretch-out line 0–0 is then divided into the same number of equal parts as was the right section, using the parallel-line method of §4.15 or with the bow dividers. With the cylinder elements located at these points, the development is begun with the shortest element. The true lengths of the elements are transferred by projecting horizontally from the front view.

22.5 DEVELOPMENT OF OBLIQUE CIRCULAR CYLINDERS

Fig. 22.7   The elements of the oblique circular cylinder connecting the two vertical duets do not appear in true length in either the front or top view. Accordingly, auxiliary view 1, showing the true lengths, is added. In this view also, any right section, such as X–X, appears in edge view. Auxiliary view 2 shows the true size of section X–X.

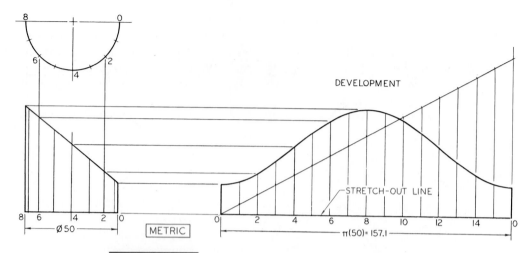

FIGURE 22.6 **Development of Right-Circular Cylinder.**

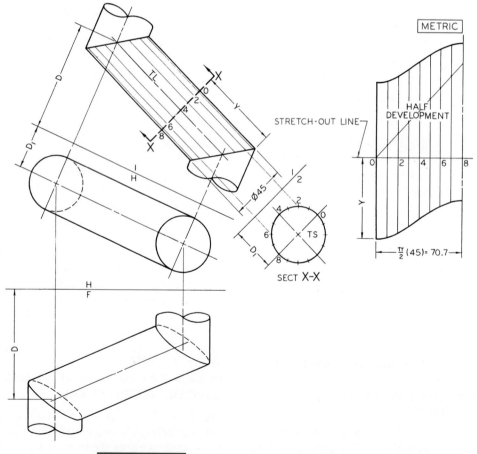

FIGURE 22.7 **Development of Oblique Cylinder.**

The development is constructed as explained in §22.4. Since the development is symmetrical about either element 0 or element 8, either point is a satisfactory starting point for the construction, thus permitting the development to be only a *half-development,* as shown. Note that half-developments are so labeled and bounded by a center line at the element of symmetry.

RADIAL-LINE DEVELOPMENTS

■ 22.6 RADIAL-LINE DEVELOPMENT

Pyramids and cones are characterized by the fact that the lateral edges or elements of each intersect at a common point or *vertex.* In their developments, Fig. 22.1 (c) and (d), this feature is retained, so that the lateral edges or elements radiate from the vertex in the development.

After the various true lengths have been secured, a radial-line development is constructed, generally as a series of triangles by the method of §4.18.

■ 22.7 DEVELOPMENT OF PYRAMID

DEVELOPMENT OF RIGHT PYRAMID *Fig. 22.8*
The true lengths of the lateral edges are found at (a) by revolution, §9.10. With point 0 in the development as center and line $0–1_R$ (distance Y) in the front view as radius, arc 1-2-3-4-1 is drawn, as shown at (b). The chords 1-2, 2-3, . . . , are then inscribed with their true lengths equal, respectively, to the sides of the base, as obtained in the top view. The lines 1-0, 2-0, . . . , are drawn and the true length lines OD_R (or $1_R–D_R$), OA, OB, . . . , respectively, are set off. The development of the lateral surface is completed by joining the points D, A, B, . . . , by straight lines.

In sheet-metal work it is not usually necessary to include bases in the development. If desired, however, they may be attached to longer edges, as shown. To transfer an irregular figure, such as the trapezoid shown here, refer to §§4.25–4.26.

DEVELOPMENT OF OBLIQUE PYRAMID *Fig. 22.9*
None of the four lateral surfaces of the pyramid is shown in true size and shape in the multiview drawing, (a). By the revolution method of §9.10, each edge is revolved until it appears in true length in the front view, as shown. Thus line 0–2 revolves to line $O–2_R$ line 0–3 revolves to line $O–3_R$; and so on. These true lengths are transferred from the front view to the develop-

ment at (b) with the compass. The true lengths of the edges of the bases are shown in the top view and are transferred directly to the development, completing the triangular lateral faces. True lengths, such as line $O–A_R$, are then found and transferred.

■ 22.8 DEVELOPMENT OF RIGHT-CIRCULAR CONE

Fig. 22.10 With the division of the base into a large number of parts (preferably equal), the development of a cone becomes similar to the development of a multisided pyramid as described in §22.7.

All elements of a right-circular cone are the same length: the so-called *slant height,* S, as shown at (a). Hence the development at (b) is a sector of a circle, with radius S. The length of the arc of the sector is made equal to the circumference of the base. This construction is most accurately and easily performed by calculation.

$$\text{Circumference} = 2\pi R$$

$$\text{Angle of sector} = \theta$$

$$\text{Length of arc of sector} = \frac{\theta}{360°} \times 2\pi S$$

Therefore

$$2\pi R = \frac{\theta}{360°} \times 2\pi S$$

or

$$\theta = \frac{R}{S} \times 360°$$

This may be simplified as shown at (a), by introducing convenient lengths proportional to R and S. Any convenient 36 units of length are set off along the slant height. The corresponding horizontal distance (12 units in this case) leads to the simple calculation shown at (b).

With the circle sector established at (b), the arc is divided into the same number of parts as the base of the cone and the elements are drawn in the development. True-length lines VA, VC′, VE′, . . . , obtained in the front view, are then set off on the respective elements to locate points on the upper curve.

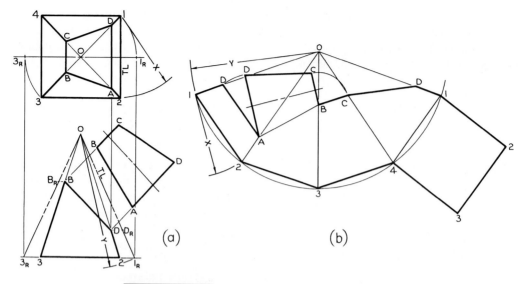

FIGURE 22.8   Development of Right Pyramid.

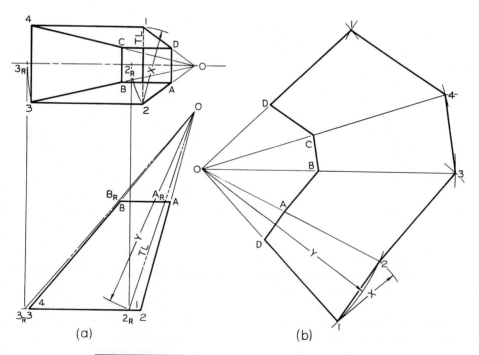

FIGURE 22.9   Development of Oblique Pyramid.

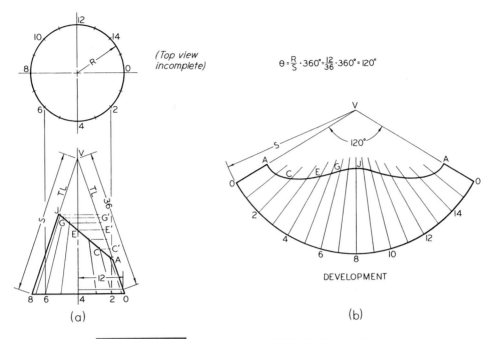

$\theta = \frac{R}{S} \cdot 360° = \frac{12}{36} \cdot 360° = 120°$

(Top view incomplete)

DEVELOPMENT

(a)

(b)

FIGURE 22.10 Development of Right-Circular Cone.

■ 22.9 DEVELOPMENT OF OBLIQUE CONE

Fig. 22.11 The form connecting the two cylinders (*transition piece*) is a frustum of an elliptical oblique cone, the vertex of which may be found by extending the contour elements to their intersection at point A. The development is similar to that of Fig. 22.9. The sides of each narrow "triangle" are the true lengths of two successive elements of the cone, and the base is the true length of the curve of the base of the cone between the two elements. This curve is not shown in its true length in the given views. The plane of the base of the frustum is therefore revolved until it is horizontal in order to find the distance from the foot of one element to the foot of the next. When the plane of the base is thus revolved, the foot of any element, such as 5, revolves to 5′ and the curve 4′–5′ (top view) is the true length of the curve of the base between elements 4 and 5. In practice the chord distance R between these points is generally used to approximate the curved distance. Relatively short chords should be used to obtain reasonable accuracy.

After the conical surface has been developed, the true lengths of the elements on the upper or imaginary section of the cone are set off from the vertex A of the development to secure points on the upper curve.

TRIANGULATION

■ 22.10 TRIANGULATION

Triangulation is simply the process of dividing a surface into a number of triangles and transferring them to a development. The radial-line developments of §§22.6–22.9 are actually examples of triangulation. However, because those constructions basically radiate from the vertices of the pyramids and cones, it is generally preferred to term them radial-line developments. In the sheet-metal industry, triangulation usually implies the method of development applied to certain forms that are not pyramids or cones.

The most common forms developed by triangulation are *transition pieces*, Fig. 22.12. A transition piece is one that connects two differently shaped, differently sized, or skew-positioned openings. In most cases, transition pieces are composed of plane surfaces and conical surfaces, which are developed by the methods applied to pyramids and cones. The oblique cone of Fig. 22.11 is also a transition piece. Sometimes such a

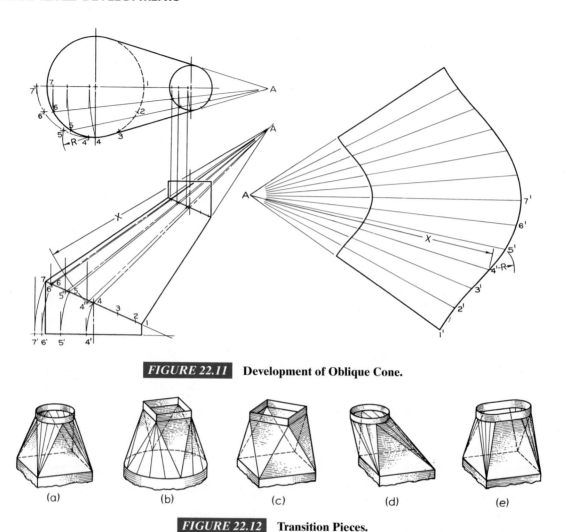

FIGURE 22.11   Development of Oblique Cone.

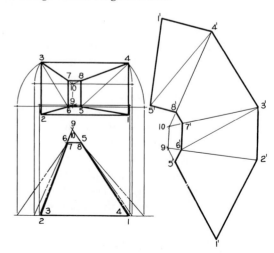

FIGURE 22.12   Transition Pieces.

piece is a frustum of a pyramid, as in Fig. 22.8. Transition pieces are extensively used in air-conditioning, heating, ventilating, and similar constructions.

■ 22.11 DEVELOPMENT OF TRANSITION PIECE CONNECTING TWO RECTANGULAR DUCTS

Fig. 22.13   In this case the lateral edges do not all intersect at the same point to form a pyramid. It is convenient to develop some of the faces by extending the sides to form triangles, as shown for faces 1–2–6–5 and 3–4–8–7. These are then developed by finding the true lengths of the sides, as in §22.7. This procedure is somewhat inconvenient for faces 1–5–8–4 and 2–6–7–3. These are best developed by triangulation, §22.10. Diagonals 4–5 and 3–6 are added to the faces

FIGURE 22.13   Development of Transition Piece Connecting Two Rectangular Ducts.

to divide them into triangles. The true lengths of the diagonals, as well as the lateral edges, are then found and assembled as shown.

As a cheek on the development, lines parallel on a surface must also be parallel on the development; for example, line 8′–5′ must be parallel to line 4′–1′ on the development.

■ 22.12 DEVELOPMENT OF TRANSITION PIECE CONNECTING TWO CIRCULAR DUCTS

Fig. 22.14 Although the transition piece resembles a cone, close inspection reveals that the elements do not intersect at a common vertex. The circular intersection with the large vertical pipe is shown true size in the top view, and the circular intersection with the small inclined pipe is shown true size in the auxiliary view. Since both intersections are true circles, and the planes containing them are not parallel, the lateral surface of the transition piece is a warped surface. It is theoretically nondevelopable, but it may be developed approximately by considering it to be made of several plane triangles, as shown. The true lengths of the sides

of the triangles are found by the method of Fig. 9.10 (d), but in a systematic manner so as to form *true-length diagrams,* as shown in Fig. 22.14.

■ 22.13 DEVELOPMENT OF TRANSITION PIECE—SQUARE TO ROUND

Fig. 22.15 The development of the transition piece is made up of four plane triangular surfaces and portions of four conical surfaces. The development is constructed with procedures resembling a combination of those used for Figs. 22.11 and 22.13. Curve 1–7 is revolved to position 1′–7′ in a manner similar to Fig. 22.11. The true-length diagrams are similar to those in Fig. 22.14. See Fig. 9.10 (d).

■ 22.14 APPROXIMATE DEVELOPMENT OF SPHERE

The surface of a sphere is not developable, §§22.1 and 22.2. The spherical surface may be developed approximately by dividing it into small sections and substituting for each section a segment of a developable surface, such as that of a cone or a cylinder.

FIGURE 22.14 **Development of Transition Piece Connecting Two Circular Ducts.**

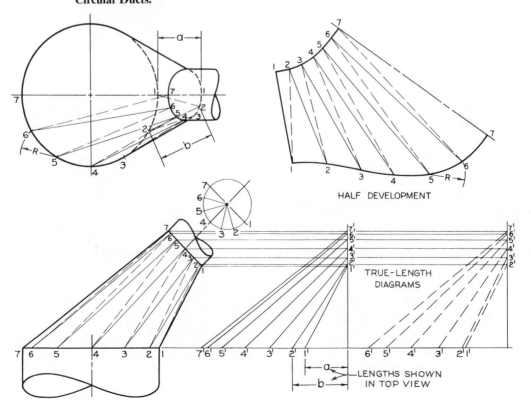

GRAPHICS SPOTLIGHT

Patents: A Computer Program that Provides Technical Help for Inventors*

Perhaps is was inevitable that someone would one day patent the act of inventing. Certainly the age of computer modeling and powerful, nimble techical software has made it possible for a group of inventors to now own what it calls an "invention that creates new inventions."

The patented software program aims to allow everyday inventors to find out what would happen if they really tried to build a device or carry out a process they have in mind. It takes its inspiration from computer-aided design programs now commonly used by engineers and designers to try out and modify new ideas.

Five inventors, all originally from the former Soviet Union, designed the software, which enables a user to create an engineering model of an invention idea, and then test the model's scientific properties. At each stage, the software offers descriptions and illustrations of the science or engineering concept involved, chooses technical properties that will work in combination, and suggests alternatives to the mechanics or the method.

Theoretically, an inventor should not need a technical background, though the software is laden with physical, geometric and chemical laws, theorems and hypotheses that only an engineer or scientist could love.

The inventors say their software goes beyond computer-assisted design and into what they call in the patent "concept engineering." The purpose, they say, is to help designers and inventors come up with new ideas, evaluate and trouble-shoot them, and consider different engineering approaches for realizing their plans.

"These systems serve to increase the designer's inventive and creative abilities in solving engineering and scientific operational or functional problems and, in the course of such problem solving, induce the designer to invent new structural and functional concepts applicable to his/her design goals," the inventors write in their patent.

The software enables a user to quickly understand the technical results that are being sought, the structure and function of engineering devices, and how such devices might produce those results if two or more are linked together. The inventor can simulate the effect of combining engineering properties, manipulating parts of those properties or adding his own customized properties. The software itself will automatically combine engineering components in an effort to refine the invention or suggest new designs and problem-solving techniques to the inventor.

The program is built around a data base of scientific and engineering concepts and their effects. Using key words, an inventor can search through 6,500 entries and pull up full-motion animation and a text description of each concept.

But first, the inventor must state an objective. The software then suggests engineering concepts that can make that design objective a reality. Each suggestion is accompanied by an explanation and illustration.

An inventor can connect the effects of two or more concepts to see what will happen, or he can simulate the results of changing parameters. He can click on a prompt that leads the software to offer other engineering or scientific facts that will improve the performance of his device or process. The inventor can bookmark his most commonly researched effects, and he can store the system's suggestions in his computer.

The software creates animated versions of the engineering effects, shows 5 to 10 seconds of full-motion video, describes the reactions and gives any examples of similar technologies that already exist.

The patent belongs to the Invention Machine Corporation. The company has already incorporated into its computer-assisted design software, which can be examined at www.invention-machine.com. The inventors, Leonid Batchilo, of Arlington, Mass., Valery Tsourikov of Boston, Vitaly Glazunov of Moscow, and Alexandre Kirkovski and Alexandre Korzoun, from Minsk, Belarus, received patent 5,901,068.

Patents may be viewed on the Web at www.uspto.gov or may be ordered through the mail, by patent number, for $3 from the Patent and Trademark Office, Washington, D.C. 20231.

---

*Adapted from "Patents—A Computer Program that Provides Technical Help for Inventors without Scientific Training," by Sabra Chartrand, *New York Times*, May 17, 1999.

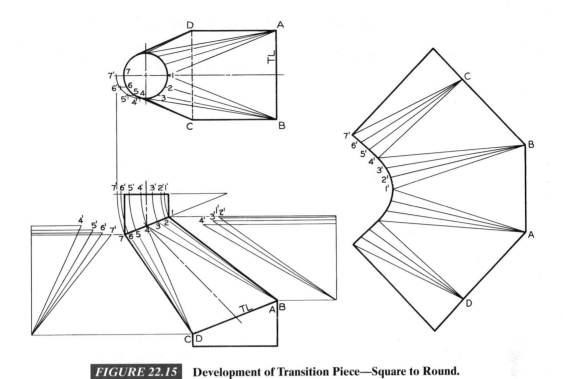

FIGURE 22.15 Development of Transition Piece—Square to Round.

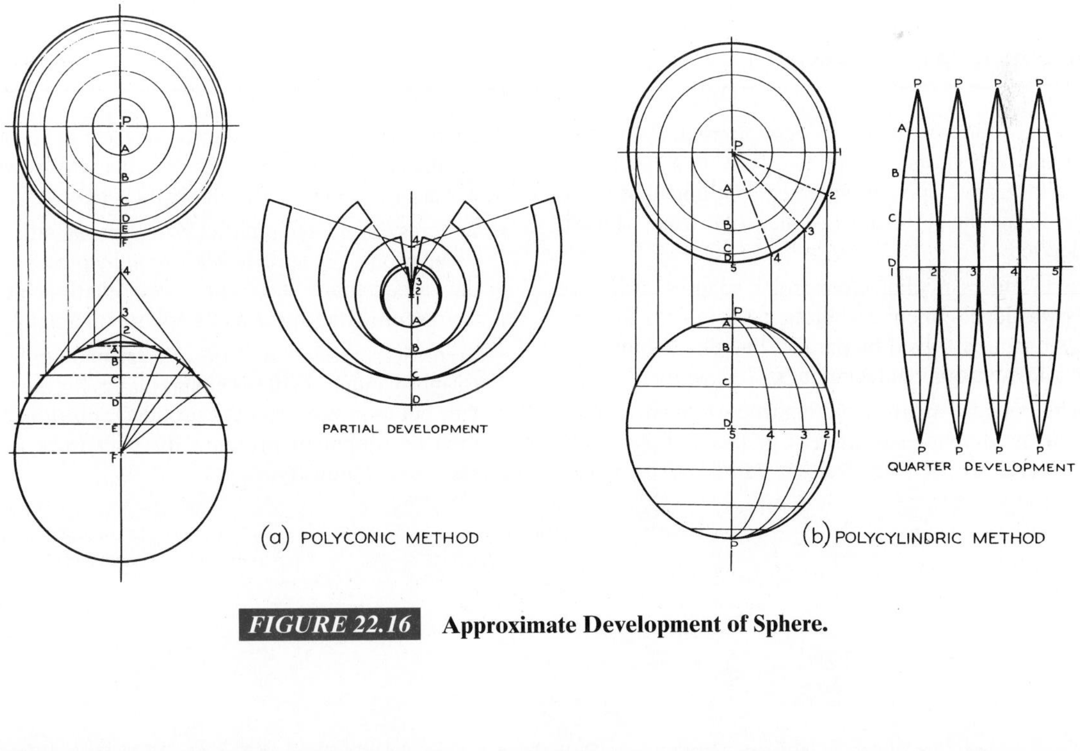

(a) POLYCONIC METHOD

PARTIAL DEVELOPMENT

(b) POLYCYLINDRIC METHOD

QUARTER DEVELOPMENT

FIGURE 22.16 Approximate Development of Sphere.

POLYCONIC METHOD *Fig. 22.16(a)* Also known as the *zone* method, the polyconic procedure is the basis for most official United States maps. See also §25.1. The sphere is divided by a series of horizontal planes, usually equally spaced, and each zone is considered to be a frustum of a right-circular cone. The development of the conical surfaces, §22.8, becomes an approximate development of the spherical surface. If the conical surfaces are inscribed in the sphere, the development will be smaller than the spherical surface. If the conical surfaces are circumscribed about the sphere, the development will be larger. If the conical surfaces are partly within and partly without the sphere, as in (a), the resulting development more closely approximates the spherical surface.

POLYCYLINDRIC METHOD *Fig. 22.16(b)* The polycylindric method, which is also known as the *gore* method, consists of cutting the sphere with *meridian* planes (planes containing the vertical axis, §25.8). Cylindrical sections are then substituted for each spherical section. The cylindrical surfaces may be inscribed within the sphere, circumscribed about it, or located partly within and partly without. The cylindrical surfaces are inscribed in the sphere in this example. The development of the series of cylindrical surfaces, §22.4, is an approximate development of the spherical surface.

■ KEY WORDS

SURFACE DEVELOPMENT	PATTERN
BEND ALLOWANCE	PARALLEL-LINE DEVELOPMENT
RADIAL-LINE DEVELOPMENT	TRIANGULATION
STRETCH OUT LINE	HALF DEVELOPMENT
TRANSITION PIECES	SLANT HEIGHT
TRUE LENGTH DIAGRAM	GORE METHOD
MERIDIAN PLANES	ZONE METHOD

■ CHAPTER SUMMARY

- Surface development is the process of creating the true size shape of a geometric solid surface on a flat plane. The surface development, when cut out and folded together, forms the three-dimensional solid from which it was derived.

- Prisms, cylinders, pyramids, cones, and spheres each have their own development techniques. All but the sphere, however, can be solved by parallel line development or radial line development techniques.

- All development involves creating true size planes of the solid on the drawing surface. For prisms and pyramids, these planes are all true size. For curved surfaces like cylinders and cones, the curved surfaces are broken up into small approximations of planes that are drawn on a flat surface to create the development.

- Spheres are the most difficult to approximate as a development on a flat surface. Cartography professionals have developed many different approximation methods for representing the surface of a sphere on a flat plane.

- Transitions require the use of a triangulation technique to approximate small triangular planes for the curved surface between two geometric solids. Transitions are a common development problem for duct work used in the sheet metal industry.

■ REVIEW QUESTIONS

1. Which types of geometric solids would use the parallel line development technique for creating a surface development?
2. Which types of geometric solids would use the radial line development technique for creating a surface development?
3. Are the plane surfaces on the development of a prism and pyramid exactly true size?
4. Are the plane surfaces on the development of a cylinder and cone exactly true size?
5. Which of the following solids is the most difficult and least accurate surface development: prism, cylinder, pyramid, cone, or sphere?

6. How many radial elements would you need to create on a cone so its surface development was perfectly accurate?
7. Are the triangles created on a transition development true size or approximations of true size?
8. When creating the surface development of a truncated pyramid, which of the radial elements extend beyond the actual surface boundaries of the pyramid?
9. Is the stretch-out line of a cylinder's surface development exact size or approximate?
10. Which development method of a sphere would have the most accurate representation of the North Pole?

■ DEVELOPMENT PROBLEMS

A variety of development problems is provided in Figs. 22.17–22.22. These problems are designed to fit on Layout B-3 or A3-3. Dimensions may be included on the given views if assigned. The student is cautioned to take special pains to obtain accuracy on these drawings and to draw smooth curves as required.

Many suitable problems, involving intersections, also may be assigned for development from Figs. 22.24–22.27.

Since many of the problems in this chapter are of a general nature, they can also be solved on most computer graphics systems. If a system is available, the instructor may choose to assign specific problems to be completed by this method.

Additional problems, in convenient form for solution, are available in *Engineering Graphics Problems*, Series 1, by Spencer, Hill, Loving, Dygdon, and Novak, designed to accompany this text and published by Prentice Hall Publishing Company.

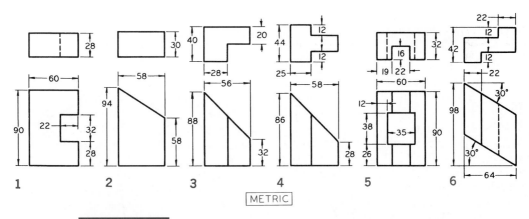

METRIC

FIGURE 22.17 Draw given views and develop lateral surface (Layout A3–3 or B–3).

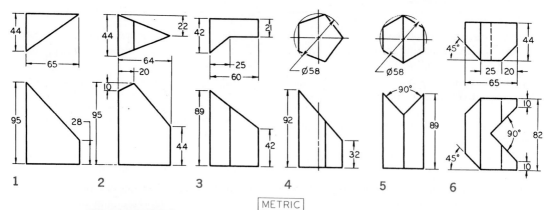

METRIC

FIGURE 22.18   Draw given views and develop lateral surface (Layout A3–3 or B–3).

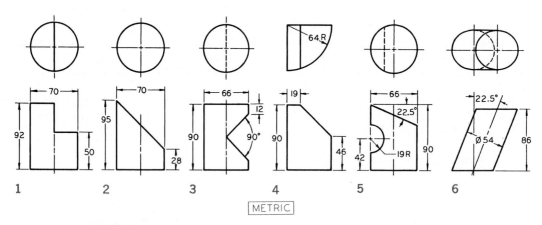

METRIC

FIGURE 22.19   Draw given views and develop lateral surface (Layout A3–3 or B–3).

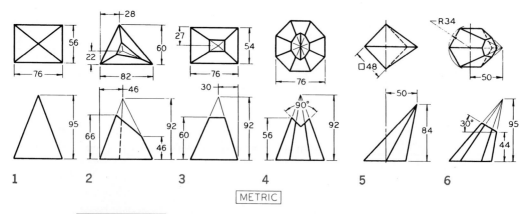

METRIC

FIGURE 22.20   Draw given views and develop lateral surface (Layout A3–3 or B–3).

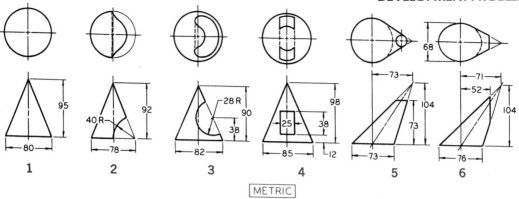

FIGURE 22.21 Draw given views and develop lateral surface (Layout A3–3 or B–3).

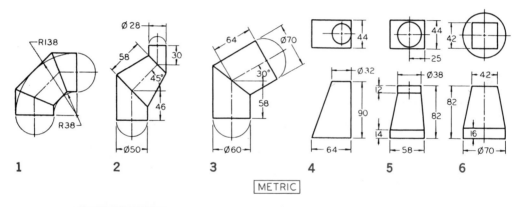

FIGURE 22.22 Draw given views and develop lateral surface (Layout A3–3 or B–3).

C H A P T E R   2 3

LINE AND PLANE TANGENCIES

OBJECTIVES

After studying the material in this chapter, you should be able to:

1. Describe the conditions that make a line or plane tangent to a three-dimensional surface.

2. Draw a plane through a point and tangent to a cone or cylinder.

3. Draw a plane parallel to a line and tangent to a cone or cylinder.

4. Draw a plane tangent to a sphere at a given point on its surface.

5. Draw a plane tangent to a sphere and parallel to a line.

6. Draw a plane tangent to a sphere and containing a given line not intersecting the sphere.

7. Draw a line at a specified angle to one or more principle planes of projection.

OVERVIEW

A plane tangent to a ruled surface such as a cone or cylinder contains only one straight-line element of that surface. A plane tangent to a double-curved surface such as a sphere contains only one point in that surface. All lines tangent to a curved surface at a particular point or at points along the same straight-line element lie in a plane tangent at the point or element. A plane tangent to a ruled surface is conveniently represented by two straight lines, one an element and the other line tangent to the surface at a point on the element. For double-curved surfaces the plane is represented by two straight lines, both tangent at the same point on the double-curved surface.

One of the most complex engineering graphics solutions is the plane tangent to a three-dimensional curved surface like a cone, cylinder, or sphere. While most CAD programs can quickly find the tangent point for plane geometry, solutions for three-dimensional tangencies can only be handled by the most sophisticated CAD program. However, any drafter with a thorough understanding of engineering graphics theory can use a drawing board or simple CAD program to solve complex tangency problems. Similar solutions can be used to solve for true angles between oblique lines and the principle planes of projection.

■ 23.1 LINE AND PLANE TANGENCIES

Line tangencies as encountered in plane geometry are discussed in Chapter 4, "Geometric Constructions." Methods such as those of §§4.33–4.35 are based on the principle that a line tangent to a circle is perpendicular to the radius drawn to the point of tangency. Thus in Fig. 23.1 (a) line 1–2 is perpendicular to radius O–1 at point 1 and is tangent to the circle O. If the assumption is made that line 1–2 and circle O are in the plane of the paper, a plane geometry construction suffices.

At (b) is given a multiview (or three-dimensional) drawing. Although the top view of line 3–4 appears to be tangent to the top view of the circle, line 3–4 is *not* tangent to the circle *in space* because line 3–4 is not in the plane of the circle, as is evident in the front view. By contrast, line 5–6 lies in the plane of the circle and *is* tangent in space to the given circle.

Because planes are easily represented by lines, §19.7, planes tangent to curved surfaces are often represented by suitable pairs of tangent lines, or by one

tangent line and a line lying in the curved surface (a straight-line *element*, §6.28). It is sometimes convenient to represent a tangent plane by a tangent edge view of the plane, Fig. 23.10 (b).

■ 23.2 PLANES TANGENT TO CONES

PLANE TANGENT TO A RIGHT-CIRCULAR CONE THROUGH POINT ON SURFACE *Fig. 23.2* Let the cone and the top view of point 1 on the surface of the cone be given as at (a).

At (b) the element through point 1 is drawn in the top view, establishing point 2 on the circular base. Point 2 is then projected to the front view, and point 1 is projected to the now-established front view of the element.

If the view of element O–2 is nearly parallel to the projectors, more dependable accuracy may be secured through the use of revolution, as at (c), which is an application of §9.6.

At (d) line 2–3 is drawn tangent to the circular base by drawing its top view perpendicular to element O–2 at point 2 and by drawing its front view coinciding with the edge view of the base. Plane 0–1–2–3 is the required tangent plane.

PLANE TANGENT TO OBLIQUE CONE THROUGH POINT OUTSIDE ITS SURFACE *Fig. 23.3* Let the cone and point be given as at (a).

Because all elements of a cone pass through its vertex, a tangent plane, which must contain an element, will contain the vertex also. Hence line V–1 lies in the tangent plane. Any line tangent to the base circle lies in the plane of the base, §23.1, and thus can intersect line V–1 only at point 2, shown in the front view at (b).

At (c) point 2 is projected to the top view of line V–1, and from the top view of point 2 line 2–3 and 2–3′ may be drawn tangent to the base circle as shown. Either plane V–2–3 or V–2–3′ meets the requirements of the problem. In practice, it is usually evident which of two optional solutions is compatible with other features of the design.

PLANE TANGENT TO CONE AND PARALLEL TO GIVEN LINE *Fig. 23.4* With the given cone and line 1–2 as shown at (a), a line is drawn through vertex V of the cone, parallel to the given line and intersecting the plane of the base at point 3, as at (b), thus establishing a line in the required tangent plane. (Also see §20.4.)

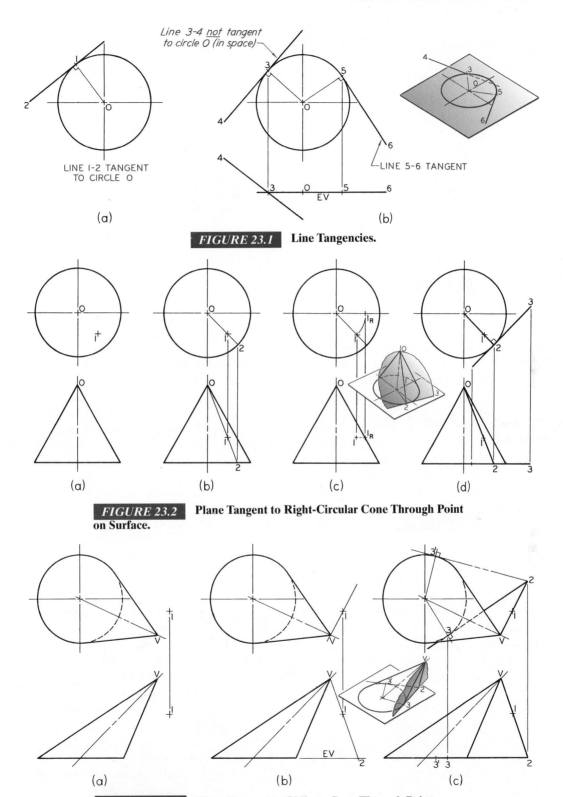

FIGURE 23.1 Line Tangencies.

FIGURE 23.2 Plane Tangent to Right-Circular Cone Through Point on Surface.

FIGURE 23.3 Plane Tangent to Oblique Cone Through Point Outside Cone.

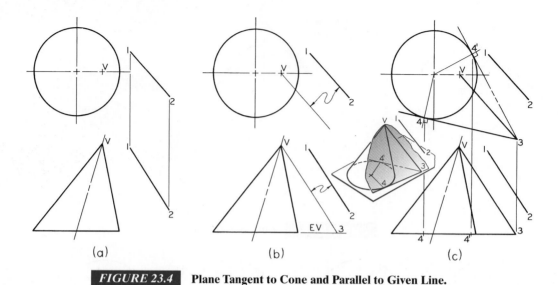

FIGURE 23.4   **Plane Tangent to Cone and Parallel to Given Line.**

At (c) point 3 is projected to the top view of the line through V and parallel to line 1–2. From the top view of point 3 lines 3–4 and 3–4′ may be drawn tangent to the base circle as shown. Either plane V–3–4 or V–3–4′ satisfies the requirements of the problem.

■ 23.3 PLANES TANGENT TO CYLINDERS

By definition, §4.7, all elements of a cylinder are parallel to each other and to the axis of the cylinder. It follows from §20.4 that any plane tangent to a cylinder,

and thus containing one element, is parallel to the remaining elements and to the axis.

PLANE TANGENT TO CYLINDER THROUGH POINT ON SURFACE *Fig. 23.5*   With one view given of point 1 on the surface of an oblique cylinder as at (a), element 1–2 is introduced as at (b). When point 2 is projected to the top view, it is observed that point 2 may fall at either position 2 or position 2′. There are thus alternative solutions, and point 1 may be at either of the locations 1 and 1′, as shown. Addition of line 2–3 tangent to the base of the cylinder at point 2 (or

FIGURE 23.5   **Plane Tangent to Cylinder Through Point Outside Surface.**

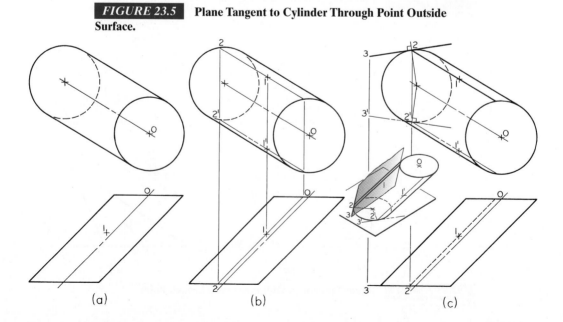

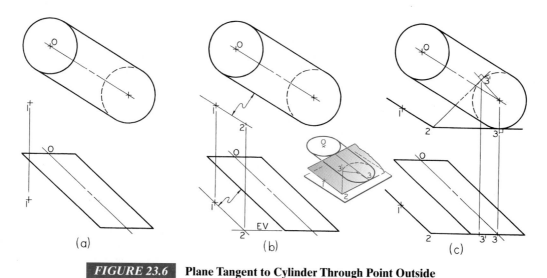

FIGURE 23.6 **Plane Tangent to Cylinder Through Point Outside Surface.**

line 2′–3′ at point 2′) completes the representation of the required tangent plane.

PLANE TANGENT TO CYLINDER THROUGH POINT OUTSIDE ITS SURFACE *Fig. 23.6* The cylinder and point 1 are given as at (a). As previously noted, any plane tangent to a cylinder must be parallel to the elements. Hence a line 1–2 drawn through point 1 and parallel to the elements, as at (b), must be common to all planes containing point 1 and parallel to the elements, §20.4. The representation of a tangent plane is completed by the addition of a line tangent to one base of the cylinder. As observed earlier, such a tangent line

must lie in the same plane as the chosen base—in this example the lower base, (c). Therefore the tangent line could intersect line 1–2 only at point 2 located in the front view at the intersection of the (extended) edge view of the base plane with line 1–2. Observe that lines could be drawn from point 2 tangent to the lower base at either point 3 or point 3′, so that again there are alternative solutions, and in an application it would normally be apparent which solution is practical.

PLANE TANGENT TO CYLINDER AND PARALLEL TO GIVEN LINE OUTSIDE THE CYLINDER *Fig. 23.7* Let the cylinder and line 1–2 be given as at (a).

FIGURE 23.7 **Plane Tangent to Cylinder and Parallel to Given Line Outside the Cylinder.**

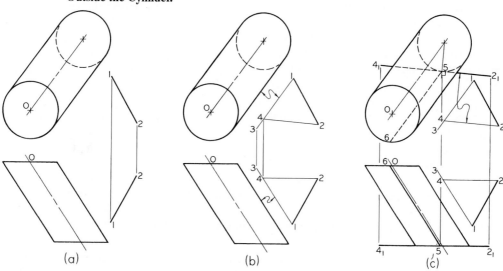

A plane is constructed parallel to a given line by the method of §20.4. However, at this stage it is not known which element will be the line of tangency. The tangent plane must be parallel to all elements of the cylinder as well as to line 1–2. By the method of §20.4 a plane can be constructed at any convenient location and parallel both to the elements and to line 1–2. The required tangent plane will then be parallel to this plane.

A convenient representation for this preliminary plane includes the given line 1–2, as at (b). With line 1–3 drawn parallel to the cylinder as shown, plane 2–1–3 is established parallel to the cylinder.

Any line tangent to either given cylinder base must be a horizontal line, §23.1. Since all horizontal lines in the same oblique plane are parallel to each other, §19.8, it follows that horizontal lines in parallel oblique planes are likewise parallel to each other. Thus, if the direction of one such horizontal line is established, such as line 2–4 at (b), the direction of horizontal lines in planes parallel to plane 2–1–3, including the required plane, is also established.

At (c) line 2_1–4_1 is drawn parallel to the top view of line 2–4 and tangent to either base—in this case the lower base—of the cylinder. The point of tangency is 5, and line 5–6 is the element of tangency. Construction of the front view completes the representation of the required tangent plane parallel to given line 1–2.

Since tangent line 2_1–4_1 could have been drawn on the opposite side of the base, there is an alternative tangent plane, which is not shown here.

■ 23.4 PLANES TANGENT TO SPHERES

A plane tangent to a *double-curved surface*, §23.1, contains one and only one point of that surface, since it follows from the definition of such a surface that it contains no straight-line elements. Hence planes tangent to double-curved surfaces are represented by appropriate combinations of lines tangent at the desired point or points of tangency. Under suitable circumstances such a tangent line may represent an edge view of the required tangent plane. An example of this appears later in this section.

The sphere, §4.7, is by far the most practical, hence most common, form of the double-curved surface. This discussion will be limited to spherical surfaces.

LINES TANGENT TO A SPHERE AT A GIVEN POINT ON ITS SURFACE   *Fig. 23.8*   Let the front and top views of a hemisphere be given, as at (a), and let the top view of a point 1 on its surface be given also. The

front view of point 1 can be located by passing a convenient cutting plane through the point and finding another view of the line (circle) of intersection. It follows that a "convenient" cutting plane is one in such position that the circle appears in one of the given views as a circle and not as an ellipse. As an example, at (b) a frontal, edge-view plane is introduced. The circle is located and constructed in the front view as shown, and point 1 is projected to it. A line tangent to this circle is tangent to the spherical surface. At (c) line 1–2 is constructed tangent to the circle at point 1 by drawing the front view of the tangent line perpendicular to radius O–1 and then drawing the top view coincident with the edge view of the cutting plane, §23.1.

At (d) point 1 is revolved, in the top view, to the frontal plane through center O. See §9.6. This amounts to revolving the edge view of a vertical cutting plane (EV), as indicated. The revolved view of the circle of intersection coincides with the circular front view of the sphere, and the revolved position 1_R of point 1 is projected to it. As shown at (e), line 1_R–3 is now drawn tangent to this circle, intersecting the vertical center line of the sphere at point 3. Since this vertical center line is also the axis of revolution, point 3 will not move as the cutting plane counterrevolves, as at (f). Line 1–3 is thus another line tangent to the spherical surface at given point 1.

At (g) a horizontal cutting plane is introduced by first drawing the top view of the circle of intersection passing through the top view of point 1. This in turn projects to the front view as shown, locating the edge of the cutting plane, to which point 1 is projected. Line 1–4, constructed tangent to the circle of intersection at point 1, is also tangent to the spherical surface.

Finally, at (h) and (i), line 1–5 is constructed tangent to the spherical surface in a variation of the method shown at (d) to (f). An edge-view cutting plane is introduced in the front view, through points 1 and O. After it revolves to horizontal, the circle of intersection coincides with the top view of the sphere, and line 1_R–5 is drawn tangent to the cut circle of intersection. Counterrevolution establishes the views of the tangent line 1–5, as shown at (i).

PLANE TANGENT TO A SPHERE AT A GIVEN POINT ON ITS SURFACE   *Fig 23.9*   In Fig. 23.8 four lines, 1–2, 1–3, 1–4, and 1–5, were constructed tangent to the spherical surface at point 1. Any two of these constitute intersecting tangent lines and thus establish the plane tangent at point 1. As an example, tangent

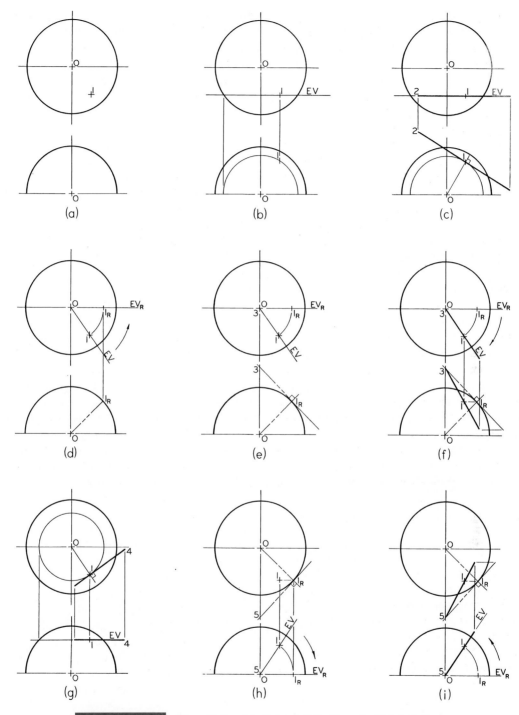

FIGURE 23.8   Lines Tangent to Spherical Surface at a Point on the Surface.

lines 1–2 and 1–4 are shown in Fig. 23.9. Plane 2–1–4 is one representation of the plane tangent to the spherical surface at point 1. Incidentally, study reveals that this is simply the construction of a plane perpen-

dicular to radius O–1 at point 1 by the given-view method of §20.6. Analogous to the plane geometry description of a line tangent to a circle, §23.1, a plane tangent to a sphere may be defined as a plane

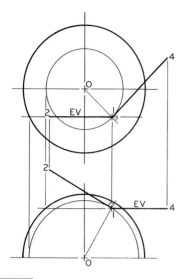

FIGURE 23.9 **Plane Tangent to Sphere at Given Point on its Surface.**

perpendicular to the radius of the sphere drawn to the point of tangency.

PLANE TANGENT TO A SPHERE AND CONTAINING A GIVEN LINE THAT DOES NOT INTERSECT THE SPHERE   *Fig 23.10*   Let sphere O and line 1–2 be given in the front and top views, as at (a). If a point view of a line is constructed, any plane containing the line appears in edge view, §19.9. Also any orthographic

projection of a sphere shows the true diameter of the sphere. Hence in the given problem, if a point view of line 1–2 is constructed, the required tangent plane will appear in edge view and tangent to the corresponding view of the sphere. Since line 1–2 appears in true length in the top view at (a), its point view may be constructed in primary auxiliary view 1 as shown.

At (h) alternative edge views of planes are drawn through point view 2, 1 and tangent to the sphere at point 3 or point 3′ as preferred. The front and top views of point 3 or 3′ are then projected as shown, completing the representation of the tangent plane. In practice, other lines of the tangent plane could, and probably would, be drawn to establish a recognizable configuration. Theoretically, additional lines are not needed.

■ 23.5 APPLICATIONS OF RIGHT-CIRCULAR CONES

All elements of a right-circular cone form the same angle with the base plane of the cone. This feature is the basis for the constructions following.

PLANE CONTAINING AN OBLIQUE LINE AND MAKING A SPECIFIED ANGLE WITH HORIZONTAL *Fig. 23.11*   Let line 1–2 be given, as at (a), and let it be required to construct a plane containing line 1–2 and forming an angle of 45° (or 135°) with horizontal.

FIGURE 23.10   **Plane Tangent to Sphere and Containing Line Outside Sphere.**

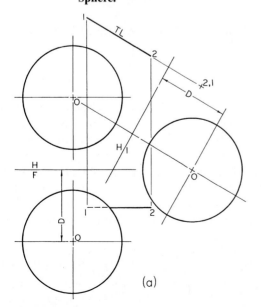

(a)

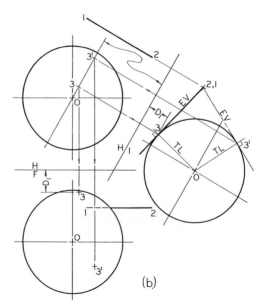

(b)

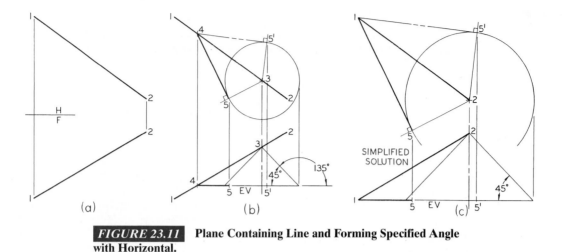

FIGURE 23.11 **Plane Containing Line and Forming Specified Angle with Horizontal.**

A plane tangent to a right-circular cone contains one element and forms the same angle as does any element with the base plane of the cone. Because the vertex is common to all elements, it must lie in any tangent plane. Hence at (b) the vertex of a cone of suitable dimensions is placed at some chosen point 3 along given line 1–2. In this case the required angle is 45° with horizontal, so the cone is placed with its base horizontal (axis vertical) and with its elements at 45° with the base. This same construction is used for a specified angle of 135°, the supplement of 45°. Line 1–2 pierces the extended edge view of the base plane at point 4. See Fig. 23.3. Lines tangent to the base may be drawn alternatively from point 4 to point 5 or from point 4 to point 5′. Either of the two resulting tangent planes may be selected according to additional specifications, if any.

As shown at (c), the foregoing construction could be somewhat simplified in detail, not in principle, by placing the cone vertex at point 2 and the base plane at the same elevation as point 1.

LINE AT SPECIFIED ANGLES WITH GIVEN PLANES
Fig. 23.12 If two right-circular cones with the same vertex intersect, the common element or elements of the two cones form the same angles with the two base planes as do the respective sets of elements. To simplify determining which elements are common, the two cones should have elements of the same length so that their base lines intersect, as exemplified by points 1 and 2 in Fig. 23.12.

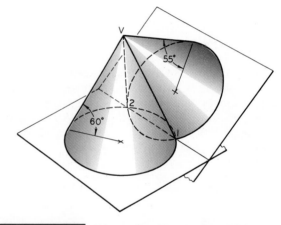

FIGURE 23.12 **Line at Specified Angles with Two Given Planes.**

As an example of an application of the foregoing, Fig. 23.13, let it be required to construct a line 1–2, with point 1 given, such that line 1–2 forms an angle of 30° with a horizontal plane and an angle of 50° with a frontal plane. At (a) a right-circular cone is constructed with its vertex at point 1 and with its elements at 30° with horizontal. The length S of the elements can be any convenient or specified length.

At (b) a second cone is introduced with its vertex at point 1 but with its elements at 50° with a frontal plane. Note that the previously selected length S must also be used for the elements of the second cone. This selection

GRAPHICS SPOTLIGHT
Creating Digital Terrain Models

3D CAD TERRAIN

You can use CAD to create 3D models of terrain. These models can be created from point data, which contains the *x* and *y* coordinates, or northings and eastings, of a point, and their *z* coordinate or elevation. The 3D point information gathered from a survey can be used to define a CAD surface model representing the terrain. Stereophotogrammetry can also be used to generate the 3D point information needed to create CAD terrain models. There are a number of CAD packages available which help you in creating 3D surface models of terrain. One company which makes a full suite of civil software applications that run inside of AutoCAD® is Softdesk.

TERRAIN MODELS INSIDE AUTOCAD

Softdesk, Inc. provides Digital Terrain Modeling (DTM) software which you can use to do site analysis and terrain modeling for sites, reservoirs, channels, roadways, landfills, slope analysis, and mining operations (see Fig. A). Softdesk's DTM runs inside AutoCAD software. The drawings it creates are standard AutoCAD drawings, which you can also edit and plot using just AutoCAD. One of the nice features of DTM is that it can interpret a variety of data including drawing or project points, AutoCAD points, 3D faces, and 3D entities such as contours to create a surface model. DTM™ also works with Softdesk's COGO (Coordinate Geometry) module which you can use to import survey locations from data recorders, and to reduce survey data.

FROM SURFACE MODEL TO CONTOUR MAP

The surface models created with DTM are called Triangulated Irregular Networks or TINs because they are made up of triangular shaped facets that describe the shape of the surface (see Fig. B).

Once you have defined a TIN, you can use it to automatically generate a contour map. You can digitize, edit, label, and smooth the contours. You can also get surface statistics such as average slope and total surface area, and can view the model based on elevation or slope. You can generate multiple surface cross sections utilizing an unlimited number of surfaces. This is extremely useful for planning road locations, or general site development.

Softdesk also has software for Advanced Design and Hydrology Tools which let you sample profiles and cross sections automatically. You can also calculate volumes between surfaces utilizing Softdesk's Earthworks™ module, which is helpful for designing roadway drainages.

FLY THROUGH SITE

The surface model describes the shape of the terrain. Once you have defined it, it can be used visually to create fly-throughs of the site. This can be very useful for presenting your design in a way that is friendly for managers and decision makers. As well as improving visualization, 3D fly-throughs involve the viewer in a way that static presentations do not. The increased involvement in the design can help to provide increased understanding and acceptance of the project.

(B) *Courtesy of Softdesk, Inc.*

(A) *Courtesy of Softdesk, Inc.*

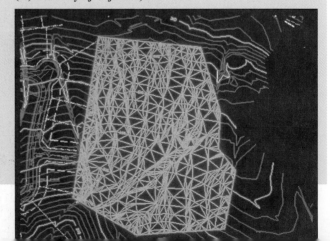

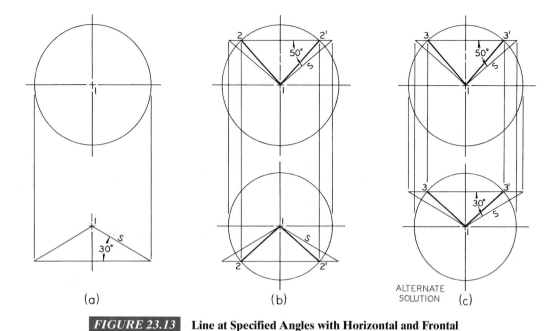

FIGURE 23.13   **Line at Specified Angles with Horizontal and Frontal Planes.**

results in the intersection of two base circles at points 2 and 2′. Thus the requirements of the problems are fulfilled by either line 1–2 or line 1–2′. There are additional alternative solutions. If we choose to reverse either cone, we find two more solutions. At (c) the cone with the 30° angle is drawn sloping upward from point 1. The two additional solutions are lines 1–3 and 1–3′.

Additional reversals of the cones produce line-segment extensions of lines 1–2 and 1–2′ or 1–3 and 1–3′—not additional alternative solutions.

It is important to realize that there are limitations on the selection of the two angles. If their sum is greater than 90°, the cones do not intersect, and there is no solution.* If the sum equals 90°, the cones are tangent and the element of tangency is the solution. (Two such single-element solutions are possible.) Only when the sum of the required angles is less than 90° do we have the four alternative possibilities shown in Fig. 23.13. For given planes that are not perpendicular, there are different but similar limitations dependent on the dihedral angle, §9.11, between the two planes. In general, the sum of the required angles must be equal to or less than the dihedral angle between the given planes.

■ **KEY WORDS**

TANGENCY	**APPARENT TANGENT POINT**
RIGHT CYLINDER	**RIGHT CONE**
OBLIQUE CYLINDER	**OBLIQUE CONE**
DOUBLE-CURVED SURFACE	**SINGLE-CURVED SURFACE**
RADIAL ELEMENTS	**DIHEDRAL ANGLE**

■ CHAPTER SUMMARY

- In plane geometry, lines that appear to intersect with a curved figure at one point are considered tangent at that point. In three-dimensional space, a line can *appear* to be tangent in one view, but in fact may not be tangent when viewed from an alternate view.

- Radial elements are defined on cones and cylinders to make the intersection of a tangent plane and the radial element coplanar. Once the geometry is coplanar, the simplified tangency rules of plane geometry apply.

- The process of constructing radial elements in one view and projecting onto an adjacent view is a common construction technique.

- When radial elements are nearly parallel to the plane of projection in which they are drawn, standard revolution techniques can be used to improve construction accuracy when solving problems with instruments on paper.

- CAD programs that do not have automatic tangency functions can still be used to solve complex tangency problems by using the software as a super-accurate drafting tool and employing descriptive geometry solution techniques.

- A plane is tangent to a cylinder or cone at only one radial element (line). A plane is tangent to a sphere at only one point.

- All lines tangent to a curved surface at one point are contained in the plane tangent to the curved surface at that point.

■ REVIEW QUESTIONS

1. What is the difference between a true tangent point and an apparent tangent point?

2. How many lines are contained in the intersection of a tangent plane and the cone or cylinder with which it is tangent?

3. How many points are contained in the intersection of a tangent plane and the cone or cylinder with which it is tangent?

4. What is the purpose of an edge view (EV) cutting plane when solving for a line tangent to a sphere at a given point on the sphere?

5. Why is the first step when solving for a plane tangent to a cylinder through a point outside the surface of the cylinder to draw a line through the point and *parallel* to the center line of the cylinder?

6. How many solutions are there for a plane tangent to a cylinder through a point outside the surface of the cylinder? of a cone? of a sphere?

7. Some television screens are called *vertically flat*. That is, a straight edge held vertically against the screen would be coincident with a vertical element of the screen, but if held horizontal to the screen would be tangent at a single point. Is this type of television screen a single-curved surface or a double-curved surface?

8. Can two non-coincident planes be tangent to the same point on the cylinder? on a cone? on a sphere?

■ LINE AND PLANE TANGENCY PROBLEMS

In Figs. 23.14, 23.15, and 23.16 are problems involving planes tangent to cones, cylinders, and spheres, and applications of right-circular cones.

Use Layout A–1 or A4–1 (adjusted) and divide the working area into four equal areas for problems to be assigned by the instructor. Some problems require two problem areas or one-half sheet. Data for most problems are given by a coordinate system using metric dimensions. For example, in Fig. 23.14. Prob. 1, point O is located by the full-scale coordinates (60 mm, 50 mm, 90 mm). The first coordinate locates the front view from the left edge of the problem area. The second coordinate locates the front view of the point from the bottom edge of the problem area. The third coordinate locates either the top view of the point from the bottom edge of the problem area or the side view of the point from the left edge of the problem area. Inspection of the given problem layout will determine which application to use.

Since many of the problems in this chapter are of a general nature, they can also be solved on most computer graphics systems. If a system is available, the instructor may choose to assign specific problems to be completed by this method.

Additional problems, in convenient form for solution, are available in *Engineering Graphics Problems*, Series 1, by Spencer, Hill, Loving, Dygdon, and Novak, designed to accompany this text and published by Prentice-Hall Publishing Company.

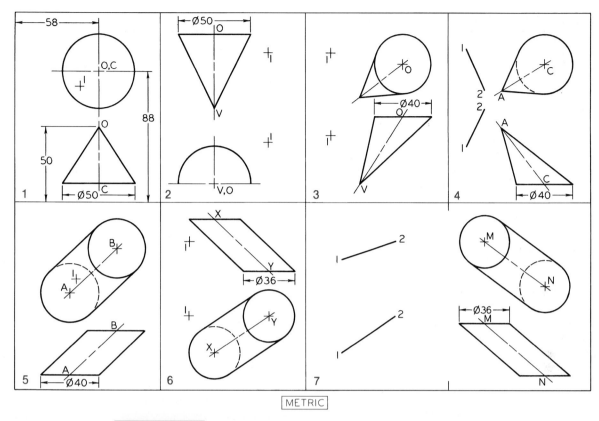

METRIC

FIGURE 23.14   Lay out and solve problems as assigned. Use Layout
A–1 or A4–1 (adjusted) divided into four equal areas.

1. Point 1 (46, –, 79) is on the surface of cone O (58, 30, 88)–C (58, 12, 88). Pass a plane tangent to the cone and containing point 1.
2. Pass a plane through point 1 (75, 40, 100) and tangent to cone V (38, 12, 63)–O (38, 12, 114).
3. Pass a plane through point 1 (18, 46, 100) and tangent to cone V (38, 12, 71)–O (69, 61, 94).
4. Pass a plane tangent to cone A (38, 50, 75)–C (68, 12, 94) and parallel to line 1 (12, 38, 104)–2 (25, 63, 75).
5. Pass a plane tangent to cylinder A (38, 10, 66)–B (71, 40, 96) and containing point 1 (43, –, 75) on the surface of the cylinder.
6. Pass a plane through point 1 (20, 50, 100) and tangent to cylinder X (38, 25, 117)–Y (75, 50, 81).
7. Pass a plane tangent to cylinder M (127, 46, 100)–N (170, 10, 70) and parallel to line 1 (25, 25, 88)–2 (63, 50, 100).

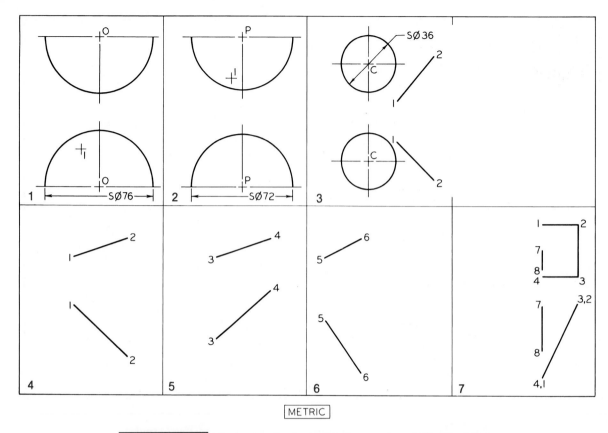

METRIC

FIGURE 23.15   **Lay out and solve problems as assigned. Use Layout
A–1 or A4–1 (adjusted) divided into four equal areas.**

1. Pass a plane tangent to sphere O (56, 12, 114) and containing point 1 (43, 38, –) on the surface of the sphere.
2. Draw three lines tangent to sphere P (56, 12, 114) and containing point 1 (48, –, 86) on the surface of the sphere.
3. Pass a plane tangent to sphere C (43, 30, 96) and containing line 1 (63, 43, 70)–2 (88, 18, 100). Show the point of tangency in all views.
4. Pass a plane through line 1 (38, 61, 94)–2 (75, 25, 107) and making an angle of 30° with a frontal plane.
5. Pass a plane through line 3 (38, 38, 94)–4 (75, 71, 107) and making an angle of 60° with horizontal.
6. Pass a plane through line 5 (12, 50, 94)–6 (38, 15, 107) and making an angle of 135° with a profile plane.
7. Pass a plane through line 7 (63, 61, 99)–8 (63, 30, 86) and making an angle of 60° with plane 1 (63, 12, 117)–2 (88, 63, 117)–3 (88, 63, 81)–4 (63, 12, 81).

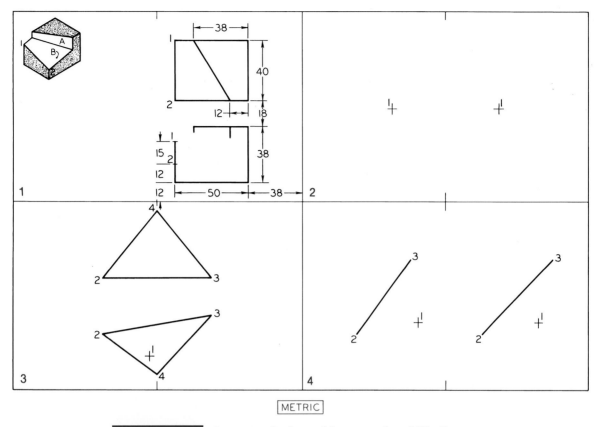

METRIC

FIGURE 23.16 **Lay out and solve problems as assigned. Use Layout A–1 or A4–1 (adjusted) divided into four equal areas.**

1. Surfaces A and B form a dihedral angle of 130°. Complete the front view. Omit the pictorial in the layout.
2. Complete the views of a line 1 (63, 63, 140)–2, which is 50 mm in length and forms angles of 45° with a profile plane and 35° with a frontal plane.
3. Point 1 (96, 23, –) is in plane 2 (63, 38, 75)–3 (140, 50, 75)–4 (100, 10, 122). Find in plane 2–3–4 a line 1–5 that forms an angle of 25° with a horizontal plane.
4. Find a line 1 (81, 45, 167)–4 which is 40 mm in length, is perpendicular to line 2 (38, 38, 127)–3 (75, 88, 178), and makes an angle of 40° with a frontal plane.

CHAPTER 24

CARTOGRAPHY, GEOLOGY, AND SPHERICAL GEOMETRY

OBJECTIVES

After studying the material in this chapter, you should be able to:

1. Understand the concept of producing and representing three-dimensional contours on a two-dimensional map drawing.

2. Define and understand the terminology used in cartography and map projections.

3. Decribe the four basic types of map projections.

4. Draw a contour map given elevation data.

5. Solve strike, dip, stratum thickness, outcrop, and cut and fill problems using descriptive geometry techniques.

6. Solve problems using spherical triangles.

OVERVIEW

Although graphical methods may be used to advantage in all fields of engineering, the nature of many problems in mining and civil engineering makes them suitable for graphical representation and solution. Highway engineers, geologists, mining engineers, and military strategists have frequent need to prepare and use *topographic maps*, which are a graphical means of representing the earth's surface and related information to a convenient scale in single views. Orthographic projection, as used in technical drawings, in which the line of sight is assumed to be perpendicular to the plane of the map, is the method most commonly used in topographic representation.

Specific applications of the principles of descriptive geometry are required to illustrate graphically the areas of topography, mining, and geology. Spherical geometry is fundamental to celestial and terrestrial navigation and inertial guidance.

Cartography uses some of the most exciting new technology developments. Inexpensive computer programs now display maps for nearly every country, and personal Global Positioning System (GPS) receivers determine your position to within 50 meters. Both technologies rely on cartography to display and triangulate the curved surface of the Earth on a two-dimensional surface. Civil engineers use cartography and engineering graphics to layout cut and fill areas for road design. Mining and geological stratum plots require descriptive geometry solutions to show strike, dip, thickness, and outcrop. Cellular telephone providers use sophisticated mapping software in conjunction with modeling software to predict and measure path loss over coverage terrain when designing new or expanding cellular networks across the globe. Some automobile manufacturers use on-board computer maps and GPS software combined with traffic condition data to direct a driver along the best route to their destination. All of these technologies require thorough understanding of the basic principles of cartography and spherical geometry.

■ 24.1 CARTOGRAPHY

Cartography is the science and art of map making. It is the process of representing the surface of a sphere, usually the earth or other heavenly body, on a plane surface. Because a sphere is not truly developable, it is impossible to represent exactly any extensive portion of a spherical surface on a plane. A sphere may be developed approximately by the polyconic or polycylindric methods of §22.14. Shown in Fig. 24.1 (a) is a modification of the polyconic method, in which the zones are not separated. Alternatively, the spherical surface may be projected on a plane or a developable surface, such as a cone or cylinder, Fig. 24.1 (b) to (d). The projection may be parallel projection, as in Fig. 24.1 (d), or central projection, as in (b) and (c). The central projection may be from the center of the Earth or any other point. Frequently, the resulting developed map is intentionally altered to improve accuracy in areas most severely distorted. All these variations, and many others, are pursued with the objective of minimizing one or another type of distortion for a particular purpose. Detailed treatment of the theory of map drawing is beyond the scope of this text.

Civil engineers, mining engineers, and geologists normally are concerned with maps of relatively small areas, and at proportionally large scales, so that the curvature of the Earth's surface is small compared to local irregularities such as hills and valleys. Under these circumstances, the curvature is ignored and the map is regarded simply as a top view in orthographic projection.

■ 24.2 CONSTRUCTION OF TOPOGRAPHIC MAPS

Maps that are designed to describe in detail local features of the earth's surface, either natural or man-made, are called *topographic maps* (or drawings). The source of information for the construction of topographic maps is the *survey*. Surveying is the actual measurement of distances, elevations, and directions on the Earth's surface. Hence maps are plotted from field data provided by the surveyor. These data are obtained with traditional surveying equipment or by *photogrammetry*.

Distances are measured by several alternative or complementary methods. The traditional measuring instrument is the steel tape, with stakes driven to mark the points of measurements. Distances may also be calculated from photographs via aircraft or satellite if the conditions under which the photographs were made are known. There are also several instrumental methods for measuring distances. The *stadia transit* is an optical instrument used in conjunction with a special *stadia rod*. A sight is made on the rod, and the in-

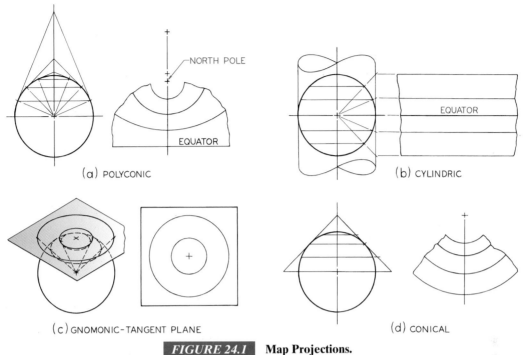

(a) POLYCONIC

(b) CYLINDRIC

(c) GNOMONIC-TANGENT PLANE

(d) CONICAL

FIGURE 24.1   **Map Projections.**

strument reading is converted to distance with a *conversion factor.*

The *level,* also an optical instrument, is equipped with a telescope containing cross hairs. This instrument is commonly used to determine differences in elevation in the field, a process called *differential leveling.* When this instrument is leveled, the line of sight of the telescope becomes horizontal. A *level rod,* graduated into feet or meters, is held vertically on selected points. Instrument readings of the rod then serve to determine the difference in elevations of the points.

Photogrammetry is now widely used for map surveying. In this method actual photographs of the earth's surface, usually aerial, are used. Originally, *aerial photogrammetry* was used for mapping enemy territory during wartime. Now this method is used for government and commercial surveying, exploration, property valuations, real estate tax records, and so on. It is easy to use in rough or inaccessible terrain, and large areas can be mapped from a relatively few clear photographs. By combining photogrammetry pictures through the use of special optical devices, it is possible to determine not only the relative position of objects in a horizontal plane but also their relative elevations. The method is also used in connection with ground surveying by including in the photographs *control points* already located on the ground by precise surveying.

Contours are lines drawn on a map to show points of equal elevation; that is, all points on a single contour line have the same elevation. A *contour interval* is the vertical distance between horizontal planes passing through successive contours. For example, in Fig. 24.2 the contour interval is 10′. The contour interval should not vary on any one map. As an aid to reading, particularly where contours are closely spaced, every fifth contour may be drawn somewhat heavier, as shown in the figure.

When extended far enough (if the map is sufficiently large), every contour line will close. At streams, contours form V's pointing upstream. Evenly spaced contours indicate uniform slope, and closely spaced contours suggest steep slopes.

A *profile*—not to be confused with a view on the profile plane in multiview drawing—is a line contained in a vertical plane that depicts the relative elevations of points along the line. Thus, if a vertical section were to be cut into the earth, the top boundary of this section would represent the ground profile. See the lower or front view of Fig. 24.2.

Locations of points on contour lines are determined by *interpolation.* In Fig. 24.2 the locations and elevations of seven control points are determined, and contour lines are drawn on the assumption that the slope of the surface of the ground is uniform between

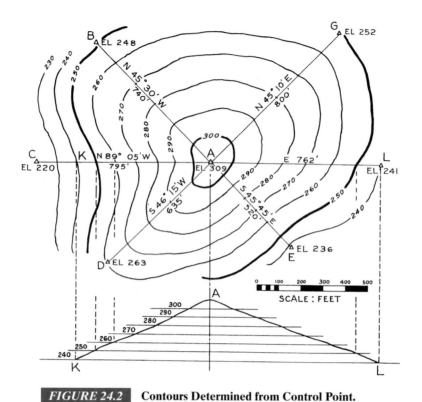

FIGURE 24.2   Contours Determined from Control Point.

station A and the six adjacent stations. A contour interval of 10′ was adopted, and the locations of the points of intersection of the contour lines with the straight lines joining the point A and the six adjacent points were calculated as follows.

The horizontal distance between stations A and B is 740′. The difference in elevation of these stations is 61′. The difference in elevation of station A and contour 300 is 9′. Therefore, assuming a uniform slope between A and B, contour 300 crosses line AB at a distance from station A of $\frac{9}{61}$ of 740, or 109.1′. Contour 290 crosses line AB at a distance from contour 300 of $\frac{10}{61}$ of 740, or 121.3′. This distance of 121.3′ between contour lines is uniform along line AB and can be set off without further calculation.

In the same way, points where the contours cross the other lines of the survey are interpolated, after which the several contour lines are drawn (freehand) through points of equal elevation, as shown.

After contours have been plotted, it is easy to construct a profile of the ground along a line in any direc-

tion. In Fig. 24.2, the profile of line KAL is shown in the front view. In civil engineering it is customary to draw the profile to a larger vertical scale than that of the map or plan to emphasize the varying slopes. However, in mining engineering and geology this practice complicates angular and certain other measurements. See §24.4.

Contours may also be plotted through the use of other patterns of distribution of points of recorded elevation. A popular pattern is the *checkerboard* or *grid survey,* as in Fig. 24.3, in which lines are established at right angles to each other, dividing the survey into squares of appropriate size, here 100′, and elevations are determined at the corners of the squares. The contour interval is selected here as 2′, and the slope of the ground between adjacent stations is assumed to be uniform.

The points where the contour lines cross the survey lines may be located approximately by eye, or more accurately by the graphical methods of Figs. 4.14–4.17, or by the numerical method explained for Fig. 24.2.

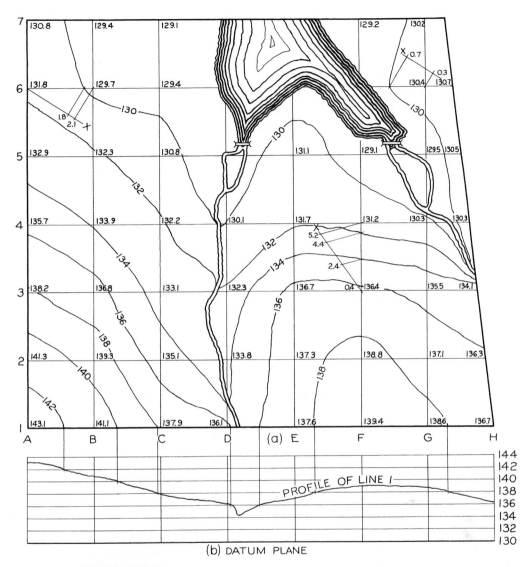

FIGURE 24.3   Contours Determined from Readings at Regular Intervals.

The points of intersection of contour lines with survey lines also may be found by constructing a profile of each line of the survey, as shown for grid line 1 in Fig. 24.3 (b). Horizontal lines are drawn at elevations at which it is desired to show contours. The points in which the profile line intersects these horizontal lines are projected upward to survey line 1 as indicated. While this procedure is more reasonable than assuming uniform slopes between measured elevations, it must be repeated for every grid line of the survey. It is therefore quite tedious and time consuming, and it is the responsibility of the engineer in charge to determine if the increased accuracy is sufficiently significant to justify the extra expense.

■ 24.3 GEOLOGY AND MINING TERMS

An extensive vocabulary of special terminology has been developed in geology and mining. For the limited

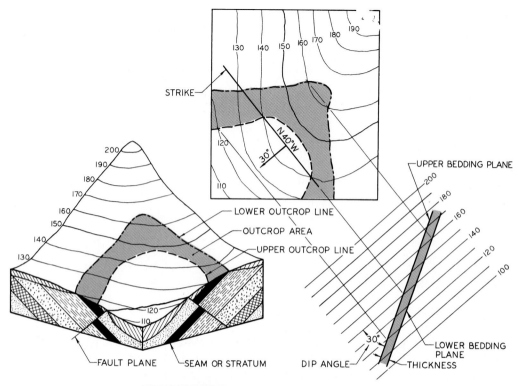

FIGURE 24.4 **Geology and Mining Terms.**

treatment here, however, only the few terms illustrated in Fig. 24.4 are needed.

1. *Strike.* The bearing, §20.5, of a horizontal line in a plane, customarily measured from north. In the illustration, the strike is N40°W.

2. *Dip.* Includes both an *angle* and a *direction*. The angle is the dihedral angle, §10.11, between a given plane and a horizontal plane. The direction is that of maximum downward slope. Because this direction is always perpendicular to the strike of the given plane, and because the strike and dip are usually given together, it is sufficient to give the quadrant in which the direction of dip falls: NE, NW, SE, or SW. In Fig. 24.4 the dip is 30°SW. Note on a map that the dip direction is given by the angle and the arrow only.

3. *Stratum, seam.* A sedimentary layer or deposit bounded by parallel *bedding planes* or bounding planes (within limits).

4. *Vein.* A deposit in a fissure or fault. May be bounded by bedding planes, or may be highly irregular in outline.

5. *Fault.* A discontinuity in a formation involving displacement of one segment with respect to another. If the displacement takes place along an essentially plane surface, that plane is called a *fault plane.*

6. *Thickness.* The *perpendicular* distance between the two bedding planes of a stratum, seam, or vein.

7. *Outcrop.* If a sloping stratum continues without faults, it eventually outcrops (becomes exposed) at the earth's surface. The area is established on a topographic map by plotting the lines of intersection, *outcrop lines,* of the two bedding planes with the surface of the earth. Outcrop lines are similar to contour lines except that they are intersections of sloping planes, as opposed to horizontal planes, with the surface of the earth.

■ 24.4 STRIKE, DIP, AND THICKNESS OF STRATUM

Strata are represented by plotting points on the bedding planes on a map and on an adjacent elevation view. Field data on such points are obtained by surveying and test drillings. Theoretically, only three points not in line are necessary to establish a plane, §19.7.

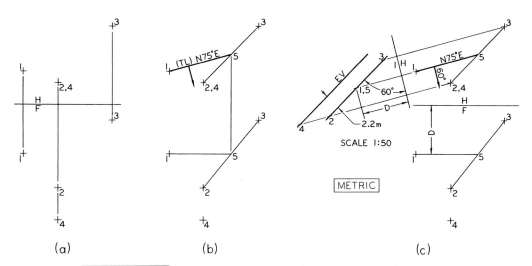

FIGURE 24.5   Strike, Dip, and Thickness of Stratum Established by Points in Bedding Planes.

Actually, because of the likelihood of concealed faults and irregularities, many field measurements are made, and discrepancies are balanced out or otherwise accounted for in the plotting room.

For simplicity, in Fig. 24.5 (a) only three points, 1, 2, and 3, are given in the upper bedding plane of a stratum and one point, point 4, in the lower bedding plane (since the bedding planes are to be assumed parallel). Let it be required to find the strike, dip, and thickness of the stratum.

Note that, because of the angle and thickness measurements, the vertical scale should be the same as the map scale. Compare with Fig. 24.2.

STRIKE   At (b) the top view of the horizontal line 1–5 is found by connecting points 2 and 3 with a line, adding horizontal line 1–5 in the front view, and projecting point 5 to the top view. The bearing of line 1–5 is N 75°E, which is the strike of plane 1–2–3. Note that strike is conventionally given from the north, not from the south.

DIP   In the front view at (b), it is observed that point 2 is below (lower than) line 1–5 in space. The general direction of the dip in the top view is thus from line 1–5 toward point 2, or southeasterly, as indicated by the arrow. The amount of dip—the actual angle with horizontal—is not known as yet, however.

Since line 1–5 is the line of intersection of a horizontal plane with plane 1–2–3, and since the top view of line 1–5 is true length, the top view of 1–5 serves to establish the direction of sight for the auxiliary view at

(c), which shows the dihedral angle between plane 1–2–3 and a horizontal plane, §19.9. This is the dip angle, which is referred to as "a dip of 60°SE"; on the map (top view), at (c), the dip is indicated by the arrow plus the number of degrees in the dip angle.

THICKNESS   Point 4 is next located in the auxiliary view at (c), and the edge view of the lower bedding plane is drawn parallel to that of the upper bedding plane. The perpendicular distance between the parallel edge views, 2.2 m in the example, is the thickness of the stratum.

■ 24.5 OUTCROP

In Fig. 24.6 an enlarged portion of Fig. 24.5 (c) is shown with a contour map superimposed on the top view. The front view is not shown here, the elevations of the various points being indicated by numbers in parentheses on the map. These numbers in turn permit the proper location in the auxiliary view of the parallel lines representing the edge views of the horizontal planes (horizontal in space) of the contours. Points common to the planes of the stratum and the planes of the contours appear at the intersections of the edge views in the auxiliary view. As an example, point A, B in the auxiliary view represents the point view of the line of intersection of the upper bedding plane 1–2–3 and a horizontal plane at the 130 m elevation. In the top view this line intersects the 130 m contour at points A and B, which are thus points on the outcrop

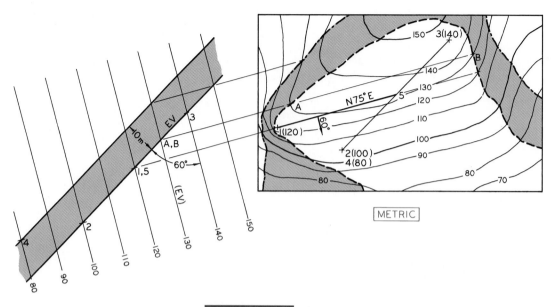

FIGURE 24.6   **Outcrop.**

line of the upper bedding plane. This process is continued for all such points, and the upper outcrop line is drawn through the points thus located. Note that the outcrop line must not intersect a contour line *except* at a point located by the foregoing construction process.

The lower outcrop line is located by the same procedures. For clarity, it is advisable to draw the outcrop lines somewhat heavier than other lines of the map. In addition, a dash line or dashdot line coding or the use of colors will help distinguish these lines. Shading the outcrop area, as shown, further emphasizes the outcrop.

■ 24.6 CUT AND FILL AREAS

In laying out a railroad or highway right of way through hilly country, material must be removed from high areas and added to low areas to keep percent grades within reasonable limits. If the material removed from the *cuts* is suitable for use as *fill*, economies can frequently be realized by equating the quantities of the two to minimize haul distances. Also, cuts that are excessively deep, and fills that are unusually high, may be too wide to fit the acquired right of way. The plotting of these areas, before the final design is approved, is therefore very important.

Like outcrop lines, cut and fill lines for relatively loose material (not solid rock) are simply the intersections of plane bounding surfaces with the surface of the earth. Figure 24.7 is an example. A level highway at

100 m elevation is shown; the cut is at the ratio of 1:1 and the fill at 1.5:1. The plotting of the cut and fill lines is the same as the plotting of outcrop lines. Note the curved construction lines concentric with the center line of the highway. Also note the points where the cut and fill lines intersect the edges of the highway. Since the cut and fill lines are lines of intersection of planes at opposing slopes, they are not continuous with each other but would intersect, or cross, if extended into the highway area. However, since the highway bed is level (a horizontal plane), there is little to be gained by showing cut or fill areas in the highway area itself. A profile along the center of the highway, §24.2, would give a better picture of the relation between cut and fill in that area.

■ 24.7 SPHERICAL GEOMETRY, SPHERICAL TRIANGLE

Attention here will be confined to what might be more accurately called *spherical trigonometry:* the study and solving of *spherical triangles.*

A plane containing the center of a sphere cuts from the surface of the sphere a circle having the same radius or diameter as the sphere. Such a maximum-sized circle on a spherical surface is called a *great circle.* Any three intersecting great circles that do not intersect at the same point on the surface of the sphere form a series of spherical triangles, Fig. 24.8. Let us confine our attention to the small spherical triangle ABC.

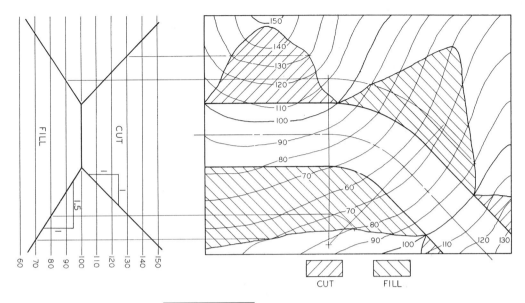

FIGURE 24.7   **Cut and Fill Areas.**

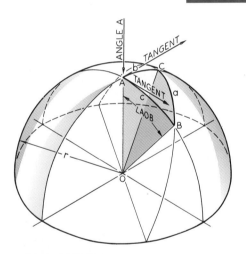

FIGURE 24.8   **Spherical Triangle ABC.**

Conventional notation is similar to that used in plane trigonometry, with the vertices (and interior angles) of a spherical triangle being denoted by capital letters, and the sides opposite each vertex with the same letter in lowercase form, as shown. The sides of the spherical triangle could be measured by the corresponding lengths of the great-circle arcs. As an example, the length of arc AB (side c) in the figure is

$$\frac{\angle AOB}{360°} \times 2\pi r$$

where r is the radius of the sphere (and all great circles). For a given sphere the quantity $2\pi r/360°$ is a

constant, so that it is conventional to give the sides of a spherical triangle in degrees, converting to a linear distance only in special cases. The interior angle of the triangle at a vertex such as A is the angle between the sides b and c where they intersect at A, which is the angle between the two tangents to the curves at A. The true size of this angle is seen in a view showing radius OA in point view. This, in turn, is a view showing the size of the dihedral angle, §19.9, between surfaces AOB and AOC of the *spherical pyramid* OABC.

Thus, just as in plane trigonometry, there are six quantities to be measured on a spherical triangle: the three sides a, b, and c and the three angles A, B, and C. Given any three of these, it is possible to solve for the other three. Only one example will be given here.

GIVEN THREE SIDES OF A SPHERICAL TRIANGLE, TO FIND THE THREE ANGLES *Fig. 24.9* Let it be given that side a = 45°, side b = 30°, and side c = 40°. The precise size of the sphere is immaterial. It needs to be large enough for clarity, yet compatible with the working area available. To keep the construction as simple as practicable, it is advisable to assume one vertex at the top of the front view (or at the front of the top view). At (a) vertex A is placed at the top of the front view. Again for simplicity, vertex C is placed in the same frontal plane as A, so that side b = 30° is laid out along the circular arc in the front view, as shown.

Although various arrangements and/or combinations of auxiliary views and revolved views may be employed for the remainder of the layout, *revolution,*

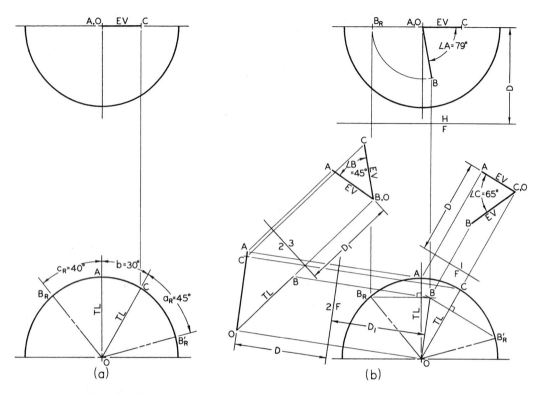

FIGURE 24.9   Solution of Spherical Triangle—Three Sides Given.

§20.4, yields a fairly simple construction in this example. Vertex B is assumed to be revolved about truelength axis OA until B is in the same frontal plane as A. Setting off side c = 40° in the front view then locates B_R, as shown at the left in the front view. Similarly, point B is assumed to be revolved about truelength axis OC until B is in the same frontal plane as C. Setting off side a = 45° in the front view then locates another revolved position of point B, denoted by B′_R at the right of the front view.

Point B is then counterrevolved about both axes OA and OC, as shown at (b). The paths (perpendicular to the axes) intersect at the front view of vertex B. The top view of vertex B is located by drawing the circular top view of the path of revolution of vertex B about axis OA and by projecting the front view of vertex B to the top view. This completes the layout of the spherical triangle ABC. It is not necessary to plot the elliptical arcs of sides a and c in the front view and side a in the top view, since these are not needed for the following solution of the triangle.

Radius OA appears in point view in the top view. Hence, as indicated at (b), angle A is measured in the top view. Radius OC appears in true length in the front view. Auxiliary view 1, showing radius OC in point view, then yields angle C. Finally, radius OB does not appear in true length in either given view. Auxiliary views 2 and 3 are constructed to show radius OB in true length and then in point view. Angle B appears in view 3, as indicated, completing the solution of the triangle ABC.

■ 24.8 SPHERICAL TRIANGLE IN NAVIGATION

Although the Earth is not a perfect sphere, being slightly flattened at the poles, the difference is imperceptible at the scale of ordinary graphical representations.

The coordinate system of *latitude* and *longitude* has been in use since ancient times for locating points on the Earth. It is natural to use the great circle midway between the North and South poles, the *equator*, as a reference. The equator is assigned the latitude of 0°. The *parallels* of latitude then proceed north and south from the equator, so that the poles are at 90° north and south. The equator is divided into 360°, each degree into 60 minutes, and each minute into 60 seconds. The great circles through these division points and the poles then become the *meridians* of longitude. Remember, parallels are *east–west lines,* and meridians are *north–south lines.* Most countries of the world use the meridian through Greenwich, England, as the zero reference of longitude—the *prime meridian.* Longitude is reckoned east and west of the prime meridian up to 180°.

In §24.7 it was mentioned that the length of an arc of a great circle of a sphere of designated size is the number of degrees in the arc multiplied by a constant. For the Earth this constant is approximately 60 nautical miles. [The international nautical mile = 1852 m (6076.1′). Because of the flattening at the poles, the length of 1 degree of latitude varies from approximately 59.5 nautical miles at the equator to about 60.4 nautical miles nearer the poles. The average figure of 60 nautical miles is accurate enough for our problem solutions. A degree of longitude varies in length from about 60 nautical miles at the equator to zero at the poles.

Any two points on the Earth's surface, plus either the North or South pole, can be considered the three vertices of a spherical triangle, if the Earth is treated as a sphere. The solution of the spherical triangle is basic to navigation over appreciable distances.

Let it be assumed that two points are given by means of the following information: point A, latitude 45°N, longitude 50°W; and point B, latitude 30°N, longitude 10°E. A fundamental navigation problem would be as follows: Assume a trip is made from point A to point B along the greatcircle course.

1. What distance is traveled?
2. What is the bearing of the course at the beginning of the trip?
3. What is the bearing at the end of the trip?

First, a layout is made, Fig. 24.10. The horizontal center line in the front view is designated as the equator, which places the North Pole N at the top of the front view as shown. To simplify the layout, the horizontal center line in the top view is assigned the longitude of one of the given points, in this illustration point

B, longitude 10°E. The dihedral angle at N then has the value 10° + 50° = 60°, which locates in the top view the meridian through point A. The front view of point B is located 30° north of the equator, as shown. Point A is considered revolved about axis ON to the frontal plane through point N, and its revolved A_R is located at 45°N. Point A is then counterrevolved to longitude 50°W to complete the layout of the given problem. The great circle from point A to point B would appear elliptical in the front and top views. It is not plotted, since this is not required in the solution.

1. The distance from point A to point B is measured by side n (opposite vertex N) of the spherical triangle NAB. The true size of side n (angle AOB) is seen in a true-size view of plane AOB. This could be obtained by successive auxiliary views as in §19.10; however, it is simpler in this case to revolve point A about true-length axis OB in the front view, since the path of revolution must be perpendicular to line OB and terminate at the circular outline of the front view. Angle $A_R'OB$ is measured as 48°, and the great-circle distance from point A to point B is therefore 48° × 60 = 2880 nautical miles.

2. The initial bearing (from north) is measured from the meridian through point A (side b of the spherical triangle) to the great-circle course arc AB (side n of the spherical triangle). This angle is the dihedral angle at A. Auxiliary views 1 and 2 are used to find this angle, and it is recorded as azimuth bearing N 87°, §19.5.

3. Since at the end of the journey the direction of travel will be through point B *away* from point A, the final bearing is measured as the *supplement* of the dihedral angle at vertex B of the spherical triangle NAB. Auxiliary view 3 is constructed to show this angle, and the final bearing is measured from the meridian NB to the extension of side AB, and is recorded as N 125°.

Of interest is the difference in the initial and final bearings. The course is initially in a northeasterly direction and finally southeasterly. Thus a great-circle course between two points of different latitudes and longitudes involves constantly changing bearings. This suggests navigation by computer. If the navigation is controlled by a human, a great-circle course is approximated by calculating a series of courses (*rhumb lines*) that have constant bearings toward points on or near the great-circle course.

G R A P H I C S   S P O T L I G H T
CAD Solutions for Industrial Piping

PROCESS DESIGN SPECIALISTS

The Harris Group is an international consulting company, whose main focus is power plant design, process design for the petroleum industry, research and development for power plants using special coal drying techniques and a wide variety of other projects. To design a large industrial plant, first the plant is broken down into systems such as heating systems, lube oil systems, and chemical reaction systems. Design teams composed of chemical process engineers, mechanical engineers, and process engineers work together to design the total plant. Kip Funk is a process design engineer with the Harris Group. He designs complex piping systems for industrial plants using CAD.

Kip says that, to start, he creates piping and instrumentation drawings (P&IDs) using Rebus's Autoplant software running inside AutoCAD. The P&ID drawings are usually a single line drawing that show all valves and equipment, like turbines, boilers, pneumatic and electrical controls, gauges, levels, and anything else that is needed in the system.

3D USED TO CHECK CLEARANCES

The completed P&ID drawings are sent to the engineers within the Harris Group to size all of the compo-nents such as pumps, pressure vessels, and pipe diameters. When the engineers have sized the equipment, the pipe designers go to work. They use Propipe software to create 3D drawings of all the equipment and piping. Using a 3D representation allows them to check and make sure that they have the necessary clearances between runs of pipe. They also check to see that all of the equipment fits into the space provided. Previously, using hand drafting methods, it was very difficult to visualize whether the piping interfered. By creating 3D drawings of the design like shown in Figure A, they are able to ensure that the piping will fit when the equipment is installed.

PIPE STRESS ANALYSIS

Once the 3D drawings are completed, Kip uses the 3D point locations of the pumps, tanks, vessels, and pipe to model the pipe stress analysis using Algor's pipe stress software. He makes sure that the stresses from thermal expansion, the weight of the pipe, and other factors do not damage the equipment when it is in place. Figures B and C show pipe stress analysis. Stress concentrations near the pumps can be particularly critical. According to Kip balancing the stresses in the system is kind of like a game. "You try to spread out the stresses so that you

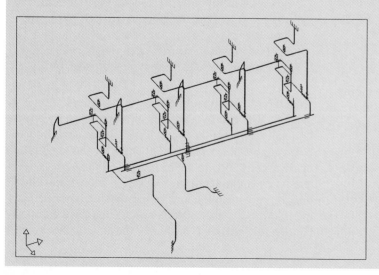

(A)

don't have critical concentrations at any one point, but as you are moving equipment and load, it has an affect in other places. You must reanalyze the system and check to see that by moving one thing you haven't created a new concentration of stress somewhere else in the system."

COMMUNICATION IS IMPORTANT TO SUCCESS

Kip also points out that they usually work under short time frames so that while the P&IDs are being designed, the engineers may be sizing equipment at the same time. Communication is important because the changes to the design that one engineer makes to improve one part of the system may affect the stress analysis or some other portion of the design. All parts of the group must work together to design an effective system in a short time frame. The use of CAD piping software helps them meet this goal.

FINAL APPROVAL

Once the stresses are checked, the design is sent back to the engineers to double check the design. They make sure the pressure drop in the system is not too large, and verify the sizes of the other equipment. Finally, the design is approved for production. Although revisions to the drawings are noted throughout the design process, after the drawings are released for production, further revisions are specially documented on the drawings.

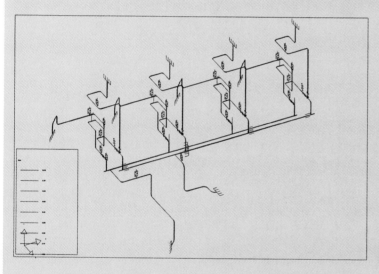

(B)

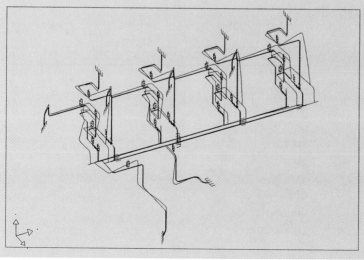

(C)

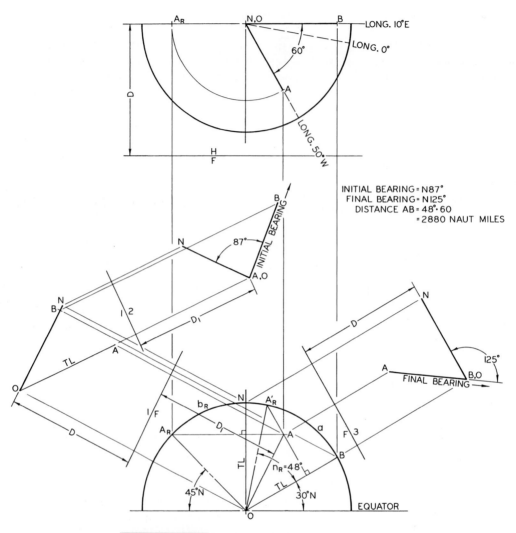

INITIAL BEARING = N87°
FINAL BEARING = N125°
DISTANCE AB = 48°·60
= 2880 NAUT MILES

FIGURE 24.10 Spherical Triangle in Navigation.

■ KEY WORDS

CARTOGRAPHY	INTERPOLATION
AERIAL PHOTOGRAMMETRY	GRID SURVEY
PROFILE	GEOLOGY
INTERVAL	STRIKE
OUTCROP	STRATUM
DIP	VEIN
SPHERICAL TRIANGLES	CUT AND FILL
LATITUDE	BEDDING PLANE
TOPOGRAPHY	LONGITUDE
CONTOUR	

■ CHAPTER SUMMARY

- Cartography is the drawing of surface maps that describe surface contour over a defined area. Studies of elevation change, surface profile along a straight line, and the outcrop of geological veins are common applications of cartography.

- Because the Earth's surface approximates a sphere, flat maps will distort the curvature of the actual surface. Several projection techniques are used to approximate the curved surface on a flat projection.

- Maps use contour lines to show changes in surface elevation. Each contour line represents points on the surface that have the same elevation. The vertical distance between the elevation of two adjacent contour lines is the contour interval.

- A profile is a graph of the changes in elevation along a straight line drawn on a map. Civil engineers use profile and contour maps to smooth out the elevation of a road using cut and fill techniques based on descriptive geometry theory.

- Geologists use strike, dip and thickness of stratum to predict the location of mineral veins. Locations for drilling and outcrop plots can be accurately defined using descriptive geometry.

- Computers can aid in the drawing of map images, but the underlying principles of descriptive geometry are the same whether the map is drawn by computer or by hand.

■ REVIEW QUESTIONS

1. How do flat maps show changes in elevation?
2. Is a profile graph an elevation drawing or a plan drawing?
3. What is meant by interpolation?
4. Are contour lines horizontal or vertical cutting planes through the earth?
5. When solving for outcrop, are you more interested in finding edge view of a vein or true size of the surface boundary of a vein.

6. What is the difference between strike and dip?
7. Describe the procedure for finding cut and fill.
8. Which type of map projection is most accurate at the equator?

■ CARTOGRAPHY, GEOLOGY, AND SPHERICAL GEOMETRY PROBLEMS

Following are problems involving mining and geology, contour maps, and spherical triangles. Use Layouts A–1 or A4–1 (adjusted), A–2 or A4–2 (adjusted), or B–3 or A3–3 as specified.

Since many of the problems in this chapter are of a general nature, they can also be solved on most computer graphics systems. If a system is available, the instructor may choose to assign specific problems to be completed by this method.

Additional problems, in convenient form for solution, are available in *Engineering Graphics Problems*, Series 1, by Spencer, Hill, Loving, Dygdon, and Novak, designed to accompany this text and published by Prentice-Hall Publishing Company.

Prob. 24.1. Assuming the slope of the ground to be uniform and assuming a horizontal scale of $1'' = 200'$ and a contour interval of $5'$, plot by interpolation the contours of Fig. 24.2 (Layout B–3 or A3–3).

Prob. 24.2. Draw, to assigned vertical scales, profiles of any three of the six lines shown in Prob. 24.1.

Prob. 24.3. Using the elevations shown in Fig. 24.3 (a) and a contour interval of 1 m, plot the contours to any convenient horizontal and vertical scales, and draw profiles of lines

3 and 5 and of any two lines perpendicular to them. Check graphically the points in which the contours cross these lines (Layout B–3 or A3–3).

Prob. 24.4. Using a contour interval of 1 m and a horizontal scale of 1/1000, plot the contours from the elevations given, Fig. 24.11, at 10 m stations; check graphically the points in which the contours cross one of the horizontal grid lines and one of the vertical grid lines, using a vertical scale of 1/100 (Layout B–3 or A3–3).

Prob. 24.5. Fig. 24.12 (a). Find the strike and dip of plane 1 (12, 38, 75)–2 (50, 38, 88)–3 (50, 25, 68).

Prob. 24.6. Fig. 24.12 (b). Find the strike and dip of plane 1 (50, 12, 63)–2 (71, 46, 94)–3 (88, 25, 63).

Prob. 24.7. Fig. 24.12 (c). Plane 1 (50, 38, 100)–2 (88, –, 106)–3 (88, –, 75) has a strike of N 75°W and a dip of 30°SW. Find the front views of points 2 and 3.

Prob. 24.8. Fig. 24.12 (d). Points 1 (25, 12, 63) and 2 (75, 50, 100) lie in a plane that has a dip of 60°NW. Find the strike of this plane. *Hint:* See §24.5.

Prob. 24.9. Fig. 24.12 (e). Points 1 (25, 38, 75), 2 (63, 12, 117), and 3 (100, 25, 75) are in the upper bedding plane of a stratum. Point 4 (25, 12, 75) is in the parallel lower bedding plane of the stratum. Find the strike, dip, and thickness of the stratum. Scale: 1/200.

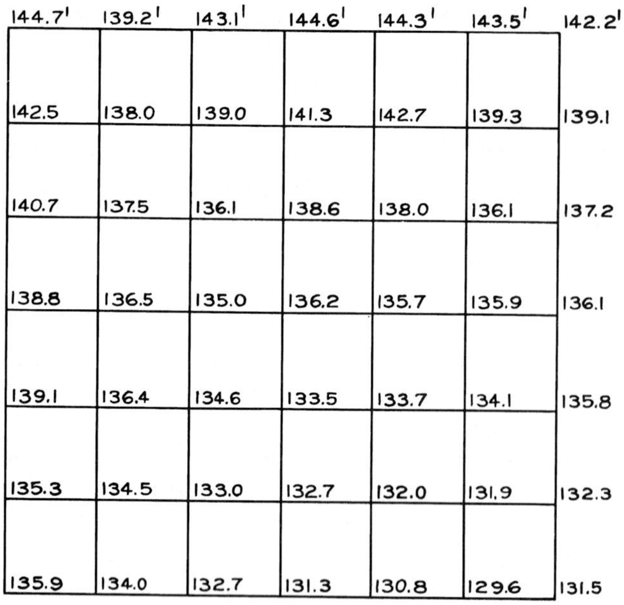

144.7'	139.2'	143.1'	144.6'	144.3'	143.5'	142.2'
142.5	138.0	139.0	141.3	142.7	139.3	139.1
140.7	137.5	136.1	138.6	138.0	136.1	137.2
138.8	136.5	135.0	136.2	135.7	135.9	136.1
139.1	136.4	134.6	133.5	133.7	134.1	135.8
135.3	134.5	133.0	132.7	132.0	131.9	132.3
135.9	134.0	132.7	131.3	130.8	129.6	131.5

FIGURE 24.11   **To Draw Contours (Prob. 24.4).**

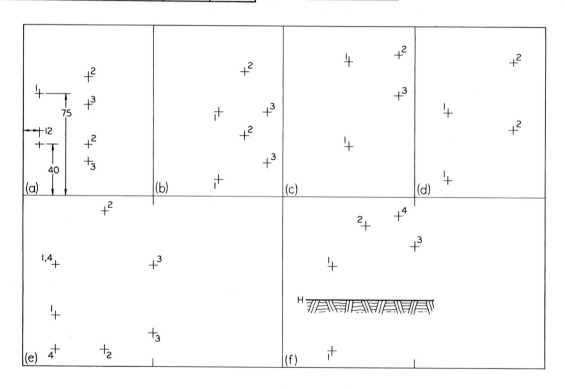

METRIC

FIGURE 24.12   **Probs. 24.5–24.10. Use Layout A–1 or A4–1 (adjusted). Divide the working area into four equal areas for problems as assigned by the instructor. Some problems require a single problem area and others require two problems areas (one half sheet). Data for the layout for each problem are given by a coordinate system. For example, in Fig. 24.12 (a), point 1 is located by the full-scale coordinates (12 mm, 40 mm, 75 mm). The first coordinate locates the front view from the left edge of the problem area. The second and third coordinates locate the front and top views of the point from the bottom edge of the problem area.**

Prob. 24.10. Fig. 24.12 (f). Point 1 (38, 12, 75) lies in the upper bedding plane of a stratum. The surface of the Earth locally is represented by the horizontal plane through H (–, 50, –). Points 2 (63, –, 106) and 3 (100, –, 91) are on the upper outcrop line of the stratum. Point 4 (88, –,114) is on the lower outcrop line. Find the strike, dip, and thickness of the stratum. Scale: 1/400.

Prob. 24.11. Fig. 24.13. Draw the profile along grid D. Use a vertical scale of 1/200.

Prob. 24.12. Fig. 24.13. Point P (at the intersection of grids 7 and C) is at an elevation of 80 m and is in a thin vein that has a strike on N 68°W and dip of 14°SW. Plot the outcrop of this vein.

Prob. 24.13. Fig. 24.13. Points R, S, and T are in the upper outcrop line of a stratum. Point U is in the lower outcrop line. Find the strike, dip, and thickness of the stratum and plot the outcrop area. Vertical scale (same as horizontal scale): 1/500.

Prob. 24.14. Fig. 24.13. Line VWXY is the center line of a level 12 m wide roadway that is at an elevation of 70 m. Plot the cut and fill lines. Shade the cut and fill areas.

Prob. 24.15. Fig. 24.14. Given that side a = 35°, side b = 40°, and side c = 60°, solve for angles A, B, and C of spherical triangle ABC.

Prob. 24.16. Fig. 24.14. Given that side b = 45°, side c = 50°, and angle A = 105°, solve for side a and angles B and C of the spherical triangle.

Prob. 24.17. Fig. 24.14. Given that side b = 40°, angle A = 75°, and angle C = 75°, solve for sides a and c and angle B.

Prob. 24.18. Fig. 24.14. Given that side b = 50°, side c = 30°, and angle C = 60°, solve for side a and angles A and B.

Prob. 24.19. Fig. 24.14. Point A is at latitude 15°N and longitude 175°W. Point B is at latitude 50°N and longitude 15°W. Determine the great-circle distance between the points. Plot the course in the given views.

Prob. 24.20. Fig. 24.14. Point A is at latitude 60°N and longitude 90°W. Point B is at latitude 30°N and longitude 30°E. Determine the great-circle distance between the points and the initial and final bearings of the great-circle course from A to B.

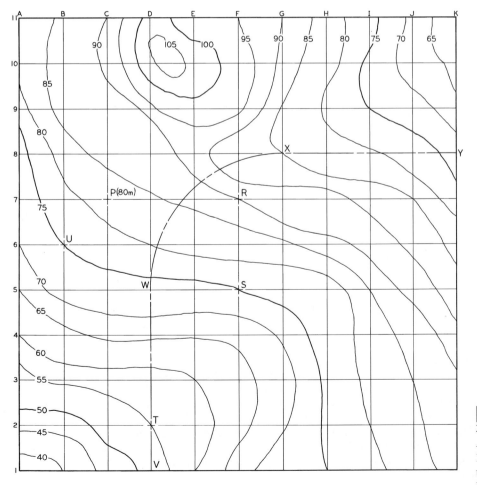

METRIC

FIGURE 24.13 Probs. 24.11–24.14. Use Layout A–2 or A4–2 (adjusted). The map scale is 1/500. Place the map toward the upper left of Layout A–2. Reproduce the grids and contour lines by tracing.

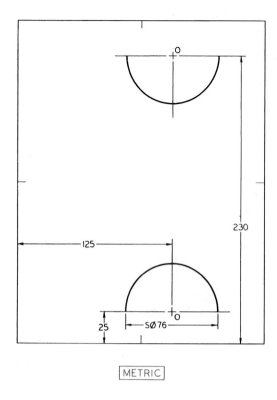

METRIC

FIGURE 24.14   Probs. 24.15–24.20. Use Layout A–1 or A4–1 (adjusted). Point O (125, 25, 230) is the center of the sphere.

C H A P T E R   2 5

GRAPHICAL
VECTOR ANALYSIS

OBJECTIVES

After studying the material in this chapter, you should be able to:

1. Notate vector diagrams using standard Bow's notation.
2. Solve concurrent and nonconcurrent coplanar vector problems.
3. Use funicular polygons for vector analysis.
4. Use vector analysis to solve for reactions and stresses for a truss.
5. Use point view of unknown force to resolve concurrent noncoplanar vectors with one unknown force.
6. Use edge view of a plane to resolve concurrent noncoplanar vectors with two unknown forces.
7. Use graphical vectors to solve velocity problems.

OVERVIEW

Vector analysis, an important branch of mathematics, is the solution of problems involving directed magnitudes such as forces, velocities, accelerations, and moments. Analytic applications of vector analysis are in wide use in advanced engineering analysis and design. Graphical solutions of vector problems are sufficiently accurate for, and are commonly used by, civil engineers, structural engineers, and architects in the form of *graphic statics*, §25.1. Structural members are selected or designed from handbooks, and extreme accuracy is impractical if not impossible in predicting loads that members can withstand. For this reason, a structural designer will apply a *factor of safety*, so that the structure will be capable of withstanding stresses much greater than any likely to occur. Extreme accuracy in predicting stresses is therefore not necessary, and the accuracy required is well within the capabilities of graphic methods. Most vector problems can be solved quickly by graphic methods, and errors are usually apparent. Thus graphic statics is in wide use for structural design.

Since most engineering data are graphical in origin, being read from scales, meters, and the like, graphic methods are suitable for solution of problems involving such data. The results, with reasonable care, are likely to be within the accuracy range of the original data. This chapter presents some of the more common graphical solutions of problems involving vectors.

Many static force problems can be solved using graphical vector analysis. Vectors are graphical representations of magnitude and direction. Vector solution techniques can be applied to both planar problems as well as three-dimensional problems in space. The magnitude component of a vector can be a distance (e.g., miles), a force (e.g., Newtons), a rate (mph), or any scalar that represents the amount applied in the direction of the vector. Three basic types of vector representations are common: the space diagram, the vector diagram, and the vector polygon. Concurrent vectors pass their action through a single point; nonconcurrent vectors apply their action through different points in space. Vector analysis is the process of graphically summing the action of multiple vectors in space. Taking one step forward, and then one step backward is a simple representation of two vectors whose graphical sum is zero. By combining common descriptive geometry constructions like point view and edge view, vector analysis can be expanded into three di-

mensional spatial analysis, summing multiple, noncoplanar vectors in space to find the resultant sum. Many CAD programs can combine graphical vector techniques with numerical precision to provide quick and accurate vector analysis solutions. While the CAD program can perform the calculations quickly, the user still needs to understand the principles behind the construction techniques in order to ensure the correctness of the result.

■ 25.1 DEFINITIONS AND NOTATIONS

1. *Vector.* A directed magnitude represented graphically by a straight-line segment that has a definite length proportional to the magnitude and an arrowhead pointing in the direction of action or *sense* of the vector. Vectors are commonly used to represent forces, velocities, accelerations, and moments.
2. *Scalar.* A magnitude without a specified action. The magnitude or length of a vector is a scalar quantity.
3. *Statics.* A branch of mechanics dealing with forces in equilibrium.
4. *Equilibrium.* A state of rest or of uniform motion.
5. *Graphic statics.* Solutions of problems in statics by graphic methods.
6. *Force.* An action causing or tending to produce motion.
7. *Resultant.* A single vector that will produce the same effect (externally, in the case of forces) as a system of vectors.
8. *Equilibrant.* A vector that is equal to a resultant but opposite in sense.
9. *Component.* One member of a system of vectors.
10. *Resolution of vectors.* Act of resolving a single vector into two or more components.
11. *Concurrent and nonconcurrent vectors.* Vectors whose lines of action pass and do not pass, respectively, through a single point.
12. *Coplanar and noncoplanar vectors.* Vectors that lie and do not lie, respectively, in the same plane.
13. *Moment.* The product of the magnitude of a vector and its perpendicular distance from a point called the *center of moments*. The perpendicular distance is called the *lever arm*. Moments tending to produce clockwise rotation are considered positive.
14. *Couple.* Two equal and opposite forces whose lines of action are parallel. The external effect on the object to which the couple is applied is a ten-

dency to produce rotation only. The *arm* of a couple is the perpendicular distance between the two forces. The *magnitude* of a couple is the magnitude of either force multiplied by the arm.

15. *Vector scale.* The scale to which the magnitudes of the vectors are drawn, as contrasted with the scale at which physical objects are represented in a drawing.

16. *Space diagram.* A graphical layout or sketch of a problem showing the spatial arrangement of physical objects or members of a structure, together with the locations, magnitudes, and senses of vectors.

17. *Vector diagram* or *stress diagram.* A diagram showing the magnitudes to scale and directions of the vectors specified by the space diagram.

18. *Reactions.* Force components representing the forces resisting externally applied loads. For static conditions, the vector sum of the reactions must equal and oppose the vector sum of the applied loads.

19. *Bow's notation.* A system for identifying members and vectors in space and vector diagrams, Fig. 25.1. In the space diagram at (a) capital letters are placed in the spaces between structural members (if any) and lines of action of vectors (usually forces for Bow's notation). Note the temporarily

assumed resultant **R** for completion for the notation. The vectors are identified by pairs of letters: AB, BC, CD, and DA. The letters are usually read in the clockwise direction, as indicated by the arrow. In the vector diagram at (b) corresponding lowercase letters are placed at the ends of the vectors. Many variations, including numerals, are used in practice, but the fundamental ideas are the same. The advantages of Bow's notation will become apparent in later examples. Bow's notation is most commonly used in structural analysis involving multiple loads and members.

20. *Vector notation.* When Bow's notation is not used, vectors are commonly identified in printed matter in boldface type, for example, **A**, **B**, **V**. For handwritten or handlettered material an identifying symbol, such as $\vec{A}$, may be used.

21. *Newton* (*N*). The unit of force, kg m/sec². A mass of 1 kilogram (kg) exerts a gravitational force of 9.8 N (theoretically 9.80665 N) at mean sea level.

■ 25.2 CONCURRENT COPLANAR VECTORS

RESULTANT OF CONCURRENT COPLANAR VECTORS   This process is called *vector addition*. In Fig. 25.2 two forces acting at point A are added vectorially. The space diagram at (a) shows the directions and specifies the magnitudes of the vectors. The vector diagram at (b) shows the application of the *parallelogram of vectors* to obtain the resultant **R**.

Note that either the lower half or the upper half of the parallelogram at Fig. 25.2 (b) could be omitted

| FIGURE 25.1 | Bow's Notation. |

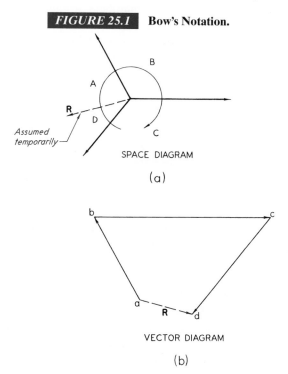

SPACE DIAGRAM

(a)

VECTOR DIAGRAM

(b)

| FIGURE 25.2 | Parallelogram of Vectors. |

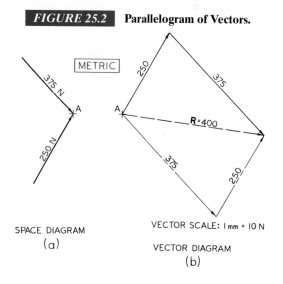

SPACE DIAGRAM

(a)

VECTOR SCALE: 1 mm = 10 N

VECTOR DIAGRAM

(b)

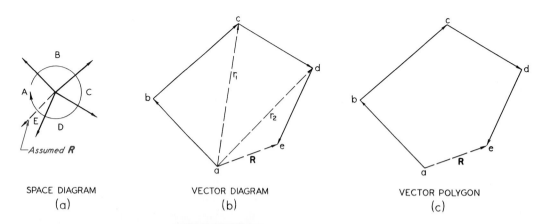

SPACE DIAGRAM
(a)

VECTOR DIAGRAM
(b)

VECTOR POLYGON
(c)

FIGURE 25.3   **Polygon of Forces.**

without altering the resultant. This simplification is of value when several vectors must be combined, as in Fig. 25.3.

In Fig. 25.3 (b) the vectors are combined two at a time to form *subresultants:* vectors ab and bc combine to form subresultant r_1, r_1 and cd to form r_2, and finally r_2 and de to form **R**, the overall resultant.

As seen at (c), the subresultants r_1 and r_2 need not be shown. This simplification results in a *vector polygon,* a great convenience when many vectors must be shown, as in a stress diagram of a truss or beam with several loads and reactions. (Note the use of Bow's notation in Fig. 25.3.)

RESOLUTION OF VECTOR INTO COMPONENTS
A very important process in vector analysis, either analytical or graphical, is the decomposition of a vector into two or more components acting in selected directions. A common practice is to resolve a vector into components along Cartesian axes as shown in Fig.

25.4. At (b) conventional x- and y-axes are introduced with their origin at A. Here, vector **V** is drawn to scale; hence (b) is a vector diagram. At (c) lines drawn through the arrowhead end of **V** parallel to the x- and y-axes resolve **V** into components V_y and V_x, respectively. In analytical vector analysis, addition of multiple vectors is usually accomplished through the separate additions of the x- and y-components.

Essentially, resolution into two components is the reversal of the parallelogram process of Fig. 25.2. A practical application is shown in Fig. 25.5, which is a simplified drawing of a crane. Members BC and BD (Bow's notation) are rigid; members AB and AD are cables.

The force polygon at (b) is started with the vertical force vector ac since the 350 kg load is the only quantity known. This is multiplied by 9.8 to obtain the 3430 N force of the force polygon. Lines parallel to AB and BC are drawn through points a and c, respectively, intersecting to locate point b. Then lines are

FIGURE 25.4   **Resolution of Vector into Components Along Axes.**

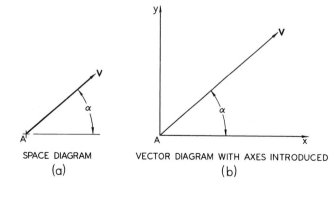

SPACE DIAGRAM
(a)

VECTOR DIAGRAM WITH AXES INTRODUCED
(b)

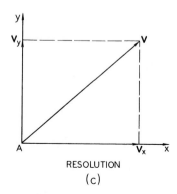

RESOLUTION
(c)

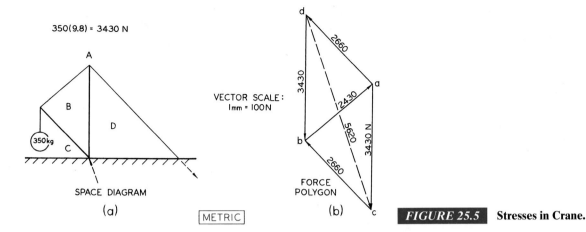

350(9.8) = 3430 N

VECTOR SCALE:
1mm = 100N

SPACE DIAGRAM

(a)

METRIC

FORCE POLYGON

(b)

FIGURE 25.5   **Stresses in Crane.**

drawn through points a and b, respectively parallel to AD and BD, intersecting at point d. The components ab, bd, ad, and bc are measured at the same scale as for component ac to determine the magnitude of the stresses in these members.

The reactions at the ground attachment points are readily determined from the vector diagram. The reaction at the foot of cable AD is simply equal to component ad but opposite in sense. The reaction at the foot of mast BD is equal and opposite to the dashed diago-

nal component cd, since it falls between members C and D of the space diagram at (a).

Two concurrent vectors are shown in multiview form in Fig. 25.6. Since the front and top views of a line segment are its projections on frontal and horizontal planes, respectively, the views shown at (a) are actually frontal and horizontal components of the space vectors **A** and **B**. The sum of two vectors is the vector sum of the components of the vectors. Hence at (b) vectors A_F and B_F are added vectorially to locate

FIGURE 25.6   **Concurrent Vectors in Multiview Projection.**

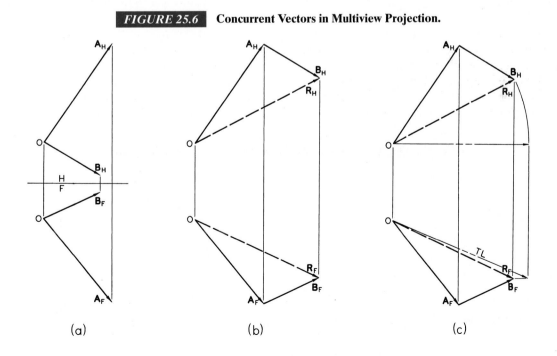

(a)   (b)   (c)

vector $\mathbf{R}_F$. Likewise, vectors $\mathbf{A}_H$ and $\mathbf{B}_H$ are combined to establish vector $\mathbf{R}_H$. That $\mathbf{R}_H$ and $\mathbf{R}_F$ are views of the same vector is substantiated by the vertical projection line connecting the right-hand ends. The true length of resultant $\mathbf{R}$ can now be determined by revolution, as shown at (c).

Two concurrent vectors are necessarily coplanar, §19.7. However, any number of concurrent vectors can be combined two at a time in the preceding manner to obtain a single resultant, even though the complete system is not coplanar. This procedure will be shown later, §25.4.

■ 25.3 NONCONCURRENT COPLANAR VECTORS—FUNICULAR POLYGON

A system of coplanar vectors that do not intersect at a common point is shown in Fig. 25.7 (a). (The student may wish to extend them to check this.) The magnitude and sense of the resultant $\mathbf{R}$ are obtained as usual, by means of the vector polygon shown at (b). Since the given vector system does not have a common point of action, the location of the resultant is yet to be determined, (c) and (d).

At (c) a *pole* o has been added to the vector polygon and connected to the ends of each vector with *strings* or *rays* oa, ob, oc, od, and oe. Each given vector may now be replaced by components along the strings. For example, vector ab is replaced by strings ao and ob, and vector bc by bo and oc. Observe that rays ob of the first triangle and bo of the second triangle are of opposite sense and thus cancel. The same observation may be made of rays oc and od. The final result, then, is that the entire set of original vectors has been replaced by the two strings ao and oe, which add vectorially to give the same resultant $\mathbf{R}$.

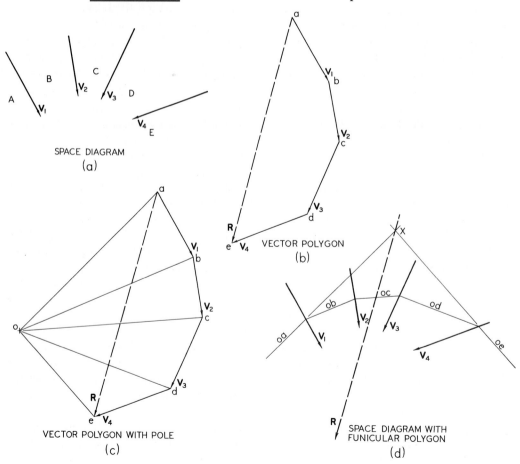

FIGURE 25.7 **Resultant of Nonconcurrent Coplanar Vectors.**

SPACE DIAGRAM
(a)

VECTOR POLYGON
(b)

VECTOR POLYGON WITH POLE
(c)

SPACE DIAGRAM WITH FUNICULAR POLYGON
(d)

At (d) the space diagram is repeated from (a), and the *funicular polygon** is constructed as follows. Any convenient point on vector V_1 is selected, and through it are drawn lines parallel to strings oa and ob. Through the intersection of string ob and vector V_2, a line is drawn parallel to string oc to intersect vector V_3. This process is continued until the last string oe is located. The vectors V_1 to V_4 have now been replaced by components (strings) proportional to the strings shown at (c). The intermediate components ob, oe, and od are canceled, leaving components oa and oe to represent the entire system. These intersect at point X, which is thus a point on the line of action of their resultant R. This resultant R, as has been demonstrated, is the same as the resultant of the original system. The magnitude and sense may therefore be transferred from (b) or (c) to complete the funicular polygon.

The thoughtful student may ponder the situation in which the resultant of a system of forces turns out to be zero. In this case the force polygon is said to *close*, as illustrated in Fig. 25.8 (a).

At (c) is shown the space diagram with a funicular polygon corresponding to (b). Note that the funicular polygon closes at point X on force CB (extended). This system is said to be in *equilibrium*, §25.1.

By contrast, at (d) the funicular polygon closes at point X, which is not on the line of action of force CB. Although the *resultant* of this system is known to be zero, the system is *not* in equilibrium. Any object subjected to this force system would tend to rotate clockwise—it would be subjected to a *couple*, §25.1. The couple can be described numerically as the product of the magnitude of vector CD and distance d. (The couple may also be represented by a vector perpendicular to the plane of the couple, but this concept is of no concern in the discussion to follow. The student may wish to investigate this idea in other texts.)

In summary, for any coplanar system of vectors to be in equilibrium, it is necessary that both the vector polygon and any funicular polygon close. This principle is employed to find the reactions at the supports of the *truss* in the following example.

Assume a truss is given with the loading[†] shown at Fig. 25.9 (a), with reactions R_1 and R_2 to be determined. Since one load, force AB, is not vertical, it has a

horizontal component as well as a vertical component that must be resisted for equilibrium. It is assumed that the connection at R_1 is *roller supported*, so that reaction R_1 must be vertical. This leaves the horizontal component to be resisted at reaction R_2, which is then assumed to be at a *pin-connected joint*[†]. Since the direction of reaction R_2 is not now known, it is conventionally represented by a wavy line, as shown.

The force polygon at (b) is now begun. Forces ab, bc, and cd are laid out from conditions given at (a). The direction of reaction ae is known to be vertical, but point e cannot be located at this stage, since the direction of force de is not known. However, force DE in the space diagram at (a) acts through the pinned joint, so the funicular polygon is started at the pinned joint at (a). Line od is drawn through the pinned joint at (a) and parallel to string od of the force polygon at (b). The procedure is then similar to Fig. 25.7 (d). Lines oc, ob, and oa are added in turn, finally locating point X at the intersection of reaction R_1 and component oa.

As was shown before, for a system to be in equilibrium the funicular polygon must close. Therefore, the direction of line oe is from point X to the pinned joint. This same direction, transferred to (b), locates point e in the force polygon. Reaction R_1 is string ea, and reaction R_2 is determined by drawing string de.

With all external forces established for the truss, the internal stresses in the members can now be determined. First, Bow's notation is extended to the inner spaces of the truss. Numbers are frequently used, as shown at (c). Each joint could be considered in turn as isolated—a so-called *free body*. A force polygon could then be drawn for each joint in a fashion similar to Fig. 25.5. This procedure is called *joint by joint analysis*. However, it saves time and space, and also builds in accuracy checks, to combine all the free-body diagrams into one *stress diagram*, or *Maxwell diagram*.

At (d) the first step is to reproduce the force polygon from (b), but without the pole and strings. The stress diagram may be started at any one of several points. A good starting point is e, since stress e-1 is known to be horizontal. A horizontal line is then drawn from point e to the left, and a second line is drawn from point a parallel to member A-1. These two lines intersect at point 1. Triangle a-e-1 is the force polygon for the roller joint, and stresses (forces)

---

* Funicle means a small chord or string.

[†] The loads would, of course, have to include the weights of the truss members. It is assumed here that these weights have been distributed among the loads shown.

[†] If both joints are pin-connected, the system becomes *indeterminant*, and the solution requires more sophisticated methods than will be presented here.

c
a
b
FORCE POLYGON
(a)
c
o
a
b
FORCE POLYGON
WITH STRINGS
(b)
C
B
A
oa
oc
ob
X
SPACE DIAGRAM AND
FUNICULAR POLYGON,
CONCURRENT FORCES
(c)
B
C
A
oa
oc
ob
X
d
SPACE DIAGRAM AND
FUNICULAR POLYGON,
NON-CONCURRENT FORCES
(d)

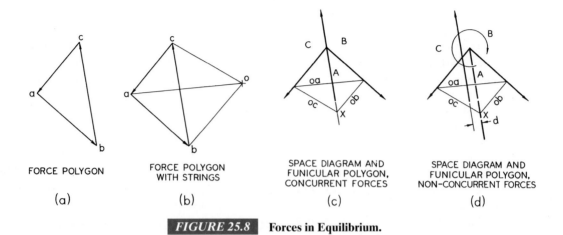

FIGURE 25.8 **Forces in Equilibrium.**

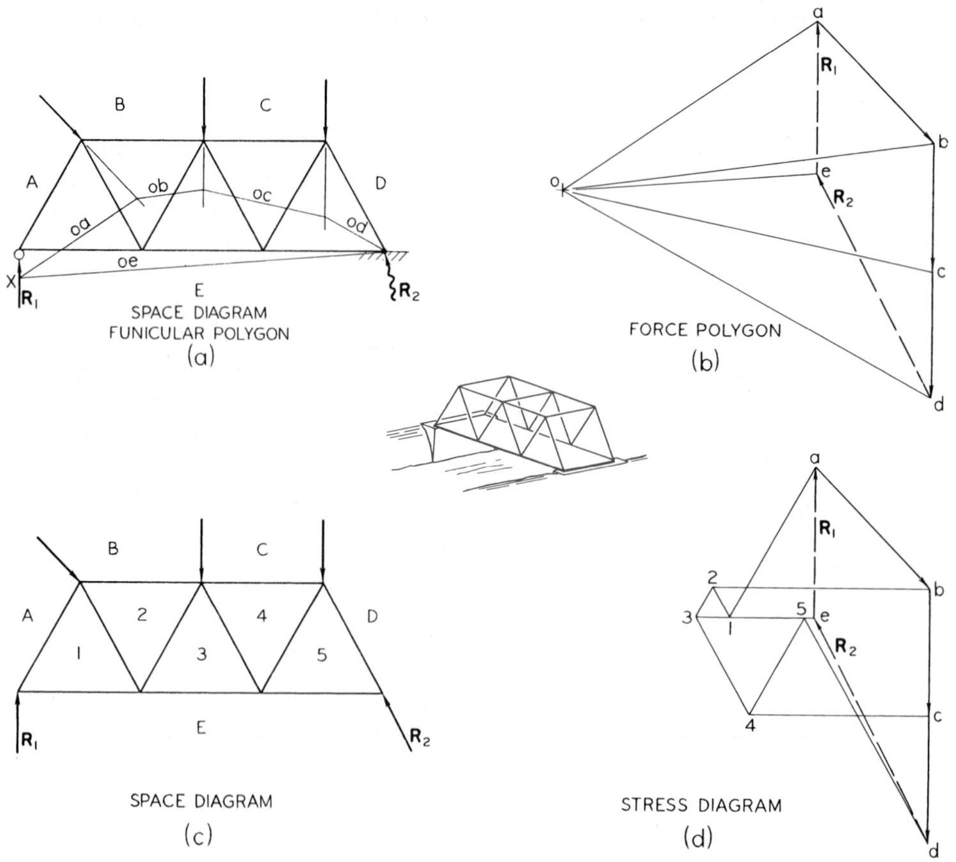

B
C
A
D
ob
oa
oc
od
oe
X
R_1
E
R_2
SPACE DIAGRAM
FUNICULAR POLYGON
(a)
a
R_1
b
o
e
R_2
c
FORCE POLYGON
(b)
d
B
C
A
2
4
D
1
3
5
E
R_1
R_2
SPACE DIAGRAM
(c)
a
R_1
2
b
3
1
5
e
R_2
4
c
STRESS DIAGRAM
(d)
d

FIGURE 25.9 **Reactions and Stresses for a Truss.**

a–1 and e–1 may now be measured at the same scale as for the force polygon.

Through point 1 a line is now added parallel to member 1–2, and through point b a horizontal line (parallel to B–2) is drawn. These intersect at point 2 of the stress diagram, establishing the force polygon for the upper left joint of the truss.

This procedure is continued from joint to joint until all are covered. Frequently, in trusses of different designs, a point on a stress vector at a joint can be located only by proceeding first to the joint at the other end of the same member and drawing a line back toward the first stress vector.

Every line of the stress diagram will be parallel to either an external force or a truss member, and, of course, all portions of the stress diagram must close, attesting to the overall accuracy.

■ 25.4 CONCURRENT NONCOPLANAR VECTORS

As pointed out in §25.2, multiview representation of vectors in space actually consists of components of the vectors—their projections on the coordinate planes. In Fig. 25.10 (a) is shown a multiview space diagram of four concurrent vectors, **A**, B, C, and **D**. (For simplicity the vectors are assumed drawn to scale.) At (b) the

horizontal and frontal components are added separately by vector polygons in the horizontal and frontal projection planes. Continual checks of vertical alignment are made to assure accuracy. The vector sums, **R**$_H$ and **R**$_F$, are the horizontal and frontal components of the resultant **R**. Its magnitude (true length) and angle with one of the coordinate planes could now be found by one of the methods of §§19.3 and 19.4.

■ 25.5 RESOLUTION INTO CONCURRENT NONCOPLANAR VECTORS

Frequently in practice it is necessary to resolve a known load or force into components along noncoplanar directions—that is, to find the stresses in the supporting members of a structure. This is a problem involving forces in equilibrium.

In any one multiview projection of a concurrent force system, *two* unknown magnitudes can be determined, as in §25.3, by closure of the force (vector) polygon. Three or more unknown magnitudes in a single view constitute an indeterminate system that cannot be solved directly by graphical means. However, if two adjacent views of a system are shown, it is possible to determine three unknown magnitudes by several methods, two of which follow.

FIGURE 25.10 **Addition of Concurrent Noncoplanar Vectors.**

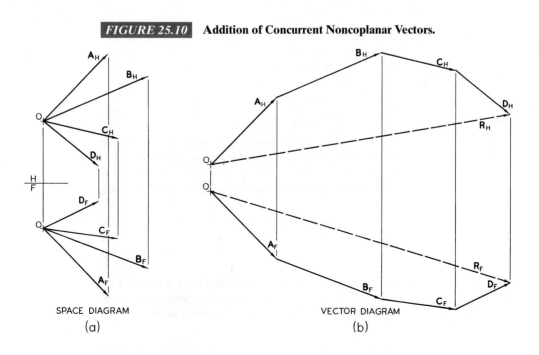

SPACE DIAGRAM
(a)

VECTOR DIAGRAM
(b)

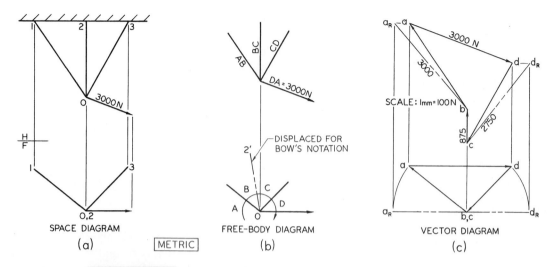

SPACE DIAGRAM
(a)

METRIC

FREE-BODY DIAGRAM
(b)

VECTOR DIAGRAM
(c)

FIGURE 25.11 Resolution into Components by Point View of Unknown Force.

POINT VIEW OF UNKNOWN FORCE If one view shows (or a new view is constructed to show) one of the unknown vectors as a point, that vector has no effect in that view. The known load or force can then be resolved into components along the other members.

In Fig. 25.11 (a), a tripod structure is shown acted upon by a 3000 N force. Member O–2 appears in the front view as a point. Thus there are actually only two unknown frontal components, and the problem can be solved by beginning the construction of the vector diagram with its front view.

At (b) a *free-body diagram* of joint O is shown. *Bow's notation is applied in only one view,* in this case the front view. Note the arbitrary bending or offset of member O–2 to position O–2′ for applying Bow's notation. This offset can be in any desired direction.

The vector diagram at (c) is now begun by drawing the top view of member αd to scale and projecting its front view. The front view of member αd is then resolved into the components bα and dc, as shown. In the top view the directions of components bα and dc are known from the space and free-body diagrams. Projection upward from point b, c in the front view locates points b and c in the top view, completing the two views of the vector diagram. The actual stresses in the tripod members are now found by constructing true-length views of members bα, cb, and dc, §20.4.

EDGE VIEW OF PLANE OF TWO UNKNOWN FORCES In Fig. 25.12 (a), a tripod is shown supporting a load of 1000 N. The preceding method could be applied by constructing a point view of one of the

members, §20.6. In this case, however, two auxiliary views would be required, complicating the construction. A line connecting points 1 and 3 is true length in the top view, as indicated. This establishes the direction of sight for auxiliary view 1, showing plane O–1–3 in edge view. With members O–1 and O–3 coinciding in view 1, there are only two unknown force components in view 1: the force in O–2 and the combined force in O–3–1. The auxiliary view also serves as a free-body diagram, and Bow's notation is applied after offsetting force O–3 to some temporary and arbitrary position such as O–3′.

The vector diagram at (b) is now constructed. The vectors are parallel to their space-diagram positions as usual. First, load vector αd is resolved in view 1 into vector components dc and cbα. (The position of point b is not yet determined.) This permits the location of point c in view H, the top view. This component, in turn, must be the equilibrant of components cb and bα in the top view, locating point b in the top view. Projection back to the auxiliary view 1 establishes point b in that view. Thus every member of the vector diagram is now shown in the two adjacent views, permitting the construction of true-length views of the vectors, §19.4. The front view of the vector diagram is not needed.

■ 25.6 RELATIVE MOTION

As was pointed out in the definitions of §25.1, velocities are frequently represented by vectors. Graphical vector

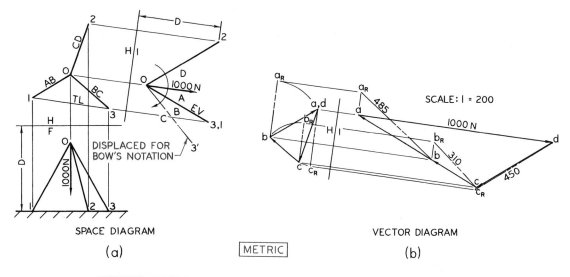

FIGURE 25.12   Resolution by Edge View of Plane of Two Unknown Forces.

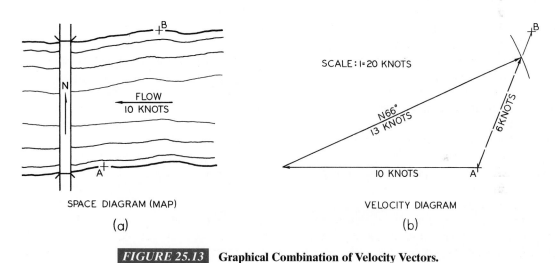

FIGURE 25.13   Graphical Combination of Velocity Vectors.

analysis is readily applicable to many velocity problems.

In Fig. 25.13 (a) let it be given that the river has a current velocity of 10 knots and that it is desired to proceed from station A to station B in a motorboat traveling at 13 knots. The problem is to determine the direction (bearing) in which the motorboat should be headed.

At (b) a suitable scale is selected, and the current-velocity vector (10 knots) is set off. The problem then becomes that of positioning the motorboat velocity vector (13 knots) so that the resultant points from station A toward station B. The required bearing is N 66°, as shown, and the net velocity of the motorboat turns out to be 6 knots.

In the preceding example all velocities were relative to the surface of the earth, and the result was obtained by simple vector arithmetic. In Fig. 25.14 (a) let it be given that two objects A and B are moving on a

GRAPHICS SPOTLIGHT

CAD Welding Symbols and Much More

EXPAND AUTOCAD'S ABILITIES FOR MECHANICAL DESIGN

Design Pacifica's SI Mechanical design and drawing annotation software expands AutoCAD's capabilities to offer a wide range of productivity tools for mechanical engineers, designers, and drafters. It supports both ANSI and ISO standards. You can use either metric or inch units to automatically create parametric 2D and 3D fasteners, hardware, holes, and shafts, among other things. It contains a wide variety of stock hardware and parts, is easy to use, and can create 2D, solid, surface, or wireframe representation for the various parts. Because the parts are parametric, they are easy to edit once they are inserted. In addition, you can write your own routines to access SI Mechanical functions because its program interface is open and documented to allow you ease of access. Its database editor also allows you to customize all aspects of how you want the program to work.

DIALOG BOX DRIVEN WELDING SYMBOL CREATION

Automated generation of welding symbols, either attached to a leader or separate, is just one of the many capabilities of SI Mechanical. The dialog box shown as

Figure A, used to create welding symbols makes it easy to add the symbol you want to your drawing. The following types of weld symbols are available for you to choose from the pull-down list in the upper left of the dialog box:

- Backing weld
- Bevel weld
- Corner flange weld
- Edge flange weld
- Fillet weld
- Flared bevel weld
- Flared V weld
- J groove weld
- Plug weld
- Scarf weld
- Seam weld
- Spot weld
- Square weld
- Stud weld
- Surface weld
- U groove weld
- V groove weld

You can also use the dialog box to input the number of welds to display with the symbol, the contour

(a)

WELDIM - Symbol Editor

Main side

Symbol: BEV   Grv ang: 0
Finish: None   Number: 0
Contour: None   Root: 0.0000
Size (throat): 0.2500   (0.0000)
Length-Pitch: 1.0000 - 3.0000

☑ All around
☐ Field weld
☐ Melt-thru
☐ Backing

Opposite side

Symbol: None   Grv ang: 0
Finish: None   Number: 0
Contour: None   Root: 0.0000
Size (throat): 0.0000   (0.0000)
Length-Pitch: 0.0000 - 0.0000

Side
⦿ Arrow
○ Other
○ Neither

Justify
⦿ Left
○ Right

Main > Opposite
Clear Opposite

Specification:

OK   Cancel   Help...

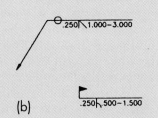

(b)

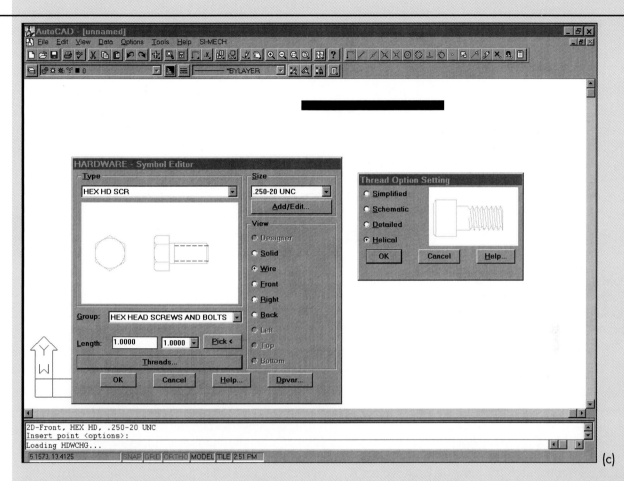

(c)

symbol that will be displayed, the root size, size and throat values, length and pitch, and the symbol for the side of the material where the weld will appear. You can also pick to set the all around designation for the symbol, whether it is a field weld, and whether melt-thru designation and/or backing designation will appear with the symbol. To transfer the settings from the main side to the opposite side to create double sided weld from a single sided weld symbol you just pick. Weld symbols can also be automatically attached to the end of a leader, as shown in Figure B as they usually are in drawings.

One of the other nice features of the SI-Mechanical software is the ease which you can edit the symbols, hardware, and other items that have been placed in your drawing. To edit a symbol, just *pick Symbols, Modify, Symchg* from the SI-Mech pull-down menu that appears on the AutoCAD pull-down menu. The same dialog box you used to create the symbol will

appear again. Make the changes you want inside the dialog box and when you pick OK, your existing symbol will be updated to show the changes.

SI-Mechanical is easy to install and also easy to uninstall. Unlike some packages that add bits and pieces of software throughout your AutoCAD and system program files, SI-Mechanical installs all of its files into a single directory. To remove it all you need to do is delete the directory and change your AutoCAD path so that it no longer points to SI-Mech. Beside the welding symbols, fasteners, hardware and other standard parts, SI-Mechanical also creates surface finish marks, title blocks and borders, balloon tags, geometric dimensioning and tolerancing symbols, revision blocks, breaks, and many other useful items, with same easy dialog approach. Hs Hardware Symbol Editor is shown in Figure C. SI-Mechanical makes your AutoCAD package into a CAD package for mechanical designers, engineers, and detail drafters.

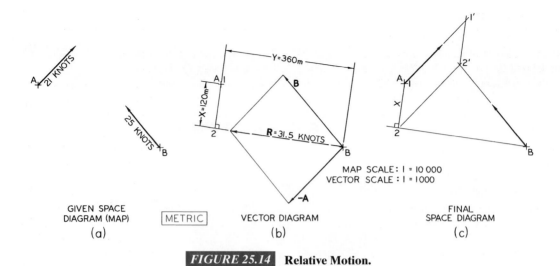

GIVEN SPACE
DIAGRAM (MAP)
(a)

METRIC

VECTOR DIAGRAM
(b)

FINAL
SPACE DIAGRAM
(c)

FIGURE 25.14 **Relative Motion.**

horizontal plane in the directions and with the velocities indicated. Will the objects collide? If not, how close will they pass? What will be the elapsed time to the point of the closest approach?

The given velocities are with respect to the plane of motion. The required answer must be in terms of the motion of one object with respect to the other. If, for example, the plane of motion is imagined to be moving in the opposite sense but with the same velocity as object A, object A becomes stationary with respect to the drawing paper; and object B moves in the direction and with the composite velocity **R** shown at (b). Vector **R** is the vector *difference* between vectors **A** and **B**. It is then seen that the closest approach of object B to object A is distance X, as indicated.

Since the scalar magnitude of vector **R** and the distance traveled to closest approach can be measured at

their individual specified scales, the elapsed time can be calculated as follows.

$$\text{Time} = \text{distance} \div \text{velocity}$$

$$= \frac{360 \ (\text{m}) \ (\text{hr}) \ (\text{n. miles}) \ 3600 \ (\text{sec})}{31.5 \ (\text{n. miles}) \ 1852 \ (\text{m}) \ (\text{hr})}$$

$$= 22.2 \ \text{sec}$$

The results can be referred back to the horizontal plane of motion (map), as at (c), by moving distance X (line 1–2) along vector **A** until point 2 falls on the line of action of vector **B** at point 2'. It is then seen that the actual map positions of objects A and B at the moment of closest approach will be at points 1' and 2', respectively. As the objects continue beyond these positions, the distance between them increases.

■ KEY WORDS

VECTOR	SCALAR
STATICS	EQUILIBRIUM
FORCE	RESULTANT
EQUILIBRANT	RESOLUTION
CONCURRENT	NONCONCURRENT
MOMENT	COUPLE
SPACE DIAGRAM	VECTOR DIAGRAM
VECTOR POLYGON	REACTIONS
BOW'S NOTATION	FUNICULAR POLYGON
RELATIVE MOTION	FREE BODY

■ CHAPTER SUMMARY

- Graphical vector analysis is a simple technique for solving spatial relations between multiple forces acting in different directions. Vectors are graphical representations that show the magnitude of an action applied in a specified direction. Vectors can be graphically summed to determine the resultant action of all the vectors.

- Specific notation, like many descriptive geometry techniques, are required to effectively share solutions with others. Vector analysis uses Bow's notation to label vectors and describe their relationship in space.

- Space diagrams show the vectors connected at a common source (tails connected). The resultant is usually not apparent in a space diagram. Vector diagrams show the vectors connected (head to tail) and are resolved sequentially, showing intermediate resultants. Vector polygons show the vectors connected sequentially (head to tail) with a single final resultant.

- One technique for resolving vectors is to break each vector up into two Cartesian axis component vectors, and summing the vectors along each axis to find the resultant vector.

- When resolving oblique vectors in space via multiview projection, the resultant must be revolved to determine its true length magnitude.

- Stress diagrams are a common and accepted method of determining stresses in civil engineering applications.

- Use the point view of an unknown force to resolve concurrent noncoplanar vectors with one unknown force. Use the edge view of a plane to resolve concurrent noncoplanar vectors with two unknown forces.

■ REVIEW QUESTIONS

1. What is the difference between a vector and a scalar? Which part of the vector is the same as a scalar?

2. The resultant vector of a fishing pole with a fish pulling on the line would be a vector with some force from the handle of the fishing pole to the fish. Give another example of a real-life two-vector problem, and describe the resultant vector.

3. A kite on a string has a resultant force of zero (the force of the kite pulling against the string is canceled by the equal and opposite direction force of the person pulling down on the string). Give another real-life example of a two-vector problem whose resultant is zero.

4. What are concurrent vectors?

5. What is the difference between a space diagram and a vector polygon?

6. In Bow's notation, how are structural members or lines of force notated?

7. What is meant by a free body in a stress diagram?

8. Give an example of a relative motion problem that could be solved using vector analysis. What information must be given in order to solve the problem?

■ GRAPHICAL VECTOR ANALYSIS PROBLEMS

In Figs. 25.15–25.20 are problems on concurrent coplanar vectors, nonconcurrent coplanar vectors, and concurrent noncoplanar vectors.

Use Layout A–1 or A4–1 (adjusted), and divide the working area into four equal areas for problems to be assigned by the instructor. Some problems require two problem areas (one-half sheet) or the entire sheet. Data for the layout for each problem are given by a coordinate system. For example, in Fig. 25.15, Prob. 1, point a is located by the full-scale coordinates (100 mm, 25 mm, –). The first coordinate locates the front view of the point from the left edge of the problem area. The second coordinate locates the front view from the bottom edge of the problem area. The third coordinate (not given in this example; see Fig. 25.19, Prob. 1) locates the top view of the point from the bottom edge of the problem area.

Since many of the problems in this chapter are of a general nature, they can also be solved on most computer graphics systems. If a system is available, the instructor may choose to assign specific problems to be completed by this method.

Additional problems, in convenient form for solution, are available in *Engineering Graphics Problems*, Series 1, by Spencer, Hill, Loving, Dygdon, and Novak, designed to accompany this text and published by Prentice-Hall Publishing Company.

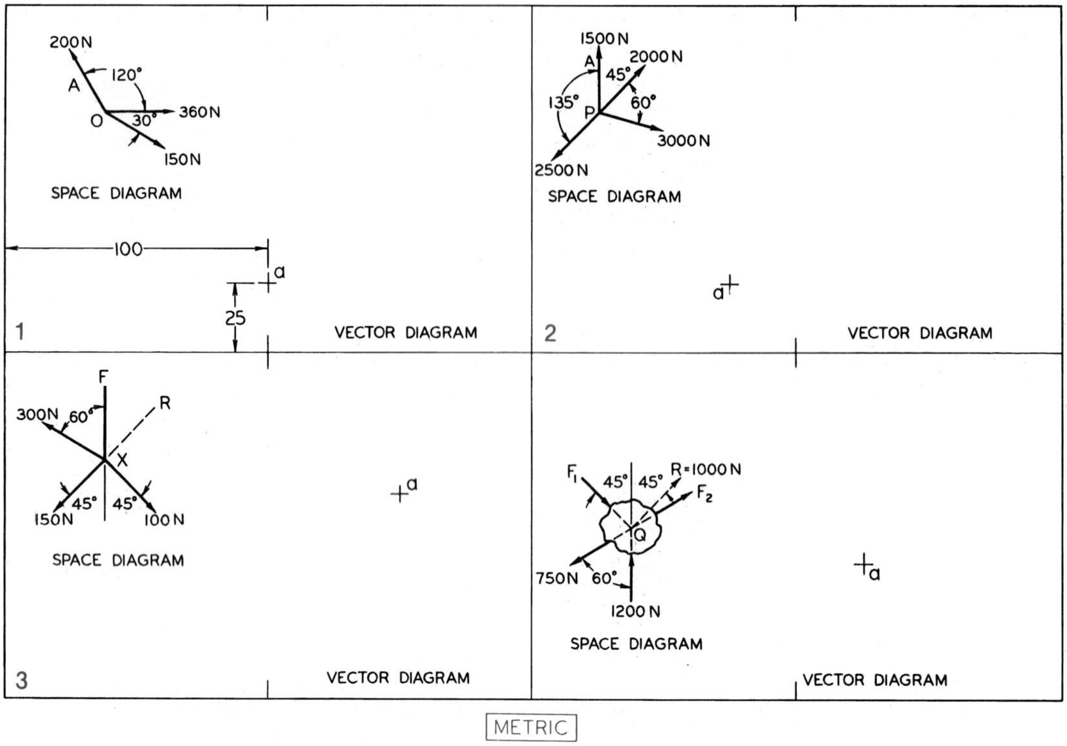

METRIC

FIGURE 25.15 **Concurrent Coplanar Forces. Lay out and solve problems as assigned. Layout A–1 or A4–1 (adjusted) divided into four equal areas.**

1. Find the resultant of the force system. Point O of the space diagram is at $(38, 88, -)$. Complete the Bow's notation beginning with A as shown, and start the vector polygon at a $(100, 25, -)$. Vector scale: 1 mm = 4 N.

2. Find the horizontal and vertical components of the resultant of the force system whose space diagram is shown at P$(25, 88, -)$. Use Bow's notation beginning with A as shown, and start the vector polygon at a $(75, 25, -)$. Vector scale: 1 mm = 40 N.

3. Determine the magnitudes and senses of force F and resultant R of the system shown at X$(38, 88, -)$. Start the force polygon at a$(152, 75, -)$. Vector scale: 1 mm = 4 N. *Hint:* Arrange Bow's notation so that all known forces are included before an unknown force is encountered.

4. Determine the magnitudes of forces F_1 and F_2 for the system shown at Q$(38, 63, -)$. The resultant is R. Begin the force polygon at a$(127, 50, -)$. Vector scale: 1 mm = 20 N. *Hint:* Translate the "pushing" forces to "pulling" forces before applying Bow's notation. Also see the hint in Prob. 3.

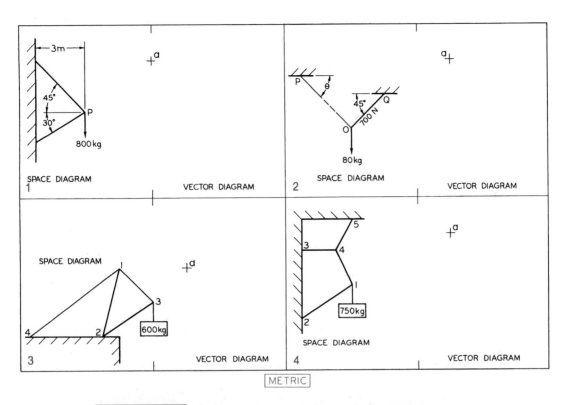

FIGURE 25.16   Concurrent Coplanar Forces. Lay out and solve problems as assigned. Use Layout A–1 or A4–1 (adjusted) divided into four equal areas.

1. Point P of the space diagram is at $(50, 63, -)$. Space scale: 1/50. Starting the force polygon at a $(100, 100, -)$, find the stresses in the two members, supporting the 800 kg load. Vector scale: 1 mm = 100 N.

2. Cables OP and OQ are supporting an 80 kg mass. The stress in and position of cable OQ are as shown. Find the stress in cable OP and angle θ. Vector scale: 1 mm = 10 N. Start the force polygon at a $(125, 100, -)$.

3. The members and coordinates of the space diagram of a loading crane are as follows: mast 1 $(75, 75, -)$–2 $(63, 25, -)$, boom 2–3 $(100, 50, -)$, cable 1–4 $(7, 25, -)$, and cable 1–3. Start the force polygon at a $(127, 75, -)$ and find the stress in each of the four members. Also find the reactions at mounting points 2 and 4. Vector scale: 1 mm = 100 N.

4. The 750 kg load is supported by four struttural members the coordinates of whose end points are as follows in the space diagram: 1 $(50, 63, -)$, 2 $(12, 38, -)$, 3 $(12, 88, -)$, 4 $(38, 88, -)$, and 5 $(50, 111, -)$. Find the stresses in the members, starting the force polygon at a $(127, 110, -)$. Vector scale: 1 mm = 100 N.

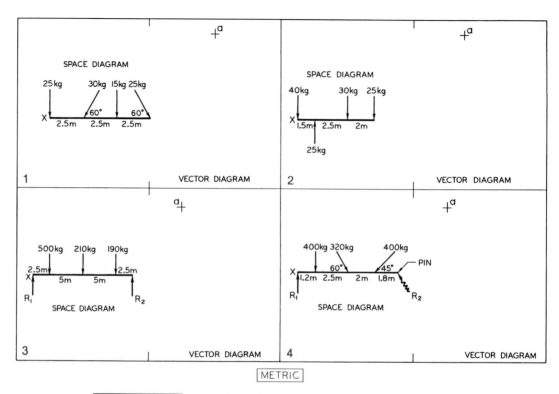

METRIC

FIGURE 25.17 Noncurrent Coplanar Forces. Lay out and solve problems as assigned. Use Layout A-1 or A4-1 (adjusted) divided into four equal areas.

1. Find the magnitude, sense, and line of action of the resultant of the force system positioned at $X(25, 50, -)$. Start the force polygon at $a(152, 114, -)$. Space scale: 1/100. Vector scale: 1 mm = 10 N.

2. Find the magnitude, sense, and line of action of the force that will place the given system located at $X(12, 50, -)$ in equilibrium. Start the force polygon at $a(140, 114, -)$. Space scale: 1/100. Vector scale: 1 mm = 10 N.

3. Find the reactions R_1 and R_2 of the space diagram at $X(12, 63, -)$. Start the force polygon at $a(127, 114, -)$. Space scale: 1/100. Vector scale: 1 mm = 100 N.

4. Find the reactions R_1 and R_2 of the space diagram at $X(12, 66, -)$. Start the force polygon at $a(127, 114, -)$. Space scale: 1/100. Vector scale: 1 mm = 100 N. *Remember:* Start the funicular polygon at the pinned joint.

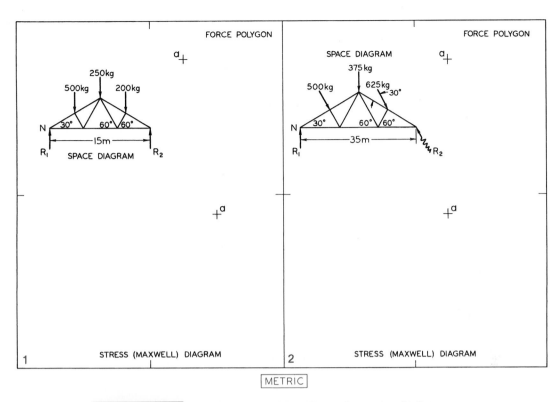

METRIC

FIGURE 25.18 Noncurrent Coplanar Forces. Lay out and solve problems as assigned. Use Layout A–1 or A4–1 (adjusted) divided into four equal areas.

1. Find the reactions R_1 and R_2 of the truss at N(25, 178, –). Start the force polygon at a(127, 228, –). Find the loads in each member of the truss starting the stress diagram at a new a(152, 114, –). Space scale: 1/200. Vector scale: 1 mm = 100 N. *Note:* In this problem it is not necessary to repeat the space diagram as was done in Fig. 26.9 (c). Simply extend Bow's notation in the original space diagram.

2. Find the reactions R_1 and R_2 of the truss at N(12, 178, –). Start the force polygon at a(127, 230, –). Find the loads in each member of the truss starting the stress diagram at a new a(127, 114, –). Space scale: 1–400. Vector scale: 1 mm = 100 N. See *Note* in Prob. 1.

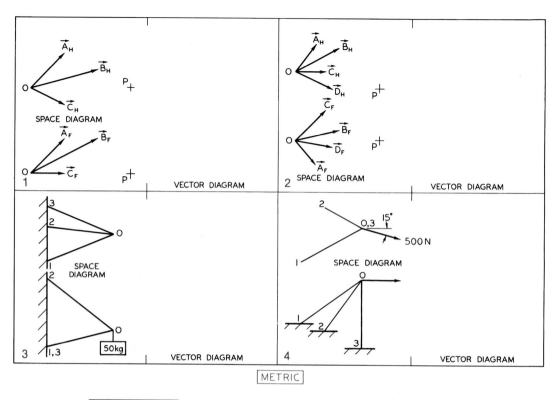

FIGURE 25.19 Concurrent Noncoplanar Vectors. Lay out and solve problems as assigned. Use Layout A–1 or A4–1 (adjusted) divided into four equal areas.

1.   Three concurrent vectors are shown to scale in multiview form. The coordinates are O(12, 12, 75). A(38, 38, 100), B(63, 38, 88), and C(38, 12, 63). Starting the vector diagram at P(88, 12, 75), find the magnitude and slope of the resultant of the vectors. Vector scale: 1 mm = 1 N.

2. The vectors are shown to scale in the space diagram. The coordinates are O(12, 38, 88), A(28, 20, 110), B(45, 45, 106), C(35, 60, 88), and D(40, 33, 75). Start the vector diagram at P(75, 38, 75). Find the magnitudes of the horizontal and vertical components of the resultant. Vector scale: 1 mm = 10 N.

3. The tripod structure shown in the space diagram is located as follows: O(75, 25, 96), 1 (25, 12, 75), 2 (25, 65, 100), and 3 (25, 12, 117). Find the loads in the three members O–1, O–2, and O–3. Vector scale: 1 mm = 10 N. *Hint:* Use the front view in the space diagram as a free-body diagram.

4. The tripod structure withstanding the 500 N horizontal force is located by the following coordinates: O (63, 63, 100), 1 (18, 30, 75), 2 (35, 25, 117), and 3 (63, 12, 100). Find the loads in the tripod legs. Vector scale: 1 mm = 10 N.

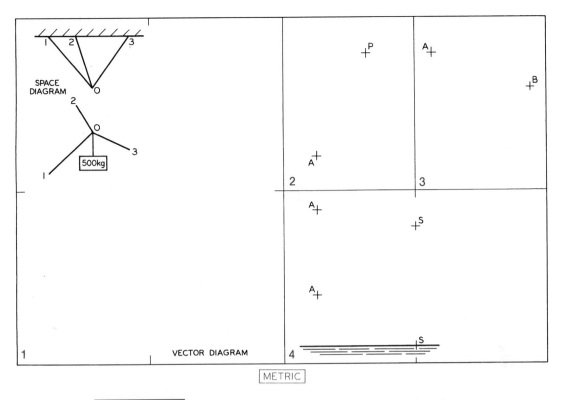

FIGURE 25.20 Concurrent Noncoplanar Vectors and Velocity Vectors.
Lay out and solve problems as assigned. Use Layout A–1 or A4–1
(adjusted) divided into four equal areas.

1. The tripod structure supports a weight of 500 kg and is located by the following coordinates: O (58, 172, 205), 1 (25, 142, 244), 2 (46, 193, 244), and 3 (86, 160, 244). Find the loads in the tripod legs. Vector scale: 1 mm = 100N.

2. An aircraft at A (25, –, 25) is in level flight on a compass course of N 45° and at an indicated air speed of 200 knots. If there is a 100 knot wind blowing from due east, what is the resulting true ground speed and course? Scale: 1 mm = 4 knots. If the aircraft is to pass over point P (63, –, 100), with the original compass course and wind velocity, what must be the indicated air speed of the craft?

3. A ship at A (12, –, 100) is sailing a course of N 135° at 16 knots. Another ship at B (88, –, 75) is on a course of N 215° at 12 knots. How close will the ships pass? Distance scale: 1/2000. Velocity scale: 1 mm = 0.5 knot. How much time will have elapsed at the moment of closest approach? (Show calculations.) Show the actual positions of the ships at this time.

4. An aircraft at A (25, 50, 114) is flying a true course of N 120° at a corrected air speed of 120 knots. A destroyer at S (100, 12, 100) is on a course of N 240° at 20 knots. Find the resulting minimum antiaircraft artillery range in meters. Distance scale: 1/50 000. Velocity scale: 1 mm = 2 knots.

APPENDIX

1 Bibliography of American National Standards

American National Standards Institute, 11 West 42nd St., New York, NY. 10036. For complete listing of standards, see ANSI catalog of American National Standards.

Abbreviations

Abbreviations for Use on Drawings and in Text, ANSI/ASME Y1.1–1989

Bolts, Screws, and Nuts

Bolts, Metric Heavy Hex, ANSI B18.2.3.6M–1979 (R1995)

Bolts, Metric Heavy Hex Structural, ANSI B18.2.3.7M–1979 (R1995)

Bolts, Metric Hex, ANSI B18.2.3.5M–1979 (R1995)

Bolts, Metric Round Head Short Square Neck, ANSI/ASME B18.2.2.1M–1981 (R1995)

Bolts, Metric Round Head Square Neck, ANSI/ASME B18.5.2.2M–1982 (R1993)

Hex Jam Nuts, Metric, ANSI B18.2.4.5M–1979 (R1990)

Hex Nuts, Heavy, Metric, ANSI B18.2.4.6M–1979 (R1990)

Hex Nuts, Slotted, Metric, ANSI/ASME B18.2.4.3M–1979 (R1995)

Hex Nuts, Style 1, Metric, ANSI/ASME B18.2.4.1M–1979 (R1995)

Hex Nuts, Style 2, Metric, ANSI/ASME B18.2.4.2M–1979 (R1995)

Hexagon Socket Flat Countersunk Head Cap Screws (Metric Series), ANSI/ASME B18.3.5M–1986 (R1993)

Mechanical Fasteners, Glossary of Terms, ANSI B18.12–1962 (R1995)

Miniature Screws, ANSI B18.11–1961 (R1992)

Nuts, Metric Hex Flange, ANSI B18.2.4.4M–1982 (R1993)

Plow Bolts, ANSI/ASME B18.9–1958 (R1995)

Round Head Bolts, Metric Round Head Short Square Neck, ANSI/ASME B18.5.2.1M–1981 (R1995)

Screws, Hexagon Socket Button Head Cap, Metric Series, ANSI/ASME B18.3.4M–1986 (R1993)

Screws, Hexagon Socket Head Shoulder, Metric Series, ANSI/ASME B18.3.3M–1986 (R1993)

Screws, Hexagon Socket Set, Metric Series, ANSI/ASME B18.3.6M–1986 (R1993)

Screws, Metric Formed Hex, ANSI/ASME B18.2.3.2M–1979 (R1995)

Screws, Metric Heavy Hex, ANSI/ASME B18.2.3.3M–1979 (R1995)

Screws, Metric Hex Cap, ANSI/ASME B18.2.3.1M–1979 (R1995)

Screws, Metric Hex Flange, ANSI/ASME B18.2.3.4M–1984 (R1995)

Screws, Metric Hex Lag, ANSI B18.2.3.8M–1981 (R1991)

Screws, Metric Machine, ANSI/ASME B18.6.7M–1985 (R1993)

Screws, Socket Head Cap, Metric Series, ANSI/ASME B18.3.1M–1986 (R1993)

Screws, Tapping and Metallic Drive, Inch Series, Thread Forming and Cutting, ANSI B18.6.4–1981 (R1991)

Slotted and Recessed Head Machine Screws and Machine Screw Nuts, ANSI B18.6.3–1972 (R1991)

Slotted Head Cap Screws, Square Head Set Screws, and Slotted Headless Set Screws, ANSI/ASME B18.6.2–1995

Socket Cap, Shoulder, and Set Screws (Inch Series) ANSI/ASME B18.3–1986 (R1995)

Square and Hex Bolts and Screws, Inch Series, ANSI B18.2.1–1981 (R1992)

Square and Hex Nuts (Inch Series) ANSI/ASME B18.2.2–1987 (R1993)

Track Bolts and Nuts, ANSI/ASME B18.10–1982 (R1992)

Wood Screws, Inch Series, ANSI B18.6.1–1981 (R1991)

Dimensioning and Surface Finish

General Tolerances for Metric Dimensioned Products, ANSI B4.3–1978 (R1994)

Preferred Limits and Fits for Cylindrical Parts, ANSI B4.1–1967 (R1994)

Preferred Metric Limits and Fits, ANSI B4.2–1978 (R1994)

Surface Texture, ANSI/ASME B46.1–1995

Drafting Manual (Y14)

Casting and Forgings, ANSI/ASME Y14.BM–1989

Decimal Inch, Drawing Sheet Size and Format, ANSI/ASME Y14.1–1995

Dimensioning and Tolerancing, ANSI/ASME Y14.5M–1994

Electrical and Electronics Diagrams, ANSI Y14.15–1966 (R1988)

Electrical and Electronics Diagrams—Supplement, ANSI Y14.15a–1971 (R1988)

Electrical and Electronics Diagrams—Supplement, ANSI Y14.15b–1973 (R1988)

Engineering Drawings, Types, and Applications, ANSI/ASME Y14.24M–1989. Revision of Engineering Drawings, ANSI/ASME Y14.35M–1992

Gear and Spline Drawing Standards—Part 2, Bevel and Hypoid Gears, ANSI Y14.7.2–1978 (R1994)

Gear Drawing Standards—Part 1, for Spur, Helical, Double Helical, and Rack, ANSI/ASME Y14.7.1–1971 (R1993)

Line Conventions and Lettering, ANSI/ASME Y14.2M–1992

Mechanical Spring Representation, ANSI/ASME Y14.13M–1981 (R1992)

Metric Drawing Sheet Size and Format, ANSI/ASME Y14.1M–1995

Multiview and Sectional View Drawings, ANSI/ASME Y14.3M–1994

Parts Lists, Data Lists, and Index Lists, ANSI/ASME Y14.34M–1990

Pictorial Drawing, ANSI/ASME Y14.4M–1989 (R1994)

Screw Thread Representation, ANSI/ASME Y14.6–1978 (R1993)

Screw Thread Representation, Metric, ANSI/ASME Y14.6aM–1981 (R1993)

Surface Texture Symbols, ANSI/ASME Y14.36M–1996

Gears

Basic Gear Geometry, ANSI/AGMA 115.01–1989

Gear Nomenclature—Terms, Definitions, Symbols, and Abbreviations, ANSI/AGMA 1012–F90

Nomenclature of Gear-Tooth Failure Modes, ANSI/AGMA 110.04–1980 (R1989)

Design Manual for Bevel Gearing, ANSI/AGMA 2005–B88

Tooth Proportions for Fine-Pitch Spur and Helical Gears, ANSI/AGMA 1003–G93

Graphic Symbols

Public Fire Safety Symbols, ANSI/NFPA 170–1994

Graphic Symbols for Electrical and Electronics Diagrams, ANSI/IEEE 315–1975 (R1994)

Graphic Symbols for Electrical Wiring and Layout Diagrams Used in Architecture and Building Construction, ANSI Y32.9–1972 (R1989)

Graphic Symbols for Fluid Power Diagrams, ANSI/ASME Y32.10–1967 (R1994)

Graphic Symbols for Grid and Mapping Used in Cable Television Systems, ANSI/IEEE 623–1976 (R1989)

Graphic Symbols for Heat-Power Apparatus, ANSI Y32.2.6M–1950 (R1993)

Graphic Symbols for Heating, Ventilating and Air Conditioning, ANSI Y32.2.4–1949 (R1993)

Graphic Symbols for Logic Functions, ANSI/IEEE 91–1984

Graphic Symbols for Pipe Fittings, Valves, and Piping, ANSI/ASME Y32.2.3–1949 (R1994)

Graphic Symbols for Plumbing Fixtures for Diagrams Used in Architecture and Building Construction, ANSI/ASME Y32.4–1977 (R1994)

Graphic Symbols for Process Flow Diagrams in the Petroleum and Chemical Industries, ANSI Y32.11–1961 (R1993)

Graphic Symbols for Railroad Maps and Profiles, ANSI/ASME Y32.7–1972 (R1994)

Instrumentation Symbols and Identification, ANSI/ISA S5.1–1984 (R1992)

Reference Designations for Electrical and Electronics Parts and Equipment, ANSI/IEEE 200–1975 (R1989)

Symbols for Mechanical and Acoustical Elements as Used in Schematic Diagrams, ANSI Y32.18–1972 (R1993)

Symbols for Welding, Brazing, and Nondestructive Examination, ANSI/AWS A2.4–93

Keys and Pins

Clevis Pins and Cotter Pins, ANSI/ASME B18.8.1–1994

Hexagon Keys and Bits (Metric Series), ANSI B18.3.2M–1979 (R1994)

Keys and Keyseats, ANSI B17.1–1967 (R1989)

Pins—Taper Pins, Dowel Pins, Straight Pins, Grooved Pins and Spring Pins (Inch Series), ANSI/ASME B18.8.2–1994

Woodruff Keys and Keyseats, ANSI B17.2–1967 (Rl990)

Piping

Cast Bronze Threaded Fittings, Class 125 and 250, ANSI/ASME B16.15–1985 (R1994)

Cast Copper Alloy Pipe Flanges and Flanged Fittings, ANSI/ASME B16.24–1991

Cast Iron Pipe Flanges and Flanged Fittings, Class 25, 125, 250 and 800, ANSI/ASME B16.1–1989

Gray Iron Threaded Fittings, ANSI/ASME B16.4–1992

Ductile Iron Pipe, Centrifugally Cast, ANSI/AWWA C151/A21.51–91

Factory-Made Wrought Steel Buttwelding Fittings, ANSI/ASME B16.9–1993

Ferrous Pipe Plugs, Bushings, and Locknuts with Pipe Threads, ANSI/ASME B16.14–1991

Flanged Ductile-Iron Pipe with Threaded Flanges, ANSI/AWWA C115/A21.15–94

Malleable-Iron Threaded Fittings, ANSI/ASME B16.3–1992

Pipe Flanges and Flanged Fittings, ANSI/ASME B16.5–1988

Stainless Steel Pipe, ANSI/ASME B36.19M–1985 (R1994)

Welded and Seamless Wrought Steel Pipe, ANSI/ASME B36.10M–1995

Rivets

Large Rivets ($\frac{1}{2}$ Inch Nominal Diameter and Larger), ANSI/ASME B18.1.2–1972 (R1995)

Small Solid Rivets ($\frac{7}{16}$ Inch Nominal Diameter and Smaller), ANSI/ASME B18.1.1–1972 (R1995)

Small Solid Rivets, Metric, ANSI/ASME B18.1.3M–1983 (R1995)

Small Tools and Machine Tool Elements

Jig Bushings, ANSI B94.33–1974 (R1994)

Machine Tapers, ANSI/ASME B5.10–1994

Milling Cutters and End Mills, ANSI/ASME B94.19–1985

Reamers, ANSI/ASME B94.2–1995

T-Slots—Their Bolts, Nuts and Tongues, ANSI/ASME B5.1M–1985 (R1992)

Twist Drills, ANSI/ASME B94.11M–1993

Threads

Acme Screw Threads, ANSI/ASME B1.5–1988 (R1994)

Buttress Inch Screw Threads, ANSI B1.9–1973 (R1992)

Class 5 Interference–Fit Thread, ANSI/ASME B1.12–1987 (R1992)

Dryseal Pipe Threads (Inch), ANSI B1.20.3–1976 (R1991)

Hose Coupling Screw Threads, ANSI/ASME B1.20.7–1991

Metric Screw Threads—M Profile, ANSI/ASME B1.13M–1995

Metric Screw Threads—MJ Profile, ANSI/ASME B1.21M–1978

Nomenclature, Definitions and Letter Symbols for Screw Threads, ANSI/ASME B1.7M–1984 (R1992)

Pipe Threads, General Purpose (Inch), ANSI/ASME B1.20.1–1983 (R1992)

Stub Acme Threads, ANSI/ASME B1.8–1988 (R1994)

Unified Screw Threads (UN and UNR Thread Form), ANSI/ASME B1.1–1989

Unified Miniature Screw Threads, ANSI B1.10–1958 (R1988)

Washers

Lock Washers, Inch, ANSI/ASME B18.21.1–1994

Lock Washers, Metric, ANSI/ASME B18.21.2M–1994

Plain Washers, ANSI B18.22.1–1965 (R1981)

Plain Washers, Metric, ANSI B18.22M–1981

Miscellaneous

Knurling, ANSI/ASME B94.6–1984 (R1995)

Preferred Metric Sizes for Flat Metal Products, ANSI/ASME B32.3M–1984 (R1994)

Preferred Metric Equivalents of Inch Sizes for Tubular Metal Products Other Than Pipe, ANSI/ASME B32.6M–1984 (R1994)

Preferred Metric Sizes for Round, Square, Rectangle and Hexagon Metal Products, ANSI B32.4M–1980 (R1994)

Preferred Metric Sizes for Tubular Metal Products Other Than Pipe, ANSI B32.5–1977 (R1994)

Preferred Thickness for Uncoated Thin Flat Metals (Under 0.250 in.), ANSI B32.1–1952 (R1994)

Surface Texture (Surface Roughness, Waviness and Lay), ANSI/ASME B46.1–1995

Technical Drawings, ISO Handbook, 12–1991

2   Technical Terms

"The beginning of wisdom is to call things by their right names."

–CHINESE PROVERB

n *means a* noun; v *means a* verb

acme (*n*)   Screw thread form, §§14.3 and 14.13

addendum (*n*)   Radial distance from pitch circle to top of gear tooth.

allen screw (*n*)   Special set screw or cap screw with hexagon socket in head, §14.31.

allowance (*n*)   Minimum clearance between mating parts, §13.12.

alloy (*n*)   Two or more metals in combination, usually a fine metal with a baser metal.

aluminum (*n*)   A lightweight but relatively strong metal. Often alloyed with copper to increase hardness and strength.

angle iron (*n*)   A structural shape whose section is a right angle, §11.20.

anneal (*v*)   To heat and cool gradually, to reduce brittleness and increase ductility, §11.22.

arc-weld (*v*)   To weld by electric arc. The work is usually the positive terminal.

babbitt (*n*)   A soft alloy for bearings, mostly of tin with small amounts of copper and antimony.

bearing (*n*)   A supporting member for a rotating shaft.

bevel (*n*)   An inclined edge, not at right angle to joining surface.

bolt circle (*n*)   A circular center line on a drawing, containing the centers of holes about a common center, §12.25.

bore (*v*)   To enlarge a hole with a boring mill, Figs. 11.15b and 11.18b.

boss (*n*)   A cylindrical projection on a casting or a forging.

BOSS

brass (*n*)   An alloy of copper and zinc.

braze (*v*)   To join with hard solder of brass or zinc.

Brinell (*n*)   A method of testing hardness of metal.

broach (*n*)   A long cutting tool with a series of teeth that gradually increase in size which is forced through a hole or over a surface to produce a desired shape, §11.15.

bronze (*n*)   An alloy of eight or nine parts of copper and one part of tin.

buff (*v*)   To finish or polish on a buffing wheel composed of fabric with abrasive powders.

burnish (*v*)   To finish or polish by pressure upon a smooth rolling or sliding tool.

burr (*n*)   A jagged edge on metal resulting from punching or cutting.

bushing (*n*)   A replaceable lining or sleeve for a bearing.

calipers (*n*)   Instrument (of several types) for measuring diameters, §11.17.

cam (*n*)   A rotating member for changing circular motion to reciprocating motion.

carburize (*v*)   To heat a low-carbon steel to approximately 2000°F in contact with material which adds carbon to the surface of the steel, and to cool slowly in preparation for heat treatment, §11.22.

caseharden (*v*)   To harden the outer surface of a carburized steel by heating and then quenching.

castellate (*v*)   To form like a castle, as a castellated shaft or nut.

casting (*n*)   A metal object produced by pouring molten metal into a mold, §11.3.

cast iron (*n*)   Iron melted and poured into molds, §11.3.

center drill (*n*)   A special drill to produce bearing holes in the ends of a workpiece to be mounted between centers. Also called a "combined drill and countersink," §12.35.

COMBINED DRILL & C SINK

chamfer (*n*)   A narrow inclined surface along the intersection of two surfaces.

CHAMFER

chase (*v*)   To cut threads with an external cutting tool.

cheek (*n*)   The middle portion of a three-piece flask used in molding, §11.3.

chill (*v*)   To harden the outer surface of cast iron by quick cooling, as in a metal mold.

chip (*v*)   To cut away metal with a cold chisel.

chuck (*n*)   A mechanism for holding a rotating tool or workpiece.

coin (*v*)   To form a part in one stamping operation.

cold-rolled steel (CRS) (*n*)   Open hearth or Bessemer steel containing 0.12–0.20% carbon that has been rolled while cold to produce a smooth, quite accurate stock.

collar (*n*)   A round flange or ring fitted on a shaft to prevent sliding.

COLLAR

colorharden (*v*)   Same as *caseharden*, except that it is done to a shallower depth, usually for appearance only.

cope (*n*)   The upper portion of a flask used in molding, §11.3.

core (*v*)   To form a hollow portion in a casting by using a dry-sand core or a green-sand core in a mold, §11.3.

coreprint (*n*)   A projection on a pattern which forms an opening in the sand to hold the end of a core, §11.3.

cotter pin (*n*)   A split pin used as a fastener, usually to prevent a nut from unscrewing, Figs. 14.31g and 14.31h and Appendix 30.

counterbore (*v*)   To enlarge an end of a hole cylindrically with a *counterbore*, § 11.16.

COUNTERBORE

countersink (*v*)   To enlarge an end of a hole conically, usually with a *countersink*, §11.16.

COUNTERSINK

crown (*n*)   A raised contour, as on the surface of a pulley.

cyanide (*v*)   To surface-harden steel by heating in contact with a cyanide salt, followed by quenching.

dedendum (*n*)   Distance from pitch circle to bottom of tooth space.

development (*n*)   Drawing of the surface of an object unfolded or rolled out on a plane.

diametral pitch (*n*)   Number of gear teeth per inch of pitch diameter.

die (*n*)   (1) Hardened metal piece shaped to cut or form a required shape in a sheet of metal by pressing it against a mating die. (2) Also used for cutting small male threads. In a sense is opposite to a tap.

die casting (*n*)   Process of forcing molten metal under pressure into metal dies or molds, producing a very accurate and smooth casting.

die stamping (*n*)   Process of cutting or forming a piece of sheet metal with a die.

dog (*n*)   A small auxiliary clamp for preventing work from rotating in relation to the face plate of a lathe.

dowel (*n*)   A cylindrical pin, commonly used to prevent sliding between two contacting flat surfaces.

DOWEL

draft (*n*)   The tapered shape of the parts of a pattern to permit it to be easily withdrawn from the sand or, on a forging, to permit it to be easily withdrawn from the dies, §§11.3 and 11.21.

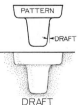

PATTERN

DRAFT

DRAFT

drag (*n*)   Lower portion of a flask used in molding, §11.3.

draw (*v*)   To stretch or otherwise to deform metal. Also to temper steel.

drill (*v*)   To cut a cylindrical hole with a drill. A *blind hole* does not go through the piece, §11.16.

drill press (*n*)   A machine for drilling and other hole-forming operations, §11.8.

drop forge (*v*)   To form a piece while hot between dies in a drop hammer or with great pressure, §11.21.

face (*v*)   To finish a surface at right angles, or nearly so, to the center line of rotation on a lathe.

FAO Finish all over, §12.17.

feather key (*n*)   A flat key, which is partly sunk in a shaft and partly in a hub, permitting the hub to slide lengthwise of the shaft, §14.34.

file (*v*)   To finish or smooth with a file.

fillet (*n*)   An interior rounded intersection between two surfaces, §§7.34 and 11.5.

fin (*n*)   A thin extrusion of metal at the intersection of dies or sand molds.

fit (*n*)   Degree of tightness or looseness between two mating parts, as a *loose fit*, a *snug fit*, or a *tight fit*. §§12.9 and 13.1–13.3.

fixture (*n*)   A special device for holding the work in a machine tool, *but not for guiding the cutting tool.*

flange (*n*)   A relatively thin rim around a piece.

FLANGE

flash (*n*)   Same as *fin.*

flask (*n*)   A box made of two or more parts for holding the sand in sand molding, §11.3.

flute (*n*)   Groove, as on twist drills, reamers, and taps.

forge (*v*)   To force metal while it is hot to take on a desired shape by hammering or pressing, §11.21.

galvanize (*v*)   To cover a surface with a thin layer of molten alloy, composed mainly of zinc, to prevent rusting.

gasket (*n*)   A thin piece of rubber, metal, or some other material, placed between surfaces to make a tight joint.

gate (*n*)   The opening in a sand mold at the bottom of the *sprue* through which the molten metal passes to enter the cavity or mold, §11.3.

graduate (*v*)   To set off accurate divisions on a scale or dial.

grind (*v*)   To remove metal by means of an abrasive wheel, often made of carborundum. Use chiefly where accuracy is required, §11.14.

harden (*v*)   To heat steel above a critical temperature and then quench in water or oil, §11.22.

heat-treat (*v*)   To change the properties of metals by heating and then cooling, §11.22.

interchangeable (*adj.*)   Refers to a part made to limit dimensions so that it will fit any mating part similarly manufactured, §13.1.

jig (*n*)   A device *for guiding a tool* in cutting a piece. Usually it holds the work in position.

journal (*n*)   Portion of a rotating shaft supported by a bearing.

kerf (*n*)   Groove or cut made by a saw.

KERF

key (*n*)   A small piece of metal sunk partly into both shaft and hub to prevent rotation, §14.34.

keyseat (*n*)   A slot or recess in a shaft to hold a key, §14.34.

KEYSEAT

keyway (*n*)   A slot in a hub or portion surrounding a shaft to receive a key. §14.34.

KEYWAY

knurl (*v*)   To impress a pattern of dents in a turned surface with a knurling tool to produce a better hand grip, §12.37.

lap (*v*)   To produce a very accurate finish by sliding contact with a *lap*, or piece of wood, leather, or soft metal impregnated with abrasive powder.

lathe (*n*)   A machine used to shape metal or other materials by rotating against a tool, §11.7.

lug (*n*)   An irregular projection of metal, but not round as in the case of a *boss*, usually with a hole in it for a bolt or screw.

malleable casting (*n*)   A casting that has been made less brittle and tougher by annealing.

mill (*v*)   To remove material by means of a rotating cutter on a milling machine, §11.9.

mold (*n*)   The mass of sand or other material that forms the cavity into which molten metal is poured, §11.3.

MS (*n*)   Machinery steel, sometimes called *mild steel* with a small percentage of carbon. Cannot be hardened.

neck (*v*)   To cut a groove called a *neck* around a cylindrical piece.

NECK

normalize (*v*) To heat steel above its critical temperature and then to cool it in air, §11.22.

pack-harden (*v*) To *carburize*, then to *caseharden*, §11.22.

pad (*n*) A slight projection, usually to provide a bearing surface around one or more holes.

PAD

pattern (*n*) A model, usually of wood, used in forming a mold for a casting. In sheet metal work a pattern is called a *development*.

peen (*v*) To hammer into shape with a ballpeen hammer.

pickle (*v*) To clean forgings or castings in dilute sulphuric acid.

pinion (*n*) The smaller of two mating gears.

pitch circle (*n*) An imaginary circle corresponding to the circumference of the friction gear from which the spur gear was derived.

plane (*v*) To remove material by means of the *planer*, §11.11.

planish (*v*) To impart a planished surface to sheet metal by hammering with a smooth-surfaced hammer.

plate (*v*) To coat a metal piece with another metal, such as chrome or nickel, by electrochemical methods.

polish (*v*) To produce a highly finished or polished surface by friction, using a very fine abrasive.

profile (*v*) To cut any desired outline by moving a small rotating cutter, usually with a master template as a guide.

punch (*v*) To cut an opening of a desired shape with a rigid tool having the same shape, by pressing the tool through the work.

quench (*v*) To immerse a heated piece of metal in water or oil to harden it.

rack (*n*) A flat bar with gear teeth in a straight line to engage with teeth in a gear.

ream (*v*) To enlarge a finished hole slightly to give it greater accuracy, with a *reamer*, §11.16.

relief (*n*) An offset of surfaces to provide clearance for machining.

RELIEF

rib (*n*) A relatively thin flat member acting as a brace or support.

RIB

rivet (*v*) To connect with rivets or to clench over the end of a pin by hammering, §14.36.

round (*n*) An exterior rounded intersection of two surfaces, §§7.34 and 11.5.

SAE Society of Automotive Engineers.

sandblast (*v*) To blow sand at high velocity with compressed air against castings or forgings to clean them.

scleroscope (*n*) An instrument for measuring hardness of metals.

scrape (*v*) To remove metal by scraping with a hand scraper, usually to fit a bearing.

shape (*v*) To remove metal from a piece with a *shaper*, §11.10.

shear (*v*) To cut metal by means of shearing with two blades in sliding contact.

sherardize (*v*) To galvanize a piece with a coating of zinc by heating it in a drum with zinc powder, to a temperature of 575–850°F.

shim (*n*) A thin piece of metal or other material used as a spacer in adjusting two parts.

solder (*v*) To join with solder, usually composed of lead and tin.

spin (*v*) To form a rotating piece of sheet metal into a desired shape by pressing it with a smooth tool against a rotating form.

spline (*n*) A keyway, usually one of a series cut around a shaft or hole.

SPLINED HOLE

spotface (*v*) To produce a round *spot* or bearing surface around a hole, usually with a *spotfacer*. The spotface may be on top of a boss or it may be sunk into the surface, §§7.33 and 11.16.

SPOTFACE

sprue (*n*) A hole in the sand leading to the *gate* which leads to the mold, through which the metal enters, §11.3.

steel casting (*n*) Like cast-iron casting except that in the furnace scrap steel has been added to the casting.

swage (*v*)   To hammer metal into shape while it is held over a *swage,* or die, which fits in a hole in the *swage block*, or anvil.

sweat (*v*)   To fasten metal together by the use of solder between the pieces and by the application of heat and pressure.

tap (*v*)   To cut relatively small internal threads with a *tap*. §11.16.

tape (*n*)   Conical form given to a shaft or a hole. Also refers to the slope of a plane surface, §12.33.

taper pin (*n*)   A small tapered pin for fastening, usually to prevent a collar or hub from rotating on a shaft.

TAPER PIN

taper reamer (*n*)   A tapered reamer for producing accurate tapered holes, as for a taper pin, §§12.33 and 14.35.

temper (*v*)   To reheat hardened steel to bring it to a desired degree of hardness, §11.22.

template or *templet* (*n*)   A guide or pattern used to mark out the work, guide the tool in cutting it, or check the finished product.

tin (*n*)   A silvery metal used in alloys and for coating other metals, such as tin plate.

tolerance (*n*)   Total amount of variation permitted in limit dimension of a part, §13.2.

trepan (*v*)   To cut a circular groove in the flat surface at one end of a hole.

tumble (*v*)   To clean rough castings or forgings in a revolving drum filled with scrap metal.

turn (*v*)   To produce, on a lathe, a cylindrical surface parallel to the center line, §11.7.

twist drill (*n*)   A drill for use in a drill press, §11.16.

undercut (*n*)   A recessed cut or a cut with inwardly sloping sides.

UNDERCUT

upset (*v*)   To form a head or enlarged end on a bar or rod by pressure or by hammering between dies.

web (*n*)   A thin flat part joining larger parts. Also known as a *rib*.

weld (*v*)   Uniting metal pieces by pressure or fusion welding processes, §11.19.

Woodruff key (*n*)   A semicircular flat key, §14.34.

WOODRUFF KEYS

wrought iron (*n*)   Iron of low carbon content useful because of its toughness, ductility, and malleability.

3   CAD/CAM Glossary*

access time (or disk access time)   One measure of system response. The time interval between the instant that data is called for from storage and the instant that delivery is completed—i.e., read time. See also *response time*.

alphanumeric (or alphameric)   A term that encompasses letters, digits, and special characters that are machine-processable.

alphanumeric display (or alphameric display)   A work-station device consisting of a CRT on which text can be viewed. An alphanumeric display is capable of showing a fixed set of letters, digits, and special characters. It allows the designer to observe entered commands and to receive messages from the system.

alphanumeric keyboard (or alphameric keyboard)   A work-station device consisting of a typewriter-like keyboard that allows the designer to communicate with the system using an English-like command language.

American Standard Code for Information Interchange (ASCII)   An industry-standard character code widely used for information interchange among data processing systems, communications systems, and associated equipment.

analog   Applied to an electrical or computer system, this denotes the capability to represent data in continuously varying physical quantities.

annotation   Process of inserting text or a special note or identification (such as a flag)   on a drawing, map, or diagram constructed on a CAD/CAM system. The text can be generated and positioned on the drawing using the system.

application program (or package)   A computer program or collection of programs to perform a task or tasks specific to a particular user's need or class of needs.

*Extracted from *The CAD/CAM Glossary*, 1983 edition, published by the Computervision Corporation, Bedford, MA 01730; reproduced with permission of the publisher.

archival storage Refers to memory (on magnetic tape, disks, printouts, or drums) used to store data on completed designs or elements outside of main memory.

array (*v*) To create automatically on a CAD system an arrangement of identical elements or components. The designer defines the element once, then indicates the starting location and spacing for automatic generation of the array. (*n*) An arrangement created in the above manner. A series of elements or sets of elements arranged in a pattern—i.e., matrix.

ASCII See *American National Standard Code for Information Exchange.*

assembler A computer program that converts (i.e., translates) programmer-written symbolic instructions, usually in mnemonic form, into machine-executable (computer or binary-coded) instructions. This conversion is typically one-to-one (one symbolic instruction converts to one machine-executable instruction). A software programming aid.

associative dimensioning A CAD capability that links dimension entities to geometric entities being dimensioned. This allows the value of a dimension to be automatically updated as the geometry changes.

attribute A nongraphic characteristic of a part, component, or entity under design on a CAD system. Examples include: dimension entities associated with geometry, text with text nodes, and nodal lines with connect nodes. Changing one entity in an association can produce automatic changes by the system in the associated entity; e.g., moving one entity can cause moving or stretching of the other entity.

automatic dimensioning A CAD capability that computes the dimensions in a displayed design, or in a designated section, and automatically places dimensions, dimensional lines, and arrowheads where required. In the case of mapping, this capability labels the linear feature with length and azimuth.

auxiliary storage Storage that supplements main memory devices such as disk or drum storage. Contrast with *archival storage.*

benchmark The program(s) used to test, compare, and evaluate in real time the performance of various CAD/CAM systems prior to selection and purchase. A *synthetic* benchmark has preestablished parameters designed to exercise a set of system features and resources. A *live* benchmark is drawn from the prospective user's workload as a model of the entire workload.

bit The smallest unit of information that can be stored and processed by a digital computer. A bit may assume only one of two values: 0 or 1 (i.e., ON/ OFF or YES/NO). Bits are organized into larger units called *words* for access by computer instructions.

Computers are often categorized by word size in bits, i.e., the maximum word size that can be processed as a unit during an instruction cycle (e.g., 16-bit computers or 32-bit computers). The number of bits in a word is an indication of the processing power of the system, especially for calculations or for high-precision data.

bit rate The speed at which bits are transmitted, usually expressed in bits per second.

bits per inch (bpi) The number of bits that can be stored per inch of a magnetic tape. A measure of the data storage capacity of a magnetic tape.

blank A CAD command that causes a predefined entity to go temporarily blank on the CRT. The reversing command is *unblank.*

blinking A CAD design aid that makes a predefined graphic entity blink on the CRT to attract the attention of the designer.

boot up Start up (a system).

B-spline A sequence of parametric polynomial curves (typically quadratic or cubic polynomials) forming a smooth fit between a sequence of points in 3D space. The piece-wise defined curve maintains a level of mathematical continuity dependent upon the polynomial degree chosen. It is used extensively in mechanical design applications in the automotive and aerospace industries.

bug A flaw in the design or implementation of a software program or hardware design that causes erroneous results or malfunctions.

bulk memory A memory device for storing a large amount of data, e.g., disk, drum, or magnetic tape. It is not randomly accessible as main memory is.

byte A sequence of adjacent bits, usually eight, representing a character that is operated on as a unit. Usually shorter than a word. A measure of the memory capacity of a system, or of an individual storage unit (as a 300-million-byte disk).

CAD See *computer-aided design.*

CAD/CAM See *computer-aided design/computer-aided manufacturing.*

CADDS® Computervision Corporation's registered trademark for its prerecorded software programs.

CAE See *computer-aided engineering.*

CAM See *computer-aided manufacturing.*

CAMACS™ (CAM Asynchronous Communications Software) Computervision's communications link, which enables users to exchange machine control data with other automation systems and devices, or to interact directly with local or remote manufacturing systems and machines. CAMACS tailors CAD/CAM data automatically for a wide range of machine tools, robots, coordinate measurement systems, and off-line storage devices.

cathode ray tube (CRT) The principal component in a CAD display device. A CRT displays graphic representations of geometric entities and designs and can be of various types: storage tube, raster scan, or refresh. These tubes create images by means of a controllable beam of electrons striking a screen. The term *CRT* is often used to denote the entire display device.

central processing unit (CPU)   The computer brain of a CAD/CAM system that controls the retrieval, decoding, and processing of information, as well as the interpretation and execution of operating instructions—the building blocks of application and other computer programs. A CPU comprises arithmetic, control, and logic elements.

character   An alphabetical, numerical, or special graphic symbol used as part of the organization, control, or representation of CAD/CAM data.

characters per second (cps)   A measure of the speed with which an alphanumeric terminal can process data.

chip   See *integrated circuit.*

code   A set of specific symbols and rules for representing data (usually instructions) so that the data can be understood and executed by a computer. A code can be in binary (machine) language, assembly language, or a high-level language. Frequently refers to an industry-standard code such as ANSI, ASCII, IPC, or Standard Code for Information Exchange. Many application codes for CAD/CAM are written in FORTRAN.

color display   A CAD/CAM display device. Color raster-scan displays offer a variety of user-selectable, contrasting colors to make it easier to discriminate among various groups of design elements on different layers of a large, complex design. Color speeds up the recognition of specific areas and subassemblies, helps the designer interpret complex surfaces, and highlights interference problems. Color displays can be of the penetration type, in which various phosphor layers give off different colors (refresh display) or the TV-type with red, blue, and green electron guns (raster-scan display).

command   A control signal or instruction to a CPU or graphics processor, commonly initiated by means of a menu/tablet and electronic pen or by an alphanumeric keyboard.

command language   A language for communicating with a CAD/CAM system in order to perform specific functions or tasks.

communication link   The physical means, such as a telephone line, for connecting one system module or peripheral to another in a different location in order to transmit and receive data. See also *data link.*

compatibility   The ability of a particular hardware module or software program, code, or language to be used in a CAD/CAM system without prior modification or special interfaces. *Upward compatible* denotes the ability of a system to interface with new hardware or software modules or enhancements (i.e., the system vendor provides with each new module a reasonable means of transferring data, programs, and operator skills from the user's present system to the new enhancements).

compiler   A computer program that converts or translates a high-level, user-written language (e.g., PASCAL, COBOL, VARPRO, or FORTRAN) or source, into a language that a computer can understand. The conversion is typically one to many (i.e., one user instruction to many machine-executable instructions). A software programming aid, the compiler allows the designer to write programs in an English-like language with relatively few statements, thus saving program development time.

component   A physical entity, or a symbol used in CAD to denote such an entity. Depending on the application, a component might refer to an IC or part of a wiring circuit (e.g., a resistor), or a valve, elbow, or vee in a plant layout, or a substation or cable in a utility map. Also applies to a subassembly or part that goes into higher level assemblies.

computer-aided design (CAD)   A process that uses a computer system to assist in the creation, modification, and display of a design.

computer-aided design/computer-aided manufacturing (CAD/CAM)   Refers to the integration of computers into the entire design-to-fabrication cycle of a product or plant.

computer-aided engineering (CAE)   Analysis of a design for basic error checking, or to optimize manufacturability, performance, and economy (for example, by comparing various possible materials or designs). Information drawn from the CAD/CAM design data base is used to analyze the functional characteristics of a part, product, or system under design and to simulate its performance under various conditions. In electronic design, CAE enables users of the Computervision Designer system to detect and correct potentially costly design flaws. CAE permits the execution of complex circuit loading analyses and simulation during the circuit definition stage. CAE can be used to determine section properties, moments of inertia, shear and bending moments, weight, volume, surface area, and center of gravity. CAE can precisely determine loads, vibration, noise, and service life early in the design cycle so that components can be optimized to meet those criteria. Perhaps the most powerful CAE technique is finite element modeling. See also *kinematics.*

computer-aided manufacturing (CAM)   The use of computer and digital technology to generate manufacturing-oriented data. Data drawn from a CAD/ CAM data base can assist in or control a portion or all of a manufacturing process, including numerically controlled machines, computer-assisted parts programming, computer-assisted process planning, robotics, and programmable logic controllers, CAM can involve production programming, manufacturing engineering, industrial engineering, facilities engineering, and reliability engineering (quality control).

CAM techniques can be used to produce process plans for fabricating a complete assembly, to program robots, and to coordinate plant operation.

computer graphics A general term encompassing any discipline or activity that uses computers to generate, process, and display graphic images. The essential technology of CAD/CAM systems. See also *computer-aided design.*

computer network An interconnected complex (arrangement or configuration) of two or more systems. See also *network.*

computer program A specific set of software commands in a form acceptable to a computer and used to achieve a desired result. Often called a *software program* or *package.*

configuration A particular combination of a computer, software and hardware modules, and peripherals at a single installation and interconnected in such a way as to support certain application(s).

connector A termination point for a signal entering or leaving a PC board or a cabling system.

convention Standardized methodology or accepted procedure for executing a computer program. In CAD, the term denotes a standard rule or mode of execution undertaken to provide consistency. For example, a drafting convention might require all dimensions to be in metric units.

core (core memory) A largely obsolete term for *main storage.*

CPU See *central processing unit.*

CRT See *cathode ray tube.*

cursor A visual tracking symbol, usually an underline or cross hairs, for indicating a location or entity selection on the CRT display. A text cursor indicates the alphanumeric input; a graphics cursor indicates the next geometric input. A cursor is guided by an electronic or light pen, joystick, keyboard, etc., and follows every movement of the input device.

cycle A preset sequence of events (hardware or software) initiated by a single command.

data base A comprehensive collection of interrelated information stored on some kind of mass data storage device, usually a disk. Generally consists of information organized into a number of fixed-format record types with logical links between associated records. Typically includes operating systems instructions, standard parts libraries, completed designs and documentation, source code, graphic and application programs, as well as current user tasks in progress.

data communication The transmission of data (usually digital) from one point (such as a CAD/CAM workstation or CPU) to another point via communication channels such as telephone lines.

data link The communication line(s), related controls, and interface(s) for the transmission of data between two or more computer systems. Can include modems, telephone lines, or dedicated transmission media such as cable or optical fiber.

data tablet A CAD/CAM input device that allows the designer to communicate with the system by placing an electronic pen or stylus on the tablet surface. There is a direct correspondence between positions on the tablet and addressable points on the display surface of the CRT. Typically used for indicating positions on the CRT, for digitizing input of drawings, or for menu selection. See also *graphic tablet.*

debug To detect, locate, and correct any bugs in a system's software or hardware.

dedicated Designed or intended for a single function or use. For example, a dedicated work station might be used exclusively for engineering calculations or plotting.

default The predetermined value of a parameter required in a CAD/CAM task or operation. It is automatically supplied by the system whenever that value (e.g., text, height, or grid size) is not specified.

density (1) A measure of the complexity of an electronic design. For example, IC density can be measured by the number of gates or transistors per unit area or by the number of square inches per component. (2) Magnetic tape storage capacity. High capacity might be 1600 bits/inch; low, 800 bits/inch.

device A system hardware module external to the CPU and designed to perform a specific function—i.e., a CRT, plotter, printer, hard-copy unit, etc. See also *peripheral.*

diagnostics Computer programs designed to test the status of a system or its key components and to detect and isolate malfunctions.

dial up To initiate station-to-station communication with a computer via a dial telephone, usually from a workstation to a computer.

digital Applied to an electrical or computer system, this denotes the capability to represent data in the form of digits.

digitize (1) General description: to convert a drawing into digital form (i.e., coordinate locations) so that it can be entered into the data base for later processing. A digitizer, available with many CAD systems, implements the conversion process. This is one of the primary ways of entering existing drawings, crude graphics, lines, and shapes into the system. (2) Computervision usage: to specify a coordinate location or entity using an electronic pen or other device; or a single coordinate value or entity pointer generated by a digitizing operation.

digitizer A CAD input device consisting of a data tablet on which is mounted the drawing or design to be digitized into the system. The designer moves a puck or electronic pen to selected points on the drawing and enters coordinate data for lines and shapes by simply pressing down the digitize button with the puck or pen.

dimensioning automatic   A CAD capability that will automatically compute and insert the dimensions of a design or drawing, or a designated section of it.

direct access (linkage)   Retrieval or storage of data in the system by reference to its location on a tape, disk, or cartridge, without the need for processing on a CPU.

direct-view storage tube (DVST)   One of the most widely used graphics display devices, DVST generates a long-lasting, flicker-tree image with high resolution and no refreshing. It handles an almost unlimited amount of data. However, display dynamics are limited since DVSTs do not permit selective erase. The image is not as bright as with refresh or raster. Also called *storage tube*.

directory   A named space on the disk or other mass storage device in which are stored the names of files and some summary information about them.

discrete components   Components with a single functional capability per package—for example, transistors and diodes.

disk (storage)   A device on which large amounts of information can be stored in the data base. Synonymous with *magnetic disk storage* or *magnetic disk memory*.

display   A CAD/CAM work station device for rapidly presenting a graphic image so that the designer can react to it, making changes interactively in real time. Usually refers to a CRT.

dot-matrix plotter   A CAD peripheral device for generating graphic plots. Consists of a combination of wire nibs (styli) spaced 100 to 200 styli per inch, which place dots where needed to generate a drawing. Because of its high speed, it is typically used in electronic design applications. Accuracy and resolution are not as great as with pen plotters. Also known as *electrostatic plotter*.

drum plotter   An electromechanical pen plotter that draws an image on paper or film mounted on a rotatable drum. In this CAD peripheral device a combination of plotting-head movement and drum rotation provides the motion.

dynamic (motion)   Simulation of movement using CAD software, so that the designer can see on the CRT screen 3D representations of the parts in a piece of machinery as they interact dynamically. Thus, any collision or interference problems are revealed at a glance.

dynamic menuing   This feature of Computervision's Instaview terminal allows a particular function or command to be initiated by touching an electronic pen to the appropriate key word displayed in the status text area on the screen.

dynamics   The capability of a CAD system to zoom, scroll, and rotate.

edit   To modify, refine, or update an emerging design or text on a CAD system. This can be done on-line interactively.

electrostatic plotter   See *dot-matrix plotter*.

element   The basic design entity in computer-aided design whose logical, positional, electrical, or mechanical function is identifiable.

enhancements   Software or hardware improvements, additions, or updates to a CAD/CAM system.

entity   A geometric primitive—the fundamental building block used in constructing a design or drawing, such as an arc, circle, line, text, point, spline, figure, or nodal line. Or a group of primitives processed as an identifiable unit. Thus, a square may be defined as a discrete entity consisting of four primitives (vectors), although each side of the square could be defined as an entity in its own right. See also *primitive*.

feedback   (1) The ability of a system to respond to an operator command in real time either visually or with a message on the alphanumeric display or CRT. This message registers the command, indicates any possible errors, and simultaneously displays the updated design on the CRT. (2) The signal or data fed back to a commanding unit from a controlled machine or process to denote its response to a command. (3) The signal representing the difference between actual response and desired response and used by the commanding unit to improve performance of the controlled machine or process. See also *prompt*.

figure   A symbol or a part that may contain primitive entities, other figures, nongraphic properties, and associations. A figure can be incorporated into other parts or figures.

file   A collection of related information in the system that may be accessed by a unique name. May be stored on a disk, tape, or other mass storage media.

file protection   A technique for preventing access to or accidental erasure of data within a file on the system.

firmware   Computer programs, instructions, or functions implemented in user-modifiable hardware, i.e., a microprocessor with read-only memory. Such programs or instructions, stored permanently in programmable read-only memories, constitute a fundamental part of system hardware. The advantage is that a frequently used program or routine can be invoked by a single command instead of multiple commands as in a software program.

flatbed plotter   A CAD/CAM peripheral device that draws an image on paper, glass, or film mounted on a flat table. The plotting head provides all the motion.

flat-pattern generation   A CAD/CAM capability for automatically unfolding a 3D design of a sheet metal part into its corresponding flat-pattern design. Calculations for material bending and stretching are performed automatically for any specified material. The reverse flat-pattern generation package automatically folds a flat-pattern design into its 3D version. Flat-pattern generation eliminates major bottlenecks for sheet metal fabricators.

flicker   An undesired visual effect on a CRT when the refresh rate is low.

font, line   Repetitive pattern used in CAD to give a displayed line appearance characteristics that make it more easily distinguishable, such as a solid, dashed, or dotted line. A line font can be applied to graphic images in order to provide meaning, either graphic (e.g., hidden lines) or functional (roads, tracks, wires, pipes, etc.). It can help a designer to identify and define specific graphic representations of entities that are view-dependent. For example, a line may be solid when drawn in the top view of an object but, when a line font is used, becomes dotted in the side view where it is not normally visible.

font, text   Sets of type faces of various styles and sizes. In CAD, fonts are used to create text for drawings, special characters such as Greek letters, and mathematical symbols.

FORTRAN   *FOR*mula *TRAN*slation, a high-level programming language used primarily for scientific or engineering applications.

fracturing   The division of IC graphics by CAD into simple trapezoidal or rectangular areas for pattern-generation purposes.

function key   A specific square on a data tablet, or a key on a function key box, used by the designer to enter a particular command or other input. See also *data tablet*.

function keyboard   An input device located at a CAD/CAM workstation and containing a number of function keys.

gap   The gap between two entities on a computer-aided design is the length of the shortest line segment that can be drawn from the boundary of one entity to the other without intersecting the boundary of the other. CAD/CAM design-rules checking programs can automatically perform gap checks.

graphic tablet   A CAD/CAM input device that enables graphic and location instruments to be entered into the system using an electronic pen on the tablet. See also *data tablet*.

gray scales   In CAD systems with a monochromatic display, variations in brightness level (gray scale) are employed to enhance the contrast among various design elements. This feature is very useful in helping the designer discriminate among complex entities on different layers displayed concurrently on the CRT.

grid   A network of uniformly spaced points or crosshatch optionally displayed on the CRT and used for exactly locating and digitizing a position, inputting components to assist in the creation of a design layout, or constructing precise angles. For example, the coordinate data supplied by digitizers is automatically calculated by the CPU from the closest grid point. The grid determines the minimum accuracy with which design entities are described or connected. In the mapping environment, a grid is used to describe the distribution network of utility resources.

hard copy   A copy on paper of an image displayed on the CRT—for example, a drawing, printed report, plot, listing, or summary. Most CAD/CAM systems can automatically generate hard copy through an on-line printer or plotter.

hardware   The physical components, modules, and peripherals comprising a system—computer disk, magnetic tape, CRT terminal(s), plotter(s), etc.

hard-wired link   A technique of physically connecting two systems by fixed circuit interconnections using digital signals.

high-level language   A problem-oriented programming language using words, symbols, and command statements that closely resemble English-language statements. Each statement typically represents a series of computer instructions. Relatively easy to learn and use, a high-level language permits the execution of a number of subroutines through a simple command. Examples are BASIC, FORTRAN, PL/I, PASCAL, and COBOL.

A high-level language must be translated or compiled into machine language before it can be understood and processed by a computer. See also *assembler: low-level language*.

host computer   The primary or controlling computer in a multicomputer network. Large-scale host computers typically are equipped with mass memory and a variety of peripheral devices, including magnetic tape, line printers, card readers, and possibly hard-copy devices. Host computers may be used to support, with their own memory and processing capabilities, not only graphics programs running on a CAD/CAM system but also related engineering analysis.

host-satellite system   A CAD/CAM system configuration characterized by a graphic workstation with its own computer (typically holding the display file) that is connected to another, usually larger, computer for more extensive computation or data manipulation. The computer local to the display is a satellite to the larger host computer, and the two comprise a host-satellite system.

IC   See *integrated circuit*.

IGES   See *Initial Graphics Exchange Specification*.

inches per second (ips)   Measure of the speed of a device (i.e., the number of inches of magnetic tape that can be processed per second, or the speed of a pen plotter).

Initial Graphics Exchange Specification (IGES)   An interim CAD/CAM data base specification until the American National Standards Institute develops its own specification. IGES attempts to standardize communication of drawing and geometric product information between computer systems.

initialize To set counters, switches, and addresses on a computer to zero or to other starting values at the beginning of, or at predetermined stages in, a program or routine.

input (data) (1) The data supplied to a computer program for processing by the system. (2) The process of entering such data into the system.

input devices A variety of devices (such as data tablets or keyboard devices) that allow the user to communicate with the CAD/CAM system, for example, to pick a function from many presented, to enter text and/or numerical data, to modify the picture shown on the CRT, or to construct the desired design.

input/output (I/O) A term used to describe a CAD/CAM communications device as well as the process by which communications take place in a CAD/CAM system. An I/O device is one that makes possible communications between a device and a workstation operator or between devices on the system (such as workstations or controllers). By extension, input/output also denotes the process by which communications takes place. Input refers to the data transmitted to the processor for manipulation, and output refers to the data transmitted from the processor to the workstation operator or to another device (i.e., the results). Contrast with the other major parts of a CAD/CAM system: the CPU or central processing unit, which performs arithmetic and logical operations, and data storage devices (such as memories, disks, or tapes).

insert To create and place entities, figures, or information on a CRT or into an emerging design on the display.

instruction set (1) All the commands to which a CAD/CAM computer will respond. (2) The repertoire of functions the computer can perform.

integrated circuit (IC) A tiny complex of electronic components and interconnections comprising a circuit that may vary in functional complexity from a simple logic gate to a microprocessor. An IC is usually packaged in a single substrate such as a slice of silicon. The complexity of most IC designs and the many repetitive elements have made computer-aided design an economic necessity. Also called a *chip*.

integrated system A CAD/CAM system that integrates the entire product development cycle—analysis, design, and fabrication—so that all processes flow smoothly from concept to production.

intelligent work station/terminal A workstation in a system that can perform certain data processing functions in a stand-alone mode, independent of another computer. Contains a built-in computer, usually a microprocessor or minicomputer, and dedicated memory.

interactive Denotes two-way communications between a CAD/CAM system or workstation and its operators. An operator can modify or terminate a program and receive feedback from the system for guidance and verification. See also *feedback*.

interactive graphics system (IGS) or interactive computer graphics (ICG) A CAD/CAM system in which the workstations are used interactively for computer-aided design and/or drafting, as well as for CAM, all under full operator control, and possibly also for text-processing, generation of charts and graphs, or computer-aided engineering. The designer (operator) can intervene to enter data and direct the course of any program, receiving immediate visual feedback via the CRT. Bilateral communication is provided between the system and the designer(s). Often used synonymously with *CAD*.

interface (*n*) (1) A hardware and/or software link that enables two systems, or a system and its peripherals, to operate as a single, integrated system. (2) The input devices and visual feedback capabilities that allow bilateral communication between the designer and the system. The interface to a large computer can be a communications link (hardware) or a combination of software and hard-wired connections. An interface might be a portion of storage accessed by two or more programs or a link between two subroutines in a program.

I/O See *input/output*.

ips See *inches per second*.

jaggies A CAD jargon term used to refer to straight or curved lines that appear to be jagged or sawtoothed on the CRT screen.

joystick A CAD data-entering device employing a hand-controlled lever to manually enter the coordinates of various points on a design being digitized into the system.

key file A disk file that provides user-defined definitions for a tablet menu. See *menu*.

kinematics A computer-aided engineering (CAE) process for plotting or animating the motion of parts in a machine or a structure under design on the system. CAE simulation programs allow the motion of mechanisms to be studied for interference, acceleration, and force determinations while still in the design stage.

layering A method of logically organizing data in a CAD/CAM data base. Functionally different classes of data (e.g., various graphic/geometric entities) are segregated on separate layers, each of which can be displayed individually or in any desired combination. Layering helps the designer distinguish among different kinds of data in creating a complex product such as a multilayered PC board or IC.

layers User-defined logical subdivisions of data in a CAD/CAM data base that may be viewed on the CRT individually or overlaid and viewed in groups.

learning curve A concept that projects the expected improvement in operator productivity over a period of time.

Usually applied in the first 1 to $1\frac{1}{2}$ years of a new CAD/CAM facility as part of a cost-justification study, or when new operators are introduced. An accepted tool of management for predicting manpower requirements and evaluating training programs.

library, graphics (or parts library)   A collection of standard, often-used symbols, components, shapes, or parts stored in the CAD data base as templates or building blocks to speed up future design work on the system. Generally an organization of files under a common library name.

light pen   A hand-held photosensitive CAD input device used on a refreshed CRT screen for identifying display elements, or for designating a location on the screen where an action is to take place.

line font   See *font, line*.

line printer   A CAD/CAM peripheral device used for rapid printing of data.

line smoothing   An automated mapping capability for the interpolation and insertion of additional points along a linear entity yielding a series of shorter linear segments to generate a smooth curved appearance to the original linear component. The additional points or segments are created only for display purposes and are interpolated from a relatively small set of stored representative points. Thus, data storage space is minimized.

low-level language   A programming language in which statements translate on a one-for-one basis. See also *machine language*.

machine   A computer, CPU, or other processor.

machine instruction   An instruction that a machine (computer) can recognize and execute.

machine language   The complete set of command instructions understandable to and used directly by a computer when it performs operations.

macro   (1) A sequence of computer instructions executable as a single command. A frequently used, multistep operation can be organized into a macro, given a new name, and remain in the system for easy use, thus shortening program development time. (2) In Computer-vision's IC design system, macro refers to macroexpansion of a cell. This system capability enables the designer to replicate the contents of a cell as primitives without the original cell grouping.

magnetic disk   A flat circular plate with a magnetic surface on which information can be stored by selective magnetization of portions of the flat surface. Commonly used for temporary working storage during computer-aided design. See also *disk*.

magnetic tape   A tape with a magnetic surface on which information can be stored by selective polarization of portions of the surface. Commonly used in CAD/CAM for off-line storage of completed design files and other archival material.

mainframe (computer)   A large central computer facility.

main memory/storage   The computer's general-purpose storage from which instructions may be executed and data loaded directly into operating registers.

mass storage   Auxiliary large-capacity memory for storing large amounts of data readily accessible by the computer. Commonly a disk or magnetic tape.

matrix   A 2D or 3D rectangular array (arrangement) of identical geometric or symbolic entities. A matrix can be generated automatically on a CAD system by specifying the building block entity and the desired locations. This process is used extensively in computer-aided electrical/electronic design.

memory   Any form of data storage where information can be read and written. Standard memories include RAM, ROM, and PROM. See also *programmable read-only memory; random access memory; read-only memory; storage*.

menu   A common CAD/CAM input device consisting of a checkerboard pattern of squares printed on a sheet of paper or plastic placed over a data tablet. These squares have been preprogrammed to represent a part of a command, a command, or a series of commands. Each square, when touched by an electronic pen, initiates the particular function or command indicated on that square. See also *data tablet, dynamic menuing*.

merge   To combine two or more sets of related data into one, usually in a specified sequence. This can be done automatically on a CAD/CAM system to generate lists and reports.

microcomputer   A smaller, lower-cost equivalent of a full-scale minicomputer. Includes a microprocessor (CPU), memory, and necessary interface circuits. Consists of one or more ICs (chips) comprising a chip set.

microprocessor   The central control element of a microcomputer, implemented in a single integrated circuit. It performs instruction sequencing and processing, as well as all required computations. It requires additional circuits to function as a microcomputer. See *microcomputer*.

minicomputer   A general-purpose, single processor computer of limited flexibility and memory performance.

mirroring   A CAD design aid that automatically creates a mirror image of a graphic entity on the CRT by flipping the entity or drawing on its *x* or *y* axis.

mnemonic symbol   An easily remembered symbol that assists the designer in communicating with the system (e.g., an abbreviation such as MPY for *multiply*).

model, geometric   A complete, geometrically accurate 3D or 2D representation of a shape, a part, a geographic area, a plant or any part of it, designed on a CAD system and stored in the data base. A mathematical or analytic model of a physical system used to determine the response of that system to a stimulus or load. See *modeling, geometric*.

modeling, geometric   Constructing a mathematical or analytic model of a physical object or system for the purpose of determining the response of that object or system to a stimulus or load. First, the designer describes the shape under design using a geometric model constructed on the system. The computer then converts this pictorial representation on the CRT into a mathematical model later used for other CAD functions such as design optimization.

modeling, solid   A type of 3D modeling in which the solid characteristics of an object under design are built into the data base so that complex internal structures and external shapes can be realistically represented. This makes computer-aided design and analysis of solid objects easier, clearer, and more accurate than with wire-frame graphics.

modem   *MO*dulator-*DEM*odulator, a device that converts digital signals to analog signals, and vice versa, for long-distance transmission over communications circuits such as telephone lines, dedicated wires, optical fiber, or microwave.

module   A separate and distinct unit of hardware or software that is part of a system.

mouse   A hand-held data-entering device used to position a cursor on a data tablet. See *cursor*.

multiprocessor   A computer whose architecture consists of more than one processing unit. See *central processing unit: microcomputer*.

network   An arrangement of two or more interconnected computer systems to facilitate the exchange of information in order to perform a specific function. For example, a CAD/CAM system might be connected to a mainframe computer to off-load heavy analytic tasks. Also refers to a piping network in computer-aided plant design.

numerical control   (NC) A technique of operating machine tools or similar equipment in which motion is developed in response to numerically coded commands. These commands may be generated by a CAD/CAM system on punched tapes or other communications media. Also, the processes involved in generating the data or tapes necessary to guide a machine tool in the manufacture of a part.

off-line   Refers to peripheral devices not currently connected to and under the direct control of the system's computer.

on-line   Refers to peripheral devices connected to and under the direct control of the system's computer, so that operator-system interaction, feedback, and output are all in real time.

operating system   A structured set of software programs that control the operation of the computer and associated peripheral devices in a CAD/CAM system, as well as the execution of computer programs and data flow to and from peripheral devices. May provide support for activities and programs such as scheduling, debugging, input/output control, accounting, editing, assembly, compilation, storage assignment, data management, and diagnostics. An operating system may assign task priority levels, support a file system, provide drives for I/O devices, support standard system commands or utilities for on-line programming, process commands, and support both networking and diagnostics.

output   The end result of a particular CAD/CAM process or series of processes. The output of a CAD cycle can be artwork and hard-copy lists and reports. The output of a total design-to-manufacturing CAD/CAM system can also include numerical control tapes for manufacturing.

overlay   A segment of code or data to be brought into the memory of a computer to replace existing code or data.

paint   To fill in a bounded graphic figure on a raster display using a combination of repetitive patterns or line fonts to add meaning or clarity. See *font, line*.

paper-tape punch/reader   A peripheral device that can read as well as punch a perforated paper tape generated by a CAD/CAM system. These tapes are the principal means of supplying data to an NC machine.

parallel processing   Executing more than one element of a single process concurrently on multiple processors in a computer system.

password protection   A security feature of certain CAD/CAM systems that prevents access to the system or to files within the system without first entering a password, i.e., a special sequence of characters.

PC board   See *printed circuit board*.

pen plotter   An electromechanical CAD output device that generates hard copy of displayed graphic data by means of a ballpoint pen or liquid ink. Used when a very accurate final drawing is required. Provides exceptional uniformity and density of lines, precise positional accuracy, as well as various user-selectable colors.

peripheral (device)   Any device, distinct from the basic system modules, that provides input to and/or output from the CPU. May include printers, keyboards, plotters, graphics display terminals, paper-tape reader/punches, analog-to-digital converters, disks, and tape drives.

permanent storage   A method or device for storing the results of a completed program outside the CPU—usually in the form of magnetic tape or punched cards.

photo plotter   A CAD output device that generates high-precision artwork masters photographically for PC board design and IC masks.

pixel   The smallest portion of a CRT screen that can be individually referenced. An individual dot on a display image. Typically, pixels are evenly spaced, horizontally and vertically, on the display.

plotter   A CAD peripheral device used to output for external use the image stored in the data base. Generally makes large, accurate drawings substantially better than what is displayed. Plotter types include pen, drum, electrostatic, and flatbed.

postprocessor   A software program or procedure that formats graphic or other data processed on the system for some other purpose. For example, a postprocessor might format cutter centerline data into a form that a machine controller can interpret.

precision   The degree of accuracy. Generally refers to the number of significant digits of information to the right of the decimal point for data represented within a computer system. Thus, the term denotes the degree of discrimination with which a design or design element can be described in the data base.

preplaced line (or bus)   A run (or line) between a set of points on a PC board layout that has been predefined by the designer and must be avoided by a CAD automatic routing program.

preprocessor   A computer program that takes a specific set of instructions from an external source and translates it into the format required by the system.

primitive   A design element at the lowest stage of complexity. A fundamental graphic entity. It can be a vector, a point, or a text string. The smallest definable object in a display processor's instruction set.

printed circuit (PC) board   A baseboard made of insulating materials and an etched copper-foil circuit pattern on which are mounted ICs and other components required to implement one or more electronic functions. PC boards plug into a rack or subassembly of electronic equipment to provide the brains or logic to control the operation of a computer, or a communications system, instrumentation, or other electronic systems. The name derives from the fact that the circuitry is connected not by wires but by copper-foil lines, paths, or traces actually etched onto the board surface. CAD/CAM is used extensively in PC board design, testing, and manufacture.

process simulation   A program utilizing a mathematical model created on the system to try out numerous process design iterations with real-time visual and numerical feedback. Designers can see on the CRT what is taking place at every stage in the manufacturing process. They can therefore optimize a process and correct problems that could affect the actual manufacturing process downstream.

processor   In CAD/CAM system hardware, any device that performs a specific function. Most often used to refer to the CPU. In software, it refers to a complex set of instructions to perform a general function. See also *central processing unit.*

productivity ratio   A widely accepted means of measuring CAD/CAM productivity (throughput per hour) by comparing the productivity of a design/engineering group before and after installation of the system or relative to some standard norm or potential maximum. The most common way of recording productivity is Actual Manual Hours/Actual CAD Hours, expressed as $4:1, 6:1$, etc.

program   (*n*) A precise sequential set of instructions that direct a computer to perform a particular task or action or to solve a problem. A complete program includes plans for the transcription of data, coding for the computer, and plans for the absorption of the results into the system. (*v*) To develop a program. See also *computer program.*

Programmable Read-Only Memory (PROM)   A memory that, once programmed with permanent data or instructions, becomes a ROM. See *read-only memory.*

PROM   See *programmable read-only memory.*

prompt   A message or symbol generated automatically by the system, and appearing on the CRT, to inform the user of (a) a procedural error or incorrect input to the program being executed or (b) the next expected action, option(s), or input. See also *tutorial.*

puck   A hand-held, manually controlled input device that allows coordinate data to be digitized into the system from a drawing placed on the data tablet or digitizer surface. A puck has a transparent window containing cross hairs.

RAM   See *random access memory.*

random access memory (RAM)   A main memory read/write storage unit that provides the CAD/CAM operator direct access to the stored information. The time required to access any word stored in the memory is the same as for any other word.

raster display   A CAD workstation display in which the entire CRT surface is scanned at a constant refresh rate. The bright, flicker-free image can be selectively written and erased. Also called a digital TV display.

raster scan (video)   Currently, the dominant technology in CAD graphic displays. Similar to conventional television, it involves a line-by-line sweep across the entire CRT surface to generate the image. Raster-scan features include good brightness, accuracy, selective erase, dynamic motion capabilities, and the opportunity for unlimited color. The device can display a large amount of information without flicker, although resolution is not as good as with storage-tube displays.

read-only memory (ROM)   A memory that cannot be modified or reprogrammed. Typically used for control and execute programs. See also *programmable read-only memory.*

real time   Refers to tasks or functions executed so rapidly by a CAD/CAM system that the feedback at various stages in the process can be used to guide the designer in completing the task. Immediate visual feedback through the CRT makes possible real time, interactive operation of a CAD/CAM system.

rectangular array   Insertion of the same entity at multiple locations on a CRT using the system's ability to copy design elements and place them at user-specified intervals to create a rectangular arrangement or matrix. A feature of PC and IC design systems.

refresh (or vector refresh)   A CAD display technology that involves frequent redrawing of an image displayed on the CRT to keep it bright, crisp, and clear. Refresh permits a high degree of movement in the displayed image as well as high resolution. Selective erase or editing is possible at any time without erasing and repainting the entire image. Although substantial amounts of high-speed memory are required, large, complex images may flicker.

refresh rate   The rate at which the graphic image on a CRT is redrawn in a refresh display, i.e., the time needed for one refresh of the displayed image.

registration   The degree of accuracy in the positioning of one layer or overlay in a CAD display or artwork, relative to another layer, as reflected by the clarity and sharpness of the resulting image.

repaint   A CAD feature that automatically redraws a design displayed on the CRT.

resolution   The smallest spacing between two display elements that will allow the elements to be distinguished visually on the CRT. The ability to define very minute detail. For example, the resolution of Computervision's IC design system is one part in 33.5 million. As applied to an electrostatic plotter, resolution means the number of dots per square inch.

response time   The elapsed time from initiation of an operation at a workstation to the receipt of the results at that workstation. Includes transmission of data to the CPU, processing, file access, and transmission of results back to the initiating workstation.

restart   To resume a computer program interrupted by operator intervention.

restore   To bring back to its original state a design currently being worked on in a CAD/CAM system after editing or modification that the designer now wants to cancel or rescind.

resume   A feature of some application programs that allows the designer to suspend the data-processing operation at some logical break point and restart it later from that point.

reticle   The photographic plate used to create an IC mask. See also *photo plotter*.

rotate   To turn a displayed 2D or 3D construction about an axis through a predefined angle relative to the original position.

robotics   The use of computer-controlled manipulators or arms to automate a variety of manufacturing processes such as welding, material handling, painting and assembly.

ROM   See *read-only memory*.

routine   A computer program, or a subroutine in the main program. The smallest separately compilable source code unit. See *computer program: source*.

rubber banding   A CAD capability that allows a component to be tracked (dragged) across the CRT screen, by means of an electronic pen, to a desired location, while simultaneously stretching all related interconnections to maintain signal continuity. During tracking the interconnections associated with the component stretch and bend, providing an excellent visual guide for optimizing the location of a component to best fit into the flow of the PC board, or other entity, minimizing total interconnect length and avoiding areas of congestion.

satellite   A remote system connected to another, usually larger, host system. A satellite differs from a remote intelligent work station in that it contains a full set of processors, memory, and mass storage resources to operate independently of the host. See *host-satellite system*.

scale   (*v*) To enlarge or diminish the size of a displayed entity without changing its shape, i.e., to bring it into a user-specified ratio to its original dimensions. Scaling can be done automatically by a CAD system. (*n*) Denotes the coordinate system for representing an object.

scissoring   The automatic erasing of all portions of a design on the CRT that lie outside user-specified boundaries.

scroll   To automatically roll up, as on a spool, a design or text message on a CRT to permit the sequential viewing of a message or drawing too large to be displayed all at once on the screen. New data appear on the CRT at one edge as other data disappear at the opposite edge. Graphics can be scrolled up, down, left, or right.

selective erase   A CAD feature for deleting portions of a display without affecting the remainder or having to repaint the entire CRT display.

shape fill   The automatic painting-in of an area, defined by user-specified boundaries, on an IC or PC board layout, for example, the area to be filled by copper when the PC board is manufactured. Can be done on-line by CAD.

smoothing   Fitting together curves and surfaces so that a smooth, continuous geometry results.

software   The collection of executable computer programs including application programs, operating systems, and languages.

source   A text file written in a high-level language and containing a computer program. It is easily read and understood by people but must be compiled or assembled to generate machine-recognizable instructions. Also known as *source code*. See also *high-level language*.

source language   A symbolic language comprised of statements and formulas used in computer processing. It is translated into object language (object code) by an assembler or compiler for execution by a computer.

spline   A subset of a B-spline where in a sequence of curves is restricted to a plane. An interpolation routine executed on a CAD/CAM system automatically adjusts a curve by design iteration until the curvature is continuous over the length of the curve. See also *B-spline*.

storage   The physical repository of all information relating to products designed on a CAD/CAM system. It is typically in the form of a magnetic tape or disk. Also called *memory*.

storage tube   A common type of CRT that retains an image continuously for a considerable period of time without redrawing (refreshing). The image will not flicker regardless of the amount of information displayed. However, the display tends to be slow relative to raster scan, the image is rather dim, and no single element by itself can be modified or deleted without redrawing. See also *direct view storage tube*.

stretch   A CAD design/editing aid that enables the designer to automatically expand a displayed entity beyond its original dimensions.

string   A linear sequence of entities, such as characters or physical elements, in a computer-aided design.

stylus   A hand-held pen used in conjunction with a data table to enter commands and coordinate input into the system. Also called an *electronic pen*.

subfigure   A part or a design element that may be extracted from a CAD library and inserted intact into another part displayed on the CRT.

surface machining   Automatic generation of NC tool paths to cut 3D shapes. Both the tool paths and the shapes may be constructed using the mechanical design capabilities of a CAD/CAM system.

symbol   Any recognizable sign, mark, shape or pattern used as a building block for designing meaningful structures. A set of primitive graphic entities (line, point, arc, circle, text, etc.) that form a construction that can be expressed as one unit and assigned a meaning. Symbols may be combined or nested to form larger symbols and/or drawings. They can be as complex as an entire PC board or as simple as a single element, such as a pad. Symbols are commonly used to represent physical things. For example, a particular graphic shape may be used to represent a complete device or a certain kind of electrical component in a schematic. To simplify the preparation of drawings of piping systems and flow diagrams, standard symbols are used to represent various types of fittings and components in common use.

Symbols are also basic units in a language. The recognizable sequence of characters END may inform a compiler that the routine it is compiling is completed. In computer-aided mapping, a symbol can be a diagram, design, letter, character, or abbreviation placed on maps and charts, that, by convention or reference to a legend, is understood to stand for or represent a specific characteristic or feature. In a CAD environment, symbol libraries contribute to the quick maintenance, placement, and interpretation of symbols.

syntax   (1) A set of rules describing the structure of statements allowed in a computer language. To make grammatical sense, commands and routines must be written in conformity to these rules. (2) The structure of a computer command language, i.e., the English-sentence structure of a CAD/CAM command language, e.g., verb, noun, modifiers.

system   An arrangement of CAD/CAM data processing, memory, display, and plotting modules—coupled with appropriate software—to achieve specific objectives. The term CAD/CAM system implies both hardware and software. See also *operating system* (a purely software term).

tablet   An input device on which a designer can digitize coordinate data or enter commands into a CAD/CAM system by means of an electronic pen. See also *data tablet*.

task   (1) A specific project that can be executed by a CAD/CAM software program. (2) A specific portion of memory assigned to the user for executing that project.

template   The pattern of a standard, commonly used component or part that serves as a design aid. Once created, it can be subsequently traced instead of redrawn whenever needed. The CAD equivalent of a designer's template might be a standard part in the data-base library that can be retrieved and inserted intact into an emerging drawing on the CRT.

temporary storage   Memory locations for storing immediate and partial results obtained during the execution of a program on the system.

terminal   See *workstation*.

text editor   An operating system program used to create and modify text files on the system.

text file   A file stored in the system in text format that can be printed and edited on-line as required.

throughput   The number of units of work performed by a CAD/CAM system or a work station during a given period of time. A quantitative measure of system productivity.

time-sharing   The use of a common CPU memory and processing capabilities by two or more CAD/CAM terminals to execute different tasks simultaneously.

tool path   Centerline of the tip of an NC cutting tool as it moves over a part produced on a CAD/CAM system. Tool paths can be created and displayed interactively or automatically by a CAD/CAM system, and reformatted into NC tapes, by means of postprocessor, to guide or control machining equipment. See also *surface machining*.

track ball   A CAD graphics input device consisting of a ball recessed into a surface. The designer can rotate it in any direction to control the position of the cursor used for entering coordinate data into the system.

tracking   Moving a predefined (tracking) symbol across the surface of the CRT with a light pen or an electronic pen.

transform   To change an image displayed on the CRT by, for example, scaling, rotating, translating, or mirroring.

transformation   The process of transforming a CAD display image. Also the matrix representation of a geometric space.

translate   (1) To convert CAD/CAM output from one language to another, for example, by means of a postprocessor such as Computervision's IPC-to-Numerics Translator program. (2) Also, by an editing command, to move a CAD display entity a specified distance in a specified direction.

trap   The area that is searched around each digitize to find a hit on a graphics entity to be edited. See also *digitize*.

turnaround time   The elapsed time between the moment a task or project is input into the CAD/CAM system and the moment the required output is obtained.

turnkey   A CAD/CAM system for which the supplier/vendor assumes total responsibility for building, installing, and testing both hardware and software, and the training of user personnel. Also, loosely, a system that comes equipped with all the hardware and software required for a specific application or applications. Usually implies a commitment by the vendor to make the system work and to provide preventive and remedial maintenance of both hardware and software. Sometimes used interchangeably with stand-alone, although stand-alone applies more to system architecture than to terms of purchase.

tutorial   A characteristic of CAD/CAM systems. If the user is not sure how to execute a task, the system will show how. A message is displayed to provide information and guidance.

utilities   Another term for system capabilities and/or features that enable the user to perform certain processes.

vector   A quantity that has magnitude and direction and that, in CAD, is commonly represented by a directed line segment.

verification   (1) A system-generated message to a work station acknowledging that a valid instruction or input has been received. (2) The process of checking the accuracy, viability, and/or manufacturability of an emerging design on the system.

view port   A user-selected, rectangular view of a part, assembly, etc., that presents the contents of a window on the CRT. See also *window*.

window   A temporary, usually rectangular, bounded area on the CRT that is user-specified to include particular entities for modification, editing, or deletion.

wire-frame graphics   A computer-aided design technique for displaying a 3D object on the CRT screen as a series of lines outlining its surface.

wiring diagram   (1) Graphic representation of all circuits and device elements of an electrical system and its associated apparatus or any clearly defined functional portion of that system. A wiring diagram may contain not only wiring system components and wires but also nongraphic information such as wire number, wire size, color, function, component label, and pin number. (2) Illustration of device elements and their interconnectivity as distinguished from their physical arrangement. (3) Drawing that shows how to hook things up.

Wiring diagrams can be constructed, annotated, and documented on a CAD system.

word   A set of bits (typically 16 to 32) that occupies a single storage location and is treated by the computer as a unit. See also *bit*.

working storage   That part of the system's internal storage reserved for intermediate results (i.e., while a computer program is still in progress). Also called *temporary storage*.

workstation   The work area and equipment used for CAD/CAM operations. It is where the designer interacts (communicates) with the computer. Frequently consists of a CRT display and an input device as well as, possibly, a digitizer and a hard-copy device. In a distributed processing system, a work station would have local processing and mass storage capabilities. Also called a *terminal* or *design terminal*.

write   To transfer information from CPU main memory to a peripheral device, such as a mass storage device.

write-protect   A security feature in a CAD/CAM data storage device that prevents new data from being written over existing data.

zero   The origin of all coordinate dimensions defined in an absolute system as the intersection of the baselines of the x, y, and z axes.

zero offset   On an NC unit, this features allows the zero point on an axis to be relocated anywhere within a specified range, thus temporarily redefining the coordinate frame of reference.

zoom   A CAD capability that proportionately enlarges or reduces a figure displayed on a CRT screen.

4 Abbreviations for Use on Drawings and in Text— American National Standard

(Selected from ANSI/ASME Y1.1–1989)

A					
absolute	ABS	assistant	ASST	both sides	BS
accelerate	ACCEL	associate	ASSOC	both ways	BW
accessory	ACCESS.	association	ASSN	bottom	BOT
account	ACCT	atomic	AT	bottom chord	BC
accumulate	ACCUM	audible	AUD	bottom face	BF
actual	ACT.	audio frequency	AF	bracket	BRKT
adapter	ADPT	authorized	AUTH	brake	BK
addendum	ADD.	automatic	AUTO	brake horsepower	BHP
addition	ADD.	auto-transformer	AUTO TR	brass	BRS
adjust	ADJ	auxiliary	AUX	brazing	BRZG
advance	ADV	avenue	AVE	break	BRK
after	AFT.	average	AVG	Brinell hardness	BH
aggregate	AGGR	aviation	AVI	British Standard	BR STD
air condition	AIR COND	azimuth	AZ	British thermal unit	BTU
airplane	APL			broach	BRO
allowance	ALLOW	*B*		bronze	BRZ
alloy	ALY	Babbitt	BAB	Brown & Sharpe (wire gage, same as AWG)	B&S
alteration	ALT	back feed	BF	building	BLDG
alternate	ALT	back pressure	BP	bulkhead	BHD
alternating current	AC	back to back	B to B	burnish	BNH
altitude	ALT	backface	BF	bushing	BUSH.
aluminum	AL	balance	BAL	button	BUT.
American National Standard	AMER NATL STD	ball bearing	BB		
		barometer	BAR	*C*	
American wire gage	AWG	base line	BL		
amount	AMT	base plate	BP	cabinet	CAB.
ampere	AMP	bearing	BRG	calculate	CALC
amplifier	AMPL	bench mark	BM	calibrate	CAL
anneal	ANL	bending moment	M	cap screw	CAP SCR
antenna	ANT.	bent	BT	capacity	CAP
apartment	APT.	bessemer	BESS	carburetor	CARB
apparatus	APP	between	BET.	carburize	CARB
appendix	APPX	between centers	BC	carriage	CRG
approved	APPD	between perpendiculars	BP	case harden	CH
approximate	APPROX	bevel	BEV	cast iron	CI
arc weld	ARC/W	bill of material	B/M	cast steel	CS
area	A	Birmingham wire gage	BWG	casting	CSTG
armature	ARM.	blank	BLK	castle nut	CAS NUT
armor plate	ARM-PL	block	BLK	catalogue	CAT.
army navy	AN	blueprint	BP	cement	CEM
arrange	ARR.	board	BD	center	CTR
artificial	ART.	boiler	BLR	center line	CL
asbestos	ASB	boiler feed	BF	center of gravity	CG
asphalt	ASPH	boiler horsepower	BHP	center of pressure	CP
assemble	ASSEM	boiling point	BP	center to center	C to C
assembly	ASSY	bolt circle	BC	centering	CTR
		both faces	BF	chamfer	CHAM

change	CHG	**D**		existing	EXIST.
channel	CHAN			exterior	EXT
check	CHK	decimal	DEC	extra heavy	X HVY
check valve	CV	dedendum	DED	extra strong	X STR
chord	CHD	deflect	DEFL	extrude	EXTR
circle	CIR	degree	(°) DEG		
circular	CIR	density	D		
circular pitch	CP	department	DEPT	**F**	
circumference	CIRC	design	DSGN		
clear	CLR	detail	DET	fabricate	FAB
clearance	CL	develop	DEV	face to face	F to F
clockwise	CW	diagonal	DIAG	Fahrenheit	F
coated	CTD	diagram	DIAG	far side	FS
cold drawn	CD	diameter	DIA	federal	FED.
cold-drawn steel	CDS	diametral pitch	DP	feed	FD
cold finish	CF	dimension	DIM.	feet	(') FT
cold punched	CP	discharge	DISCH	figure	FIG.
cold rolled	CR	distance	DIST	fillet	FIL
cold-rolled steel	CRS	division	DIV	fillister	FIL
combination	COMB.	double	DBL	finish	FIN.
combustion	COMB	dovetail	DVTL	finish all over	FAO
commercial	COML	dowel	DWL	flange	FLG
company	CO	down	DN	flat	F
complete	COMPL	dozen	DOZ	flat head	FH
compress	COMP	drafting	DFTG	floor	FL
concentric	CONC	drawing	DWG	fluid	FL
concrete	CONC	drill or drill rod	DR	focus	FOC
condition	COND	drive	DR	foot	(') FT
connect	CONN	drive fit	DF	force	F
constant	CONST	drop	D	forged steel	FST
construction	CONST	drop forge	DF	forging	FORG
contact	CONT	duplicate	DUP	forward	FWD
continue	CONT			foundry	FDRY
copper	COP.	**E**		frequency	FREQ
corner	COR			front	FR
corporation	CORP	each	EA	furnish	FURN
correct	CORR	east	E		
corrugate	CORR	eccentric	ECC		
cotter	COT	effective	EFF	**G**	
counter	CTR	elbow	ELL	gage or gauge	GA
counterbore	CBORE	electric	ELEC	gallon	GAL
counter clockwise	CCW	elementary	ELEM	galvanize	GALV
counterdrill	CDRILL	elevate	ELEV	galvanized iron	GI
counterpunch	CPUNCH	elevation	EL	galvanized steel	GS
countersink	CSK	engine	ENG	gasket	GSKT
coupling	CPLG	engineer	ENGR	general	GEN
cover	COV	engineering	ENGRG	glass	GL
cross section	XSECT	entrance	ENT	government	GOVT
cubic	CU	equal	EQ	governor	GOV
cubic foot	CU FT	equation	EQ	grade	GR
cubic inch	CU IN.	equipment	EQUIP	graduation	GRAD
current	CUR	equivalent	EQUIV	graphite	GPH
customer	CUST	estimate	EST	grind	GRD
cyanide	CYN	exchange	EXCH	groove	GRV
		exhaust	EXH	ground	GRD

H

half-round	½ RD
handle	HDL
hanger	HGR
hard	H
harden	HDN
hardware	HDW
head	HD
headless	HDLS
heat	HT
heat-treat	HT TR
heavy	HVY
hexagon	HEX
high-pressure	HP
high-speed	HS
horizontal	HOR
horsepower	HP
hot rolled	HR
hot-rolled steel	HRS
hour	HR
housing	HSG
hydraulic	HYD

I

illustrate	ILLUS
inboard	INBD
inch	(") IN.
inches per second	IPS
inclosure	INCL
include	INCL
inside diameter	ID
instrument	INST
interior	INT
internal	INT
intersect	INT
iron	I
irregular	IREG

J

joint	JT
joint army-navy	JAN
journal	JNL
junction	JCT

K

key	K
keyseat	KST
Keyway	KWY

L

laboratory	LAB
laminate	LAM
lateral	LAT
left	L
left hand	LH
length	LG
length over all	LOA
letter	LTR
light	LT
line	L
locate	LOC
logarithm	LOG.
long	LG
lubricate	LUB
lumber	LBR

M

machine	MACH
machine steel	MS
maintenance	MAINT
malleable	MALL
malleable iron	MI
manual	MAN.
manufacture	MFR
manufactured	MFD
manufacturing	MFG
material	MATL
maximum	MAX
mechanical	MECH
mechanism	MECH
median	MED
metal	MET.
meter	M
miles	MI
miles per hour	MPH
millimeter	MM
minimum	MIN
minute	(') MIN
miscellaneous	MISC
month	MO
Morse taper	MOR T
motor	MOT
mounted	MTD
mounting	MTG
multiple	MULT
music wire gage	MWG

N

national	NATL
natural	NAT
near face	NF
near side	NS
negative	NEG
neutral	NEUT
nominal	NOM
normal	NOR
north	N
not to scale	NTS
number	NO.

O

obsolete	OBS
octagon	OCT
office	OFF.
on center	OC
opposite	OPP
optical	OPT
original	ORIG
outlet	OUT.
outside diameter	OD
outside face	OF
outside radius	OR
overall	OA

P

pack	PK
packing	PKG
page	P
paragraph	PAR.
part	PT
patent	PAT.
pattern	PATT
permanent	PERM
perpendicular	PERP
piece	PC
piece mark	PC MK
pint	PT
pitch	P
pitch circle	PC
pitch diameter	PD
plastic	PLSTC
plate	PL
plumbing	PLMB
point	PT
point of curve	PC
point of intersection	PI
point of tangent	PT
polish	POL
position	POS
potential	POT.
pound	LB
pounds per square inch	PSI
power	PWR
prefabricated	PREFAB
preferred	PFD
prepare	PREP
pressure	PRESS.

process	PROC	screw	SCR
production	PROD	second	SEC
profile	PF	section	SECT
propeller	PROP	semi-steel	SS
publication	PUB	separate	SEP
push button	PB	set screw	SS
		shaft	SFT
Q		sheet	SH
		shoulder	SHLD
quadrant	QUAD	side	S
quality	QUAL	single	S
quarter	QTR	sketch	SK
		sleeve	SLV
R		slide	SL
		slotted	SLOT.
radial	RAD	small	SM
radius	R	socket	SOC
railroad	RR	space	SP
ream	RM	special	SPL
received	RECD	specific	SP
record	REC	spot faced	SF
rectangle	RECT	spring	SPG
reduce	RED.	square	SQ
reference line	REF L	standard	STD
reinforce	REINF	station	STA
release	REL	stationary	STA
relief	REL	steel	STL
remove	REM	stock	STK
require	REQ	straight	STR
required	REQD	street	ST
return	RET.	structural	STR
reverse	REV	substitute	SUB
revolution	REV	summary	SUM.
revolutions per minute	RPM	support	SUP.
right	R	surface	SUR
right hand	RH	symbol	SYM
rivet	RIV	system	SYS
Rockwell hardness	RH		
roller bearing	RB	**T**	
room	RM		
root diameter	RD	tangent	TAN.
root mean square	RMS	taper	TPR
rough	RGH	technical	TECH
round	RD	template	TEMP
		tension	TENS.
S		terminal	TERM.
		thick	THK
schedule	SCH	thousand	M
schematic	SCHEM		
scleroscope hardness	SH		

thread	THD	wall	W
threads per inch	TPI	washer	WASH.
through	THRU	watt	W
time	T	week	WK
tolerance	TOL	weight	WT
tongue & groove	T & G	west	W
tool steel	TS	width	W
tooth	T	wood	WD
total	TOT	Woodruff	WDF
transfer	TRANS	working point	WP
typical	TYP	working pressure	WP
		wrought	WRT
U		wrought iron	WI
ultimate	ULT		
unit	U	**Y**	
universal	UNIV		
		yard	YD
V		year	YR
vacuum	VAC		
valve	V		
variable	VAR		
versus	VS		
vertical	VERT		
volt	V		
volume	VOL		
W			

5   Running and Sliding Fits[a]—American National Standard

RC 1   *Close sliding fits* are intended for the accurate location of parts which must assemble without perceptible play.

RC 2   *Sliding fits* are intended for accurate location, but with greater maximum clearance than class RC 1. Parts made to this fit move and turn easily but are not intended to run freely, and in the larger sizes may seize with small temperature changes.

RC 3   *Precision running fits* are about the closest fits which can be expected to run freely and are intended for precision work at slow speeds and light journal pressures, but they are not suitable where appreciable temperature differences are likely to be encountered.

RC 4   *Close running fits* are intended chiefly for running fits on accurate machinery with moderate surface speeds and journal pressures, where accurate location and minimum play are desired.

Basic hole system. Limits are in thousandths of an inch. See §12.8.
Limits for hole and shaft are applied algebraically to the basic size to obtain the limits of size for the parts.
Data in **boldface** are in accordance with ABC agreements.
Symbols H5, g5, etc., are hole and shaft designations used in ABC System.

Nominal Size Range, inches Over To	Class RC 1 Limits of Clearance	Class RC 1 Standard Limits Hole H5	Class RC 1 Standard Limits Shaft g4	Class RC 2 Limits of Clearance	Class RC 2 Standard Limits Hole H6	Class RC 2 Standard Limits Shaft g5	Class RC 3 Limits of Clearance	Class RC 3 Standard Limits Hole H7	Class RC 3 Standard Limits Shaft f6	Class RC 4 Limits of Clearance	Class RC 4 Standard Limits Hole H8	Class RC 4 Standard Limits Shaft f7
0–0.12	0.1 0.45	+0.2 −0	−0.1 −0.25	0.1 0.55	+0.25 −0	−0.1 −0.3	0.3 0.95	+0.4 −0	−0.3 −0.55	0.3 1.3	+0.6 −0	−0.3 −0.7
0.12–0.24	0.15 0.5	+0.2 −0	−0.15 −0.3	0.15 0.65	+0.3 −0	−0.15 −0.35	0.4 1.12	+0.5 −0	−0.4 −0.7	0.4 1.6	+0.7 −0	−0.4 −0.9
0.24–0.40	0.2 0.6	+0.25 −0	−0.2 −0.35	0.2 0.85	+0.4 −0	−0.2 −0.45	0.5 1.5	+0.6 −0	−0.5 −0.9	0.5 2.0	+0.9 −0	−0.5 −1.1
0.40–0.71	0.25 0.75	+0.3 −0	−0.25 −0.45	0.25 0.95	+0.4 −0	−0.25 −0.55	0.6 1.7	+0.7 −0	−0.6 −1.0	0.6 2.3	+1.0 −0	−0.6 −1.3
0.71–1.19	0.3 0.95	+0.4 −0	−0.3. −0.55	0.3 1.2	+0.5 −0	−0.3 −0.7	0.8 2.1	+0.8 −0	−0.8 −1.3	0.8 2.8	+1.2 −0	−0.8 −1.6
1.19–1.97	0.4 1.1	+0.4 −0	−0.4 −0.7	0.4 1.4	+0.6 −0	−0.4 −0.8	1.0 2.6	+1.0 −0	−1.0 −1.6	1.0 3.6	+1.6 −0	−1.0 −2.0
1.97–3.15	0.4 1.2	+0.5 −0	−0.4 −0.7	0.4 1.6	+0.7 −0	−0.4 −0.9	1.2 3.1	+1.2 −0	−1.2 −1.9	1.2 4.2	+1.8 −0	−1.2 −2.4
3.15–4.73	0.5 1.5	+0.6 −0	−0.5 −0.9	0.5 2.0	+0.9 −0	−0.5 −1.1	1.4 3.7	+1.4 −0	−1.4 −2.3	1.4 5.0	+2.2 −0	−1.4 −2.8
4.73–7.09	0.6 1.8	+0.7 −0	−0.6 −1.1	0.6 2.3	+1.0 −0	−0.6 −1.3	1.6 4.2	+1.6 −0	−1.6 −2.6	1.6 5.7	+2.5 −0	−1.6 −3.2
7.09–9.85	0.6 2.0	+0.8 −0	−0.6 −1.2	0.6 2.6	+1.2 −0	−0.6 −1.4	2.0 5.0	+1.8 −0	−2.0 −3.2	2.0 6.6	+2.8 −0	−2.0 −3.8
9.85–12.41	0.8 2.3	+0.9 −0	−0.8 −1.4	0.8 2.9	+1.2 −0	−0.8 −1.7	2.5 5.7	+2.0 −0	−2.5 −3.7	2.5 7.5	+3.0 −0	−2.5 −4.5
12.41–15.75	1.0 2.7	+1.0 −0	−1.0 −1.7	1.0 3.4	+1.4 −0	−1.0 −2.0	3.0 6.6	+2.2 −0	−3.0 −4.4	3.0 8.7	+3.5 −0	−3.0 −5.2

[a] From ANSI B4.1–1967 (RI994). For larger diameters, see the standard.

5 Running and Sliding Fits[a]—American National Standard (continued)

RC 5
RC 6 } *Medium running fits* are intended for higher running speeds, or heavy journal pressures, or both.

RC 7 } *Free running fits* are intended for use where accuracy is not essential, or where large temperature variations are likely to be encountered, or under both these conditions.

RC 8
RC 9 } *Loose running fits* are intended for use where wide commercial tolerances may be necessary, together with an allowance, on the external member.

Nominal Size Range, inches	Class RC 5			Class RC 6			Class RC 7			Class RC 8			Class RC 9		
	Limits of Clearance	Standard Limits		Limits of Clearance	Standard Limits		Limits of Clearance	Standard Limits		Limits of Clearance	Standard Limits		Limits of Clearance	Standard Limits	
Over To		Hole H8	Shaft e7		Hole H9	Shaft e8		Hole H9	Shaft d8		Hole H10	Shaft c9		Hole H11	Shaft
0–0.12	0.6	+0.6	−0.6	0.6	+1.0	−0.6	1.0	+1.0	−1.0	2.5	+1.6	−2.5	4.0	+2.5	−4.0
	1.6	−0	−1.0	2.2	−0	−1.2	2.6	−0	−1.6	5.1	−0	−3.5	8.1	−0	−5.6
0.12–0.24	0.8	+0.7	−0.8	0.8	+1.2	−0.8	1.2	+1.2	−1.2	2.8	+1.8	−2.8	4.5	+3.0	−4.5
	2.0	−0	−1.3	2.7	−0	−1.5	3.1	−0	−1.9	5.8	−0	−4.0	9.0	−0	−6.0
0.24–0.40	1.0	+0.9	−1.0	1.0	+1.4	−1.0	1.6	+1.4	−1.6	3.0	+2.2	−3.0	5.0	+3.5	−5.0
	2.5	−0	−1.6	3.3	−0	−1.9	3.9	−0	−2.5	6.6	−0	−4.4	10.7	−0	−7.2
0.40–0.71	1.2	+1.0	−1.2	1.2	+1.6	−1.2	2.0	+1.6	−2.0	3.5	+2.8	−3.5	6.0	+4.0	−6.0
	2.9	−0	−1.9	3.8	−0	−2.2	4.6	−0	−3.0	7.9	−0	−5.1	12.8	−0	−8.8
0.71–1.19	1.6	+1.2	−1.6	1.6	+2.0	−1.6	2.5	+2.0	−2.5	4.5	+3.5	−4.5	7.0	+5.0	−7.0
	3.6	−0	−2.4	4.8	−0	−2.8	5.7	−0	−3.7	10.0	−0	−6.5	15.5	−0	−10.5
1.19–1.97	2.0	+1.6	−2.0	2.0	+2.5	−2.0	3.0	+2.5	−3.0	5.0	+4.0	−5.0	8.0	+6.0	−8.0
	4.6	−0	−3.0	6.1	−0	−3.6	7.1	−0	−4.6	11.5	−0	−7.5	18.0	−0	−12.0
1.97–3.15	2.5	+1.8	−2.5	2.5	+3.0	−2.5	4.0	+3.0	−4.0	6.0	+4.5	−6.0	9.0	+7.0	−9.0
	5.5	−0	−3.7	7.3	−0	−4.3	8.8	−0	−5.8	13.5	−0	−9.0	20.5	−0	−13.5
3.15–4.73	3.0	+2.2	−3.0	3.0	+3.5	−3.0	5.0	+3.5	−5.0	7.0	+5.0	−7.0	10.0	+9.0	−10.0
	6.6	−0	−4.4	8.7	−0	−5.2	10.7	−0	−7.2	15.5	−0	−10.5	24.0	−0	−15.0
4.73–7.09	3.5	+2.5	−3.5	3.5	+4.0	−3.5	6.0	+4.0	−6.0	8.0	+6.0	−8.0	12.0	+10.0	−12.0
	7.6	−0	−5.1	10.0	−0	−6.0	12.5	−0	−8.5	18.0	−0	−12.0	28.0	−0	−18.0
7.09–9.85	4.0	+2.8	−4.0	4.0	+4.5	−4.0	7.0	+4.5	−7.0	10.0	+7.0	−10.0	15.0	+12.0	−15.0
	8.6	−0	−5.8	11.3	−0	−6.8	14.3	−0	−9.8	21.5	−0	−14.5	34.0	−0	−22.0
9.85–12.41	5.0	+3.0	−5.0	5.0	+5.0	−5.0	8.0	+5.0	−8.0	12.0	+8.0	−12.0	18.0	+12.0	−18.0
	10.0	−0	−7.0	13.0	−0	−8.0	16.0	−0	−11.0	25.0	−0	−17.0	38.0	−0	−26.0
12.41–15.75	6.0	+3.5	−6.0	6.0	+6.0	−6.0	10.0	+6.0	−10.0	14.0	+9.0	−14.0	22.0	+14.0	−22.0
	11.7	−0	−8.2	15.5	−0	−9.5	19.5	−0	13.5	29.0	−0	−20.0	45.0	−0	−31.0

[a] From ANSI B4.1–1967 (R1994). For larger diameters, see the standard.

6 Clearance Locational Fits[a]—American National Standard

LC *Locational clearance fits* are intended for parts which are normally stationary but which can be freely assembled or disassembled They run from snug fits for parts requiring accuracy of location, through the medium clearance fits for parts such as spigots, to the looser fastener fits, where freedom of assembly is of prime importance.

Basic hole system. Limits are in thousandths of an inch. See §12.8.
Limits for hole and shaft are applied algebraically to the basic size to obtain the limits of size for the parts.
Data in **boldface** are in accordance with ABC agreements.
Symbols H6, H5, etc., are hole and shaft designations used in ABC System.

Nominal Size Range, inches Over To	Class LC 1 Limits of Clearance	Standard Limits Hole H6	Shaft h5	Class LC 2 Limits of Clearance	Standard Limits Hole H7	Shaft h6	Class LC 3 Limits of Clearance	Standard Limits Hole H8	Shaft h7	Class LC 4 Limits of Clearance	Standard Limits Hole H10	Shaft h9	Class LC 5 Limits of Clearance	Standard Limits Hole H7	Shaft g6
0–0.12	0 0.45	+0.25 −0	+0 −0.2	0 0.65	+0.4 −0	+0 −0.25	0 1	+0.6 −0	+0 −0.4	0 2.6	+1.6 −0	+0 −1.0	0.1 0.75	+0.4 −0	−0.1 −0.35
0.12–0.24	0 0.5	+0.3 −0	+0 −0.2	0 0.8	+0.5 −0	+0 −0.3	0 1.2	+0.7 −0	+0 −0.5	0 3.0	+1.8 −0	+0 −1.2	0.15 0.95	+0.5 −0	−0.15 −0.45
0.24–0.40	0 0.65	+0.4 −0	+0 −0.25	0 1.0	+0.6 −0	+0 −0.4	0 1.5	+0.9 −0	+0 −0.6	0 3.6	+2.2 −0	+0 −1.4	0.2 1.2	+0.6 −0	−0.2 −0.6
0.40–0.71	0 0.7	+0.4 −0	+0 −0.3	0 1.1	+0.7 −0	+0 −0.4	0 1.7	+1.0 −0	+0 −0.7	0 4.4	+2.8 −0	+0 −1.6	0.25 1.35	+0.7 −0	−0.25 −0.65
0.71–1.19	0 0.9	+0.5 −0	+0 −0.4	0 1.3	+0.8 −0	+0 −0.5	0 2	+1.2 −0	+0 −0.8	0 5.5	+3.5 −0	+0 −2.0	0.3 1.6	+0.8 −0	−0.3 −0.8
1.19–1.97	0 1.0	+0.6 −0	+0 −0.4	0 1.6	+1.0 −0	+0 −0.6	0 2.6	+1.6 −0	+0 −1	0 6.5	+4.0 −0	+0 −2.5	0.4 2.0	+1.0 −0	−0.4 −1.0
1.97–3.15	0 1.2	+0.7 −0	+0 −0.5	0 1.9	+1.2 −0	+0 −0.7	0 3	+1.8 −0	+0 −1.2	0 7.5	+4.5 −0	+0 −3	0.4 2.3	+1.2 −0	−0.4 −1.1
3.15–4.73	0 1.5	+0.9 −0	+0 −0.6	0 2.3	+1.4 −0	+0 −0.9	0 3.6	+2.2 −0	+0 −1.4	0 8.5	+5.0 −0	+0 −3.5	0.5 2.8	+1.4 −0	−0.5 −1.4
4.73–7.09	0 1.7	+1.0 −0	+0 −0.7	0 2.6	+1.6 −0	+0 −1.0	0 4.1	+2.5 −0	+0 −1.6	0 10	+6.0 −0	+0 −4	0.6 3.2	+1.6 −0	−0.6 −1.6
7.09–9.85	0 2.0	+1.2 −0	+0 −0.8	0 3.0	+1.8 −0	+0 −1.2	0 4.6	+2.8 −0	+0 −1.8	0 11.5	+7.0 −0	+0 −4.5	0.6 3.6	+1.8 −0	−0.6 −1.8
9.85–12.41	0 2.1	+1.2 −0	+0 −0.9	0 3.2	+2.0 −0	+0 −1.2	0 5	+3.0 −0	+0 −2.0	0 13	+8.0 −0	+0 −5	0.7 3.9	+2.0 −0	−0.7 −1.9
12.41–15.75	0 2.4	+1.4 −0	+0 −1.0	0 3.6	+2.2 −0	+0 −1.4	0 5.7	+3.5 −0	+0 −2.2	0 15	+9.0 −0	+0 −6	0.7 4.3	+2.2 −0	−0.7 −2.1

[a] From ANSI B4.1–1967 (R1994). For larger diameters, see the standard.

Nominal Size Range, inches (Over To)	LC 6 Limits of Clearance	LC 6 Hole H9	LC 6 Shaft f8	LC 7 Limits of Clearance	LC 7 Hole H10	LC 7 Shaft e9	LC 8 Limits of Clearance	LC 8 Hole H10	LC 8 Shaft d9	LC 9 Limits of Clearance	LC 9 Hole H11	LC 9 Shaft c10	LC 10 Limits of Clearance	LC 10 Hole H12	LC 10 Shaft	LC 11 Limits of Clearance	LC 11 Hole H13	LC 11 Shaft
0–0.12	0.3 / 1.9	+1.0 / −0	−0.3 / −0.9	0.6 / 3.2	+1.6 / −0	−0.6 / −1.6	1.0 / 3.6	+1.6 / −0	−1.0 / −2.0	2.5 / 6.6	+2.5 / −0	−2.5 / −4.1	4 / 12	+4 / −0	−4 / −8	5 / 17	+6 / −0	−5 / −11
0.12–0.24	0.4 / 2.3	+1.2 / −0	−0.4 / −1.1	0.8 / 3.8	+1.8 / −0	−0.8 / −2.0	1.2 / 4.2	+1.8 / −0	−1.2 / −2.4	2.8 / 7.6	+3.0 / −0	−2.8 / −4.6	4.5 / 14.5	+5 / −0	−4.5 / −9.5	6 / 20	+7 / −0	−6 / −13
0.24–0.40	0.5 / 2.8	+1.4 / −0	−0.5 / −1.4	1.0 / 4.6	+2.2 / −0	−1.0 / −2.4	1.6 / 5.2	+2.2 / −0	−1.6 / −3.0	3.0 / 8.7	+3.5 / −0	−3.0 / −5.2	5 / 17	+6 / −0	−5 / −11	7 / 25	+9 / −0	−7 / −16
0.40–0.71	0.6 / 3.2	+1.6 / −0	−0.6 / −1.6	1.2 / 5.6	+2.8 / −0	−1.2 / −2.8	2.0 / 6.4	+2.8 / −0	−2.0 / −3.6	3.5 / 10.3	+4.0 / −0	−3.5 / −6.3	6 / 20	+7 / −0	−6 / −13	8 / 28	+10 / −0	−8 / −18
071–1.19	0.8 / 4.0	+2.0 / −0	−0.8 / −2.0	1.6 / 7.1	+3.5 / −0	−1.6 / −3.6	2.5 / 8.0	+3.5 / −0	−2.5 / −4.5	4.5 / 13.0	+5.0 / −0	−4.5 / −8.0	7 / 23	+8 / −0	−7 / −15	10 / 34	+12 / −0	−10 / −22
1.19–1.97	1.0 / 5.1	+2.5 / −0	−1.0 / −2.6	2.0 / 8.5	+4.0 / −0	−2.0 / −4.5	3.0 / 9.5	+4.0 / −0	−3.0 / −5.5	5 / 15	+6 / −0	−5 / −9	8 / 28	+10 / −0	−8 / −18	12 / 44	+16 / −0	−12 / −28
1.97–3.15	1.2 / 6.0	+3.0 / −0	−1.2 / −3.0	2.5 / 10.0	+4.5 / −0	−2.5 / −5.5	4.0 / 11.5	+4.5 / −0	−4.0 / −7.0	6 / 17.5	+7 / −0	−6 / −10.5	10 / 34	+12 / −0	−10 / −22	14 / 50	+18 / −0	−14 / −32
3.15–4.73	1.4 / 7.1	+3.5 / −0	−1.4 / −3.6	3.0 / 11.5	+5.0 / −0	−3.0 / −6.5	5.0 / 13.5	+5.0 / −0	−5.0 / −8.5	7 / 21	+9 / −0	−7 / −12	11 / 39	+14 / −0	−11 / −25	16 / 60	+22 / −0	−16 / −38
4.73–7.09	1.6 / 8.1	+4.0 / −0	−1.6 / −4.1	3.5 / 13.5	+6.0 / −0	−3.5 / −7.5	6 / 16	+6 / −0	−6 / −10	8 / 24	+10 / −0	−8 / −14	12 / 44	+16 / −0	−12 / −28	18 / 68	+25 / −0	−18 / −43
7.09–9.85	2.0 / 9.3	+4.5 / −0	−2.0 / −4.8	4.0 / 15.5	+7.0 / −0	−4.0 / −8.5	7 / 18.5	+7 / −0	−7 / −11.5	10 / 29	+12 / −0	−10 / −17	16 / 52	+18 / −0	−16 / −34	22 / 78	+28 / −0	−22 / −50
9.85–12.41	2.2 / 10.2	+5.0 / −0	−2.2 / −5.2	4.5 / 17.5	+8.0 / −0	−4.5 / −9.5	7 / 20	+8 / −0	−7 / −12	12 / 32	+12 / −0	−12 / −20	20 / 60	+20 / −0	−20 / −40	28 / 88	+30 / −0	−28 / −58
12.41–15.75	2.5 / 12.0	+6.0 / −0	−2.5 / −6.0	5.0 / 20.0	+9.0 / −0	−5 / −11	8 / 23	+9 / −0	−8 / −14	14 / 37	+14 / −0	−14 / −23	22 / 66	+22 / −0	−22 / −44	30 / 100	+35 / −0	−30 / −65

^a From ANSI B4.1-1967 (R1994). For larger diameters, see the standard.

LT *Transition fits* are a compromise between clearance and interference fits, for application where accuracy of location is important, but either a small amount of clearance or interference is permissible.

Basic hole system. Limits are in thousandths of an inch. See §12.8.
Limits for hole and shaft are applied algebraically to the basic size to obtain the limits of size for the mating parts.
Data in **boldface** are in accordance with ABC agreements.
"Fit" represents the maximum interference (minus values) and the maximum clearance (plus values).
Symbols H7, js6, etc., are hole and shaft designations used in ABC System.

Nominal Size Range, inches Over–To	Class LT 1 Fit	Hole H7	Shaft js6	Class LT 2 Fit	Hole H8	Shaft js7	Class LT 3 Fit	Hole H7	Shaft k6	Class LT 4 Fit	Hole H8	Shaft k7	Class LT 5 Fit	Hole H7	Shaft n6	Class LT 6 Fit	Hole H7	Shaft n7
0–0.12	−0.10	+0.4	+0.10	−0.2	+0.6	+0.2							−0.5	+0.4	+0.5	−0.65	+0.4	+0.65
	+0.50	−0	−0.10	+0.8	−0	−0.2							+0.15	−0	+0.25	+0.15	−0	+0.25
0.12–0.24	−0.15	+0.5	+0.15	−0.25	+0.7	+0.25							−0.6	+0.5	+0.6	−0.8	+0.5	+0.8
	+0.65	−0	−0.15	+0.95	−0	−0.25							+0.2	−0	+0.3	+0.2	−0	+0.3
0.24–0.40	−0.2	+0.6	+0.2	−0.3	+0.9	+0.3	−0.5	+0.6	+0.5	−0.7	+0.9	+0.7	−0.8	+0.6	+0.8	−1.0	+0.6	+1.0
	+0.8	−0	−0.2	+1.2	−0	−0.3	+0.5	−0	+0.1	+0.8	−0	+0.1	+0.2	−0	+0.4	+0.2	−0	+0.4
0.40–0.71	−0.2	+0.7	+0.2	−0.35	+1.0	+0.35	−0.5	+0.7	+0.5	−0.8	+1.0	+0.8	−0.9	+0.7	+0.9	−1.2	+0.7	+1.2
	+0.9	−0	−0.2	+1.35	−0	−0.35	+0.6	−0	+0.1	+0.9	−0	+0.1	+0.2	−0	+0.5	+0.2	−0	+0.5
0.71–1.19	−0.25	+0.8	+0.25	−0.4	+1.2	+0.4	−0.6	+0.8	+0.6	−0.9	+1.2	+0.9	−1.1	+0.8	+1.1	−1.4	+0.8	+1.4
	+1.05	−0	−0.25	+1.6	−0	−0.4	+0.7	−0	+0.1	+1.1	−0	+0.1	+0.2	−0	+0.6	+0.2	−0	+0.6
1.19–1.97	−0.3	+1.0	+0.3	−0.5	+1.6	+0.5	−0.7	+1.0	+0.7	−1.1	+1.6	+1.1	−1.3	+1.0	+1.3	−1.7	+1.0	+1.7
	+1.3	−0	−0.3	+2.1	−0	−0.5	+0.9	−0	+0.1	+1.5	−0	+0.1	+0.3	−0	+0.7	+0.3	−0	+0.7
1.97–3.15	−0.3	+1.2	+0.3	−0.6	+1.8	+0.6	−0.8	+1.2	+0.8	−1.3	+1.8	+1.3	−1.5	+1.2	+1.5	−2.0	+1.2	+2.0
	+1.5	−0	−0.3	+2.4	−0	−0.6	+1.1	−0	+0.1	+1.7	−0	+0.1	+0.4	−0	+0.8	+0.4	−0	+0.8
3.15–4.73	−0.4	+1.4	+0.4	−0.7	+2.2	+0.7	−1.0	+1.4	+1.0	−1.5	+2.2	+1.5	−1.9	+1.4	+1.9	−2.4	+1.4	+2.4
	+1.8	−0	−0.4	+2.9	−0	−0.7	+1.3	−0	+0.1	+2.1	−0	+0.1	+0.4	−0	+1.0	+0.4	−0	+1.0
4.73–7.09	−0.5	+1.6	+0.5	−0.8	+2.5	+0.8	−1.1	+1.6	+1.1	−1.7	+2.5	+1.7	−2.2	+1.6	+2.2	−2.8	+1.6	+2.8
	+2.1	−0	−0.5	+3.3	−0	−0.8	+1.5	−0	+0.1	+2.4	−0	+0.1	+0.4	−0	+1.2	+0.4	−0	+1.2
7.09–9.85	−0.6	+1.8	+0.6	−0.9	+2.8	+0.9	−1.4	+1.8	+1.4	−2.0	+2.8	+2.0	−2.6	+1.8	+2.6	−3.2	+1.8	+3.2
	+2.4	−0	−0.6	+3.7	−0	−0.9	+1.6	−0	+0.2	+2.6	−0	+0.2	+0.4	−0	+1.4	+0.4	−0	+1.4
9.85–12.41	−0.6	+2.0	+0.6	−1.0	+3.0	+1.0	−1.4	+2.0	+1.4	−2.2	+3.0	+2.2	−2.6	+2.0	+2.6	−3.4	+2.0	+3.4
	+2.6	−0	−0.6	+4.0	−0	−1.0	+1.8	−0	+0.2	+2.8	−0	+0.2	+0.6	−0	+1.4	+0.6	−0	+1.4
12.41–15.75	−0.7	+2.2	+0.7	−1.0	+3.5	+1.0	−1.6	+2.2	+1.6	−2.4	+3.5	+2.4	−3.0	+2.2	+3.0	−3.8	+2.2	+3.8
	+2.9	−0	−0.7	+4.5	−0	−1.0	+2.0	−0	+0.2	+3.3	−0	+0.2	+0.6	−0	+1.6	+0.6	−0	+1.6

[a] From ANSI B4.1-1967 (R1994). For larger diameters, see the standard.

8   Interference Locational Fits[a]—American National Standard

LN   *Locational interference fits* are used where accuracy of location is of prime importance and for parts requiring rigidity and alignment with no special requirements for bore pressure. Such fits are not intended for parts designed to transmit frictional loads from one part to another by virtue of the tightness of fit, as these conditions are covered by force fits.

Basic hole system. Limits are in thousandths of an inch. See §12.8.
Limits for hole and shaft are applied algebraically to the basic size to obtain the limits of size for the parts.
Data in **boldface** are in accordance with ABC agreements.
Symbols H7, p6, etc., are hole and shaft designations used in ABC System.

Nominal Size Range, inches Over To	Class LN 1 Limits of Interference	Class LN 1 Standard Limits Hole H6	Class LN 1 Standard Limits Shaft n5	Class LN 2 Limits of Interference	Class LN 2 Standard Limits Hole H7	Class LN 2 Standard Limits Shaft p6	Class LN 3 Limits of Interference	Class LN 3 Standard Limits Hole H7	Class LN 3 Standard Limits Shaft r6
0–0.12	0 0.45	+0.25 −0	+0.45 +0.25	0 0.65	+0.4 −0	+0.65 +0.4	0.1 0.75	+0.4 −0	+0.75 +0.5
0.12–0.24	0 0.5	+0.3 −0	+0.5 +0.3	0 0.8	+0.5 −0	+0.8 +0.5	0.1 0.9	+0.5 0	+0.9 +0.6
0.24–0.40	0 0.65	+0.4 −0	+0.65 +0.4	0 1.0	+0.6 −0	+1.0 +0.6	0.2 1.2	+0.6 −0	+1.2 +0.8
0.40–0.71	0 0.8	+0.4 −0	+0.8 +0.4	0 1.1	+0.7 −0	+1.1 +0.7	0.3 1.4	+0.7 −0	+1.4 +1.0
0.71–1.19	0 1.0	+0.5 −0	+1.0 +0.5	0 1.3	+0.8 −0	+1.3 +0.8	0.4 1.7	+0.8 −0	+1.7 +1.2
1.19–1.97	0 1.1	+0.6 −0	+1.1 +0.6	0 1.6	+1.0 −0	+1.6 +1.0	0.4 2.0	+1.0 −0	+2.0 +1.4
1.97–3.15	0.1 1.3	+0.7 −0	+1.3 +0.7	0.2 2.1	+1.2 −0	+2.1 +1.4	0.4 2.3	+1.2 −0	+2.3 +1.6
3.15–4.73	0.1 1.6	+0.9 −0	+1.6 +1.0	0.2 2.5	+1.4 −0	+2.5 +1.6	0.6 2.9	+1.4 −0	+2.9 +2.0
4.73–7.09	0.2 1.9	+1.0 −0	+1.9 +1.2	0.2 2.8	+1.6 −0	+2.8 +1.8	0.9 3.5	+1.6 −0	+3.5 +2.5
7.09–9.85	0.2 2.2	+1.2 −0	+2.2 +1.4	0.2 3.2	+1.8 −0	+3.2 +2.0	1.2 4.2	+1.8 −0	+4.2 +3.0
9.85–12.41	0.2 2.3	+1.2 −0	+2.3 +1.4	0.2 3.4	+2.0 −0	+3.4 +2.2	1.5 4.7	+2.0 −0	+4.7 +3.5

[a] From ANSI B4.1–1967 (R1994). For larger diameters, see the standard.

FN 1   Light *drive fits* are those requiring light assembly pressures, and produce more or less permanent assemblies. They are suitable for thin sections or long fits, or in cast-iron external members.

FN 2   Medium *drive fits* are suitable for ordinary steel parts, or for shrink fits on light sections. They are about the tightest fits that can be used with high-grade cast-iron external members.

FN 3   Heavy *drive fits* are suitable for heavier steel parts or for shrink fits in medium sections.

FN 4   }
FN 5   }   *Force fits* are suitable for parts which can be highly stressed, or for shrink fits where the heavy pressing forces required are impractical.

Basic hole system. Limits are in thousandths of an inch. See §12.8.
Limits for hole and shaft are applied algebraically to the basic size to obtain the limits of size for the parts.
Data in **boldface** are in accordance with ABC agreements.
Symbols H7, s6, etc., are hole and shaft designations used in ABC System.

Nominal Size Range, inches		Class FN 1			Class FN 2			Class FN 3			Class FN 4			Class FN 5		
		Limits of Interference	Standard Limits		Limits of Interference	Standard Limits		Limits of Interference	Standard Limits		Limits of Interference	Standard Limits		Limits of Interference	Standard Limits	
Over	To		Hole H6	Shaft		Hole H7	Shaft s6		Hole H7	Shaft t6		Hole H7	Shaft u6		Hole H8	Shaft x7
0	0.12	0.05 / 0.5	+0.25 / −0	+0.5 / +0.3	**0.2 / 0.85**	**+0.4 / −0**	**+0.85 / +0.6**				**0.3 / 0.95**	**+0.4 / −0**	**+0.95 / +0.7**	**0.3 / 1.3**	**+0.6 / −0**	**+1.3 / +0.9**
0.12	0.24	0.1 / 0.6	+0.3 / −0	+0.6 / +0.4	**0.2 / 1.0**	**+0.5 / −0**	**+1.0 / +0.7**				**0.4 / 1.2**	**+0.5 / −0**	**+1.2 / +0.9**	**0.5 / 1.7**	**+0.7 / −0**	**+1.7 / +1.2**
0.24	0.40	0.1 / 0.75	+0.4 / −0	+0.75 / +0.5	**0.4 / 1.4**	**+0.6 / −0**	**+1.4 / +1.0**				**0.6 / 1.6**	**+0.6 / −0**	**+1.6 / +1.2**	**0.5 / 2.0**	**+0.9 / −0**	**+2.0 / +1.4**
0.40	0.56	0.1 / 0.8	+0.4 / −0	+0.8 / +0.5	**0.5 / 1.6**	**+0.7 / −0**	**+1.6 / +1.2**				**0.7 / 1.8**	**+0.7 / −0**	**+1.8 / +1.4**	**0.6 / 2.3**	**+1.0 / −0**	**+2.3 / +1.6**
0.56	0.71	0.2 / 0.9	+0.4 / −0	+0.9 / +0.6	**0.5 / 1.6**	**+0.7 / −0**	**+1.6 / +1.2**				**0.7 / 1.8**	**+0.7 / −0**	**+1.8 / +1.4**	**0.8 / 2.5**	**+1.0 / −0**	**+2.5 / +1.8**
0.71	0.95	0.2 / 1.1	+0.5 / −0	+1.1 / +0.7	**0.6 / 1.9**	**+0.8 / −0**	**+1.9 / +1.4**				**0.8 / 2.1**	**+0.8 / −0**	**+2.1 / +1.6**	**1.0 / 3.0**	**+1.2 / −0**	**+3.0 / +2.2**
0.95	1.19	0.3 / 1.2	+0.5 / −0	+1.2 / +0.8	**0.6 / 1.9**	**+0.8 / −0**	**+1.9 / +1.4**	**0.8 / 2.1**	**+0.8 / −0**	**+2.1 / +1.6**	**1.0 / 2.3**	**+0.8 / −0**	**+2.3 / +1.8**	**1.3 / 3.3**	**+1.2 / −0**	**+3.3 / +2.5**
1.19	1.58	0.3 / 1.3	+0.6 / −0	+1.3 / +0.9	**0.8 / 2.4**	**+1.0 / −0**	**+2.4 / +1.8**	**1.0 / 2.6**	**+1.0 / −0**	**+2.6 / +2.0**	**1.5 / 3.1**	**+1.0 / −0**	**+3.1 / +2.5**	**1.4 / 4.0**	**+1.6 / −0**	**+4.0 / +3.0**

[a] ANSI B4.1-1967 (R1994).

Nominal Size Range, inches Over–To	Class FN 1 Limits of Interference	Class FN 1 Standard Limits Hole H6	Class FN 1 Standard Limits Shaft	Class FN 2 Limits of Interference	Class FN 2 Standard Limits Hole H7	Class FN 2 Standard Limits Shaft s6	Class FN 3 Limits of Interference	Class FN 3 Standard Limits Hole H7	Class FN 3 Standard Limits Shaft t6	Class FN 4 Limits of Interference	Class FN 4 Standard Limits Hole H7	Class FN 4 Standard Limits Shaft u6	Class FN 5 Limits of Interference	Class FN 5 Standard Limits Hole H8	Class FN 5 Standard Limits Shaft x7
1.58–1.97	0.4 / 1.4	+0.6 / −0	+1.4 / −1.0	0.8 / 2.4	+1.0 / −0	+2.4 / +1.8	1.2 / 2.8	+1.0 / −0	+2.8 / +2.2	1.8 / 3.4	+1.0 / −0	+3.4 / +2.8	2.4 / 5.0	+1.6 / −0	+5.0 / +4.0
1.97–2.56	0.6 / 1.8	+0.7 / −0	+1.8 / +1.3	0.8 / 2.7	+1.2 / −0	+2.7 / +2.0	1.3 / 3.2	+1.2 / −0	+3.2 / +2.5	2.3 / 4.2	+1.2 / −0	+4.2 / +3.5	3.2 / 6.2	+1.8 / −0	+6.2 / +5.0
2.56–3.15	0.7 / 1.9	+0.7 / −0	+1.9 / +1.4	1.0 / 2.9	+1.2 / −0	+2.9 / +2.2	1.8 / 3.7	+1.2 / −0	+3.7 / +3.0	2.8 / 4.7	+1.2 / −0	+4.7 / +4.0	4.2 / 7.2	+1.8 / −0	+7.2 / +6.0
3.15–3.94	0.9 / 24	+0.9 / −0	+2.4 / +1.8	1.4 / 3.7	+1.4 / −0	+3.7 / +2.8	2.1 / 4.4	+1.4 / −0	+4.4 / +3.5	3.6 / 5.9	+1.4 / −0	+5.9 / +5.0	4.8 / 8.4	+2.2 / −0	+8.4 / +7.0
3.94–4.73	1.1 / 2.6	+0.9 / −0	+2.6 / +2.0	1.6 / 3.9	+1.4 / −0	+3.9 / +3.0	2.6 / 4.9	+1.4 / −0	+4.9 / +4.0	4.6 / 6.9	+1.4 / −0	+6.9 / +6.0	5.8 / 9.4	+2.2 / −0	+9.4 / +8.0
4.73–5.52	1.2 / 2.9	+1.0 / −0	+2.9 / +2.2	1.9 / 4.5	+1.6 / −0	+4.5 / +3.5	3.4 / 6.0	+1.6 / −0	+6.0 / +5.0	5.4 / 8.0	+1.6 / −0	+8.0 / +7.0	7.5 / 11.6	+2.5 / −0	+11.6 / +10.0
5.52–6.30	1.5 / 3.2	+1.0 / −0	+3.2 / +2.5	2.4 / 5.0	+1.6 / −0	+5.0 / +4.0	3.4 / 6.0	+1.6 / −0	+6.0 / +5.0	5.4 / 8.0	+1.6 / −0	+8.0 / +7.0	9.5 / 13.6	+2.5 / −0	+13.6 / +12.0
6.30–7.09	1.8 / 3.5	+1.0 / −0	+3.5 / +2.8	2.9 / 5.5	+1.6 / −0	+5.5 / +4.5	4.4 / 7.0	+1.6 / −0	+7.0 / +6.0	6.4 / 9.0	+1.6 / −0	+9.0 / +8.0	9.5 / 13.6	+2.5 / −0	+13.6 / +12.0
7.09–7.88	1.8 / 3.8	+1.2 / −0	+3.8 / +3.0	3.2 / 6.2	+1.8 / −0	+6.2 / +5.0	5.2 / 8.2	+1.8 / −0	+8.2 / +7.0	7.2 / 10.2	+1.8 / −0	+10.2 / +9.0	11.2 / 15.8	+2.8 / −0	+15.8 / +14.0
7.88–8.86	2.3 / 4.3	+1.2 / −0	+4.3 / +3.5	3.2 / 6.2	+1.8 / −0	+6.2 / +5.0	5.2 / 8.2	+1.8 / −0	+8.2 / +7.0	8.2 / 11.2	+1.8 / −0	+11.2 / +10.0	13.2 / 17.8	+2.8 / −0	+17.8 / +16.0
8.86–9.85	2.3 / 4.3	+1.2 / −0	+4.3 / +3.5	4.2 / 7.2	+1.8 / −0	+7.2 / +6.0	6.2 / 9.2	+1.8 / −0	+9.2 / +8.0	10.2 / 13.2	+1.8 / −0	+13.2 / +12.0	13.2 / 17.8	+2.8 / −0	+17.8 / +16.0
9.85–11.03	2.8 / 4.9	+1.2 / −0	+4.9 / +4.0	4.0 / 7.2	+2.0 / −0	+7.2 / +6.0	7.0 / 10.2	+2.0 / −0	+10.2 / +9.0	10.0 / 13.2	+2.0 / −0	+13.2 / +12.0	15.0 / 20.0	+3.0 / −0	+20.0 / +18.0
11.03–12.41	2.8 / 4.9	+1.2 / −0	+4.9 / +4.0	5.0 / 8.2	+2.0 / −0	+8.2 / +7.0	7.0 / 10.2	+2.0 / −0	+10.2 / +9.0	12.0 / 15.2	+2.0 / −0	+15.2 / +14.0	17.0 / 22.0	+3.0 / −0	+22.0 / +20.0
12.41–13.98	3.1 / 5.5	+1.4 / −0	+5.5 / +4.5	5.8 / 9.4	+2.2 / −0	+9.4 / +8.0	7.8 / 11.4	+2.2 / −0	+11.4 / +10.0	13.8 / 17.4	+2.2 / −0	+17.4 / +16.0	18.5 / 24.2	+3.5 / +0	+24.2 / +22.0

[a] From ANSI B4.1-1967 (R1994). For larger diameters, see the standard.

10 International Tolerance Grades[a]

Dimensions are in millimeters.

Basic sizes		Tolerance grades[b]																		
Over	Up to and Including	IT01	IT0	IT1	IT2	IT3	IT4	IT5	IT6	IT7	IT8	IT9	IT10	IT11	IT12	IT13	IT14	IT15	IT16
0	3	0.0003	0.0005	0.0008	0.0012	0.002	0.003	0.004	0.006	0.010	0.014	0.025	0.040	0.060	0.100	0.140	0.250	0.400	0.600
3	6	0.0004	0.0006	0.001	0.0015	0.0025	0.004	0.005	0.008	0.012	0.018	0.030	0.048	0.075	0.120	0.180	0.300	0.480	0.750
6	10	0.0004	0.0006	0.001	0.0015	0.0025	0.004	0.006	0.009	0.015	0.022	0.036	0.058	0.090	0.150	0.220	0.360	0.580	0.900
10	18	0.0005	0.0008	0.0012	0.002	0.003	0.005	0.008	0.011	0.018	0.027	0.043	0.070	0.110	0.180	0.270	0.430	0.700	1.100
18	30	0.0006	0.001	0.0015	0.0025	0.004	0.006	0.009	0.013	0.021	0.033	0.052	0.084	0.130	0.210	0.330	0.520	0.840	1.300
30	50	0.0006	0.001	0.0015	0.0025	0.004	0.007	0.011	0.016	0.025	0.039	0.062	0.100	0.160	0.250	0.390	0.620	1.000	1.600
50	80	0.0008	0.0012	0.002	0.003	0.005	0.008	0.013	0.019	0.030	0.046	0.074	0.120	0.190	0.300	0.460	0.740	1.200	1.900
80	120	0.001	0.0015	0.0025	0.004	0.006	0.010	0.015	0.022	0.035	0.054	0.087	0.140	0.220	0.350	0.540	0.870	1.400	2.200
120	180	0.0012	0.002	0.0035	0.005	0.008	0.012	0.018	0.025	0.040	0.063	0.100	0.160	0.250	0.400	0.630	1.000	1.600	2.500
180	250	0.002	0.003	0.0045	0.007	0.010	0.014	0.020	0.029	0.046	0.072	0.115	0.185	0.290	0.460	0.720	1.150	1.850	2.900
250	315	0.0025	0.004	0.006	0.008	0.012	0.016	0.023	0.032	0.052	0.081	0.130	0.210	0.320	0.520	0.810	1.300	2.100	3.200
315	400	0.003	0.005	0.007	0.009	0.013	0.018	0.025	0.036	0.057	0.089	0.140	0.230	0.360	0.570	0.890	1.400	2.300	3.600
400	500	0.004	0.006	0.008	0.010	0.015	0.020	0.027	0.040	0.063	0.097	0.155	0.250	0.400	0.630	0.970	1.550	2.500	4.000
500	630	0.0045	0.006	0.009	0.011	0.016	0.022	0.030	0.044	0.070	0.110	0.175	0.280	0.440	0.700	1.100	1.750	2.800	4.400
630	800	0.005	0.007	0.010	0.013	0.018	0.025	0.035	0.050	0.080	0.125	0.200	0.320	0.500	0.800	1.250	2.000	3.200	5.000
800	1000	0.0055	0.008	0.011	0.015	0.021	0.029	0.040	0.056	0.090	0.140	0.230	0.360	0.560	0.900	1.400	2.300	3.600	5.600
1000	1250	0.0065	0.009	0.013	0.018	0.024	0.034	0.046	0.066	0.105	0.165	0.260	0.420	0.660	1.050	1.650	2.600	4.200	6.600
1250	1600	0.008	0.011	0.015	0.021	0.029	0.040	0.054	0.078	0.125	0.195	0.310	0.500	0.780	1.250	1.950	3.100	5.000	7.800
1600	2000	0.009	0.013	0.018	0.025	0.035	0.048	0.065	0.092	0.150	0.230	0.370	0.600	0.920	1.500	2.300	3.700	6.000	9.200
2000	2500	0.011	0.015	0.022	0.030	0.041	0.057	0.077	0.110	0.175	0.280	0.440	0.700	1.100	1.750	2.800	4.400	7.000	11.000
2500	3150	0.013	0.018	0.026	0.036	0.050	0.069	0.093	0.135	0.210	0.330	0.540	0.860	1.350	2.100	3.300	5.400	8.600	13.500

[a] From ANSI B4.2–1978 (R1994).
[b] IT Values for tolerance grades larger than IT16 can be calculated by using the formulas: IT17 = IT × 10, IT18 = IT13 × 10, etc.

Dimensions are in millimeters.

Basic Size		Loose Running			Free Running			Close Running			Sliding			Locational Clearance		
		Hole H11	Shaft c11	Fit	Hole H9	Shaft d9	Fit	Hole H8	f7	Fit	Hole H7	Shaft g6	Fit	Hole H7	Shaft h6	Fit
1	Max	1.060	0.940	0.180	1.025	0.980	0.070	1.014	0.994	0.030	1.010	0.998	0.018	1.010	1.000	0.016
	Min	1.060	0.880	0.060	1.000	0.955	0.020	1.000	0.984	0.006	1.000	0.992	0.002	1.000	0.994	0.000
1.2	Max	1.260	1.140	0.180	1.225	1.180	0.070	1.214	1.194	0.030	1.210	1.198	0.018	1.210	1.200	0.016
	Min	1.200	1.080	0.060	1.200	1.155	0.020	1.200	1.184	0.006	1.200	1.192	0.002	1.200	1.194	0.000
1.6	Max	1.660	1.540	0.180	1.625	1.580	0.070	1.614	1.594	0.030	1.610	1.598	0.018	1.610	1.600	0.016
	Min	1.600	1.480	0.060	1.600	1.555	0.020	1.600	1.584	0.006	1.600	1.592	0.002	1.600	1.594	0.000
2	Max	2.060	1.940	0.180	2.025	1.980	0.070	2.014	1.994	0.030	2.010	1.998	0.018	2.010	2.000	0.016
	Min	2.000	1.880	0.060	2.000	1.955	0.020	2.000	1.984	0.006	2.000	1.992	0.002	2.000	1.994	0.000
2.5	Max	2.560	2.440	0.180	2.525	2.480	0.070	2.514	2.494	0.030	2.510	2.498	0.018	2.510	2.500	0.016
	Min	2.500	2.380	0.060	2.500	2.455	0.020	2.500	2.484	0.006	2.500	2.492	0.002	2.500	2.494	0.000
3	Max	3.060	2.940	0.180	3.025	2.980	0.070	3.014	2.994	0.030	3.010	2.998	0.018	3.010	3.000	0.016
	Min	3.000	2.880	0.060	3.000	2.955	0.020	3.000	2.984	0.006	3.000	2.992	0.002	3.000	2.994	0.000
4	Max	4.075	3.930	0.220	4.030	3.970	0.090	4.018	3.990	0.040	4.012	3.996	0.024	4.012	4.000	0.020
	Min	4.000	3.855	0.070	4.000	3.940	0.030	4.000	3.978	0.010	4.000	3.988	0.004	4.000	3.992	0.000
5	Max	5.075	4.930	0.220	5.030	4.970	0.090	5.018	4.990	0.040	5.012	4.996	0.024	5.012	5.000	0.020
	Min	5.000	4.855	0.070	5.000	4.940	0.030	5.000	4.978	0.010	5.000	4.988	0.004	5.000	4.992	0.000
6	Max	6.075	5.930	0.220	6.030	5.970	0.090	6.018	5.990	0.040	6.012	5.996	0.024	6.012	6.000	0.020
	Min	6.000	5.855	0.070	6.000	5.940	0.030	6.000	5.978	0.010	6.000	5.988	0.004	6.000	5.992	0.000
8	Max	8.090	7.920	0.260	8.036	7.960	0.112	8.022	7.987	0.050	8.015	7.995	0.029	8.015	8.000	0.024
	Min	8.000	7.830	0.080	8.000	7.924	0.040	8.000	7.972	0.013	8.000	7.986	0.005	8.000	7.991	0.000
10	Max	10.090	9.920	0.260	10.036	9.960	0.112	10.022	9.987	0.050	10.015	9.995	0.029	10.015	10.000	0.024
	Min	10.000	9.830	0.080	10.000	9.924	0.040	10.000	9.972	0.013	10.000	9.986	0.005	10.000	9.991	0.000
12	Max	12.110	11.905	0.315	12.043	11.950	0.136	12.027	11.984	0.061	12.018	11.994	0.035	12.018	12.000	0.029
	Min	12.000	11.795	0.095	12.000	11.907	0.050	12.000	11.966	0.016	12.000	11.983	0.006	12.000	11.989	0.000
16	Max	16.110	15.905	0.315	16.043	15.950	0.136	16.027	15.984	0.061	16.018	15.994	0.035	16.018	16.000	0.029
	Min	16.000	15.795	0.095	16.000	15.907	0.050	16.000	15.966	0.016	16.000	15.983	0.006	16.000	15.989	0.000
20	Max	20.130	19.890	0.370	20.052	19.935	0.169	20.033	19.980	0.074	20.021	19.993	0.041	20.021	20.000	0.034
	Min	20.000	19.760	0.110	20.000	19.883	0.065	20.000	19.959	0.020	20.000	19.980	0.007	20.000	19.987	0.000
25	Max	25.130	24.890	0.370	25.052	24.935	0.169	25.033	24.980	0.074	25.021	24.993	0.041	25.021	25.000	0.034
	Min	25.000	24.760	0.110	25.000	24.883	0.065	25.000	24.959	0.020	25.000	24.980	0.007	25.000	24.987	0.000
30	Max	30.130	29.890	0.370	30.052	29.935	0.169	30.033	29.980	0.074	30.021	29.993	0.041	30.021	30.000	0.034
	Min	30.000	29.760	0.110	30.000	29.883	0.065	30.000	29.959	0.020	30.000	29.980	0.007	30.000	29.987	0.000

[a] From ANSI B4.2-1978 (R1994). For description of preferred fits, see Table 12.2.

Dimensions are in millimeters.

Basic Size		Loose Running			Free Running			Close Running			Sliding			Locational Clearance		
		Hole H11	Shaft c11	Fit	Hole H9	Shaft d9	Fit	Hole H8	Shaft f7	Fit	Hole H7	Shaft g6	Fit	Hole H7	Shaft h6	Fit
40	Max	40.160	39.880	0.440	40.062	39.920	0.204	40.039	39.975	0.089	40.025	39.991	0.050	40.025	40.000	0.041
	Min	40.000	39.720	0.120	40.000	39.858	0.080	40.000	39.950	0.025	40.000	39.975	0.009	40.000	39.984	0.000
50	Max	50.160	49.870	0.450	50.062	49.920	0.204	50.039	49.975	0.089	50.025	49.991	0.050	50.025	50.000	0.041
	Min	50.000	49.710	0.130	50.000	49.858	0.080	50.000	49.950	0.025	50.000	49.975	0.009	50.000	49.984	0.000
60	Max	60.190	59.860	0.520	60.074	59.900	0.248	60.046	59.970	0.106	60.030	59.990	0.059	60.030	60.000	0.049
	Min	60.000	59.670	0.140	60.000	59.826	0.100	60.000	59.940	0.030	60.000	59.971	0.010	60.000	59.981	0.000
80	Max	80.190	79.950	0.530	80.074	79.900	0.248	80.046	79.970	0.106	80.030	79.990	0.059	80.030	80.000	0.049
	Min	80.000	79.660	0.150	80.000	79.826	0.100	80.000	79.940	0.030	80.000	79.971·	0.010	80.000	79.981	0.000
100	Max	100.220	99.830	0.610	100.087	99.880	0.294	100.054	99.964	0.125	100.035	99.988	0.069	100.035	100.000	0.057
	Min	100.000	99.610	0.170	100.000	99.793	0.120	100.000	99.929	0.036	100.000	99.966	0.012	100.000	99.978	0.000
120	Max	120.220	119.820	0.620	120.087	119.880	0.294	120.054	119.964	0.125	120.035	119.988	0.069	120.035	120.000	0.057
	Min	120.000	119.600	0.180	120.000	119.793	0.120	120.000	119.929	0.036	120.000	119.966	0.012	120.000	119.978	0.000
160	Max	160.250	159.790	0.710	160.100	159.855	0.345	160.063	159.957	0.146	160.040	159.986	0.079	160.040	160.000	0.065
	Min	160.000	159.540	0.210	160.000	159.755	0.145	160.000	159.917	0.043	160.000	159.961	0.014	160.000	159.975	0.000
200	Max	200.290	199.760	0.820	200.115	199.830	0.400	200.072	199.950	0.168	200.046	199.985	0.090	200.046	200.000	0.075
	Min	200.000	199.470	0.240	200.000	199.715	0.170	200.000	199.904	0.050	200.000	199.956	0.015	200.000	199.971	0.000
250	Max	250.290	249.720	0.860	250.115	249.830	0.400	250.072	249.950	0.168	250.046	249.985	0.090	250.046	250.000	0.075
	Min	250.000	249.430	0.280	250.000	249.715	0.170	250.000	249.904	0.050	250.000	249.956	0.015	250.000	249.971	0.000
300	Max	300.320	299.670	0.970	300.130	299.810	0.450	300.081	299.944	0.189	300.052	299.983	0.101	300.052	300.000	0.084
	Min	300.000	299.350	0.330	300.000	299.680	0.190	300.000	299.892	0.056	300.000	299.951	0.017	300.000	299.968	0.000
400	Max	400.360	399.600	1.120	400.140	399.790	0.490	400.089	399.938	0.208	400.057	399.982	0.111	400.057	400.000	0.093
	Min	400.000	399.240	0.400	400.000	399.650	0.210	400.000	399.881	0.062	400.000	399.946	0.018	400.000	399.964	0.000
500	Max	500.400	499.520	1.280	500.155	499.770	0.540	500.097	499.932	0.228	500.063	499.980	0.123	500.063	500.000	0.103
	Min	500.000	499.120	0.480	500.000	499.615	0.230	500.000	499.869	0.068	500.000	499.940	0.020	500.000	499.960	0.000

[a] From ANSI B4.2–1978 (R1994). For description of preferred fits, see Table 12.2.

12 Preferred Metric Hole Basis Transition and Interference Fits[a]—American National Standard

Dimensions are in millimeters.

Basic Size		Locational Transn. Hole H7	Locational Transn. Shaft k6	Locational Transn. Fit	Locational Transn. Hole H7	Locational Transn. Shaft n6	Locational Transn. Fit	Locational Interf. Hole H7	Locational Interf. Shaft p6	Locational Interf. Fit	Medium Drive Hole H7	Medium Drive Shaft s6	Medium Drive Fit	Force Hole H7	Force Shaft u6	Force Fit
1	Max	1.010	1.006	0.010	1.010	1.010	0.006	1.010	1.012	0.004	1.010	1.020	−0.004	1.010	1.024	−0.008
	Min	1.000	1.000	−0.006	1.000	1.004	−0.010	1.000	1.006	−0.012	1.000	1.014	−0.020	1.000	1.018	−0.024
1.2	Max	1.210	1.206	0.010	1.210	1.210	0.006	1.210	1.212	0.004	1.210	1.220	−0.004	1.210	1.224	−0.008
	Min	1.200	1.200	−0.006	1.200	1.204	−0.010	1.200	1.206	−0.012	1.200	1.214	−0.020	1.200	1.218	−0.024
1.6	Max	1.610	1.606	0.010	1.610	1.610	0.006	1.610	1.612	0.004	1.610	1.620	−0.004	1.610	1.624	−0.008
	Min	1.600	1.600	−0.006	1.600	1.604	−0.010	1.600	1.606	−0.012	1.600	1.614	−0.020	1.600	1.618	−0.024
2	Max	2.010	2.006	0.010	2.010	2.010	0.006	2.010	2.012	0.004	2.010	2.020	−0.004	2.010	2.024	−0.008
	Min	2.000	2.000	−0.006	2.000	2.004	−0.010	2.000	2.006	−0.012	2.000	2.014	−0.020	2.000	2.018	−0.024
2.5	Max	2.510	2.506	0.010	2.510	2.510	0.006	2.510	2.512	0.004	2.510	2.520	−0.004	2.510	2.524	−0.008
	Min	2.500	2.500	−0.006	2.500	2.504	−0.010	2.500	2.506	−0.012	2.500	2.514	−0.020	2.500	2.518	−0.024
3	Max	3.010	3.006	0.010	3.010	3.010	0.006	3.010	3.012	0.004	3.010	3.020	−0.004	3.010	3.024	−0.008
	Min	3.000	3.000	−0.006	3.000	3.004	−0.010	3.000	3.006	−0.012	3.000	3.014	−0.020	3.000	3.018	−0.024
4	Max	4.012	4.009	0.011	4.012	4.016	0.004	4.012	4.020	0.000	4.012	4.027	−0.007	4.012	4.031	−0.011
	Min	4.000	4.001	−0.009	4.000	4.008	−0.016	4.000	4.012	−0.020	4.000	4.019	−0.027	4.000	4.023	−0.031
5	Max	5.012	5.009	0.011	5.012	5.016	0.004	5.012	5.020	0.000	5.012	5.027	−0.007	5.012	5.031	−0.011
	Min	5.000	5.001	−0.009	5.000	5.008	−0.016	5.000	5.012	−0.020	5.000	5.019	−0.027	5.000	5.023	−0.031
6	Max	6.012	6.009	0.011	6.012	6.016	0.004	6.012	6.020	0.000	6.012	6.027	−0.007	6.012	6.031	−0.011
	Min	6.000	6.001	−0.009	6.000	6.008	−0.016	6.000	6.012	−0.020	6.000	6.019	−0.027	6.000	6.023	−0.031
8	Max	8.015	8.010	0.014	8.015	8.019	0.005	8.015	8.024	0.000	8.015	8.032	−0.008	8.015	8.037	−0.013
	Min	8.000	8.001	−0.010	8.000	8.010	−0.019	8.000	8.015	−0.024	8.000	8.023	−0.032	8.000	8.028	−0.037
10	Max	10.015	10.010	0.014	10.015	10.019	0.005	10.015	10.024	0.000	10.015	10.032	−0.008	10.015	10.037	−0.013
	Min	10.000	10.001	−0.010	10.000	10.010	−0.019	10.000	10.015	−0.024	10.000	10.023	−0.032	10.000	10.028	−0.037
12	Max	12.018	12.012	0.017	12.018	12.023	0.006	12.018	12.029	0.000	12.018	12.039	−0.010	12.018	12.044	−0.015
	Min	12.000	12.001	−0.012	12.000	12.012	−0.023	12.000	12.018	−0.029	12.000	12.028	−0.039	12.000	12.033	−0.044
16	Max	16.018	16.012	0.017	16.018	16.023	0.006	16.018	16.029	0.000	16.018	16.039	−0.010	16.018	16.044	−0.015
	Min	16.000	16.001	−0.012	16.000	16.012	−0.023	16.000	16.018	−0.029	16.000	16.028	−0.039	16.000	16.033	−0.044
20	Max	20.021	20.015	0.019	20.021	20.028	0.006	20.021	20.035	−0.001	20.021	20.048	−0.014	20.021	20.054	−0.020
	Min	20.000	20.002	−0.015	20.000	20.015	−0.028	20.000	20.022	−0.035	20.000	20.035	−0.048	20.000	20.041	−0.054
25	Max	25.021	25.015	0.019	25.021	25.028	0.006	25.021	25.035	−0.001	25.021	25.048	−0.014	25.021	25.061	−0.027
	Min	25.000	25.002	−0.015	25.000	25.015	−0.028	25.000	25.022	−0.035	25.000	25.035	−0.048	25.000	25.048	−0.061
30	Max	30.021	30.015	0.019	30.021	30.028	0.006	30.021	30.035	−0.001	30.021	30.048	−0.014	30.021	30.061	−0.027
	Min	30.000	30.002	−0.015	30.000	30.015	−0.028	30.000	30.022	−0.035	30.000	30.035	−0.048	30.000	30.048	−0.061

[a] From ANSI B4.2-1978 (R1994).

Dimensions are in millimeters.

Basic Size		Locational Transn.			Locational Transn.			Locational Interf.			Medium Drive			Force		
		Hole H7	Shaft k6	Fit	Hole H7	Shaft n6	Fit	Hole H7	Shaft p6	Fit	Hole H7	Shaft s6	Fit	Hole H7	Shaft u6	Fit
40	Max	40.025	40.018	0.023	40.025	40.033	0.08	40.025	40.042	−0.001	40.025	40.059	−0.018	40.025	40.076	−0.035
	Min	40.000	40.002	−0.018	40.000	40.017	−0.033	40.000	40.026	−0.042	40.000	40.043	−0.059	40.000	40.060	−0.076
50	Max	50.025	50.018	0.023	50.025	50.033	0.008	50.025	50.042	−0.001	50.025	50.059	−0.018	50.025	50.086	−0.045
	Min	50.000	50.002	−0.018	50.000	50.017	−0.033	50.000	50.026	−0.042	50.000	50.043	−0.059	50.000	50.070	−0.086
60	Max	60.030	60.021	0.028	60.030	60.039	0.010	60.030	60.051	−0.002	60.030	60.072	−0.023	60.030	60.106	−0.057
	Min	60.000	60.002	−0.021	60.000	60.020	−0.039	60.000	60.032	−0.051	60.000	60.053	−0.072	60.000	60.087	−0.106
80	Max	80.030	80.021	0.028	80.030	80.039	0.010	80.030	80.051	−0.002	80.030	80.078	−0.029	80.030	80.121	−0.072
	Min	80.000	80.002	−0.021	80.000	80.020	−0.039	80.000	80.032	−0.051	80.000	80.059	−0.078	80.000	80.102	−0.121
100	Max	100.035	100.025	0.032	100.035	100.045	0.012	100.035	100.059	−0.002	100.035	100.093	−0.036	100.035	100.146	−0.089
	Min	100.000	100.003	−0.025	100.000	100.023	−0.045	100.000	100.037	−0.059	100.000	100.071	−0.093	100.000	100.124	−0.146
120	Max	120.035	120.025	0.032	120.035	120.045	0.012	120.035	120.059	−0.002	120.035	120.101	−0.044	120.035	120.166	−0.109
	Min	120.000	120.003	−0.025	120.000	120.023	−0.045	120.000	120.037	−0.059	120.000	120.079	−0.101	120.000	120.144	−0.166
160	Max	160.040	160.028	0.037	160.040	160.052	0.013	160.040	160.068	−0.003	160.040	160.125	−0.060	160.040	160.215	−0.150
	Min	160.000	160.003	−0.028	160.000	160.027	−0.052	160.000	160.043	−0.068	160.000	160.100	−0.125	160.000	160.190	−0.215
200	Max	200.046	200.033	0.042	200.046	200.060	0.015	200.046	200.079	−0.004	200.046	200.151	−0.076	200.046	200.265	−0.190
	Min	200.000	200.004	−0.033	200.000	200.031	−0.060	200.000	200.050	−0.079	200.000	200.122	−0.151	200.000	200.236	−0.265
250	Max	250.046	250.033	0.042	250.046	250.060	0.015	250.046	250.079	−0.004	250.046	250.169	−0.094	250.046	250.313	−0.238
	Min	250.000	250.004	−0.033	250.000	250.031	−0.060	250.000	250.050	−0.079	250.000	250.140	−0.169	250.000	250.284	−0.313
300	Max	300.052	300.036	0.048	300.052	300.066	0.018	300.052	300.088	−0.004	300.052	300.202	−0.118	300.052	300.382	−0.298
	Min	300.000	300.004	−0.036	300.000	300.034	−0.066	300.000	300.056	−0.088	300.000	300.170	−0.202	300.000	300.350	−0.382
400	Max	400.057	400.040	0.053	400.057	400.073	0.020	400.057	400.098	−0.005	400.057	400.244	−0.151	400.057	400.471	−0.378
	Min	400.000	400.004	−0.040	400.000	400.037	−0.073	400.000	400.062	−0.098	400.000	400.208	−0.244	400.000	400.435	−0.471
500	Max	500.063	500.045	0.058	500.063	500.080	0.023	500.063	500.108	−0.005	500.063	500.292	−0.189	500.063	500.580	−0.477
	Min	500.000	500.005	−0.045	500.000	500.040	−0.080	500.000	500.068	−0.108	500.000	500.252	−0.292	500.000	500.540	−0.580

[a] From ANSI B4.2–1978 (R1994).

Dimensions are in millimeters.

Basic Size		Loose Running			Free Running			Close Running			Sliding			Locational Clearance		
		Hole C11	Shaft h11	Fit	Hole D9	Shaft h9	Fit	Hole F8	Shaft h7	Fit	Hole G7	Shaft h6	Fit	Hole H7	Shaft h6	Fit
1	Max	1.120	1.000	0.180	1.045	1.000	0.070	1.020	1.000	0.030	1.012	1.000	0.018	1.010	1.000	0.016
	Min	1.060	0.940	0.060	1.020	0.975	0.020	1.006	0.990	0.006	1.002	0.994	0.002	1.000	0.994	0.000
1.2	Max	1.320	1.200	0.180	1.245	1.200	0.070	1.220	1.200	0.030	1.212	1.200	0.018	1.210	1.200	0.016
	Min	1.260	1.140	0.060	1.220	1.175	0.020	1.206	1.190	0.006	1.202	1.194	0.002	1.200	1.194	0.000
1.6	Max	1.720	1.600	0.180	1.645	1.600	0.070	1.620	1.600	0.030	1.612	1.600	0.018	1.610	1.600	0.016
	Min	1.660	1.540	0.060	1.620	1.575	0.020	1.606	1.590	0.006	1.602	1.594	0.002	1.600	1.594	0.000
2	Max	2.120	2.000	0.180	2.045	2.000	0.070	2.020	2.000	0.030	2.012	2.000	0.018	2.010	2.000	0.016
	Min	2.060	1.940	0.060	2.020	1.975	0.020	2.006	1.990	0.006	2.002	1.994	0.002	2.000	1.994	0.000
2.5	Max	2.620	2.500	0.180	2.545	2.500	0.070	2.520	2.500	0.030	2.512	2.500	0.018	2.510	2.500	0.016
	Min	2.560	2.440	0.060	2.520	2.475	0.020	2.506	2.490	0.006	2.502	2.494	0.002	2.500	2.494	0.000
3	Max	3.120	3.000	0.180	3.045	3.000	0.070	3.020	3.000	0.030	3.012	3.000	0.018	3.010	3.000	0.016
	Min	3.060	2.940	0.060	3.020	2.975	0.020	3.006	2.990	0.006	3.002	2.994	0.002	3.000	2.994	0.000
4	Max	4.145	4.000	0.220	4.060	4.000	0.090	4.028	4.000	0.040	4.016	4.000	0.024	4.012	4.000	0.020
	Min	4.070	3.925	0.070	4.030	3.970	0.030	4.010	3.988	0.010	4.004	3.992	0.004	4.000	3.992	0.000
5	Max	5.145	5.000	0.220	5.060	5.000	0.090	5.028	5.000	0.040	5.016	5.000	0.024	5.012	5.000	0.020
	Min	5.070	4.925	0.070	5.030	4.970	0.030	5.010	4.988	0.010	5.004	4.992	0.004	5.000	4.992	0.000
6	Max	6.145	6.000	0.220	6.060	6.000	0.090	6.028	6.000	0.040	6.016	6.000	0.024	6.012	6.000	0.020
	Min	6.070	5.925	0.070	6.030	5.970	0.030	6.010	5.988	0.010	6.004	5.992	0.004	6.000	5.992	0.000
8	Max	8.170	8.000	0.260	8.076	8.000	0.112	8.035	8.000	0.050	8.020	8.000	0.029	8.015	8.000	0.024
	Min	8.080	7.910	0.080	8.040	7.964	0.040	8.013	7.985	0.013	8.005	7.991	0.005	8.000	7.991	0.000
10	Max	10.170	10.000	0.260	10.076	10.000	0.112	10.035	10.000	0.050	10.020	10.000	0.029	10.015	10.000	0.024
	Min	10.080	9.910	0.080	10.040	9.964	0.040	10.013	9.985	0.013	10.005	9.991	0.005	10.000	9.991	0.000
12	Max	12.205	12.000	0.315	12.093	12.000	0.136	12.043	12.000	0.061	12.024	12.000	0.035	12.018	12.000	0.029
	Min	12.095	11.890	0.095	12.050	11.957	0.050	12.016	11.982	0.016	12.006	11.989	0.006	12.000	11.989	0.000
16	Max	16.205	16.000	0.315	16.093	16.000	0.136	16.043	16.000	0.061	16.024	16.000	0.035	16.018	16.000	0.029
	Min	16.095	15.890	0.095	16.050	15.957	0.050	16.016	15.982	0.016	16.006	15.989	0.006	16.000	15.989	0.000
20	Max	20.240	20.000	0.370	20.117	20.000	0.169	20.053	20.000	0.074	20.028	20.000	0.041	20.021	20.000	0.034
	Min	20.110	19.870	0.110	20.065	19.948	0.065	20.020	19.979	0.020	20.007	19.987	0.007	20.000	19.987	0.000
25	Max	25.240	25.000	0.370	25.117	25.000	0.169	25.053	25.000	0.074	25.028	25.000	0.041	25.021	25.000	0.034
	Min	25.110	24.870	0.110	25.065	24.948	0.065	25.020	24.979	0.020	25.007	24.987	0.007	25.000	24.987	0.000
30	Max	30.240	30.000	0.370	30.117	30.000	0.169	30.053	30.000	0.074	30.028	30.000	0.041	30.021	30.000	0.034
	Min	30.110	29.870	0.110	30.065	29.948	0.065	30.020	29.979	0.020	30.007	29.987	0.007	30.000	29.987	0.000

[a] From ANSI B4.2-1978 (R1994).

Dimensions are in millimeters.

Basic Size		Loose Running			Free Running			Close Running			Sliding			Locational Clearance		
		Hole C11	Shaft h11	Fit	Hole D9	Shaft h9	Fit	Hole F8	Shaft h7	Fit	Hole G7	Shaft h6	Fit	Hole H7	Shaft h6	Fit
40	Max	40.280	40.000	0.440	40.142	40.000	0.204	40.064	40.000	0.089	40.034	40.000	0.050	40.025	40.000	0.041
	Min	40.120	39.840	0.120	40.080	39.938	0.080	40.025	39.975	0.025	40.009	39.984	0.009	40.000	39.984	0.000
50	Max	50.290	50.000	0.450	50.142	50.000	0.204	50.064	50.000	0.089	50.034	50.000	0.050	50.025	50.000	0.041
	Min	50.130	49.840	0.130	50.080	49.938	0.080	50.025	49.975	0.025	50.009	49.984	0.009	50.000	49.984	0.000
60	Max	60.330	60.000	0.520	60.174	60.000	0.248	60.076	60.000	0.106	60.040	60.000	0.059	60.030	60.000	0.049
	Min	60.140	59.810	0.140	60.100	59.926	0.100	60.030	59.970	0.030	60.010	59.981	0.010	60.000	59.981	0.000
80	Max	80.340	80.000	0.530	80.174	80.000	0.248	80.076	80.000	0.106	80.040	80.000	0.059	80.030	80.000	0.049
	Min	80.150	79.810	0.150	80.100	79.926	0.100	80.030	79.970	0.030	80.010	79.981	0.010	80.000	79.981	0.000
100	Max	100.390	100.000	0.610	100.207	100.000	0.294	100.090	100.000	0.125	100.047	100.000	0.069	100.035	100.000	0.057
	Min	100.170	99.780	0.170	100.120	99.913	0.120	100.036	99.965	0.036	100.012	99.978	0.012	100.000	99.978	0.000
120	Max	120.400	120.000	0.620	120.207	120.000	0.294	120.090	120.000	0.125	120.047	120.000	0.069	120.035	120.000	0.057
	Min	120.180	119.780	0.180	120.120	119.913	0.120	120.036	119.965	0.036	120.012	119.978	0.012	120.000	119.978	0.000
160	Max	160.460	160.000	0.710	160.245	160.000	0.345	160.106	160.000	0.146	160.054	160.000	0.079	160.040	160.000	0.065
	Min	160.210	159.750	0.210	160.145	159.900	0.145	160.043	159.960	0.043	160.014	159.975	0.014	160.000	159.975	0.000
200	Max	200.530	200.000	0.820	200.285	200.000	0.400	200.122	200.000	0.168	200.061	200.000	0.090	200.046	200.000	0.075
	Min	200.240	199.710	0.240	200.170	199.885	0.170	200.050	199.954	0.050	200.015	199.971	0.015	200.000	199.971	0.000
250	Max	250.570	250.000	0.860	250.285	250.000	0.400	250.122	250.000	0.168	250.061	250.000	0.090	250.046	250.000	0.075
	Min	250.280	249.710	0.280	250.170	249.885	0.170	250.050	249.954	0.050	250.015	249.971	0.015	250.000	249.971	0.000
300	Max	300.650	300.000	0.970	300.320	300.000	0.450	300.137	300.000	0.189	300.069	300.000	0.101	300.052	300.000	0.084
	Min	300.330	299.680	0.330	300.190	299.870	0.190	300.056	299.948	0.056	300.017	299.968	0.017	300.000	299.968	0.000
400	Max	400.760	400.000	1.120	400.350	400.000	0.490	400.151	400.000	0.208	400.075	400.000	0.111	400.057	400.000	0.093
	Min	400.400	399.640	0.400	400.210	399.860	0.210	400.062	399.943	0.062	400.018	399.964	0.018	400.000	399.964	0.000
500	Max	500.880	500.000	1.280	500.385	500.000	0.540	500.165	500.000	0.228	500.083	500.000	0.123	500.063	500.000	0.103
	Min	500.480	499.600	0.480	500.230	499.845	0.230	500.068	499.937	0.068	500.020	499.960	5.020	500.000	499.960	0.000

[a] From ANSI B4.2-1978 (R1994).

Dimensions are in millimeters.

Basic Size		Locational Transn. Hole K7	Shaft h6	Fit	Locational Transn. Hole N7	Shaft h6	Fit	Locational Interf. Hole P7	Shaft h6	Fit	Medium Drive Hole S7	Shaft h6	Fit	Force Hole U7	Shaft h6	Fit
1	Max	1.000	1.000	0.006	0.996	1.000	0.002	0.994	1.000	0.000	0.986	1.000	−0.008	0.982	1.000	−0.012
	Min	0.990	0.994	−0.010	0.986	0.994	−0.014	0.984	0.994	−0.016	0.976	0.994	−0.024	0.972	0.994	−0.028
1.2	Max	1.200	1.200	0.006	1.196	1.200	0.002	1.194	1.200	0.000	1.186	1.200	−0.008	1.182	1.200	−0.012
	Min	1.190	1.194	−0.010	1.186	1.194	−0.014	1.184	1.194	−0.016	1.176	1.194	−0.024	1.172	1.194	−0.028
1.6	Max	1.600	1.600	0.006	1.596	1.600	0.002	1.594	1.600	0.000	1.586	1.600	−0.008	1.582	1.600	−0.012
	Min	1.590	1.594	−0.010	1.586	1.594	−0.014	1.584	1.594	−0.016	1.576	1.594	−0.024	1.572	1.594	−0.028
2	Max	2.000	2.000	0.006	1.996	2.000	0.002	1.994	2.000	0.000	1.986	2.000	−0.008	1.982	2.000	−0.012
	Min	1.990	1.994	−0.010	1.986	1.994	−0.014	1.984	1.994	−0.016	1.976	1.994	−0.024	1.972	1.994	−0.028
2.5	Max	2.500	2.500	0.006	2.496	2.500	0.002	2.494	2.500	0.000	2.486	2.500	−0.008	2.482	2.500	−0.012
	Min	2.490	2.494	−0.010	2.486	2.494	−0.014	2.484	2.494	−0.016	2.476	2.494	−0.024	2.472	2.494	−0.028
3	Max	3.000	3.000	0.006	2.996	3.000	0.002	2.994	3.000	0.000	2.986	3.000	−0.008	2.982	3.000	−0.012
	Min	2.990	2.994	−0.010	2.986	2.994	−0.014	2.984	2.994	−0.016	2.976	2.994	−0.024	2.972	2.994	−0.028
4	Max	4.003	4.000	0.011	3.996	4.000	0.004	3.992	4.000	0.000	3.985	4.000	−0.007	3.981	4.000	−0.011
	Min	3.991	3.992	−0.009	3.984	3.992	−0.016	3.980	3.992	−0.020	3.973	3.992	−0.027	3.969	3.992	−0.031
5	Max	5.003	5.000	0.011	4.996	5.000	0.004	4.992	5.000	0.000	4.985	5.000	−0.007	4.981	5.000	−0.011
	Min	4.991	4.992	−0.009	4.984	4.992	−0.016	4.980	4.992	−0.020	4.973	4.992	−0.027	4.969	4.992	−0.031
6	Max	6.003	6.000	0.011	5.996	6.000	0.004	5.992	6.000	0.000	5.985	6.000	−0.007	5.981	6.000	−0.011
	Min	5.991	5.992	−0.009	5.984	5.992	−0.016	5.980	5.992	−0.020	5.973	5.992	−0.027	5.969	5.992	−0.031
8	Max	8.005	8.000	0.014	7.996	8.000	0.005	7.991	8.000	0.000	7.983	8.000	−0.008	7.978	8.000	−0.013
	Min	7.990	7.991	−0.010	7.981	7.991	−0.019	7.976	7.991	−0.024	7.968	7.991	−0.032	7.963	7.991	−0.037
10	Max	10.005	10.000	0.014	9.996	10.000	0.005	9.991	10.000	0.000	9.983	10.000	−0.008	9.978	10.000	−0.013
	Min	9.990	9.991	−0.010	9.981	9.991	−0.019	9.976	9.991	−0.024	9.968	9.991	−0.032	9.963	9.991	−0.037
12	Max	12.006	12.000	0.017	11.995	12.000	0.006	11.989	12.000	0.000	11.979	12.000	−0.010	11.974	12.000	−0.015
	Min	11.988	11.989	−0.012	11.977	11.989	−0.023	11.971	11.989	−0.029	11.961	11.989	−0.039	11.956	11.989	−0.044
16	Max	16.006	16.000	0.017	15.995	16.000	0.006	15.989	16.000	0.000	15.979	16.000	−0.010	15.974	16.000	−0.015
	Min	15.988	15.989	−0.012	15.977	15.989	−0.023	15.971	15.989	−0.029	15.961	15.989	−0.039	15.956	15.989	−0.044
20	Max	20.006	20.000	0.019	19.993	20.000	0.006	19.986	20.000	−0.001	19.973	20.000	−0.014	19.967	20.000	−0.020
	Min	19.985	19.987	−0.015	19.972	19.987	−0.028	19.965	19.987	−0.035	19.952	19.987	−0.048	19.946	19.987	−0.054
25	Max	25.006	25.000	0.019	24.993	25.000	0.006	24.986	25.000	−0.001	24.973	25.000	−0.014	24.960	25.000	−0.027
	Min	24.985	24.987	−0.015	24.972	24.987	−0.028	24.965	24.987	−0.035	24.952	24.987	−0.048	24.939	24.987	−0.061
30	Max	30.006	30.000	0.019	29.993	30.000	0.006	29.986	30.000	−0.001	29.973	30.000	−0.014	29.960	30.000	−0.027
	Min	29.985	29.987	−0.015	29.972	29.987	−0.028	29.965	29.987	−0.035	29.952	29.987	−0.048	29.939	29.987	−0.061

[a] From ANSI B4.2–1978 (R1994). For description of preferred fits, see Table 12.2.

Dimensions are in millimeters.

Basic Size		Locational Transn. Hole K7	Shaft h6	Fit	Locational Transn. Hole N7	Shaft h6	Fit	Locational Interf. Hole P7	Shaft h6	Fit	Medium Drive Hole S7	Shaft h6	Fit	Force Hole U7	Shaft h6	Fit
40	Max	40.007	40.000	0.023	39.992	40.000	0.008	39.983	40.000	−0.001	39.966	40.000	−0.018	39.949	40.000	−0.035
	Min	39.982	39.984	−0.018	39.967	39.984	−0.033	39.958	39.984	−0.042	39.941	39.984	−0.059	39.924	39.984	−0.076
50	Max	50.007	50.000	0.023	49.992	50.000	0.008	49.983	50.000	−0.001	49.966	50.000	−0.018	49.939	50.000	−0.045
	Min	49.982	49.984	−0.018	49.967	49.984	−0.033	49.958	49.984	−0.042	49.941	49.984	−0.059	49.914	49.984	−0.086
60	Max	60.009	60.000	0.028	59.991	60.000	0.010	59.979	60.000	−0.002	59.958	60.000	−0.023	59.924	60.000	−0.057
	Min	59.979	59.981	−0.021	59.961	59.981	−0.039	59.949	59.981	−0.051	59.928	59.981	−0.072	59.894	59.981	−0.106
80	Max	80.009	80.000	0.028	79.991	80.000	0.010	79.979	80.000	−0.002	79.952	80.000	−0.029	79.909	80.000	−0.072
	Min	79.979	79.981	−0.021	79.961	79.981	−0.039	79.949	79.981	−0.051	79.922	79.981	−0.078	79.879	79.981	−0.121
100	Max	100.010	100.000	0.032	99.990	100.000	0.012	99.976	100.000	−0.002	99.942	100.000	−0.036	99.889	100.000	−0.089
	Min	99.975	99.978	−0.025	99.955	99.978	−0.045	99.941	99.978	−0.059	99.907	99.978	−0.093	99.854	99.978	−0.146
120	Max	120.010	120.000	0.032	119.990	120.000	0.012	119.976	120.000	−0.002	119.934	120.000	−0.044	119.869	120.000	−0.109
	Min	119.975	119.978	−0.025	119.955	119.978	−0.045	119.941	119.978	−0.059	119.899	119.978	−0.101	119.834	119.978	−0.166
160	Max	160.012	160.000	0.037	159.988	160.000	0.013	159.972	160.000	−0.003	159.915	160.000	−0.060	159.825	160.000	−0.150
	Min	159.972	159.975	−0.028	159.948	159.975	−0.052	159.932	159.975	−0.068	159.875	159.975	−0.125	159.785	159.975	−0.215
200	Max	200.013	200.000	0.042	199.986	200.000	0.015	199.967	200.000	−0.004	199.895	200.000	−0.076	199.781	200.000	−0.190
	Min	199.967	199.971	−0.033	199.940	199.971	−0.060	199.921	199.971	−0.079	199.849	199.971	−0.151	199.735	199.971	−0.265
250	Max	250.013	250.000	0.042	249.986	250.000	0.015	249.967	250.000	−0.004	249.877	250.000	−0.094	249.733	250.000	−0.238
	Min	249.967	249.971	−0.033	249.940	249.971	−0.060	249.921	249.971	−0.079	249.831	249.971	−0.169	249.687	249.971	−0.313
300	Max	300.016	300.000	0.048	299.986	300.000	0.018	299.964	300.000	−0.004	299.850	300.000	−0.118	299.670	300.000	−0.298
	Min	299.964	299.968	−0.036	299.934	299.968	−0.066	299.912	299.968	−0.088	299.798	299.968	−0.202	299.618	299.968	−0.382
400	Max	400.017	400.000	0.053	399.984	400.000	0.020	399.959	400.000	−0.005	399.813	400.000	−0.151	399.586	400.000	−0.378
	Min	399.960	399.964	−0.040	399.927	399.964	−0.073	399.902	399.964	−0.098	399.756	399.964	−0.244	399.529	399.964	−0.471
500	Max	500.018	500.000	0.058	499.983	500.000	0.023	499.955	500.000	−0.005	499.771	500.000	−0.189	499.483	500.000	−0.477
	Min	499.955	499.960	−0.045	499.920	499.960	−0.080	499.892	499.960	−0.108	499.708	499.960	−0.292	499.420	499.960	−0.580

[a] From ANSI B4.2–1978 (R1994). For description of preferred fits, see Table 12.2.

15 Screw Threads, American National, Unified, and Metric

AMERICAN NATIONAL STANDARD UNIFIED AND AMERICAN NATIONAL SCREW THREADS[a]

Nominal Diameter	Coarse[b] NC UNC Thds. per Inch	Tap Drill[d]	Fine[b] NF UNF Thds. per Inch	Tap Drill[d]	Extra Fine[c] NEF UNEF Thds. per Inch	Tap Drill[d]
0 (.060)			80	$\frac{3}{64}$		
1 (.073)	64	No. 53	72	No. 53	...	...
2 (.086)	56	No. 50	64	No. 50	...	...
3 (.099)	48	No. 47	56	No. 45	...	...
4 (.112)	40	No. 43	48	No. 42	...	...
5 (.125)	40	No. 38	44	No. 37	...	...
6 (.138)	32	No. 36	40	No. 33	...	...
8 (.164)	32	No. 29	36	No. 29	...	...
10 (.190)	24	No. 25	32	No. 21	...	...
12 (.216)	24	No. 16	28	No. 14	32	No. 13
$\frac{1}{4}$	20	No. 7	28	No. 3	32	$\frac{7}{32}$
$\frac{5}{16}$	18	F	24	I	32	$\frac{9}{32}$
$\frac{3}{8}$	16	$\frac{5}{16}$	24	Q	32	$\frac{11}{32}$
$\frac{7}{16}$	14	U	20	$\frac{25}{64}$	28	$\frac{13}{32}$
$\frac{1}{2}$	13	$\frac{27}{64}$	20	$\frac{29}{64}$	28	$\frac{15}{32}$
$\frac{9}{16}$	12	$\frac{31}{64}$	18	$\frac{33}{64}$	24	$\frac{33}{64}$
$\frac{5}{8}$	11	$\frac{17}{32}$	18	$\frac{37}{64}$	24	$\frac{37}{64}$
$\frac{11}{16}$	...	...	...	...	24	$\frac{41}{64}$
$\frac{3}{4}$	10	$\frac{21}{32}$	16	$\frac{11}{16}$	20	$\frac{45}{64}$
$\frac{13}{16}$	...	...	...	...	20	$\frac{49}{64}$
$\frac{7}{8}$	9	$\frac{49}{64}$	14	$\frac{13}{16}$	20	$\frac{53}{64}$
$\frac{15}{16}$	...	...	...	...	20	$\frac{57}{64}$
1	8	$\frac{7}{8}$	12	$\frac{59}{64}$	20	$\frac{61}{64}$
$1\frac{1}{16}$	...	...	...	...	18	1
$1\frac{1}{8}$	7	$\frac{63}{64}$	12	$1\frac{3}{64}$	18	$1\frac{5}{64}$
$1\frac{3}{16}$	...	...	...	...	18	$1\frac{9}{64}$
$1\frac{1}{4}$	7	$1\frac{7}{64}$	12	$1\frac{11}{64}$	18	$1\frac{3}{16}$
$1\frac{5}{16}$	...	...	...	...	18	$1\frac{17}{64}$
$1\frac{3}{8}$	6	$1\frac{7}{32}$	12	$1\frac{19}{64}$	18	$1\frac{5}{16}$
$1\frac{7}{16}$	...	...	...	...	18	$1\frac{3}{8}$
$1\frac{1}{2}$	6	$1\frac{11}{32}$	12	$1\frac{27}{64}$	18	$1\frac{7}{16}$
$1\frac{9}{16}$	...	...	...	...	18	$1\frac{1}{2}$
$1\frac{5}{8}$	...	...	...	...	18	$1\frac{9}{16}$
$1\frac{11}{16}$	...	...	...	...	18	$1\frac{5}{8}$
$1\frac{3}{4}$	5	$1\frac{9}{16}$	...	...	...	...
2	$4\frac{1}{2}$	$1\frac{25}{32}$	...	...	...	...
$2\frac{1}{4}$	$4\frac{1}{2}$	$2\frac{1}{32}$	...	...	...	...
$2\frac{1}{2}$	4	$2\frac{1}{4}$	...	...	...	...
$2\frac{3}{4}$	4	$2\frac{1}{2}$	...	...	...	...
3	4	$2\frac{3}{4}$	...	...	...	...
$3\frac{1}{4}$	4	...	...	...	...	...
$3\frac{1}{2}$	4	...	...	...	...	...
$3\frac{3}{4}$	4	...	...	...	...	...
4	4	...	...	...	...	...

[a] ANSI/ASME B1.1–1989. For 8-, 12-, and 16-pitch thread series, see next page.
[b] Classes 1A, 2A, 3A, 1B, 2B, 3B, 2, and 3.
[c] Classes 2A, 2B, 2, and 3.
[d] For approximate 75% full depth of thread. For decimal sizes of numbered and lettered drills, see Appendix 16.

15 Screw Threads, American National, Unified, and Metric (continued)

AMERICAN NATIONAL STANDARD UNIFIED AND AMERICAN NATIONAL SCREW THREADS[a] (continued)

Nominal Diameter	8-Pitch[b] Series 8N and 8UN		12-Pitch[b] Series 12N and 12UN		16-Pitch[b] Series 16N and 16UN		Nominal Diameter	8-Pitch[b] Series 8N and 8UN		12-Pitch[b] Series 12N and 12UN		16-Pitch[c] Series 16N and 16UN	
	Thds. per Inch	Tap Drill[c]	Thds. per Inch	Tap Drill[c]	Thds. per Inch	Tap Drill[c]		Thds. per Inch	Tap Drill[c]	Thds. per Inch	Tap Drill[c]	Thds. per Inch	Tap Drill[c]
$\frac{1}{2}$	...	...	12	$\frac{27}{64}$	...	...	$2\frac{1}{16}$	...	...	...	...	16	2
$\frac{9}{16}$	...	...	12[e]	$\frac{31}{64}$	...	...	$2\frac{1}{8}$	...	...	12	$2\frac{3}{64}$	16	$2\frac{1}{16}$
$\frac{5}{8}$	...	...	12	$\frac{35}{64}$	...	...	$2\frac{3}{16}$	...	...	...	...	16	$2\frac{1}{8}$
$\frac{11}{16}$	...	...	12	$\frac{39}{64}$	...	...	$2\frac{1}{4}$	8	$2\frac{1}{8}$	12	$2\frac{17}{64}$	16	$2\frac{3}{16}$
$\frac{3}{4}$	...	...	12	$\frac{43}{64}$	16[e]	$\frac{11}{16}$	$2\frac{5}{16}$	...	...	...	...	16	$2\frac{1}{4}$
$\frac{13}{16}$	...	...	12	$\frac{47}{64}$	16	$\frac{3}{4}$	$2\frac{3}{8}$	...	...	12	$2\frac{19}{64}$	16	$2\frac{5}{16}$
$\frac{7}{8}$	...	...	12	$\frac{51}{64}$	16	$\frac{13}{16}$	$2\frac{7}{16}$	...	...	...	...	16	$2\frac{3}{8}$
$\frac{15}{16}$	...	...	12	$\frac{55}{64}$	16	$\frac{7}{8}$	$2\frac{1}{2}$	8	$2\frac{3}{8}$	12	$2\frac{27}{64}$	16	$2\frac{7}{16}$
1	8[e]	$\frac{7}{8}$	12	$\frac{59}{64}$	16	$\frac{15}{16}$	$2\frac{5}{8}$	...	...	12	$2\frac{35}{64}$	16	$2\frac{9}{16}$
$1\frac{1}{16}$	...	...	12	$\frac{63}{64}$	16	1	$2\frac{3}{4}$	8	$2\frac{5}{8}$	12	$2\frac{43}{64}$	16	$2\frac{11}{16}$
$1\frac{1}{8}$	8	1	12[e]	$1\frac{3}{64}$	16	$1\frac{1}{16}$	$2\frac{7}{8}$	...	...	12	...	16	...
$1\frac{3}{16}$	...	...	12	$1\frac{7}{64}$	16	$1\frac{1}{8}$	3	8	$2\frac{7}{8}$	12	...	16	...
$1\frac{1}{4}$	8	$1\frac{1}{8}$	12	$1\frac{11}{64}$	16	$1\frac{3}{16}$	$3\frac{1}{8}$	...	...	12	...	16	...
$1\frac{5}{16}$	...	...	12	$1\frac{15}{64}$	16	$1\frac{1}{4}$	$3\frac{1}{4}$	8	...	12	...	16	...
$1\frac{3}{8}$	8	$1\frac{1}{4}$	12[e]	$1\frac{19}{64}$	16	$1\frac{5}{16}$	$3\frac{3}{8}$	...	...	12	...	16	...
$1\frac{7}{16}$	...	...	12	$1\frac{23}{64}$	16	$1\frac{3}{8}$	$3\frac{1}{2}$	8	...	12	...	16	...
$1\frac{1}{2}$	8	$1\frac{3}{8}$	12[e]	$1\frac{27}{64}$	16	$1\frac{7}{16}$	$3\frac{5}{8}$	...	...	12	...	16	...
$1\frac{9}{16}$	...	...	...	...	16	$1\frac{1}{2}$	$3\frac{3}{4}$	8	...	12	...	16	...
$1\frac{5}{8}$	8	$1\frac{1}{2}$	12	$1\frac{35}{64}$	16	$1\frac{9}{16}$	$3\frac{7}{8}$	...	...	12	...	16	...
$1\frac{11}{16}$	...	...	...	...	16	$1\frac{5}{8}$	4	8	...	12	...	16	...
$1\frac{3}{4}$	8	$1\frac{5}{8}$	12	$1\frac{43}{64}$	16[e]	$1\frac{11}{16}$	$4\frac{1}{4}$	8	...	12	...	16	...
$1\frac{13}{16}$	...	...	...	...	16	$1\frac{3}{4}$	$4\frac{1}{2}$	8	...	12	...	16	...
$1\frac{7}{8}$	8	$1\frac{3}{4}$	12	$1\frac{51}{64}$	16	$1\frac{13}{16}$	$4\frac{3}{4}$	8	...	12	...	16	...
$1\frac{15}{16}$	...	...	...	...	16	$1\frac{7}{8}$	5	8	...	12	...	16	...
2	8	$1\frac{7}{8}$	12	$1\frac{59}{64}$	16[e]	$1\frac{15}{16}$	$5\frac{1}{4}$	8	...	12	...	16	...

[a] ANSI/ASME B1.1–1989.
[b] Classes 2A, 3A, 2B, 3B, 2, and 3.
[c] For approximate 75% full depth of thread.
[d] Boldface type indicates Amrican National Threads only.
[e] This is a standard size of the Unified or American National threads of the coarse, fine, or extra fine series. See preceding page.

15 Screw Threads, American National, Unified, and Metric (continued)

METRIC SCREW THREADS[a]

Preferred sizes for commercial threads and fasteners are shown in **boldface** type.

Coarse (general purpose)		Fine	
Nominal Size & Thd Pitch	Tap Drill Diameter, mm	Nominal Size & Thd Pitch	Tap Drill Diameter, mm
M1.6 × 0.35	1.25	—	—
M1.8 × 0.35	1.45	—	—
M2 × 0.4	1.6	—	—
M2.2 × 0.45	1.75	—	—
M2.5 × 0.45	2.05	—	—
M3 × 0.5	2.5	—	—
M3.5 × 0.6	2.9	—	—
M4 × 0.7	3.3	—	—
M4.5 × 0.75	3.75	—	—
M5 × 0.8	4.2	—	—
M6 × 1	5.0	—	—
M7 × 1	6.0		
M8 × 1.25	6.8	**M8 × 1**	7.0
M9 × 1.25	7.75	—	—
M10 × 1.5	8.5	**M10 × 1.25**	8.75
M11 × 1.5	9.50	—	—
M12 × 1.75	10.30	**M12 × 1.25**	10.5
M14 × 2	12.00	**M14 × 1.5**	12.5
M16 × 2	14.00	**M16 × 1.5**	14.5
M18 × 2.5	15.50	**M18 × 1.5**	16.5
M20 × 2.5	17.5	**M20 × 1.5**	18.5
M22 × 25[b]	19.5	**M22 × 1.5**	20.5
M24 × 3	21.0	**M24 × 2**	22.0
M27 × 3[b]	24.0	**M27 × 2**	25.0
M30 × 3.5	26.5	**M30 × 2**	28.0
M33 × 3.5	29.5	**M30 × 2**	31.0
M36 × 4	32.0	**M36 × 2**	33.0
M39 × 4	35.0	M39 × 2	36.0
M42 × 4.5	37.5	**M42 × 2**	39.0
M45 × 4.5	40.5	M45 × 1.5	42.0
M48 × 5	43.0	**M48 × 2**	45.0
M52 × 5	47.0	M52 × 2	49.0
M56 × 5.5	50.5	**M56 × 2**	52.0
M60 × 5.5	54.5	M60 × 1.5	56.0
M64 × 6	58.0	**M64 × 2**	60.0
M68 × 6	62.0	M68 × 2	64.0
M72 × 6	66.0	**M72 × 2**	68.0
M80 × 6	74.0	**M80 × 2**	76.0
M90 × 6	84.0	**M90 × 2**	86.0
M100 × 6	94.0	**M100 × 2**	96.0

[a] ANSI/ASME B1.13M–1995.
[b] Only for high strength structural steel fasteners.

16 Twist Drill Sizes—American National Standard and Metric

AMERICAN NATIONAL STANDARD DRILL SIZES[a]

All dimensions are in inches.
Drills designated in common fractions are available in diameters $\frac{1}{64}''$ to $1\frac{3}{4}''$ in $\frac{1}{64}''$ increments, $1\frac{3}{4}''$ to $2\frac{1}{4}''$ in $\frac{1}{32}''$ increments. $2\frac{1}{4}''$ to $3''$ in $\frac{1}{16}''$ increments and $3''$ to $3\frac{1}{2}''$ in $\frac{1}{8}''$ increments. Drills larger than $3\frac{1}{2}''$ are seldom used, and are regarded as special drills.

Size	Drill Diameter	Size	Drill Diameter	Size	Drill Diameter	Size	Drill Diameter	Size	Drill Diameter	Size	Drill Diameter
1	.2280	17	.1730	33	.1130	49	.0730	65	.0350	81	.0130
2	.2210	18	.1695	34	.1110	50	.0700	66	.0330	82	.0125
3	.2130	19	.1660	35	.1100	51	.0670	67	.0320	83	.0120
4	.2090	20	.1610	36	.1065	52	.0635	68	.0310	84	.0115
5	.2055	21	.1590	37	.1040	53	.0595	69	.0292	85	.0110
6	.2040	22	.1570	38	.1015	54	.0550	70	.0280	86	.0105
7	.2010	23	.1540	39	.0995	55	.0520	71	.0260	87	.0100
8	.1990	24	.1520	40	.0980	56	.0465	72	.0250	88	.0095
9	.1960	25	.1495	41	.0960	57	.0430	73	.0240	89	.0091
10	.1935	26	.1470	42	.0935	58	.0420	74	.0225	90	.0087
11	.1910	27	.1440	43	.0890	59	.0410	75	.0210	91	.0083
12	.1890	28	.1405	44	.0860	60	.0400	76	.0200	92	.0079
13	.1850	29	.1360	45	.0820	61	.0390	77	.0180	93	.0075
14	.1820	30	.1285	46	.0810	62	.0380	78	.0160	94	.0071
15	.1800	31	.1200	47	.0785	63	.0370	79	.0145	95	.0067
16	.1770	32	.1160	48	.0760	64	.0360	80	.0135	96	.0063
										97	.0059

LETTER SIZES

A	.234	G	.261	L	.290	Q	.332	V	.377
B	.238	H	.266	M	.295	R	.339	W	.386
C	.242	I	.272	N	.302	S	.348	X	.397
D	.246	J	.277	O	.316	T	.358	Y	.404
E	.250	K	.281	P	.323	U	.368	Z	.413
F	.257								

[a] ANSI/ASME B94.11M–1993.

16 Twist Drill Sizes—American National Standard and Metric (continued)

METRIC DRILL SIZES

Decimal-inch equivalents are for reference only.

Drill Diameter		Drill Diameter		Drill Diameter		Drill Diameter		Drill Diameter		Drill Diameter	
mm	in.	mm	in.	mm	in.	mm	in.	mm	in.	mm	in.
0.40	.0157	1.95	.0768	4.70	.1850	8.00	.3150	13.20	.5197	25.50	1.0039
0.42	.0165	2.00	.0787	4.80	.1890	8.10	.3189	13.50	.5315	26.00	1.0236
0.45	.0177	2.05	.0807	4.90	.1929	8.20	.3228	13.80	.5433	26.50	1.0433
0.48	.0189	2.10	.0827	5.00	.1969	8.30	.3268	14.00	.5512	27.00	1.0630
0.50	.0197	2.15	.0846	5.10	.2008	8.40	.3307	14.25	.5610	27.50	1.0827
0.55	.0217	2.20	.0866	5.20	.2047	8.50	.3346	14.50	.5709	28.00	1.1024
0.60	.0236	2.25	.0886	5.30	.2087	8.60	.3386	14.75	.5807	28.50	1.1220
0.65	.0256	2.30	.0906	5.40	.2126	8.70	.3425	15.00	.5906	29.00	1.1417
0.70	.0276	2.35	.0925	5.50	.2165	8.80	.3465	15.25	.6004	29.50	1.1614
0.75	.0295	2.40	.0945	5.60	.2205	8.90	.3504	15.50	.6102	30.00	1.1811
0.80	.0315	2.45	.0965	5.70	.2244	9.00	.3543	15.75	.6201	30.50	1.2008
0.85	.0335	2.50	.0984	5.80	.2283	9.10	.3583	16.00	.6299	31.00	1.2205
0.90	.0354	2.60	.1024	5.90	.2323	9.20	.3622	16.25	.6398	31.50	1.2402
0.95	.0374	2.70	.1063	6.00	.2362	9.30	.3661	16.50	.6496	32.00	1.2598
1.00	.0394	2.80	.1102	6.10	.2402	9.40	.3701	16.75	.6594	32.50	1.2795
1.05	.0413	2.90	.1142	6.20	.2441	9.50	.3740	17.00	.6693	33.00	1.2992
1.10	.0433	3.00	.1181	6.30	.2480	9.60	.3780	17.25	.6791	33.50	1.3189
1.15	.0453	3.10	.1220	6.40	.2520	9.70	.3819	17.50	.6890	34.00	1.3386
1.20	.0472	3.20	.1260	6.50	.2559	9.80	.3858	18.00	.7087	34.50	1.3583
1.25	.0492	3.30	.1299	6.60	.2598	9.90	.3898	18.50	.7283	35.00	1.3780
1.30	.0512	3.40	.1339	6.70	.2638	10.00	.3937	19.00	.7480	35.50	1.3976
1.35	.0531	3.50	.1378	6.80	.2677	10.20	.4016	19.50	.7677	36.00	1.4173
1.40	.0551	3.60	.1417	6.90	.2717	10.50	.4134	20.00	.7874	36.50	1.4370
1.45	.0571	3.70	.1457	7.00	.2756	10.80	.4252	20.50	.8071	37.00	1.4567
1.50	.0591	3.80	.1496	7.10	.2795	11.00	.4331	21.00	.8268	37.50	1.4764
1.55	.0610	3.90	.1535	7.20	.2835	11.20	.4409	21.50	.8465	38.00	1.4961
1.60	.0630	4.00	.1575	7.30	.2874	11.50	.4528	22.00	.8661	40.00	1.5748
1.65	.0650	4.10	.1614	7.40	.2913	11.80	.4646	22.50	.8858	42.00	1.6535
1.70	.0669	4.20	.1654	7.50	.2953	12.00	.4724	23.00	.9055	44.00	1.7323
1.75	.0689	4.30	.1693	7.60	.2992	12.20	.4803	23.50	.9252	46.00	1.8110
1.80	.0709	4.40	.1732	7.70	.3031	12.50	.4921	24.00	.9449	48.00	1.8898
1.85	.0728	4.50	.1772	7.80	.3071	12.50	.5039	24.50	.9646	50.00	1.9685
1.90	.0748	4.60	.1811	7.90	.3110	13.00	.5118	25.00	.9843		

17 Acme Threads, General Purpose[a]

Size	Threads per Inch	Size	Threads per Inch	Size	Threads per Inch	Size	Threads per Inch
$\frac{1}{4}$	16	$\frac{3}{4}$	6	$1\frac{1}{2}$	4	3	2
$\frac{5}{16}$	14	$\frac{7}{8}$	6	$1\frac{3}{4}$	4	$3\frac{1}{2}$	2
$\frac{3}{8}$	12	1	5	2	4	4	2
$\frac{7}{16}$	12	$1\frac{1}{8}$	5	$2\frac{1}{4}$	3	$4\frac{1}{2}$	2
$\frac{1}{2}$	10	$1\frac{1}{4}$	5	$2\frac{1}{2}$	3	5	2
$\frac{5}{8}$	8	$1\frac{3}{8}$	4	$2\frac{3}{4}$	3	...	...

[a] ANSI/ASME B1.5–1988 (R1994).

18 Bolts, Nuts, and Cap Screws—Square and Hexagon— American National Standard and Metric

AMERICAN NATIONAL STANDARD SQUARE AND HEXAGON BOLTS[a] AND NUTS[b]
AND HEXAGON CAP SCREWS[c]

Boldface type indicates product features unified dimensionally with British and Canadian standards.
All dimensions are in inches.
For thread series, minimum thread lengths, and bolt lengths, see §13.25.

Nominal Size D Body Diameter of Bolt		Regular Bolts					Heavy Bolts		
		Width Across Flats W		Height H			Width Across Flats W	Height H	
		Sq.	Hex.	Sq. (Unfin.)	Hex (Unfin.)	Hex Cap Scr.[c] (Fin.)		Hex. (Unfin.)	Hex Screw (Fin.)
$\frac{1}{4}$	0.2500	$\frac{3}{8}$	$\frac{7}{16}$	$\frac{11}{64}$	$\frac{11}{64}$	$\frac{5}{32}$	...	...	...
$\frac{5}{16}$	0.3125	$\frac{1}{2}$	$\frac{1}{2}$	$\frac{13}{64}$	$\frac{7}{32}$	$\frac{13}{64}$	...	...	...
$\frac{3}{8}$	0.3750	$\frac{9}{16}$	$\frac{9}{16}$	$\frac{1}{4}$	$\frac{1}{4}$	$\frac{15}{64}$	...	...	...
$\frac{7}{16}$	0.4375	$\frac{5}{8}$	$\frac{5}{8}$	$\frac{19}{64}$	$\frac{19}{64}$	$\frac{9}{32}$	...	...	...
$\frac{1}{2}$	0.5000	$\frac{3}{4}$	$\frac{3}{4}$	$\frac{21}{64}$	$\frac{11}{32}$	$\frac{5}{16}$	$\frac{7}{8}$	$\frac{11}{32}$	$\frac{5}{16}$
$\frac{9}{16}$	0.5625	...	$\frac{13}{16}$	...	...	$\frac{23}{64}$	...	...	...
$\frac{5}{8}$	0.6250	$\frac{15}{16}$	$\frac{15}{16}$	$\frac{27}{64}$	$\frac{27}{64}$	$\frac{25}{64}$	$1\frac{1}{16}$	$\frac{27}{64}$	$\frac{25}{64}$
$\frac{3}{4}$	0.7500	$1\frac{1}{8}$	$1\frac{1}{8}$	$\frac{1}{2}$	$\frac{1}{2}$	$\frac{15}{32}$	$1\frac{1}{4}$	$\frac{1}{2}$	$\frac{15}{32}$
$\frac{7}{8}$	0.8750	$1\frac{5}{16}$	$1\frac{5}{16}$	$\frac{19}{32}$	$\frac{37}{64}$	$\frac{35}{64}$	$1\frac{7}{16}$	$\frac{37}{64}$	$\frac{35}{64}$
1	1.000	$1\frac{1}{2}$	$1\frac{1}{2}$	$\frac{21}{32}$	$\frac{43}{64}$	$\frac{39}{64}$	$1\frac{5}{8}$	$\frac{43}{64}$	$\frac{39}{64}$
$1\frac{1}{8}$	1.1250	$1\frac{11}{16}$	$1\frac{11}{16}$	$\frac{3}{4}$	$\frac{3}{4}$	$\frac{11}{16}$	$1\frac{13}{16}$	$\frac{3}{4}$	$\frac{11}{16}$
$1\frac{1}{4}$	1.2500	$1\frac{7}{8}$	$1\frac{7}{8}$	$\frac{27}{32}$	$\frac{27}{32}$	$\frac{25}{32}$	2	$\frac{27}{32}$	$\frac{25}{32}$
$1\frac{3}{8}$	1.3750	$2\frac{1}{16}$	$2\frac{1}{16}$	$\frac{29}{32}$	$\frac{29}{32}$	$\frac{27}{32}$	$2\frac{3}{16}$	$\frac{29}{32}$	$\frac{27}{32}$
$1\frac{1}{2}$	1.5000	$2\frac{1}{4}$	$2\frac{1}{4}$	1	1	$\frac{15}{16}$	$2\frac{3}{8}$	1	$\frac{15}{16}$
$1\frac{3}{4}$	1.7500	...	$2\frac{5}{8}$	...	$1\frac{5}{32}$	$1\frac{3}{32}$	$2\frac{3}{4}$	$1\frac{5}{32}$	$1\frac{3}{32}$
2	2.0000	...	3	...	$1\frac{11}{32}$	$1\frac{7}{32}$	$3\frac{1}{8}$	$1\frac{11}{32}$	$1\frac{7}{32}$
$2\frac{1}{4}$	2.2500	...	$3\frac{3}{8}$	...	$1\frac{1}{2}$	$1\frac{3}{8}$	$3\frac{1}{2}$	$1\frac{1}{2}$	$1\frac{3}{8}$
$2\frac{1}{2}$	2.5000	...	$3\frac{3}{4}$	...	$1\frac{21}{32}$	$1\frac{17}{32}$	$3\frac{7}{8}$	$1\frac{21}{32}$	$1\frac{17}{32}$
$2\frac{3}{4}$	2.7500	...	$4\frac{1}{8}$	...	$1\frac{13}{16}$	$1\frac{11}{16}$	$4\frac{1}{4}$	$1\frac{13}{16}$	$1\frac{11}{16}$
3	3.0000	...	$4\frac{1}{2}$	...	2	$1\frac{7}{8}$	$4\frac{5}{8}$	2	$1\frac{7}{8}$
$3\frac{1}{4}$	3.2500	...	$4\frac{7}{8}$	...	$2\frac{3}{16}$	...	...	...	...
$3\frac{1}{2}$	3.5000	...	$5\frac{1}{4}$	...	$2\frac{5}{16}$	...	...	...	...
$3\frac{3}{4}$	3.7500	...	$5\frac{5}{8}$	...	$2\frac{1}{2}$	...	...	...	...
4	4.0000	...	6	...	$2\frac{11}{16}$	...	...	...	...

[a] ANSI B18.2.1–1981 (R1992).
[b] ANSI/ASME B18.2.2.–1987 (R1993).
[c] Hexagon cap screws and finished hexagon bolts are combined as a single product.

18　Bolts, Nuts, and Cap Screws—Square and Hexagon— American National Standard and Metric (continued)

AMERICAN NATIONAL STANDARD SQUARE AND HEXAGON BOLTS AND NUTS AND HEXAGON CAP SCREWS (continued)

See ANSI B18.2.2 for jam nuts, slotted nuts, thick nuts, thick slotted nuts, and castle nuts. For methods of drawing bolts and nuts and hexagon-head cap screws, see Figs. 13.29, 13.30, and 13.32.

Nominal Size D Body Diameter of Bolt		Regular Bolts					Heavy Nuts			
		Width Across Flats W		Thickness T			Width Across Flats W	Thickness T		
		Sq.	Hex.	Sq. (Unfin.)	Hex. Flat (Unfin.)	Hex. (Fin.)		Sq. (Unfin.)	Hex. Flat (Unfin.)	Hex. (Fin.)
$\frac{1}{4}$	0.2500	$\frac{7}{16}$	$\frac{7}{16}$	$\frac{7}{32}$	$\frac{7}{32}$	$\frac{7}{32}$	$\frac{1}{2}$	$\frac{1}{4}$	$\frac{15}{64}$	$\frac{15}{64}$
$\frac{5}{16}$	0.3125	$\frac{9}{16}$	$\frac{1}{2}$	$\frac{17}{64}$	$\frac{17}{64}$	$\frac{17}{64}$	$\frac{9}{16}$	$\frac{5}{16}$	$\frac{19}{64}$	$\frac{19}{64}$
$\frac{3}{8}$	0.3750	$\frac{5}{8}$	$\frac{9}{16}$	$\frac{21}{64}$		$\frac{21}{64}$	$\frac{11}{16}$	$\frac{3}{8}$	$\frac{23}{64}$	$\frac{23}{64}$
$\frac{7}{16}$	0.4375	$\frac{3}{4}$	$\frac{11}{16}$	$\frac{3}{8}$	$\frac{3}{8}$	$\frac{3}{8}$	$\frac{3}{4}$	$\frac{7}{16}$	$\frac{27}{64}$	$\frac{27}{64}$
$\frac{1}{2}$	0.5000	$\frac{13}{16}$	$\frac{3}{4}$	$\frac{7}{16}$	$\frac{7}{16}$	$\frac{7}{16}$	$\frac{7}{8}$ a	$\frac{1}{2}$	$\frac{31}{64}$	$\frac{31}{64}$
$\frac{9}{16}$	0.5625	...	$\frac{7}{8}$	...	$\frac{31}{64}$	$\frac{31}{64}$	$\frac{15}{16}$	...	$\frac{35}{64}$	$\frac{35}{64}$
$\frac{5}{8}$	0.6250	1	$\frac{15}{16}$	$\frac{35}{64}$	$\frac{35}{64}$	$\frac{35}{64}$	$1\frac{1}{16}$ a	$\frac{5}{8}$	$\frac{39}{64}$	$\frac{39}{64}$
$\frac{3}{4}$	0.7500	$1\frac{1}{8}$	$1\frac{1}{8}$	$\frac{21}{32}$	$\frac{41}{64}$	$\frac{41}{64}$	$1\frac{1}{4}$ a	$\frac{3}{4}$	$\frac{47}{64}$	$\frac{47}{64}$
$\frac{7}{8}$	0.8750	$1\frac{5}{16}$	$1\frac{5}{16}$	$\frac{49}{64}$	$\frac{3}{4}$	$\frac{3}{4}$	$1\frac{7}{16}$ a	$\frac{7}{8}$	$\frac{55}{64}$	$\frac{55}{64}$
1	1.0000	$1\frac{1}{2}$	$1\frac{1}{2}$	$\frac{7}{8}$	$\frac{55}{64}$	$\frac{55}{64}$	$1\frac{5}{8}$ a	1	$\frac{63}{64}$	$\frac{63}{64}$
$1\frac{1}{8}$	1.1250	$1\frac{11}{16}$	$1\frac{11}{16}$	1	1	$\frac{31}{32}$	$1\frac{13}{16}$ a	$1\frac{1}{8}$	$1\frac{1}{8}$	$1\frac{7}{64}$
$1\frac{1}{4}$	1.2500	$1\frac{7}{8}$	$1\frac{7}{8}$	$1\frac{3}{32}$	$1\frac{3}{32}$	$1\frac{1}{16}$	2 a	$1\frac{1}{4}$	$1\frac{1}{4}$	$1\frac{7}{32}$
$1\frac{3}{8}$	1.3750	$2\frac{1}{16}$	$2\frac{1}{16}$	$1\frac{13}{64}$	$1\frac{13}{64}$	$1\frac{11}{64}$	$2\frac{3}{16}$ a	$1\frac{3}{8}$	$1\frac{3}{8}$	$1\frac{11}{32}$
$1\frac{1}{2}$	1.5000	$2\frac{1}{4}$	$2\frac{1}{4}$	$1\frac{5}{16}$	$1\frac{5}{16}$	$1\frac{9}{32}$	$2\frac{3}{8}$ a	$1\frac{1}{2}$	$1\frac{1}{2}$	$1\frac{15}{32}$
$1\frac{5}{8}$	1.6250	...	...	...	...	...	$2\frac{9}{16}$	...	...	$1\frac{19}{32}$
$1\frac{3}{4}$	1.7500	...	...	...	...	...	$2\frac{3}{4}$	...	$1\frac{3}{4}$	$1\frac{23}{32}$
$1\frac{7}{8}$	1.8750	...	...	...	...	...	$2\frac{15}{16}$	...	...	$1\frac{27}{32}$
2	2.0000	...	...	...	...	...	$3\frac{1}{8}$	...	2	$1\frac{31}{32}$
$2\frac{1}{4}$	2.2500	...	...	...	...	...	$3\frac{1}{2}$	...	$2\frac{1}{4}$	$2\frac{13}{64}$
$2\frac{1}{2}$	2.5000	...	...	...	...	...	$3\frac{7}{8}$	...	$2\frac{1}{2}$	$2\frac{29}{64}$
$2\frac{3}{4}$	2.7500	...	...	...	...	...	$4\frac{1}{4}$	...	$2\frac{3}{4}$	$2\frac{45}{64}$
3	3.0000	...	...	...	...	...	$4\frac{5}{8}$	...	3	$2\frac{61}{64}$
$3\frac{1}{4}$	3.2500	...	...	...	...	...	5	...	$3\frac{1}{4}$	$3\frac{3}{16}$
$3\frac{1}{2}$	3.5000	...	...	...	...	...	$5\frac{3}{8}$	...	$3\frac{1}{2}$	$3\frac{7}{16}$
$3\frac{3}{4}$	3.7500	...	...	...	...	...	$5\frac{3}{4}$	...	$3\frac{3}{4}$	$3\frac{11}{16}$
4	4.0000	...	...	...	...	...	$6\frac{1}{8}$	...	4	$3\frac{15}{16}$

a Product feature not unified for heavy square nut.

18 Bolts, Nuts, and Cap Screws—Square and Hexagon—American National Standard and Metric (continued)

METRIC HEXAGON BOLTS, HEXAGON CAP SCREWS, HEXAGON STRUCTURAL BOLTS, AND HEXAGON NUTS

Nominal Size D, mm	Width Across Flats W (max)		Thickness T (max)			
Body Dia and Thd Pitch	Bolts,[a] Cap Screws,[b] and Nuts[c]	Heavy Hex & Hex Structural Bolts[a] & Nuts[c]	Bolts (Unfin.)	Cap Screw (Fin.)	Nut (Fin. or Unfin.)	
					Style 1	Style 2
M5 × 0.8	8.0		3.88	3.65	4.7	5.1
M6 × 1	10.0		4.38	4.47	5.2	5.7
M8 × 1.25	13.0		5.68	5.50	6.8	7.5
M10 × 1.5	16.0		6.85	6.63	8.4	9.3
M12 × 1.75	18.0	21.0	7.95	7.76	10.8	12.0
M14 × 2	21.0	24.0	9.25	9.09	12.8	14.1
M16 × 2	24.0	27.0	10.75	10.32	14.8	16.4
M20 × 2.5	30.0	34.0	13.40	12.88	18.0	20.3
M24 × 3	36.0	41.0	15.90	15.44	21.5	23.9
M30 × 3.5	46.0	50.0	19.75	19.48	25.6	28.6
M36 × 4	55.0	60.0	23.55	23.38	31.0	34.7
M42 × 4.5	65.0		27.05	26.97	…	…
M48 × 5	75.0		31.07	31.07	…	…
M56 × 5.5	85.0		36.20	36.20	…	…
M64 × 6	95.0		41.32	41.32	…	…
M72 × 6	105.0		46.45	46.45	…	…
M80 × 6	115.0		51.58	51.58	…	…
M90 × 6	130.0		57.74	57.74	…	…
M100 × 6	145.0		63.90	63.90	…	…

HIGH STRENGTH STRUCTURAL HEXAGON BOLTS[a] (FIN.) AND HEXAGON NUTS[c]

M16 × 2	27.0	…	10.75	…	…	17.1
M20 × 2.5	34.0	…	13.40	…	…	20.7
M22 × 2.5	36.0	…	14.9	…	…	23.6
M24 × 3	41.0	…	15.9	…	…	24.2
M27 × 3	46.0	…	17.9	…	…	27.6
M30 × 3.5	50.0	…	19.75	…	…	31.7
M36 × 4	60.0	…	23.55	…	…	36.6

[a] ANSI/ASME B18.2.3.5M–1979 (R1995), B18.2.3.6M–1979 (R1995), B18.2.3.7M–1979 (R1995).
[b] ANSI/ASME B18.2.3.1M–1979 (R1995).
[c] ANSI/ASME B18.2.4.1M–1979 (R1995), B18.2.4.2M–1979 (R1995).

19 Cap Screws, Slotted[a] and Socket Head[b]— American National Standard and Metric

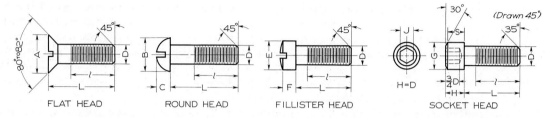

FLAT HEAD   ROUND HEAD   FILLISTER HEAD   SOCKET HEAD

For methods of drawing cap screws, screw lengths, and thread data, see Fig. 13.32.

Nominal Size D	Flat Head[a]	Round Head[a]		Fillister Head[a]		Socket Head[b]		
	A	B	C	E	F	G	J	S
0 (.060)	...	...	...	...	...	.096	.05	.054
1 (.073)	...	...	...	...	...	.118	$\frac{1}{16}$	.066
2 (.086)	...	...	...	...	...	.140	$\frac{5}{64}$	.077
3 (.099)	...	...	...	...	...	.161	$\frac{5}{64}$	.089
4 (.112)	...	...	...	...	...	.183	$\frac{3}{32}$	.101
5 (.125)	...	...	...	...	...	.205	$\frac{3}{32}$	.112
6 (.138)	...	...	...	...	...	.226	$\frac{7}{64}$	.124
8 (.164)	...	...	...	...	...	.270	$\frac{9}{64}$	.148
10 (.190)	...	...	...	...	...	.312	$\frac{5}{32}$	.171
$\frac{1}{4}$	.500	.437	.191	.375	.172	.375	$\frac{3}{16}$	.225
$\frac{5}{16}$	.625	.562	.245	.437	.203	.469	$\frac{1}{4}$	.281
$\frac{3}{8}$	.750	.675	.273	.562	.250	.562	$\frac{5}{16}$	.337
$\frac{7}{16}$	.812	.750	.328	.625	.297	.656	$\frac{3}{8}$	.394
$\frac{1}{2}$	.875	.812	.354	.750	.328	.750	$\frac{3}{8}$	.450
$\frac{9}{16}$	1.000	.937	.409	.812	.375	...	...	...
$\frac{5}{8}$	1.125	1.000	.437	.875	.422	.938	$\frac{1}{2}$	.562
$\frac{3}{4}$	1.375	1.250	.546	1.000	.500	1.125	$\frac{5}{8}$	.675
$\frac{7}{8}$	1.625	...	...	1.125	.594	1.312	$\frac{3}{4}$	.787
1	1.875	...	...	1.312	.656	1.500	$\frac{3}{4}$	.900
$1\frac{1}{8}$	2.062	...	...	...	...	1.688	$\frac{7}{8}$	1.012
$1\frac{1}{4}$	2.312	...	...	...	...	1.875	$\frac{7}{8}$	1.125
$1\frac{3}{8}$	2.562	...	...	...	...	2.062	1	1.237
$1\frac{1}{2}$	2.812	...	...	...	...	2.250	1	1.350

[a] ANSI/ASME B18.6.2–1995.
[b] ANSI/ASME B18.3–1986 (R1995). For hexagon-head screws, see Appendix 18.

19 Cap Screws, Slotted[a] and Socket Head[b]—American National Standard and Metric (continued)

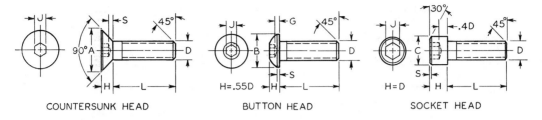

COUNTERSUNK HEAD   BUTTON HEAD   SOCKET HEAD

For methods of drawing cap screws, screw lengths, and thread data, see Fig. 13.32.

Nominal Size D	Countersunk Head[a]			Button Head[a]			Socket Head[b]		Hex Socket Size
	A (max)	H	S	B	S	G	C	S	J
M1.6 × 0.35	...	...	...	...	...	...	3.0	0.16	1.5
M2 × 0.4	...	...	...	...	...	...	3.8	0.2	1.5
M2.5 × 0.45	...	...	...	...	...	...	4.5	0.25	2.0
M3 × 0.5	6.72	1.86	0.25	5.70	0.38	0.2	5.5	0.3	2.5
M4 × 0.7	8.96	2.48	0.45	7.6	0.38	0.3	7.0	0.4	3.0
M5 × 0.8	11.2	3.1	0.66	9.5	0.5	0.38	8.5	0.5	4.0
M6 × 1	13.44	3.72	0.7	10.5	0.8	0.74	10.0	0.6	5.0
M8 × 1.25	17.92	4.96	1.16	14.0	0.8	1.05	13.0	0.8	6.0
M10 × 1.5	22.4	6.2	1.62	17.5	0.8	1.45	16.0	1.0	8.0
M12 × 1.75	26.88	7.44	1.8	21.0	0.8	1.63	18.0	1.2	10.0
M14 × 2	30.24	8.12	2.0	...	...	...	21.0	1.4	12.0
M16 × 2	33.6	8.8	2.2	28.0	1.5	2.25	24.0	1.6	14.0
M20 × 2.5	19.67	10.16	2.2	...	...	...	30.0	2.0	17.0
M24 × 3	...	...	...	...	...	...	36.0	2.4	19.0
M30 × 3.5	...	...	...	...	...	...	45.0	3.0	22.0
M36 × 4	...	...	...	...	...	...	54.0	3.6	27.0
M42 × 4.5	...	...	...	...	...	...	63.0	4.2	32.0
M48 × 5	...	...	...	...	...	...	72.0	4.8	36.0

Metric Socket Head Cap Screws

[a] ANSI/ASME B18.3.4M–1986 (R1993).
[b] ANSI/ASME B18.3.1M–1986 (R1993).

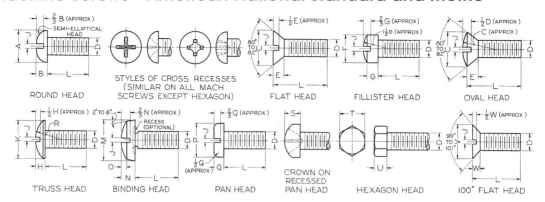

ROUND HEAD · STYLES OF CROSS RECESSES (SIMILAR ON ALL MACH SCREWS EXCEPT HEXAGON) · FLAT HEAD · FILLISTER HEAD · OVAL HEAD · TRUSS HEAD · BINDING HEAD · PAN HEAD · CROWN ON RECESSED PAN HEAD · HEXAGON HEAD · 100° FLAT HEAD

AMERICAN NATIONAL STANDARD MACHINE SCREWS[a]

Length of Thread: On screws 2″ long and shorter, the threads extend to within two threads of the head and closer if practicable; longer screws have minimum thread length of $1\frac{3}{4}″$.

Points: Machine screws are regularly made with plain sheared ends, not chamfered.

Threads: Either Coarse or Fine Thread Series, Class 2 fit.

Recessed Heads: Two styles of cross recesses are available on all screws except hexagon head.

Nominal Size	Max Diameter D	Round Head		Flat Heads & Oval Head		Fillister Head		Truss Head			Slot Width
		A	B	C	E	F	G	K	H	R	J
0	0.060	0.113	0.053	0.119	0.035	0.096	0.045	0.131	0.037	0.087	0.023
1	0.073	0.138	0.061	0.146	0.043	0.118	0.053	0.164	0.045	0.107	0.026
2	0.086	0.162	0.069	0.172	0.051	0.140	0.062	0.194	0.053	0.129	0.031
3	0.099	0.187	0.078	0.199	0.059	0.161	0.070	0.226	0.061	0.151	0.035
4	0.112	0.211	0.086	0.225	0.067	0.183	0.079	0.257	0.069	0.169	0.039
5	0.125	0.236	0.095	0.252	0.075	0.205	0.088	0.289	0.078	0.191	0.043
6	0.138	0.260	0.103	0.279	0.083	0.226	0.096	0.321	0.086	0.211	0.048
8	0.164	0.309	0.120	0.332	0.100	0.270	0.113	0.384	0.102	0.254	0.054
10	0.190	0.359	0.137	0.385	0.116	0.313	0.130	0.448	0.118	0.283	0.060
12	0.216	0.408	0.153	0.438	0.132	0.357	0.148	0.511	0.134	0.336	0.067
$\frac{1}{4}$	0.250	0.472	0.175	0.507	0.153	0.414	0.170	0.573	0.150	0.375	0.075
$\frac{5}{16}$	0.3125	0.590	0.216	0.635	0.191	0.518	0.211	0.698	0.183	0.457	0.084
$\frac{3}{8}$	0.375	0.708	0.256	0.762	0.230	0.622	0.253	0.823	0.215	0.538	0.094
$\frac{7}{16}$	0.4375	0.750	0.328	0.812	0.223	0.625	0.265	0.948	0.248	0.619	0.094
$\frac{1}{2}$	0.500	0.813	0.355	0.875	0.223	0.750	0.297	1.073	0.280	0.701	0.106
$\frac{9}{16}$	0.5625	0.938	0.410	1.000	0.260	0.812	0.336	1.198	0.312	0.783	0.118
$\frac{5}{8}$	0.625	1.000	0.438	1.125	0.298	0.875	0.375	1.323	0.345	0.863	0.133
$\frac{3}{4}$	0.750	1.250	0.547	1.375	0.372	1.000	0.441	1.573	0.410	1.024	0.149

Nominal Size	Max Diameter D	Binding Head			Pan Head			Hexagon Head		100° Flat Head		Slot Width
		M	N	O	P	Q	S	T	U	V	W	J
2	0.086	0.181	0.050	0.018	0.167	0.053	0.062	0.125	0.050	...	...	0.031
3	0.099	0.208	0.059	0.022	0.193	0.060	0.071	0.187	0.055	...	...	0.035
4	0.112	0.235	0.068	0.025	0.219	0.068	0.080	0.187	0.060	0.225	0.049	0.039
5	0.125	0.263	0.078	0.029	0.245	0.075	0.089	0.187	0.070	...	...	0.043
6	0.138	0.290	0.087	0.032	0.270	0.082	0.097	0.250	0.080	0.279	0.060	0.048
8	0.164	0.344	0.105	0.039	0.322	0.096	0.115	0.250	0.110	0.332	0.072	0.054
10	0.190	0.399	0.123	0.045	0.373	0.110	0.133	0.312	0.120	0.385	0.083	0.060
12	0.216	0.454	0.141	0.052	0.425	0.125	0.151	0.312	0.155	...	...	0.067
$\frac{1}{4}$	0.250	0.513	0.165	0.061	0.492	0.144	0.175	0.375	0.190	0.507	0.110	0.075
$\frac{5}{16}$	0.3125	0.641	0.209	0.077	0.615	0.178	0.218	0.500	0.230	0.635	0.138	0.084
$\frac{3}{8}$	0.375	0.769	0.253	0.094	0.740	0.212	0.261	0.562	0.295	0.762	0.165	0.094
$\frac{7}{16}$	.4375	...	...	...	.865	.247	.305	...	...	...	...	.094
$\frac{1}{2}$	.500	...	...	...	.987	.281	.348	...	...	...	...	.106
$\frac{9}{16}$	.5625	...	...	...	1.041	.315	.391	...	...	...	...	.118
$\frac{5}{8}$	.625	...	...	...	1.172	.350	.434	...	...	...	...	.133
$\frac{3}{4}$	.750	...	...	...	1.435	.419	.521	...	...	...	...	.149

20 Machine Screws—American National Standard and Metric (continued)

METRIC MACHINE SCREWS

Length of Thread: On screws 36 mm long or shorter, the threads extend to within one thread of the head: on longer screws the thread extends to within two threads of the head.

Points: Machine screws are regularly made with sheared ends, not chamfered.

Threads: Coarse (genera purpose) threads series are given.

Recessed Heads: Two styles of cross-recesses are available on all screws except hexagon head.

Nominal Size & Thd Pitch	Max. Dia. D. mm	Flat Heads & Oval Head		Pan Heads			Hex Head		Slot Width
		C	E	P	Q	S	T	U	J
M2 × M	2.0	3.5	1.2	4.0	1.3	1.6	3.2	1.6	0.7
M2.5 × 0.45	2.5	4.4	1.5	5.0	1.5	2.1	4.0	2.1	0.8
M3 × 0.5	3.0	5.2	1.7	5.6	1.8	2.4	5.0	2.3	1.0
M3.5 × 0.6	3.5	6.9	2.3	7.0	2.1	2.6	5.5	2.6	1.2
M4 × 0.7	4.0	8.0	2.7	8.0	2.4	3.1	7.0	3.0	1.5
M5 × 0.8	5.0	8.9	2.7	9.5	3.0	3.7	8.0	3.8	1.5
M6 × 1	6.0	10.9	3.3	12.0	3.6	4.6	10.0	4.7	1.9
M8 × 1.25	8.0	15.14	4.6	16.0	4.8	6.0	13.0	6.0	2.3
M10 × 1.5	10.0	17.8	5.0	20.0	6.0	7.5	15.0	7.5	2.8
M12 × 1.75	12.0	...	...	...	...	...	18.0	9.0	...

Nominal Size	Metric Machine Screw Lengths—L																				
	2.5	3	4	5	6	8	10	13	16	20	25	30	35	40	45	50	55	60	65	70	80	90
M2 × 0.4	PH	A	A	A	A	A	A	A	A	A												
M2.5 × 0.45		PH	A	A	A	A	A	A	A	A	A	Min. Thd Length—28 mm										
M3 × 0.5			PH	A	A	A	A	A	A	A	A	A										
M3.5 × 0.6				PH	A	A	A	A	A	A	A	A	A	Min. Thd Length—38 mm								
M4 × 0.7				PH	A	A	A	A	A	A	A	A										
M5 × 0.8					PH	A	A	A	A	A	A	A	A	A	A	A						
M6 × 1						A	A	A	A	A	A	A	A	A	A	A	A					
M8 × 1.25						A	A	A	A	A	A	A	A	A	A	A	A	A	A	A	A	
M10 × 1.5						A	A	A	A	A	A	A	A	A	A	A	A	A	A	A	A	A
M12 × 1.75							A	A	A	A	A	A	A	A	A	A	A	A	A	A	A	A

[a] PH = recommended lengths for only pan and hex head metric screws.
A = recommended lengths for all metric screw head-styles.

21   Keys—Square, Flat, Plain Taper,[a] and Gib Head

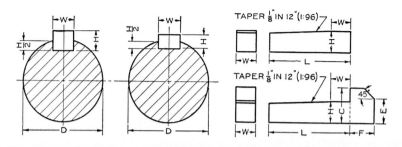

Shaft Diameters	Square Stock Key	Flat Stock Key	Gib Head Taper Stock Key					
			Square			Flat		
			Height	Length	Height to Chamfer	Height	Length	Height to Chamfer
D	$W = H$	$W \times H$	C	F	E	C	F	E
$\frac{1}{2}$ to $\frac{9}{16}$	$\frac{1}{8}$	$\frac{1}{8} \times \frac{3}{32}$	$\frac{1}{4}$	$\frac{7}{32}$	$\frac{5}{32}$	$\frac{3}{16}$	$\frac{1}{8}$	$\frac{1}{8}$
$\frac{5}{8}$ to $\frac{7}{8}$	$\frac{3}{16}$	$\frac{3}{16} \times \frac{1}{8}$	$\frac{5}{16}$	$\frac{9}{32}$	$\frac{7}{32}$	$\frac{1}{4}$	$\frac{3}{16}$	$\frac{5}{32}$
$\frac{15}{16}$ to $1\frac{1}{4}$	$\frac{1}{4}$	$\frac{1}{4} \times \frac{3}{16}$	$\frac{7}{16}$	$\frac{11}{32}$	$\frac{11}{32}$	$\frac{5}{16}$	$\frac{1}{4}$	$\frac{3}{16}$
$1\frac{5}{16}$ to $1\frac{3}{8}$	$\frac{5}{16}$	$\frac{5}{16} \times \frac{1}{4}$	$\frac{9}{16}$	$\frac{13}{32}$	$\frac{13}{32}$	$\frac{3}{8}$	$\frac{5}{16}$	$\frac{1}{4}$
$1\frac{7}{16}$ to $1\frac{3}{4}$	$\frac{3}{8}$	$\frac{3}{8} \times \frac{1}{4}$	$\frac{11}{16}$	$\frac{15}{32}$	$\frac{15}{32}$	$\frac{7}{16}$	$\frac{3}{8}$	$\frac{5}{16}$
$1\frac{13}{16}$ to $2\frac{1}{4}$	$\frac{1}{2}$	$\frac{1}{2} \times \frac{3}{8}$	$\frac{7}{8}$	$\frac{19}{32}$	$\frac{5}{8}$	$\frac{5}{8}$	$\frac{1}{2}$	$\frac{7}{16}$
$2\frac{5}{16}$ to $2\frac{3}{4}$	$\frac{5}{8}$	$\frac{5}{8} \times \frac{7}{16}$	$1\frac{1}{16}$	$\frac{23}{32}$	$\frac{3}{4}$	$\frac{3}{4}$	$\frac{5}{8}$	$\frac{1}{2}$
$2\frac{7}{8}$ to $3\frac{1}{4}$	$\frac{3}{4}$	$\frac{3}{4} \times \frac{1}{2}$	$1\frac{1}{4}$	$\frac{7}{8}$	$\frac{7}{8}$	$\frac{7}{8}$	$\frac{3}{4}$	$\frac{5}{8}$
$3\frac{3}{8}$ to $3\frac{3}{4}$	$\frac{7}{8}$	$\frac{7}{8} \times \frac{5}{8}$	$1\frac{1}{2}$	1	1	$1\frac{1}{16}$	$\frac{7}{8}$	$\frac{3}{4}$
$3\frac{7}{8}$ to $4\frac{1}{2}$	1	$1 \times \frac{3}{4}$	$1\frac{3}{4}$	$1\frac{3}{16}$	$1\frac{3}{16}$	$1\frac{1}{4}$	1	$\frac{13}{16}$
$4\frac{3}{4}$ to $5\frac{1}{2}$	$1\frac{1}{4}$	$1\frac{1}{4} \times \frac{7}{8}$	2	$1\frac{7}{16}$	$1\frac{7}{16}$	$1\frac{1}{2}$	$1\frac{1}{4}$	1
$5\frac{3}{4}$ to 6	$1\frac{1}{2}$	$1\frac{1}{2} \times 1$	$2\frac{1}{2}$	$1\frac{3}{4}$	$1\frac{3}{4}$	$1\frac{3}{4}$	$1\frac{1}{2}$	1

[a] Plain taper square and flat keys have the same dimensions as the plain parallel stock keys, with the addition of the taper on top. Gib head taper square and flat keys have the same dimensions as the plain taper keys, with the addition of the gib head.

Stock lengths for plain taper and gib head taper keys: The minimum stock length equals 4W, and the maximum equals 16W. The increments of increase of length equal 2W.

22   Screw Threads,[a] Square and Acme

Size	Threads per Inch	Size	Threads per Inch	Size	Threads per Inch	Size	Threads per Inch
$\frac{3}{8}$	12	$\frac{7}{8}$	5	2	$2\frac{1}{2}$	$3\frac{1}{2}$	$1\frac{1}{3}$
$\frac{7}{16}$	10	1	5	$2\frac{1}{4}$	2	$3\frac{3}{4}$	$1\frac{1}{3}$
$\frac{1}{2}$	10	$1\frac{1}{8}$	4	$2\frac{1}{2}$	2	4	$1\frac{1}{3}$
$\frac{9}{16}$	8	$1\frac{1}{4}$	4	$2\frac{3}{4}$	2	$4\frac{1}{4}$	$1\frac{1}{3}$
$\frac{5}{8}$	8	$1\frac{1}{2}$	3	3	$1\frac{1}{2}$	$4\frac{1}{2}$	1
$\frac{3}{4}$	6	$1\frac{3}{4}$	$2\frac{1}{2}$	$3\frac{1}{4}$	$1\frac{1}{2}$	over $4\frac{1}{2}$	1

[a] See Appendix 17 for General-Purpose Acme Threads.

23 Woodruff Keys[a]—American National Standard

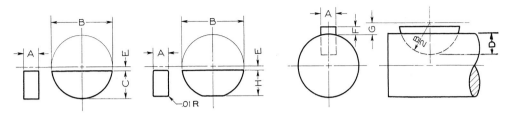

Key No.[b]	Nominal Sizes $A \times B$	E	F	G	Maximum Sizes H	D	C	Key No.[b]	Nominal Sizes $A \times B$	E	F	G	Maximum Sizes H	D	C
204	$\frac{1}{16} \times \frac{1}{2}$	$\frac{3}{64}$	$\frac{1}{32}$	$\frac{5}{64}$	.194	.1718	.203	808	$\frac{1}{4} \times 1$	$\frac{1}{16}$	$\frac{1}{8}$	$\frac{3}{16}$	.428	.3130	.438
304	$\frac{3}{32} \times \frac{1}{2}$	$\frac{3}{64}$	$\frac{3}{64}$	$\frac{3}{32}$	.194	.1561	.203	809	$\frac{1}{4} \times 1\frac{1}{8}$	$\frac{5}{64}$	$\frac{1}{8}$	$\frac{13}{64}$	.475	.3590	.484
305	$\frac{3}{32} \times \frac{5}{8}$	$\frac{1}{16}$	$\frac{3}{64}$	$\frac{7}{64}$	.240	.2031	.250	810	$\frac{1}{4} \times 1\frac{1}{4}$	$\frac{5}{64}$	$\frac{1}{8}$	$\frac{13}{64}$	.537	.4220	.547
404	$\frac{1}{8} \times \frac{1}{2}$	$\frac{3}{64}$	$\frac{1}{16}$	$\frac{7}{64}$	.194	.1405	.203	811	$\frac{1}{4} \times 1\frac{3}{8}$	$\frac{3}{32}$	$\frac{1}{8}$	$\frac{7}{32}$	.584	.4690	.594
405	$\frac{1}{8} \times \frac{5}{8}$	$\frac{1}{16}$	$\frac{1}{16}$	$\frac{1}{8}$	.240	.1875	.250	812	$\frac{1}{4} \times 1\frac{1}{2}$	$\frac{7}{64}$	$\frac{1}{8}$	$\frac{15}{64}$	.631	.5160	.641
406	$\frac{1}{8} \times \frac{3}{4}$	$\frac{1}{16}$	$\frac{1}{16}$	$\frac{1}{8}$	.303	.2505	.313	1008	$\frac{5}{16} \times 1$	$\frac{1}{16}$	$\frac{5}{32}$	$\frac{7}{32}$	.428	.2818	.438
505	$\frac{5}{32} \times \frac{5}{8}$	$\frac{1}{16}$	$\frac{5}{64}$	$\frac{9}{64}$	.240	.1719	.250	1009	$\frac{5}{16} \times 1\frac{1}{8}$	$\frac{5}{64}$	$\frac{5}{32}$	$\frac{15}{64}$	.475	.3278	.484
506	$\frac{5}{32} \times \frac{3}{4}$	$\frac{1}{16}$	$\frac{5}{64}$	$\frac{9}{64}$	.303	.2349	.313	1010	$\frac{5}{16} \times 1\frac{1}{4}$	$\frac{5}{64}$	$\frac{5}{32}$	$\frac{15}{64}$	.537	.3908	.547
507	$\frac{5}{32} \times \frac{7}{8}$	$\frac{1}{16}$	$\frac{5}{64}$	$\frac{9}{64}$	.365	.2969	.375	1011	$\frac{5}{16} \times 1\frac{3}{8}$	$\frac{3}{32}$	$\frac{5}{32}$	$\frac{8}{32}$	.584	.4378	.594
606	$\frac{3}{16} \times \frac{3}{4}$	$\frac{1}{16}$	$\frac{3}{32}$	$\frac{5}{32}$	.303	.2193	.313	1012	$\frac{5}{16} \times 1\frac{1}{2}$	$\frac{7}{64}$	$\frac{5}{32}$	$\frac{17}{64}$	.631	.4848	.641
607	$\frac{3}{16} \times \frac{7}{8}$	$\frac{1}{16}$	$\frac{3}{32}$	$\frac{5}{32}$	.365	.2813	.375	1210	$\frac{3}{8} \times 1\frac{1}{4}$	$\frac{5}{64}$	$\frac{3}{16}$	$\frac{17}{64}$	.537	.3595	.547
608	$\frac{3}{16} \times 1$	$\frac{1}{16}$	$\frac{3}{32}$	$\frac{5}{32}$	.428	.3443	.438	1211	$\frac{3}{8} \times 1\frac{3}{8}$	$\frac{3}{32}$	$\frac{3}{16}$	$\frac{9}{32}$	.584	.4065	.594
609	$\frac{3}{16} \times 1\frac{1}{8}$	$\frac{5}{64}$	$\frac{3}{32}$	$\frac{11}{64}$	.475	.3903	.484	1212	$\frac{3}{8} \times 1\frac{1}{2}$	$\frac{7}{64}$	$\frac{3}{16}$	$\frac{19}{64}$	.631	.4535	.641
807	$\frac{1}{4} \times \frac{7}{8}$	$\frac{1}{16}$	$\frac{1}{8}$	$\frac{3}{16}$	.365	.2500	.375	...	....	...	...	...	...	...	...

[a] ANSI B17.2–1967 (R1990).
[b] Key numbers indicate nominal key dimensions. The last two digits give the nominal diameter B in eighths of an inch, and the digits before the last two give the nominal width A in thirty-seconds of an inch.

24 Woodruff Key Sizes for Different Shaft Diameters[a]

Shaft Diameter	$\frac{5}{16}$ to $\frac{3}{8}$	$\frac{7}{16}$ to $\frac{1}{2}$	$\frac{9}{16}$ to $\frac{3}{4}$	$\frac{13}{16}$ to $\frac{15}{16}$	1 to $1\frac{3}{16}$	$1\frac{1}{4}$ to $1\frac{7}{16}$	$1\frac{1}{2}$ to $1\frac{3}{4}$	$1\frac{13}{16}$ to $2\frac{1}{8}$	$2\frac{3}{16}$ to $2\frac{1}{2}$
Key Numbers	204	304	404	505	606	807	810	1011	1211
		305	405	506	607	808	811	1012	1212
			406	507	608	809	812		
					609				

[a] Suggested sizes; not standard.

25 Pratt and Whitney Round-End Keys

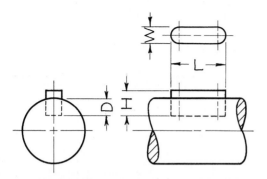

KEYS MADE WITH ROUND
ENDS AND KEYWAYS CUT
IN SPLINE MILLER

Maximum length of slot is 4″ + W. Note that key is sunk two-thirds into shaft in all cases.

Key No.	L^a	W or D	H	Key No.	L^a	W or D	H
1	$\frac{1}{2}$	$\frac{1}{16}$	$\frac{3}{32}$	22	$1\frac{3}{8}$	$\frac{1}{4}$	$\frac{3}{8}$
2	$\frac{1}{2}$	$\frac{3}{32}$	$\frac{9}{64}$	23	$1\frac{1}{38}$	$\frac{5}{16}$	$\frac{15}{32}$
3	$\frac{1}{2}$	$\frac{1}{8}$	$\frac{3}{16}$	F	$1\frac{3}{8}$	$\frac{3}{8}$	$\frac{9}{16}$
4	$\frac{5}{8}$	$\frac{3}{32}$	$\frac{9}{64}$	24	$1\frac{1}{2}$	$\frac{1}{4}$	$\frac{3}{8}$
5	$\frac{5}{8}$	$\frac{1}{8}$	$\frac{3}{16}$	25	$1\frac{1}{2}$	$\frac{5}{16}$	$\frac{15}{32}$
6	$\frac{5}{8}$	$\frac{5}{32}$	$\frac{15}{64}$	G	$1\frac{1}{2}$	$\frac{3}{8}$	$\frac{9}{16}$
7	$\frac{3}{4}$	$\frac{1}{8}$	$\frac{3}{16}$	51	$1\frac{3}{4}$	$\frac{1}{4}$	$\frac{3}{8}$
8	$\frac{3}{4}$	$\frac{5}{32}$	$\frac{15}{64}$	52	$1\frac{3}{4}$	$\frac{5}{16}$	$\frac{15}{32}$
9	$\frac{3}{4}$	$\frac{3}{16}$	$\frac{9}{32}$	53	$1\frac{3}{4}$	$\frac{3}{8}$	$\frac{9}{16}$
10	$\frac{7}{8}$	$\frac{5}{32}$	$\frac{15}{64}$	26	2	$\frac{3}{16}$	$\frac{9}{32}$
11	$\frac{7}{8}$	$\frac{3}{16}$	$\frac{9}{32}$	27	2	$\frac{1}{4}$	$\frac{3}{8}$
12	$\frac{7}{8}$	$\frac{7}{32}$	$\frac{21}{64}$	28	2	$\frac{5}{16}$	$\frac{15}{32}$
A	$\frac{7}{8}$	$\frac{1}{4}$	$\frac{3}{8}$	29	2	$\frac{3}{8}$	$\frac{9}{16}$
13	1	$\frac{3}{16}$	$\frac{9}{32}$	54	$2\frac{1}{4}$	$\frac{1}{4}$	$\frac{3}{8}$
14	1	$\frac{7}{32}$	$\frac{21}{64}$	55	$2\frac{1}{4}$	$\frac{5}{16}$	$\frac{15}{32}$
15	1	$\frac{1}{4}$	$\frac{3}{8}$	56	$2\frac{1}{4}$	$\frac{3}{8}$	$\frac{9}{16}$
B	1	$\frac{5}{16}$	$\frac{15}{32}$	57	$2\frac{1}{4}$	$\frac{7}{16}$	$\frac{21}{32}$
16	$1\frac{1}{8}$	$\frac{3}{16}$	$\frac{9}{32}$	58	$2\frac{1}{2}$	$\frac{5}{16}$	$\frac{15}{32}$
17	$1\frac{1}{8}$	$\frac{7}{32}$	$\frac{21}{64}$	59	$2\frac{1}{2}$	$\frac{3}{8}$	$\frac{9}{16}$
18	$1\frac{1}{8}$	$\frac{1}{4}$	$\frac{3}{8}$	60	$2\frac{1}{2}$	$\frac{7}{16}$	$\frac{21}{32}$
C	$1\frac{1}{8}$	$\frac{5}{16}$	$\frac{15}{32}$	61	$2\frac{1}{2}$	$\frac{1}{2}$	$\frac{3}{4}$
19	$1\frac{1}{4}$	$\frac{3}{16}$	$\frac{9}{32}$	30	3	$\frac{3}{8}$	$\frac{9}{16}$
20	$1\frac{1}{4}$	$\frac{7}{32}$	$\frac{21}{64}$	31	3	$\frac{7}{16}$	$\frac{21}{32}$
21	$1\frac{1}{4}$	$\frac{1}{4}$	$\frac{3}{8}$	32	3	$\frac{1}{2}$	$\frac{3}{4}$
D	$1\frac{1}{4}$	$\frac{5}{16}$	$\frac{15}{32}$	33	3	$\frac{9}{16}$	$\frac{27}{32}$
E	$1\frac{1}{4}$	$\frac{3}{8}$	$\frac{9}{16}$	34	3	$\frac{5}{8}$	$\frac{15}{16}$

^{a} The length L may vary from the table, but equals at least 2W.

For parts lists, etc., give inside diameter, outside diameter, and the thickness; for example, .344 × .688 × .065 TYPE A PLAIN WASHER.

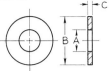

PREFERRED SIZES OF TYPE A PLAIN WASHERS[b]

Nominal Washer Size[c]			Inside Diameter	Outside Diameter	Nominal Thickness
			A	B	C
…	…		0.078	0.188	0.020
…	…		0.094	0.250	0.020
…	…		0.125	0.312	0.032
No. 6	0.138		0.156	0.375	0.049
No. 8	0.164		0.188	0.438	0.049
No. 10	0.190		0.219	0.500	0.049
$\frac{3}{16}$	0.188		0.250	0.562	0.049
No. 12	0.216		0.250	0.562	0.065
$\frac{1}{4}$	0.250	N	0.281	0.625	0.065
$\frac{1}{4}$	0.250	W	0.312	0.734	0.065
$\frac{5}{16}$	0.312	N	0.344	0.688	0.065
$\frac{5}{16}$	0.312	W	0.375	0.875	0.083
$\frac{3}{8}$	0.375	N	0.406	0.812	0.065
$\frac{3}{8}$	0.375	W	0.438	1.000	0.083
$\frac{7}{16}$	0.438	N	0.469	0.922	0.065
$\frac{7}{16}$	0.438	W	0.500	1.250	0.083
$\frac{1}{2}$	0.500	N	0.531	1.062	0.095
$\frac{1}{2}$	0.500	W	0.562	1.375	0.109
$\frac{9}{16}$	0.562	N	0.594	1.156	0.095
$\frac{9}{16}$	0.562	W	0.625	1.469	0.109
$\frac{5}{8}$	0.625	N	0.656	1.312	0.095
$\frac{5}{8}$	0.625	W	0.688	1.750	0.134
$\frac{3}{4}$	0.750	N	0.812	1.469	0.134
$\frac{3}{4}$	0.750	W	0.812	2.000	0.148
$\frac{7}{8}$	0.875	N	0.938	1.750	0.134
$\frac{7}{8}$	0.875	W	0.938	2.250	0.165
1	1.000	N	1.062	2.000	0.134
1	1.000	W	1.062	2.500	0.165
$1\frac{1}{8}$	1.125	N	1.250	2.250	0.134
$1\frac{1}{8}$	1.125	W	1.250	2.750	0.165
$1\frac{1}{4}$	1.250	N	1.375	2.500	0.165
$1\frac{1}{4}$	1.250	W	1.375	3.000	0.165
$1\frac{3}{8}$	1.375	N	1.500	2.750	0.165
$1\frac{3}{8}$	1.375	W	1.500	3.250	0.180
$1\frac{1}{2}$	1.500	N	1.625	3.000	0.165
$1\frac{1}{2}$	1.500	W	1.625	3.500	0.180
$1\frac{5}{8}$	1.625		1.750	3.750	0.180
$1\frac{3}{4}$	1.750		1.875	4.000	0.180
$1\frac{7}{8}$	1.875		2.000	4.250	0.180
2	2.000		2.125	4.500	0.180
$2\frac{1}{4}$	2.250		2.375	4.750	0.220
$2\frac{1}{2}$	2.500		2.625	5.000	0.238
$2\frac{3}{4}$	2.750		2.875	5.250	0.259
3	3000		3.125	5.500	0.284

[a] From ANSI B18.22.1–1965 (R1981). For complete listings, see the standard.
[b] Preferred sizes are for the most part from series previously designated "Standard Plate" and "SAE." Where common sizes existed in the two series, the SAE size is designated "N" (narrow) and the Standard Plate "W" (wide).
[c] Nominal washer sizes are intended for use with comparable nominal screw or bolt sizes.

27 Washers,[a] Lock—American National Standard

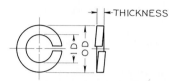

For parts lists, etc., give nominal size and series; for example, $\frac{1}{4}$ REGULAR LOCK WASHER

PREFERRED SERIES

Nominal Washer Size[b]		Inside Diameter, Min.	Regular		Extra Duty		Hi-Collar	
			Outside Diameter, Max.	Thickness, Min.	Outside Diameter, Max.	Thickness, Min.	Outside Diameter, Max.	Thickness, Min.
No. 2	0.086	0.088	0.172	0.020	0.208	0.027	. . .	. . .
No. 3	0.099	0.101	0.195	0.025	0.239	0.034	. . .	. . .
No. 4	0.112	0.115	0.209	0.025	0.253	0.034	0.173	0.022
No. 5	0.125	0.128	0.236	0.031	0.300	0.045	0.202	0.030
No. 6	0.138	0.141	0.250	0.031	0.314	0.045	0.216	0.030
No. 8	0.164	0.168	0.293	0.040	0.375	0.057	0.267	0.047
No. 10	0.190	0.194	0.334	0.047	0.434	0.068	0.294	0.047
No. 12	0.216	0.221	0.377	0.056	0.497	0.080	. . .	. . .
$\frac{1}{4}$	0.250	0.255	0.489	0.062	0.535	0.084	0.365	0.078
$\frac{5}{16}$	0.312	0.318	0.586	0.078	0.622	0.108	0.460	0.093
$\frac{3}{8}$	0.375	0.382	0.683	0.094	0.741	0.123	0.553	0.125
$\frac{7}{16}$	0.438	0.446	0.779	0.109	0.839	0.143	0.647	0.140
$\frac{1}{2}$	0.500	0.509	0.873	0.125	0.939	0.162	0.737	0.172
$\frac{9}{16}$	0.562	0.572	0.971	0.141	1.041	0.182	. . .	. . .
$\frac{5}{8}$	0.625	0.636	1.079	0.156	1.157	0.202	0.923	0.203
$\frac{11}{16}$	0.688	0.700	1.176	0.172	1.258	0.221	. . .	. . .
$\frac{3}{4}$	0.750	0.763	1.271	0.188	1.361	0.241	1.111	0.218
$\frac{13}{16}$	0.812	0.826	1.367	0.203	1.463	0.261	. . .	. . .
$\frac{7}{8}$	0.875	0.890	1.464	0.219	1.576	0.285	1.296	0.234
$\frac{15}{16}$	0.938	0.954	1.560	0.234	1.688	0.308	. . .	. . .
1	1.000	1.017	1.661	0.250	1.799	0.330	1.483	0.250
$1\frac{1}{16}$	1.062	1.080	1.756	0.266	1.910	0.352	. . .	. . .
$1\frac{1}{8}$	1.125	1.144	1.853	0.281	2.019	0.375	1.669	0.313
$1\frac{3}{16}$	1.188	1.208	1.950	0.297	2.124	0.396	. . .	. . .
$1\frac{1}{4}$	1.250	1.271	2.045	0.312	2.231	0.417	1.799	0.313
$1\frac{5}{16}$	1.312	1.334	2.141	0.328	2.335	0.438	. . .	. . .
$1\frac{3}{8}$	1.375	1.398	2.239	0.344	2.439	0.458	2.041	0.375
$1\frac{7}{16}$	1.438	1.462	2.334	0.359	2.540	0.478	. . .	. . .
$1\frac{1}{2}$	1.500	1.525	2.430	0.375	2.638	0.496	2.170	0.375

[a] From ANSI/ASME B18.21.1–1994. For complete listing, see the standard.
[b] Nominal washer sizes are intended for use with comparable nominal screw or bolt sizes.

28 Wire Gage Standards[a]

Dimensions of sizes in decimal parts of an inch.[b]

No. of Wire	American or Brown & Sharpe for Non-ferrous Metals	Birmingham, or Stubs' Iron Wire[c]	American S. & W. Co.'s (Washburn & Moen) Std. Steel Wire	American S. & W. Co.'s Music Wire	Imperial Wire	Stubs' Steel Wire[c]	Steel Manufacturers' Sheet Gage[b]	No. of Wire
7–0's	.651354	...	.4900	...	.500	...	...	7–0's
6–0's	.580049	...	.4615	.004	.464	...	...	6–0's
5–0's	.516549	.500	.4305	.005	.432	...	...	5–0's
4–0's	.460	.454	.3938	.006	.400	...	...	4–0's
000	.40964	.425	.3625	.007	.372	...	...	000
00	.3648	.380	.3310	.008	.348	...	...	00
0	.32486	.340	.3065	.009	.324	...	...	0
1	.2893	.300	.2830	.010	.300	.227	...	1
2	.25763	.284	.2625	.011	.276	.219	...	2
3	.22942	.259	.2437	.012	.252	.212	.2391	3
4	.20431	.238	.2253	.013	.232	.207	.2242	4
6	.16202	.203	.1920	.016	.192	.201	.1943	6
7	.14428	.180	.1770	.018	.176	.199	.1793	7
8	.12849	.165	.1620	.020	.160	.197	.1644	8
9	.11443	.148	.1483	.022	.144	.194	.1495	9
10	.10189	.134	.1350	.024	.128	.191	.1345	10
11	.090742	.120	.1205	.026	.116	.188	.1196	11
12	.080808	.109	.1055	.029	.104	.185	.1046	12
13	.071961	.095	.0915	.031	.092	.182	.0897	13
14	.064084	.083	.0800	.033	.080	.180	.0747	14
15	.057068	.072	.0720	.035	.072	.178	.0763	15
16	.05082	.065	.0625	.037	.064	.175	.0598	16
17	.045257	.058	.0540	.039	.056	.172	.0538	17
18	.040303	.049	.0475	.041	.048	.168	.0478	18
19	.03589	.042	.0410	.043	.040	.164	.0418	19
20	.031961	.035	.0348	.045	.036	.161	.0359	20
21	.028462	.032	.0317	.047	.032	.157	.0329	21
22	.025347	.028	.0286	.049	.028	.155	.0299	22
23	.022571	.025	.0258	.051	.024	.153	.0269	23
24	.0201	.022	.0230	.055	.022	.151	.0239	24
25	.0179	.020	.0204	.059	.020	.148	.0209	25
26	.01594	.018	.0181	.063	.018	.146	.0179	26
27	.014195	.016	.0173	.067	.0164	.143	.0164	27
28	.012641	.014	.0162	.071	.0149	.139	.0149	28
29	.011257	.013	.0150	.075	.0136	.134	.0135	29
30	.010025	.012	.0140	.080	.0124	.127	.0120	30
31	.008928	.010	.0132	.085	.0116	.120	.0105	31
32	.00795	.009	.0128	.090	.0108	.115	.0097	32
33	.00708	.008	.0118	.095	.0100	.112	.0090	33
34	.006304	.007	.0104	...	.0092	.110	.0082	34
35	.005614	.005	.0095	...	.0084	.108	.0075	35
36	.005	.004	.0090	...	.0076	.106	.0067	36
37	.004453	...	.0085	...	.0068	.103	.0064	37
38	.003965	...	.0080	...	.0060	.101	.0060	38
39	.003531	...	.0075	...	.0052	.099	...	39
40	.003144	...	.0070	...	.0048	.097	...	40

[a] Courtesy Brown & Sharpe Mfg. Co.
[b] Now used by steel manufacturers in place of old U.S. Standard Gage.
[c] The difference between the Stubs' Iron Wire Gage and the Stubs' Steel Wire Gage should be noted, the first being commonly known as the English Standard Wire, or Birmingham Gage, which designates the Stubs' soft wire sizes and the second being used in measuring drawn steel wire or drill rods of Stubs' make.

29 Taper Pins[a]—American National Standard

D — L (MAX) — TAPER .25 PER FT

To find small diameter of pin, multiply the length by .02083 and subtract the result from the larger diameter. All dimensions are given in inches.
Standard reamers are available for pins given above the heavy line.

Number	7/0	6/0	5/0	4/0	3/0	2/0	0	1	2	3	4	5	6	7	8
Size (Large End)	.0625	.0780	.0940	.1090	.1250	.1410	.1560	.1720	.1930	.2190	.2500	.2890	.3410	.4090	.4920
Shaft Diameter (Approx)[b]		$\frac{7}{32}$	$\frac{1}{4}$	$\frac{5}{16}$	$\frac{3}{8}$	$\frac{7}{16}$	$\frac{1}{2}$	$\frac{9}{16}$	$\frac{5}{8}$	$\frac{3}{4}$	$\frac{13}{16}$	$\frac{7}{8}$	1	$1\frac{1}{4}$	$1\frac{1}{2}$
Drill Size (Before Reamer)[b]	.0312	.0312	.0625	.0625	.0781	.0938	.0938	.1094	.1250	.1250	.1562	.1562	.2188	.2344	.3125
Length L															
.250	×	×	×	×	×										
.375	×	×	×	×	×	×	×								
.500	×	×	×	×	×	×	×	×							
.625	×	×	×	×	×	×	×	×	×						
.750	×	×	×	×	×	×	×	×	×	×					
.875	×	×	×	×	×	×	×	×	×	×	×	×			
1.000	×	×	×	×	×	×	×	×	×	×	×	×	×		
1.250		×	×	×	×	×	×	×	×	×	×	×	×	×	
1.500		×	×	×	×	×	×	×	×	×	×	×	×	×	×
1.750				×	×	×	×	×	×	×	×	×	×	×	×
2.000					×	×	×	×	×	×	×	×	×	×	×
2.250						×	×	×	×	×	×	×	×	×	×
2.500								×	×	×	×	×	×	×	×
2.750									×	×	×	×	×	×	×
3.000										×	×	×	×	×	×
3.250											×	×	×	×	×
3.500											×	×	×	×	×
3.750											×	×	×	×	×
4.000												×	×	×	×
4.250													×	×	×
4.500														×	×

[a] ANSI/ASME B18.8.2–1994. For Nos. 9 and 10, see the standard. Pins Nos 11 (size .8600), 12 (size 1.032), 13 (size 1.241), and 14 (size 1.523) are special sizes; hence their lengths are special.
[b] Suggested sizes; not American National Standard.

30 Cotter Pins[a]—American National Standard

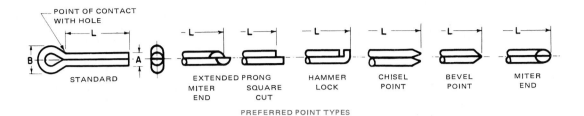

PREFERRED POINT TYPES

All dimensions are given in inches.

Nominal Size or Pin Diameter		Diameter A		Outside Eye Diameter B Min.	Extended Prong Length Min.	Hole Sizes Recommended
		Max.	Min.			
$\frac{1}{32}$	.031	.032	.028	.06	.01	.047
$\frac{3}{64}$	.047	.048	.044	.09	.02	.062
$\frac{1}{16}$	.062	.060	.056	.12	.03	.078
$\frac{5}{64}$	.078	.076	.072	.16	.04	.094
$\frac{3}{32}$	.094	.090	.086	.19	.04	.109
$\frac{7}{64}$	.109	.104	.100	.22	.05	.125
$\frac{1}{8}$	.125	.120	.116	.25	.06	.141
$\frac{9}{64}$	.141	.134	.130	.28	.06	.156
$\frac{5}{32}$	.156	.150	.146	.31	.07	.172
$\frac{3}{16}$	.188	.176	.172	.38	.09	.203
$\frac{7}{32}$	.219	.207	.202	.44	.10	.234
$\frac{1}{4}$	.250	.225	.220	.50	.11	.266
$\frac{5}{16}$	.312	.280	.275	.62	.14	.312
$\frac{3}{8}$	.375	.335	.329	.75	.16	.375
$\frac{7}{16}$	.438	.406	.400	.88	.20	.438
$\frac{1}{2}$	.500	.473	.467	1.00	.23	.500
$\frac{5}{8}$	.625	.598	.590	1.25	.30	.625
$\frac{3}{4}$	.750	.723	.715	1.50	.36	.750

[a] ANSI/ASME B18.8.1–1994.

31   Metric Equivalents

Length	
U.S. to Metric	**Metric to U.S.**
1 inch = 2.540 centimeters 1 foot = .305 meter 1 yard = .914 meter 1 mile = 1.609 kilometers	1 millimeter = .039 inch 1 centimeter = .394 inch 1 meter = 3.281 feet or 1.094 yards 1 kilometer = .621 mile
Area	
$1\ \text{inch}^2 = 6.451\ \text{centimeter}^2$ $1\ \text{foot}^2 = .093\ \text{meter}^2$ $1\ \text{yard}^2 = .836\ \text{meter}^2$ $1\ \text{acre}^2 = 4{,}046.873\ \text{meter}^2$	$1\ \text{millimeter}^2 = .00155\ \text{inch}^2$ $1\ \text{centimeter}^2 = .155\ \text{inch}^2$ $1\ \text{meter}^2 = 10.764\ \text{foot}^2$ or $1.196\ \text{yard}^2$ $1\ \text{kilometer}^2 = .386\ \text{mile}^2$ or $247.04\ \text{acre}^2$
Volume	
$1\ \text{inch}^3 = 16.387\ \text{centimeter}^3$ $1\ \text{foot}^3 = .028\ \text{meter}^3$ $1\ \text{yard}^3 = .764\ \text{meter}^3$ 1 quart = 0.946 liter $1\ \text{gallon} = .003785\ \text{meter}^3$	$1\ \text{centimeter}^3 = .061\ \text{inch}^3$ $1\ \text{meter}^3 = 35.314\ \text{foot}^3$ or $1.308\ \text{yard}^3$ 1 liter = .2642 gallons 1 liter = 1.057 quarts $1\ \text{meter}^3 = 264.02\ \text{gallons}$
Weight	
1 ounce = 28.349 grams 1 pound = .454 kilogram 1 ton = .907 metric ton	1 gram = .035 ounce 1 kilogram = 2.205 pounds 1 metric ton = 1.102 tons
Velocity	
1 foot/second = .305 meter/second 1 mile/hour = .447 meter/second	1 meter/second = 3.281 feet/second 1 kilometer/hour = .621 mile/second
Acceleration	
$1\ \text{inch/second}^2 = .0254\ \text{meter/second}^2$ $1\ \text{foot/second}^2 = .305\ \text{meter/second}^2$	$1\ \text{meter/second}^2 = 3.278\ \text{feet/second}^2$
Force	

N (newton) = basic unit of force, kg-m/s^2. A mass of one kilogram (1 kg) exerts a gravitational force of 9.8 N (theoretically 9.80665 N) at mean sea level.

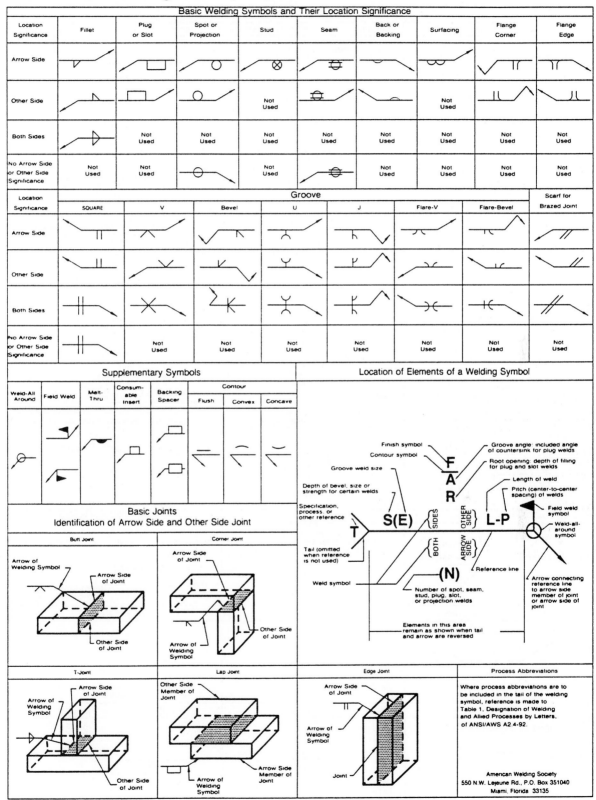

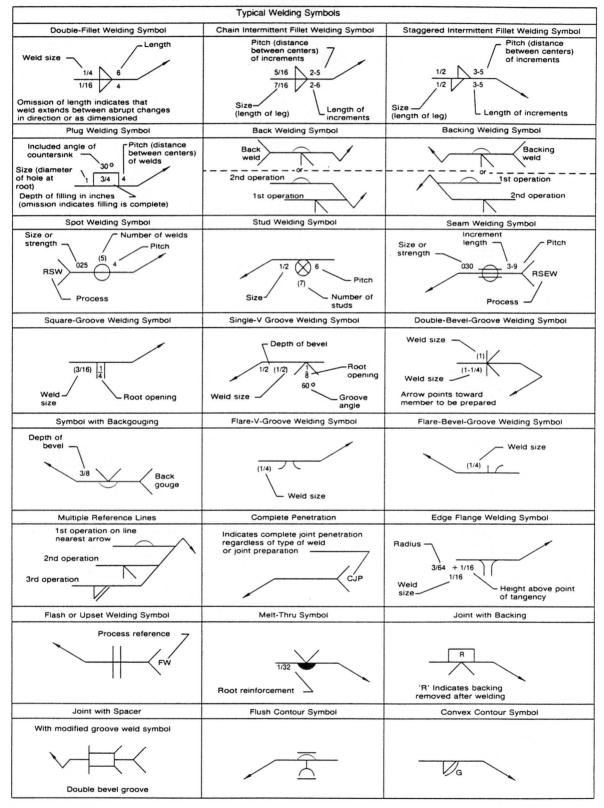

* It should be understood that these charts are intended only as shop akis. The only complete and official presentation of the standard welding symbols is in A2.4.

32 Welding Symbols and Processes—
American Welding Society Standard (continued)

MASTER CHART OF WELDING AND ALLIED PROCESSES

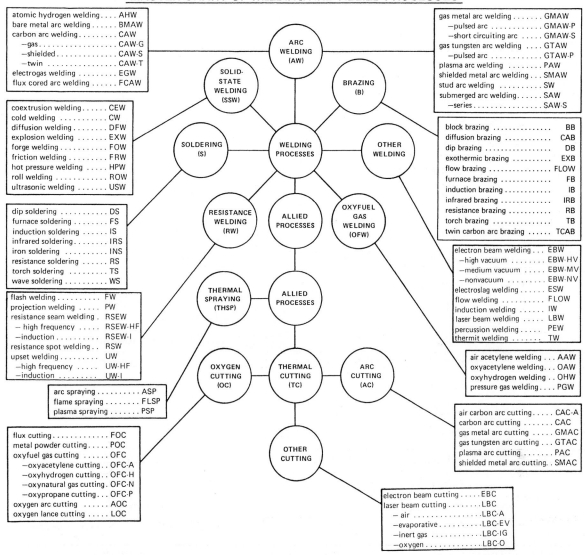

atomic hydrogen welding.... AHW	gas metal arc welding....... GMAW
bare metal arc welding...... BMAW	—pulsed arc............ GMAW-P
carbon arc welding......... CAW	—short circuiting arc..... GMAW-S
—gas................. CAW-G	gas tungsten arc welding.... GTAW
—shielded............. CAW-S	—pulsed arc............ GTAW-P
—twin............... CAW-T	plasma arc welding....... PAW
electrogas welding......... EGW	shielded metal arc welding... SMAW
flux cored arc welding...... FCAW	stud arc welding.......... SW
	submerged arc welding...... SAW
	—series.............. SAW-S

ARC WELDING (AW)
SOLID-STATE WELDING (SSW)
BRAZING (B)

coextrusion welding...... CEW	block brazing.............. BB
cold welding.......... CW	diffusion brazing............ CAB
diffusion welding....... DFW	dip brazing............... DB
explosion welding...... EXW	exothermic brazing.......... EXB
forge welding.......... FOW	flow brazing.............. FLOW
friction welding....... FRW	furnace brazing............ FB
hot pressure welding..... HPW	induction brazing............ IB
roll welding........... ROW	infrared brazing............ IRB
ultrasonic welding....... USW	resistance brazing.......... RB
	torch brazing.............. TB
	twin carbon arc brazing...... TCAB

SOLDERING (S)
WELDING PROCESSES
OTHER WELDING

dip soldering.......... DS	electron beam welding... EBW
furnace soldering....... FS	—high vacuum........ EBW-HV
induction soldering...... IS	—medium vacuum..... EBW-MV
infrared soldering...... IRS	—nonvacuum........ EBW-NV
iron soldering.......... INS	electroslag welding...... ESW
resistance soldering...... RS	flow welding.......... FLOW
torch soldering......... TS	induction welding...... IW
wave soldering......... WS	laser beam welding..... LBW
	percussion welding..... PEW
	thermit welding....... TW

RESISTANCE WELDING (RW)
ALLIED PROCESSES
OXYFUEL GAS WELDING (OFW)

flash welding......... FW	air acetylene welding... AAW
projection welding..... PW	oxyacetylene welding... OAW
resistance seam welding. RSEW	oxyhydrogen welding.. OHW
— high frequency..... RSEW-HF	pressure gas welding.... PGW
—induction.......... RSEW-I	
resistance spot welding.. RSW	
upset welding........ UW	air carbon arc cutting..... CAC-A
—high frequency..... UW-HF	carbon arc cutting....... CAC
—induction.......... UW-I	gas metal arc cutting..... GMAC

THERMAL SPRAYING (THSP)
ALLIED PROCESSES

arc spraying......... ASP
flame spraying....... FLSP
plasma spraying....... PSP

OXYGEN CUTTING (OC)
THERMAL CUTTING (TC)
ARC CUTTING (AC)

flux cutting............ FOC	gas tungsten arc cutting... GTAC
metal powder cutting..... POC	plasma arc cutting....... PAC
oxyfuel gas cutting...... OFC	shielded metal arc cutting.. SMAC
—oxyacetylene cutting.. OFC-A	
—oxyhydrogen cutting.. OFC-H	
—oxynatural gas cutting. OFC-N	
—oxypropane cutting... OFC-P	
oxygen arc cutting..... AOC	
oxygen lance cutting..... LOC	

OTHER CUTTING

electron beam cutting..... EBC
laser beam cutting....... LBC
— air................. LBC-A
—evaporative.......... LBC-EV
—inert gas............ LBC-IG
—oxygen.............. LBC-O

[a] ANSI/AWS A3.0–94.

33 Topographic Symbols

Symbol	Name	Symbol	Name
	Highway		National or State Line
	Railroad		County Line
	Highway Bridge		Township or District Line
	Railroad Bridge		City or Village Line
	Drawbridges		Triangulation Statio
	Suspension Bridge	BM X 1232	Bench Mark and Elevation
	Dam		Any Location Station (WITH EXPLANATORY NOTE)
	Telegraph or Telephone Line		Streams in General
	Power-Transmission Line		Lake or Pond
	Buildings in General		Falls and Rapids
	Capital		Contours
	County Seat		Hachures
	Other Towns		Sand and Sand Dunes
	Barbed Wire Fence		Marsh
	Smooth Wire Fence		Woodland of Any Kind
	Hedge		Orchard
	Oil or Gas Wells		Grassland in General
	Windmill		Cultivated Fields
	Tanks		Commercial or Municipal Field
	Canal or Ditch		Airplane Landing Field Marked or Emergency
	Canal Lock		Mooring Mast
	Canal Lock (POINT UPSTREAM)		Airway Light Beacon, (arrow indicates course lights)
	Aqueduct or Water Pipe		Auxiliary Airway Light Beacon, flashing

34 Piping Symbols—American National Standard

	FLANGED	SCREWED	BELL & SPIGOT	WELDED	SOLDERED
1. Joint					
2. Elbow—90°					
3. Elbow—45°					
4. Elbow—Turned Up					
5. Elbow—Turned Down					
6. Elbow—Long Radius					
7. Reducing Elbow					
8. Tee					
9. Tee—Outlet Up					
10. Tee—Outlet Down					
11. Side Outlet Tee—Outlet Up					
12. Cross					
13. Reducer—Concentric					
14. Reducer—Eccentric					
15. Lateral					
16. Gate Valve—Elev.					
17. Globe Valve—Elev.					
18. Check Valve					
19. Stop Cock					
20. Safety Valve					
21. Expansion Joint					
22. Union					
23. Sleeve					
24. Bushing					

[a] ANSI/ASME Y32.2.3–1949 (R1994).

35 Heating, Ventilating, and Ductwork Symbols[a]— American National Standard

Symbol	Description	Symbol	Description
	High Pressure Steam		Soil, Waste or Leader (Above Grade)
	Medium Pressure Return		Cold Water
FOF	Fuel Oil Flow		Hot Water
A	Compressed Air		Hot Water Return
RD	Refrigerant Discharge	F——F	Fire Line
RS	Refrigerant Suction	G——G	Gas
B	Brine Supply	S	Sprinklers—Main Supplies

	Wall Radiator, Plan		Volume Damper
	Wall Radiator on Ceiling, Plan	*Elev.*	
	Unit Heater (Propeller), Plan		Deflecting Damper
	Unit Heater (Centrifugal Fan), Plan		Turning Vanes
	Thermostatic Trap		
	Thermostatic Float		Automatic Dampers
	Thermometer		
Ⓣ	Thermostat		
20×12	Duct Plan (1st Figure, Width; 2nd Depth)		Canvas Connections
D	Inclined Drop in Respect to Air Flow		
S ◄—12×20	Supply Duct Section		Fan and Motor with Belt Guard
E ◄—12×20	Exhaust Duct Section		
R ◄—12×20	Recirculation Duct Section		
A F ◄—12×20	Fresh Air Duct Section		
	Supply Outlet		Intake Louvres and Screen
	Exhaust Inlet		
Plan	Volume Damper		

[a] ANSI/ASME Y32.2.3–1949 (R1994) and ANSI Y32.2.4–1949 (R1993).

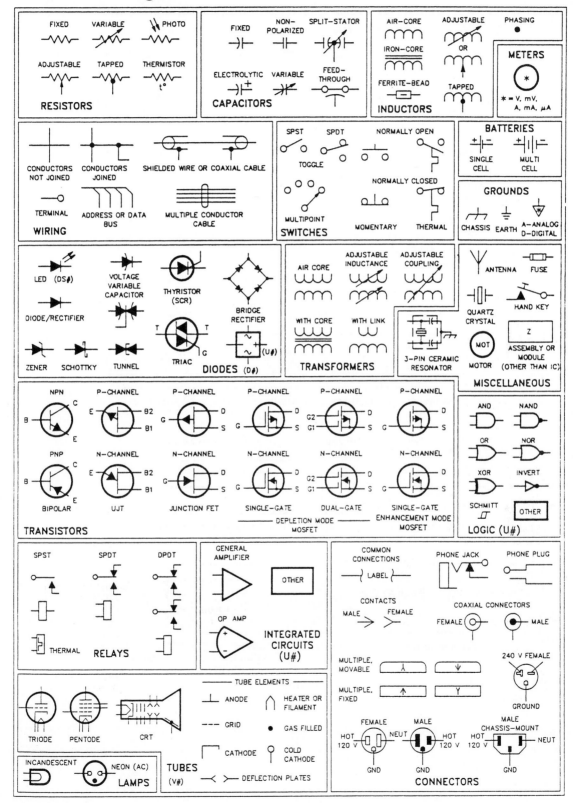

37   Form and Proportion of Geometric Tolerancing Symbols[a]

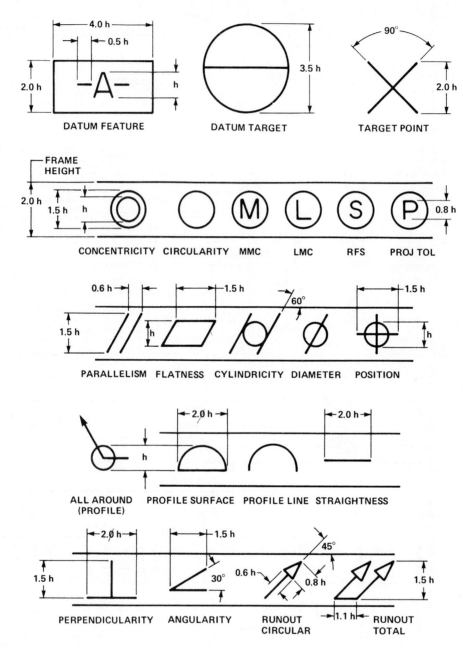

[a] ANSI/ASME Y14.5M–1994.

All dimensions are in inches except those in last two columns.

| Nominal Pipe Size | D Outside Diameter of Pipe | Threads per Inch | L₁[c] Normal Engagement by Hand Between External and Internal Threads | L₂[c] Length of Effective Thread | Nominal Wall Thickness | | | | | | | | | | Length of Pipe, Feet, per Square Foot External Surface[f] | Length of Standard Weight Pipe, Feet, Containing 1 cu. ft.[f] |
					Sched. 10	Sched. 20[d]	Sched. 30[d]	Sched. 40[d]	Sched. 60[d]	Sched. 80[d]	Sched. 100	Sched. 120	Sched. 140	Sched. 160		
⅛	.405	27	.1615	.2639	...	...	...	.068	...	.095	...	...	...	...	9.431	2,533.8
¼	.540	18	.2278	.4018	...	...	...	.088	...	.119	...	...	...	...	7.073	1,383.8
⅜	.675	18	.240	.4078	...	...	...	.091	...	.126	...	...	...	...	5.658	754.36
½	.840	14	.320	.5337	...	...	...	.109	...	.147	...	...	...	.188	4.547	473.91
¾	1.050	14	.339	.5457	...	...	...	.113	...	.154	...	...	...	.219	3.637	270.03
1	1.315	11.5	.400	.6828	...	...	...	.133	...	.179	...	...	...	.250	2.904	166.62
1¼	1.660	11.5	.420	.7068	...	...	...	.140	...	.191	...	...	...	.250	2.301	96.275
1½	1.900	11.5	.420	.7235	...	...	...	.145	...	.200	...	...	...	.281	2.010	70.733
2	2.375	11.5	.436	.7565	...	...	...	.154	...	.218	...	...	...	.344	1.608	42.913
2½	2.875	8	.682	1.1375	...	...	...	.203	...	.276	...	...	...	.375	1.328	30.077
3	3.500	8	.766	1.2000	...	...	...	.216	...	.300	...	...	...	.438	1.091	19.479
3½	4.000	8	.821	1.2500	...	...	...	.226	...	.318	...	...	...	...	.954	14.565
4	4.500	8	.844	1.3000	...	...	...	.237	...	.337	...	.438	...	.531	.848	11.312
5	5.563	8	.937	1.4063	...	...	...	.258	...	.375	...	.500	...	.625	.686	7.199
6	6.625	8	.958	1.5125	...	...	...	.280	...	.432	...	.562	...	.719	.576	4.984
8	8.625	8	1.063	1.7125	...	.250	.277	.322	.406	.500	.594	.719	.812	.906	.443	2.878
10	10.750	8	1.210	1.9250	...	.250	.307	.365	.500	.594	.719	.844	1.000	1.125	.355	1.826
12	12.750	8	1.360	2.1250	...	.250	.330	.406	.562	.688	.844	1.000	1.125	1.312	.299	1.273
14 OD	14.000	8	1.562	2.2500	.250	.312	.375	.438	.594	.750	.938	1.094	1.250	1.406	.273	1.065
16 OD	16.000	8	1.812	2.4500	.250	.312	.375	.500	.656	.844	1.031	1.219	1.438	1.594	.239	.815
18 OD	18.000	8	2.000	2.6500	.250	.312	.438	.562	.750	.938	1.156	1.375	1.562	1.781	.212	.644
20 OD	20.000	8	2.125	2.8500	.250	.375	.500	.594	.812	1.031	1.281	1.500	1.750	1.969	.191	.518
24 OD	24.000	8	2.375	3.2500	.250	.375	.562	.688	.969	1.219	1.531	1.812	2.062	2.344	.159	.358

a ANSI/ASME B36.10M–1995.
b ANSI/ASME B1.20.1–1983 (R1992).
c Refer to §13.22 and Fig. 13.20.
d Boldface figures correspond to "standard" pipe.
e Boldface figures correspond to "extra strong" pipe
f Calculated values for Schedule 40 pipe.

Size, inches	Thickness, inches	Outside Diameter, inches	16 ft Laying Length Avg. per Foot[b] Weight (lb)	Per Length Based on
Class 50: 50 psi Pressure—115 ft Head				
3	.32	3.96	12.4	195
4	.35	4.80	16.5	265
6	.38	6.90	25.9	415
8	.41	9.05	37.0	590
10	.44	11.10	49.1	785
12	.48	13.20	63.7	1,020
14	.48	15.30	74.6	1,195
16	.54	17.40	95.2	1,525
18	.54	19.50	107.6	1,720
20	.57	21.60	125.9	2,015
24	.63	25.80	166.0	2,655
30	.79	32.00	257.6	4,120
36	.87	38.30	340.9	5,455
42	.97	44.50	442.0	7,070
48	1.06	50.80	551.6	8,825
Class 100: 100 psi Pressure—231 ft Head				
3	.32	3.96	12.4	195
4	.35	4.80	16.5	265
6	.38	6.90	25.9	415
8	.41	9.05	37.0	590
10	.44	11.10	49.1	785
12	.48	13.20	63.7	1,020
14	.51	15.30	78.8	1,260
16	.54	17.40	95.2	1,525
18	.58	19.50	114.8	1,835
20	.62	21.60	135.9	2,175
24	.68	25.80	178.1	2,850
30	.79	32.00	257.6	4,120
36	.87	38.30	340.9	5,455
42	.97	44.50	442.0	7,070
48	1.06	50.80	551.6	8,825
Class 150: 150 psi Pressure—346 ft Head				
3	.32	3.96	12.4	195
4	.35	4.80	16.5	265
6	.38	6.90	25.9	415
8	.41	9.05	37.0	590
10	.44	11.10	49.1	785
12	.48	13.20	63.7	1,020
14	.51	15.30	78.8	1,260
16	.54	17.40	95.2	1,525
18	.58	19.50	114.8	1,835
20	.62	21.60	135.9	2,175
24	.73	25.80	190.1	3,040
30	.85	32.00	275.4	4,405
36	.94	38.30	365.9	5,855
42	1.05	44.50	475.3	7,605
48	1.14	50.80	589.6	9,435
Class 200: 200 psi Pressure—462 ft Head				
3	.32	3.96	12.4	195
4	.35	4.80	16.5	265
6	.38	6.90	25.9	415

Size, inches	Thickness, inches	Outside Diameter, inches	16 ft Laying Length Avg. per Foot[b] Weight (lb)	Per Length Based on
Class 200: 200 psi Pressure—462 ft Head				
8	.41	9.05	37.0	590
10	.44	11.10	49.1	785
12	.48	13.20	63.7	1,020
14	.55	15.30	84.4	1,350
16	.58	17.40	101.6	1,625
18	.63	19.50	123.7	1,980
20	.67	21.60	145.9	2,335
24	.79	25.80	205.6	3,290
30	.92	32.00	297.8	4,765
36	1.02	38.30	397.1	6,355
42	1.13	44.50	512.3	8,195
48	1.23	50.80	637.2	10,195
Class 250: 250 psi Pressure—577 ft Head				
3	.32	3.96	12.4	195
4	.35	4.80	16.5	265
6	.38	6.90	25.9	415
8	.41	9.05	37.0	590
10	.44	11.10	49.1	785
12	.52	13.20	68.5	1,095
14	.59	15.30	90.6	1,450
16	.63	17.40	110.4	1,765
18	.68	19.50	133.4	2,135
20	.72	21.60	156.7	2,505
24	.79	25.80	205.6	3,290
30	.99	32.00	318.4	5,095
36	1.10	38.30	425.5	6,810
42	1.22	44.50	549.5	8,790
48	1.33	50.80	684.5	10,950
Class 300: 300 psi Pressure—693 ft Head				
3	.32	3.96	12.4	195
4	.35	4.80	16.5	265
6	.38	6.90	25.9	415
8	.41	9.05	37.0	590
10	.48	11.10	53.1	850
12	.52	13.20	68.5	1,095
14	.59	15.30	90.6	1,450
16	.68	17.40	118.2	1,890
18	.73	19.50	142.3	2,275
20	.78	21.60	168.5	2,695
24	.85	25.80	219.8	3,515
Class 350: 350 psi Pressure—808 ft Head				
3	.32	3.96	12.4	195
4	.35	4.80	16.5	265
6	.38	6.90	25.9	415
8	.41	9.05	37.0	590
10	.52	11.10	57.4	920
12	.56	13.20	73.8	1,180
14	.64	15.30	97.5	1,605
16	.68	17.40	118.2	1,945
18	.79	19.50	152.9	2,520
20	.84	21.60	180.2	2,970
24	.92	25.80	236.3	3,895

[a] Average weight per foot based on calculated weight of pipe before rounding.

40 Cast-Iron Pipe Screwed Fittings,[a] 125 lb—American National Standard

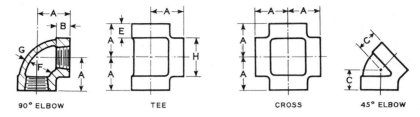

90° ELBOW   TEE   CROSS   45° ELBOW

DIMENSIONS OF 90° AND 45° ELBOWS, TEES, AND CROSSES (STRAIGHT SIZES)

All dimensions given in inches.

Fittings having right- and left-hand threads shall have four or more ribs or the letter "L" cast on the band at end with left-hand thread.

Nominal Pipe Size	Center to End, Elbows, Tees, and Crosses A	Center to End, 45° Elbows C	Length of Thread, Min. B	Width of Band. Min. E	Inside Diameter of Fitting F Max.	Min.	Metal Thickness G	Diameter of Band, Min. H
$\frac{1}{4}$	.81	.73	.32	.38	.58	.54	.11	.93
$\frac{3}{8}$	.95	.80	.36	.44	.72	.67	.12	1.12
$\frac{1}{2}$	1.12	.88	.43	.50	.90	.84	.13	1.34
$\frac{3}{4}$	1.31	.98	.50	.56	1.11	1.05	.15	1.63
1	1.50	1.12	.58	.62	1.38	1.31	.17	1.95
$1\frac{1}{4}$	1.75	1.29	.67	.69	1.73	1.66	.18	2.39
$1\frac{1}{2}$	1.94	1.43	.70	.75	1.97	1.90	.20	2.68
2	2.25	1.68	.75	.84	2.44	2.37	.22	3.28
$2\frac{1}{2}$	2.70	1.95	.92	.94	2.97	2.87	.24	3.86
3	3.08	2.17	.98	1.00	3.60	3.50	.26	4.62
$3\frac{1}{2}$	3.42	2.39	1.03	1.06	4.10	4.00	.28	5.20
4	3.79	2.61	1.08	1.12	4.60	4.50	.31	5.79
5	4.50	3.05	1.18	1.18	5.66	5.56	.38	7.05
6	5.13	3.46	1.28	1.28	6.72	6.62	.43	8.28
8	6.56	4.28	1.47	1.47	8.72	8.62	.55	10.63
10	8.08[b]	5.16	1.68	1.68	10.85	10.75	.69	13.12
12	9.50[b]	5.97	1.88	1.88	12.85	12.75	.80	15.47

[a] From ANSI/ASME B16.4–1992.
[b] This applies to elbows and tees only.

41   Cast-Iron Pipe Screwed Fittings,[a] 250 lb—American National Standard

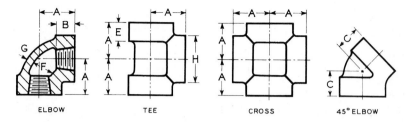

ELBOW   TEE   CROSS   45° ELBOW

DIMENSIONS OF 90° AND 45° ELBOWS, TEES, AND CROSSES (STRAIGHT SIZES)

All dimensions given in inches.

The 250-lb standard for screwed fittings covers only the straight sizes of 90° and 45° elbows, tees, and crosses.

Nominal Pipe Size	Center to End, Elbows, Tees, and Crosses A	Center to End, 45° Elbows C	Length of Thread, Min. B	Width of Band. Min. E	Inside Diameter of Fitting F Max.	Min.	Metal Thickness G	Diameter of Band, Min. H
$\frac{1}{4}$	.94	.81	.43	.49	.58	.54	.18	1.17
$\frac{3}{8}$	1.06	.88	.47	.55	.72	.67	.18	1.36
$\frac{1}{2}$	1.25	1.00	.57	.60	.90	.84	.20	1.59
$\frac{3}{4}$	1.44	1.13	.64	.68	1.11	1.05	.23	1.88
1	1.63	1.31	.75	.76	1.38	1.31	.28	2.24
$1\frac{1}{4}$	1.94	1.50	.84	.88	1.73	1.66	.33	2.73
$1\frac{1}{2}$	2.13	1.69	.87	.97	1.97	1.90	.35	3.07
2	2.50	2.00	1.00	1.12	2.44	2.37	.39	3.74
$2\frac{1}{2}$	2.94	2.25	1.17	1.30	2.97	2.87	.43	4.60
3	3.38	2.50	1.23	1.40	3.60	3.50	.48	5.36
$3\frac{1}{2}$	3.75	2.63	1.28	1.49	4.10	4.00	.52	5.98
4	4.13	2.81	1.33	1.57	4.60	4.50	.56	6.61
5	4.88	3.19	1.43	1.74	5.66	5.56	.66	7.92
6	5.63	3.50	1.53	1.91	6.72	6.62	.74	9.24
8	7.00	4.31	1.72	2.24	8.72	8.62	.90	11.73
10	8.63	5.19	1.93	2.58	10.85	10.75	1.08	14.37
12	10.00	6.00	2.13	2.91	12.85	12.75	1.24	16.84

[a] From ANSI/ASME B16.4–1992.

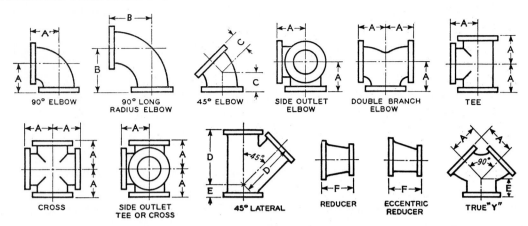

90° ELBOW 90° LONG RADIUS ELBOW 45° ELBOW SIDE OUTLET ELBOW DOUBLE BRANCH ELBOW TEE

CROSS SIDE OUTLET TEE OR CROSS 45° LATERAL REDUCER ECCENTRIC REDUCER TRUE "Y"

DIMENSIONS OF ELBOWS, DOUBLE BRANCH ELBOWS, TEES, CROSSES, LATERALS, TRUE Y'S (STRAIGHT SIZES), AND REDUCERS

All dimensions in inches.

Nominal Pipe Size	Inside Diameter of Fittings	Center to Face 90° Elbow, Tees, Crosses True "Y" and Double Branch Elbow A	Center to Face, 90° Long Radius Elbow B	Center to Face 45° Elbow C	Center to Face Lateral D	Short Center to Face True "Y" and Lateral E	Face to Face Reducer F	Diameter of Flange	Thickness of Flange, Min.	Wall Thickness
1	1.00	3.50	5.00	1.75	5.75	1.75	...	4.25	.44	.31
1¼	1.25	3.75	5.50	2.00	6.25	1.75	...	4.62	.50	.31
1½	1.50	4.00	6.00	2.25	7.00	2.00	...	5.00	.56	.31
2	2.00	4.50	6.50	2.50	8.00	2.50	5.0	6.00	.62	.31
2½	2.50	5.00	7.00	3.00	9.50	2.50	5.5	7.00	.69	.31
3	3.00	5.50	7.75	3.00	10.00	3.00	6.0	7.50	.75	.38
3½	3.50	6.00	8.50	3.50	11.50	3.00	6.5	8.50	.81	.44
4	4.00	6.50	9.00	4.00	12.00	3.00	7.0	9.00	.94	.50
5	5.00	7.50	10.25	4.50	13.50	3.50	8.0	10.00	.94	.50
6	6.00	8.00	11.50	5.00	14.50	3.50	9.0	11.00	1.00	.56
8	8.00	9.00	14.00	5.50	17.50	4.50	11.0	13.50	1.12	.62
10	10.00	11.00	16.50	6.50	20.50	5.00	12.0	16.00	1.19	.75
12	12.00	12.00	19.00	7.50	24.50	5.50	14.0	19.00	1.25	.81
14 OD	14.00	14.00	21.50	7.50	27.00	6.00	16.0	21.00	1.38	.88
16 OD	16.00	15.00	24.00	8.00	30.00	6.50	18.0	23.50	1.44	1.00
18 OD	18.00	16.50	26.50	8.50	32.00	7.00	19.0	25.00	1.56	1.06
20 OD	20.00	18.00	29.00	9.50	35.00	8.00	20.0	27.50	1.69	1.12
24 OD	24.00	22.00	34.00	11.00	40.50	9.00	24.0	32.00	1.88	1.25
30 OD	30.00	25.00	41.50	15.00	49.00	10.00	30.0	38.75	2.12	1.44
36 OD	36.00	28.00	49.00	18.00	...	...	36.0	46.00	2.38	1.62
42 OD	42.00	31.00	56.50	21.00	...	...	42.0	53.00	2.62	1.81
48 OD	48.00	34.00	64.00	24.00	...	...	48.0	59.50	2.75	2.00

[a] ANSI/ASME B16.1–1989.

43 Cast-Iron Pipe Flanges, Drilling for Bolts and Their Lengths,[a] 125 lb—American National Standard

Nominal Pipe Size	Diameter of Flange	Thickness of Flange, Min.	Diameter of Bolt Circle	Number of Bolts	Diameter of Bolts	Diameter of Bolt Holes	Length of Bolts
1	4.25	.44	3.12	4	.50	.62	1.75
$1\frac{1}{4}$	4.62	.50	3.50	4	.50	.62	2.00
$1\frac{1}{2}$	5.00	.56	3.88	4	.50	.62	2.00
2	6.00	.62	4.75	4	.62	.75	2.25
$2\frac{1}{2}$	7.00	.69	5.50	4	.62	.75	2.50
3	7.50	.75	6.00	4	.62	.75	2.50
$3\frac{1}{2}$	8.50	.81	7.00	8	.62	.75	2.75
4	9.00	.94	7.50	8	.62	.75	3.00
5	10.00	.94	8.50	8	.75	.88	3.00
6	11.00	1.00	9.50	8	.75	.88	3.25
8	13.50	1.12	11.75	8	.75	.88	3.50
10	16.00	1.19	14.25	12	.88	1.00	3.75
12	19.00	1.25	17.00	12	.88	1.00	3.75
14 OD	21.00	1.38	18.75	12	1.00	1.12	4.25
16 OD	23.50	1.44	21.25	16	1.00	1.12	4.50
18 OD	25.00	1.56	22.75	16	1.12	1.25	4.75
20 OD	27.50	1.69	25.00	20	1.12	1.25	5.00
24 OD	32.00	1.88	29.50	20	1.25	1.38	5.50
30 OD	38.75	2.12	36.00	28	1.25	1.38	6.25
36 OD	46.00	2.38	42.75	32	1.50	1.62	7.00
42 OD	53.00	2.62	49.50	36	1.50	1.62	7.50
48 OD	59.50	2.75	56.00	44	1.50	1.62	7.75

[a] ANSI B16.1–1989.

44 Shaft Center Sizes

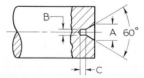

Shaft Diameter D	A	B	C	Shaft Diameter D	A	B	C
$\frac{3}{16}$ to $\frac{7}{32}$	$\frac{5}{64}$	$\frac{3}{64}$	$\frac{1}{16}$	$1\frac{1}{8}$ to $1\frac{15}{32}$	$\frac{5}{16}$	$\frac{5}{32}$	$\frac{5}{32}$
$\frac{1}{4}$ to $\frac{11}{32}$	$\frac{3}{32}$	$\frac{3}{64}$	$\frac{1}{16}$	$1\frac{1}{2}$ to $1\frac{31}{32}$	$\frac{3}{8}$	$\frac{3}{32}$	$\frac{5}{32}$
$\frac{3}{8}$ to $\frac{17}{32}$	$\frac{1}{8}$	$\frac{1}{16}$	$\frac{5}{64}$	2 to $2\frac{31}{32}$	$\frac{7}{16}$	$\frac{7}{32}$	$\frac{3}{16}$
$\frac{9}{16}$ to $\frac{25}{32}$	$\frac{3}{16}$	$\frac{5}{64}$	$\frac{3}{32}$	3 to $3\frac{31}{32}$	$\frac{1}{2}$	$\frac{7}{32}$	$\frac{7}{32}$
$\frac{13}{16}$ to $1\frac{3}{32}$	$\frac{1}{4}$	$\frac{3}{32}$	$\frac{3}{32}$	4 and over	$\frac{9}{16}$	$\frac{7}{32}$	$\frac{7}{32}$

45 Cast-Iron Pipe Flanges and Fittings,[a] 250 lb— American National Standard

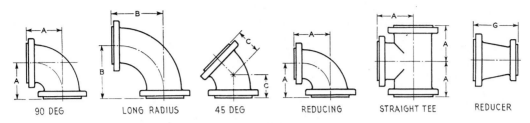

90 DEG   LONG RADIUS   45 DEG   REDUCING   STRAIGHT TEE   REDUCER

DIMENSIONS OF ELBOWS, TEES, AND REDUCERS

All dimensions are given in inches.

Nominal Pipe Size	Inside Diameter of Fitting, Min.	Wall Thickness of Body	Diameter of Flange	Thickness of Flange Min.	Diameter of Raised Face	Center-to-Face Elbow and Tee A	Center-to-Face Long Radius Elbow B	Center-to-Face 45° Elbow C	Face-to-Face Reducer G
1	1.00	.44	4.88	.69	2.69	4.00	5.00	2.00	. . .
1¼	1.25	.44	5.25	.75	3.06	4.25	5.50	2.50	. . .
1½	1.50	.44	6.12	.81	3.56	4.50	6.00	2.75	. . .
2	2.00	.44	6.50	.88	4.19	5.00	6.50	3.00	5.00
2½	2.50	.50	7.50	1.00	4.94	5.50	7.00	3.50	5.50
3	3.00	.56	8.25	1.12	5.69	6.00	7.75	3.50	6.00
3½	3.50	.56	9.00	1.19	6.31	6.50	8.50	4.00	6.50
4	4.00	.62	10.00	1.25	6.94	7.00	9.00	4.50	7.00
5	5.00	.69	11.00	1.38	8.31	8.00	10.25	5.00	8.00
6	6.00	.75	12.50	1.44	9.69	8.50	11.50	5.50	9.00
8	8.00	.81	15.00	1.62	11.94	10.00	14.00	6.00	11.00
10	10.00	.94	17.50	1.88	14.06	11.50	16.50	7.00	12.00
12	12.00	1.00	20.50	2.00	16.44	13.00	19.00	8.00	14.00
14 OD	13.25	1.12	23.00	2.12	18.94	15.00	21.50	8.50	16.00
16 OD	15.25	1.25	25.50	2.25	21.06	16.50	24.00	9.50	18.00
18 OD	17.00	1.38	28.00	2.38	23.31	18.00	26.50	10.00	19.00
20 OD	19.00	1.50	30.50	2.50	25.56	19.50	29.00	10.50	20.00
24 OD	23.00	1.62	36.00	2.75	30.31	22.50	34.00	12.00	24.00
30 OD	29.00	2.00	43.00	3.00	37.19	27.50	41.50	15.00	30.00

[a] ANSI B16.1–1989.

46   Cast-Iron Pipe Flanges, Drilling for Bolts and Their Lengths,[a] 250 lb—American National Standar

Nominal Pipe Size	Diameter of Flange	Thickness of Flange, Min.	Diameter of Raised Face	Diameter of Bolt Circle	Diameter of Bolt Holes	Number of Bolts	Size of Bolts	Length of Bolts	Length of Bolt Studs with Two Nuts
1	4.88	.69	2.69	3.50	.75	4	.62	2.50	...
$1\frac{1}{4}$	5.25	.75	3.06	3.88	.75	4	.62	2.50	...
$1\frac{1}{2}$	6.12	.81	3.56	4.50	.88	4	.75	2.75	...
2	6.50	.88	4.19	5.00	.75	8	.62	2.75	...
$2\frac{1}{2}$	7.50	1.00	4.94	5.88	.88	8	.75	3.25	...
3	8.25	1.12	6.69	6.62	.88	8	.75	3.50	...
$3\frac{1}{2}$	9.00	1.19	6.31	7.25	.88	8	.75	3.50	...
4	10.00	1.25	6.94	7.88	.88	8	.75	3.75	...
5	11.00	1.38	8.31	9.25	.88	8	.75	4.00	...
6	12.50	1.44	9.69	10.62	.88	12	.75	4.00	...
8	15.00	1.62	11.94	13.00	1.00	12	.88	4.50	...
10	17.50	1.88	14.06	5.25	1.12	16	1.00	5.25	...
12	20.50	2.00	16.44	17.75	1.25	16	1.12	5.50	...
14 OD	23.00	2.12	18.94	20.25	1.25	20	1.12	6.00	...
16 OD	25.50	2.25	21.06	22.50	1.38	20	1.25	6.25	...
18 OD	28.00	2.38	23.31	24.75	1.38	24	1.25	6.50	...
20 OD	30.50	2.50	25.56	27.00	1.38	24	1.25	6.75	...
24 OD	36.00	2.75	30.31	32.00	1.62	24	1.50	7.50	9.50
30 OD	43.00	3.00	37.19	39.25	2.00	28	1.75	8.50	10.50

[a] ANSI B16.1–1989.